Worthington George Smith

SYNO PSIS
of the
BRITISH
BASIDIOM YCETES

Elibron Classics
www.elibron.com

Elibron Classics series.

ISBN 0-543-91388-0 (paperback)
ISBN 0-543-91387-2 (hardcover)

This Elibron Classics Replica Edition is an unabridged facsimile of the edition published in 1908 by the British Museum, London.

SYNOPSIS OF THE
BRITISH BASIDIOMYCETES

A DESCRIPTIVE CATALOGUE OF THE DRAWINGS AND SPECIMENS IN THE DEPARTMENT OF BOTANY BRITISH MUSEUM

BY

WORTHINGTON GEORGE SMITH, F.L.S.

LONDON
PRINTED BY ORDER OF THE TRUSTEES OF THE
BRITISH MUSEUM
1908

LONDON:
PRINTED BY WILLIAM CLOWES AND SONS, LIMITED,
DUKE STREET, STAMFORD STREET, S.E., AND GREAT WINDMILL STREET, W.

PREFACE

In 1905 the Trustees acquired the manuscript descriptions drawn up by Mr. Worthington G. Smith, when preparing the fine series of coloured drawings of British Fungi which are exhibited in the Public Gallery of the Department of Botany. The descriptions were accompanied by line drawings illustrating the characters of each genus. In 1907 the Trustees gave permission for their publication in the form of the present Handbook, which it is hoped will prove a useful introduction to the study in the field of the larger British Fungi. Of the numerous introduced species all that have appeared in the open air are included; but those occurring in greenhouses and stoves are omitted, except those which continually reappear, and those which stray into gardens.

Mr. Smith himself and Mr. W. E. St. John Brooks, Assistant in the Department, have given valuable help in the preparation of the book for the press.

A. B. Rendle.

Department of Botany,
British Museum.

October, 1908.

SYNOPSIS

OF

BRITISH BASIDIOMYCETES

INTRODUCTION

THE British Basidiomycetes include nearly all the larger British Fungi. They comprise all the Mushroom-like Fungi, practically all the larger woody corky and gelatinous fungi of trees stumps and fallen branches (*Hymenomycetes*), the Phalli, Puff-balls and their allies, and a small number of Truffle-like subterranean species (*Gasteromycetes*). The term Basidiomycetes is used in the sense in which

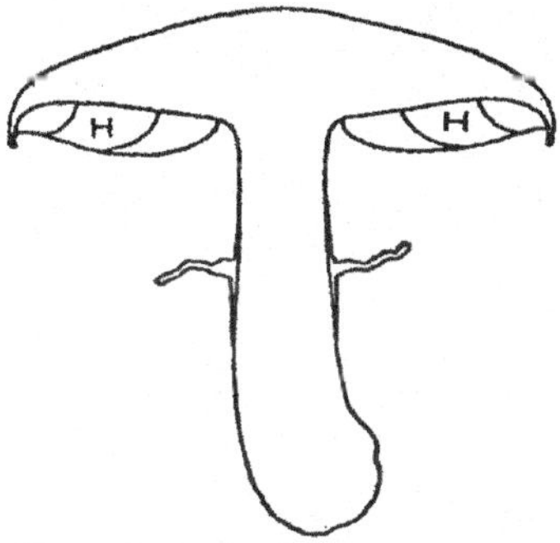

Fig. 1.—Section through Mushroom, showing exposed hymenium (H).
One-half natural size.

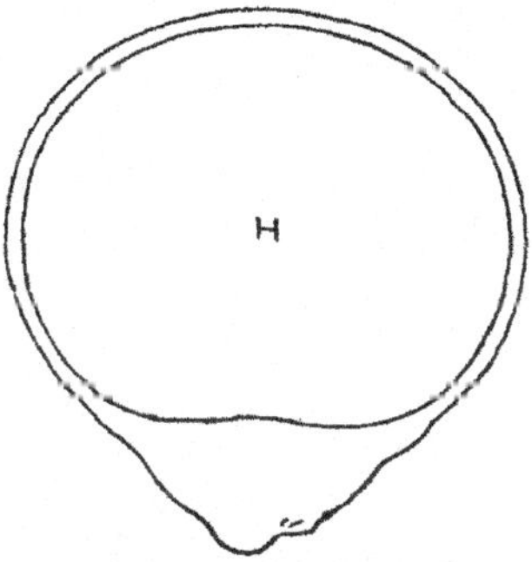

Fig. 2.—Section through Puff-ball, showing hymenium (H) closed in.
One-quarter natural size.

it was employed by De Bary, and not with the larger meaning assigned to it by Brefeld, which includes the Rust-fungi or *Uredineæ*.

In the *Hymenomycetes* the spore-bearing surface or *hymenium* is exposed during development, as in the Mushroom (fig. 1); in the *Gasteromycetes* the hymenium is closed or hidden within the substance of the fungus until maturity, as in the Puff-ball (fig. 2).

B

The larger Fungi not described in this work are a few of the Ascomycetes, including the Morel and its allies, the true ascus-bearing Truffles and a limited number of Cup-fungi.

The microscope is unnecessary for the determination of the greater number of the Basidiomycetes; nearly all are large and can be satisfactorily examined by the unaided eye or with the assistance of a hand-lens. A few forms found under Family IV *Thelephoraceæ*, as *Solenia* and *Cyphella*, superficially resemble certain of the Ascomycetes, as *Peziza;* but with a little experience even obscure forms may be easily determined with the aid of a simple lens. In some genera of the *Thelephoraceæ* a microscopic examination of the hymenium is sometimes desirable.

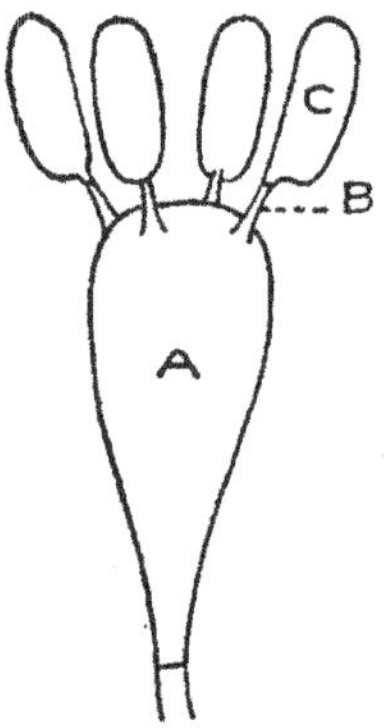

Fig. 3.—*Pleurotus ostreatus* Quél.
A, basidium; B, sterigma; C, spore. × 750.

The Basidiomycetes are highly plastic and variable. No one species is constant in all its characters, and a single example seldom wholly accords with any other single example of the same species. Examples which appear to be intermediate between allied, and sometimes between not allied, species are frequently met with. About one species in ten is perhaps fairly well and distinctly marked, but all species will at times present aberrant characters. Any one character is liable to fail; in the determination of species, therefore, all the characters must be studied together.

The Basidiomycetes are so named on account of the spores being borne on more or less club-shaped cells named *basidia* (fig. 3). They are commonly borne in fours, on slender usually short threads or supports named *sterigmata*. In rare instances, however, the basidium bears two spores or perhaps only one; in these cases the normal number of four is sometimes reverted to in well-developed examples. In some species of the *Tremellinaceæ* the spores are septate, and in some *Gasteromycetes* six or even more spores are produced on a basidium.

The spores are usually smooth and simple or composed of one cell; they are sometimes echinulate or warted. The commonest form is oval or round; a few are multiangular. They are of microscopic size, varying from about 18μ to 3μ in length, and of various colours.

The character of the basidium with its four naked spores is one of great importance, as it separates the Basidiomycetes from the Ascomycetes. In the latter the spores are borne usually in a series of eight, within microscopic transparent elongated colourless sacs named *asci* (fig. 4); the asci when ripe open at the apex and the spores escape.

In an examination of the spore-bearing surface or hymenium of the Basidiomycetes three kinds of cells are met with,—the ordinary barren cells of the hymenium, the basidia with the spores, and certain other large cells named *cystidia.* The last-named are sometimes very small or practically absent, as in *Clavaria;* in other genera they are very large, as in *Coprinus* (fig. 5), *Hymenochæte* (fig. 6) and *Peniophora.*

The more highly developed Basidiomycetes as a rule bear white or slightly tinted spores; the lower or dung-borne species usually produce black spores. In most instances the spore-colour is permanent, but in some species it vanishes and leaves the spores hyaline. The colour of the spores is often used as a generic or subgeneric character; the tint often, but not invariably, gives the distinctive hue to the mature hymenial surface. It is readily seen by placing examples with the hymenium downwards upon a piece of glass on which the spores will be deposited.

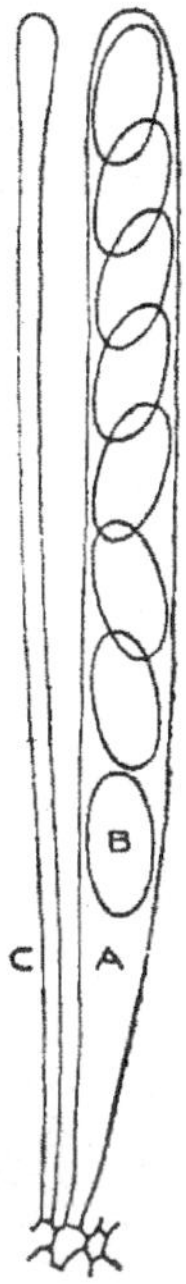

Fig. 4.—*Morchella esculenta* Pers. A, ascus; B, spore; C, paraphysis. × 500.

At maturity, the spores of the *Hymenomycetes* drop from the sterigmata and are free in the air, but in the *Gasteromycetes* the outer coats become naturally ruptured before the spores can escape. In the *Phalloidaceæ* the sporiferous surface is at first enclosed in a veil (*volva*), and even after the elongation of the stem, the cap (*pileus*) is covered by a very thin, almost invisible membrane; when the spores are ripe they are enveloped in mucus which runs down from the upper parts of the fungus to the lower, where there is a gelatinous stratum; the spores may possibly germinate in the decomposing gelatine. In the fetid species the odour attracts swarms of flies and sometimes small beetles to the sporiferous mucus, which they eagerly devour. On flying away the insects disseminate the spores, which have been seen to germinate after having passed through flies. In the *Lycoperdaceæ* the spores are at first enclosed in a peridium, and it is only when this is ruptured that they are set free. The *Hymenogastraceæ* are subterranean, or nearly so, and the spores must either germinate in the decaying substance of the mother plant or be carried away by larvæ, insects, etc.

In germination the spores open at one or both ends; a transparent germinal tube emerges (fig. 7) and forms strands, threads or hyphæ of cellular mycelium. The strands branch and coalesce with other hyphæ belonging to other spores of the same species. In rare instances, as in the *Tremellinaceæ*, the spores germinate and

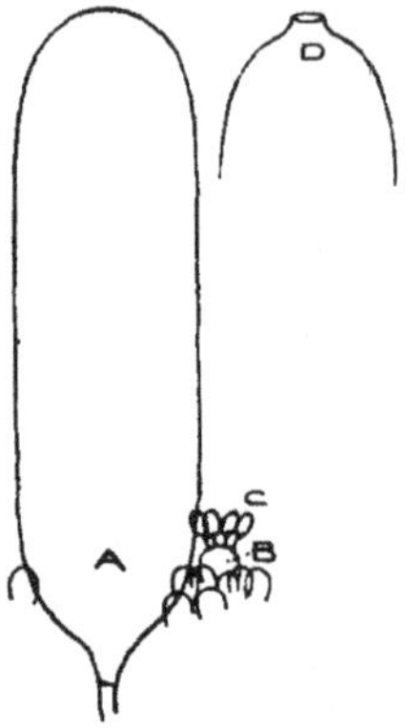

Fig. 5.—*Coprinus atramentarius* Fr.
A, cystidium, orifice open at D; B, basidium; C, spores. × 200.

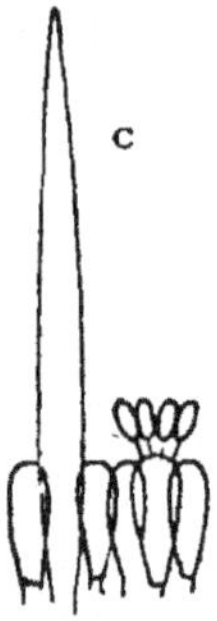

Fig. 6.—*Hymenochæte nigrescens* Cooke.
Part of hymenium with long cystidium, C. × 300.

produce a rudimentary mycelium (*promycelium*) bearing spores, from which only is a true mycelium produced. The mycelium or finely matted growth produced by germinating spores is the "spawn" of gardeners; it is usually white in colour, but is sometimes tinted; in one or two instances it is blood-red. If the mycelium or spawn be placed upon a suitable matrix the fungus will, under favourable circumstances, be reproduced. The spores of some Basidiomycetes very speedily die after falling from the hymenium. No sexual organs are known in the Basidiomycetes.

Fig. 7.—*Psathyrella prona* Gill.
Spores germinating. × 350.

The season of growth is usually the autumn, but a considerable number appear in the spring and summer, and some continue to grow till late in the autumn or even early winter. Many species of *Marasmius* will, after being quite parched in dry weather, revive in rain, and several of the late-growing species of *Clitocybe* and *Collybia* revive after being frozen. There are two crops of some Basidiomycetes every year, one vernal and the other autumnal; this is the case with *Coprinus atramentarius*, of which the first crop appears after the rains of April and May, the second in September or October. If vernal examples are gathered and buried with decayed wood or beside decaying posts a crop of perfect fungi will appear in September or October. There are sometimes three crops of *Coprinus micaceus* in one year. If living plants of *Coprinus radiatus* are placed upon manure heaps the spores will germinate at once and reproduce the plant in a day or two.

Some of the perennial *Polyporaceæ* continue to increase in size for many years. The years of growth can be counted in the annual flesh-strata, which are easily seen when sections are made. Mycelium will continue to live in wood or in the ground for an apparently unlimited time. I have known a "fairy-ring" of *Clitocybe geotropa* on Dunstable Downs for forty or more years, which under favourable conditions of light can be seen at a distance of more than a mile. The diameter has not altered much during the time I have known it, for sometimes it grows inwardly for several years and then again outwardly.

Most species are terrestrial and grow on or just below the ground; a large number grow on dead leaves and twigs; others on dead wood, bark, branches and trunks. It is unusual to see fungi on healthy trees; the majority of the larger fungi so commonly seen growing on tree-trunks in parks, woods and forests are "wound parasites," which germinate upon a damp injured place and grow parasitically upon the living host. Rootlets are often attacked by mycelium which at length develops as a perfect fungus at the tree's base, as in *Fomes annosus*, or in other cases travels further up the trunk as in *F. igniarius*, and other species. *Polyporus Schweinitzii* is well known to be highly destructive to conifers; *Pinus Strobus*, *P. sylvestris* and *Larix europæa* especially suffer. Many other species of *Polyporus* are more or less destructive. Amongst the *Agaricaceæ*, *Armillaria mellea* is one of the most destructive; it often grows upon living roots, not uncommonly under the shade of large trees, the growth of the fungus being limited to the extent of the root-growth of the trees. One form of its mycelium, named *Rhizomorpha*, lives perennially beneath the bark of trees. It destroys tree-roots, chiefly those of conifers, but sometimes attacks plum and other trees; it also appears on squared timber, when used for bridges, in mines, etc.

Certain species grow parasitically upon other Basidiomycetes. *Volvaria Loveiana* grows upon the pileus of *Clitocybe nebularis*; two species of *Nyctalis* grow upon the pileus or stem of species of *Russula* and *Lactarius*; and *Boletus parasiticus* grows from the base or side of *Scleroderma vulgare.* Some species are almost peculiar to human habitations: *Coprinus radians* grows upon plastered walls, *C. aphthosus* commonly appears in kitchens and cellars, *Merulius lacrymans* (Dry Rot), *Poria hybrida* and *Polyporus destructor* are almost entirely confined to the interior of houses and ships, and *Merulius corium* often grows in dwelling-houses. *Coprinus domesticus* is usually seen in or close to human dwellings; *C. comatus* and *Cortinarius urbicus* commonly grow close to houses or to places where human beings resort. Many species grow on burnt wood and earth, as *Clitocybe fumosa*, *Collybia atrata*, *C. ambusta*, *Flammula decipiens*, *F. sapinea*, *F. carbonaria*, *Lepiota Friesii*, *Cantharellus radicosus*, *Polystictus cinnamomeus*, and many others.

The taste of the majority of the Basidiomycetes when raw is watery-insipid; many are practically tasteless or mild. Some when uncooked have a pleasant nutty flavour, as *Psalliota campestris* (mushroom) and *Lepiota procera*. Several are bitter, as *Boletus felleus;* others are bitter-nauseous, as *Hypholoma fasciculare*, and many species of *Lactarius* and *Russula* are very acrid or bitter acrid.

The odours are most diverse. *Clitocybe fragrans* is very sweet and recalls Melilot, as does also *Hydnum graveolens; C. odora* is fragrant of Woodruff or Vernal Grass; *Trametes suaveolens, T. odora, Lactarius glyciosmus* and *Clavaria stricta* are also very sweet-scented. One variety of *Cantharellus cibarius* smells strongly of apricots, and *Clitocybe geotropa* is almond-scented. Many smell strongly of onions or garlic, the best known examples being different species of *Marasmius*. Fetid and disgusting odours are common; a familiar instance is that of *Ithyphallus impudicus*. The odours possibly serve some purpose at present unknown. The carrion-scented species attract swarms of carrion-feeding insects which greedily devour the highly fetid, soft, sporiferous material of the *Phalloidaceæ*.

The exudation of fluid, the so-called milk, when the stem is broken, is a remarkable character of some species. *Mycena galopus* and *M. lactea* contain a white, *M. crocata* and *M. chelidonia* a yellow, *M. hæmatopus* a dark purple-red, and *M. sanguinolenta* a red juice; *Lactarius deliciosus* exudes an orange-coloured and *L. sanguifluus* a deep blood-red juice which, on exposure to the air, quickly becomes green. The milk of *L. chrysorrheus* and *L. theiogalus* is sulphur-yellow, that of *L. acris* is at first white, then reddish. The milk of *L. uvidus* is white and quickly changes to violet; that of *L. scrobiculatus* is first white, then sulphur; that of *L. fuliginosus* is white, then saffron. The stem of *Marasmius varicosus* is filled with dark blood-red juice, which flows when the stem is bruised or broken.

Some species of *Agaricaceæ*, as *Hypholoma lacrymabundum* and *H. velutinum*, have "weeping gills"; in mature examples drops of fluid may be seen sprinkled all over the surface of the gills which, when examined under the microscope, are seen to be charged with spores and cystidia. The hymenium of the dry rot fungus, *Merulius lacrymans*, is usually covered with globules of exuded moisture; *Polyporus dryadeus* is often seen in the same condition.

The Basidiomycetes as a rule do not exhibit brilliant colouring, but there are remarkable exceptions, the most striking being scarlet and crimson, as in *Amanita muscaria*. Yellow, orange, blue, purple, and white also occur, also rarely green and black, but the majority are pallid, watery-brown, brown, greyish or buff. The brown and buff colours of Agarics often cause the fungi to be overlooked when growing amongst dead leaves.

The Mushroom is a highly nitrogenous plant and contains a considerable amount of fat. The following are its chief constituents, as given by Professor A. H. Church:—

	In 100 parts.	In 1 lb.
Water	90·0	14 oz. 175 gr.
Albuminoids etc.	5·0	0 oz. 350 gr.
Carbohydrates etc.	3·8	0 oz. 266 gr.
Fat	·7	0 oz. 49 gr.
Mineral matter	·5	0 oz. 35 gr.

The chemical composition of the Champignon, *Marasmius oreades*, is nearly the same. Several species are known to be highly poisonous in all stages of growth, both cooked and uncooked; those without unpleasant taste are the most dangerous, as *Amanita phalloides*.

Specimens of fungi for determination must be quite perfect and fresh; examples past maturity are useless. Young specimens should be secured whenever possible, as some characters which are then very obvious vanish as the plant advances to full growth.

In the descriptions which follow the sequence is almost identical with that of Fries's *Hymenomycetes Europæi*, 1874. This sequence has been almost universally approved, and has been followed in this country by Berkeley, Cooke, Stevenson, and in my supplementary volume to Berkeley's *Outlines of British Fungology*. The large collection of drawings, prints and dried examples of British Basidiomycetes in the Herbarium of the British Museum is arranged on the same plan.

The number before each species corresponds with the number in my series of coloured drawings of British Fungi exhibited in the Public Gallery of the Department of Botany, British Museum.

The letters *a b c* used in the text refer to the drawings and specimens in the British Museum. The letter *a* indicates that there is a drawing of the species in the Public Gallery, *b* that there is a drawing and *c* that there is a specimen in the Herbarium.

The specific descriptions contain the salient distinctive characters, which must be used with the generic and sectional characters. If fuller descriptions of the Hymenomycetes are desired, Fries's *Hymenomycetes Europæi* (1874) and his *Monographia Hymenomycetum Sueciæ* (1857) may be consulted; the Gasteromycetes are described in greater detail in Massee's *Monograph of the British Gasteromycetes* (1889) and the same author's *British Fungus Flora*, vol. i. (1892). Saccardo's *Sylloge Fungorum* (1882–1908) may also be consulted.

The names of species obviously introduced are printed in small capitals; introduced genera are distinguished typographically.

The text was prepared and checked at the time when the drawings were made, when living and dried examples, drawings, plates, printed descriptions, and notes were before me.

The dimensions give the typical size in inches of well-developed examples as found in Britain, for which those given by Fries are sometimes very misleading. The numerals represent the diameter of the pileus, the height from the base of the stem to the top of the pileus, and the diameter of the stem near the middle. The stem in most of the stem-bearing Basidiomycetes is cylindrical or slightly attenuated upwards; this character is not reiterated in the descriptions, where departures from the type only are given. In stemless species, which are generally excentric, the longer diameter of the pileus is given.

The notes as to taste and odour are generally original, supplemented by the observations of others; those as to colour have in most instances been drawn up from living examples, supplemented by printed descriptions and by reference to characteristic coloured drawings and plates.

The habitats and dates of appearance are from my own notes, and from observations of other British botanists; in a few instances no habitat or date is on record.

Spore measurements have been omitted as untrustworthy; in many cases authors have measured immature spores from young plants, in others it is certain that authors have misnamed the fungus from which the spores have been taken.

The following abbreviations are employed:—

HYMENOMYCETES.

A.	Annulus.	Pa.	Papillæ.
Br.	Bristles.	Pi.	Pileoli.
Cl.	Clubs.	Po.	Pores.
Co.	Cortina.	Scl.	Sclerotium.
G.	Gills.	Se.	Setæ.
Gr.	Granules,	Sp.	Spines.
Hym.	Hymenium.	St.	Stem.
H.	Hymenophore.	Sub.	Subiculum.
Myc.	Mycelium.	T.	Tubes.
P.	Pileus.	Te.	Teeth.
circ.	circumference	Tub.	Tubercles.
gr.	granules	V.	Veil.
marg.	margin	Ve.	Veins.
mid.	middle	Vo.	Volva.
sc.	scales	Z.	Zone.
sq.	squamules		
um.	umbo		
z.	zones.		

The stem is supposed to be cylindrical or nearly so unless otherwise described, and simple if without volva and annulus.

GASTEROMYCETES.

Cap.	Capillitium.	Ex.	Exoperidium.	Per.	Peristome.
Col.	Columella.	Fun.	Funiculus.	Peri.	Peridiolum.
Cor.	Cortex.	Gl.	Gleba.	Rec.	Receptacle.
End.	Endoperidium.	Pe.	Peridium.	Vo.	Volva.
		Ped.	Pedicel.		

ORDER I. HYMENOMYCETES

Basidia and spores exposed in the mature plant (fig. 8).
Families I–VI.

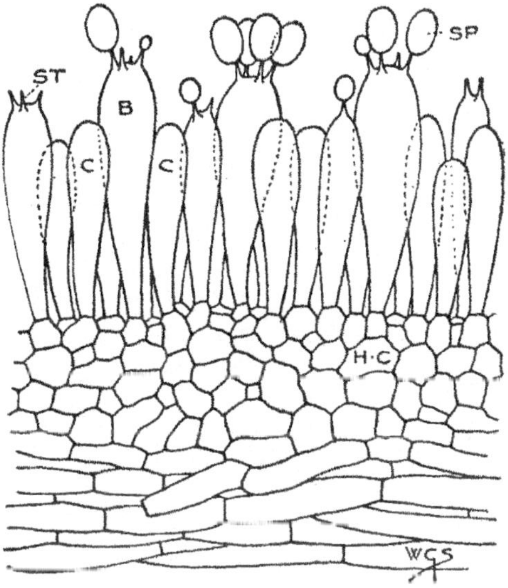

Fig. 8.—Basidia and spores of Mushroom (*Psalliota campestris* Quél.). B, basidia; SP, spores; C, cystidia or barren basidia; ST, sterigmata; H.C, hymenial cells. × 1000.

I. AGARICACEÆ. Hymenium spread over the surface of radiating gills or lamellæ. Genera **1–58**

II. POLYPORACEÆ. Hymenium lining the surface of small pores or tubes. Genera **59–69**

III. HYDNACEÆ. Hymenium spread over spines or protuberances. Genera **70–79**

IV. THELEPHORACEÆ. Hymenium unilateral, spread over a firm smooth or corrugate under or upper face. Genera **80–90**

V. CLAVARIACEÆ. Hymenium spread over the surface of smooth, simple or branched clubs; in *Sparassis* over a large sponge-like mass of coalescing branches. Genera **91–96**

VI. TREMELLINACEÆ. Hymenium spread over a smooth gelatinous surface. Genera **97–106**

Order II. GASTEROMYCETES

Hymenium concealed within a peridium until the spores are mature (fig. 9). Families VII–XI

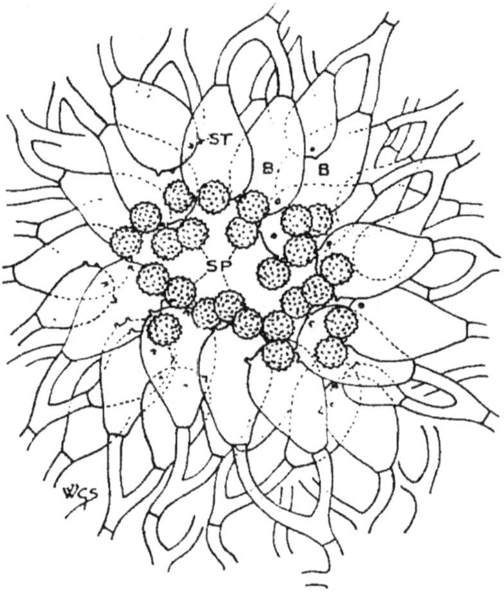

Fig. 9.—Basidia and spores of *Scleroderma vulgare* Fr. B, basidia; SP, spores; ST, sterigmata. × 400.

VII. Phalloidaceæ. Pileus and stem or latticed receptacle at first enclosed in a globular peridium with a gelatinous stratum. Peridium at length opening at the apex and the gleba extended on a stem or lattice. Genera **107–110**

VIII. Lycoperdaceæ. Peridium globular, with a capillitium; at length opening for the dispersion of the spores. Genera **111–116**

IX. Sclerodermaceæ. Peridium globose, thick, without a capillitium, dehiscing irregularly. Genera **117–118**

X. Nidulariaceæ. Peridium bell- or cup-shaped, containing compressed or globose peridiola which become free at maturity. Genera **119–122**

XI. Hymenogastraceæ. Subterranean, peridium without a capillitium, indehiscent. Genera **123–128**

Order I. HYMENOMYCETES

Fam. I. AGARICACEÆ

Hymenium spread over the surface of radiating gills, the axis a stem, rarely a point, usually central, sometimes lateral. In some forms the young fungus is entirely enclosed in a *primary* or *universal veil* (fig. 10 B). During growth, as the stem elongates, the veil is ruptured, the lower portion called the *volva* (Vo.) remaining as a sheath round the base of the stem, while the upper portion, still attached to the pileus, is torn into fragments as the latter expands (U.V.). In some species a *secondary* or *partial veil* is also present in the earlier stages, spreading from the upper part of the stem to the edge of the pileus. This veil is finally ruptured, and partly persists as a *ring* or *annulus* (A), encircling the stem. Contains two-thirds of the whole Basidiomycetes. Species **1—1455**

Key to the Genera.

In using the key for the determination of the genus the following characters must be ascertained:—1. Colour of the spores. 2. Nature of the substance, whether fleshy, putrescent, deliquescent or coriaceous. 3. Whether the hymenophore is distinct from the stem, or confluent and homogeneous with it. 4. Nature of the pileus, whether dry, viscid, silky, etc. 5. Nature of the stem, whether simple, annulate or volvate, and whether it is homogeneous with the pileus, or heterogeneous with a cartilaginous bark. 6. Nature of the gills and their attachment to the stem.

I. Spores white.
  A. Plant fleshy, more or less firm, not deliquescent or coriaceous.
    Hymenophore distinct from the fleshy stem.
      Pileus bearing warts or patches free from the cuticle.
        Stem volvate and annulate .... 1 Amanita.
        Stem volvate.. 2 Amanitopsis.
      Pileus pulverulent, smooth or scaly; scales, when present, concrete with the cuticle.
        Stem annulate .... 3 Lepiota.
        Stem simple .... 4 Hiatula.
    Hymenophore confluent and homogeneous with the fleshy stem.
      Stem without cartilaginous bark.
        Stem central (some of the plants under 50 are stemless).
          Annulate .... 5 Armillaria.

Simple.		
Gills sinuate	6	TRICHOLOMA.
Gills decurrent		
edge acute	7	CLITOCYBE.
edge swollen	50	CANTHARELLUS.
separating from hymenophore	46	PAXILLUS. (Group I. Lepista.)
Gills adnate.		
Parasitic on Agarics	51	NYCTALIS.
Not parasitic.		
Milky	48	LACTARIUS.
Not milky.		
Rigid and brittle	49	RUSSULA.
Waxy	47	HYGROPHORUS.
Stem lateral or absent	11	PLEUROTUS.
Stem with cartilaginous bark.		
Gills usually adnate	8	COLLYBIA.
Gills sinuate	9	MYCENA.
Gills decurrent	10	OMPHALIA.
B. Plant tough, coriaceous or woody.		
Stem central.		
Gills simple	52	MARASMIUS.
Gills branched	55	XEROTUS.
Stem lateral or absent.		
Gills toothed	53	LENTINUS.
Gills not toothed	54	PANUS.
Gills channelled longitudinally or crisped	56	TROGIA.
Gills splitting longitudinally	57	SCHIZOPHYLLUM.
Gills anastomosing	58	LENZITES.
II. Spores rosy or salmon-colour.		
A. Stem without cartilaginous bark.		
Hymenophore distinct from the fleshy stem.		
Stem volvate	12	VOLVARIA.
Stem annulate	13	ANNULARIA.
Stem simple	14	PLUTEUS.
Hymenophore confluent and homogeneous with the fleshy stem.		
Stem central.		
Gills adnate or sinuate	15	ENTOLOMA.
Gills decurrent	16	CLITOPILUS.
Stem lateral or absent	20	CLAUDOPUS.
B. Stem with cartilaginous bark.		
Gills decurrent	19	ECCILIA.
Gills not decurrent.		
Pileus scaly	17	LEPTONIA.
Pileus papillose, subcampanulate.		
Gills membranous, persistent	18	NOLANEA.
Gills subdeliquescent	43	BOLBITIUS.
III. Spores brown, reddish- or yellowish-brown.		
Stem without cartilaginous bark.		
Stem central.		
Stem annulate.		
Annulus membranous.		
Terrestrial	22	TOGARIA.
Growing on wood	24	PHOLIOTA.
Annulus arachnoid, filamentous or evanescent.		
Gills adnate, terrestrial	44	CORTINARIUS.
Gills decurrent or adnate, growing on wood	27	FLAMMULA.

Stem simple.
With a volva ... 21 ACETABULARIA.
Without a volva.
Gills adhering to the hymenophore and sinuate.
Cuticle fibrillose or silky ... 25 INOCYBE.
Cuticle smooth, viscid ... 26 HEBELOMA.
Gills separating from the hymenophore and decurrent, stem sometimes obsolete ... 46 PAXILLUS. (Group 2. Tapinia.)
Gills decurrent ... 30 TUBARIA.
Gills not decurrent.
Margin of pileus at first incurved ... 28 NAUCORIA.
Margin of pileus always straight.
Hymenophore free from the stem ... 23 PLUTEOLUS.
Hymenophore confluent with the stem ... 29 GALERA.
Stem lateral or absent ... 31 CREPIDOTUS.

IV. Spores purple, brownish-purple, dark purple, or dark brown.
A. Stem without cartilaginous bark.
Hymenophore distinct from the fleshy stem.
Stem volvate ... 32 CHITONIA.
Stem without a volva.
Stem annulate ... 33 PSALLIOTA.
Stem simple ... 34 PILOSACE.
Hymenophore confluent with the stem.
Veil normally ring-like on the stem ... 35 STROPHARIA.
Veil normally fibrillose and adhering to the margin of the pileus ... 36 HYPHOLOMA.
B. Stem with cartilaginous bark.
Margin of pileus at first incurved ... 37 PSILOCYBE.
Margin of pileus at first straight ... 38 PSATHYRA.

V. Spores black or nearly so.
Gills deliquescent ... 42 COPRINUS.
Gills not deliquescent.
Gills decurrent ... 45 GOMPHIDIUS.
Gills not decurrent.
Stem annulate ... 39 ANELLARIA.
Stem simple ... 40 PANÆOLUS.
Pileus striate ... 41 PSATHYRELLA.

Genera **1–41** represent the genus *Agaricus* as understood by Fries who regarded the modern genera as of subgeneric rank only. They include species **1–920**, and are arranged in five series defined by the colour of the spores as illustrated in the five diagrams (Plates I.–V.).

The genera which are marked in corresponding positions on the different diagrams are structurally identical; the only difference is in the colour of the spores. For instance, *Amanitopsis*, on the first diagram, has white spores; it corresponds in structure with *Volvaria*, with pink spores, on the second diagram, with *Acetabularia*, with brown spores, on the third, and with *Chitonia*, with purple spores, on the fourth. There is no genus structurally identical with *Amanitopsis* in Britain with black spores, consequently this position on the fifth diagram is left vacant.

White-spored species represent the highest types and black-spored species the lowest, the pink-, brown-, and purple-spored species are intermediate. *Amanita* is the highest type of the white-spored Agarics. The majority of the white-spored species are terrestrial and more or less persistent, whilst the black-spored species commonly grow on dung and are evanescent.

Series I. LEUCOSPORI Fr.

(From the white spores; Gr. *leukos*, white, *sporos*, a seed.)

Spores white, rarely with a faint shade of ashy-grey, a faint tint of ochre, pale brown or pale greenish; usually oval or round and smooth, rarely slightly spinulose. Genera I—XI.

I. **AMANITA** Pers.

(Galen's name for certain fungi.)

Veil universal, clothy, distinct from the cuticle of the pileus, after rupture, as maturity is reached, remaining as a volva at the base of the stem, with broken patches of its substance frequent on the pileus. *Hymenophore* distinct from the fleshy stem. *Pileus* convex,

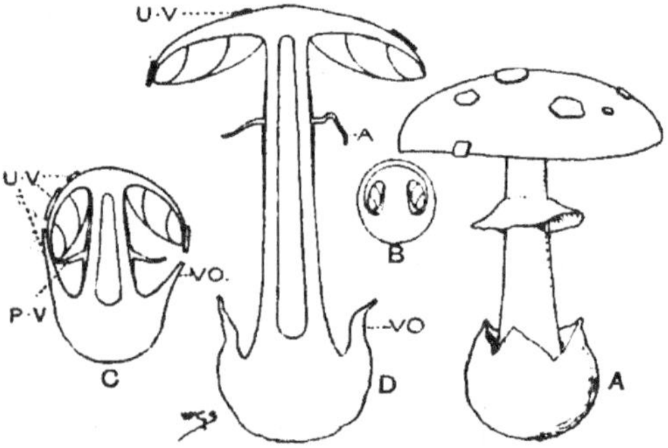

Fig. 10.—*Amanita phalloides* Link. One-third natural size. A, perfect plant; B, section of young plant showing universal veil; C, section of half-grown example showing rupture of universal veil; D, section of mature example; U.V, universal veil; P.V, partial veil, becoming annulus; A, annulus; VO, volva.

then expanded, not decidedly fleshy. *Stem* central, annulate and volvate. *Gills* free or nearly so, white or whitish. (Fig. 10.)

All the species grow on the ground in woods and uncultivated places. A few are edible, others are poisonous or suspicious. Species 1—17

Volva bulb-like, large, exhibiting a free expanded border in splitting. *Pileus* naked or nearly so, and viscid, except **4a**, **4b**, and **4c**; margin striate, except **4a**, **4b**, and **4c**. *Gills* free or adnexed. **1—4**

Volva splitting regularly, its margin closely adpressed to the stem and the detached upper part forming patches or warts on the pileus. *Gills* free. **5—9**

Volva friable, broken into small scales or warts, not persistent at the somewhat bulbous base of the stem. **10—13**

Volva rudimentary, floccose, disappearing. *Stem* slightly enlarged at the base, not truly bulbous. **14—17**

1. **A. virosa** Quél. (*virus*, poison) *a b c*. Wholly white.
P. marg. lobed, floccose with V. *St.* somewhat scaly. *A.* large. *Vo.* lax, wide.

Poisonous, fetid. Woods; uncommon. Aug.–Oct. 5 × 7½ × ⅝ in. Compare **3**. This and the next when young greatly resemble the edible mushroom.

2. **A. phalloides** Link (from the volva as in *Phalloidaceæ*) *a b c*.
P. white, yellowish or light, or dark olivaceous. *St.* somewhat smooth, white.

Poisonous, narcotic, acrid. Odour when young, strong, not disagreeable, at length fetid. Woods; frequent. Apr.–Nov. 3½ × 4½ × ⅝ in. Many fatal cases of poisoning on record.

3. **A. verna** Quél. (*ver*, spring) *a b*. Wholly white.
P. marg. not lobed, not floccose. *St.* floccose. *A.* not large. *Vo.* somewhat adpressed to stem.

Probably poisonous. Odour not strong, but unpleasant. Woods; uncommon. May–Oct. 3 × 5 × ⅝ in. Probably a var. of **2**. Compare **1**.

4. **A. mappa** Quél. (from the resemblance of the volva to a napkin, *mappa*) *a b c*.
P. dry, patched with remains of Vo., white to pale yellowish-buff. *St.* somewhat smooth, white.

Poisonous, fetid. Woods; frequent. Usually small. Aug.–Nov. 3 × 3⅝ × ½ in. Compare **12**.

4a. **A. porphyria** A. & S. (from the colour of the pileus) *a b*.
P. warm brown or purplish-brown; marg. smooth. *St.* slightly scaly below, whitish. *A.* distant.

Odour none. Woods, pine. Sept. 9½ × 3½ × ⅝ in.

4b. **A. recutita** Gill. (from the smoothly cut edge of the volva) *a*.
P. sometimes umbonate, dry, glabrous, pale or vinous-umber; marg. smooth. *St.* slightly scaly below, white, becoming faintly brownish. *A.* distant. *Vo.* more or less adpressed.

Woods, pine. Sept. 6¼ × 4⅝ × ⅝ in.

4c. **A. aureola** Gill. (from the golden colour of the pileus) *a b c*.
P. slightly umbonate, bright yellow or orange-yellow; marg.

smooth. *St.* floccoso-squamose, white. *A.* superior, large, spreading.

Woods, under birches. Sept. 3⅜ × 7¼ × ½ in.

4d. A. citrina Pers. (from the lemon-coloured pileus) *b.*

P. convex, obtuse, brassy-yellow, with white warts. *St.* hollow, white. *A.* large. *Vo.* imperfect. *G.* free, broad, white.

Woods. Oct. 4⅜ × 4½ × 1.

5. A. muscaria Pers. (from its former use as a fly-poison; *musca*, a fly) *a b c.*

P. scarlet or orange, rarely yellow, brown or liver-colour; marg. slightly striate. *St.* white, rarely yellowish. *Vo.* bulb-like, large.

Poisonous, narcotic, acrid. Woods, birch, fir. July–Nov. 7 × 9⅝ × 1 in. Compare brown form with **6.**

6. A. pantherina Quél. (from the markings of the pileus) *a b c.*

P. viscid, pale or dark brown; marg. striate. *St.* whitish.

Poisonous. Woods and pastures; frequent. July–Oct. 4 × 5 × ½ in. Compare **11.**

7. A. excelsa Gonn. & Rab. (*excelsus*, tall) *a b c.*

P. viscid, papilloso-wrinkled, brownish; marg. striate. *St.* brownish-white.

Solitary. Poisonous. Taste not unpleasant. Woods, chiefly beech and birch; frequent. July–Oct. 4¾ × 5½ × ¾ in.

8. A. strobiliformis Quél. (from the resemblance of the scales of the pileus to the scales of certain fir-cones; *strobilus*, a pine cone) *a b c.*

P. white, then ashy-brown, covered with large sc.; marg. even. *St.* whitish.

Said to be edible. Taste and odour at first slight, becoming disagreeable. Open places in and near woods; rare. Pileus sometimes 12 in. in diam. July–Sept. 8 × 6¼ × 1⅝ in.

9. A. solitaria Secr. (from its solitary habit) *a b.* White or buff-white.

P. marg. even. *St.* brownish-white. *Vo.* forming an obconic bulb; base rooting.

Solitary. Taste insipid. On the ground, damp places. Sept. 6 × 5½ × 1⅛ in.

10. A. rubescens Pers. (from the flesh becoming sienna-red when broken or bruised) *a b c.*

P. reddish-sienna, light to dark. *St.* pale sienna-red.

Edible; *Golmelle* of continental markets. Taste not unpleasant; odour strong. Woods; common. Commonly much eaten by larvæ. There is a white variety. June–Nov. 5½ × 5 × 1 in. Compare **14.**

11. A. spissa Opiz (from the small crowded warts; *spissus*, crowded) *a b.*

P. smoky-brown or grey, with ash-coloured warts. *St.* concentrico-squamulose, obconic at base, rooting, whitish.

Woods, mixed; uncommon. July–Oct. 4 × 4¼ × ⅝ in. Compare **6.**

11a. A. cariosa Gill. (from the carious stem) *a b.*

P. soft, even, brown, whitish-brown, or dark grey, unequally clad with thin, mealy, white patches and clouds. *St.* fragile, somewhat enlarged below, white, pale umber and brown, scaly at base. *G.* adnate then seceding and free.

Suspected poisonous. Woods. Sept. $4\frac{7}{8} \times 5\frac{7}{8} \times \frac{3}{4}$ in. Closely allied to **11**.

12. A. nitida Fr. (from the shining pileus; *nitidus*, shining) *a b*.
P. whitish, cream or whitish-buff, rarely yellowish or olive, warts brownish. *St.* squamulose, white to buff-white.

Woods. Aug. $4 \times 4\frac{1}{4} \times \frac{5}{8}$ in. Compare **4**.

13. A. aspera Pers. (from the rough pileus) *a b c*.
P. whitish- or dusky-olive, crowded with small, pointed warts. *St.* white.

Poisonous. Taste not unpleasant; odour strong. Woods, beech, fir; uncommon. June–Nov. $3 \times 3\frac{1}{2} \times \frac{5}{8}$ in.

14. A. magnifica Sacc. (*magnificus*, splendid) *a*.
P. whitish, pallid-tan or sienna-brown, floccoso-spotted; marg. sulcate. *St.* reddish-white. *G.* adnato-decurrent. *Flesh* white, tawny when broken.

Woods, fir, beech; rare. July–Oct. $4 \times 4\frac{1}{2} \times \frac{1}{2}$ in. Compare **10**.

15. A. arida Gill. (*aridus*, dry) *a*. Greyish.
P. soon plane; marg. sulcate. *St.* subglabrous, white. *A.* distant. *G.* adnexo-free or adnexed. *Flesh* white.

Plantations, birch. Sept.–Oct. $3 \times 3\frac{1}{4} \times \frac{5}{8}$ in.

15a. A. junquillea Quél. (from a fancied resemblance in the colour of the pileus in some examples to the flowers of the jonquil) *b*.
P. campanulato-convex, viscid, buff with paler marg., clad with the greyish remains of the V.; marg. deeply striate, white. *St.* hollow, villous-scaly, white, whitish or pale yellowish, pale olive at base. *A.* superior, membranous, partly appendiculate at marg. of P. in young examples. *Vo.* marginate, at length torn, white, becoming brown at the edge. *G.* crowded, adnate with a tooth or adnexed, broader in front, white to faint yellowish-ivory.

Woods. Spring and autumn. $4\frac{3}{4} \times 5\frac{1}{4} \times \frac{5}{8}$. Compare **20**.

16. A. megalodactyla Sacc. (from its tall stem; Gr. *megas*, great, *dactulos*, a finger) *a c*.
P. reddish-grey or brownish; marg. not sulcate. *St.* solid, white. *A.* large. *G.* free, at length red, not olive-tinged.

Odour strong. Woods. Oct.–Nov. $4 \times 5\frac{1}{2} \times \frac{5}{8}$ in. Allied to **17** and to *Lepiota*.

17. A. lenticularis W. G. Sm., *Agaricus lenticularis* Lasch. (from its lentil-shaped stem) *a*.
P. tan or buff; marg. not sulcate. *St.* solid or slightly stuffed, tan-white. *A.* large. *G.* free, sometimes faintly olive.

Odour mouldy. Plantations, mixed. Oct. $4\frac{1}{2} \times 6 \times \frac{7}{8}$ in. Allied to **16** and to *Lepiota*.

II. **AMANITOPSIS** Roze

(From its relationship to *Amanita*.)

Veil universal, clothy, distinct from the cuticle of the pileus. After its rupture, as maturity is reached, it remains as a volva at the base of the stem and as broken patches on the pileus. *Hymenophore*

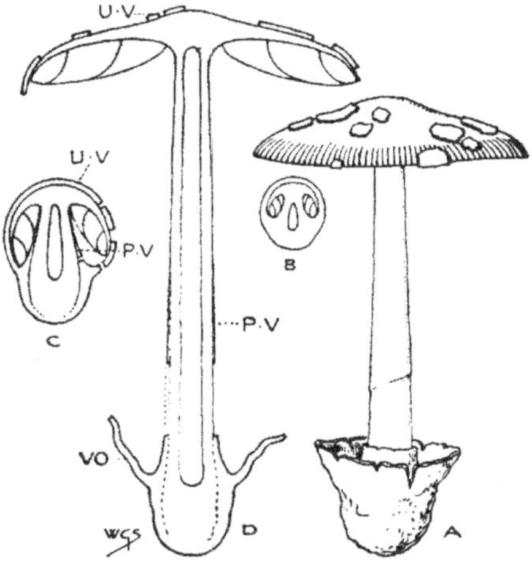

Fig. 11.—*Amanitopsis vaginata* Roze. One-third natural size. A, perfect plant; B, section of young example showing universal veil; C, section of half-grown example showing rupture of universal veil; D, section of mature example; U.V, universal veil; P.V, partial veil, adnate to stem; VO, volva.

distinct from the fleshy stem. *Pileus* convex, then expanded, not decidedly fleshy. *Stem* central, volvate, simple. *Gills* free, adnate in **20**, white or whitish. (Fig. 11.)

All the species grow on the ground. Species **18—20**

18. A. vaginata Roze (from the volva; *vagina*, a sheath) *a b c*.

P. brown; marg. sulcate. *St.* white or pale brown. *Vo.* large, free, lax, sheathing within.

Said to be edible, but sometimes produces perspiration and illness. Tasteless or insipid when raw, but cooking brings out a strong meat-like flavour; odour none. Woods; common. Sept.–Nov. $4\frac{1}{2} \times 6\frac{1}{2} \times \frac{3}{8}$ in. There are several varieties; *nivalis* W. G. Sm. (*Amanita nivalis* Grev.) is white.

18a. A. fulva W. G. Sm., *Agaricus fulvus* Schæff. (from the bright reddish-tawny colour) *b*.

P. sienna; mid. dark, sometimes almost black; marg. striate or sulcate. *St.* squamulose, pale sienna. *Vo.* large, free, lax, sienna.

Poisonous. Woods, birch. June–Sept. $3\frac{1}{4} \times 6\frac{1}{4} \times \frac{3}{8}$ in. *Amanita spadicea* Pers. is a brown form of this.

19. A. strangulata Roze (from the transverse constrictions of the stem) *a b c.*

P. mouse-colour, brown or yellowish-brown; marg. sulcate. *St.* brownish-white. *Vo.* not large, adpressed to St. *Flesh* whitish, tawny when broken.

Taste sweet; odour none. Woods; uncommon. May–Oct. $6\frac{1}{2} \times 8 \times \frac{7}{8}$ in.

20. A. adnata Sacc. (from the adnate gills) *a b c.*

P. firm, not brittle, buff-yellow; marg. not sulcate. *St.* minutely fibrillose, buff-white. *G.* crowded. *Vo.* not large, lax, irregularly torn, pubescent.

Woods and plantations, oak, holly; uncommon. Oct. $3 \times 3\frac{3}{4} \times \frac{1}{2}$ in. Compare **15a.**

III. **LEPIOTA** Quél.

(From the scaly pileus and stem; Gr. *lepis*, a scale.)

Veil universal, concrete with the cuticle of the pileus, breaking up in the form of scales, granules or silkiness. *Hymenophore* distinct from the fleshy stem. *Pileus* never compact, often very thin, at first plane,

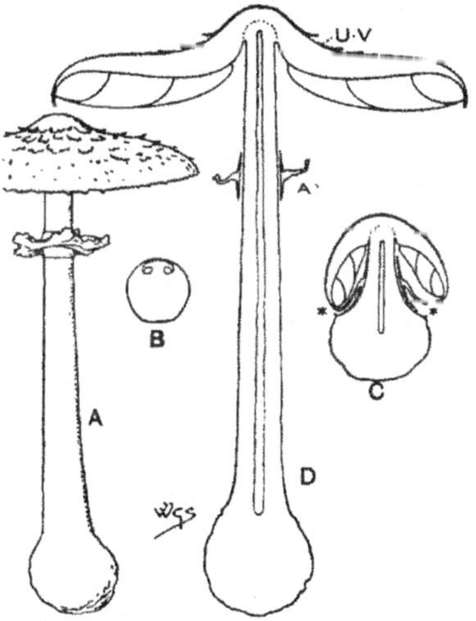

Fig 12.—*Lepiota procera* Quél. One-quarter nat. size.
A, perfect plant; B, section of young example showing universal veil; C, section of half-grown example showing at ** rupture of veil; D, section of mature example showing at A annulus free from the stem; U.V, remains of universal veil.

then scaly, granular, silky, smooth or viscid. *Stem* central, annulate, somewhat cartilaginous, firmer in texture than the pileus. *Gills* free or nearly so, except **46, 47,** and **50.** *Spores* normally white, sometimes yellowish, greenish or green. (Fig. 12.)

The species grow mostly in rich grassy places, more often in fields than woods; some occur in stoves and greenhouses. Nearly all are autumnal. Some are edible. Species **21—59**

A. Epidermis dry.

a. Proceræ. Scales of pileus brown or brownish on a white ground. *Stem* bulbous. *Annulus* more or less movable. *Gills* remote. Edible. The species are nearly allied; intermediate forms occur between all. **21—25**

b. Clypeolariæ. *Pileus* and usually the *stem* scaly or covered with granules. *Annulus* fixed. *Gills* free. Odour, except **29,** strong, disagreeable, pungent, sometimes weak. **26—37**

c. Annulosæ. *Annulus* superior, fixed, somewhat persistent; fugacious in **41** and **42.** **38—43**

d. Granulosæ. *Pileus* and lower part of *stem* covered with granules. *Annulus* inferior. *Gills* approaching stem, adnate in **47.** **44—48**

e. Mesomorphæ. *Pileus* dry, entire, not torn, not usually granular, small, slender. **48a—54**

B. Pileus smooth, viscid.

f. Viscidæ. **55—59**

a. Proceræ.

21. L. procera Quél. (*procerus*, tall) *a b c.*

P. umbonate, finely scaly. *St.* minutely scaly. *G.* subremote. *Flesh* unchanging white.

Solitary or in troops. Edible. Taste and odour pleasant, delicate. Pastures, parks, gardens, heaths, among bracken, etc.; common. July–Oct. 8 × 10⅝ × ⅝ in.

22. L. rachodes Quél. (from the ragged pileus; Gr. *rakos*, a ragged garment) *a b c.*

P. subglobose, coarsely scaly. *G.* subremote, sometimes reddening when bruised. *Flesh* changing to saffron-reddish when broken.

Gregarious or cæspitose. Edible. Taste and odour pleasant, strong, varying to rank. Woods and hedges, fir-plantations, greenhouses; common. July–Dec. 5½ × 6 × 1 in. The form *puellaris* is smaller and shining white. Bears the same relation to **21** as the horse-mushroom does to the pasture-mushroom. There is a white variety.

22a. L. prominens Sacc. (from the prominent umbo) *a b.*

P. sphærico-expanded, strongly and subacutely umbonate; sc. imbricate on um.; brownish or dull pale ochreous. *St.* abruptly bulbous at the base.

Edible. Fields. Oct. 8⅝ × 4⅜ × ½ in.

22b. L. permixta Barla (from its mixed characters) *a.*

P. campanulato-convex, slightly subumbonate, then expanded, smooth then fibrillose; sc. opaque, somewhat pale brown on a brownish-ivory ground; marg. sublacerate, white. *St.* hollow, enlarged at base, irregularly scaly, colour as P. *Flesh* becoming faintly rufous. *A*. membranous, fibrillose, lax. *G*. free, ventricose, crowded, ivory or faintly flesh-coloured.

Odour weak. Hilly pastures. Autumn. $7 \times 4\frac{1}{8} \times \frac{1}{2}$ in. With the mixed characters of **21**, **22**, and **23**.

23. **L. excoriata** Quél. (from the cracked membrane of the pileus) *a b c*.
P. semiglobose, slightly scaly, pale buff. *A*. less movable than in **21** and **22**. *Flesh* unchanging white.

Said to be edible. Taste and odour usually delicate and agreeable, but varying to rank and unpleasant. Pastures and hedge-sides; less common than **21** and **22**. May–Oct. $4 \times 4 \times \frac{5}{8}$ in.

24. **L. gracilenta** Quél. (from its slender stem) *a b c*.
P. subumbonate, slightly scaly; sc. broad, adpressed, pale brown. *A*. slightly movable. *Flesh* unchanging white.

Often cæspitose. Edible. Pastures, under beeches; uncommon. June–Nov. $4 \times 7 \times \frac{3}{8}$ in.

25. **L. mastoidea** Quél. (from the resemblance of the pileus to a breast; Gr. *mastos*) *a b c*.
P. acutely umbonate, whitish; sc. minute, dark brown, varying whitish. *St*. brownish-white.

Woods. Sept.–Oct. $2 \times 3\frac{1}{4} \times \frac{3}{16}$ in.

b. Clypeolariæ.

26. **L. Friesii** Quél. (after Elias Fries) *a b c*.
P. tomentoso-scaly, rich dark brown. *St*. pale brown. *A*. superior. *G*. crowded, branched.

Odour heavy, strong. Pastures, grassy places, gardens, sawdust, tree-roots, cinder-heaps. Aug.–Oct. $4\frac{1}{4} \times 4 \times \frac{5}{8}$ in.

27. **L. acutesquamosa** Gill. (from the sharp pointed scales of the pileus) *a b c*.
P. lobed, ochreous buff-brown; sc. dark brown. *St*. pale brown. *A*. large. *G*. free, but close to stem, branched.

Sometimes cæspitose. Said to be edible. Taste mild, insipid; odour slight, mephitic, disagreeable. Gardens, yards, hot-beds, melon-frames, churchyards, greenhouses. Sept.–Nov. $4\frac{1}{2} \times 4\frac{3}{4} \times \frac{3}{4}$ in.

28. **L. Badhami** Quél. (after C. D. Badham) *a b c*.
P. pale crimson-brown; sc. small velvety, dark purple-brown. *St*. crimson-brown or deep purple-brown below, paler above. *G*. at length black edged. *Flesh* changing to saffron-crimson when broken.

Taste none or insipid; odour strong, usually unpleasant. Under yews, hawthorns, hedges. Sept.–Nov. $3\frac{3}{4} \times 5 \times \frac{3}{8}$ in.

29. **L. emplastrum** Sacc. (from the plaister-like scales) *a b*.
P. yellow-buff; sc. large, smooth, brown. *St*. dull pale buff-yellow. *Flesh* pink when broken.

Gregarious. Taste and odour none. Amongst grass under trees. Oct. $3 \times 3\frac{1}{4} \times \frac{1}{2}$ in.

30. L. MELEAGRIS Quél. (from its markings, as of the guinea-fowl, *meleagris*) *a b c.*

P. pale salmon; sc. minute, almost black. *St.* bulbous-rooting, black-spotted, blackish at base. *G.* white or lemon-yellow. *Flesh* bright red when broken, reddish when dry.

Hothouses, greenhouses, on spent tan; rare. May–Oct. $2\frac{1}{4} \times 2\frac{1}{2} \times \frac{3}{16}$ in.

31. L. BIORNATA Sacc. (from its twofold ornamentation) *a c.*

P. whitish; sc. dot-like, dark red. *St.* enlarged below, rooting, red-spotted. *A.* red-spotted at marg. *Flesh* white, reddish in St., dark when dry.

In hot-beds, stoves, melon and cucumber frames. July. $2\frac{1}{4} \times 4\frac{1}{4} \times \frac{5}{16}$ in.

32. L. hispida Gill. (from the rough pileus; *hispidus*, rough) *a b.*

P. velvety, rich brown. *St.* woolly-scaly, brown below A. *Flesh* unchanging white.

Odour disagreeable, somewhat pungent of radishes. Woods, chiefly pine. July–Oct. $3 \times 3 \times \frac{3}{8}$ in.

33. L. clypeolaria Quél. (from the shield-like pileus; *clypeus*, a shield) *a b c.*

P. floccoso-scaly, sometimes pale buff, rosy or rufescent. *St.* scaly, whitish. *Flesh* white.

Odour weak or none. Woods, stoves, greenhouses. Oct.–Nov. $2\frac{3}{4} \times 3\frac{3}{4} \times \frac{1}{4}$ in. Colour variable, sometimes wholly white. Sometimes very small.

Var. **alba** Bres. (from the dead-white colour) *b.*

P. fleshy, convex, then expanded, broadly umbonate, silky-white; mid. glabrous; marg. fibrilloso-flocculose, then squamulose, becoming yellowish with age. *St.* equal or attenuate downwards, white, somewhat fuscous at base, white-floccose and yellow-spotted below the distant, floccose, fugacious white A. *G.* free, crowded, white then ochreous. *Flesh* white, somewhat yellowish in stem. Often forming a spurious second A.

Edible. On the ground. Nov.

33a. L. felina Karst. (from the markings on the pileus; *felis*, a cat) *a b.*

P. subumbonate, black when young, then black-scaly on a white ground. *St.* white, often black-scaly below. *A.* large, often black-dotted.

Woods, fir, amongst moss. May–Sept. $1\frac{3}{4} \times 3 \times \frac{3}{16}$ in.

33b. L. nigromarginata Mass. (from the black-edged gills) *a b.*

P. campanulato-expanded, subumbonate; sc. umber on a pale sienna-ochre ground. *St.* hollow, attenuate upwards, smooth, salmon-white. *A.* persistent, membranous, white. *G.* broader in front, whitish; edge bordered with dark umber. *Flesh* thin.

Amongst grass. May. $2 \times 2\frac{3}{4} \times \frac{1}{10}$ in. Allied to **33**, and especially to **33a**, but differing from both in the dark-edged gills.

34. L. metulæspora Sacc. (from its obelisk-shaped spores; *metula*, an obelisk) *a b c*.

P. umbonate, white; um. and sc. pallid; marg. sulcate. *V.* appendiculate. *St.* pallid or lemon-colour within and without.

Odour weak or none. Woods, fir, amongst moss and leaves; rare. Sept.-Oct. $1\frac{1}{2} \times 2\frac{1}{2} \times \frac{3}{16}$ in.

35. L. cristata Quél. (from the scaly-crested pileus) *a b c*.

P. white; sc. and mid. bright dark brown. *St.* not scaly, brown at base. Sometimes red when broken.

Suspected poisonous. Taste strong, disagreeable; odour pungent, mephitic, sulphurous, or like radishes, but noxious. Fields, orchards, gardens, lawns, etc.; common. Aug.–Nov. $2\frac{1}{2} \times 3\frac{1}{2} \times \frac{3}{16}$ in. Sometimes red when broken.

36. L. erminea Gill. (from the ermine-like pileus) *a b c*. Wholly white. Very fragile.

P. silky.

Taste radish-like or none; odour slight. Grassy places. Sept.–Nov. $2 \times 3\frac{1}{8} \times \frac{3}{16}$ in.

37. L. MICROPHOLIS Sacc. (from the minute scales on the pileus; Gr. *mikros*, small, *pholis*, a scale).

P white; sc. radiating, erect, dark grey or black; marg. substriate. *St.* white.

Stoves on coco-nut fibre. $\frac{3}{4} \times 1 \times \frac{1}{16}$ in.

c. Annulosæ.

38. L. Vittadinii Quél. (after Carlo Vittadini, Italian mycologist) *a b*. Wholly white or whitish.

P. rough with pointed sc. *St.* concentrically scaly. *A.* large.

Suspected poisonous. Groves, pastures, downs. June–Aug. $5\frac{1}{2} \times 6\frac{3}{4} \times \frac{7}{8}$ in. Allied to *Amanita.*

38a. L. nympharum Karst. (from its beauty, as of a wood-nymph) *a b*.

P. umbonate, covered with large revolute scales, white. *St.* white, or faintly buff-stained below. *G.* crowded, ventricose, white.

Grassy places. Oct. $4 \times 4\frac{3}{4} \times \frac{1}{2}$ in.

39. L. holosericea Gill. (from the silky pileus; Gr. *holos*, entire, *serikos*, silken) *a b c*. Buff-white. Substance very soft.

P. fibrilloso-silky, then even. *St.* longitudinally fibrillose, then even. *A.* large.

Edible. Odour none. Gardens, cultivated places. Sept.–Oct. $3\frac{3}{4} \times 2\frac{3}{4} \times \frac{5}{8}$ in.

40. L. naucina Quél. (from the nut-like flavour; *nucinus*, of a nut) *a b c*. Buff-white.

P. granular; marg. not plicate, appendiculate with V. *A.* large. *G.* becoming faintly rosy. *Flesh* white, usually becoming pale salmon or salmon-brownish when cut.

Somewhat cæspitose. Edible. Taste mild, pleasant; odour strong. Fields; rare. July–Oct. $5 \times 3\frac{1}{2} \times \frac{5}{8}$ in. Compare 790.

41. L. CEPÆSTIPES Quél. (from the onion- or chive-shaped stem; *cepa*, an onion, *stipes*, a stem) *a b c*. White, pale lemon, or sulphur.

P. floccose; marg. plicate. *St*. floccose. *A*. fugacious.

Gregarious. Stoves, frames, pine-pits, on tan and leaves, melon-beds, etc.; common. Mar.–Sept. $2\frac{3}{4} \times 4\frac{3}{4} \times \frac{3}{8}$ in. Sometimes becomes vinous or claret-colour in drying.

Var. *cretacea* Sacc. *P*. white; sc. darker.

42. L. citrophylla Sacc. (from the lemon-tinted gills) *a c*. Lemon-yellow.

P. sc. rufous; um. rich-brown; marg. not crenate or sulcate. *St*. squamulose. *A*. fugacious. *Flesh* and *G*. sometimes change to green.

On the ground; rare. Oct. $1\frac{1}{4} \times 1\frac{3}{4} \times \frac{1}{8}$ in.

43. L. LICMOPHORA Sacc. (from the sulcate pileus: a fancied resemblance to a winnowing fan, Gr. *likmos*) *a b c*.

P. lemon-yellow. *A*. small. *G*. remote.

Hothouses, conservatories, etc., on the ground, on coco-nut fibre. Aug.–Oct. $2\frac{1}{8} \times 4\frac{3}{4} \times \frac{3}{16}$ in. Resembles a *Bolbitius* or a yellow *Hygrophorus*. Sometimes wholly sulphur-white. At one time regarded as a var. of **41**.

d. Granulosæ.

44. L. carcharias Karst. (from a fancied resemblance in the granules of the pileus to the teeth of a dog-fish, Gr. *karcharias*) *a b c*.

P. Flesh-colour; gr. darker. *A*. granulose.

Taste disagreeable, bitter or none; odour faint to strong of meal. Woods, fir. May–Nov. $2 \times 1\frac{3}{4} \times \frac{3}{16}$ in. Sometimes very small.

45. L. cinnabarina Karst. (from the cinnabar or vermilion colour) *a b c*.

P. granular, often coarsely so; marg. fimbriate with V. *St*. granular below A., orange-vermilion.

Taste mild. Woods, fir, grassy places. Sept.–Nov. $2\frac{1}{2} \times 2\frac{1}{2} \times \frac{5}{16}$ in. *Agaricus Terreyi* B. & Br. is a form of this species.

46. L. granulosa Quél. (from the granular pileus) *a b c*.

P. floccoso-granular, red-brown. *St*. granular-scaly, red-brown below. *G*. slightly adnexed, sometimes adnate.

Woods, fir, grassy places; very common. July–Nov. $2 \times 2\frac{1}{2} \times \frac{3}{16}$ in. There is a white var. *rufescens* Sacc., which becomes red in drying.

46a. L. atrocrocea Mass. (*ater*, black, *croceus*, saffron) *a b*.

P. expanded, slightly depressed, bright salmon-orange, more or less covered with purple-brown almost black granular flocci. *St*. hollow, attenuate upwards, bright salmon-orange; salmon-brown-squamulose. *A*. fugacious. *G*. broadly adnate, salmon-white. *Flesh* thin; salmon-orange-brown.

Oct. $1\frac{1}{2} \times 1\frac{1}{4} \times \frac{1}{8}$ in. Allied to **46** and **47**.

47. L. amianthina Karst. (from its pure colour; Gr. *amiantos*, undefiled) *a b*.

P. ochraceous; mid. orange. *St*. squamulose, yellow. *A*. fugacious. *G*. adnate or slightly decurrent, becoming light yellowish. *Flesh* wholly or partially yellow.

Woods; uncommon. Aug.–Nov. $1\frac{1}{2} \times 2 \times \frac{1}{8}$ in. A form occurs intermediate between this and **46**.

Var. *Broadwoodiæ* Sacc. *P*. hemispherical, tomentose, yellow; *G*. white.

48. L. polysticta Gill. (from the many-spotted pileus; Gr. *polus*, many, *stiktos*, spotted) *a b c*.

P. fleshy, livid-brownish; sc. small, red-brown. *A*. fugacious. *G*. faintly yellowish.

Taste insipid; odour none or mephitic, like **35**. Grassy places, roadsides, etc.; uncommon. Sept.–Nov. $2 \times 1\frac{3}{4} \times \frac{3}{8}$ in.

c. Mesomorphæ.

48a. L. parvannulata Gill. (from the small annulus) *a*.

P. umbonate or gibbous, at first slightly pruinose, white to faintly yellowish. *St*. fibrillose below A., white. *A*. spreading, subpersistent.

Pastures. Autumn. $\frac{7}{8} \times 1\frac{1}{2} \times \frac{1}{10}$ in.

49. L. sistrata Quél. (from a fancied resemblance to a rattle, *sistrum*) *a b*.

P. shining-atomate, whitish, becoming light yellowish or flesh-colour; marg. appendiculate with V. *St*. pruinoso-silky, white. *A*. fibrillose.

Sandy ground amongst grass, gardens, etc. Oct. $1 \times 2 \times \frac{1}{8}$ in.

50. L. seminuda Gill. (from the partially mealy pileus and stem; *semi*, half, *nudus*, naked) *a b c*. Whitish or flesh-coloured.

P. floccoso-mealy then naked, usually white. *St*. mealy and flesh-coloured below, naked and white above. *G*. adnate

Odour none. Woods. May–Nov. $\frac{5}{8} \times 1\frac{1}{2} \times \frac{1}{16}$ in.

51. L. Bucknalli Sacc. (after Cedric Bucknall) *a c*. White.

P. and *St*. minutely granular, lilac.

Odour strong of gas-tar. Oct. $1\frac{1}{8} \times 2\frac{1}{8} \times \frac{1}{8}$ in.

52. L. mesomorpha Gill. (from its intermediate characters; Gr. *mesos*, middle, *morphe*, form) *a b c*. Whitish, yellowish, brownish or brown.

P. and *St*. smooth. *A*. entire, persistent.

Woods. Oct.–Dec. $1\frac{1}{8} \times 1\frac{3}{4} \times \frac{1}{16}$ in.

53. L. MARTIALIS Sacc. (from its red colour) *a b*.

P. fibrillose, ochreous-red; marg. striate. *St*. ochreous above, rufous below.

Conservatories, tree-fern stems. March. $1\frac{1}{8} \times 1\frac{3}{8} \times \frac{1}{16}$ in.

54. L. IANTHINA Sacc. (from the violet pileus; Gr. *ion*, a violet) *a*.

P. fibrillose; mid. darker; marg. not striate.

Stoves. March. $1 \times 1\frac{1}{8} \times \frac{1}{8}$ in.

54a. L. submarasmioides Sacc. (from its slight resemblance to a species of *Marasmius*) *a b.*

P. umbonate, tough, arachnoid-squamulose, white to pale buff; um. rufous; marg. subplicate, cracking. *St.* at first arachnoid-pruinose, white. *A.* superior, movable. *G.* membranous-tough. *Flesh* tough, rufous within towards base.

Low ground at Worcester. Sept. 1 × 2⅛ × ⅛ in.

f. Viscidæ.

55. L. medullata Gill. (from its pith-bearing stem; *medulla*, pith) *a b.* White or grey-white.

P. umbonate; marg. appendiculate with V. *St.* with a removable pith.

Odour of radish or mephitic like **35.** Woods, fir. Aug.–Oct. 2 × 3½ × ¼ in. Compare **58.** Resembles **1188** in stature, colour and viscidity of pileus.

56. L. glioderma Gill. (from the clammy pileus; Gr. *gloios*, clammy, *derma*, skin) *a c.*

P. brownish-red. *St.* dry, floccoso-scaly below, white or buff-white.

Woods, thickets, pine; uncommon. July–Sept. 1⅝ × 3⅛ × ¼ in.

57. L. delicata Gill. (delicate) *a b.*

P. thin, subumbonate, rufescent, yellowish, pale rose or brownish. *St.* dry, floccoso-scaly below, whitish.

Taste and odour sometimes rank, like *Polyporus squamosus.* Hothouses, woods, stumps, etc. Sept. 1¼ × 1 × ⅛ in.

58. L. illinita Quél. (from the glutinous pileus; *illinitus*, besmeared) *a b.*

P. umbonate white, varying to ochreous or clay-colour with a brown um. *St.* glutinous below. *A.* well developed on St., or appendiculate at edge of P.

Woods. Aug.–Sept. 2¼ × 3⅜ × $\frac{5}{16}$ in. Compare **55** and **1188.**

59. L. Georginæ Sacc. (after Georgina E. Johnstone) *a b.*

White, all parts becoming bright crimson when touched. *P.* viscido-pruinose. Brown when dry.

Taste and odour none. On sphagnum in greenhouses. May–Nov. 1¼ × 2⅛ × ⅛ in.

IV. HIATULA Mont.

(From the furrows on the pileus; Gr. *hio*, to gape.)

Hymenophore distinct from the stem. *Veil* universal, manifest in the granular or powdery covering, and sometimes in the appendiculate margin of the pileus. *Pileus* fleshy to membranous. *Stem* central, simple. *Gills* free. *Spores* normally white, sometimes greenish or green as in *Lepiota.* (Fig. 13.)

Differing from *Lepiota* in the absence of an annulus, and agreeing in structure with *Pluteus*, *Pluteolus*, and *Pilosace.*

Pileus somewhat fleshy. **60, 60a.**
Pileus membranous, sulcate. **60b, 60c.**

60. H. Eyrei W. G. Sm., *Schulzeria Eyrei* Mass. (after Rev. W. L. W. Eyre) *a b c*.

P. broadly umbonate, smooth, pallid; mid. minutely granular ochreous; marg. incurved, appendiculate with V. *St.* smooth,

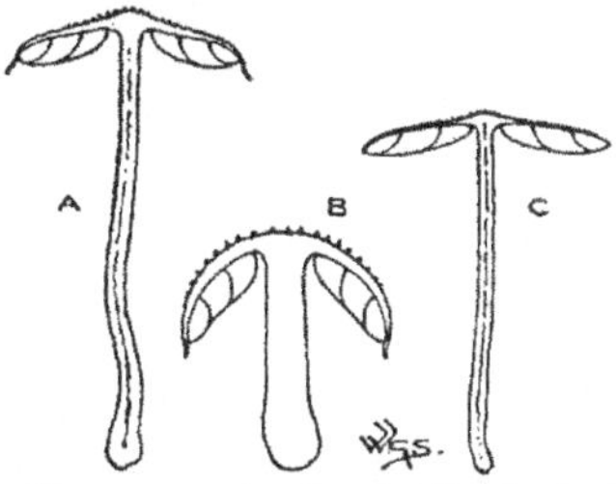

Fig. 13.—A, section through *Hiatula Eyrei* W. G. Sm.; B, ditto, *H. lycoperdoides* W. G. Sm.; C, ditto, *H. Wynniæ* B. & Br. One-half natural size.

even, glabrous, minutely mealy above, pallid. *G.* free, narrowed behind, thin, crowded then distant, pale green then deep bluish-green. *Spores* same colour as G.

Gregarious. Taste and odour none. Under spruce-firs. Autumn. 1½ × 2⅝ × ⅛ in.

60a. H. lycoperdoides W. G. Sm., *Schulzeria lycoperdoides* Cooke & Mass. (from a resemblance in the pileus to certain species of *Lycoperdon* as **2063** and **2064**) *a b*.

P. convex, then expanded, soft, buff-ivory, beset with pyramidal silky-fibrous dark brown warts; marg. appendiculate. *St.* solid, equal, minutely fibrillose, whitish; faintly buff-ivory below. *G.* somewhat crowded, white. *Flesh* white, pale buff-brown under cuticle of P.

Taste and odour none. Under cedars. Autumn. 1¼ × 1½ × 1¼ in.

60b. H. Wynneæ B. & Br. (after Mrs. Lloyd Wynne).

P. campanulate then plane, subumbonate, pulverulent, white; mid. brownish. *St.* hollow. *G.* free, at first very slightly adnexed, somewhat distant, with a shrivelled appearance as in **41.** Has been observed emitting a greenish phosphorescent light.

Stoves, on wood.

60c. H. grangei W. G. Sm., *Schulzeria grangei* Eyre (from the original locality, Grange Park, Hants).

P. flatly umbonate, cracking into dark green fibrous scales on a white ground. *St.* squamose, sq. tipped with colour of P. *G.* widest in front, minutely denticulate.

Cæspitose. On soil amongst beech-leaves. Nov. 1⅜ × 2⅜ × ⅛ in.

V. **ARMILLARIA** Quél.

(From the annulus; *armilla*, a ring.)

Veil universal, imperfect, remaining as minute, concrete warts or flocci on the pileus and as the scales of the stem. *Hymenophore* confluent and homogeneous with the fleshy stem. *Pileus* usually

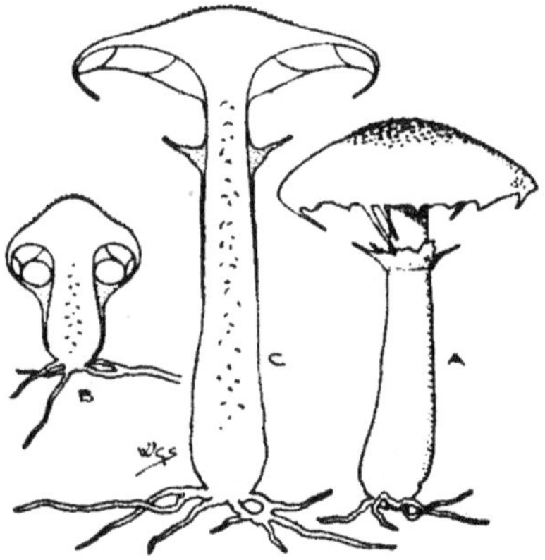

Fig. 14.—*Armillaria mellea* Quél. One-third natural size. A, perfect plant; B, section of young example showing universal veil; C, section of mature example.

fleshy, the surface broken up into small scurfy warts or floccose scales. *Stem* central, of the same substance with the pileus, annulate. *Gills* sinuato-adnexed, adnate or decurrent. *Spores* white, elliptical, smooth. (Fig. 14.)

The species grow on the ground or on or about stumps.

Must not be confounded with the last section of *Amanita*.

Species **61—72**

Stem furnished with a fugitive annulus. *Gills* sinuato-adnexed. **61—68**

Stem furnished with a sometimes fugitive annulus. *Gills* more or less decurrent. **69—71**

Stem annulate. *Gills* adnate, varying rounded, sinuate or sinuato-decurrent. **72**

61. A. bulbiger Quél. (from the bulbous stem) *a b c*.

P. subumbonate, nearly smooth, pale yellowish or buff. *St.* strongly bulbous, whitish, longitudinally black fibrillose. *G.* becoming pale yellowish. *Flesh* whitish.

Taste and odour none. Woods, pine. Oct.-Nov. $3\frac{1}{2} \times 3\frac{1}{2} \times \frac{1}{2}$ in.

62. A. focalis Gill. (from the thick clothy annulus; *focale*, a neck-cloth) *a b*.

P. umbonate, silky-fibrillose, rich yellowish-red, ochre or livid yellow. *Flesh* pale yellowish-brown.

Woods, pine, heaths, bare ground under old laurels. Aug.–Oct. $4\frac{1}{2} \times 3\frac{1}{2} \times \frac{7}{8}$ in. The dimensions of the variety called *Goliath* by Fries are $6\frac{3}{4} \times 6\frac{3}{4} \times 1\frac{1}{2}$ in.

63. A. robusta Gill. (from its sturdy growth) *a b c*.

P. convex, nearly smooth, rich brown. *St.* attenuate downwards, not scaly, brownish-white. *Flesh* hard.

Odour of new meal, but nauseous. Woods and plantations, pine, maple; rare. Oct. $2\frac{1}{2} \times 2 \times \frac{3}{4}$ in. The form *major* is twice the size of *minor*, the form illustrated.

63a. A. caligata Gill. (from a fancied resemblance in the stem to a leg with a soldier's shoe, *caliga*) *b*.

P. compact, convex, then plane, tawny, weasel-colour, spotted with adpressed silky sq. of the same colour. *St.* solid, zoned with brown sq. below the persistent membranous A. *G.* emarginate, white.

Odour strong. On the ground in woods, pine. Autumn. $3\frac{3}{4} \times 3\frac{3}{8} \times 1\frac{1}{8}$ in. *Agaricus causetta* Barla, is a form of this species.

64. A. aurantia Quél. (from its orange colour) *a b*.

P. convex, innato-squamulose, viscid. *St.* very scaly. *Flesh* yellowish.

Woods, pine. Oct. $2\frac{3}{4} \times 3\frac{3}{4} \times \frac{1}{2}$ in.

65. A. ramentacea Quél. (from a fancied resemblance in the pileus to wood-shavings, *ramenta*) *a c*.

P. floccose becoming revolute, whitish, then yellowish or brownish; sc. and punctate mid. darker. *St.* white, clad with adpressed brown sc. below A. *G.* becoming yellowish. *Flesh* white.

Odour unpleasant. Under pines; uncommon. Sept. $2\frac{1}{2} \times 2\frac{1}{2} \times \frac{5}{16}$ in. Allied to *Lepiota*.

66. A. hæmatites Sacc. (from the colour of the pileus; Gr. *haima*, blood) *a*.

P. hemispherical, red liver-colour. *St.* liver-colour. *Flesh* pale liver-colour, darker at base of St.

Amongst fir-leaves. Nov. $1\frac{3}{8} \times 1\frac{5}{8} \times \frac{1}{8}$ in. Allied to *Lepiota*.

67. A. Jasonis Sacc. (from a fancied resemblance in the pileus to the golden fleece of Jason) *a b*.

P. campanulate, granular, golden-yellow; mid. reddish. *St.* golden-yellow.

Cæspitose. Taste none; odour usually none, but sometimes strong. Stumps. Sept. $2 \times 2\frac{1}{4} \times \frac{5}{16}$ in. Resembling *Pholiota*.

68. A. constricta Gill. (from the compact substance; *constrictus*, compact) *a*. White.

P. convex, silky-smooth. *Flesh* becoming pale yellow when bruised.

Odour of new meal. Pastures, sunny and urine-scorched places. Sept.–Oct. $2 \times 1\frac{3}{4} \times \frac{3}{8}$ in.

69. A. mellea Quél. (from the honey-coloured pileus; *mel*, honey) *a b c*.
P. smoky yellow-olivaceous, usually dark-scurfy or squamulose; marg. striate. *St.* often sulphur-yellow at base. *A.* usually large and white-floccose, sometimes small or absent. *G.* usually yellowish or pale salmon.

Solitary, gregarious, or cæspitose. Taste acrid and disagreeable; odour strong, foxy. Sold as an esculent in Continental markets; said by Mrs. Hussey to be poisonous. On or near stumps, tree-roots or root-fibres, squared timber. July–Nov. $4\frac{1}{2} \times 5 \times \frac{5}{8}$ in. Extremely variable. Compare **222.** *Collybia tabescens* Sacc. is a ringless form.

Var. **versicolor** W. G. Sm., *Agaricus versicolor* With. (from the change of colour in the stem and gills).
St. bulbous, white then brown. *G.* yellow-white, then deep red-brown.

Amongst grass. July. $1–4 \times 2 \times \frac{1}{4}$ in.

Var. **laricina** W. G. Sm., *Agaricus laricinus* Bolt.
P. flesh-colour, glabrous; marg. smooth. *G.* white. On larch.

69a. A. denigrata Gill. (from the very dark or blackish pileus; *denigro*, to colour black) *a*.
P. subviscid, guttato-punctate; marg. even. *St.* solid, pallid-fuscous, whitish above. *A.* superior, entire fugitive, white. *G.* pale brown, becoming darker.

Solitary or gregarious. On the ground, damp, shady places, bases of trunks. $2\frac{1}{2} \times 2\frac{1}{2} \times \frac{5}{8}$ in. Sometimes confused with **560**, which has brown, not white spores.

70. A. citri Gill. (from the citron-colour of pileus) *a b*.
P. subumbonate, smooth, sulphur-yellow; marg. crenulate, becoming white. *St.* whitish, reddish-brown below.

Cæspitose. Odour of new meal. Stumps. Sept. $2 \times 3 \times \frac{3}{16}$ in.

71. A. subcava Gill. (from the partially hollow stem; *cavus*, hollow) *a*. White.
P. umbonate, thin; mid. brownish.

Woods, pine. Nov. $2 \times 3\frac{1}{2} \times \frac{3}{16}$ in. Allied to *Lepiota*.

72. A. mucida Quél. (from the slimy pileus) *a b c*. White.
P. hemispherical, radiato-wrinkled; mid. sometimes faintly brownish. *St.* externally cartilaginous. *G.* broad, mucid.

On beeches. Aug.–Nov. $3\frac{1}{2} \times 6 \times \frac{1}{4}$ in.

VI. TRICHOLOMA Quél.

(From the hairy, silky, or downy covering of the pileus, best seen in young examples; Gr. *thrix*, a hair, *loma*, a fringe.)

Veil universal, imperfect, manifest in scales, flocci, fibrils or down, which adhere to the margin of the pileus in infancy, sometimes obsolete or nearly so. *Hymenophore* confluent and homogeneous with the fleshy stem. *Pileus* usually convex and fleshy, never umbilicate. *Stem* central, fleshy, simple, with cartilaginous bark. *Gills* sinuate, sometimes separating from the hymeno-phore. (Fig. 15.)

The species, except **91, 98** and **149,** grow on the ground, the larger usually grow in upland woods, the smaller in pastures. Mostly autumnal; some late; section *Guttatæ* is usually vernal. Many have a pleasant odour of meal, a few have a disagreeable odour; some are edible, a few suspicious or poisonous. Species **73—152**

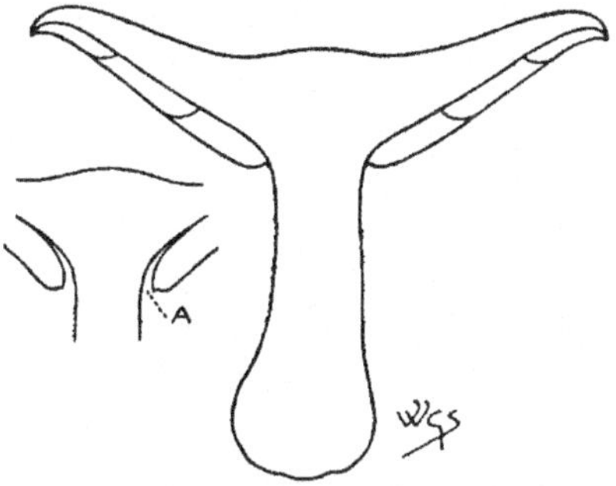

Fig. 15.—Section of *Tricholoma personatum* Quél. One-third natural size. The separation of the stem from the hymenophore is shown at A.

A. *Pileus* viscid when damp, pubescent, fibrillose or scaly, not scaly-torn. *Stem* fibrillose. *Flesh*, except **77**, not hygrophanous.

a. Limacinæ. Pileus fleshy. *Stem* solid, except **73a, 76**, and **79**. *Gills* not changing colour or becoming rufescent. **73—79**

Gills changing colour usually to rufescent, commonly red-spotted. **80—89**

b. Genuinæ. Pileus torn into scales or fibrils, not viscid or hygrophanous, moist in **94**. *Taste* sometimes bitter and disagreeable. *Odour*, except **93**, generally not unpleasant or none.

Gills, except **94**, not changing colour, or becoming rufous or black-spotted. **90—95**

Gills rufescent or becoming grey, usually marked with rufous or blackish spots. **96—101**

c. Rigidæ. Pileus, except **106**, punctato-granulate or broken up into smooth squamules, not viscid, scaly-floccose or fibrillose. *Flesh* in stout species cartilaginous-hard, in thin species fragile. *Odour* of the majority of the species unpleasant.

Gills unchanging white or becoming hyaline-whitish, except **103**, which often becomes reddish. **102—108**

Gills changing colour, rufescent, or becoming ashy or spotted. **109—112** *d. Sericellæ.* *Pileus* smooth, dry, silky, never scaly, moist or viscid, flocculose in **121.**

Gills broad, somewhat thick, subdistant, except **115.** *Odour* strong, fetid. **113—116**

Gills narrow, thin, crowded. *Odour* none, except **119.** **117—121**

B. *Pileus* smooth, moist in wet weather, not viscid, downy or scaly. *Flesh* soft, spongy, watery, hygrophanous.

e. Guttatæ. *Pileus* thick, fleshy. *Stem* obese, solid. *Odour* pleasant. Usually vernal; often growing in troops, or cæspitose or forming large rings.

Gills whitish. **122—124**

Gills changing colour, rufescent or smoky. **125—127**

f. Spongiosæ. *Pileus* obtuse, smooth, moist, except **133**; not hygrophanous. *Gills* sinuato-decurrent. Often late autumnal. Usually growing in troops. Must not, on account of the sinuato-decurrent gills, be confounded with *Clitocybe.*

Gills whitish. **128—134**

Gills changing colour. **135—142**

g. Hygrophanæ. *Pileus* hygrophanous, changing colour on parting from its absorbed water.

Gills white or faintly brownish, unspotted. **143—148**

Gills becoming violet, grey, or smoky. **149—152**

a. Limacinæ.

73. T. equestre Quél. (from its being the handsomest of the group; *eques*, a knight) *a b c.*

P. gibbous, then expanded, sulphur-olive or ochraceous, streaked olive-brown or olive-purplish. *St.* sulphur-olive, sulphur-white within. *G.* sulphur-olive.

Taste pleasant or none; odour none. Woods, fir; uncommon. Aug.-Nov. $4 \times 2\frac{3}{4} \times 1$.

73a. T. coryphæum Gill. (from its position as type of group; Gr. *koruphaios*, a leader).

P. white, shaded sulphur-white or sulphur, punctato-virgate with dull purplish scales. *St.* equal to obclavate, white, shaded sulphur-white. *G.* crowded, white, edged or clouded pale sulphur.

Under beeches. Nov. $2\frac{1}{2} \times 4 \times \frac{5}{8}$ in.

74. T. sejunctum Quél. (from the gills, separated from the stem; *sejunctus*, separated) *a b c.*

P. at first umbonate, buff or yellow, streaked brownish or purple-sooty. *St.* white, clouded pale yellow. *G.* white.

Taste disagreeable, bitter; odour strong of rancid meal. Woods, chiefly pine; uncommon. Sept.-Nov. $4\frac{1}{4} \times 4\frac{3}{4} \times \frac{7}{8}$ in.

75. T. portentosum Quél. (from its variability; *portentosus*, strange) *a b c.*

P. unequal, slaty-brown, black-fibrillose; mid. darker. *St.* white or greyish. *G.* somewhat distant, white, pale grey or yellowish.

Solitary, gregarious or cæspitose. Taste mild or somewhat nauseous; odour none. Woods, fir; frequent. Aug.–Nov. $4\frac{1}{2} \times 3\frac{1}{2} \times \frac{7}{8}$ in. Pileus sometimes 8 in. in diameter. Sometimes springing from an involved mass of thick cord-like mycelium.

76. T. fucatum Gill. (from the dyed appearance of the pileus; *fuco*, to dye) *a b c.*

P. yellowish or palest yellow, streaked brownish-buff or pale slate. *St.* yellowish, sometimes white above and below, grey- or blackish-squamulose or fibrillose. *G.* whitish.

Woods, chiefly pine; frequent. Aug.–Nov. $3\frac{1}{4} \times 3\frac{3}{4} \times \frac{5}{8}$ in.

77. T. quinquepartitum Gill. (from the pileus being supposititiously five-lobed; the character does not however belong to this species) *a b.* Somewhat hygrophanous.

P. yellowish-buff; mid. darker, unstreaked. *St.* ivory-white, not squamulose. *G.* white.

Taste mild; odour none. Woods, pine; rare. Oct. $3\frac{3}{4} \times 4\frac{1}{4} \times \frac{5}{8}$ in.

78. T. resplendens Karst. (from its brightly shining appearance) *a b.* Shining white.

P. sometimes very faintly stained yellowish or brownish.

Gregarious. Taste and odour pleasant. Woods, beech, hazel, etc.; frequent. Sept.–Oct. $2\frac{3}{4} \times 2\frac{5}{8} \times \frac{1}{2}$ in. Resembles **1188** in general appearance.

79. T. spermaticum Gill. (derivation obscure) *a b.* White.

P. sometimes cream or faintest ochre-white. *G.* coarsely notched.

Poisonous. Odour strong like tallow, varying to cucumber, but stinking. Woods, fir. Oct. $3\frac{3}{4} \times 2\frac{1}{4} \times \frac{3}{4}$ in. Resembling **78.**

80. T. colossum Quél. (from its great size) *a b.*

P. rich brown, paler or whitish at marg. *St.* bulbous, reddish-brown with a white floccose apex. *G.* white or dull flesh-colour. *Flesh* hard, changing to tawny when broken.

Odour none or cheese-like. Under Scotch firs, elms. Oct. $8 \times 4\frac{1}{2} \times 1\frac{3}{4}$ in. Resembling **1490** in size and colour. A form occurs apparently between this and **83.** Boudier considers this species to belong to *Armillaria.*

81. T. nictitans Gill. (from its sporadic appearance; *nictitans*, winking) *a b.*

P. rich brown; mid. darker. *St.* enlarged below. *G.* yellowish.

Taste usually mild or sweet, slightly of mushrooms, sometimes disagreeable, but not bitter; odour none or sweet. Woods; rare. Sept. $5\frac{1}{4} \times 3\frac{1}{2} \times \frac{5}{8}$ in. Resembling **1458** in form, colour and viscidity.

82. T. fulvellum Gill. (from the tawny pileus; *fulvus*, tawny) *a c.*

P. dotted-wrinkled, rich tawny-brown; mid. darker. *St.* not enlarged below. *G.* white.

Odour none. Woods, beech. Oct.–Nov. $3\frac{1}{2} \times 3 \times \frac{1}{2}$ in.

83. T. flavobrunneum Quél. (*flavus*, light yellow, *brunneus*, brown) *a b c.*

P. streaked, tawny or rich brown; mid. darker. *St.* slightly enlarged below, rich brown or brownish. *G.* light yellowish. *Flesh* white or partially pale yellowish.

Gregarious, cæspitose or in troops. Odour strong of rank meal or cheese. Deciduous woods, especially birch, beech; amongst gorse. Sept.–Nov. $3\frac{1}{2} \times 2\frac{1}{4} \times \frac{5}{8}$ in. A form occurs apparently between this and **80.**

84. T. albobrunneum Quél. (*albus*, white, *brunneus*, brown) *a b c.*

P. streaked, rich ochreous to pale brown; mid. darker. *St.* not enlarged below, yellowish-brown, white above and below. *G.* white. *Flesh* permanently white.

Said to be edible. Taste usually mild, sometimes bitter; odour none. Plantations, fir. Sept.–Dec. $4 \times 2 \times \frac{3}{4}$ in.

85. T. ustale Quél. (from its scorched appearance; *ustus*, burnt) *a b c.*

P. dotted-wrinkled, rich- or purple-brown, becoming black-brown, unstreaked. *St.* slightly attenuate downwards, brownish-white. *G.* white. *Flesh* sometimes changing to faintest rose.

Odour none. Woods, chiefly pine; rare. Aug.–Nov. $3 \times 2\frac{3}{8} \times \frac{3}{8}$ in. Intermediate between **83** and **86.**

86. T. pessundatum Quél. (from the foot-trodden appearance in Sterbeek's illustration) *a b c.*

P. guttato-spotted, not streaked, brown. *St.* equal, pale ochre-brown. *G.* white.

Said to be edible, but suspected poisonous. Taste mild; odour strong of new meal. Woods, fir; rare. Sept.–Nov. $4\frac{1}{8} \times 2\frac{1}{2} \times 1$ in.

87. T. stans Gill. (from its standing distinct from **86**) *a b c.*

P. reddish-brown. *St.* equal, white. *G.* white. *Flesh* reddish under cuticle.

Woods. Sept.–Oct. $3\frac{5}{8} \times 4 \times 1$. Compare **106.**

88. A. Russula Gill. (from its resemblance to certain of the red *Russulæ*) *a b c.*

P. rose, or crimson-granulate, or rose-liver-purplish, variegated rose and ivory-white. *G.* white, palest rose or clouded as P.

Eaten in Austria. Taste and odour mild, pleasant. Under trees. Oct. $4\frac{5}{8} \times 2\frac{1}{2} \times \frac{7}{8}$ in. Perhaps allied to **1192.**

89. T. frumentaceum Quél. (from its odour of meal; *frumentum*, corn) *a c.*

P. whitish-buff; mid. brownish-buff. *St.* equal. *G.* white. *Flesh* white.

Odour strong of new meal. Sept.–Oct. $3\frac{1}{2} \times 1\frac{7}{8} \times \frac{5}{8}$ in.

b. Genuinæ.

90. T. rutilans Quél. (*rutilo*, to be reddish) *a b c.*

P. dark purple on a yellow ground. *St.* enlarged below, light sulphur, purple-floccose. *G.* bright yellow. *Flesh* sulphur to yellow.

Suspected poisonous. Odour none. Woods, pine, birch; base of trees; common. Aug.–Nov. $5\frac{1}{2} \times 4\frac{1}{4} \times 1$ in. Pileus sometimes 7 in. or more in diam.

91. T. **variegatum** Gill. (from the variegated pileus) *a b*.
P. reddish-purple on a yellow ground. *St.* slightly enlarged below, reddish-purple-floccose on a yellowish-white ground. *G.* yellowish-white. *Flesh* whitish to palest sulphur.

Rotten wood. June–Oct. $2\frac{1}{2} \times 2\frac{3}{4} \times \frac{3}{8}$ in. Resembling a small, attenuate, dull-coloured state of **90**.

92. T. **luridum** Quél. (from its lurid colour) *a b c*.
P. whitish-brown-livid, or ashy-yellowish; mid. darker. *St.* white. *G.* very crowded, whitish.

Taste mild; odour of new meal. Woods, pine, etc.; common. Sept.–Oct. $2\frac{1}{2} \times 2\frac{1}{2} \times \frac{1}{2}$ in. Compare **74** with distant gills and **103** with odour of soap.

93. T. **guttatum** Gill. (from its spotted pileus; *gutta*, a drop) *a*.
P. pale buff, mid. darker, clad with large brown granular sq.; marg. remotely sulcate, white or faint rose-floccose. *G.* white. *Flesh* white, pale rose under cuticle and at base of St.

Somewhat cæspitose. Taste and odour acid-bitter. Woods. Oct. $3 \times 2 \times \frac{5}{8}$ in. Resembling **99**.

94. T. **columbetta** Quél. (from its dove-colour; *columba*, a dove) *a b c*.
P. moist, white or faintly clouded buff-white; mid. sometimes brownish-grey. *G.* separating from stem.

Solitary. Said to be edible. Taste mild; odour none. Pastures, woods, birch, fir, beech. Aug.–Nov. $4 \times 3 \times \frac{3}{4}$ in. Sometimes red- or blue-spotted; sometimes exuding a deep bright permanent violet stain.

95. T. **scalpturatum** Quél. (from its scratched appearance; *scalpto*, to scratch) *a c*.
P. livid- or grey-brown on a whitish ground. *St.* and *Flesh* whitish. *G.* white to yellowish.

Woods, fir; uncommon. Sept.–Nov. $3\frac{1}{4} \times 3\frac{1}{4} \times \frac{3}{8}$ in. Differs from **101** in the yellowish, not grey gills.

96. T. **imbricatum** Quél. (from the scales of the pileus; *imbricatus*, overlapping) *a b c*.
P. rich reddish- or crimson-brown. *St.* crimson-brown, white-pulverulent above. *G.* crowded, at first white.

Scattered or in troops. Said to be edible. Taste when raw disagreeable. Woods, pine, oak. Sept.–Nov. $4\frac{1}{2} \times 2\frac{3}{4} \times \frac{3}{4}$ in.

97. T. **vaccinum** Quél. (from the reddish-brown colour; *vacca*, a cow) *a b c*.
P. warm-rose to crimson-brown. *St.* crimson-brown, not white above. *G.* somewhat distant, separating from St.

In troops. Taste bitter, disagreeable; odour strong, not unpleasant. Woods, fir. Aug.–Oct. $2\frac{3}{4} \times 2\frac{1}{4} \times \frac{5}{8}$ in.

98. T. **immundum** Quél. (*immundus*, discoloured) *a b c*.
P. grey-buff. *St.* greyish-buff. *G.* pearly-grey.

Cæspitose. Taste bitter, disagreeable; odour mealy to offensive. On sheep-dung; rare. Sept.–Oct. $3 \times 2\frac{3}{4} \times \frac{3}{8}$ in.

99. T. inodermum Gill. (from the fibrils of the pileus; Gr. *is*, *inos*, a nerve, *derma*, the skin) *a b*.

P. pink-buff; mid. brownish, brown-scaly. *St*. white, then partially reddish, mealy above. *G*. white, red-spotted when bruised. *Flesh* pale pink-buff.

Damp, dense pine-woods. 2¾ × 2¼ × ½ in. Resembling **93**.

100. T. gausapatum Quél. (from the shaggy pileus; *gausapa*, a shaggy woollen cloth) *a b*.

P. bullate, obtuse, wavy, woolly-fibrillose, somewhat dark-ashy-grey; edge at first incurved, white-woolly. *St*. solid, white, fibrillose. *G*. adnate, pale grey.

Taste mild; odour none. Woods; grassy places. Sept.-Oct. 3 × 2 × ⅝ in.

101. T. terreum Quél. (from its earthy colour; *terra*, the earth) *a b c*.

P. umbonate, dark purplish-grey or pale slate. *St*. grey, white-pruinose above. *G*. white to grey.

Solitary, gregarious or cæspitose. Taste slightly rancid; odour none. Woods, chiefly beech, pine; common. Aug.–Nov. 3¼ × 2⅝ × ⅝ in. Often becoming black in drying. Sometimes springing from rose-coloured mycelium or from rhizomorphoid threads. Var. *orirubens* Mass. *G*. red-edged. Var. *argyraceum* Mass. Wholly silvery-whitish. Var. *virescens* W. G. Sm., *Agaricus scalpturatus* Fr. var. *virescens* Cooke. Becoming yellowish-green. Var. *atrosquamosum* Mass. Black, scaly. Var. *chrysites* Mass. Becoming yellow.

101a. T. squarrulosum Bres. (*squarrulosus*, scaly) *a*.

P. becoming striato-hairy, fuscous to lurid tan; mid. black-squamulose. *St*. broadly striato-rough, dark grey-umber on a pale dull salmon ground. *G*. grey-white as in **101**.

Under oaks, hazel. Sept. 3⅛ × 3½ × ⅝ in.

101b. T. horribile Rea (*horribilis*, rough) *a*.

P. dark-fuscous-squamulose, the ground pink at maturity. *St*. enlarged below, white, smooth, hollow. *G*. emarginate, crowded; marg. uneven, whitish, becoming pinkish. *Flesh* white, becoming pinkish and fuscous-spotted.

Under beech. 4⅝ × 3¾ × 1 in.

c. Rigidæ.

102. T. macrorhizum Sacc. (Gr. *makros*, long, *rhiza*, a root) *a b*.

P. smooth, then tessulato-cracked, ochreous-buff. *St*. whitish-ochreous. *Flesh* slightly yellowish.

Taste mushroom-like or unpleasant; odour strong like **113**, cadaverous, like *Lilium auratum*, like carrion or fetid-sweet, varying pleasant to unpleasant. Nov. 4½ × 5¾ × 1¼ in. Pileus sometimes 8 in. or more in diam. with a long rooting stem in proportion.

103. T. saponaceum Quél. (from its soapy odour) *a b c*.

P. smooth, then olive-brown scaly. *St*. pale yellowish-grey, often minutely black scaly. *G*. distant, yellowish-grey-white.

Woods, hedgesides, etc., amongst leaves; common. Aug.-Nov. 3½ × 2¾ × ⅝ in. Often reddish when bruised. Must not be confounded with the scentless **112b**.

104. T. cartilagineum Quél. (from its gristly substance) *a c*.

P. black-dotted on a brownish ground; mid. darker. *St.* smooth, whitish. *Flesh* white. *G.* crowded, white or pale greyish.

Odour none. Grassy places in woods and pastures, amongst pines; uncommon. Aug.–Nov. 3½ × 2¼ × ⅜ in. Compare **112a.** An abnormal growth of this, 4 ft. 1 in. × 2 ft. 1 in., with many pilei, raised the pavement of Goswell Road, London, in 1865.

105. T. tenuiceps Mass. (from the thin flesh of the pileus; *tenuiceps*, thin-headed) *a*.

P. conico-hemispherical, granular, dark smoky-brown. *St.* punctato-granulose, ochreous-white. *G.* subdistant, ventricose, white.

Solitary or in clusters of two or three, springing from thick cord-like mycelium. Amongst grass under trees. July. 2½ × 1¾ × ⅜ in. Resembling **227** (var. *repens*).

106. T. loricatum Gill. (from the leathery cuticle; *lorica*, a leather cuirass) *a c*.

P. unusually lobed, moist or subviscid, the thick cuticle peeling off, rich brown; marg. paler. *St.* reddish-buff. *G.* whitish straw-colour.

Odour strong, pleasant or nauseous. Woods, mixed. Sept. 2⅜ × 2¼ × ½ in. Compare with **87.**

107. T. atrocinereum Quél. (*ater*, black, *cinereus*, ash-coloured) *a*.

P. ashy-brown or blackish; mid. darker. *St.* whitish. *Flesh* hygrophanous.

Odour of new meal. Grassy ground. Oct. 1¾ × 1¾ × ¼ in.

108. T. cuneifolium Gill. (from the wedge-shaped gills; *cuneus*, a wedge, *folium*, a leaf) *a b c*. Fragile.

P. smooth, brittle, often concentrically cracked, rich brown to livid. *St.* slightly attenuate downwards, splitting, pale brownish, white above. *G.* distant.

Said to be edible. Taste disagreeable, rancid; odour strong of new meal. Pastures; frequent. Sept.–Nov. 1¾ × 1¼ × ⅜ in. *Agaricus cinereo-rimosum* Batsch, a form of this, is British.

109. T. crassifolium Sacc. (from its thick gills; *crassus*, thick, *folium*, a leaf) *a*.

P. ochraceous or somewhat buff; mid. umber. *St.* pale ochreous.

Gregarious or subcæspitose. Odour strong. Woods, fir; uncommon. Oct. 2½ × 1¾ × ⅜ in.

109a. T. sudum Quél. (*sudus*, dry) *a*.

P. ashy-rufous or blackish-umber; marg. paler. *St.* punctate with minute squamules, pallid, shaded rufous. *G.* whitish, edge rufescent.

Woods. Amongst grass. 3¾ × 3½ × ⅜ in. Compare **131.**

110. T. tumidum Gill. (*tumidus*, swollen) *a b*.

P. irregular, often cracked, ashy-livid; mid. darker. *St*. often swollen and attenuate below, white, often red-tinged within.

Taste and odour weak, not unpleasant. Woods, pine, moist places. Oct.–Nov. $3\frac{1}{4} \times 3\frac{3}{4} \times \frac{5}{8}$ in. Var. *Keithii* Sacc. (after the Rev. Dr. J. Keith) the whole plant becomes partially rufescent.

111. T. murinaceum Gill. (from its mouse-colour; *mus*, a mouse) *a b c*. Ashy- or lilac-greyish.

P. cracked, streaked grey-silky-scaly; sc. sometimes blackish. *St*. more or less scaly like P. *G*. grey.

Taste very disagreeable, bitter; odour strong, unpleasant, sometimes nitrous. Open woods and pastures; rare. Aug.–Nov. $4\frac{1}{8} \times 3\frac{1}{2} \times \frac{3}{4}$ in. Must not be confounded with **1244**.

111a. T. hordum Quél. (perhaps from *hordeum*, barley, on account of the hard substance).

P. subumbonate, soon cracking, in dry weather squarrose, mouse-grey; mid. darker. *St*. glabrous, whitish, shaded greyish. *G*. subdistant, white to greyish.

Taste mild; odour none. Under beeches. $3\frac{1}{2} \times 4\frac{3}{8} \times \frac{5}{8}$ in.

112. T. virgatum Gill. (from the streaked pileus; *virga*, a stripe) *a c*.

P. pale ashy-grey, finely-streaked black-fibrillose. *St*. subbulbous. *G*. crowded, becoming hoary. *Flesh* cinereous-whitish.

Usually solitary. Taste when young very bitter, becoming tasteless; odour usually none, sometimes sweet like melilot. Woods, mixed, oak, pine. Aug.–Oct. $3\frac{1}{4} \times 3\frac{5}{8} \times \frac{1}{2}$ in.

112a. T. elytroides Karst. (from the appearance of the pileus, like the wing-cases of certain beetles; Gr. *elutron*, a beetle's wing-case, *eidos*, appearance) *a b*.

P. expanded, obtuse, scabrous, with minute linear black-brown scales on a yellowish or pale brown ground, caused by the cracking of the cuticle. *St*. soft, densely brownish-fibrillose. *G*. ashy-grey or brownish-white. *Flesh* white.

Odour faint of new meal. Oct. $3\frac{1}{2} \times 3\frac{1}{4} \times \frac{5}{8}$ in. Compare **104**.

112b. T. opicum Gill. (*opicus*, clownish) *a*.

P. somewhat thin, obtusely umbonate, even, soon squamulose, at length unusually upturned and split, very dry, becoming minutely squamulose, grey. *St*. stuffed, fibrillose, becoming almost glabrous, pallid to greyish. *G*. arcuato-adfixed, somewhat thick, hoary. *Flesh* greyish.

Odour none. Woods, pine, amongst moss. $1\frac{1}{2} \times 3 \times \frac{1}{4}$ in. Must not be confounded with **103**, which smells strongly of soap.

d. Sericellæ.

113. T. sulphureum Quél. (from its sulphur-colour) *a b c*.

P. sulphur-yellow, sometimes rufescent-clouded. *St*. striate, sulphur. *G*. light sulphur. *Flesh* sulphur.

Gregarious. Probably poisonous. Odour strong and penetrating of gas-tar, fetid, like *Hemerocallis flava*, sometimes pleasant. Woods, mixed; common. Aug.–Nov. $3\frac{1}{2} \times 4 \times \frac{1}{2}$ in.

114. T. bufonium Gill. (from a fancied resemblance in the pileus to a toad's back; *bufo*, a toad) *a b c*.

P. dotted-wrinkled, yellowish-tan or buff; mid. rufescent. *St*. flocculose, yellow or yellow rufescent. *G*. yellow-tan.

Odour same as **113**, but fainter, sometimes not unpleasant. Woods, under pines. Sept.–Nov. $3 \times 2\frac{3}{4} \times \frac{1}{2}$ in.

115. T. lascivum Gill. (from its many affinities; *lascivus*, playful, wanton) *a b*.

P. not umbonate, becoming depressed, cream to pale buff; mid. darker. *St*. ivory-white. *G*. separating from stem, crowded, white.

Taste strong, disagreeable, pungent, somewhat acid; odour like gas-tar, varying to fried mushrooms, but disagreeable, more or less like **113**. Mixed woods; frequent. May–Nov. $3\frac{1}{4} \times 2\frac{3}{4} \times \frac{1}{2}$ in. Not unlike **1188** in general appearance.

116. T. inamœnum Gill. (from its unpleasant odour; *inamœnus*, unpleasant) *a c*.

P. subumbonate, ivory-white. *G*. white or buff-white. *Spores* ochre-white.

Suspected poisonous. Odour, especially when young, as in **113**, but stronger and more fetid. Woods, amongst pine-leaves. Sept.–Nov. $2\frac{1}{4} \times 3\frac{1}{2} \times \frac{3}{8}$ in. Sometimes confounded with **115**. Resembling **1188** in general appearance.

117. T. cerinum Quél. (from its colour, like yellow wax; *cera*, wax) *a c*.

P. yellow, then fuscous-brown. *St*. yellow, base often brownish, or wholly fuscous. *G*. deep yellow.

Pine-woods, lawns. June–July. $1\frac{5}{8} \times 2\frac{1}{8} \times \frac{1}{4}$ in.

118. T. fallax Sacc. (from its resemblance to a dwarf form of **114** or **117**; *fallax*, deceitful) *a*.

P. yellow; mid. sometimes rufous. *St*. yellowish. *G*. white, then yellowish.

Under firs. Autumn. $1\frac{1}{8} \times 1 \times \frac{3}{16}$ in.

118a. T. onychinum Gill. (*onychinus*, of the colour of the human nail) *a*.

P. fleshy, subumbonate, convexo-plane, glabrous, purplish or sienna-umber; marg. striate, orange sienna or silky-purple. *St*. solid, almost equal, fibrilloso-silky, pallid or dull orange-sulphur, reddish or sienna pulverulent above, brownish below. *G*. rounded, at length free, crowded, yellow. *Flesh* yellowish-white, or pale olive-sulphur.

Amongst pines, mossy places. Autumn. $2\frac{3}{4} \times 3 \times \frac{1}{4}$ in.

119. T. ionides Quél. (from its violet colour; Gr. *ion*) *a c*.

P. lilac or violaceous, sometimes livid reddish. *St*. pale lilac. *Flesh* lilac under membrane of P., and at base of St.

Odour faint, agreeable, like **1280**. Woods, moist places, open pastures, under firs, cedars. Aug.–Nov. $2\frac{1}{4} \times 2\frac{1}{8} \times \frac{1}{4}$ in. Resembles **298** in general appearance.

120. T. carneum Quél. (from its flesh-colour; *caro*) *a b c*.

P. becoming whitish. *St.* attenuate downwards, paler than P. *G.* white. *Flesh* white.

Fields and waysides in woods; frequent. May–Oct. $1\frac{3}{4} \times 1\frac{1}{2} \times \frac{1}{4}$ in.

121. T. cælatum Gill. (from its umbilicus, chiselled out; *cælum*, a chisel) *a b c*.

P. becoming flocculose, brownish, then pale grey. *St.* pale brown. *G.* dull white or grey.

Woods, downs. April–Sept. $1\frac{1}{4} \times 1\frac{1}{4} \times \frac{3}{16}$ in. Resembling certain species of *Omphalia*.

c. Guttatæ.

122. T. gambosum Gill. (from the hoof-like pileus; *gamba*, a hoof) *a b c*. Buff-white.

P. hemispherical. *G.* crowded.

Solitary or 2–3 connate; in large rings or troops. Edible if gathered in dry weather and not rank smelling. Odour strong of new meal, pleasant, varying to rank as in *Polyporus squamosus*. Fields and downs. April–July. $3\frac{5}{8} \times 2 \times 1\frac{1}{8}$ in. "St. George's mushroom." Must not be confounded with **133** or the solitary form of **123**.

123. T. albellum Quél. (*albus*, white) *a b c*.

P. conical, then convex, pale dull yellowish-white. *St.* ovato-bulbous, colour as P. *G.* very crowded.

Solitary or connate. Taste pleasant; odour weak. Woods. April–Oct. Solitary—$3 \times 2 \times 1$ in. Connate—$1\frac{3}{8} \times 1\frac{1}{2} \times \frac{1}{2}$ in. The connate form must not be confounded with connate **153**.

124. T. boreale Karst. (from its growing chiefly in northern countries; *borealis*, northern) *a b c*.

P. subumbonate, flesh-colour. *St.* attenuate downwards, pale flesh-colour. *G.* crowded.

Said to be edible. Odour of fresh meal. Grassy places. $2\frac{3}{4} \times 2\frac{1}{4} \times \frac{1}{2}$ in. Not unlike **648** in general appearance.

125. T. amethystinum Gill. (from its colour) *a*.

P. livid lilac or grey-purple. *St.* livid.

Woods, pine. Sept. $2\frac{3}{8} \times 2 \times \frac{1}{2}$ in.

126. T. tigrinum Quél. (from the markings on the pileus) *a b*.

P. pallid brown, spotted-brown. *St.* white.

Solitary or cæspitose. Odour usually none, sometimes fetid. Woods, fir, open places; rare. June–July. Allied to **129** and **142**.

127. T. pes-capræ Quél. (from a fancied resemblance in the stem to a goat's foot, *pes-capræ*) *a c*.

P. umbonate, fragile, grey or brownish; marg. cracked. *St.* brownish-white. *G.* becoming ashy, rarely olive.

Odour of meal. Open places, under oaks. Oct.–Nov. $1\frac{7}{8} \times 2\frac{3}{4} \times \frac{3}{8}$ in. Var. *multiforme* Mass. Smaller and flesh of pileus thinner than type, gregarious.

f. Spongiosæ.

128. T. Schumacheri Gill. (after Christian Friedrich Schumacher) *a b*. *P.* livid grey or faint brownish-lilac; mid. darker; marg. exceeding G. *St.* ivory-greyish or white, base villous. *G.* subdecurrent, very crowded, white or brownish-white.

Woods, hothouses. Autumn. 3⅜ × 3⅜ × ⅜ in. Not unlike **153.**

128a. T. amicum Gill. (from its being well known; *amicus*, a friend) *a*. *P.* fleshy, convexo-expanded, broadly umbonate, even, glabrous, dusky- or olive-brown; membrane of P. extending beyond G. *St.* solid, firm, bulbous, white or slightly suffused with colour of P. *G.* rounded, almost free, broad, distant, white. *Flesh* white.

Solitary. Odour none. Amongst pines, mossy places. Autumn. 3 × 3½ × ½ in.

129. T. circumtectum Sacc. (from the encircling margin of the pileus, *circumtectus*) *a b*.
P. olive or dusky; mid. tawny. *St.* white. *G.* emarginate, distant.

Taste mild; odour none. Sept. 2½ × 1⅜ × ⅜ in. Elastic. Compare **126** and **142.**

130. T. patulum Quél. (from the spread out pileus, *patulus*) *a b c*.
P. ivory-brown to pale, dull lavender-brown or whitish, becoming pale yellowish. *St.* white. *G.* rounded, almost free to distant, whitish.

Solitary, cæspitose or in troops. Odour none. Woods, under birches. Only appearing in very wet seasons. Oct. 4 × 3 × 1⅛ in. The cæspitose form must not be confounded with **179.**

131. T. arcuatum Quél. (from its arched gills) *a c*.
P. dark-brown; mid. almost black, becoming lighter. *St.* subbulbous, white, brown at base. *G.* sinuate, white or buff-white. *Flesh* becoming yellow or brownish.

Gregarious. Grassy ground. Oct.–Nov. 2¾ × 1½ × $\frac{7}{16}$ in. Sometimes confounded with **109a** and **144.**

132. T. oreinum Gill. (from its usual habitat; Gr. *oreinos*, hilly) *a*.
P. livid brown; marg. exceeding G. *St.* subbulbous, white above, ochreous below. *G.* rounded, free. *Flesh* white.

Solitary. Odour none. Heaths and open ground. Autumn. 1½ × 1⅛ × ¼ in. Sometimes confounded with **144.**

133. T. album Quél. (*albus*, white) *a b c*. Ivory-white, sometimes becoming faintly yellowish or slightly buff, fleshy.
P. dry. *St.* attenuate upwards. *G.* sinuate.

Perhaps poisonous. Taste unpleasant, bitter, acrid; odour weak or none. Woods; frequent. Aug.–Nov. 4½ × 3¾ × ⅝ in. Var. *cæsariatum* Quél. *P.* thin. *St.* thin. G. almost free. Must be carefully distinguished from other white species.

134. T. leucocephalum Quél. (Gr. *leukos*, white, *cephale*, head) *a b c*. Unchanging white.

P. silky-downy. *St.* attenuate downwards. *G.* rounded.

Odour strong and pleasant of new meal. Woods, amongst leaves, lawns, numerous specimens growing through the flattened remains of a dead hedgehog. Aug.–Oct. 2¼ × 2⅜ × ⅜ in.

135. T. acerbum Quél. (*acerbus*, bitter) *a b c*.

P. viscid when moist, buff or brown; marg. involute brown-wrinkled. *St.* solid, squamulose, light buff, darker below. *G.* crowded, pallid, then spotted or rufescent.

Taste usually bitter, sometimes none; odour strong and unpleasant, sometimes pleasant or none. Woods, shady places; frequent. Aug.–Dec. 6 × 3¼ × 1¼ in.

136. T. militare Gill. (from its appearance as contrasted with **137**; *miles*, a soldier) *a*.

P. very fleshy, viscid, cinnamon; marg. white-floccose. *St.* pallid. *G.* somewhat crowded, becoming livid-spotted.

Cæspitose or in rows. Taste and odour unpleasant. Woods. Oct. 5⅛ × 3¼ × 1 in.

137. T. civile Gill. (from its appearance as contrasted with **136**; *civis*, a citizen) *a c*.

P. subfleshy, subviscid, pale yellowish; mid. darker. *St.* whitish. *G.* becoming pale yellow, rarely spotted.

Odour none. Woods, pine. Oct.–Nov. 3 × 2¼ × ½ in. Resembling an exannulate **17**.

138. T. personatum Quél. (from its mask—the velvety edge of the pileus and the scales on the stem as contrasted with **139**; *persona*, a mask) *a b c*.

P. very fleshy, somewhat pale livid-brown. *St.* brownish, bright purple above, brittle below. *G.* livid, sometimes becoming purplish.

Solitary or gregarious. Edible in dry weather, soddened and unwholesome in wet. Pastures and downs, sometimes woods; common. June–Nov. 5⅛ × 3¼ × 1 in. The "bluette," "blewitt," or "blue-leg." Often confounded with **139** and **1039**.

138a. T. glaucocanum Bres. (*glaucus*, bluish-grey, *canus*, grey-whitish) *a b*.

P. fleshy, somewhat soft, convexo-expanded, moist; marg. involute, somewhat flocculoso-pruinose. *St.* solid, bulbous, fibrilloso-striate, subsquamulose above, whitish-grey. *G.* emarginate, very crowded, separating from hymenophore, purple-greyish.

Edible. Taste mild; odour strong of new meal. Under conifers. Autumn. 2⅜ × 3 × ¾ in. At length wholly hoary. Intermediate between **138** and **139**.

138b. T. sævum Gill. (from its aspect; *sævus*, sinister) *a*.

P. convex, then plane, pallid brownish. *St.* somewhat shaggy-squamulose, bright lilac-purple, lighter above, white at base. *G.* crowded, pallid whitish.

On the earth, in woods, pasturages. Nov. 3⅝ × 2⅜ × ⅝ in.

139. T. nudum Quél. (from the naked edge of the pileus and smooth stem when contrasted with **138**) *a b c*.

P. somewhat thin, purple, rufescent. *St.* purple. *G.* purple to rufescent.

Odour acid, pleasant. Woods, often amongst firs; frequent. Sept.-Dec. 3⅛ × 2⅝ × ½ in. Often confounded with **138**, **149**, **160**, and **1039**. There is a large form twice the size of type.

140. T. cinerascens Gill. (from its ashy colour; *cinis*, ashes) *a b*.

P. white to ashy-grey. *St.* whitish. *G.* rounded, free, white, then darker than P.

Gregarious, cæspitose. Odour somewhat disagreeable, pungent, sometimes like new meal. Woods; frequent. Sometimes becomes blackish when bruised. Aug.-Oct. 3 × 3¼ × ½ in.

141. T. panæolum Quél. (Gr. *panaiolos*, variegated) *a b*.

P. convex, then flat, sooty-grey pruinose, or brown-yellow blotched different shades of umber. *St.* naked, whitish-grey. *G.* sinuato-rounded, then subdecurrent, white, then grey to dusky or brownish.

Said to be edible and delicious. Odour not unpleasant. Grassy places. Sept.-Nov. 3¼ × 1⅞ × ½ in.

141a. T. cnista Gill. (perhaps from Gr. *knisa*, odour of roasting fat) *a*.

P. convex, obtuse, plane, even, glabrous, moist, pale tan or whitish; disc darker; marg. incurved, even, naked. *St.* solid, equal, even, glabrous, tough, white. *G.* adnexo-rounded, crowded, transversely veined, inclined to separate from hymenophore, white, pallid yellow or reddish when bruised. *Flesh* rather thick, soft, white.

Grassy open places. Autumn. 3⅞ × 3¼ × ⅜ in. Similar to **141**, but white, not becoming grey.

142. T. duracinum Sacc. (*durus*, hard, *acinus*, a berry) *a*. Ashy-grey.

P. very fleshy, broadly umbonate. *St.* attenuate upwards, reticulato-squamose above, paler than P. or white. *G.* cinereous.

Under cedars. Oct. 2⅝ × 2¼ × 1⅛ in. Allied to **126**.

g. Hygrophanæ.

143. T. grammopodium Quél. (from the striate stem; Gr. *gramme*, a line, *pous*, a foot) *a b c*.

P. depressed, umbonate, livid, opaque, brownish. *St.* whitish brown, enlarged and white-villous below. *G.* whitish or livid.

Taste unpleasant; odour mouldy, unpleasant. Pastures and grassy places in woods; frequent. June-Nov. 5¼ × 4 × ⅝ in. The pileus is frequently clouded with white mycelium as in **153**. There is a wholly white variety.

144. T. melaleucum Quél. (from the dark pileus and white stem; Gr. *melas*, black, *leukos*, white) *a b c*.

P. flat, subumbonate, dark, smoky, olive- or slate-brown, varying to lavender. *St.* white, often buff below. *G.* white, sometimes becoming yellowish. *Flesh* unchanging white.

Taste strong and disagreeable or none; odour pleasant or none. Grassy places, orchards, gardens; uncommon. March-Oct. 3⅝ × 2½ × ½ in.

Sometimes confounded with **131** and **132**. Var. *polioleucum* Gill. *P.* livid, then grey. Var. *porphyroleucum* Gill. Firmer. *P.* fuliginous or fuscous, then rufescent, with evanescent um.

145. T. brevipes Quél. (from its short stem; *brevis*, short, *pes*, a foot) *a b c*.

P. broadly subumbonate, livid pale- or red-brown or dull lavender. *St.* bulbous, white to brownish. *G.* whitish to fuscous. *Flesh* brown, white when dry.

Solitary or somewhat connate. Odour none. Woods, open places, fields, etc. June–Nov. 3⅞ × 1⅜ × ½ in.

146. T. humile Quél. (*humilis*, low, dwarf) *a b c*.

P. livid ashy- or warm-brown, pale dull buff or pale slate, sometimes faintly zoned darker. *St.* pulverulento-villous, ashy-white. *G.* whitish. *Flesh* whitish or ashy-brown.

Gregarious or cæspitose. Taste not unpleasant; odour none. Woods, open places, gardens, cinder-heaps; frequent. Aug.–Nov. 2¼ × 1⅜ × ¼ in.

147. T. excissum Quél. (from the linear gills, as if cut away; *excindo*, to tear out) *a b c*.

P. umbonate, brownish-white, pale mouse-colour or somewhat buff, varying dark brown. *St.* smooth, white. *Flesh* white.

Pastures. May–Nov. 3 × 2½ × ⅜ in.

148. T. subpulverulentum Karst. (from the minute hoary powder frequent on the pileus; *pulvis*, dust) *a b*.

P. broadly subumbonate, pale livid-brown, greyish or whitish. *St.* smooth, slightly striate, white. *G.* white. *Flesh* white.

Sometimes in large rings, twenty or more feet in diameter. Taste at first agreeable, at length astringent. Woods, pastures, gardens. Aug.–Oct. 2⅛ × 1⅝ × ⅜ in.

149. T. sordidum Quél. (*sordidus*, dirty) *a b c*.

P. broadly subumbonate, pale livid to purplish pale brown. *St.* colour as P. *G.* sinuato- or adnato-decurrent, becoming distant, purplish or pale brownish.

Gregarious, somewhat cæspitose. Odour none. Pastures, manure, dung heaps. July–Nov. 3⅛ × 2⅜ × ⅜ in. Sometimes mistaken for a pale form of **139**.

150. T. pædidum Quél. (*pædidus*, nasty, mean) *a*.

P. tough, flaccid, umbonate, smoky-grey or dull brown, at first radiato-streaked with fibrils. *St.* paler than P. *G.* crowded, rounded, whitish, then grey.

Odour none. Woods, grassy places, gardens, on dunghills. Aug. 1⅞ × 1¼ × ¼ in.

151. T. lixivium Karst. (from its watery-ashy colour; *lixivium*, ashes and water) *a b c*.

P. umbonate, umber, ashy-brown when dry. *St.* floccoso-pruinose, white, then as P. *G.* distant, adnexo-rounded, grey.

Woods, pine. Nov. 3 × 2⅜ × ½ in.

152. T. putidum Karst. (*putidus*, stinking) *a c.*

P. subumbonate, olive-grey or brown, hoary or sprinkled white-silky. *St.* grey, white pruinose. *G.* crowded, adnexo-free, ashy-grey.

Odour strong, rancid. Woods, pine. Oct.-Nov. $2 \times 2\frac{1}{8} \times \frac{1}{4}$ in.

VII. CLITOCYBE Quél.

(From the decurrent gills; Gr. *klitos*, a declivity, *kube*, a head.)

Veil universal, imperfect or obsolete, manifest as pruina, flock, silkiness or squamules on the pileus and stem. *Hymenophore* confluent and homogeneous with the fleshy stem. *Pileus* varying

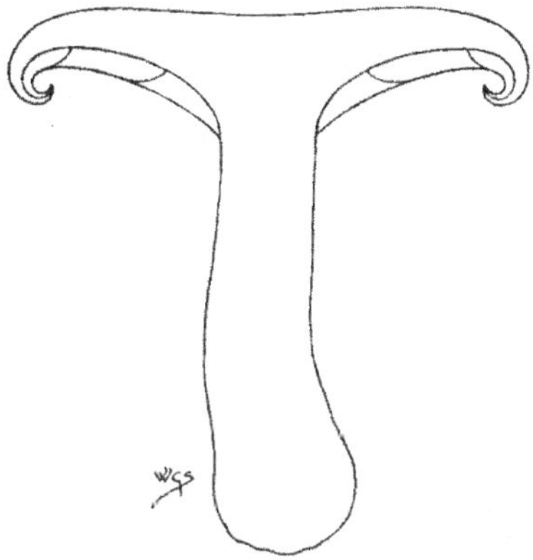

Fig. 16.—Section of *Clitocybe nebularis* Quél.
One-third natural size.

infundibuliform to plano-depressed, usually fleshy at the disc, margin at first involute. *Stem* central, simple, without cartilaginous bark. *Gills* decurrent, rarely adnate, with an acute edge. *Spores* elliptical or subglobose, smooth. (Fig. 16.)

The species usually grow on the ground, but exceptions occur in **185, 201, 208, 212,** and **223,** which rarely grow on logs, stumps or rotten wood; they commonly grow in clusters, many are fragrant and appear in the late autumn or early winter; a few are edible.

Must not be confounded with *Hygrophorus*, where the gills are thick and often of a wax-like substance, or with *Cantharellus*, where the gills are very thick and fold-like, or *Lentinus*, *Panus* or *Xerotus*, where the substance is leathery and the habitat usually stumps.

Species **153—224**

A. Pileus fleshy, not hygrophanous, often becoming pale and silky in drying.

a. Disciformes. Pileus convex, then plane or depressed, regular. *Gills* adnate or adnato decurrent.

Pileus ashy or brownish. **153—159**
Pileus violaceous, rufescent. **160—163**
Pileus becoming yellow. **164—167**
Pileus green, except **170**, then pale; odour sweet. **168—170**
Pileus white. **171—177**

b. Difformes. Pileus at first umbonate, then expanded or depressed. *Gills* unequally decurrent. Generally cæspitose, sometimes solitary. **178—189**

c. Infundibuliformes. Pileus at length deeply depressed. *Gills* truly decurrent. Often losing colour in drying.

Pileus coloured, becoming hyaline, minutely flocculose or silky. Hygrophanous but not externally moist. **190—195**

Pileus varied in colour. Moist in rainy weather. **196—204**

Pileus shining whitish, slightly flocculoso-sprinkled or smooth. **205—207**

B. Hygrophanous.

d. Cyathiformes. Pileus cup-shaped. *Gills* at first adnate, then decurrent. Colours dull when moist. (Must not be confounded with the *Hydrogrammæ*, under *Omphalia.*) **208—214**

e. Orbiformes. Pileus convex, then plane or depressed, shining, not mealy or squamulose. *Gills* adnate or adnato-toothed. Colours dull or watery-hyaline.

Pileus at first dark. *Gills* becoming ashy. **215—217**
Pileus becoming pale. *Gills* whitish. **218—221**

f. Versiformes. Pileus tough, more or less squamulose or furfuraceous. *Gills* adnate or decurrent, somewhat thick, distant. *Stem* tough.

Pileus dull or dusky with darker squamules. **222, 223**
Pileus bright in colour. **224**

a. Disciformes.

153. C. nebularis Quél. (from the frequent clouds of white mycelium on the pileus; *nebula*, a cloud) *a b c.*

P. smoky-brown, grey, or pale slate. *St.* spongy-elastic, whitish or grey. *G.* crowded, shortly decurrent, white. *Flesh* opaque white.

Laxly gregarious, rarely connate as in **123.** Edible. Taste somewhat peppery; odour of curd cheese. Woods, fir, beech, hedge-bottoms, etc., amongst leaves; common. Aug.–Dec. 6 × 4½ × 1½ in. There is a white variety. Must not be confounded with **128.** The mycelium on pileus gives rise to **451.** A similar mycelium occurs on **143.**

154. C. clavipes Gill. (from the club-shaped stem; *clava*, a club, *pes*, a foot) *a b c*.

P. flat, sooty-brown or grey; marg. often white. *St.* smoky-livid, somewhat fibrillose. *G.* subdistant, deeply decurrent, white. *Flesh* hyaline-whitish.

Odour mild, pleasant. Woods, spruce-fir, pine, beech; common. July–Nov. $2\frac{5}{8} \times 2\frac{3}{8} \times \frac{1}{2}$ in. There is a white variety. The pileus is sometimes 4 in. or more in diam.

154a. C. comitialis Gill. (perhaps from its being one of a number of allied forms which meet at this position in *Clitocybe*) *a*.

P. obtuse, soon plane, even, glabrous, somewhat moist, not hygrophanous, sooty-brown, almost black. *St.* stuffed, attenuate upwards, obconic, glabrous, sooty-brown. *G.* subdecurrent, horizontal, thin, crowded, white. *Flesh* white or whitish.

Odour none. Pine-woods, damp places amongst moss. Autumn. $1\frac{3}{8} \times 3\frac{1}{4} \times \frac{5}{16}$ in. Firmer and smaller than **154**.

155. C. gangrænosa Gill. (from its ulcerous appearance; *gangræna*, an ulcer) *a b c*.

P. livid, at first white-pulverulent. *St.* slender to stout, sooty-grey. *G.* sub- or sinuato-decurrent.

Taste insipid, unpleasant; odour sometimes strongly fetid, meal-like but nauseous, sometimes strong and not unpleasant. Woods and plantations, larch; rare. Oct.–Nov. $3\frac{3}{8} \times 3\frac{1}{4} \times \frac{5}{8}$ in. More or less changing within and without to slate-colour or black; the flesh changes from livid to indigo, then black; sepia-black when dry. Must not be confounded with **1168**.

155a. C. polia Karst. (Gr. *polios*, grey) *a b*.

P. glabrous, not hygrophanous, pale grey or pale olive-brownish; marg. whitish. *St.* solid, slightly attenuate upwards, white, pale brownish below. *G.* somewhat deeply decurrent, closely crowded, very narrow, white. *Flesh* white, pale brown in P.

Cæspitose, usually in small clusters, but sometimes 1 ft. 4 in. in diam. Woods. Autumn. $2 \times 3\frac{1}{4} \times \frac{3}{8}$ in.

156. C. inornata Gill. (*inornatus*, unadorned) *a b c*.

P. with a separable livid-brownish pellicle, which frequently cracks from the marg. upwards and peels up, unchanging livid. *St.* greyish or brownish. *G.* decurrent, colour as St. *Flesh* grey.

Odour mushroom-like. Fields, woods; under cedars; rare. Oct.–Nov. $4 \times 2\frac{1}{4} \times \frac{1}{2}$ in.

156a. C. luscina Karst. (from the often small eye-like pileus and the solitary habit; *luscinus*, one-eyed) *a*.

P. somewhat fleshy, even, edge spreading, glabrous, grey on a pale reddish or buff ground, purple or slate-brown when young; marg. striate. *St.* solid, attenuate downwards, greyish or faintly brownish, white pulverulent. *G.* subdecurrent, crowded, yellowish- or salmon-whitish, shaded grey.

In pastures. Autumn. $2 \times 2 \times \frac{3}{16}$ in.

157. C. curtipes Gill. (from the short stem; *curto*, to shorten, *pes*, a foot) *a*.

P. expanded, oblique, obtuse, olive-ochre to olive- or flesh-brown. *St*. rigid, thickened upwards, olive-brown, subpruinose. *G*. adnate, closely crowded, ivory-white.

Grassy places. Autumn. $1\frac{3}{8} \times 1 \times \frac{1}{8}$ in.

158. C. hirneola Quél. (from a fancied jug-like shape; *hirnea*, a jug) *a b c*.

P. umbilicate, minutely silky-hoary, whitish, grey, tan or olive shaded. *St*. colour as P., hoary above. *G*. decurrent, crowded, lighter than P., becoming hoary.

Amongst moss and grass, roadsides. Oct. $1\frac{1}{2} \times 1\frac{3}{8} \times \frac{1}{8}$ in. Must not be confounded with members of *Omphalia*.

159. C. zygophylla Sacc. (from the vein-connected gills; Gr. *zygon*, a yoke, *phullon*, a leaf) *a b*.

P. whitish; mid. pale buff; marg. rugose or plicate, becoming white. *St*. whitish, downy at base. *G*. decurrent, subdistant, pale clay-colour.

Amongst leaves. Oct.–Nov. $4 \times 1\frac{3}{4} \times \frac{1}{2}$ in. Resembling in general appearance **202** and **1278**. In the whitish-tawny pileus and clay-coloured gills it resembles **570**.

160. C. cyanophæa Sacc. (Gr. *kuanos*, blue, *phaios*, dusky) *a*.

P. pale slate-blue; mid. brown. *St*. paler than P., becoming yellowish, abruptly white above. *G*. deeply decurrent, pale slate-blue.

Woods. $2\frac{3}{4} \times 2\frac{3}{8} \times \frac{1}{2}$ in. Must not be confounded with **139**.

161. C. opipara Gill. (*opiparus*, splendid) *a b*.

P. shining, bronzy flesh-colour. *St*. pale brownish. *G*. adnato-decurrent, white to pale yellowish.

Taste and odour pleasant. Mossy places under trees, beech. Oct. $3 \times 3 \times \frac{7}{8}$ in.

162. C. amara Quél. (*amarus*, bitter) *a b c*.

P. flat, dull buff, then whitish. *St*. white. *G*. adnato-decurrent, crowded, white.

Odour of new meal. Woody places. Oct. $3\frac{1}{8} \times 1\frac{7}{8} \times \frac{3}{8}$ in.

163. C. socialis Gill. (from its gregarious habit) *a*.

P. acutely umbonate, reddish-buff. *St*. brownish, strigose at base. *G*. sinuato-decurrent, scarcely crowded, becoming yellowish.

Odour none. Amongst fir-leaves. Sept.–Oct. $1\frac{5}{8} \times 1\frac{3}{8} \times \frac{1}{4}$ in. Must not be confounded with **165**.

164. C. amarella Quél. (*amarus*, bitter) *a b*.

P. subumbonate, pallid fawn to whitish. *St*. solid, colour as P. *G*. subdecurrent, crowded, pallid.

Taste very fetid-bitter; odour strong of prussic acid or very disagreeable. Woods. Oct. $2\frac{1}{8} \times 1\frac{3}{4} \times \frac{1}{4}$ in.

165. C. vernicosa Gill. (from the shining pileus; *vernis*, varnish) *a b*.
P. yellowish or buff. *St.* stuffed, yellow. *G.* adnato-decurrent, subdistant, sulphur-yellow. *Flesh* whitish to pale sulphur.

Odour none. Woods, fir, larch; rare. Autumn. $3\frac{1}{8} \times 1\frac{3}{4} \times \frac{3}{8}$ in. Somewhat resembles a yellow *Hygrophorus*.

166. C. venustissima Karst. (*venustus*, beautiful) *a b*.
P. thin, subumbilicate, orange-vermilion; marg. crenate. *St.* orange-yellow, base white. *G.* decurrent, subdistant, yellow.

Odour none. Amongst larch and pine leaves. Autumn. $2 \times 1\frac{1}{2} \times \frac{1}{4}$ in. Resembles a yellow *Hygrophorus*.

167. C. subalutacea Gill. (from its colour; *aluta*, soft leather) *a c*.
P. thick, subumbonate or depressed, pale tan. *St.* pale straw. *G.* adnato-decurrent, distant, pallid white.

Odour as in **168**, or weak of anise, sometimes strong and disagreeable of rancid meal or none. Under trees, hollies. Nov. $2 \times 1\frac{3}{4} \times \frac{3}{8}$ in.

168. C. odora Quél. (*odorus*, fragrant) *a b c*.
P. subumbonate, dull greenish or bluish. *St.* greyish, whitish, greenish or pale brownish. *G.* paler than P.

Single or subcæspitose. Odour strong and pleasant of melilot, anise, almonds, meadow-sweet or new-mown hay, stronger in recently dried examples. Woods, grassy places, roadsides. Aug.–Nov. $3 \times 1\frac{3}{4} \times \frac{3}{8}$ in.

169. C. Trogii Sacc. (after Jacob Gabriel Trog, Swedish botanist) *a*.
P. subumbonate, ashy-white to buff-grey. *St.* solid, white. *G.* white.

Said to be edible. Odour fragrant and spicy as in **168**. Woods, amongst leaves. Oct.–Nov. $3\frac{1}{4} \times 1\frac{1}{4} \times \frac{1}{2}$ in. The colour approaches **215**.

170. C. rivulosa Quél. (from the rivulose cracks on the pileus) *a b*.
P. whitish, very pale or brownish flesh-colour or brownish-ochre with no shade of green, appearing at first as if whitish-varnished, the cracking of this coat giving the rivulose appearance. *St.* colour as P. *G.* flesh-colour.

Solitary, gregarious or cæspitose. Taste mild; odour pleasant. On leaves, amongst moss, pastures, roadsides, etc. Oct.–Nov. $1\frac{3}{4} \times 1\frac{1}{8} \times \frac{3}{16}$ in. Some small forms, as the var. *neptunea* Mass. approach **171**.

171. C. cerussata Quél. (from its colour; *cerussa*, white-lead) *a b c*.
St. solid. *G.* adnato-decurrent, very crowded, never yellow.

Often in troops. Taste mild, pleasant, sweet, or like **1394**, sometimes rancid; odour pleasant or none. Woods, mixed, fir; frequent. April–Oct. $3\frac{3}{8} \times 2 \times \frac{3}{8}$ in. There are three forms:—1. Stout; like **122** and **123**. 2. Slender; like **172**. 3. Large, stout, irregular. The var. *difformis* Gill. is British. Sometimes springs from dense, creeping mycelium. A close ally of **188**.

172. C. phyllophila Quél. (Gr. *phullon*, a leaf, *philos*, loving) *a b c*.
P. tan or white. *St.* stuffed or hollow, sometimes pale brownish. *G.* slightly decurrent, subcrowded, sometimes buff-white or yellowish.

Cæspitose and larger, solitary and smaller. Odour mealy or none. Woods, chiefly amongst beech-leaves, one form on pine-leaves. Sept.–Dec. $3 \times 2\frac{1}{2} \times \frac{3}{8}$ in. Must not be confounded with **367**.

173. C. pithyophila Gill. (Gr. *pitus*, pine, *philos*, loving) *a b c*. Persistently white.

P. plano-umbilicate, thin, hygrophanous. *St*. somewhat hollow, tomentose at base. *G*. adnato-decurrent, very crowded.

Gregarious or subcæspitose. Odour pleasant. Pine-woods; frequent. Sept.-Nov. $2\frac{3}{8} \times \frac{5}{8} \times \frac{3}{16}$ in. Must not be confounded with **206**.

174. C. tornata Quél. (from its shape; *torno*, to turn on a lathe) *a*.

P. gibbous; mid. fleshy. *St*. stuffed, pubescent at base. *G*. adnate with a decurrent tooth, very crowded.

Odour none. Woods, amongst grass, by decayed stumps, elm. Oct. $2 \times 2 \times \frac{1}{4}$ in. Always small.

175. C. candicans Quél. (*candicans*, shining-white) *a b c*.

P. mid. not fleshy. *St*. somewhat hollow. *G*. adnato-decurrent, crowded.

Odour of new meal. Woods, amongst damp leaves; common. July–Nov. $1\frac{7}{8} \times 1\frac{1}{2} \times \frac{1}{8}$ in. Must not be confounded with **176**.

176. C. dealbata Gill. (*dealbatus*, whitewashed) *a b c*.

P. not fleshy. *St*. stuffed or hollow. *G*. adnate or adnato-decurrent, crowded.

Odour sweet, weak, pleasant, mealy but fungoid. Woods and pastures, sometimes on old mushroom-beds; common. July–Nov. $2\frac{1}{2} \times 1\frac{3}{8} \times \frac{3}{8}$ in. Some small, delicate forms, var. minor Sacc., are at times confounded with **175**.

177. C. gallinacea Gill. (from its colour, like a hen's egg; *gallina*, a hen) *a b c*.

St. solid. *G*. adnato-decurrent, crowded.

In troops, sometimes in rings, amongst grass. Taste slightly fetid-acrid; odour strong, earthy. Woods and pastures, amongst grass and moss; common. Sept.–Nov. $1\frac{3}{4} \times 1\frac{7}{8} \times \frac{1}{4}$ in.

b. Difformes.

178. C. ampla Gill. (*amplus*, large) *a b*.

P. fleshy, somewhat silky, sooty or ochreous-brown. *St*. white or brownish-white. *G*. decurrent or sinuato-adnate and toothed. *Flesh* fibrous.

Tasteless; odour none. Woods, amongst moss. Sept.–Nov. $6\frac{3}{4} \times 3\frac{3}{4} \times 1\frac{3}{8}$ in.

178a. C. molybdina Gill. (from the dark, lead-coloured pileus; Gr. *molubdaina*, lead-ore) *a*.

P. fleshy, campanulate, expanded, umbonate, even, glabrous; marg. thin, somewhat dark sooty-umber; mid. sooty-buff. *St*. solid, stout, fibroso-striate, mealy-squamulose above, pallid or yellowish salmon-white, white at base. *G*. adnate, broad, pale yellowish-salmon.

Cæspitose. Woods. Autumn. $5\frac{1}{4} \times 6\frac{3}{4} \times \frac{3}{4}$ in. With the general appearance of **458** (var. *eximius*), but the adnate gills and white spores very different.

179. **C. decastes** Quél. (from its growing in clusters of about ten; Gr. *dekas*, a company of ten) *a b*.

P. very thin, of two easily separable strata, tan-grey or tan. *St.* whitish. *G.* adnato-decurrent, subdistant, white.

Cæspitose, in clusters of about one foot in diam. Woods, oak, beech, by-paths, plantations, gardens, orchards, on sawdust. Oct.–Nov. $9 \times 4\frac{3}{4} \times \frac{1}{2}$ in. Must not be confounded with **130.**

180. **C. subdecastes** Sacc. (from its affinity with **179**) *a b*.

P. very thin, campanulate, smooth, even, ivory-white; mid. ochreous. *St.* silky-fibrous, shining-white. *G.* rounded, crowded, thin, white.

Cæspitose. Taste and odour none. On the ground. Sept. $3\frac{1}{4} \times 5\frac{1}{2} \times \frac{1}{2}$ in.

181. **C. aggregata** Gill. (from its growing in clusters; *aggregatus*, collected together) *a c*.

P. gibbous, then depressed, not fleshy, often excentric, grey or dull ochreous. *St.* whitish. *G.* subdecurrent, crowded, pale ochreous or flesh-colour.

Odour strong. Woods, oak, on sawdust, mushroom-beds. July–Oct. $3\frac{1}{2} \times 3\frac{1}{2} \times \frac{5}{8}$ in. Becoming rufous-stained.

182. **C. elixa** Karst. (from its sodden condition in wet weather; *elixus*, soaked) *a b*.

P. gibbous, very fleshy, hygrophanous, sooty-brownish or greyish, becoming pale. *St.* pallid brownish. *G.* decurrent, distant, white.

In troops. Woods; frequent. Oct.–Nov. $4\frac{3}{4} \times 2\frac{1}{2} \times \frac{3}{4}$ in. Very bibulous, becoming saturated with moisture; heavy and brittle in wet or frosty weather, breaking with its own weight.

183. **C. fumosa** Quél. (from its smoky colour; *fumus*, smoke) *a b c*.

P. subgibbous, then plane, greasy-looking brown to livid-whitish; mid. sometimes slightly scurfy as in **69.** *St.* whitish or brownish. *G.* rounded, adnate, or decurrent, crowded, greyish- or brownish-white, or yellowish-olive-white.

Gregarious or cæspitose in groups a foot or more in diam. Suspected poisonous. Tasteless or insipid; odour strong, fungoid, or none. Woods, often on wood-ashes, old sawdust, amongst grass, on rubbish-heaps, on earth where trees have been felled; uncommon. June–Nov. $4\frac{1}{4} \times 2\frac{3}{4} \times \frac{5}{8}$ in. Often in company with **277** and **1374.**

184. **C. tumulosa** Sacc. (from the mound-like appearance of the connate groups; *tumulus*, a mound) *a b*.

P. gibbous, umber becoming pale. *St.* floccoso-pruinose, white. *G.* subdecurrent, crowded, white to livid grey.

Larger forms laxly cæspitose, smaller, densely cæspitose. Woods. Sept. $3\frac{1}{8} \times 3\frac{1}{2} \times \frac{3}{4}$ in.

185. **C. pergamena** Mass. (from its colour, like parchment, *pergamena*) *a*.

P. umbonate, at first brown. *St.* solid, with a brittle cartilagenous bark. *G.* broadly sinuato-decurrent.

Cæspitose. Stumps. Oct. $2 \times 5 \times \frac{5}{16}$ in. The habitat and cartilaginous bark point to *Collybia* rather than *Clitocybe*.

186. C. cryptarum Mass. (from its occasional appearance in cellars; *crypta*, a vault) *a b*.

P. subconical, then depressed, floccose, pale buff or brown, spotted. *St*. hollow, paler than P., mottled within. *G*. subdecurrent, white.

Densely cæspitose. Taste insipid; odour none. Sawdust. Oct. $2\frac{1}{8} \times 3\frac{1}{2} \times \frac{7}{8}$ in. With the varying habit and size of **184**.

187. C. monstrosa Gill. (*monstrosus*, strange) *a b*. White.

P. convexo-umbonate. *St*. solid, white-farinose above. *G*. rounded, crowded.

Often densely cæspitose. Probably edible. On the ground. Sept.–Oct. $2\frac{7}{8} \times 1\frac{1}{2} \times \frac{7}{8}$ in. May be an autumnal form of **122** or a variety of **133**, or a form of **94**. It may be a *Tricholoma*, as suggested by Berkeley, but cannot be a form of **124** as suggested by Fries. It is more probably a form of **123**; the spores are the same in size. Stem sometimes 6 in. long.

188. C. opaca Gill. (from its opaque colour) *a b c*. White with a silvery glare.

P. convexo-plane, subumbonate then repand, floccose. *St*. stuffed. *G*. adnato-decurrent, very crowded. *Flesh* pale brownish.

Single or connate. Woods. Oct.–Nov. $3 \times 2\frac{3}{8} \times \frac{5}{16}$ in. A close ally of **171**.

189. C. occulta Mass. (from its doubtful characters; *occultus*, hidden) *a*. Cartilaginous, tough.

P. convexo-plane, then slightly depressed, innato-virgate; mid. livid-smoky; marg. whitish. *St*. solid, paler than P. or white. *G*. subdecurrent, subdistant, white.

Gregarious. Charred ground. Nov. $3 \times 2\frac{1}{8} \times \frac{3}{8}$ in.

c. Infundibuliformes.

190. C. maxima Quél. (*maximus*, the greatest) *a b*. Tan-whitish.

P. thin, umbonate; marg. even. *St*. solid. *G*. subcrowded, not branched. *Flesh* white.

Odour weak and pleasant, sometimes strong. Grassy places, hedge-banks, woods, pastures; frequent. July–Nov. $10\frac{3}{4} \times 4\frac{7}{8} \times 1\frac{3}{8}$ in. Sometimes a foot or more high and broad. Often confounded with **1172**; the latter has no umbo, has a sulcate margin, is less infundibuliform, has branched gills, and a short stem.

191. C. infundibuliformis Quél. (*infundibulum*, a funnel, *forma*, form) *a b c*.

P. thin, typically umbonate, pale buff then white. *St*. spongy-stuffed, ivory-white. *G*. subcrowded, ivory-white.

Odour usually weak and pleasant, often like anise, sometimes none. Woods, grassy places, near trees, beeches, rarely on rotten wood; common. June–Dec. $3\frac{1}{8} \times 1\frac{7}{8} \times \frac{1}{4}$ in. A variable species. There is a wholly white variety. Var. *membranacea* Gill. *P*. membranous; brick-red. Woods, pine.

192. C. trullæformis Karst. (*trulla*, a ladle, *forma*, form) *a c.*

P. thin, flocculoso-villous, dry, brown-grey to brown. *St.* spongy-stuffed, ashy-grey. *G.* distant, connected by veins, white.

Borders of fir-woods, amongst grass and bushes. Oct. $2\frac{1}{4} \times 1\frac{7}{8} \times \frac{1}{4}$ in. Must not be confounded with **208.**

193. C. incilis Gill. (from its incised or crenate pileus; *incilis*, incised) *a b.*

P. plano-depressed, silky-flocculose, reddish. *St.* hollow, attenuate downwards, colour as P. *G.* distant, veined, white.

Odour strong of meal, sometimes none. Woods, under firs. Nov. $2\frac{1}{8} \times \frac{3}{4} \times \frac{1}{4}$ in. Resembles a small **1297.**

194. C. sinopica Gill. (from Sinope in Patagonia, where red-lead is found—from the vermilion-brownish pileus) *a b c.*

P. slightly depressed, floccoso-rivulose, dry. *St.* stuffed, colour as P. or buff. *G.* subdecurrent, very crowded, white, then yellowish.

Woods, chiefly on burnt places. Oct. $1\frac{3}{8} \times 1\frac{1}{4} \times \frac{3}{16}$ in. Resembling a small **1300.**

195. C. parilis Gill. (from its symmetrical growth) *a b c.*

P. subdepressed, atomate or flocculose, brownish, mottled darker, then white-grey. *St.* stuffed, pale yellowish-brown or brown-grey. *G.* very crowded, ivory-grey.

Woods, sides of plantations. Sept.–Oct. $2\frac{1}{8} \times 1\frac{1}{2} \times \frac{1}{4}$ in. Resembling a *Lactarius.* Must not be confounded with *Tricholoma* or *Omphalia.*

196. C. gilva Quél. (*gilvus*, pale yellowish-brown) *a b c.*

P. subdepressed, sometimes reddish, guttate-spotted. *St.* paler than P. *G.* pallid, then ochreous.

Woods, pine. Oct.–Dec. $3 \times 2\frac{1}{2} \times \frac{1}{2}$ in. Staining paper umber when dried. Resembles in form *Paxillus* and *Lactarius.*

197. C. subinvoluta W. G. Sm., *Agaricus subinvolutus* W. G. Sm. (from the subinvolute margin of the pileus) *a b.* Wholly pale buff, at length slightly spotted or clouded buff or yellow.

P. fleshy, plane, subdepressed; mid. darker guttate-spotted; marg. at first subinvolute. *St.* solid, subbulbous, darker below. *G.* decurrent, broad. *Flesh* buff-white, darker below.

Under firs. Oct. $3\frac{5}{8} \times 2\frac{1}{2} \times \frac{3}{4}$ in. Not *Agaricus subinvolutus* Batsch, which is a form of **1181.**

198. C. geotropa Quél. (from the often strongly deflected margin of the pileus; Gr. *ge*, the earth, *trepo*, to turn) *a b c.*

P. depressed, typically umbonate, pale brown, at first guttate-spotted. *St.* attenuate upwards, tan-brownish. *G.* white.

Sometimes in fairy-rings of great size. Odour of almonds. Woods, pastures. Sept.–Dec. $6\frac{1}{4} \times 6 \times \frac{7}{8}$ in. Pileus sometimes 9 in. in diam. Varies with a short stem and exumbonate. Intermediate forms occur between this and **197.**

199. C. spinulosa Sacc. (from a mistaken reference to the supposed spinulose spores) *a b*.

P. plane, subdepressed, umbonate, leathery, pale buff to rufescent flesh-colour. *St.* red-buff above, paler below, concentrically cracked white, as in **21**. *G.* white, then light yellowish. *Flesh* white.

In troops or two or three connate. Odour sweet, pleasant. Amongst grass by roadsides. Nov. $4 \times 3\frac{3}{4} \times \frac{3}{4}$ in.

200. C. splendens Gill. (*splendens*, shining) *a b*.

P. depressed, pale shining metallic-yellowish, shaded reddish-brown. *St.* solid, attenuate upwards, colour as P. *G.* white, then light yellow.

Solitary. Woods, amongst dead pine-leaves, etc. Sept. $4 \times 1\frac{7}{8} \times \frac{5}{8}$ in. Intermediate between **196** and **202**.

201. C. inversa Quél. (from the down-turned margin of the pileus) *a b c*.

P. depressed, somewhat fragile, oily-smooth in appearance, not guttate, yellowish-buff to dull crimson-brownish. *St.* stuffed or solid, lighter than P. *G.* whitish or yellowish. *Flesh* as P. but lighter.

Solitary, subgregarious or cæspitose. Taste same as mushroom; odour acid. Woods, chiefly fir, sometimes on logs; frequent. Aug.–Dec. $3\frac{1}{2} \times 3 \times \frac{5}{16}$ in. Not unlike **1299** in general appearance. A luxuriant form is *Agaricus lobatus* Sowerby.

202. C. flaccida Quél. (from the limp pileus; *flaccidus*, limp) *a b c*.

P. broadly umbilicate, sienna or buff. *St.* subhollow, pale sienna. *G.* obconico-decurrent, crowded, whitish, becoming yellow. *Flesh* white to brownish.

Solitary, sometimes gregarious, forming rings. Taste insipid. Plantations, fir, hills. Sept.–Jan. $3 \times 1\frac{1}{2} \times \frac{3}{8}$ in. **201** and **202** are sometimes undistinguishable; the colours vary yellowish-brown to brick-red. Must not be confounded with **159**. Resembles **1278**.

203. C. vermicularis Quél. (from its colour; *vermis*, a worm) *a b*.

P. broadly umbilicate, flesh-colour or light brownish. *St.* yellow to white. *G.* arcuato-decurrent, colour as St. *Flesh* lighter than P.

Fir-woods. Sept.–Oct. $2\frac{1}{8} \times 2\frac{1}{8} \times \frac{3}{8}$ in. Resembling a small **1288**.

204. C. senilis Gill. (from the wrinkled aspect of the pileus; *senilis*, old) *a b*.

P. infundibuliform, concentrically cracked, olive-brown or tan-brownish. *St.* solid, paler than P. *G.* decurrent in a straight line, paler than St.

Gregarious. Odour none. Woods, pine, lawns, etc. Autumn. $3\frac{1}{4} \times 1\frac{3}{4} \times \frac{5}{16}$ in.

205. C. catinus Quél. (from its bowl-shaped pileus; *catinus*, a bowl) *a*.

P. depressed, becoming flesh-colour. *St.* spongy-stuffed, attenuate upwards. *G.* colour as P.

Odour pleasant. Amongst dead leaves. Autumn. $2\frac{3}{4} \times 2 \times \frac{3}{8}$ in.

206. C. tuba Gill. (from its trumpet-shaped pileus; *tuba*, a trumpet) *a c.*

P. narrowly umbilicate, sometimes pale blue-grey. *St.* stuffed or hollow, colour as P. *G.* deeply decurrent, very crowded, white.

Gregarious. Taste mild of radish; odour of meal or horseradish or none. Woods, pine, on dead leaves. Oct.–Nov. 2¼ × 2¼ × $\frac{3}{16}$ in. Must not be confounded with **173.**

207. C. ericetorum Quél. (from its habitat; *ericetum*, a heath) *a b c.*

P. deeply depressed, arid. *St.* solid, attenuate downwards. *G.* subdecurrent, subdistant.

Odour pleasant. Sept.–Nov. 1¼ × 1 × ¼ in. Resembling **1214** in appearance.

d. Cyathiformes.

208. C. cyathiformis Quél. (*cyathus*, a cup, *forma*, form) *a b c.*

P. broadly umbilicate, deep brown, pale when dry. *St.* attenuate upwards, paler than P. *G.* distant, pale brownish-white. *Flesh* colour as G.

Woods, pastures, etc., rarely on rotten wood and tan-heaps; common. Aug.–Feb. 2¼ × 2¼ × ⅜ in. Must not be confounded with **192.**

209. C. cinerascens W. G. Sm., *Agaricus cinerascens* Batsch (from its colour; *cinis*, ashes) *a.*

P. plano-depressed, lavender-brownish. *St.* equal, striate, pale greyish or brownish. *G.* adnato-decurrent, white, then yellowish or grey.

Amongst moss. Autumn. 1¼ × 1⅛ × $\frac{3}{16}$ in.

210. C. expallens Quél. (*expallens*, growing pale) *a b.*

P. plano-depressed, very watery, white or pale brown. *St.* hollow-flocculose, paler than P. *G.* adnato-decurrent, white or greyish.

Woods, mixed, pastures, etc. Aug.–Dec. 2 × 1⅛ × $\frac{3}{16}$ in. Smaller, usually earlier, paler, less depressed and with less distant gills than **208.**

211. C. obbata Quél. (from the cup-shaped pileus; *obba*, a cup) *a c.*

P. deeply depressed, submembranous, blackish-brown or blackish-grey; marg. striate. *St.* hollow, ashy-brown. *G.* distant, dull grey.

Odour none. Woods, fir, downs. Oct.–Nov. 1¼ × 2 × $\frac{3}{16}$ in.

212. C. pruinosa Quél. (from its pruinose appearance) *a c.*

P. broadly and somewhat deeply depressed, sometimes squamulose, olive-brown, then grey. *St.* paler than P. *G.* crowded, white, then dull yellowish.

Odour none. Pine-woods amongst moss, sometimes on rotten wood. Nov.–Dec. 3 × 1⅝ × ¼ in.

213. C. concava Gill. (from the concave pileus) *a c.*

P. deeply umbilicato-depressed, very thin, tough, grey- or olive-brown; marg. not striate. *St.* ashy-grey. *G.* very crowded, colour as P.

Often gregarious. Odour none. Plantations, woods, fir, pastures; rare. Sept. $2\frac{1}{8} \times 1 \times \frac{3}{16}$ in. Must not be confounded with *Omphalia*.

214. C. brumalis Quél. (from its usual time of appearance; *bruma*, winter) *a b c.*

P. broadly plano-umbilicate, brown or whitish. *St.* hollow, paler than P. *G.* crowded, livid to yellowish-white.

Taste sometimes slightly nauseous; odour weak, earthy, not unpleasant, evanescent or none. Woods, on pine-leaves, amongst heather, on commons, on dirt-heaps; common. Oct.–Jan. $2 \times 2\frac{1}{4} \times \frac{1}{4}$ in.

e. Orbiformes.

214a. C. orbiformis Gill. (*orbis*, a circle, *forma*, form) *a c.*

P. convex, plane, obtuse, not truly depressed, glabrous, hygrophanous, smoky-grey; marg. spreading, even. *St.* stuffed, elastic, somewhat striate, attenuate and naked upwards, grey, base downy. *G.* adnato-decurrent, somewhat distant, greyish-white.

Odour none. Grassy places in pine-woods. Autumn. $2 \times 3 \times \frac{3}{16}$ in. Analogous with, but not allied to **208.**

215. C. metachroa Quél. (Gr. *metachroos*, changing colour) *a b c.*

P. umbonate, then plano-depressed, ochreous-whitish to brownish-grey; marg. at length striate. *St.* white. *G.* adnato-decurrent, ashy-whitish.

Odour none. Woods, pine, dry places, grassy woods, on leaves, rotting fern, turf, earth, dirt-heaps. May–Nov. $1\frac{3}{4} \times 2\frac{1}{8} \times \frac{1}{4}$ in.

216. C. pausiaca Gill. (from its olive-coloured gills; *pausia*, an olive) *a c.*

P. plano-depressed, subumbonate, tan, brown or grey. *St.* ashy-grey, white above. *G.* obtusely adnato-rounded.

Odour weak of meal. Woods, pine. Oct. $1\frac{3}{8} \times 1\frac{3}{4} \times \frac{1}{8}$ in.

217. C. ditopus Gill. (from the habit, often growing in twos; Gr. *dittos*, twofold, *pous*, a foot) *a b c.*

P. broadly umbilicate, horny, dark ashy-grey, or dull brown- or fuscous-ochre; marg. clouded-sooty. *St.* hollow, pale ashy-grey. *G.* adnate, crowded, colour as P.

Odour strong of new meal. Woods, pine, dead leaves. Oct. $3 \times 2 \times \frac{1}{4}$ in.

218. C. diatreta Quél. (from its perforate stem; Gr. *diatretos*, perforate) *a.*

P. plano-depressed, flesh-colour, then tan. *St.* pallid flesh-colour. *G.* adnato-decurrent, colour as St.

Odour almost obsolete. Woods, pine. Autumn. $1\frac{1}{4} \times 1\frac{1}{4} \times \frac{1}{8}$ in.

219. C. fragrans Quél. (*fragrans*, sweet-scented) *a b c.*

P. broadly depressed, pallid white; marg. substriate. *St.* colour as P. *G.* adnato-decurrent, somewhat crowded.

Odour strong of anise. Woods, pastures. July–Jan. $2\frac{1}{4} \times 3 \times \frac{3}{16}$ in. Often attacked by the mould, *Sporodinia Aspergillus.*

220. C. angustissima Gill. (from its very narrow gills; *angustus*, narrow) *a b.* Pallid white.

P. broadly umbilicate, very thin, firm. *St.* fibrous, flexuous. *G.* decurrent, very crowded.

Odour none. Woods, amongst leaves. Oct. $2 \times 2\frac{1}{2} \times \frac{1}{8}$ in.

221. C. obsoleta Quél. (from its pallid tint; *obsoletus*, faint) *a c.* Hygrophanous.

P. plano-depressed, pale grey or flesh-colour, soon white. *St.* not flexuous. *G.* adnato-decurrent, somewhat broad, pale grey.

Gregarious. Slightly fragrant of bitter almonds or anise. Amongst grass and leaves. Oct.–Nov. $1\frac{1}{4} \times 1\frac{3}{4} \times \frac{1}{8}$ in.

f. Versiformes.

222. C. ectypa Gill. (from the projecting sooty squamules in the middle of the pileus; Gr. *ektupos*, worked in relief) *a b.*

P. fleshy, plano-depressed, yellowish or brownish; marg. striate. *St.* attenuate upwards, dull light yellow, becoming black at base. *G.* decurrent, white, then rufous-spotted.

Gregarious, often cæspitose. Odour pleasant of anise, becoming fetid. Meadows, damp places. Dec. $2\frac{3}{4} \times 5\frac{1}{4} \times \frac{5}{16}$ in. Often greatly resembles **69.**

223. C. bella Gill. (*bellus*, beautiful) *a c.*

P. convex, umbilicato-depressed, orange; sq. rufescent. *St.* pale orange. *G.* adnato-toothed, broad, yellow, rufescent or pale flesh-colour.

Subcæspitose. Fetid. Plantations, fir, sometimes on rotten fir-wood; uncommon. Sept. $1\frac{5}{8} \times 1\frac{3}{4} \times \frac{3}{16}$ in.

224. C. laccata Quél. (from the red colour—as of gum-lac—of the pileus) *a b c.*

P. convex, plano-umbilicate, then repand, mealy-squamulose, rufescent, becoming ochreous. *St.* usually attenuate downwards, fibrillose, colour as P. *G.* adnato-toothed, flesh-colour.

Odour sometimes of garlic. Woods, forests, commons; common. May–Dec. $2 \times 3 \times \frac{3}{8}$ in. The pileus is sometimes brilliant or blackish-purple, becoming lavender, with stem and gills bright purple (*Agaricus amethystinus* Bolt.). A very small form with a hair-like stem occurs.

Agaricus Sadleri B. & Br. placed in this position by Stevenson is probably the same as **824.** Berkeley has proposed a new genus, *Laccaria*, for the reception of **224** and its allies.

224a. C. proxima Boud. (from its being like the brown form of **224**) *a.*

P. convex, then plane, glabrous, yellowish-brown. *St.* even, smooth, paler than P. *G.* adnato-decurrent, somewhat pink.

Woods. Oct. $1\frac{3}{4} \times 2\frac{1}{4} \times \frac{3}{16}$ in. Differs from **224** in the glabrous P. and warted spores.

VIII. **COLLYBIA** Quél.

(From the pileus, which is frequently small and regular in form like a small coin; Gr. *kollubos*.)

Veil almost obsolete, when present represented by a fibrillose, floccose, or pruinose covering. *Hymenophore* confluent with, but heterogeneous from the cartilaginous stem. *Pileus* slightly fleshy to

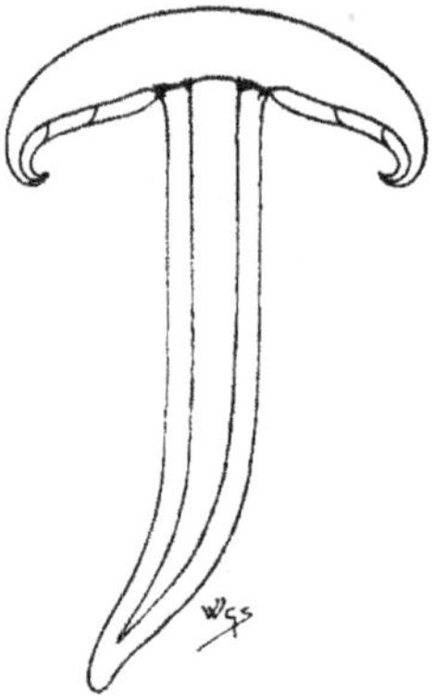

Fig. 17.—Section of *Collybia maculata* Quél. One-third natural size.

thin, margin at first involute. *Stem* central, wholly cartilaginous, or with a cartilaginous bark and stuffed with a pith; often rooting, simple. *Gills* membranous, soft, free or adnexed. *Spores* smooth. (Fig. 17.)

Often small, frequently somewhat tough, maturing more slowly and lasting longer than other Agarics. The species grow on stumps, branches and leaves, sometimes on other fungi or on the ground. Must be carefully distinguished from members of *Marasmius*, where the species are less fleshy, less putrescent, more leathery, and revive with moisture after becoming dry. Species **225—286**

A. *Gills* white or brightly coloured, not ashy. *Flesh* white.

a. *Striæpedes*. *Stem* stout, hollow, or with a spongy pith, longitudinally striate, except **225a.**

Gills broad, except **228,** somewhat distant. **225—229**

Gills crowded, narrow. **230—236**

b. *Vestipedes*. *Stem* thin, equal, fistulose or with a pith, even, velvety, floccose, or pruinose.

Gills broad, somewhat distant. **237—242**

Gills very narrow, very crowded. **243—249**

c. Lævipedes. *Stem* thin, equal, fistulose, naked, smooth except at base, not conspicuously striate.

Gills usually broad, lax, commonly more or less distant. **250—260**

Gills narrow, crowded except **270.** **261—270**

B. *Gills* becoming cinereous. Hygrophanous.

d. Tephrophanæ. Colour fuscous, becoming cinereous.

Gills crowded, somewhat narrow. **271—278**

Gills very broad, except **285,** more or less distant. **279—286**

a. Striæpedes.

225. C. radicata Quél. (from the rooting stem; *radix*, a root) *a b c.*

P. glutinous, umber; mid. radiato-rugose. *St.* rigid, brittle, attenuate downwards, deeply rooting, white. *G.* adnexed or adnate, sometimes with a tooth.

Solitary. On and about stumps, amongst fir-leaves; common. June–Nov. $4\frac{3}{4} \times 1\text{–}2\frac{1}{2} \times \frac{3}{8}$ in. There is a white variety.

225a. C. Henriettæ Sacc. (after Henrietta Smith, who first found it) *a b.*

P. dry, even, somewhat downy, somewhat yellowish-umber. *St.* attenuate upwards, even, slightly rooting, subpruinose, pale, pallid yellowish-brown within and without, somewhat darker below. *G.* broad, broadly adnate, distant, slightly rounded near St. *Flesh* very thin.

On and about trees, stumps, etc. Sept. $4 \times 7\frac{3}{8} \times \frac{1}{4}$ in. Intermediate between **225** and **226**.

225b. C. retigera Bres. (from the net-like veining of the pileus; *rete*, a net, *gero*, to carry) *a b.*

P. thin, campanulate, expanded, subumbonate, dry, glabrous, with a network of anastomosing veins best seen when old, pallid; mid. shaded tawny-brown; marg. striate. *St.* equal, solid to stuffed, pallid, minutely downy at base, somewhat rooting. *G.* somewhat broad, ventricose; edge fimbriate, pallid.

Bases of stumps, ash, beech. Dec. $2 \times 3\frac{1}{4} \times \frac{1}{4}$ in.

226. C. longipes Quél. (from the long stem; *longus*, long, *pes*, a foot) *a b c.*

P. dry, velvety, livid brown. *St.* attenuate downwards, deeply rooting, villous, dark brown, varying paler. *G.* rounded, adnexed, white.

Stumps, etc.; uncommon. Aug.–Oct. $2\frac{1}{4} \times 9\frac{1}{2} \times \frac{1}{2}$ in.

226a. C. eriocephala Rea (from the velvety pileus; Gr. *erion*, wool, *kephale*, a head) *a.*

P. convex, expanded, velvety, fulvous-tawny. *St.* stuffed, then hollow, fusiform, extended into abundant long branched rhizomorphoid proliferous brown mycelium, striate; mid.

slightly velvety, colour as P. above but paler, brown below. *G.* sinuato-adnate, distant, pale to deep ochre. *Flesh* thick at mid., thin at marg., yellow-white, then yellowish, rufous externally at base.

Cæspitose. Interior rotten elm. Jan. $2\frac{1}{4} \times 3\frac{3}{4} \times \frac{3}{8}$ in. Rooting base $4\frac{1}{2}$ in.

227. C. platyphylla Quél. (from the broad gills; Gr. *platus*, broad, *phullon*, a leaf) *a b c.*

P. flat, ashy- or livid-yellow-brown to sepia, streaked fibrillose. *St.* not attenuate or rooting, colour as P. *G.* adnate or adnexed.

Woods, amongst leaves, on rotten logs; rare. May–Oct. $4\frac{3}{4} \times 4\frac{3}{4} \times \frac{5}{8}$ in. Frequently springs from a villous cord-like mycelium. The var. *repens* Gill. resembles **105.**

228. C. semitalis Quél. (from its growing in footpaths, *semita*) *a b c.*

P. plane, hygrophanous, sooty-brown, yellowish or grey when dry. *St.* attenuate upwards, greyish. *G.* sometimes narrow, adnato-rounded, sometimes with a tooth, white, brown-spotted.

Sometimes cæspitose. On the ground after much rain. Oct.–Nov. $3\frac{1}{4} \times 4\frac{1}{4} \times \frac{1}{2}$ in. Sometimes turns black. Compare **227.**

229. C. fusipes Quél. (from the spindle-shaped stem; *fusus*, a spindle, *pes*, a foot) *a b c.*

P. convex, umbonate, brown. *St.* attenuate downwards, rooting, often split, yellowish. *G.* adnexed, pale brown, pinkish, faintly brown-spotted.

Cæspitose. Considered edible by some persons. Stumps; common. June–Nov. $2\frac{1}{4} \times 4\frac{1}{4} \times \frac{1}{2}$ in. Var. *œdematopus* has a stout ventricose St.

229a. C. lancipes Gill. (from a fancied resemblance in a longitudinal section of the stem to the blade of a spear; *lancea*, a spear, *pes*, a foot) *a b.*

P. convex, expanded, umbonate, radiately rugose from mid., dry, glabrous, pale flesh-colour, becoming pallid; marg. striate. *St.* solid or stuffed with twisted filaments, striate, gradually attenuate to the rooting, downy base, whitish or pale flesh-colour. *G.* adnexed, emarginate, broad behind, distant, thick, firm, connected with veins, pale flesh-colour. *Flesh* thick, firm, not watery.

Solitary, scattered, rarely clustered. On the ground. Autumn. $2\frac{1}{2} \times 5 \times \frac{3}{4}$ in. Every part firm and rigid.

230. C. maculata Quél. (from its foxy stains; *macula*, a spot). *a b c.*

White or pale brown, spotted—sometimes wholly—rufescent. *P.* convexo-plane. *St.* hard, subventricose, attenuate downwards, sometimes deeply rooting. *G.* rounded free.

Taste slightly acid, unpleasant, odour not unpleasant. Woods, chiefly pine, beech, alder, moist places. May–Nov. $3\frac{1}{2} \times 4\frac{1}{2} \times \frac{3}{4}$ in. There is a permanently whitish or white var. *immaculata* Mass., with *G.* sometimes abnormal, labyrinthiform, merulioid. Var. *scorzonerea* Gill. Small, yellow; *St.* long-rooting; *G.* yellow.

231. C. fodiens W. G. Sm., *Agaricus fodiens* Kalch. (from the rooting stem; *fodiens*, digging) *a*.

P. convex, faintly yellowish; mid. ochreous-flesh-colour. *St*. ventricose, attenuate downwards, rooting, white. *G*. rounded or emarginate-free, yellowish-white.

Odour of meal. Grass-borders. Oct. $3\frac{1}{8} \times 5\frac{1}{4} \times 1$ in.

232. C. prolixa Gill. (from the stretched out appearance of the pileus; *prolixus*, stretched out) *a b*.

P. campanulate, subumbonate, rufescent-brown; mid. darker. *St*. solid or imperfectly hollow, often contorted, paler than P. *G*. rounded-free, whitish, unspotted.

In dense clusters on leaf-heaps. Aug.–Sept. $4\frac{3}{4} \times 4\frac{1}{4} \times \frac{1}{2}$ in.

233. C. distorta Quél. (from the twisted stem) *a b c*.

P. convex, subumbonate, buff-brown; mid. darker, usually with darker blotches. *St*. attenuate upwards, pale brown. *G*. rounded-free, livid white, stained brown if bruised.

Gregarious or fasciculate. Taste insipid; odour strong, mephitic, like **35**. On tree-roots, pine, rotten pine-trunks, heaps of leaves. Oct.–Nov. $2\frac{3}{4} \times 4 \times \frac{3}{8}$ in. Somewhat resembling **230**.

234. C. butyracea Quél. (from the pileus, buttery to the touch; *butyrum*, butter) *a b c*.

P. plane, subumbonate, pale brown with a darker marginal zone; mid. almost black. *St*. attenuate upwards, pale rufescent or ochreous-brown. *G*. adnexed, white. *Flesh* pale brown.

Woods, fir. Jan.–Dec. $3\frac{1}{4} \times 2\frac{1}{2} \times \frac{1}{2}$ in. Becomes whitish when dry.

234a. C. bibulosa Mass. (from the bibulous substance) *a b*.

P. fleshy, subglobose, obtuse, expanded, moist, very smooth, even, deep olive- or blackish-green when moist, pale grey to whitish when dry. *St*. cartilaginous, spongy within, stuffed to imperfectly hollow, attenuate upwards, minutely striate, pale clear warm-brown, often rooting into the wood. *G*. adnexo-rounded, thin, edge subundulate, dingy.

In groups of three or four. Stumps. Autumn. $2 \times 3 \times \frac{3}{8}$ in.

235. C. stridula Quél. (*strideo*, to creak, derivation obscure), *a b*.

P. broadly umbonato-plane, dark brown. *St*. fibrous, paler than P. *G*. adnexed, pale brown.

On the ground. Oct. $2\frac{1}{4} \times 2\frac{1}{2} \times \frac{1}{4}$ in.

235a. C. pulla Gill. (from *pullus*, dark-coloured) *a b*.

P. fleshy, thin, fragile, campanulate, then expanded, obtuse, smooth, even, glabrous, hygrophanous, shining dark rich brown, purplish-bay, livid purplish-slate or nearly black, paler when dry. *St*. hollow, twisted, somewhat striate, soft, naked, sometimes slightly rooting, whitish. *G*. adnexed, somewhat broad, crowded, transversely pellucid-striate, whitish. *Flesh* white.

Under birch-trees, amongst *Polytrichum*. Oct. $2\frac{1}{2} \times 3\frac{7}{8} \times \frac{1}{4}$ in.

236. C. xylophila Gill. (Gr. *xulon*, timber, *philos*, loving) *a*.

P. campanulate, subpapillate, very thin, ivory-white; mid. ivory-buff. *St*. ivory-buff, pale yellowish-brown within. *G*. adnate with a tooth, white.

Gregarious, cæspitose. Stumps and rotten trunks of elm, etc. Sept. $3\frac{1}{2} \times 3 \times \frac{1}{4}$ in.

b. Vestipedes.

237. C. velutipes Quél. (from the velvety stem; *vellus*, fleece, *pes*, a foot) *a b c*.

P. plane, fleshy, viscid, yellow; mid. brown. *St*. yellow above, brown and almost black towards base, or wholly brown. *G*. adnate or adnexed, pallid white or yellowish.

Cæspitose. Taste agreeable. Trunks, stumps, sometimes on timber; common. Aug.–April. $2\frac{5}{8} \times 2\frac{3}{4} \times \frac{5}{16}$ in. Sometimes springs from a luxuriant golden byssoid mycelium (*Ozoneum auricomum* Link), sometimes from thick cord-like, branching and deeply rooting mycelium or from a sclerotium. *See* **1569**. The var. *rubescens* Sacc. becomes foxy-spotted.

238. C. laxipes Quél. (from the lax stem; *laxus*, lax, *pes*, a foot) *a b*.

P. convexo-plane, moist, white. *St*. white above, rich brown below. *G*. adnexo-free, white.

On wood, chips, twigs. Feb. $\frac{7}{8} \times 3\frac{1}{2} \times \frac{1}{16}$ in.

239. C. mimica Sacc. (from its resemblance to **709**) *a b*.

P. flat, thin, smooth, with a separable cuticle, pale yellow-buff; mid. buff-brownish. *St*. solid, apex subpruinose, with brownish bark, attenuate downwards, fibrilloso-strigose at base, yellow-buff above, deep brown below. *G*. adnate with a tooth.

Taste and odour strong of fish. On deal-shavings. Nov. $1\frac{5}{8} \times 2 \times \frac{1}{8}$ in.

240. C. floccipes, Gill. (from the floccose-woolly stem; *floccosus*, floccose, *pes*, a foot) *a b*.

P. plane, umbonate, deep umber. *St*. hollow, black punctato-floccose on a white ground. *G*. adnexed, white.

On the ground and about trunks. Sept. $1\frac{1}{8} \times 2 \times \frac{1}{16}$ in.

241. C. vertirugis Sacc. (from the rugose pileus; *vertex*, the top, *ruga*, a wrinkle) *a b c*.

P. convexo-plane, radiato-wrinkled, minutely pulverulent, brown-ashy or indigo-lead colour. *St*. strigose at base, brown, paler above. *G*. adnate, whitish.

Rootstocks of dead ferns, twigs, stumps, etc. Sept.–Oct. $\frac{7}{8} \times 3\frac{1}{8} \times \frac{1}{16}$ in.

242. C. stipitaria Gill. (from its habitat; *stipes*, a stem) *a b c*.

P. umbilicato-plane, velvety or fibrillose; mid. pale ochre to sienna; marg. whitish. *St*. tough, fibrilloso-shaggy, light to dark sienna. *G*. adnexo-free, white.

Frequent. July–Nov. $\frac{5}{8} \times 1\frac{3}{8} \times \frac{1}{16}$ in.

243. C. hariolorum Quél. (from its fancied use by soothsayers; *hariolus*, a soothsayer) *a b.*

P. campanulato-convex, smooth, pale ochreous-white; mid. ochreous. *St.* colour as P. or gradually dark fuscous below, naked and pallid above. *G.* adnato-free, whitish.

Densely gregarious, somewhat cæspitose. Odour strong. Woods, amongst dead leaves, beech, sometimes on rotten wood. Sept. $1\frac{3}{4} \times 2\frac{3}{8} \times \frac{1}{8}$ in. Closely allied to *Marasmius.*

244. C. confluens Quél. (from the confluent stems) *a b c.*

P. convexo-plane or subumbonate, tough, white-buff. *St.* compressed, pulverulento-woolly, white or pale brownish-lavender or purplish.

In troops, or connate confluent rows. Amongst fir-leaves; frequent. June–Oct. $2 \times 3\frac{1}{8} \times \frac{1}{4}$ in. Closely allied to *Marasmius.*

245. C. ingrata Quél. (from the unpleasant odour; *ingratus*, unpleasant) *a b.*

P. umbonato-expanded, very thin, pale livid buff. *St.* brownish-purple, white mealy above. *G.* free, colour as P.

Cæspitose. Taste none, or slightly sweet; odour slight, mouldy, fungoid, sweetish-unpleasant. Woods, pine, damp places; uncommon. Aug.–Oct. $2\frac{3}{4} \times 3\frac{1}{2} \times \frac{1}{4}$ in. There is a smaller form one-half the size of type.

246. C. conigena Quél. (from its habitat; *conus*, a cone, *gigno*, to bear) *a b c.*

P. convexo-plane or slightly umbonate, smooth, whitish-ochre or brownish. *St.* strigose at base, white. *G.* adnexo-free, white.

Usually gregarious. Under conifers on the cones and amongst the leaves. June–Dec. $1 \times 2\frac{1}{4} \times \frac{1}{8}$ in. Must not be confounded with **259** and **260.**

247. C. cirrhata Quél. (from the twisted base of the stem; *cirrus*, a curl) *a b c.*

P. convexo-plane, slightly silky, white; mid. pale sienna. *St.* flexuous, attenuate and twisted at the smooth or strigoso-fibrillose base. *G.* adnate, white.

In troops. Amongst leaves of fir, etc., and decaying fungi. Aug.–Nov. $\frac{5}{8} \times 3 \times \frac{1}{16}$ in. Must not be confounded with **1424a.**

248. C. tuberosa Quél. (from the tuberous base of the stem) *a b c.*

P. convexo-plane, subumbonate, slightly silky, white; mid. faintly ochreous. *St.* slightly pulverulent, base smooth, sometimes fibrillose, occasionally proliferous, attached to a sclerotium. *G.* adnate, white.

Gregarious. On dead Agarics, *Russulæ* (**1306, 1307**), *Lactarii*, on *Polyporus squamosus*, fir-leaves, rotten fir-twigs, amongst moss, etc. Aug.–Nov. $2\frac{3}{4} \times 4\frac{1}{4} \times \frac{9}{16}$ in. Fries says the summer form has no sclerotium; this is very doubtful. The autumnal forms may have no sclerotium as the floccose mycelium condenses in winter and forms a summer sclerotum.

249. C. racemosa Quél. (from the glandular hairs round the stem; *racemus*, a cluster) *a b.*

P. expanded, papillate, subtomentose, white or faintly greyish. *St.* racemose, apparently proliferous with minute hairs bearing glands or abortive pilei, colour as P. Attached to *Sclerotium lacunosum*. *G.* adnate, white.

Terrestrial and on putrid fungi, chiefly Agarics; very rare. Autumn. $\frac{3}{8} \times 1\frac{3}{4} \times \frac{1}{32}$ in. Turns black in drying.

c. Lævipedes.

250. C. collina Quél. (from its frequent habitat; *collinus*, a hill) *a b c*.
P. expanded, subumbonate, somewhat viscid, pale yellowish-brown. *St.* colour as P. *G.* adnexo-toothed, white.

Usually gregarious. Grassy slopes, about beech-stumps; uncommon. Oct. $2 \times 4\frac{1}{4} \times \frac{3}{16}$ in.

251. C. thelephora Sacc. (from its bearing a small umbo on the pileus; Gr. *thele*, a nipple, *phero*, to bear) *a b*.
P. campanulate, brown or yellow-brown, paler at marg. *St.* brown above and whitish below, or the reverse. *G.* adnato-toothed, white to yellowish.

Gregarious. Inodorous. Peat-bogs, partly dry sphagnum-bogs. $1\frac{5}{8} \times 3\frac{1}{2} \times \frac{1}{8}$ in. A near ally of **250.**

252. C. ventricosa Gill. (from the enlarged lower part of stem; *venter*, the belly) *a b c*. Pale warm-brownish.
P. campanulato-convex, slightly umbonate. *St.* attenuate upwards and attenuato-rooting, darker below. *G.* broadly adnato-toothed.

Woods. Oct. $2\frac{3}{8} \times 3\frac{5}{8} \times \frac{3}{16}$ in.

253. C. Stevensonii Sacc. (after the Rev. Dr. Stevenson, mycologist) *a*.
P. semi-ovate, viscid, pallid yellow. *St.* attenuate downwards, deeply rooting, whitish-yellow above, brownish below. *G.* very broad, adnato-toothed, white.

Old pastures. Aug. $\frac{5}{8} \times 2\frac{3}{8} \times \frac{1}{16}$ in. Resembling in habit members of *Psathyra*.

254. C. psathyroides Sacc. (from its resemblance to members of *Psathyra*) *a*. Ivory-white.
P. semi-ovate or conic, subviscid. *St.* not rooting. *G.* broadly adnato-toothed.

Oct. $\frac{7}{8} \times 3\frac{3}{4} \times \frac{1}{8}$ in. Greatly resembles a *Mycena*. Allied to **253.**

255. C. xanthopus Quél. (from the yellow stem; Gr. *xanthos*, yellow, *pous*, a foot) *a b c*.
P. campanulato-convex, subumbonate, whitish to deep biscuit; um. sienna. *St.* strigose at base, yellow to sienna-biscuit. *G.* adnexo-free, whitish.

Pine-woods, amongst cones, leaves and twigs. July–Nov. $1\frac{3}{4} \times 2\frac{1}{2} \times \frac{1}{8}$ in. Not unlike **1394** in general appearance.

256. C. nitellina Gill. (from its colour; *nitella*, a dormouse) *a b*.
P. convexo-plane, smooth, reddish-tan or brown, sometimes pale. *St.* yellow above, brown below, or tan. *G.* adnate, attenuate in front, tan.

Solitary or gregarious. Odour none. By roads and paths in woods. May–Oct. $2\frac{1}{4} \times 3\frac{1}{4} \times \frac{3}{16}$ in. There is a very small umbonate variety. Allied to *Clitocybe*.

257. C. succinea Quél. (from its colour; *succinum*, amber) *a b c*.
P. plano-convex, smooth, deep rich brown. *St.* colour as P. *G.* adnate, white.

Solitary. Odour none. Woods, mixed, grassy places, under trees. May–Dec. $1\frac{5}{8} \times 1\frac{5}{8} \times \frac{1}{4}$ in.

258. C. nummularia Gill. (from a fancied resemblance in the pileus to a small coin, *nummulus*) *a c*. White, sometimes faintly shaded ochreous or reddish.
P. expanded, slightly depressed, somewhat umbonate. *St.* attenuate downwards. *G.* adnexo-free.

In troops, cæspitose. Woods, mixed, amongst leaves and twigs. July–Nov. $1\frac{3}{4} \times 2\frac{1}{8} \times \frac{1}{8}$ in.

259. C. esculenta Quél. (from its edible qualities) *a b c*.
P. plane, dull ivory-ochreous to umber. *St.* attenuate downwards, deeply rooting, yellow to pale reddish-brown. *G.* adnexed to adnato-toothed, whitish.

Gregarious. Edible. Taste bitter, agreeable to some, unpleasant to others. Pastures, grassy places near plantations, pine-woods, sometimes on hotbeds; common. April–May. $1 \times 5 \times \frac{1}{16}$ in. Sold in Austrian markets as food under the name Nagelschwämme.

260. C. tenacella Quél. (*tenax*, tough) *a b c*.
P. convexo-plane, subumbonate, brown or brownish to white. *St.* attenuate and fibrillose at base, whitish above, rufescent below. *G.* adnexo-toothed, whitish.

Solitary or in troops. Taste pleasant. On leaves and cones in pine-woods, and under cedars. April–Dec. $1\frac{1}{8} \times 3\frac{3}{4} \times \frac{1}{8}$ in. The var. *stolonifer* Quél. roots for four or more inches. It is difficult to distinguish **260** from **259**, and neither must be confused with **246**.

260a. C. planipes Sacc. (from the plane stem; *planus*, plane, *pes*, a foot) *a*.
P. orbicular, expanded, somewhat viscid, bay. *St.* rooting, colour of P. *G.* free.

Cæspitose. Woods. Sept. $\frac{7}{8} \times 1\frac{3}{8} \times \frac{1}{10}$ in.

261. C. acervata Gill. (from its growing in dense clusters; *acervus*, a heap) *a b*.
P. hemispherical, then flat, tan or brown. *St.* tan- to deep rufous-brown. *G.* adnexo-free, salmon.

Woods, fir, dead wood; frequent. Sept.–Oct. $2\frac{1}{2} \times 3\frac{1}{2} \times \frac{1}{4}$ in.

262. C. dryophila Quél. (Gr. *drus*, oak, *philos*, loving) *a b c*.
P. plane, tan, brown or brownish when soddened; mid. faintly ochreous. *St.* villous at base, tan to brown. *G.* adnexo-free, whitish. *Flesh* white.

Dangerous when cooked according to a case cited by Dr. Badham (*Esculent Funguses*, 1863, p. 107). The taste when raw leaves a burning sensation on the fauces and a sense of suffocation. Amongst leaves, woods, hedge-bottoms, gardens, grassy places; common. May–Oct. $3\frac{1}{4} \times 4\frac{3}{4} \times \frac{1}{4}$ in. Variable, gills sometimes sterile. Var. *funicularis* Karst. Large, tufted, decumbent; *G.* sulphur.

263. C. aquosa Gill. (from its watery substance ; *aqua*, water) *a b c*.
P. plane to subdepressed, tan-ochreous to ochreous. *St*. fibrillose at base, ochreous-brown. *G*. adnexo-free, pale yellowish.
Woods, damp places, amongst moss. Autumn. $1\frac{3}{4} \times 2\frac{3}{4} \times \frac{1}{8}$ in.

264. C. extuberans Quél. (from the broad umbo ; *extuberans*, swelling) *a b c*.
P. olive-brownish. *St*. attenuate downwards, pale grey-brownish. *G*. adnexo-toothed, white.
Usually solitary when on the ground, cæspitose and smaller when on trunks. Odour none. Amongst grass, hillsides. May–Sept. $1\frac{3}{4} \times 3\frac{1}{4} \times \frac{3}{16}$ in.

265. C. exsculpta Gill. (from the cut-out, sinuato-rounded gills) *a b c*.
P. plane, subumbilicate, dull ochreous-orange or umber, pale whitish-buff when dry ; mid. darker. *St*. equal or subbulbous, ochreous or sulphur, darker below. *G*. adnato- or adnexo-toothed, subdecurrent, sulphury.
Gregarious. Taste and odour strong, mushroom-like, but with a suggestion of tar. Trunks, dead wood, on the ground. May–Oct. $1\frac{1}{2} \times 2 \times \frac{1}{8}$ in. Dry, tough, reviving with moisture after the manner of a *Marasmius*.

266. C. macilenta Gill. (*macies*, leanness) *a*.
P. convexo-plane, subumbonate, deep yellow. *St*. yellow. *G*. adnexo-free, bright yellow. *Flesh* pallid yellowish.
Woods, pine, on leaves. Autumn. $1\frac{1}{8} \times 2\frac{1}{8} \times \frac{1}{16}$ in. Resembles some of the slender yellow *Hygrophori*.

267. C. clavus Quél. (from the nail-like shape, *clavus*) *a c*.
P. hemispherical, orange-vermilion. *St*. substrigose at base, whitish, brownish below. *G*. adnexed, whitish.
On fir, twigs, leaves, moss ; rare. Oct. $\frac{1}{4} \times \frac{3}{4} \times \frac{1}{32}$ in.

268. C. ocellata Quél. (from the eye-like umbonate umbilicus ; *ocellus*, a little eye) *a b c*.
P. white ; mid. pale sienna. *St*. attenuate downwards, rooting, fibrillose, brownish-white, or buff above and white below. *G*. adnate, white.
Pine-woods, amongst leaves, grassy places, on *Jungermannia* ; uncommon. Sept.–Oct. $\frac{5}{8} \times 2 \times \frac{1}{16}$ in.

269. C. muscigena Quél. (from its habitat ; *muscus*, moss, *gigno*, to bear) *a*. White.
P. hemispherical, pellucid, thin. *G*. adnate.
Amongst moss and grass, near trunks. Aug. $\frac{3}{8} \times 1\frac{1}{2} \times \frac{1}{16}$ in.

270. C. leucomyosotis Sacc. (from its resemblance to a pale form of **736,** *Naucoria Myosotis* ; Gr. *leukos*, white) *a b*.
P. convex, broadly umbonate, pale translucent umber ; marg. striate. *St*. brittle, subpruinose above, colour as P. *G*. adnato-sinuate, thick, somewhat distant, whitish.
Strong-scented, somewhat fragrant. On *Sphagnum* in wet places in bogs. May. $1\frac{3}{8} \times 3\frac{7}{8} \times \frac{1}{4}$ in. Whitish when dry.

d. Tephrophanæ.

271. **C. rancida** Quél. (from its rancid odour) *a b c.*
P. convex, umbonate, sometimes viscid, dark umber, slate or slate-grey, or slate with mid. umber. *St.* subfusiform, rooting, sometimes deeply, colour as P. *G.* adnexo-free, slaty flesh-colour, grey-white or almost white.
Taste disagreeable; odour strong and offensive when cut or broken, like rancid meal. Woods, amongst trunks. Oct.–Nov. $1\frac{1}{2} \times 4 \times \frac{1}{8}$ in.

272. **C. eustygia** Sacc. (Stygian, from its becoming black) *a.*
P. convexo-plane, subdepressed, white or white-brownish. *St.* attenuate downwards, rooting, white above, grey and punctato-scaly below. *G.* adnexo-rounded, dark grey.
Odour of rancid meal. On the ground. Oct. $2\frac{1}{8} \times 3\frac{1}{8} \times \frac{5}{16}$ in. Black when dry.

273. **C. coracina** Gill. (from its colour; Gr. *korax*, a raven) *a c.*
P. convexo-expanded, black to dark- or grey-umber. *St.* attenuate upwards, not rooting, white above, brown below. *G.* adnate, greyish or yellowish.
Odour strong of new meal. Grassy places, fir-plantations. Nov. $1\frac{3}{8} \times 1\frac{1}{8} \times \frac{3}{16}$ in.

274. **C. ozes** Karst. (Gr. *ozo*, to smell) *a.*
P. convexo-plane, subumbonate, grey- or clay-brown. *St.* slightly attenuate upwards, white above, sooty-grey below. *G.* adnate, smoky-olivaceous.
Odour of new meal. On the ground and on pine-leaves. Feb. $1\frac{1}{8} \times 3\frac{1}{4} \times \frac{1}{8}$ in.

274a. **C. mephitica** Karst. (from the offensive mephitic odour, sometimes, however, like new meal) *a.*
P. fleshy, convex, subumbonate, hygrophanous, smooth, dry, silky, greyish-ochre, becoming whitish. *St.* equal, enlarged below, stuffed, grey, white-floccose or pruinose-velvety. *G.* obtusely adnate, separating from stem, attenuate in front, crowded, grey. *Flesh* yellowish.
Amongst fir-needles. Oct. $1\frac{3}{8} \times 2\frac{3}{8} \times \frac{3}{16}$ in.

275. **C. inolens** Quél. (*inolens*, scentless) *a.*
P. convexo-plane, umbonate, tan, greyish or rufous. *St.* attenuate upwards, slightly rooting at the white, strigose base. *G.* adfixed, almost free, whitish- or ochreous-grey.
Woods, chiefly pine. Sept.–Nov. $2\frac{1}{8} \times 2\frac{3}{4} \times \frac{1}{4}$ in.

276. **C. plexipes** Gill. (from the twisted stem; *plecto*, to twist, *pes*, a foot) *a b c.*
P. hemispherical, umbonate, pale umber or greyish; mid. darker. *St.* attenuate downwards, rooting, pale umber. *G.* narrowly adnate, whitish-grey. *Flesh* pale umber.
Odour none. Woods, beech, amongst grass; uncommon. Sept.–Nov. $1\frac{1}{4} \times 6 \times \frac{1}{8}$ in.

277. C. atrata Quél. (*ater*, black) *a b c*.

P. plano-depressed, greasy feeling, blackish-umber. *St.* fuscous within and without. *G.* adnato-decurrent, livid-brownish.

Scattered or in groups. Odour strong, disagreeable. On burnt ground often in company with **183**, **685** and **1374**. Sept.–Dec. $1\frac{1}{2} \times 1\frac{1}{2} \times \frac{1}{8}$ in. Must not be confounded with **369**.

278. C. ambusta Quél. (from its habitat, burnt ground; *ambustus*, burnt) *a b c*.

P. convexo-plane, papillate, umber. *St.* paler than P. *G.* adnate with a tooth, pale umber. *Flesh* colour as G.

Odour of meal. July–Nov. $\frac{3}{4} \times 1\frac{1}{8} \times \frac{1}{16}$ in.

279. C. lacerata Gill. (from the torn pileus) *a*. White-grey or slightly ochreous-brown.

P. convex, shining, often cracked, sometimes brown-fibrillose. *St.* twisted. *G.* adnexed, thick.

Somewhat cæspitose. Woods, pine; rare. Autumn. $2 \times 3 \times \frac{1}{4}$ in.

280. C. murina Gill. (from its mouse-colour; *mus*, a mouse) *a*.

P. convex, subdepressed, brown, becoming paler. *G.* attenuato-adnexed.

Woods, under oaks. Oct. $1\frac{1}{8} \times 2\frac{3}{8} \times \frac{1}{8}$ in.

281. C. protracta Gill. (from the long stem; *protractus*, drawn out) *a b*.

P. convexo-plane, slightly umbonato-depressed, grey-brown or slaty flesh-colour. *St.* attenuate downwards, rooting, strigose, colour as P., dark slate-colour below. *G.* adfixed, ventricose, grey, white-pruinose.

Mossy ground, near stumps of pine, etc., amongst dead leaves. Aug.–Nov. $1\frac{1}{2} \times 8 \times \frac{1}{8}$ in.

282. C. tesquorum Gill. (from its growing in waste places, *tesqua*) *a*.

P. convex or umbonate, fuscous-black or olive-brown. *St.* white above, brown below. *G.* adnexo-free, ashy-fuscous or pale olive-brown.

Scattered. Amongst grass, sunny banks. $\frac{3}{4} \times 1\frac{1}{2} \times \frac{1}{16}$ in.

283. C. clusilis Gill. (*clusilis*, easily closing, derivation obscure) *a b c*.

P. convexo-plane, depressed, whitish tan or greyish; mid. faintly ochreous. *St.* shining, colour as P., or paler. *G.* adnato-toothed, colour as St.

Amongst moss, grass, often with **597**. Sept.–Oct. $1\frac{3}{8} \times 2\frac{1}{2} \times \frac{1}{8}$ in.

284. C. tylicolor Gill. (from its colour, that of a small worm; Gr. *tulos*) *a b c*.

P. plano-umbonate, ashy-grey or brown; mid. darker. *St.* wavy, whitish-pulverulent on a grey-brown ground. *G.* adnato-free, varying in attachment, livid-brownish or grey. *Flesh* pale brown.

Taste slightly rank; odour weak or none. Shady woods amongst grass. Oct. $1\frac{1}{4} \times 2\frac{1}{8} \times \frac{1}{8}$ in.

285. C. DOROTHEÆ Sacc. (after Lady Dorothy Neville) *a*.
P. globose, then flat, slightly umbilicate, very thin, dark brown; marg. denticulate. *St.* slightly swollen at base, minutely velvety, white, yellowish or rufous. *G.* adnexed, narrow, white.

On dead fern-stems in hothouses. $1\frac{1}{8} \times 3\frac{1}{4} \times \frac{1}{16}$ in.

286. C. CALDARII Sacc. (from its habitat, a caldarium or hot-house) *a*.
P. hemispherical, slightly umbonate, brown. *St.* paler than P. *G.* adnato-decurrent, somewhat ashy.

On *Sphagnum* in orchid-pot. $\frac{5}{8} \times 2\frac{1}{8} \times \frac{1}{16}$ in.

IX. **MYCENA** Quél.

(Gr. *mukes*, a fungus.)

Veil almost obsolete, only seen in fibrillæ, pile, flock, tomentum or gluten. In Section *h*, *Basipedes*, the edge of the pileus is connected with the circumference of the basal disc in infancy, and

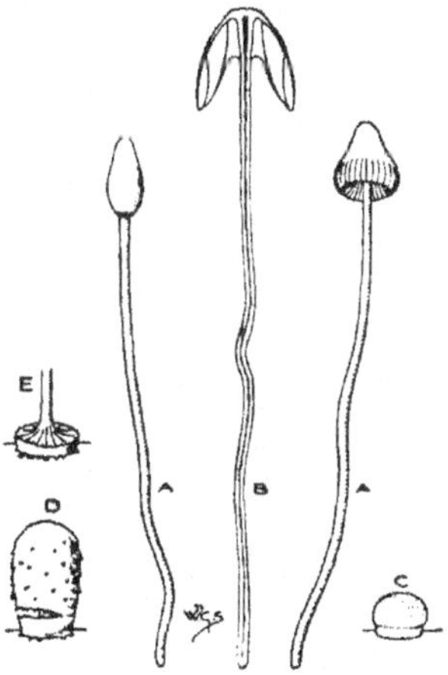

Fig. 18.—A, A, *Mycena epipterygia* Quél.; B, ditto in section. One-half natural size. C, *M. tenerrima* Quél., young state showing basal disc and veil; D, *M. stylobates* Quél., young state, showing basal disc and veil; E, basal disc; C, D, E, enlarged.

so forms a kind of universal veil. *Hymenophore* confluent with but heterogeneous from the cartilaginous stem. *Pileus* at first conico-cylindrical, becoming more or less campanulate, submembranous, not cartilaginous, never truly umbilicate, margin more or less striate, at first adpressed to the stem, never involute. A few species as the next genus is approached have the pileus slightly depressed. *Stem* simple, slender, central, fistulose (except **310**) and cartilaginous, sometimes hair-like, sometimes rooting amongst dead leaves. *Gills* adnate or adnexed, often sinuate, varied in attachment, not truly decurrent, often minutely denticulate or fringed at the edge. (Fig. 18.)

Most of the species grow on twigs, stumps, dead branches and dead leaves; some grow on the ground. They are usually slender, small, beautiful and scentless, some have a strong alkaline or mephitic odour; some contain a coloured juice termed milk. They appear after rain in summer and autumn; a few remain till early winter. Species **287—366**

a. Calodontes. Stem juiceless, base not dilated into a disc. *Gills* saw-edged or fimbriate, darker edged. **287—297**

b. Adonideæ. Colour pure and bright, not becoming fuscous or ashy, except **300**. *Stem* juiceless, base not dilated into a disc. *Gills* of one colour, including the edge. Single, terrestrial, except **306** and sometimes **305**. **298—306**

c. Rigidipedes. Stem firm, tough, juiceless, somewhat strigose and rooting. *Gills* white, changing to grey or reddish. Persistent, odourless, usually growing on stumps and very cæspitose, but single examples occur on the ground. **307—316**

d. Fragilipedes. Pileus hygrophanous. *Stem* fragile, juiceless, not dilated into a disc at the base. *Gills* changing colour. Normally terrestrial and single, a few cæspitose and growing on wood. Strong smelling. **317—334**

e. Filipedes. Not hygrophanous. *Stem* juiceless. *Gills* changing colour. Terrestrial amongst grass; usually single. **335—342**

f. Lactipedes. Whole plant exuding a white or coloured juice. **343—349**

g. Glutinipedes. Stem glutinous. Must not be confounded with species with smooth, moist or slippery stems. **350—356**

h. Basipedes. A small dilated base, bulb or disc at the base of the dry, equal, stem. Usually solitary. **357—361**

i. Insititiæ. Stem apparently inserted or grafted into dead leaves, etc. Soon flaccid, very tender. **362—366**

a. Calodontes.

287. M. pelianthina Quél. (from the livid pileus; Gr. *peliaino*, to make livid) *a b.*

P. umbonato-convex, pale brown, dull flesh-colour or lavender, whitish when dry. *St.* ochreous-whitish or whitish-lavender.

G. adnate, purplish or brownish-white, minutely black-dotted at edge. *Flesh* white to faintly purplish.

Woods, beech, oak, stumps and dead leaves. Aug.–Oct. $1\frac{5}{8} \times 3\frac{5}{8} \times \frac{3}{16}$ in. Somewhat resembles **224.**

287a. M. carneosanguinea Rea (from the change of colour in the flesh—on section—from white to blood-red; *caro*, flesh, *sanguis*, blood) *a*.

P. convex, subumbonate, smooth, livid-grey; mid. tinging to rufous, paler when old. *St*. grey, yellowish below, base clad with short mycelial threads. *G*. adnate, dull purplish-brown with a crimson-red edge. *Flesh* moderately thick at mid. of P. and apex of St., white, changing above to bright rose or blood-red.

Woods. Aug. $1\frac{1}{8} \times 1\frac{3}{4} \times \frac{1}{4}$ in.

288. M. balanina Karst. (from its frequent habitat, beech-mast; Gr. *balanos*) *a b*.

P. convexo-umbonate, granulose, umber-ochreous. *St*. attenuate downwards, satiny, white above, deep brown below. *G*. rounded, free, sprinkled and fringed with dull purple spiculæ.

Beech-mast, acorns, oak-leaves; *rare*. Sept. $1\frac{7}{8} \times 3\frac{7}{8} \times \frac{1}{8}$ in.

289. M. iris Quél. (from its rainbow tints; Gr. *iris*, the rainbow) *a b c*.

P. hemispherical, evanescent blue-fibrillose on a pale umber, lead-coloured or greenish ground; marg. sometimes cobalt. *St*. pilose, pale umber above, bright blue-fibrillose at base. *G*. adnexo-free, forming a collar behind, sometimes denticulate, ashy.

On decayed twigs, stumps, chips, sticks and cones of fir, often deeply rooting amongst pine-leaves. Sept.–Nov. $\frac{5}{8} \times 3\frac{1}{4} \times \frac{1}{16}$ in.

290. M. mirabilis Quél. (*mirabilis*, wonderful) *a*.

P. conical, dull lavender-blue; mid. ochreous. *St*. floccose, subrooting, bluish below. *G*. adnexed, white, edge dotted with dark particles.

Fir-trunks, amongst *Hypnum*. Aug. $\frac{1}{2} \times 3 \times \frac{1}{16}$ in. Perhaps the same as **291.**

291. M. marginella Quél. (from the minute fringe at the margin of the gills) *a*.

P. conico-convex, floccose, pallid; marg. dull bluish. *St*. tomentose and bluish below. *G*. adnate or adnexed, white, edge bluish- or reddish-flocculose.

Fir-trunks amongst *Hypnum cupressiforme*. Aug.–Sept. $\frac{3}{4} \times 2\frac{3}{4} \times \frac{1}{16}$ in.

292. M. aurantiomarginata Quél. (from the orange margin of the gills) *a*.

P. hemispherical, umbonate, olive-brown. *St*. attenuate upwards, pale yellowish. *G*. attenuato-adnexed with a tooth, whitish, floccose at edge.

Odour strong. Woods, silver fir. Nov. $1\frac{5}{8} \times 1\frac{5}{8} \times \frac{3}{16}$ in.

293. M. elegans Quél. (*elegans*, neat) *a b c.*

P. campanulate, livid yellowish; mid. fuscous. *St.* floccose at base, livid-yellowish. *G.* adnato-decurrent, whitish or yellowish, edge orange.

Odour weak of fennel or none. Woods, chiefly pine, fir-leaves, stems of *Rubus;* frequent. Aug.–Nov. $\frac{3}{4} \times 2\frac{5}{8} \times \frac{1}{16}$ in.

294. M. rubromarginata Gill. (from the red margin of the gills) *a c.*

P. hemispherical, subumbonate, white or pale reddish. *St.* white. *G.* adnato-decurrent, whitish or grey, edge rose or brown-purple.

Odour none. On branches, leaves and rotten twigs in woods, chiefly pine and larch; frequent. July–Nov. $\frac{5}{8} \times 1\frac{1}{2} \times \frac{1}{16}$ in. A variety, *fusco-purpurea* Mass., with a dusky purple pileus, grows on willows.

295. M. olivaceomarginata Mass. (from the olive margin of the gills) *a b.*

P. conico-hemispherical, subumbonate, honey-colour. *St.* colour as P., base tinged rufous. *G.* uncinato-adnexed, broad, subdistant, pallid with a brownish-olive edge.

Woods, lawns. Oct. $\frac{3}{4} \times 1\frac{7}{8} \times \frac{1}{16}$ in.

296. M. strobilina Gill. (from its usual habitat, fir-cones, *strobilus*) *a b.*

P. campanulate, acutely umbonate, scarlet, paler at marg. *St.* scarlet, white-strigose at base. *G.* adnato-decurrent, paler than P., edge blackish-scarlet. *Flesh* crimson.

Cones, leaves and twigs of Scotch fir, sometimes in beech-woods; rare. Oct. $\frac{3}{4} \times 2 \times \frac{1}{16}$ in.

297. M. rosella Quél. (*roseus*, rose-coloured) *a b c.*

P. campanulate, subumbonate, deep or pale rose. *St.* colour as P., seldom strigose at base. *G.* adnate, pale rose, edge dark purple.

Laxly gregarious. Amongst leaves, twigs, and decaying cones in fir-woods; uncommon. June–Dec. $\frac{5}{8} \times 1\frac{3}{4} \times \frac{1}{16}$ in.

b. Adonideæ.

298. M. pura Quél. (from its pure colour) *a b c.*

P. campanulate, broadly umbonate, brittle, flesh-colour, variable in shade. *St.* often rooting, flesh-colour, pale ochreous below. *G.* sinuato-adnate, veined, whitish to pale flesh-colour.

Taste and odour strong, hot, disagreeable, pungent of radish or cabbage-stalk. Woods, meadows; common. June–Nov. $2 \times 4 \times \frac{1}{4}$ in. There is a white variety. Varies larger and smaller, sometimes very dark when dry. Sometimes attacked by *Mucor macrocarpus.*

299. M. pseudopura Sacc. (from its resemblance to **298**; Gr. *pseudos*, false) *a.*

Structure and colour same as **298**, of which it is probably one of the smaller forms. Colour rosy. *G.* whitish.

Odour none. Woods. Autumn. $1 \times 2\frac{3}{4} \times \frac{1}{16}$ in.

300. M. zephira Quél. (from its coming with the rain-wind or west wind Zephyros) *a c.*

P. hemispherical, diaphanous, white flesh-colour; mid. often fuscous. *St.* white to flesh-colour. *G.* adnate, veined, white.

Decayed fir, amongst dead leaves in woods. Sept.–Dec. $1\frac{3}{8} \times 3\frac{1}{8} \times \frac{1}{8}$ in. *G.* and *P.* sometimes foxy-stained.

301. M. Adonis Quél. (from its beauty—like Adonis) *a b.*

P. conico-campanulate, white, yellow, greenish, orange or crimson. *St.* lighter than P. *G.* ascending, adnexo-toothed or adnate, colour as St.

Woods, pastures; uncommon. Sept.–Nov. $\frac{5}{8} \times 2\frac{1}{4} \times \frac{1}{16}$ in.

302. M. lineata Quél. (from the lined or sulcate pileus) *a b.*

P. plane, subumbonate, membranous, pale yellowish. *St.* colour as P. *G.* adnate, white. *Flesh* pale brownish.

Amongst moss. Sept.–Oct. $1 \times 2\frac{1}{4} \times \frac{1}{16}$ in. Colour variable.

302a. M. farrea Quél. (from the pileus, the colour of ripe wheat and powdered as if with flour; *farreus*, made of spelt) *a.*

P. very thin, campanulato-expanded, subumbonate, pale buff-yellowish, becoming pale, shining white pulverulent; marg. sulcate, crenulate. *St.* somewhat rooting, silky-striate, pale steel-grey, white pulverulent above. *G.* adnate, thin, connected with veins, somewhat fimbriate, white.

Mossy and grassy places. $1 \times 3\frac{1}{8} \times \frac{1}{16}$ in. This and **336a** are the only species of the genus with pulverulent pilei.

303. M. luteoalba Quél. (*luteus*, yellow, *albus*, white) *a b c.*

P. expanded, umbonate, pale yellow. *St.* white, faint yellowish at base. *G.* adnato-toothed, shining white.

Woods, fir, amongst moss; uncommon. Aug.–Nov. $\frac{5}{8} \times 1\frac{3}{4} \times \frac{1}{16}$ in.

304. M. flavoalba Quél. (*flavus*, light yellow, *albus*, white) *a b c.*

P. expanded, broadly umbonate, ochre-biscuit, paler at marg. *St.* faint yellowish at base. *G.* adnexed or adnate, distant, white.

In troops. Woods and meadows, under firs; frequent. July–Dec. $1 \times 1\frac{3}{8} \times \frac{1}{16}$ in.

305. M. lactea Quél. (from the milk-white colour; *lac*, milk) *a b c.*

P. campanulate, umbonate, varying faint biscuit. *St.* attenuate upwards, fibrillose at base. *G.* adnato-rounded or sinuate, crowded.

Scattered, rarely cæspitose. Woods, chiefly amongst and on pine-leaves; common. July–Dec. $1 \times 2\frac{3}{4} \times \frac{1}{16}$ in. Must be carefully distinguished from other species of the genus which become white.

306. M. gypsea Quél. (from its colour, like gypsum) *a b.*

P. conico-campanulate, umbonate; mid. faintly biscuit. *St.* attenuate upwards, strigose at the faintly biscuit base. *G.* adnexo-ascending or adnate.

Cæspitose. Tasteless; odour alkaline, varying strong to weak. Stumps. June–Nov. $\frac{7}{8} \times 3\frac{3}{4} \times \frac{1}{8}$ in.

c. Rigidipedes.

307. M. cohærens Gill. (from the adherent stems; *cohæro*, to stick together) *a c.*

P. campanulate, broadly umbonate, pale umber, cinnamon or orange-brown; mid. darker; marg. striate. *St.* even, smooth, shining whitish above, downy and ochreous-whitish to dark brown below. *G.* rounded-free, distant, white to yellowish or greyish. *Flesh* brown.

Sometimes solitary. Brambles, pine-leaves, etc., on the ground. Oct.–Feb. $1\frac{5}{8} \times 4\frac{1}{4} \times \frac{1}{8}$ in. Resembles **313** (var. *calopus*).

307a. M. Berkeleyi Mass. (after the Rev. M. J. Berkeley) *a.*

P. campanulate, expanded, subumbonate, hygrophanous, dull brown; um. darker and brighter, paler when dry; marg. striate. *St.* with a tapering, rooting base, somewhat striate, dull pale brownish with a purple tinge. *G.* adnato-sinuate with a decurrent tooth, subdistant, ventricose, thin, tinged purplish or flesh-colour. *Flesh* pale flesh-colour, tinged purplish.

Solitary or subcæspitose. Tree-trunks. $3\frac{1}{4} \times 5 \times \frac{1}{4}$ in. Rooting base, $2\frac{1}{4}$ in. May be the same as **309**.

308. M. prolifera Gill. (from the stem, sometimes densely proliferous with small individuals at the base) *a b.*

P. campanulato-expanded, broadly umbonate, pale umber; marg. sulcate. *St.* ventricose, splitting, white. *G.* adnexed or adnate, white.

Inodorous, at length nauseous. Woods and gardens, on gravel, near bushes. Sept.–Oct. $1\frac{7}{8} \times 6 \times \frac{3}{16}$ in. There is a white variety.

309. M. excisa Gill. (from the cut-out gills) *a c.*

P. expanded, umbonate, brown; mid. darker. *St.* as in **308**, pale brown, white above. *G.* adnexed or sinuato-free, thick, distant, hoary, veined, white, pale brown or purplish.

Woods, trunks, sometimes on the ground. Sept.–Oct. $2\frac{3}{8} \times 4 \times \frac{1}{4}$ in. Var. *fagetorum* Karst. Small, amongst beech-leaves.

310. M. psammicola Sacc. (from its habitat; Gr. *psammos*, sand) *a.*

P. hemispherical, sprinkled with minute particles, brown; marg. whitish. *St.* solid, rooting, white pulverulént on a brownish ground. *G.* adnate, white.

Sandy banks amongst moss. Sept. $\frac{3}{8} \times 1\frac{1}{2} \times \frac{1}{32}$ in.

311. M. rugosa Quél. (from the wrinkled pileus; *ruga*, a wrinkle) *a b.*

P. expanded, subumbonate, pale grey-brown; mid. brownish. *St.* subventricose, colour as P. *G.* adnato-toothed, whitish.

Tasteless or at length slightly bitter. On and near stumps; common. July–Dec. $2\frac{1}{2} \times 3\frac{3}{8} \times \frac{1}{8}$ in. Very tough and dry.

312. M. sudora Gill. (from the viscid pileus; *sudor*, sweat) *a b c.*

P. campanulate, umbonate, whitish, livid-biscuit or brownish. *St.* colour as P. *G.* adnate, white, flesh-colour or yellowish-white.

Usually solitary. Stumps, beech. Aug.–Nov. $1\frac{7}{8} \times 5\frac{1}{4} \times \frac{1}{8}$ in. Colour often like **350** or **1228**.

313. M. galericulata Quél. (from the shape of the pileus; *galericulum*, a little cap) *a b c*.

P. expanded, broadly umbonate, pale brown; mid. darker. *St.* colour as P. *G.* adnato-toothed, veined, white to pale rose.

Usually tasteless. Stumps, bramble, willow, and on the ground; common. July–Jan. $2\frac{3}{8} \times 4\frac{1}{2} \times \frac{3}{16}$ in. Var. *calopus* Karst. has an umber pileus and dark sienna stem. *Mucor macrocarpus* is frequent on this species.

314. M. polygramma Quél. (from the longitudinally striate stem; Gr. *polus*, many; *gramme*, a stroke) *a b c*.

P. campanulate, umbonate, pale to dark brown. *St.* steel-grey. *G.* adnato-toothed, attenuate near St., pale salmon.

Gregarious. Odour disagreeable, or none. Stumps; common. Sept.–Jan. $2 \times 5\frac{5}{8} \times \frac{1}{8}$ in.

315. M. parabolica Quél. (from the parabolic curves of the pileus) *a b c*.

P. conico-campanulate, sides incurved, slate; mid. umber. *St.* attenuate upwards; colour as P., paler below. *G.* adnato-ascending, white or faint grey. *Flesh* brownish-grey.

Gregarious or cæspitose. Rotten wood, stumps, especially fir, willow, fern; uncommon. Sept.–Dec. $1\frac{7}{8} \times 2\frac{1}{2} \times \frac{1}{8}$ in.

316. M. tintinnabulum Quél. (from the bell-shaped pileus; *tintinnabulum*, a bell) *a b*.

P. plane, subumbonate, umber, varying bluish, yellow-brownish and whitish. *St.* white. *G.* adnate with a tooth, white or pale flesh-colour.

Gregarious, sometimes cæspitose. Fallen trunks, beech. April–Dec. $1\frac{3}{8} \times 1\frac{1}{8} \times \frac{1}{8}$ in.

d. Fragilipedes.

317. M. CODONICEPS Sacc. (from the bell-shaped pileus; Gr. *kodon*, a bell; *kephale*, a head) *a b*.

P. campanulate, obscurely sulcate and rugulose, hispid, umber; marg. recurved. *St.* attenuate downwards, moist, smooth, white above, umber below. *G.* vertically ascending, somewhat thick, distant, adnate with a tooth, white.

Single or cæspitose. Tree-fern stems. June. $\frac{1}{16} \times \frac{1}{2} \times \frac{1}{64}$ in.

318. M. flavipes Quél. (from the yellow stem; *flavus*, yellow, *pes*, a foot) *a b*.

P. hemispherical, rose or flesh-colour. *St.* tough, villous at base. *G.* adnate, paler than P.

Odour of radishes. Stumps. Autumn. $\frac{3}{4} \times 2\frac{3}{4} \times \frac{1}{16}$ in.

319. M. atroalba Gill. (*ater*, black, *albus*, white) *a b c*.

P. broadly campanulate, dark slate-grey; marg. whitish. *St.* with a large strigose bulb, apt to split as in **320**, grey and white. *G.* free, white, then glaucous.

Woods, amongst moss. July–Nov. $1\frac{3}{8} \times 3\frac{1}{4} \times \frac{1}{8}$ in. Shape and colour of **315**, habit different.

320. M. dissiliens Quél. (from the stem splitting when broken; *dissilio*, to burst asunder) *a b.*

P. campanulato-flat, sulcate, ashy-brown; marg. whitish. *St.* ashy. *G.* rounded-free, white.

Odour insipid, sometimes strong, rancid rather than alkaline. Trunks, branches, on the ground amongst grass; frequent. July–Nov. $1\frac{3}{4} \times 2\frac{3}{4} \times \frac{1}{8}$ in.

321. M. atrocyanea Gill. (from the blackish-indigo pileus; *ater*, black, *cyaneus*, dark blue) *a b c.*

P. campanulato-conical, umbonate, sulcate. *St.* slightly attenuate upwards, colour as P. *G.* ascending, attenuato-adnate. *Flesh* pale slate.

Solitary or cæspitose. Odour none. Amongst fir-leaves, on the ground. Sept.–Nov. $\frac{5}{8} \times 3\frac{3}{8} \times \frac{1}{16}$ in. Becoming black when dry.

322. M. pullata Sacc. (*pullus*, dark-coloured) *a.*

P. campanulate, umbonate, sulcate, dark umber, becoming paler with age. *St.* slightly attenuate upwards, colour as P., base white-floccose. *G.* adnexed, white.

Odour slightly nitrous. Amongst dead leaves. Oct.–Nov. $\frac{7}{8} \times 3\frac{5}{8} \times \frac{1}{16}$ in.

323. M. paupercula Sacc. (from its appearance; *pauper*, poor) *a b.*

P. hemispherical, white or faint tawny. *St.* attenuato-rooting, white. *G.* adnexo-free, white.

Odour of fresh meal, or sweet. Inside decayed stumps, oak; uncommon. July–Sept. $\frac{3}{16} \times \frac{7}{8} \times \frac{1}{32}$ in.

324. M. leptocephala Gill. (from the delicate pileus; Gr. *leptos*, fine, *kephale*, a head) *a.*

P. expanded, umbonate, sulcate, ashy; mid. darker. *St.* colour as P. *G.* emarginate, ashy-white.

Solitary. Odour strong of nitre. On the ground and on trunks. Autumn. $\frac{1}{4} \times 2\frac{1}{8} \times \frac{1}{16}$ in.

325. M. alkalina Quél. (from its alkaline odour) *a b c.*

P. campanulate, umbonate, brown or ashy; mid. darker. *St.* brown or yellow. *G.* adnato-ascending, white, or ashy with a white edge.

Solitary or cæspitose. Odour strong, disagreeable. Stumps, etc., bramble, privet, fir; uncommon. Jan.–Dec. $1\frac{5}{8} \times 5 \times \frac{1}{8}$ in. Variable in colour, often exhibiting shades of yellow or pale rose. There is a white variety.

326. M. ammoniaca Quél. (from its ammoniacal odour) *a b.*

P. conico-campanulate, papillate, deep brown to greyish; marg. ashy. *St.* whitish-grey or brownish. *G.* adnate, whitish or grey.

Solitary. Odour strong. On the ground, amongst grass. Sept.–Nov. $1 \times 3\frac{3}{8} \times \frac{1}{8}$ in.

327. M. metata Quél. (from the pileus, like a *meta*, the conical column at the end of the Roman Circus) *a b c.*

P. conico-campanulate, cream-grey, flesh-colour or livid. *St.* colour as P. *G.* adnato-ascending, whitish.

Gregarious. Odour alkaline, strong to weak or none. Woods, amongst moss. Oct.–Nov. $\frac{7}{8} \times 3\frac{1}{4} \times \frac{1}{8}$ in.

328. M. plicosa Gill. (from the sulcate pileus; *plico*, to fold) *a b c*.
P. expanded, umbonate, often split, pale tan; mid. sienna; marg. greyish. *St.* colour as P., white at base. *G.* adnate, greyish.

Woods, on the ground; rare. Sept.–Oct. $1\frac{1}{4} \times 1\frac{7}{8} \times \frac{1}{8}$ in.

329. M. peltata Gill. (from the shield-like pileus; *pelta*, a small shield) *a b*.
P. plane, with a central projecting ring, fuscous-blackish or greyish. *St.* pale umber. *G.* sinuate, pale grey.

Woods, commons, etc., amongst moss; rare. Aug.–Oct. $1\frac{1}{4} \times 2\frac{3}{4} \times \frac{1}{16}$ in.

330. M. consimilis Sacc. (from its likeness to various other species of the genus) *a*.
P. campanulate, umbonate, splitting, pale grey; mid. umber. *St.* pale umber. *G.* adnato-ascending, pale grey.

Odour none. Amongst grass. Autumn. $1\frac{3}{8} \times 2 \times \frac{1}{8}$ in.

331. M. ætites Quél. (from the colour; Gr. *ætites*, iron ore) *a b c*.
P. expanded, umbonate, sulcate, opaque brown-ochre. *St.* paler than P. or pale livid-umber. *G.* adnato-sinuate, veined, white or whitish-grey. *Flesh* pale brown-ochre.

Scattered or loosely gregarious. Odour bitter or none. Woods, amongst moss and grass. June–Oct. $1\frac{1}{4} \times 1\frac{7}{8} \times \frac{1}{8}$ in.

332. M. stannea Quél. (from the colour; *stannum*, tin) *a b*.
P. expanded glistening, grey, sometimes umber. *St.* pale tan. *G.* adnato-toothed, whitish-grey.

Scattered or loosely gregarious. Odour none. Woods, amongst grass, on the ground. Sept.–Oct. $\frac{3}{4} \times 3\frac{1}{8} \times \frac{1}{16}$ in.

333. M. vitrea Quél. (from its fragility; *vitrum*, glass) *a b*.
P. campanulate, subumbonate, umber; mid. darker. *St.* glistening, paler than P. *G.* sinuato-adnate, whitish.

In troops. Odour none. Woods, fir, moist places; uncommon. Sept.–Oct. $1\frac{3}{8} \times 4\frac{1}{4} \times \frac{1}{8}$ in.

334. M. tenuis Gill. (*tenuis*, thin) *a b c*.
P. campanulato-convex, white; mid. pale brown; marg. toothed. *St.* white. *G.* adnate, whitish.

Woods, shady moist places, often on the root-fibres of trees; rare. Sept. $\frac{3}{4} \times 5 \times \frac{1}{16}$ in. Watery, delicate, fragile, wavering.

e. Filipedes.

335. M. filopes Quél. (from the thread-like stem; *filum*, a thread, *pes*, a foot) *a b c*.
P. expanded, pale livid-brown or whitish. *St.* rooting, pilose at base, white. *G.* adnexo-free, white or greyish.

Singly or in troops. Woods, amongst leaves. Aug.–Nov. $1 \times 6\frac{1}{4} \times \frac{1}{32}$ in.

336. M. amicta Quél. (from the pulverulent pileus and stem; *amicio*, to clothe) *a*.

P. conico-campanulate, subumbonate, greenish, pale slate or livid. *St*. rooting, livid. *G*. adnexo-free, pale grey with lighter edge.

Gregarious. Odour none. Amongst moss and leaves, about roots of *Nardus stricta*. Sept. $\frac{3}{8} \times 3\frac{1}{2} \times \frac{1}{32}$ in.

336a. M. plumbea Karst. (from the colour of the pileus; *plumbeus*, leaden).

P. membranous, convexo-flattened, obtuse, sulcate, sprinkled white-pruinose. *St*. fragile, ashy-pulverulent, white-hyaline above, strigose below. *G*. adnate, horizontal, colour as P.

Odour none. Mossy pastures. Autumn. $\frac{5}{8} \times 4$ in.

336b. M. cinerea Mass. & Crossl. (from its colour; *cinis*, ashes). Wholly grey.

P. slightly gibbous or obtuse, then expanded; marg. striate, paler and silky when dry. *St*. glabrous, base white-downy. *G*. adnate, subdistant; edge paler.

Amongst short grass. Odour strong of radishes. $1 \times 2\frac{3}{4} \times \frac{1}{16}$ in. Differs from **324** in sulcate P.; from **327** in white G.; from **336a** in absence of odour.

337. M. debilis Quél. (*debilis*, weak) *a c*.

P. conico-campanulate, subumbonate, livid-whitish or brownish-salmon. *St*. colour as P., fibrillose at rooting base. *G*. broadly adnate, whitish.

Gregarious. Woods, chestnut, fir, amongst moss. Oct. $\frac{3}{8} \times 2\frac{1}{2} \times \frac{1}{32}$ in.

338. M. vitilis Quél. (from the deep striæ of the pileus; *vitilis*, plaited) *a b*.

P. conico-campanulate, sometimes papillate, brownish, becoming whitish. *St*. shining, livid. *G*. attenuato-adnate or slightly sinuate, whitish or grey, or dark ashy with a white edge.

Solitary or gregarious. Odour none. Woods, mixed amongst leaves; frequent. Sept.–Nov. $\frac{1}{2} \times 4\frac{3}{8} \times \frac{1}{32}$ in.

339. M. collariata Quél. (from the union of the gills in a minute collar at the top of the stem) *a c*.

P. campanulato-convex, subumbonate, greyish-white, rose-tinged; mid. fuscous. *St*. smooth, whitish-brown. *G*. adnate, hoary white or faintly salmon.

Woods, amongst grass and oak-bark. Oct.–Nov. $\frac{5}{8} \times 2\frac{1}{2} \times \frac{1}{16}$ in.

340. M. speirea Gill. (from the twisted base of the stem; Gr. *speira*, a coil) *a*.

P. plane, umbonate, pallid grey or whitish; mid. dark fuscous. *St*. attenuato-rooting, fibrillose and brown below, white above. *G*. adnate, thin, deeply decurrent, white.

In troops during very rainy weather. Woods, mixed, mossy trunks; uncommon. Aug.–Oct. $\frac{1}{2} \times 1\frac{7}{8} \times \frac{1}{32}$ in.

341. M. tenella Quél. (*tener*, tender) *a*. White or livid flesh-colour.
P. campanulato-convex, pellucid. *G.* adnato-decurrent.

Cæspitose. Decayed trees. Autumn. $\frac{1}{2} \times \frac{7}{8} \times \frac{1}{32}$ in.

342. M. acicula Quél. (from its resemblance to a small pin, *acicula*) *a b c.*

P. conico-campanulate, papillate, orange-vermilion. *St.* attenuate and fibrilloso-rooting, deep yellow. *G.* adnate or adnexo-rounded, yellow with a whitish edge or wholly white.

Wood, leaves, twigs, etc.; frequent. July–Oct. $\frac{3}{8} \times 1\frac{3}{4} \times \frac{1}{32}$ in.

f. Lactipedes.

343. M. hæmatopus Quél. (from the blood-red juice of the stem; Gr. *haima*, blood, *pous*, a foot) *a b c.*

P. conico-campanulate, subumbonate, whitish flesh-colour. *St.* pulverulent, colour as P. *G.* adnate, whitish. *Milk* deep brown-crimson.

Cæspitose. Taste slightly disagreeable. Stumps, old timber, alder, birch; frequent. Aug.–Nov. $1\frac{5}{8} \times 3 \times \frac{1}{8}$ in.

344. M. cruenta Quél. (from the blood-like juice; *cruor*, gore) *a b.*
P. conico-campanulate, subumbonate, brown. *St.* tense, villous-rooting, paler than P. *G.* adnate, whitish. *Milk* deep brownish-crimson.

Solitary or subcæspitose. Woods, pine-cones, etc.; rare. Sept. $1 \times 3\frac{1}{4} \times \frac{1}{16}$ in.

345. M. sanguinolenta Quél. (from the blood-red juice; *sanguinolentus*, bloody) *a b c.*

P. hemispherical, brown or purplish-brown. *St.* flaccid, fibrillose at base, colour as P. *G.* pale purplish-brown; edge sometimes dark purple. *Milk* brownish-crimson.

Gregarious. Woods, chips, twigs, fir-leaves, moss, etc., tree-fern stems in greenhouses. May–Nov. $\frac{1}{2} \times 3\frac{5}{8} \times \frac{1}{32}$ in. Sometimes wholly pale brown and varying $\frac{1}{4}$ to $\frac{1}{2}$ in. high.

346. M. crocata Gill. (from the saffron-coloured juice; *crocus*, saffron) *a b.*

P. convex, olivaceous, ashy, yellowish or white; mid. reddish. *St.* tense, deep yellow, orange-crimson at fibrillose base. *G.* adnato-ascending or adnexed, white.

Scattered. Woods, beech, amongst leaves; uncommon. Autumn. $\frac{7}{8} \times 4\frac{3}{4} \times \frac{1}{16}$ in. A small var. appears on beech-trunks.

347. M. chelidonia Quél. (from the yellow juice, like that of *Chelidonium majus*) *a b c.*

P. hemispherical, yellowish flesh-colour; mid. ochreous. *St.* firm, yellow at the villous-rooting base. *G.* adnate, white or yellow-white. *Milk*, saffron to vinous-red.

Subcæspitose. Odour somewhat alkaline. Stumps, beech, alder; rare. Oct. $\frac{5}{8} \times 2\frac{7}{8} \times \frac{1}{16}$ in.

348. M. galopus Quél. (from the white juice; Gr. *gala*, milk, *pous*, a foot) *a b c*.

P. hemispherical, subumbonate, dark then pale purplish-flesh-colour; mid. blackish. *St*. colour as P., base villous. *G*. adnexed or adnate, white or glaucous. *Milk* white.

In troops. Almost tasteless or like radishes. Woods, amongst moss and leaves; common. July–Nov. $\frac{5}{8} \times 2\frac{7}{8} \times \frac{1}{16}$ in. Sometimes wholly dark in colour.

349. M. leucogala Sacc. (from the milk-white juice; Gr. *leukos*, white, *gala*, milk).

P. campanulate, umbonate, purple-brown. *St*. lighter than P., white tomentose at base. *G*. adnate with a tooth, ashy.

Cæspitose. Amongst grass, rotten stumps. Oct. $\frac{7}{8} \times 3 \times \frac{1}{16}$ in.

g. Glutinipedes.

350. M. epipterygia Quél. (from its frequent growth on brake-fern, *Pteris aquilina*) *a b c*.

P. conical with a separable viscid pellicle, dark olivaceous then pale livid-sulphur. *St*. greenish-sulphur. *G*. sinuato-ascending, whitish or greyish. *Flesh* pale sulphur.

Solitary, gregarious or cæspitose. Odour none. Woods, amongst grass, leaves, dead bracken, on manure-heaps; common. Aug.–Dec. $\frac{3}{4} \times 5\frac{1}{2} \times \frac{1}{16}$ in. There is a white var.

351. M. clavicularis Gill. (from its shape; *clavus*, a nail) *a b*.

P. expanded, subumbonate, dry, pelliculose, white or light yellowish. *St*. tough, slightly viscid, whitish. *G*. adnato-decurrent, whitish.

Woods, fir, grassy places. Oct. $\frac{3}{4} \times 2\frac{1}{8} \times \frac{1}{16}$ in.

352. M. pelliculosa Quél. (from the thin separable pellicle of the pileus) *a b c*.

P. at first umbonate then plane, viscid, brownish; mid. darker. *St*. rigid, livid grey-whitish. *G*. adnato-decurrent, thick, white.

In troops during very rainy weather. Taste none. Heaths, heathy woods; frequent. Sept.–Dec. $\frac{7}{8} \times 2\frac{1}{2} \times \frac{1}{16}$ in.

353. M. vulgaris Quél. (*vulgaris*, common) *a b c*.

P. convex, depressed, papillate, subviscid, brownish or ashy-grey. *St*. attenuate and slightly strigose below, ashy-grey. *G*. adnato-decurrent, white.

Gregarious. Odour strong of new meal, or none. Woods, fir, amongst moss, on dead twigs, larch and pine leaves; common. July–Nov. $\frac{1}{2} \times 1\frac{3}{4} \times \frac{1}{32}$ in.

354. M. citrinella Quél. (from its colour; *citrus*, lemon) *a b c*.

P. hemispherical, subviscid. *St*. pale brown villous at base. *G*. adnato-ascending, at first white.

Gregarious. Woods, pine, rotten trunks; uncommon. Oct. $\frac{5}{16} \times \frac{1}{8} \times \frac{1}{32}$ in. The var. *candida* Gill. is at first white, sulphur in drying.

355. **M. plicatocrenata** Gill. (from the folded and margin-notched pileus; *plica*, a fold, *crena*, a notch) *a*.

P. conic, subumbonate, white then light yellow. *St.* pale pallid-reddish. *G.* adnato-ascending, white.

Amongst heath and moss, under larches. Oct. $\frac{5}{8} \times 1\frac{3}{4} \times \frac{1}{32}$ in.

356. **M. rorida** Quél. (from the dripping jelly of the stem; *roridus*, dewy) *a b c*.

P. convex, whitish-ochre, pale buff, faintly ashy or white. *St.* white. *G.* arcuato-decurrent, white.

Taste none; odour somewhat rank, strong. Dead sticks, twigs, bramble; rare. June–Sept. $\frac{1}{2} \times 1\frac{5}{8} \times \frac{1}{32}$ in.

h. Basipedes.

357. **M. stylobates** Quél. (from the shape of the stem; Gr. *stulos*, a pillar, *basis*, a base) *a b c*. White or faint grey.

P. campanulate, sprinkled with minute hairs. *St.* pruinose above. *G.* free.

Dead grass, leaves, rushes, fern-stems, twigs, bark, chips, in wet places; uncommon. June–Oct. $\frac{3}{8} \times 2\frac{1}{8} \times \frac{1}{32}$ in. In infancy the edge of the pileus is attached to the edge of the basal disc and forms a veil.

358. **M. tenerrima** Quél. (*tenerrimus*, very tender) *a b c*. White.

P. hemispherical, pruinose. *St.* pubescent below. *G.* free.

Gregarious. Mossy bark of trees, willow, dead bramble, fir-cones, dead wood, sticks, chips, etc.; uncommon. Aug.–Mar. $\frac{1}{16} \times \frac{3}{8} \times \frac{1}{64}$ in. Edge of pileus and disc as in 357.

359. **M. discopus** Gill. (from the basal disc of the stem; Gr. *diskos*, a disc, *pous*, a foot) *a b*. White.

P. conical, subumbonate, pulverulent. *G.* adnate.

Twigs, acorns, dead wood, etc., in moist places. Autumn. $\frac{1}{16} \times \frac{3}{8} \times \frac{1}{64}$ in.

360. **M. saccharifera** Gill. (from its sugary appearance under a lens; *saccharon*, sugar, *fero*, to bear) *a b c*. Whitish.

P. hemispherical, covered with minute subglobose granules. *G.* subdecurrent.

Dead bramble, rose, furze, nettle. Nov.–Mar. $\frac{1}{16} \times \frac{3}{16} \times \frac{1}{64}$ in. In var. *electica* Mass. the *P.* is sulcate and *G.* adnate.

361. **M. pterigena** Quél. (from its habitat; *pteris*, a fern, *gigno*, to bear) *a b c*.

P. hemispherico-oblong, pale rose with orange marg. *St.* paler than P. *G.* adnate, pallid with bright orange edge.

Gregarious, sometimes fasciculate. Dead *Lastrea*, veins of dead leaves, etc.; uncommon. Sept.–Nov. $\frac{3}{16} \times 1\frac{3}{4} \times \frac{1}{32}$ in.

i. Insititiæ.

362. **M. corticola** Quél. (from its habitat; *cortex*, bark, *colo*, to inhabit) *a b c*.

P. hemispherical, depressed, brownish, varying grey. *St.* slightly furfuraceous, paler than P. *G.* sinuate, colour as St.

Amongst moss, on living ash, bramble, oak, willow, apple, pear; common. June–Jan. $\frac{5}{16} \times 1\frac{1}{2} \times \frac{1}{32}$ in.

363. M. hiemalis Quél. (from its time of growth; *hiems*, winter) *a c*. Whitish or flesh-colour, rarely brownish or slate.

P. hemispherical. *St.* slightly downy below. *G.* adnato-ascending, narrow.

Scattered. Trunks, oak, willow; rare. Sept.–Mar. $\frac{5}{8} \times 1\frac{1}{8} \times \frac{1}{16}$ in.

364. M. setosa Gill. (from the hairy stem; *seta*, a bristle) *a b*. White, becoming brownish.

P. hemispherical. *G.* adnate.

Densely gregarious. Dead beech-leaves; rare. Oct.-Nov. $\frac{1}{16} \times \frac{3}{4} \times \frac{1}{128}$ in.

365. M. capillaris Quél. (from the hair-like stem; *capillus*, a hair) *a b c*. White.

P. plane, slightly depressed. *St.* smooth, attenuate downwards. *G.* adnato-decurrent.

Woods, on herbaceous stems, moss and dead leaves, chiefly beech, oak. Sept.–Nov. $\frac{1}{16} \times \frac{3}{4} \times \frac{1}{128}$ in. Chiefly seen in very rainy weather.

366. M. juncicola Gill. (from its habitat; *juncus*, a rush, *colo*, to inhabit) *a*.

P. acutely conic, rose-colour, rufescent, blood-red or tawny. *St.* brownish. *G.* adnate, white or pale yellowish.

Rushes in bogs, twigs; rare. June–July. $\frac{1}{16} \times \frac{5}{8} \times \frac{1}{128}$ in.

X. OMPHALIA Quél.

(From the umbilicus; Gr. *omphalos*.)

Veil almost obsolete, only seen in pruina, squamules or tomentum of the pileus and stem, and rarely in a basal disc to which the edge

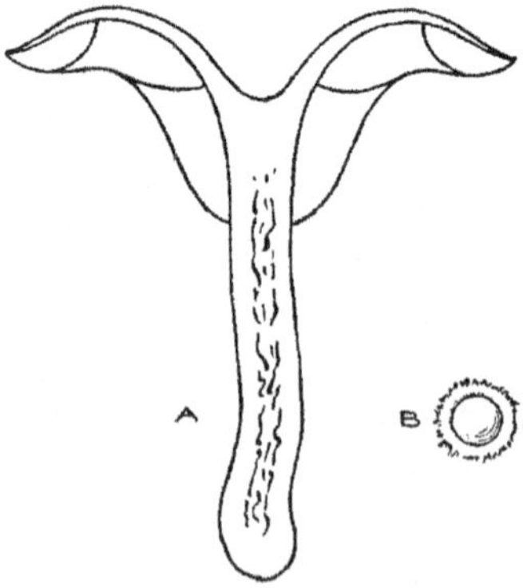

Fig. 19.—A, section of *Omphalia muralis* Quél. × 3. B, *O. pseudodirecta* W. G. Sm.; young state seen from above to show volva-like base.

of the pileus is attached in infancy. *Hymenophore* confluent with, but heterogeneous from the cartilaginous stem. *Pileus* somewhat membranous, umbilico-infundibuliform, hygrophanous. *Stem* simple, central, cartilaginous, fistulose or stuffed, more or less lustrous, sometimes hair-like. *Gills* decurrent. *Spores* more or less elliptical, smooth. (Fig. 19.)

Many of the species are small, beautiful and generally epiphytal. Odour obsolete or nearly so. Mostly peculiar to hilly positions, preferring damp, woody situations and rainy seasons.

The species must not be confounded with the thick-gilled species of similar structure found amongst the *Canthareli*.

Species **367—406**

COLLYBIARIÆ (from the resemblance to *Collybia*). *Pileus* with a deflexed margin.

a. Hydrogrammæ. Comparatively large. *Gills* narrow, very crowded, arcuate, from the first truly decurrent. **367–370**

b. Pyxidatæ. Intermediate in size. *Gills* narrow, slightly distant, attenuate at both ends. **371–382**

c. Umbelliferæ. *Gills* broad, commonly thick, very distant. Must not be confounded with *Xerotus*. **383–392**

MYCENARIÆ (from the resemblance to *Mycena*). *Pileus* at first companulate, margin straight, pressed to the stem.

d. Campanellæ. *Gills* broad, perfect, unequal. **393–403**

e. Integrellæ. *Gills* fold-like, narrow. **404–406**

COLLYBIARIÆ.

a. Hydrogrammæ.

367. O. hydrogramma Quél. (from its watery streakings; Gr. *hudor*, water, *gramme*, a line) *a b.*

P. subconvex, striate, broadly umbilicate, tan. *St.* subattenuate and villous below, tan. *G.* livid whitish or tan.

Somewhat cæspitose. In woods amongst dead leaves, beech. Oct. $2\frac{5}{8} \times 3\frac{3}{8} \times \frac{1}{4}$ in. Resembles **172** in appearance.

367a. O. detrusa Gill. (from the umbilicus—as if pushed in; *detrudo*, to push down) *a b.*

P. somewhat fleshy, convex, expanded, at length umbilicate, even, glabrous, indistinctly zoned, dark grey or whitish-umber. *St.* stuffed to hollow, glabrous, dark grey or dull lavender. *G.* subdecurrent with a tooth, thin, crowded, whitish or somewhat pale ochreous.

Woods, amongst moss. Autumn. $1\frac{5}{8} \times 2\frac{1}{4} \times \frac{3}{16}$ in.

368. O. umbilicata Gill. (from the umbilicate pileus) *a.*

P. plane, deeply umbilicate, tan to fuscous. *St.* silky striate above, villous below, whitish-buff. *G.* whitish.

Somewhat cæspitose. Woods, chiefly pine, amongst moss. Autumn. $1\frac{5}{8} \times 1\frac{1}{2} \times \frac{3}{16}$ in.

369. O. maura Gill. (from its swarthy appearance; *maurus*, a Moor) *a b*.

P. plane, striate, deeply umbilicate, sooty-brown. *St.* sometimes rooting, livid to dark brown. *G.* white.

Odour of new meal. Lawns and moist places. Oct.–Nov. $1\frac{3}{4} \times 1\frac{1}{2} \times \frac{1}{8}$ in. Must not be confounded with 277.

370. O. offuciata Gill. (from its appearance; *offucia*, paint) *a b*.

P. deeply depressed, rarely plane, with a small papilla, tan to brownish or flesh-colour. *St.* colour as P., brown within at base. *G.* colour as P.

Solitary or slightly fasciculate. Under beeches. Oct.–Nov. $1\frac{3}{8} \times 1\frac{3}{8} \times \frac{1}{8}$ in. The salmon-colour of the gills must not cause this to be confounded with Series II. *Hyporhodii.*

b. Pyxidatæ.

371. O. chrysophylla Gill. (from the golden-yellow gills; Gr. *chrusos*, gold, *phullon*, a leaf) *a b*.

P. plane, broadly umbilicate, ochreous-rufescent, tan or grey-hoary when dry. *St.* colour as G.

Gregarious. Decayed pine, stumps, chips, sawdust. Aug.–Oct. $2\frac{1}{4} \times 2 \times \frac{3}{16}$ in. The colour of gills and stem must not cause this to be confounded with *Flammula.*

372. O. Postii Karst. (after H. von Post) *a b*.

P. plane, broadly umbilicate, striate, bright orange or rufescent. *St.* light yellow to whitish. *G.* whitish to grey-white.

Charcoal beds, swampy places. July–Oct. $2\frac{1}{8} \times 2\frac{1}{4} \times \frac{1}{8}$ in. The var. *aurea* Mass. bright golden-yellow, amongst *Sphagnum* in swamps.

373. O. pyxidata Quél. (from the shape of the pileus; *pyxis*, a small box) *a b c*.

P. infundibuliform, striate, rufous-fuscous to umber or pallid. *St.* colour as P. *G.* white to brown. *Flesh* pale brown.

Lawns and woods amongst grass and moss; frequent. July–Nov. $1\frac{1}{2} \times 1\frac{3}{8} \times \frac{1}{8}$ in. Sometimes $\frac{1}{8} \times \frac{1}{4}$ in.

374. O. leucophylla Gill. (from the white gills; Gr. *leukos*, white, *phullon*, a leaf) *a b c*.

P. plane, broadly depressed, at length subrimose, dark ashy to pale or dark umber. *St.* colour as P., usually white at top and bottom.

Taste insipid; odour somewhat strong. Woods, moist places, amongst short grass. April–Sept. $1\frac{5}{8} \times 1\frac{3}{4} \times \frac{1}{8}$ in.

375. O. strisepileus Gill. (from the striate pileus) *a b c*.

P. plane, depressed, rarely with a small papilla, livid olivaceous or brownish; mid. brown. *St.* pale livid umber, becoming fuscous. *G.* whitish.

Woods, amongst moss and grass. Oct.–Nov. $1\frac{3}{8} \times 1\frac{1}{2} \times \frac{1}{8}$ in.

376. O. telmatiæa Sacc. (from its habitat; Gr. *telmatiaios*, occurring in marshy places) *a b*.

P. infundibuliform, minutely virgate, dark brown or fuscous to livid. *St*. slate or tan flesh-colour. *G*. livid or pale brown.

On *Sphagnum*. Aug. 2⅜ × 1½ × ⅛ in. (*Agaricus affricatus* B. & Br., not Fr.)

376a. O. epichysium Quél. (from a fancied resemblance to an ancient Greek wine-vessel, *epichusis*).

P. membranous, plane, umbilicate, livid umber, pale livid-reddish-umber, or mid. slate-lavender and marg. dull salmon, silky floccoso-squamose and pallid when dry; marg. striate when moist. *St*. somewhat hollow, glabrous, ashy, grey or slate, dull salmon above and below. *G*. plano-decurrent, ashy or salmon-brown-white.

On wet rotten wood, rarely in bogs. Autumn. 1⅛ × 1⅝ × $\frac{1}{16}$ in. Wholly soft and watery.

377. O. sphagnicola Karst. (from its habitat; *sphagnum*, bog-moss, *colo*, to inhabit) *a b c*.

P. infundibuliform, substriate, brownish-tan or dull pale ochreous. *St*. tan-white or ashy. *G*. thick, colour as P. or whitish.

In swamps, on *Sphagnum acutifolium*; uncommon. June–Sept. 1¼ × 1⅜ × ⅛ in.

378. O. philonotis Quél. (Gr. *philos*, loving, *notis*, moisture) *a*. Smoky-ashy or brown-white.

P. infundibuliform, fragile. *St*. floccose at base.

On *Sphagnum*, swampy places in fir-woods. Aug. 1¼ × 2⅜ × ⅛ in. Fries says this is one of the vernal species, with **377** and **380**.

379. O. onisca Gill. (from its colour, like a codfish; Gr. *oniskos*) *a b c*. Dark ashy to pale umber, becoming grey-hoary.

P. broadly umbilicate, striate. *G*. shortly decurrent.

Not cæspitose. Woods, swampy places; rare. Dec. 1⅛ × ⅞ × ⅛ in.

379a. O. Luffii Mass. (after Mr. John Luff).

P. convex, undulate, then depressed, even, glabrous; marg. usually upturned at extreme edge, pallid, white when dry. *St*. solid, thickened above, cartilaginous, often crooked, glabrous, polished, pallid. *G*. decurrent, crowded, colour as St. *Flesh* very thin, white.

Gregarious. Odour fragrant, spicy, like **219**. 1¼ × ¾ × $\frac{1}{12}$ in. Allied to **379**.

380. O. cæspitosa Sacc. (from its cæspitose habit) *a b c*.

P. hygrophanous, plano-depressed, livid buff, yellowish-white or olive-shaded, at length white; marg. crenate, sulcate. *St*. slightly enlarged below, colour as P. *G*. very distant, sometimes branched, whitish or yellowish-buff.

Moors, by boggy water-courses under pines, damp peaty soil. May–Oct. 1¼ × 1 × ⅛ in. Whole plant sometimes becomes bright yellow in drying.

381. O. glaucophylla Gill. (from the olive-grey gills; Gr. *glaukos*, the bluish-grey of the olive, *phullon*, a leaf) *a c.*

P. plano-umbilicate, mouse-colour; marg. striate. *St.* colour as P.

Woods, mossy places. Sept. $\frac{1}{2} \times \frac{5}{8} \times \frac{1}{16}$ in. Pileus sometimes $\frac{7}{8}$ in. in diameter.

382. O. rustica Quél. (from its habitat; *rus*, the country) *a b.*

P. plano-umbilicate, fuscous to grey, or whitish-umber, shaded lavender; marg. striate. *St.* colour as P., or dull rather pale lavender. *G.* thick, olivaceous or olive-grey.

Woods, grassy avenues, heathy ground, charcoal heaps. Aug.–Sept. $\frac{1}{2} \times \frac{1}{2} \times \frac{1}{16}$ in.

382a. O. alutacea Sacc. (from its colour; *aluta*, tanned leather) *a b.* Tan-colour, or dull pale ochreous.

P. membranous, convex, then umbilicate, smooth; marg. incurved, even. *St.* hollow, smooth, paler than P. *G.* somewhat narrow, crowded, arcuate, colour as St.

Amongst grass and moss. Sept. $\frac{3}{4} \times 1\frac{3}{4} \times \frac{1}{16}$ in.

c. Umbelliferæ.

383. O. demissa Karst. (*demissus*, drooping) *a b.*

P. infundibuliform, fuscous-rufescent to tan-brown. *St.* liver-rufescent to pale tan-brown. *G.* becoming purple.

Waste places, banks in woods, burnt wood; rare. Aug.–Oct. $\frac{5}{8} \times \frac{1}{2} \times \frac{1}{16}$ in. *Agaricus rufulus* B. & Br. Not unlike some forms of **224**.

384. O. hepatica Gill. (from its colour; Gr. *hepar*, the liver) *a b c.*

P. infundibuliform, tough, rufous flesh-colour to tawny and tan. *St.* fuscous flesh-colour. *G.* whitish.

Lawns, mossy pathsides in woods, etc.; rare. Nov.–Dec. $1\frac{1}{4} \times 1\frac{3}{8} \times \frac{1}{8}$ in. Often in company with **373**.

385. O. muralis Quél. (from one of its habitats; *murus*, a wall) *a b c.*

P. infundibuliform, rufous-brown to tan-umber; marg. striate. *St.* paler than P. *G.* brownish-white.

Old mossy walls, amongst moss in woods, on turf walls, sandy banks, etc. Jan.–Nov. $1 \times \frac{5}{8} \times \frac{1}{8}$ in. Sometimes only $\frac{1}{8} \times \frac{1}{4}$ in. There is a white var. Several other species of *Omphalia* occur on old mossy walls.

386. O. umbellifera Quél. (from the umbrella-like pileus; *umbella*, an umbel or umbrella, *fero*, to bear) *a b c.*

P. plano-depressed, deep brown to ochreous-whitish; marg. striate. *St.* slightly attenuate downwards, colour as P. *G.* grey-whitish, umber or dusky flesh-colour, edge whitish.

Taste none; odour at first somewhat disagreeable, evanescent. Boggy ground, stumps, rotten wood, old walls, heathy banks, etc. Mar.–Dec. $\frac{7}{8} \times \frac{7}{8} \times \frac{1}{16}$ in. Var. *abiegna* Mass. grows on firwood. Var. *viridis* Sacc. is verdigris-green. There is also a wholly deep yellow variety.

386a. O. velutina Quél. (*velutinus*, velvety) *a*.

P. convex, umbilicate, striate, greyish or yellowish-grey. *St*. solid, colour as P., white at base. *G*. arcuate, wide, distant, colour as P. *Flesh* dark grey.

In heaths. Sept. $\frac{1}{2} \times 1\frac{1}{8} \times \frac{1}{24}$ in. Has cbaracters between **386** and **391**.

387. O. infumata Sacc. (from its colour; *infumo*, to smoke) *a b*.

P. obtuse, at first green. *St*. dilated at base, tomentose, yellow. *G*. yellow.

Moss on bark. Autumn. $\frac{5}{16} \times 1\frac{3}{8} \times \frac{1}{2}$ in. Colours of **350**.

388. O. retosta Gill. (from its appearance; *retostus*, scorched) *a c*. Umber within and without.

P. plano-depressed. *St*. equal or attenuate below. *G*. somewhat darker than flesh.

Lawns, etc., amongst dead leaves. Oct. $1\frac{1}{8} \times 1 \times \frac{1}{8}$ in. Firmer than **386**.

389. O. abhorrens Sacc. (from its fetid odour) *a*.

P. plano-depressed, fuscous to warm ochreous. *St*. brown-ochre above, dark brown below. *G*. colour as P.

Cæspitose. Odour dung-like. On lawns, under yews. Oct. $\frac{1}{2} \times \frac{3}{4} \times \frac{1}{32}$ in. A close ally of **388**; much like **1411**.

390. O. pseudoandrosacea Gill. (from its resemblance to **1420** *Marasmius androsaceus* Fr.; Gr. *pseudos*, false) *a c*. White, whitish-grey, or umber-white.

P. plano-depressed; marg. striato-plicate, crenulate.

Mossy lawns. Oct.–Nov. $\frac{1}{2} \times 1\frac{1}{8} \times \frac{1}{16}$ in.

391. O. griseopallida Quél. (from its colour; *griseus*, grey, *pallidus*, pallid) *a b c*.

P. infundibuliform, squamulose, fuscous-grey or pale greyish-umber; marg. sometimes slightly striate. *St*. paler than P. *G*. fuscous- or slate-grey.

On the ground, gardens. Sept.–Oct. $\frac{7}{8} \times \frac{5}{8} \times \frac{1}{8}$ in.

392. O. stellata Quél. (from the star-like, radiate base) *a b c*. White.

P. plano-depressed, diaphanous-striate.

In woods, on sticks, decayed stems of herbaceous plants, wood; uncommon. Feb.–Nov. $\frac{5}{16} \times 1 \times \frac{1}{64}$ in. Must not be confounded with the *Basipedes* section of *Mycena*.

MYCENARIÆ.

d. Campanellæ.

393. O. campanella Quél. (from the shape of the pileus; *campanella*, a little bell) *a b c*.

P. campanulato-plane, umbilicate, yellow-ferruginous, orange or brown. *St*. brown, usually dark strigose at base. *G*. dull ochre.

Cæspitose. Woods, pine, trunks. Aug.–Nov. $\frac{3}{4} \times 1\frac{1}{2} \times \frac{1}{16}$ in. Sometimes completely covering stumps. There is a var. with a small acute papilla in the umbilicus.

394. O. picta Gill. (from its varied colours; *pictus*, painted) *a*.

P. hood-shaped, subdepressed, fuscous; marg. striate; mid. yellow or brown. *St.* horny, brown; blackish at strigose base. *G.* adnate, broader than long, subdecurrent, whitish or yellowish.

Woods, mixed, on sticks, chips, dead wood, sawdust, etc. Aug.–Oct. $\frac{3}{8} \times 1\frac{1}{4} \times \frac{1}{16}$ in.

395. O. camptophylla Sacc. (from the curved gills; Gr. *kampto*, to curve, *phullon*, a leaf) *a b*.

P. convexo-depressed, brown or brown-white; marg. striate, grey. *St.* yellow to white above, rufescent to pale brown below. *G.* adnate, subdecurrent, white.

Sticks, chips, etc.; rare. Aug.–Oct. $\frac{1}{2} \times 2 \times \frac{1}{32}$ in.

396. O. umbratilis Gill. (from its growing in the shade; *umbratilis*, remaining in shade) *a c*.

P. convexo-plane, broadly depressed, black-fuscous, hoary when dry. *St.* fuscous-dead-black. *G.* adnato-decurrent, becoming fuscous-white.

Gregarious. Amongst grass. Autumn. $\frac{7}{8} \times \frac{7}{8} \times \frac{1}{16}$ in.

397. O. grisea Quél. (*griseus*, grey) *a b c*.

P. conico-campanulate, subumbonate, then expanded and umbilicate, livid grey to umber, becoming hoary. *St.* attenuate downwards, longitudinally cracking, ashy-whitish. *G.* sinuato-decurrent, whitish-grey, dull lavender with a white edge or livid umber.

Taste and odour somewhat unpleasant. Woods, pine, mixed, dead sticks, leaves; rare. Sept.–Nov. $\frac{7}{8} \times 1\frac{3}{4} \times \frac{1}{16}$ in.

398. O. fibula Quél. (from its shape; *fibula*, a pin) *a b c*.

P. plano-depressed, sometimes papillate, orange-yellow or vermilion. *St.* colour as P. *G.* decurrent, yellow or orange. *Flesh* orange.

Mossy ground, shady banks, moist woods, moss, sedge, grass, stumps, etc. Mar.–Dec. $\frac{1}{2} \times 1\frac{3}{4} \times \frac{1}{32}$ in. Var. *Swartzii* Karst. *P.* tan or white-purple; mid. deep purple-brown. The variation from orange to purple agrees with **224** and its purple form. There is a white variety.

399. O. directa Sacc. (from the long hairs at right angles to the stem; *directus*, straight) *a c*. White.

P. plano-depressed. *St.* sometimes faintly rufous. *G.* subdistant, decurrent.

Dead leaves. May–Nov. $\frac{1}{16} \times \frac{3}{4} \times \frac{1}{128}$ in.

400. O. PSEUDODIRECTA W. G. Sm. (from its being mistaken by Dr. M. C. Cooke for **399**; Gr. *pseudos*, false) *a b*.

P. plano-depressed, white-pruinose. *St.* white, mealy-granular below, springing from a white floccose disc or volva which becomes obsolete as the stem lengthens. *G.* few, adnate, white then saffron, pruinoso-sparkling.

On Encephalartos cone at Messrs. Veitch's nursery, Chelsea. May. $\frac{1}{16} \times \frac{1}{2} \times \frac{1}{128}$ in. Stems bristle-like, often non-pileate.

401. **O. Bellisæ** Karst. (after the Misses Bell of Coldstream) *a b*.
P. infundibuliform, pale livid-brown; marg. waved and furrowed. *St.* whitish above, pale brown below, springing from a slightly enlarged, floccose base. *G.* wrinkled, veined, white.
Dead reeds. Oct. $\frac{1}{2} \times 1\frac{3}{4} \times \frac{1}{16}$ in.

401a. **O. gracilis** Quél. (*gracilis*, slender) *a*. Wholly white.
P. membranous, campanulate, papillate, striate, smooth, transparent. *St.* filiform, fistulose, transparent, pruinose, fibrillose at base. *G.* distant, very decurrent, thin.
Dead grass. Oct. $\frac{1}{8} \times 1\frac{1}{2} \times \frac{1}{30}$ in.

402. **O. gracillima** Quel. (*gracillimus*, very slender) *a b*. White.
P. convex, subumbonate then plane and umbilicate. *St.* springing from a floccose base. *G.* fold-like, disappearing short of marg. of P.
Scattered. Dead sticks, herbaceous stems. Aug. $\frac{5}{16} \times 1 \times \frac{1}{32}$ in.

403. **O. bullula** Sacc. (from its watery appearance; *bullula*, a watery vesicle) *a*. White.
P. hemispherical, diaphanous.
Scattered. Dead sticks, twigs. Autumn. $\frac{1}{4} \times \frac{5}{8} \times \frac{1}{32}$ in.

e. Integrellæ.

404. **O. integrella** Quél. (from its perfect structure; *integer*, whole) *a b c*. White.
P. hemispherical, depressed or plane. *St.* pilose downwards to a minute villous disc at base. *G.* adnato-decurrent disappearing short of marg. of P.
Commonly fasciculate. Shady places, woods, on decayed sticks, rotten wood, grass, twigs, etc. May–Oct. $\frac{3}{16} \times \frac{7}{8} \times \frac{1}{24}$ in.

405. **O. buccinalis** Sacc. (from its trumpet-shaped pileus; *buccina*, a trumpet) *a*. White.
P. fleshy, plano-depressed. *St.* with a stellate disc at base. *G.* vein-like as in *Cantharellus*.
Twigs, stems, etc.; common. $\frac{1}{4} \times \frac{7}{8} \times \frac{1}{32}$ in. Sometimes branching as in 1437 on a small scale.

406. O. NEVILLÆ Sacc. (after Lady Dorothy Neville) *a*.
P. hemispherico-depressed, granulate, striate, brown becoming pale at marg. *St.* very thin, brown, rough with black gr., villous, subdilated at base. *G.* veined at sides and interstices.
On *Sphagnum* in orchid-pot. $\frac{1}{2} \times 1 \times \frac{1}{16}$ in.

XI. PLEUROTUS Quél.

(From a fancied resemblance in many of the species to an ear; Gr. *pleuron*, a side, *ous*, an ear.)

Veil universal, seen in the squamules, scurf and appendiculate margin of the pileus and in the fugitive annulus, sometimes developed as gluten, or obsolete or nearly so. *Hymenophore* confluent and homogeneous with the fleshy stem, when present. *Pileus* usually excentric, compact, spongy to fleshy or membranous, sometimes resupinate. *Stem* usually excentric, lateral, annulate, simple or

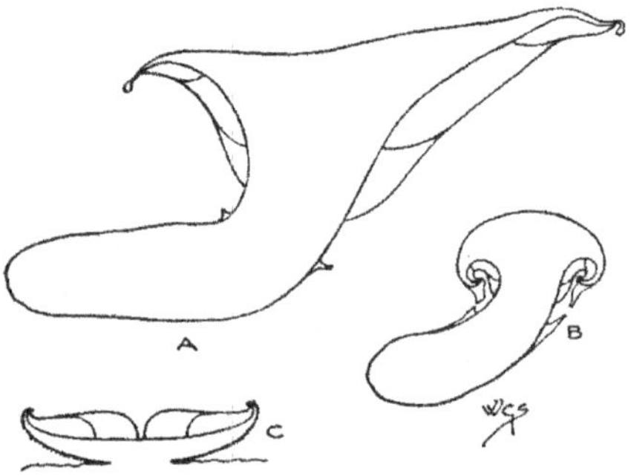

Fig. 20.—A, section of *Pleurotus dryinus* Quél. One-third natural size. B, young state, section, showing veil; C, section of *P. applicatus* Quél. × 2.

none; when present without cartilaginous bark; when the larger-stemmed species grow vertically on prostrate trunks, the stem is sometimes central or nearly so. *Gills* sinuate, adnate or decurrent, in some of the *Resupinatæ* often starting from an excentric point. *Spores* more or less elliptical, smooth. (Fig. 20.)

The majority of the species grow on trunks, branches, twigs, moss, etc., a few are terrestrial; many are large, handsome and polymorphic; some are small and resupinate. They generally appear late in the year, **421** and **422** commonly grow in the spring and reappear year after year on the same habitat. All are harmless, and a few are considered by some to be edible. All become putrid when old, and none are leathery or woody—in this they differ from members of *Lentinus*, *Panus*, *Xerotus*, *Trogia*, *Schizophyllum* and *Lenzites*. Species **407—448**

a. Excentricæ. *Pileus* laterally extended; *stem* excentric.
- *Veil* forming an annulus to the stem and hanging as a ragged fringe at the margin of the pileus. **407—409**
- *Veil* obsolete. *Gills* sinuate or obtusely adnate. **410—417**
- *Veil* obsolete. *Stem* distinct, somewhat vertical. *Gills* deeply decurrent. **418—420**
- *Veil* obsolete. *Pileus* lateral, sessile or extended behind into a short, oblique, stem-like base. *Gills* deeply decurrent. **421—425**

b. Dimidiatæ. *Pileus* definitely lateral, immarginate behind, not at first resupinate. **426—435** *c. Resupinatæ.* *Pileus* at length reflexed, sessile. *Gills* starting from an excentric point.

Pileus fleshy, uniform. **436, 437**

Pileus fleshy, striate, with an upper gelatinous stratum or viscid pellicle. **438—444**

Pileus membranous, not viscid. **445—448**

a. Excentricæ.

407. P. corticatus Quél. (from the thick covering of the pileus and stem; *cortex*, bark) *a b.*

P. fleshy, floccose, pale buff and grey. *St.* squamuloso-fibrillose white to pale buff. *A.* silky-floccose. *G.* decurrent, anasto mosing near St., yellowish when old. *Flesh* white or faintly yellowish.

Solitary or cæspitose. Tasteless; odour pleasant, rather strong. Trunks, elm, ash, apple, etc. Sept.–Oct. $9 \times 3\frac{1}{2} \times 1\frac{3}{8}$ in.

408. P. dryinus Quél. (from its frequent habitat; Gr. *drus*, oak) *a b c.*

P. fleshy, plane, whitish with fuscous sc.; mid. darker. *St.* white or yellowish-white, fuscous at base. *A.* floccose. *G.* decurrent, white, yellowish when old.

Taste pleasant, mushroom-like. Trunks, oak, elm, apple, willow, walnut, ash, poplar. Sept.–Feb. $6 \times 4 \times 1\frac{1}{8}$ in.

409. P. spongiosus Sacc. (from its spongy substance) *a b c.*

P. pulvinate, subumbonate, downy, white, tan or ashy. *St.* white-tomentose, sometimes almost obsolete. *A.* evanescent. *G.* decurrent, anastomosing on stem, whitish.

Stumps, beech, apple, poplar, hornbeam; rare. Oct.–Nov. $3\frac{1}{2} \times 1\frac{1}{4} \times \frac{5}{8}$ in. Soft and spongy, shrinking to almost half its original size in drying.

410. P. ulmarius Quél. (from its usual habitat; *ulmus*, elm) *a b c.* Whitish.

P. pulvinate, broadly umbonate, smooth. *St.* villous at base, rarely central. *G.* adnate.

Solitary or cæspitose. Said to be edible. Odour mealy or slightly acid. Trunks, elm, wych elm, beech, oak, poplar, willow, sometimes parasitic on **1587**. Sept.–Dec. $6\frac{1}{2} \times 4 \times 1\frac{3}{4}$ in. Sometimes proliferous, with many small growths at base of stem.

411. P. tessulatus Gill. (from the tesselated markings of the pileus) *a.*

P. convex, subumbonate, pale tawny, brown-tesselated. *St.* smooth, white. *G.* sinuate, white or faintly brownish.

Solitary or cæspitose. Odour of new meal. Trunks. $4\frac{1}{2} \times 2 \times \frac{5}{8}$ in.

412. P. subpalmatus Gill. (from a fancied resemblance to the shape of the hand) *a b c.*

P. convex, with a thick, gelatinous, reticulato-corrugate pellicle, rufescent-salmon. *St.* whitish above, salmon below. *G.* sinuate with a decurrent tooth or adnato-rounded, sometimes adnate with a collar, paler than P. *Flesh* salmon-streaky.

Taste bitter-aromatic. Squared timber, trunks, etc., willow, elm, oak; rare. Sept.–Nov. $4\frac{1}{2} \times 2\frac{1}{2} \times \frac{1}{2}$ in.

413. P. craspedius Gill. (from the thick stem; *crassus*, thick, *pes*, a foot) *a b*.

P. pulvinate or subumbonate, pale buff; marg. crenato-lobed. *St.* and *Flesh* white. *G.* rounded or adnate.

Cæspitose. Strong odour of cucumber. Trunks, rotten wood, poplar. Sept.–Oct. $4\frac{3}{8} \times 2 \times 1\frac{3}{8}$ in.

414. P. fimbriatus Gill. (from the fringed margin of pileus; *fimbriæ*, fringe) *a*.

P. infundibuliform, hygrophanous, whitish; marg. lobed and reflexed. *St.* central or lateral, somewhat attenuate downwards, whitish. *G.* adnato-ascending, narrow, thin, crowded, white. *Flesh* remarkably thin.

Gregarious, somewhat cæspitose. Trunks, dead; rare. Aug.–Jan. $4\frac{3}{8} \times 1\frac{5}{8} \times \frac{1}{4}$ in. Habit not unlike *Lentinus*.

415. P. Ruthæ Sacc. (after Ruth Berkeley) *a*.

P. plano-depressed, fan-shaped, tan-whitish; marg. striate. *St.* lateral or excentric, salmon-reddish. *G.* adnato-decurrent, veined, paler than St.

Sawdust. Oct. $2\frac{3}{8} \times \frac{5}{8} \times \frac{1}{4}$ in. Sometimes twisted and contorted like **1158**, for the white variety of which with similar habitat it might be readily mistaken.

416. P. lignatilis Gill. (from its habitat and hardness; *lignum*, wood) *a b c*. Whitish-tan.

P. convexo-plane, slightly depressed, flocculoso-pruinate; mid. pale buff. *St.* stuffed, woody, pruinato-villous. *G.* adnate or sinuate. *Flesh* whitish, faintly sienna at base of stem.

Cæspitose or single. Odour strong of meal. Beech, growing on **1532** and **1593**. $3 \times 2\frac{1}{2} \times \frac{1}{2}$ in. A variety has *P.* ashy with mid. black and marg. white.

417. P. circinatus Gill. (from the orbicular pileus; *circino*, to make round) *a b c*. White.

P. plane, silky-lustrous. *St.* smooth, attenuate downwards. *G.* subdecurrent.

Odour weak and pleasant, not mealy. Rotting stumps, birch. Sept.–Oct. $2\frac{5}{8} \times 1\frac{3}{4} \times \frac{3}{16}$ in.

418. P. sapidus Sacc. (from its taste; *sapidus*, savoury) *a b c*.

P. plano-depressed, fleshy, smooth, white, pale ochreous, brown or lavender shaded. *St.* connato-branched, smooth, white. *G.* subdistant, reticulate on stem, white. *Spores* pale lilac.

Cæspitose. Trunks, elm. July–Aug. $4\frac{1}{4} \times 2\frac{1}{4} \times \frac{5}{8}$ in.

419. P. pantoleucus Gill. (from its colour; Gr. *pas*, all, *leukos*, white) *a b c*.

P. convex, dimidiate, spathulate, smooth. *St.* excentric. *G.* broad.

Solitary. Trunks, branches, willow. Oct. $4\frac{1}{4} \times 2\frac{1}{4} \times \frac{5}{8}$ in. Sometimes small with a depressed pileus.

420. P. mutilus Gill. (from its being changeable in form; *mutilus*, maimed) *a c*. White.

P. excentric, lateral, reniform or spathulate, depressed, silky-smooth. *St.* excentric or lateral with a villous base.

Dead wood, rotting elm, fir-plantations, on a grassy hillside. Aug.–Dec. $1\frac{5}{8} \times \frac{7}{8} \times \frac{1}{8}$ in. Might be mistaken for a small *Clitocybe*. Sometimes appears on mushroom-beds.

421. P. ostreatus Quél. (from the shape and colour of the pileus; *ostrea*, an oyster) *a b c*.

P. convexo-depressed, subexcentric, slate or brownish, when young sometimes almost black. *St.* connato-branched, excentric, rarely central, sometimes obsolete, white. *G.* reticulate below, whitish.

Cæspitose-imbricate. Said to be edible. Odour strong. Trunks and logs, elm, poplar, walnut, beech, laburnum, apple, *Wistaria*; once on dry cerebral matter in skull of a stranded whale. Jan.–Dec. $4\frac{5}{8} \times 2 \times \frac{7}{8}$ in. Var. *columbinus* Quél. has a purplish pileus and bluish or greenish marg. Var. *glandulosus* Mass. has tear-like drops on the gills.

422. P. euosmus Sacc. (Gr. *euosmus*, sweet-smelling) *a b c*.

P. convexo-depressed, tan to olivaceous-brown. *St.* varying long, short or obsolete. *G.* reticulate below. *Spores* tan, lilac or salmon-shaded.

Cæspitose-imbricate. Said to be poisonous. Odour of tarragon. Stumps, logs, poplar, elm. April–Oct. $5\frac{1}{4} \times 3\frac{1}{4} \times 1$ in. Intermediate forms between this and 421 occur.

423. P. revolutus Gill. (from the ultimately revolute margin of the pileus) *a*.

P. plane, smoky, lead- or mouse-colour; mid. darker. *St.* whitish. *G.* serrulate, whitish.

Solitary or subcæspitose. Trunks, poplar, beech. Autumn. $10\frac{1}{2} \times 2\frac{1}{4} \times 2$ in.

424. P. salignus Quél. (from one of its habitats; *Salix*, willow) *a b c*.

P. pulvinate, plane, ochreous, smoky slate-colour or almost black. *St.* almost obsolete, tan. *Spores* dull whitish.

Solitary, rarely cæspitose or imbricate. Trunks, willow, ash, elm, sycamore, alder. Sept.–Feb. $5\frac{1}{2} \times \frac{1}{2} \times 1\frac{3}{4}$ in.

425. P. acerinus Gill. (from one of its habitats; *Acer*, maple) *a b*. White.

P. tough, fleshy, thin, silky-villous, pale rufous when dry. *St.* sublateral or obsolete, villous. *G.* very crowded, thin, white then yellow; livid-vinous when dry.

Trunks, maple, ash, plane. Autumn. $3 \times 1\frac{1}{8} \times 1$ in.

b. Dimidiatæ.

426. P. petaloides Quél. (from a fancied resemblance to a leaf; Gr. *petalon*, a leaf, *eidos*, appearance) *a b c*.

P. plane, subumbonate, fuscous, brown, ochreous or purplish. *St.* sublateral, villous, whitish. *G.* decurrent, white, yellowish or ashy.

Gregarious, cæspitose or imbricate. Taste bitter. Terrestrial and on stumps, pine; rare. Aug.–Jan. $3\frac{3}{4} \times 1 \times 1$ in.

427. P. pulmonarius Quél. (from its lung-like texture; *pulmo*, a lung) *a c*.

P. plane, smooth, ashy or rufous. *St.* lateral, whitish. *G.* decurrent or adnato-decurrent, whitish or ashy.

Solitary, rarely cæspitose. Trunks, beech. Sept. $2\frac{5}{8} \times \frac{3}{4} \times \frac{1}{2}$ in. Var. *juglandis* Karst. grows on walnut.

428. P. serotinus Gill. (from its late appearance; *serotinus*, late) *a b c*.

P. reniform, olivaceous. *St.* lateral, brilliant deep yellow, squamulose, spotted near gills, somewhat rich brown. *G.* crowded, sinuato-decurrent, cream ochre to bright yellow or orange.

Taste insipid; odour apple-like. Trunks, posts, ash, birch, hornbeam, alder, beech; rails. Oct.–Dec. $2 \times \frac{3}{4} \times \frac{1}{2}$ in.

429. P. mitis Quél. (from its taste; *mitis*, mild) *a b c*.

P. reniform, gristly, without a viscid pellicle, pale flesh-colour or tan. *St.* lateral, minutely scaly, whitish flesh-colour or tan. *G.* somewhat rounded, white.

Taste and odour almost obsolete. Dead trunks, sticks, fir, larch. Oct.–Dec. $\frac{7}{8} \times \frac{3}{8} \times \frac{3}{8}$ in. Sometimes $1\frac{3}{4}$ in. in diam. Somewhat like **1444** in general appearance.

429a. P. rufipes Sacc. & D. Sacc. (from the reddish stem; *rufus*, red, *pes*, a foot) *a b*.

P. dimidiate or reniform, convex, membranous, very glutinous when moist, white; mid. salmon; marg. incurved. *St.* becoming recurved, vinous-reddish, salmon-red within, white in centre, base white-downy. *G.* adnate, broad, distant, white, interstices pale salmon.

On wood. Autumn. $\frac{1}{8} \times \frac{3}{16} \times \frac{1}{40}$ in.

430. P. GADINIOIDES Sacc. (from its resemblance to the bivalve *Gadinia*; Gr. *eidos*, appearance) *a b c*. White.

P. convex, hygrophanous, smooth or clothed with adpressed flocci, no gelatinous upper stratum. *St.* minute, lateral or none. *G.* adnate, slightly branched.

Tree-fern stems. May. Diam. $\frac{3}{4}$ in. Reniform as seen from below.

431. P. limpidus Gill. (from its pellucid substance; *limpidus*, clear) *a b c*.

P. reniform, plane, subumbonate, whitish. *St.* sublateral, rudimentary or obsolete. *G.* sinuato-decurrent, white or straw-whitish.

Taste mild. Stumps, ash, willow, beech. Dec. Diam. $1\frac{1}{2}$ in.

432. P. reniformis Karst. (from the kidney-shaped pileus; *renis*, a kidney, *forma*, form) *a*.

P. plane, whitish, or greyish-brown. *St.* lateral, rudimentary. *G.* sinuato-decurrent, paler than P.

Branches, silver fir, on the ground amongst moss. Aug.–Oct. Diam. 1 in.

433. **P. Laurocerasi** Sacc. (from its habitat, stems of cherry-laurel, *Prunus Laurocerasus*) *a b*.

P. convex, oyster-shaped, sulcate, pale brownish-salmon. *St.* obsolete. *G.* very broad, adnate, veined, salmon-white.

Oct. Diam. 1⅜ in.

434. **P. tremulus** Quél. (from its tremulous habit) *a b c*.

P. reniform, depressed behind, fuscous-grey, pale slate or brown. *St.* attenuate downwards, grey. *G.* adnato-decurrent, whitish-umber or greyish.

Solitary. On the ground, on moss, fungi, e.g. **1790**, when growing on moss; rare. Aug.–Dec. 1¼ × 3/16 × ¼ in.

435. **P. acerosus** Quél. (from its habitat, amongst dead pine-leaves; *acersosus*, full of chaff or husk) *a b c*.

P. plane, reniform, membranous, whitish, brownish or greyish. *St.* faintly umber-whitish. *G.* adnate, branched, colour as P. or paler.

Taste insipid, somewhat disagreeable. Lawns, wood, gravel, leaves, pine-wood, pine-leaves, *Sphagnum*; rare. Aug.–Dec. 1⅞ × ¼ × ⅛ in. Resembles **1388**.

c. Resupinatæ.

436. **P. porrigens** Gill. (from its stretched out habit; *porrigo*, to stretch out) *a b c*. White.

P. at length plano-depressed, undulate. *G.* thick, sometimes forming pores. *Flesh* thin, subcartilaginous.

Imbricate. Odour pleasant or insipid. Stumps, chiefly pine. June–Nov. Diam. 6¼ in. Base villous.

437. **P. septicus** Quél. (from its habitat, rotten wood; *septicus*, putrifying) *a b c*. White.

P. convex, villous. *St.* lateral, villous, at length disappearing. *G.* rounded-adnate, subdistant.

Decayed wood, branches, twigs, pine, bramble, straw, dung, fungi. Mar.–Nov. ⅜ × 1/16 × 1/16 in.

438. **P. mastrucatus** Sacc. (from its somewhat woolly coat; *mastruca*, a sheepskin) *a b c*.

P. ear-shaped, plane, squarrose-scaly, greyish- or whitish-brown. *G.* adnato-decurrent. *Flesh* mouse-colour.

Imbricate. Old trunks, logs, beech, amongst moss and lichens; rare. Diam. 3⅜ in.

439. **P. atrocæruleus** Gill. (from the colour of the pileus; *ater*, black, *cæruleus*, azure-blue) *a b c*.

P. plane, reniform; marg. ochreous-sienna. *G.* whitish straw-colour. *Flesh* with an olive middle stratum.

Gregarious, subimbricate. Odour pleasant. Trunks, birch, elm, poplar, beech, etc. Oct.–Dec. Diam. 1⅜ in.

440. **P. Leightonii** Sacc. (after the Rev. W. A. Leighton) *a b c*.

P. at first obliquely conical, furfuraceous, umber or lead-colour. *G.* thick, distant, forked at base, pallid tan.

Wood, rotten rails. Dec. Diam. ½ in. Might be mistaken for **444**.

441. P. algidus Quél. (from its frequent growth in northern countries; *algidus*, cold) *a b c.*

P. plane, subreniform, pruinose when young, viscid, brown. *G.* adnate, becoming pallid yellow.

Usually cæspitose and imbricate. Trunks, mountain-ash, willow. Aug.–Oct. Diam. 2⅛ in.

442. P. fluxilis Gill. (from the gelatinous, upper stratum of the pileus; *fluxilis*, fluid) *a.*

P. plane, reniform, umber. *G.* adnato-rounded, whitish. *Flesh* a gelatinous stratum without a membrane.

Solitary. Wood, amongst moss on trunks, sawdust, mosses. Oct. Diam. ⅛ in. Not usually resupinate.

443. P. cyphellæformis Sacc. (from its resemblance to a *Cyphella*) *a b c.*

P. cup-shaped then conico-dependent, greyish or brownish. *G.* radiating from a central point, white. *Flesh*, upper stratum dark brown, lower white.

Gregarious. Dead herbaceous stems, *Fœniculum*, *Epilobium*; rare. Feb.–Oct. Diam. 3/16–⅜ in.

444. P. applicatus Quél. (from its close attachment to branches, etc.; *applicatus*, closely attached) *a b c.*

P. cup-shaped, then expanded or reflexed, dark grey. *G.* radiating from a central point, lighter than P.

Dead branches, twigs, bramble, oak, ash, pine, *Ribes*, *Vitis*. Sept.–June. Diam. ⅜ in. Sometimes proliferous.

445. P. Hobsonii Sacc. (after Lieut. Julian C. Hobson) *a.*

P. plano-convex, reniform, downy, pale grey. *G.* subdistant, adnate, pallid.

Stumps, larch. Sept. Diam. ⅜ in.

446. P. striatulus Quél. (from the slightly striate pileus) *a c.*

P. cup-shaped, not usually reflexed, fuscous or grey. *G.* few, radiating from a central point, lighter than P.

Gregarious. Stumps, twigs, etc., fir, hazel, elm. May–Dec. Diam. ¼ in.

447. P. hypnophilus Sacc. (from its frequent habitat, *Hypnum*; Gr. *philos*, loving) *a c.* White.

P. at length plane, subreniform. *St.* villous, at length vanishing. *G.* narrow, radiating from a central point.

On the larger mosses, leaves, leaves and stems of grass. Sept.–Dec. Diam. ⅜ in. Resembles 553 in habit.

448. P. chioneus Gill. (from its colour; Gr. *chion*, snow) *a b c.*

P. subresupinate, convex, villous. *St.* lateral, villous, at length vanishing. *G.* very distant, broad, radiating from a central point.

Wood, decorticated elder, leaves, dung, bone; rare. Sept.–Dec. ¼ × 1/32 × 1/32 in.

Series II. HYPORHODII Fr.

(From the rose-coloured gills ; Gr. *hupo*, beneath, *rhodon*, a rose.)

Spores white when young, then pale rose-colour, salmon, yellowish-salmon, salmon-whitish, nankeen, whitish-vinous or salmon-brownish, never lilac, elliptical and smooth or globose or nodulose and angularly warted.

There is no analogue of *Amanita* amongst the British *Hyporhodii*, but an analogous genus, *Metraria*, occurs in Australia.

It is remarkable that Sowerby has modelled an example of **450** with an ample annulus ; if correct this plant would be a *Metraria*.

Genera XII–XX.

XII. VOLVARIA Quél.

(From the volva.)

Veil universal, entirely enveloping the young plant. *Hymenophore* distinct from the fleshy stem. *Stem* central, simple, solid, volvate,

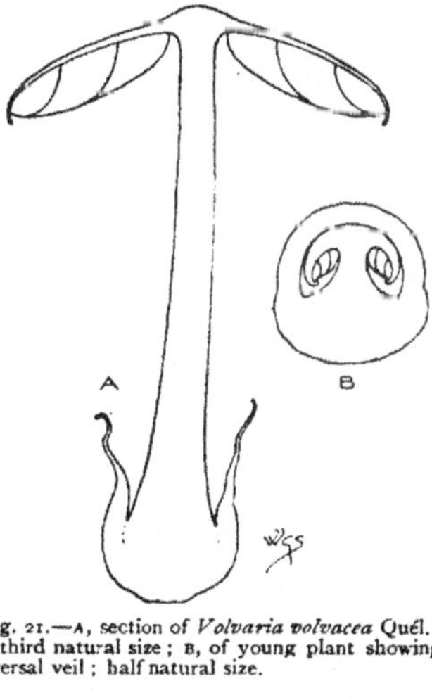

Fig. 21.—A, section of *Volvaria volvacea* Quél. ; one-third natural size ; B, of young plant showing universal veil ; half natural size.

without cartilaginous bark. *Gills* free, at first white, then rose or salmon, sometimes becoming brownish. *Spores* smooth, salmon or rose-colour. (Fig. 21.)

All the species are beautiful, some appear in spring and early summer. They grow in gardens, hothouses and pastures ; and on dung, manured ground and rotten wood ; **451** grows on the pileus of **153**. They are almost all tasteless or insipid; **455** is fetid and said by Letellier to be poisonous. None are known to be edible.

Volvaria agrees in structure with *Amanitopsis*, *Acetabularia* and *Chitonia*. Species **449—457**

Pileus dry, silky or fibrillose. **449—453**

Pileus more or less viscid, smooth. **454—457**

449. V. bombycina Quél. (from the silky pileus; *bombyx*, silk) *a b c*.

P. broadly umbonate, white or pale brownish. *St*. colour as P. *Vo*. large, lax, ochre-sienna or umber. *G*. becoming rufous.

Solitary or cæspitose. Decayed wood, stumps, hollow trees, ash, elm, birch. June–Sept. $8\frac{1}{8} \times 7 \times \frac{3}{4}$ in.

450. V. volvacea Quél. (from the large volva) *a b c*.

P. campanulate, subumbonate, black-fibrillose on a brown ground or brown-fibrillose on a white ground. *St*. white. *Vo*. large, lax, umber.

Gregarious. Reputed poisonous. Stoves, on tan, roadsides. $5 \times 7 \times \frac{1}{2}$ in. Sowerby has modelled an example of this with a large annulus.

451. V. Loveiana Gill. (after the Rev. R. T. Lowe) *a b c*.

P. convex, white-silky. *St*. white. *Vo*. lax, whitish. *G*. thick.

Gregarious, subcæspitose. On **153** and other species of *Clitocybe*. $2\frac{7}{8} \times 2 \times \frac{3}{8}$ in.

452. V. Taylorii Gill. (after Michael A. Taylor) *a b c*.

P. expanded, subumbonate, rimose, whitish; marg. faint umber. *St*. white. *Vo*. lax, small, white or umber. *G*. sinuato-free, irregular, attenuate near St.

Odour none. On the ground, gardens. July–Oct. $2\frac{3}{8} \times 2 \times \frac{1}{4}$ in. Variable in size, sometimes as small as **453** and smaller than **457**.

453. V. TEMPERATA Sacc. (from its growing in temperate greenhouses) *a b*.

P. campanulate, umbonate, pulverulent, pale ochreous-sienna; marg. striate. *St*. white. *Vo*. ample, faintly ochreous.

Greenhouses. Feb. $\frac{5}{8} \times \frac{7}{8} \times \frac{1}{16}$ in.

454. V. speciosa Gill. (*speciosus*, handsome) *a b c*.

P. campanulate, subumbonate, whitish, greyish or brownish; mid. darker. *St*. attenuate upwards, white. *Vo*. bulb-like, white.

Odour strong. Roadsides, rubbish heaps, manure heaps; rare. June–Oct. $4 \times 6\frac{1}{2} \times \frac{1}{2}$ in.

455. V. gloiocephala Gill. (from its glutinous pileus; Gr. *gloios*, sticky, *kephale*, a head) *a b c*.

P. convex, subumbonate, whitish; marg. striate, greyish or brownish; mid. darker. *St*. subfibrillose, pale brownish or tawny. *Vo*. bulb-like, adpressed, brown. *G*. serrulate, becoming reddish.

Probably poisonous. Taste disagreeable; odour strong, offensive, like putrid broad beans. On the ground. June–Nov. $4\frac{1}{2} \times 6\frac{3}{4} \times \frac{1}{2}$ in. Fragments of *Vo*. often left on *P*.

456. V. media Gill. (from its intermediate position between **455** and **457**) *a b c*.

P. convexo-plane, subumbonate, white; mid. pale brownish; marg. pale rose. *St.* white. *Vo.* small, lax, spreading; whitish.

Woods, on the ground. Aug.–Oct. $2\frac{1}{4} \times 3 \times \frac{1}{4}$ in.

457. V. parvula Quél. (*parvulus*, small) *a b c*. Whitish.

P. convexo-plane, subumbonate. *Vo.* large, lax.

Gregarious. Pastures, gardens, fir-woods, dead wood; frequent. May–Oct. $1\frac{1}{8} \times 1\frac{1}{4} \times \frac{1}{8}$ in. A minute form of this occurs in stoves.

XIII. ANNULARIA Gill.

(From the annulate stem.)

Veil universal, concrete with the cuticle of the pileus. *Hymenophore* distinct from the fleshy stem. *Pileus* plane or broken up into minute squamules. *Stem* central, annulate, hollow. *Gills* free. (Fig. 22.)

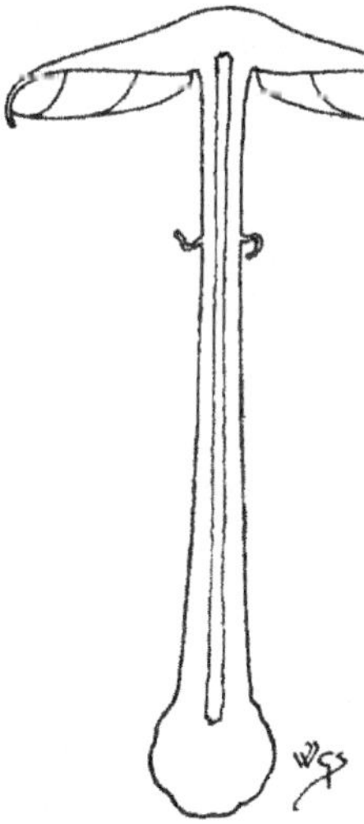

Fig. 22.—Section of *Annularia lævis* Gill. One-half natural size.

Terrestrial. *Annularia* agrees in structure with *Lepiota*, *Togaria* and *Psalliota*.

457a. A. lævis Gill. (from the usually smooth pileus; *lævis*, smooth) *a b*.

P. convex, expanded, obtuse to subumbonate, smooth, glabrous or minutely squamulose, white or pallid. *St.* slightly attenuate upwards, even, smooth, silky, bulbous, colour as P. *A.* somewhat distant, large, becoming free. *G.* free to distant, thin, somewhat crowded, white, then salmon. *Flesh* somewhat firm white.

Bushy places, amongst grass. Aug.–Oct. $3\frac{1}{4} \times 5\frac{3}{4} \times \frac{5}{8}$ in.

XIV. PLUTEUS Quél.

(From a fancied resemblance in the pileus to the roof of a turret or sentry box, *pluteus*.)

Veil represented by the fibrils, flock or pruina of the pileus. *Hymenophore* distinct from the fleshy stem. *Stem* central, simple. *Gills* free, white, then usually salmon, or with a shade of yellow.

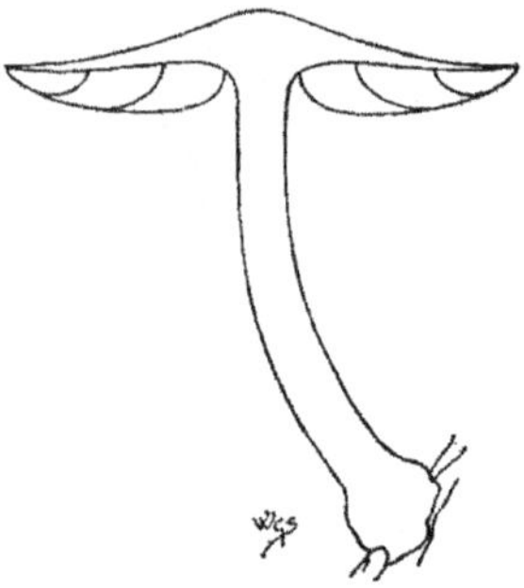

Fig. 23.—Section of *Pluteus cervinus* Quél. One-third natural size.

Spores smooth or warted, pale rose, salmon or somewhat reddish. (Fig. 23.)

Mostly growing on or close to tree-trunks, appearing from the spring to late autumn. They are all beautiful and tasteless; none are edible.

Pluteus agrees in structure with *Hiatula*, *Pluteolus* and *Pilosace*.

	Species **458—472**
Cuticle of pileus separating into fibrils or flocci.	**458—463**
Pileus pruinate, atomate, somewhat pulverulent.	**464—467**
Pileus naked, smooth.	**468—472**

458. P. cervinus Quél. (from the fawn-brown pileus; *cervus*, a deer) *a b c*.

P. expanded, broadly umbonate; marg. entire. *St.* solid, whitish, black-fibrillose, lighter above. *Flesh* silky-white.

Tasteless, insipid, or subacid. Trunks, fir, pear, sawdust, earth, straw, dung; common. April-Nov. $4\frac{3}{4} \times 4\frac{3}{4} \times \frac{1}{2}$ in. Sometimes mistaken for the mushroom. Var. *eximius* Mass. rufous-black. $6\frac{1}{2} \times 5 \times 1\frac{1}{2}$ in. Sometimes *P.* 12 in. in diam. Var. *petasatus* Mass., *P.* greyish-white. 6×8×⅜ in. Var. *patricius* Mass., *P.* scaly. Var. *Bullii* Mass., *P.* deep-brown. 6½ in. in diam.

459. P. umbrosus Gill. (from its habitat; *umbrosus*, shady) *a b.*

P. convex, broadly subumbonate, wrinkled, umber; marg. fimbriate with V. *St.* solid, villous-scaly; umber-whitish. *G.* fuliginous-fimbriate at marg.

Dead wood, fir, oak. Sept. 2⅝ × 2½ × $\frac{5}{16}$ in. This has less claim to specific rank than var. *eximius* of **458.**

460. P. ephebius Gill. (from the down on the pileus; *ephebius*, like a youth) *a b.*

P. convex, tomentose, pale slate-blue or lilac, sometimes slate. *St.* stuffed, whitish or bluish. *G.* salmon.

Rotten wood, trunks. 2⅝ × 2½ × ¼ in.

461. P. salicinus Quél. (from its usual habitat, willow, *Salix*) *a b.*

P. convex, subumbonate, subpilose, pale blue-grey; mid. slightly wrinkled. *St.* colour as P., or clouded greenish.

Willow, bramble, alder. Feb.–Dec. 1⅝ × 2½ × ¼ in. Var. *beryllus* Karst., *P.* greenish-streaked; alder. Var. *floccosa* Sacc., two or three times smaller than **400**, to which it is allied.

462. P. hispidulus Gill. (from the slightly hispid pileus) *a b.*

P. convex, grey or purplish-grey. *St.* hollow, silvery-white.

Stumps, beech, earth in hothouses. ⅝ × 1½ × $\frac{1}{16}$ in.

463. P. pellitus Quél. (from the separable membrane of the pileus; *pellitus*, clad in skins) *a b.*

P. hemispherical, silky-white. *St.* stuffed, white.

Odour disagreeable. Trunks, earth, rotten branches, twigs, chips, roots of trees, amongst grass in pastures. July–Sept. 1⅝ × 1⅝ × ⅛ in.

464. P. nanus Quél. (from its small size; *nanus*, dwarf) *a b c.*

P. convex, umber or slate-brown; mid. wrinkled. *St.* solid, whitish.

Thatch, rotten wood, rotten sawdust, earth, sticks, leaves and in gardens. Aug.–Oct. 1⅝ × 2⅞ × $\frac{3}{16}$ in. Var. *lutescens* Karst. has an orange-yellow St., sulphur-yellowish flesh and salmon or sulphur G. Var. *major* Mass. is larger than the type.

465. P. spilopus Sacc. (from the spotted stem; Gr. *spilos*, a spot, *pous*, a foot) *a b c.*

P. convex, subumbonate, rugulose, brown. *St.* stuffed, tan-white or greyish, black-dotted.

Stumps, wood. Oct. 2½ × 1⅞ × ¼ in. Sometimes small, ⅝ × 1 in. or less.

466. P. semibulbosus Gill. (from the slightly enlarged base of stem) *a b.*

P. hemispherical, sulcate, mealy, white. *St.* hollow, white, base sienna and white cottony. *G.* distant; marg. entire.

Wood, branches, rotten wood. ⅝ × ⅞ × ⅛ in.

467. P. violarius Sacc. (from its bright violet exudation) *a b*.

P. expanded, smooth, even, spongy, dry, purple, crimson-purple or sienna. *St.* hollow, silky-white, black-fibrillose on a pale umber ground below. *G.* serrulate, crowded. *Flesh* grey.

Gregarious. Taste pleasant; odour none. Stumps, rotten wood. $1\frac{1}{8} \times 1 \times \frac{1}{8}$ in.

468. P. roseoalbus Gill. (from the rosy-white or salmon pileus) *a b*.

P. expanded, subumbonate. *St.* solid, pale rose or white. *Flesh* pale salmon.

Trunks, elm. Sept.–Nov. $3\frac{1}{8} \times 3\frac{1}{4} \times \frac{5}{16}$ in.

469. P. leoninus Quél. (from the colour of the pileus, often lion-tawny) *a b c*.

P. convex, striate, lemon, yellow, orange, tawny or scarlet; mid. darker; or orange with sulphur marg. *St.* solid, yellowish. *G.* often with a yellowish edge.

Cæspitose Wood; rare. Sept.–Jan. $2\frac{3}{4} \times 2\frac{1}{2} \times \frac{1}{4}$ in. Var. *coccineus* Mass., *P.* crimson-orange or vermilion.

470. P. chrysophæus Quél. (from the dusky-golden stem; Gr. *chrusos*, gold, *phaios*, dusky) *a b c*.

P. plane, cinnamon; marg. striate. *St.* sulphur, sulphur-white or yellowish. *G.* colour as St., then salmon.

Taste insipid; odour none or subacid. Rotten wood, rotten sawdust, rotten ash, beech-stumps, boards, earth, hollow trees. May–Nov. $2\frac{7}{8} \times 2\frac{1}{8} \times \frac{3}{8}$ in.

471. P. phlebophorus Gill. (from the strongly veined pileus; Gr. *phleps*, a vein, *phero*, to carry) *a b c*.

P. convex, subumbonate, umber; marg. plain. *St.* hollow, white. *Flesh* greyish-white.

Dead stumps, rotten wood, sawdust, sticks; rare. June–Oct. $2 \times 2\frac{3}{4} \times \frac{1}{4}$ in.

472. P. umbrinellus Gill. (from its umber pileus) *a b*.

P. submembranous, expanded, sometimes subumbonate; marg. fimbriate. *St.* hollow, fibrous, shining, whitish becoming brownish.

On the ground. Oct. $1\frac{1}{4} \times 2\frac{7}{8} \times \frac{1}{8}$ in.

XV. **ENTOLOMA** Quél.

(From the potential, rather than definite veil; Gr. *entos*, within, *loma*, a fringe.)

Veil universal, faintly indicated by silkiness, flock, or scaliness. *Hymenophore* confluent and homogeneous with the fleshy stem. *Pileus* somewhat fleshy, margin at first incurved. *Stem* central, simple, fleshy, fibrous, soft or sometimes waxy. *Gills* sinuato-adnexed, sinuate or adnate often separating from stem, at first white then usually salmon. *Spores* elliptical and smooth or subglobose and coarsely warted, rosy or salmon. (Fig. 24.)

All the species grow on the ground in fields and woods chiefly after heavy rain, several appear in spring and summer; none are known to be edible, some are poisonous, many smell of meal.

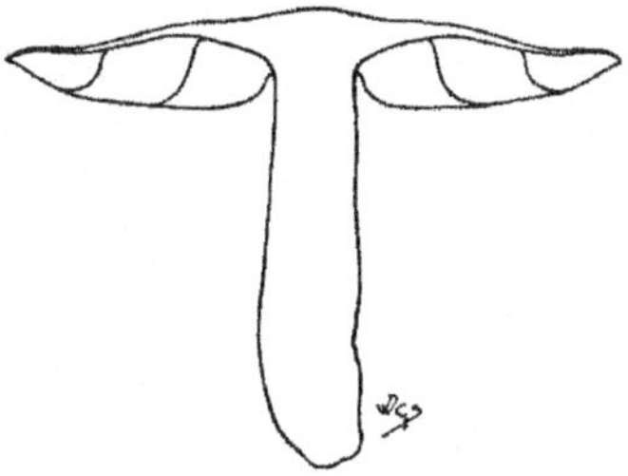

Fig. 24.—Section of *Entoloma rhodopolium* Quél. One-quarter natural size.

Entoloma agrees in structure and habit with *Tricholoma* and *Hebeloma*, and to a great extent with *Hypholoma*. It is necessary to carefully distinguish the rose- and salmon-coloured spores of *Entoloma* from the sometimes pale clay-coloured spores of *Hebeloma*.

Species **473—502**

a. Genuinæ. *Pileus* fleshy, when full grown; smooth, often viscid but not hygrophanous, not innato-floccose or squamulose. **473—483**

b. Leptonideæ. *Pileus* dry, except **484**, flocculose, somewhat scaly. **484—490**

c. Nolanideæ. *Pileus* thin, commonly irregular and repand, for the most part scissile, hygrophanous, smooth, with a silky appearance when dry. **491—502**

a. Genuinæ.

473. E. sinuatum Quel. (from the sinuate margin of the pileus) *a b c*. *P.* expanded, then repand, white or shaded tan, lavender, pale yellowish or brownish. *St.* solid, whitish. *G.* at first livid-whitish, then pinkish, at length brownish-salmon.

Gregarious. Poisonous. Odour strong, pleasant, somewhat of burnt sugar. Woods, mixed. July.–Oct. 7 × 5 × 1 in. *P.* sometimes 8 to 10 in. in diam. Young examples are sometimes mistaken for the mushroom.

474. E. lividum Quél. (from the livid colour of the pileus) *a b*. *P.* plane, broadly subumbonate, livid tan; mid. pale yellowish-buff. *St.* subhollow, whitish. *G.* adnexo-rounded, salmon.

Poisonous. Odour of new meal or none. Woods. April–Oct. $4\frac{3}{4} \times 3\frac{1}{4} \times \frac{7}{8}$ in. Sometimes mistaken for the mushroom. Var. *roseum* Sacc., *P.* slightly rose-tinted; logs.

475. E. prunuloides Quél. (from its odour being like that of **503** *Clitopilus prunulus* Quél.) *a b c.*

P. convex, broadly subumbonate, whitish, yellowish, greyish or brown; marg. striate. *St.* solid, white. *G.* emarginate, salmon.

Scattered. Odour of new meal, after maturity like decaying turnips. Woods, amongst moss and grass; uncommon. May–Sept. $2\frac{3}{4} \times 2 \times \frac{1}{2}$ in.

475a. E. porphyrophæum Karst. (Gr. *porphureos*, purple, *phaios*, dusky) *a.*

P. campanulate, expanded, subumbonate, not hygrophanous, even, almost glabrous, opaque sooty-brown or deep umber, becoming paler and mouse-colour when dry; marg. wavy, often incised and lobed. *St.* solid, attenuate upwards, fibrous, soft, opaque sooty-purple or brownish; base white-downy. *G.* truncate behind, almost free, subdistant, ventricose, greyish-white to reddish-grey or salmon. *Flesh* white.

Amongst grass. Autumn. $3\frac{3}{8} \times 4\frac{1}{4} \times \frac{5}{16}$ in. Must not be confused with **486.**

476. E. repandum Gill. (from the pileus being occasionally repand), *a b c.*

P. expanded, umbonate, whitish or faintly ochreous; marg. lobed. *St.* solid, colour as P. *G.* emarginate, salmon or faintly ochreous.

Odour of new meal. Amongst grass; rare. June–Oct. $3 \times 2\frac{1}{4} \times \frac{3}{8}$ in. Not unlike some forms of the mushroom.

476a. E. erophilum Karst. (from its frequent appearance in spring; Gr. *eros*, spring, *philos*, loving) *a b.*

P. convex, obtuse to umbonate, venoso-striate, dry, pale ashy or pale olive-brownish; mid. somewhat darker. *St.* solid, equal or attenuate downwards, farinose, white, shaded ivory. *G.* rounded-adnate, broad, subdistant, pallid or pale salmon. *Flesh* white.

Woods, amongst grass, often in large circles. Sept. $1\frac{3}{4} \times 2\frac{1}{4} \times \frac{5}{16}$ in.

477. E. placenta Quél. (from the shape of the pileus; *placenta*, a flat cake) *a b.*

P. flat, umbonate, brownish or brown. *St.* solid, tan-brownish. *G.* emarginate, pallid flesh-colour. *Flesh* thin.

Solitary. Taste at length acrid; odour none. Hedge-rows, damp places; rare. April–Oct. $2\frac{1}{4} \times 3 \times \frac{1}{4}$ in.

478. E. helodes Gill. (from its frequent habitat; Gr. *helos*, a marsh) *a.*

P. umbonate, ashy or sooty-purplish. *St.* hollow, ashy or greyish. *G.* adnate, salmon. *Flesh* thin, pallid.

Taste watery; odour of new meal. Grassy places, heathy pastures, turfy swamps, moors. Autumn. $2\frac{3}{8} \times 1\frac{1}{2} \times \frac{3}{8}$ in.

479. E. Batschianum Karst. (after Aug. Jo. Georg. Car. Batsch) *a c.*

P. convex, subumbonate, dark fuscous. *St.* hollow, grey. *G.* adnate, subsinuate, ashy or sooty. *Flesh* colour as P., white when dry.

Odour none. Woods, moist places, amongst grass in damp places. Oct. $1\frac{3}{8} \times 3\frac{3}{8} \times \frac{1}{4}$ in. Resembles one of the *Pratelli* as **804.**

480. **E. Bloxamii** Sacc. (after the Rev. A. Bloxam) *a b c*.
P. convex, umbonate or papillate, waved, splitting, dull dark purple. *St.* solid, lighter than P. *G.* adnexed, pale salmon.

Downs and exposed pastures. Sept.–Nov. 4¼ × 2⅝ × ⅝ in. Compare **482a.**

480a. **E. Farrahii** Mass. & Crossl. (after Mr. John Farrah) *a*.
P. cylindrico-ovate, then campanulate and somewhat repand, umbonate, smooth, fibrilloso-silky, blackish-blue; marg. pallid. *St.* solid, ventricose, glabrous, colour as P. with a white base. *G.* sinuato-adnexed, broad, ventricose, somewhat distant, salmon-colour.

Gregarious. Taste mild; odour none. Amongst short grass. 2¼ × 2½ × ⅜ in. Differs from its allies in its smooth, elliptical spores.

481. **E. ardosiacum** Quél. (from its slate-coloured pileus; *ardosia*, a slate) *a c*.
P. convex, steel-blue, shaded greenish, brownish or purplish. *St.* hollow, steel-blue, white at base. *G.* free, grey then salmon. *Flesh* dull pale purplish.

Odour none. Meadows, pastures, moist grassy places and under pines. 3⅜ × 4 × $\frac{5}{16}$ in. Might be mistaken for one of the *Pratelli* as **802.**

482. **E. liquescens** Sacc. (from the gills becoming liquescent) *a b*.
P. campanulate, broadly subumbonate, ochreous; mid. sienna. *St.* hollow, white. *G.* free, crowded, pale slate.

Gregarious. Under trees. April. 2⅝ × 2⅝ × ¼ in. Deliquescent like a *Coprinus*, which character with the dull pink spores and time of growth points to *Bolbitius*. Resembles one of the *Pratelli*.

482a. **E. madidum** Gill. (from the moist pileus; *madidus*, moist) *a b*.
P. campanulate, convex, obtuse, even, glabrous, moist, often viscid in rainy weather, shining when dry, not hygrophanous; when young blackish-violet, purple-slate, dull smalt or dull deep cobalt; sooty when old; mid. umber and whitish-yellow; marg. thin, inflexed, substriate. *St.* hollow, sometimes solid, fleshy-fibrous, slightly attenuate upwards, fibrillose, apex naked, colour as P., base whitish. *G.* slightly adnexed, ventricose, greyish-white to salmon. *Flesh* white.

Suspected poisonous. Odour strong, pungent, like **1341.** Amongst moss, leaf-heaps, etc. 2⅜ × 2⅜ × ⅜ in. Much like **480** in appearance.

483. **E. ameides** Sacc. (from its odour of starch) *a b*.
P. convex, broadly subumbonate, pale reddish-grey; mid. darker. *St.* stuffed, paler than P. or almost white. *G.* emarginato-adnexed, pale pink.

Tasteless; odour of hay, apples, meal, linseed, orange-flower-water and starch. Pastures. Sept.–Nov. 2 × 2¼ × ½ in. Changing to yellowish or reddish when bruised. Must not be mistaken for **1222.**

b. Leptonideæ.

484. E. Saundersii Sacc. (after W. Wilson Saunders) *a b.*

P. expanded, broadly subumbonate, then repand, glutinous, white, becoming fuscous. *St.* solid, silky-fibrous, whitish. *G.* emarginato-adnexed, salmon.

Cæspitose. On earth, river-sand, sawdust; rare. June–Oct. $3\frac{1}{2} \times 4 \times \frac{5}{8}$ in.

485. E. fertile Gill. (from its large, fruitful appearance) *a b.*

P. expanded, broadly subumbonate, pinkish-buff. *St.* solid, fibrillose, whitish. *G.* emarginato-adnexed, salmon.

Woods; rare. Aug.–Sept. $6\frac{1}{4} \times 4\frac{1}{4} \times 1\frac{1}{4}$ in.

485a. E. Rozei Quél. (after M. E. Roze, French mycologist) *a c.*

P. convexo-plane, thin, pallid grey, or somewhat pale, dull slate-grey, paler when dry, minutely white-velvety or silky shining. *St.* solid, tough, silky-fibrillose, white. *G.* adnate, crowded, narrow, whitish flesh-colour.

Amongst pines. Oct. $1\frac{1}{2} \times 2\frac{3}{8} \times \frac{3}{16}$ in. Wholly dark umber when dry.

486. E. jubatum Karst. (from the fibrillose pileus; *jubatus*, having a mane) *a b.*

P. expanded, umbonate, mouse-colour. *St.* hollow, colour as P. *G.* adnexed, salmon or dusky purplish.

Single or cæspitose. Taste watery, disagreeable. Old pastures and mossy places. May–Dec. $3\frac{1}{4} \times 4\frac{1}{4} \times \frac{5}{16}$ in. Must not be confused with 475a.

487. E. resutum Karst. (from the fibrillose pileus; *resutus*, ripped open) *a.*

P. convex, subumbonate, fuscous. *St.* stuffed, smooth, greyish. *G.* adnexed, grey or reddish-grey.

Odour none. Old pastures, woods. Oct. $1\frac{7}{8} \times 3\frac{3}{4} \times \frac{1}{8}$ in.

488. E. griseocyaneum Quél. (*griseus*, grey, *cyaneus*, dark blue) *a c.*

P. campanulate, steel-grey, lilac shaded. *St.* hollow, fibrillose, colour as P. or whitish. *G.* adnexed or adnate, ventricose, salmon.

Sunny pastures, grassy places, open woods; rare. Oct. $1\frac{3}{4} \times 3 \times \frac{3}{16}$ in.

488a. E. pulvereum Rea (from the mealy stem; *pulvereus*, dusty) *a.*

P. campanulate then flat, minutely scaly, fuliginous; marg. at first striate. *St.* stuffed then hollow, velvety, covered with reddish meal. *G.* thick, sinuato-adnate, veined, somewhat distant, exceeding marg. of P., edge irregular, pale or pinkish, then deep rose.

Amongst oak-leaves, under holly. Oct. $1\frac{1}{4} \times 2\frac{1}{8} \times \frac{1}{16}$ in.

489. E. sericellum Quél. (from the slightly silky pileus; *sericeus*, silky) *a b c.*

P. plane, whitish. *St.* subhollow, smooth, white. *G.* adnate, salmon.

Gregarious. Odour none. Woods, grassy places; frequent. July–Oct. $\frac{7}{8} \times 1\frac{5}{8} \times \frac{1}{16}$ in.

489a. E. Cookei Ch. Richon (after Dr. M. C. Cooke) *a.*

P. convex, expanded, pale reddish- or sienna-orange or brick-red, reticulate all over with raised pale yellowish ribs; marg. incurved. *St.* solid, often slightly incurved, almost central, whitish, shaded pale yellowish. *G.* adnexed, subdistant, thickish, pale salmon. *Flesh* thick at disc of P., thin elsewhere, colour as G.

Stumps. Autumn. $1\frac{1}{2} \times 1 \times \frac{3}{16}$ in. Cooke described this as a form of **471**, but Ch. Richon and Massee have made it an *Entoloma*. I have followed them with great doubt, thinking it more probably an abnormal form of **412**.

490. E. Thomsonii Sacc. (after Dr. Thomson) *a.*

P. flat, umbonate, velvety, grey; mid. reticulato-ribbed. *St.* sub-hollow, tomentose, paler than P. *G.* adnexed, salmon.

Amongst grass in plantations. Autumn. $1\frac{7}{8} \times 1\frac{7}{8} \times \frac{3}{16}$ in.

c. Nolanideæ.

491. E. clypeatum Quél. (from the shape of the pileus; *clypeus*, a shield) *a b c.*

P. flat, umbonate, umber or grey-brown, streaked darker. *St.* stuffed or hollow, paler than P. or almost white. *G.* adnexo-rounded, salmon or shaded greyish or brownish.

Solitary or cæspitose. Woods, gardens, waste places; frequent. April–Oct. $3\frac{1}{4} \times 4\frac{3}{4} \times \frac{3}{8}$ in.

492. E. nigrocinnamomeum Sacc. (from the blackish-cinnamon colour of the pileus) *a b.*

P. flat, umbonate, moist, shining, silky when dry. *St.* hollow, fibrillose, grey-brown. *G.* emarginato-rounded, reddish-salmon.

Gregarious. Odour of new meal. Pastures, amongst grass and heather. Oct. $3\frac{1}{8} \times 2\frac{1}{2} \times \frac{3}{8}$ in.

493. E. rhodopolium Quél. (from its occasionally rosy-grey colour; Gr. *rhodon*, a rose, *polios*, grey) *a b c.*

P. expanded, subumbonate, varying pale to dark fuscous, livid, brown or purplish-brown. *St.* stuffed or hollow, smooth, whitish. *G.* sinuate, rose.

Odour rank of new meal, of **1527**, of nitric acid or none. Woods; frequent. May–Oct. $2\frac{1}{8} \times 3\frac{3}{4} \times \frac{1}{4}$ in.

494. E. pluteoides Karst. (from its resemblance to *Pluteus*) *a b.*

P. convexo-plane, subumbonate, pale purplish-slate. *St.* hollow, fibrillose, whitish. *G.* emarginato-adnexed, crowded, salmon.

Stumps. $2\frac{1}{8} \times 3\frac{3}{4} \times \frac{1}{4}$ in.

495. E. majale Karst. (from its appearance in May) *a.*

P. campanulate, subumbonate, dark ochreous or brown, becoming pale. *St.* hollow, whitish. *G.* free, salmon. *Flesh* very thin.

Subcæspitose. Meadows, open woods, mossy places in fir-woods; rare. April–May. $3\frac{1}{4} \times 3\frac{3}{8} \times \frac{3}{8}$ in.

496. E. Wynnei Sacc. (after Mrs. Lloyd Wynne) *a b.*

P. plane, smooth, sometimes umbilicate, pale umber or smoky; marg. crisped, striate, at first smooth, then breaking up into minute sc. *St.* hollow, dusky grey. *G.* sinuate, veined, pallid salmon.

Taste insipid; odour of bugs or like **1278.** Woods, fir. Sept.–Nov. $2\frac{1}{4} \times 3 \times \frac{3}{16}$ in. Allied to **497,** but with G. less veined, less undulate and less distant.

497. E. costatum Gill. (from the transverse ribs or veins on the gills; *costa*, a rib) *a b c.*

P. plane, fuscous-livid or greyish. *St.* hollow, colour as P. *G.* very broad, almost free, salmon.

Solitary or cæspitose. Odour none. Damp meadows, pastures, swampy ground, grass-covered stumps; common. Aug.–Nov. $2\frac{3}{8} \times 2 \times \frac{5}{16}$ in.

498. E. sericeum Quél. (from the silky pileus; *sericeus*, silky) *a b c.*

P. plane, umber; mid. darker. *St.* hollow, pale greyish-umber. *G.* emarginate, greyish then rufescent.

Gregarious. Odour strong of new meal. Meadows, open woods; frequent. May–Oct. $2 \times 2\frac{1}{8} \times \frac{3}{16}$ in. This and **499** may be one species.

499. E. Persoonii Sacc. (after C. H. Persoon) *a b c.*

P. campanulate, subumbonate, silky-fibrous, splitting, membranous, white; mid. faintly ochreous; marg. striate to disc. *St.* hollow above, abruptly bulbous, cartilaginous, white. *G.* adnexo-adnate, pale rose or flesh-colour, then darker.

Grassy places. Autumn. $2 \times 2\frac{7}{8} \times \frac{3}{16}$ in.

500. E. nidorosum Quél. (from its strong odour; *nidor*, reeking) *a b c.*

P. plane, ashy fawn-colour, whitish or purple-brown. *St.* stuffed, whitish. *G.* emarginate, salmon.

Odour nitrous, alkaline, sometimes obsolete. Lawns, open woods; common. Aug.–Nov. $3 \times 3\frac{3}{4} \times \frac{5}{16}$ in.

501. E. speculum Quél. (from the shining pileus; *speculum*, a mirror) *a c.*

P. plane or depressed, umbonate, whitish-straw or whitish; marg. striate. *St.* stuffed or hollow, paler than P. *G.* adnexo-free, ventricose, salmon.

Solitary or subcæspitose. Odour none. Amongst grass and twigs. Oct. $2\frac{1}{4} \times 2\frac{3}{4} \times \frac{1}{4}$ in.

502. E. tortipes Mass. (from the twisted stem; *tortus*, twisted, *pes*, a foot) *a b.*

P. hygrophanous, thin, splitting, plane, depressed, umbonate, livid brown. *St.* solid, with a bark, striate above, pale brown within and without. *G.* sinuate, salmon-white.

On dung. Autumn. $3\frac{5}{8} \times 2\frac{1}{2} \times \frac{5}{16}$ in.

XVI. **CLITOPILUS** Quél.

(From the decurrent hymenophore; Gr. *klitos*, a declivity, *pilos*, a cap.)

Veil indicated by silkiness and flock. *Hymenophore* confluent and homogeneous with the fleshy stem. *Pileus* more or less depressed or umbilicate, margin at first involute. *Stem* central, simple, fleshy

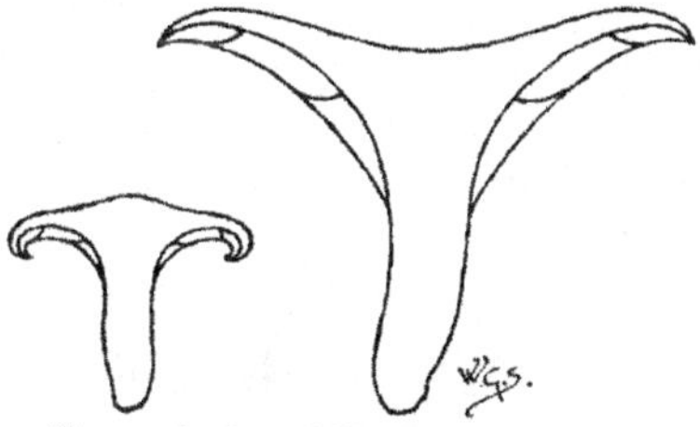

Fig. 25.—Sections of *Clitopilus prunulus* Quél. One-half natural size.

or fibrous. *Gills* decurrent, white, except **506**, then salmon. *Spores* smooth or warted, salmon, sometimes very pale. (Fig. 25.)

All the species grow on the ground. They have a more or less mealy odour. Some are tasteless, others edible.

Clitopilus agrees in structure with *Clitocybe* and in part with *Flammula*. Species **503**—**513**

a. Orcellæ. *Pileus* irregular, somewhat excentric, flexuous, somewhat hygrophanous, margin at first flocculose. *Gills* deeply decurrent. **503**—**509**

b. Sericellæ. *Pileus* regular, silky or hygrophanous-silky, margin involute, naked. *Gills* adnate, slightly decurrent. **510**—**513**

a. Orcellæ.

503. C. prunulus Quél. (from the pruinose appearance of the pileus) *a b c.*

P. plane or convex, broadly subumbonate; when young faintly grey, then whitish. *St.* solid, frequently central, paler than P. *G.* subdistant, flesh-colour.

Edible. Odour strong of new meal, or like **1527**. Woods and open places; common. June–Oct. $3\frac{1}{4} \times 2\frac{3}{4} \times \frac{1}{2}$ in. This and **504** are not insect eaten.

504. P. orcella Quél. (from the old Italian popular name; *orgella*, like an ear) *a b c.*

P. plano depressed, lobed, undulate, white. *St.* solid, flocculose, often excentric, colour as P. *G.* crowded, whitish flesh-colour.

In troops. Edible. Odour as in **503** or of Syringa leaf or cucumber, or fresh meal and cucumber. Open grassy places, sometimes in rings 20 feet in diam.; frequent. June–Oct. $4\frac{3}{8} \times \frac{7}{8} \times \frac{1}{2}$ in. Intermediate forms occur between this and **503**.

505. C. mundulus Gill. (*mundus*, neat) *a b c*.

P. convex, then depressed, white. *St.* stuffed, sometimes excentric, white, becoming black internally at base. *G.* crowded, whitish salmon. *Flesh* white.

Taste bitter; odour none. Woods, amongst leaves; uncommon. Autumn. $1\frac{3}{8} \times \frac{5}{8} \times \frac{3}{16}$ in. In var. *nigrescens* Sacc. the flesh becomes wholly black.

505a. C. Sarnicus Mass. (after the place of finding—Sarnia, the Roman name for Guernsey) *a*.

P. campanulate, plane, subumbonate, often more or less depressed round the um., slightly striate when moist, even when dry, minutely silky-flocculose, mouse-colour or grey, paler with a ruddy tinge when dry. *St.* minutely fistulose, equal, subflexuous, even, glabrous, white. *G.* subcrowded, plane nearly to the stem, then suddenly decurrent, pinkish-salmon. *Flesh* very thin.

On the ground, in Guernsey. $1\frac{1}{4} \times 1\frac{1}{4} \times \frac{1}{10}$ in. Allied to **505**.

506. C. popinalis Gill. (from its edible qualities; *popina*, a cookshop) *a b c*.

P. convex, subumbonate or depressed, ashy or dull slate. *St.* stuffed, paler than P. *G.* crowded, grey or brown, becoming reddish.

Solitary or gregarious. Odour of new meal or rank and strong. Downs, fields, grassy roadsides. Aug.–Oct. $2\frac{3}{4} \times 1\frac{7}{8} \times \frac{1}{2}$ in.

507. C. undatus Gill. (from the undulate pileus) *a b c*.

P. expanded, depressed, sometimes much lobed, sometimes with a papilla, dull ashy or tan-ochreous. *St.* hollow, colour as P. or paler, faintly salmon-shaded. *G.* not crowded, salmon, sometimes reddish.

Odour none. Downs, etc. Oct. $1\frac{5}{8} \times 1\frac{3}{4} \times \frac{1}{8}$ in.

508. C. cancrinus Quél. (from the frequently cracked pileus; *cancer*, a lattice) *a b c*.

P. hemispherical, depressed, dry, at length shining and silky fibrillose, whitish. *St.* stuffed or fistulose, colour as P. *G.* distant, flesh-colour.

Odour none. Pastures. July–Oct. $1\frac{1}{8} \times 1\frac{3}{8} \times \frac{1}{8}$.

509. C. cretatus Sacc. (from the colour of the pileus; *creta*, chalk) *a b c*.

P. lobed, umbilicato-depressed. *St.* excentric, tomentose, colour as P. *G.* rose.

Solitary or gregarious. Woods, pastures; uncommon. Oct. $\frac{3}{4} \times \frac{1}{4} \times \frac{1}{16}$ in. Resembles a small **504**.

b. Sericellæ.

510. C. carneoalbus Gill. (from the flesh-coloured gills and white pileus; *caro*, flesh, *albus*, white) *a b c*.

P. convexo-plane, subdepressed, white, faintly greyish or rufescent. *St.* stuffed or hollow, white. *G.* subdistant, flesh-colour.

Gregarious. Odour none. Woods, heathy places. Sept. $1\frac{3}{8} \times 2 \times \frac{3}{16}$ in.

511. C. vilis Gill. (*vilis*, insignificant) *a b c*.
P. convex, depressed, grey. *St*. fistulose, colour as P. *G*. crowded, pale salmon.
Downs, amongst moss. July–Aug. $1\frac{1}{2} \times 2\frac{3}{8} \times \frac{1}{8}$ in.

512. C. stilbocephalus Sacc. (from the sparkling pileus; Gr. *stilbo*, to glisten, *kephale*, a head) *a b*.
P. campanulate, obtuse or umbonate, whitish or greyish; um and marg. sometimes ochreous. *St*. hollow, pale greyish, pallid below. *G*. sinuate, veined, salmon.
Odour of new meal. Aug.–Oct. $1\frac{3}{8} \times 3\frac{1}{8} \times \frac{1}{8}$ in.

513. C. straminipes Sacc. (from the straw-coloured stem; *stramen*, a straw, *pes*, a foot) *a b*.
P. expanded, depressed, subumbonate, glabrous, shining-silky, whitish or faintly ochreous. *St*. hollow, white-farinose above. *G*. salmon.
Among grass. Sept. $2 \times 2 \times \frac{1}{8}$ in. Very fragile.

XVII. LEPTONIA Quél.

(From the slender growth; Gr. *lepos*, slender.)

Veil indicated by fibrillæ, scales, dots on stem, etc. *Hymenophore* confluent with but heterogeneous from the cartilaginous stem.

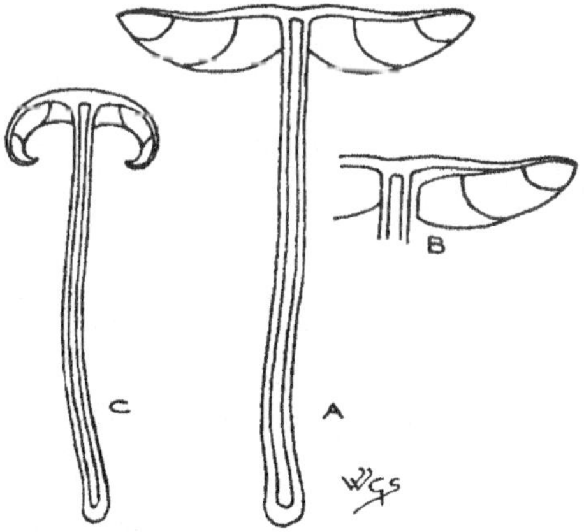

Fig. 26.—A, section of *Leptonia lampropus* Quél., showing at B separation of gills from hymenophore; C, section of *L. chalybæa* Quél. Natural size.

Pileus thin, umbilicate or with a dark middle, cuticle fibrillose or separating into darker scales; margin at first incurved. *Stem* central, simple, cartilaginous, tubular or stuffed, glabrous, sublustrous. *Gills* adnexed or adnate, readily separating from their attachment to the stem, not decurrent, variable in colour. *Spores* smooth or rough, salmon. (Fig. 26.)

Many of the species are gregarious and grow in troops, they are mostly small, inodorous, except **517** and **524**, and often beautiful in colour, shades of light or dark blue or green being frequent.

All grow on the ground except **521**; they chiefly grow in mossy pastures and marshy places. One non-British species—*L. parasitica* Quél.—grows on the hymenium of **1370**. They are most common in rainy seasons. The species generally are suspected to be poisonous.

Leptonia agrees in structure with *Collybia*, *Naucoria*, *Psilocybe* and *Panæolus*. Species **514—528**

Pileus slightly fleshy, without striæ, hygrophanous. *Gills* whitish. **514—519**

Gills light or dark blue or bluish. **520—523**

Gills pallid, becoming yellow or green. **524—526**

Pileus somewhat striate, hygrophanous. *Gills* grey or glaucous. **527, 528**

514. L. placida Karst. (from its pleasing appearance) *a*.

P. campanulate, broadly subumbonate, dusky brown or dusky-slate-squamulose on a tan ground. *St.* stuffed, slaty-blue, black-dotted above. *G.* adnexed or somewhat adnate, salmon-white. *Flesh* brownish or lavender.

Usually solitary. By beech-trunks. Autumn. $2 \times 2\frac{1}{2} \times \frac{3}{16}$ in.

515. L. anatina Quél. (from the colours; *anatinus*, pertaining to a duck) *a b c*.

P. campanulate, broadly subumbonate, umber-squamulose. *St.* stuffed, purple-bluish or smalt above, brown and whitish below. *G.* adnexed or somewhat adnate, pale salmon.

Fields. Oct. $1\frac{5}{8} \times 1\frac{1}{2} \times \frac{1}{4}$ in.

516. L. lappula Quél. (from the prickly pileus; *lappa*, a burr) *a*.

P. convex, umbilicate, grey or purple-slate; mid. black-punctate. *St.* hollow, lilac, shaded brownish. *G.* subadnate, broad, white, faint greyish or whitish-salmon.

Odour none. Amongst beech-leaves. Autumn. $1\frac{1}{2} \times 2\frac{5}{8} \times \frac{3}{16}$ in.

517. L. lampropus Quél. (from the lustrous stem; Gr. *lampros*, shining, *pous*, a foot) *a b c*.

P. expanded, depressed, mouse-colour, steel-blue, pallid yellowish, brownish or brown. *St.* fistulose, steel-blue, pale slate or purplish, brownish below, white at base. *G.* adnexed or subadnate, broad, pale salmon.

Taste and odour somewhat strong, disagreeable. Pastures; common. July–Nov. $1\frac{3}{8} \times 2\frac{1}{4} \times \frac{1}{8}$ in.

518. L. æthiops Gill. (from the black pileus and stem; Gr. *aithiops*, an Ethiopian) *a*.

P. plano-depressed, smooth. *St.* stuffed, fuscous-blackish. *G.* subadnate or adnexed, narrow, pale salmon-white.

Woods, grassy places. Sept. $1\frac{1}{2} \times 2\frac{1}{2} \times \frac{1}{8}$ in.

519. L. solstitialis Gill. (from its sometimes appearing in the summer solstice—June) *a*.

P. plane, umbonate, fuscous, brown- or olive-shaded. *St.* fistulose, dusky, whitish above or as P. *G.* emarginate, pale salmon.

Amongst stones, grassy ground, woods. Aug.–Sept. $1\frac{1}{8} \times 2 \times \frac{1}{8}$ in.

520. L. serrulata Quél. (from the black-toothed edge of the gills, contrasted with **528**; *serrula*, a saw) *a b c*.

P. convex, depressed, blackish-blue. *St.* fistulose, paler than P. *G.* adnexed or adnate, slate-whitish, then grey-salmon, varying slate-purplish. *Flesh* slate-white.

Sometimes cæspitose. Pastures and grassy woods. June–Oct. $1\frac{5}{8} \times 2\frac{1}{2} \times \frac{1}{8}$ in.

521. L. euchroa Gill. (from its fine colour, Gr. *euchroos*) *a b c*.

P. convex, subumbonate, purple-blue or bright smalt-purple. *St.* stuffed, colour as P. or lighter, or bluish flesh-colour. *G.* adnexed or subadnate, pale grey-blue, blue, purple-blue or slate.

Solitary, gregarious or cæspitose. Probably poisonous. Woods, trunks, stumps, branches, alder, hazel; rare. Aug.–Oct. $1\frac{3}{4} \times 2 \times \frac{1}{8}$ in. Pileus sometimes $\frac{1}{2}$ in. or less in diam.

522. L. chalybæa Quél. (from its steel-blue colour; Gr. *chalups*, steel) *a b c*.

P. plane, subumbonate, separable, dark violaceous, brownish-slate or slate with satin-like threads from mid. to marg. *St.* stuffed, colour as P. *G.* adnate, bluish-grey-whitish, pale olive-salmon, salmon or purplish.

Taste somewhat mushroom-like, but strong and unpleasant. Pastures; frequent. July–Oct. $1\frac{3}{8} \times 1\frac{7}{8} \times \frac{1}{8}$ in.

523. L. lazulina Quél. (from the lapis lazuli colour of the stem) *a b*.

P. expanded, dusky-black, steel-slate, dark-brown-streaky. *St.* fistulose, light or dark blue. *G.* adnate, colour as St., then shaded salmon. *Flesh* slate-steel.

Gregarious. Amongst moss and grass. Sept. $\frac{7}{8} \times 2\frac{1}{8} \times \frac{1}{16}$ in.

524. L. incana Gill. (*incanus*, hoary) *a b c*.

P. plane, umbilicate, striate, olive- or bronzy-green, streaked. *St.* fistulose, colour as P. or verdigris or greenish-blue. *G.* adnexed or adnato-toothed, whitish-green then salmon.

Gregarious. Odour strong of mice. Woods, pastures, downs; frequent. July–Oct. $1\frac{3}{8} \times 2 \times \frac{1}{16}$ in.

525. L. formosa Gill. (*formosus*, beautiful) *a b*.

P. plane, broadly umbilicate, yellow or brownish, minutely sooty-squamulose. *St.* fistulose, yellow. *G.* adnato-toothed, light salmon-yellow, then pale flesh-colour.

Amongst *Equisetum*, grass, commons, pine woods. Sept. $2\frac{1}{2} \times 3\frac{3}{4} \times \frac{1}{8}$ in. Var. *suavis* Sacc. *St.* steel-bluish.

526. L. chloropolia Gill. (from the pale green-grey pileus; Gr. *chloros*, greenish-yellow, *polios*, hoary) *a*.

P. flat, slightly subumbonate, livid or sulphur-livid; mid. black-squamulose. *St.* fistulose, colour as P. above, pale grey-bluish below. *G.* adnate, salmon-white.

Grassy places, woods. Oct. $1\frac{5}{8} \times 2\frac{5}{8} \times \frac{1}{6}$ in.

527. L. asprella Quél. (*asper*, rough) *a b c*.

P. expanded, broadly umbilicate, striate, smoky or mouse-colour; mid. villous-squamulose. *St.* fistulose, grey-white. *G.* adnato-rounded, at length grey-salmon.

Open pastures, grassy places; uncommon. Oct. $1\frac{3}{8} \times 1\frac{3}{4} \times \frac{1}{8}$ in.

528. L. nefrens Quél. (from the edge of the gills being without teeth, contrasted with **520**; *ne*, not, *frendo*, to bite) *a*.

P. flat, depressed, smoky, yellowish-umber or olivaceous. *St.* fistulose, lighter than P., or fuscous-livid. *G.* adnexed or adnate, at length light red with a blackish edge.

Grassy places. Autumn. $1\frac{5}{8} \times 1\frac{3}{8} \times \frac{3}{16}$ in.

XVIII. NOLANEA Quél.

(From a fancied resemblance in the pileus to a little bell, *nola*.)

Veil indicated by fibrils, squamules, silkiness or flock; Mr. George Massee has observed a rudimentary annulus in **534**. *Hymenophore*

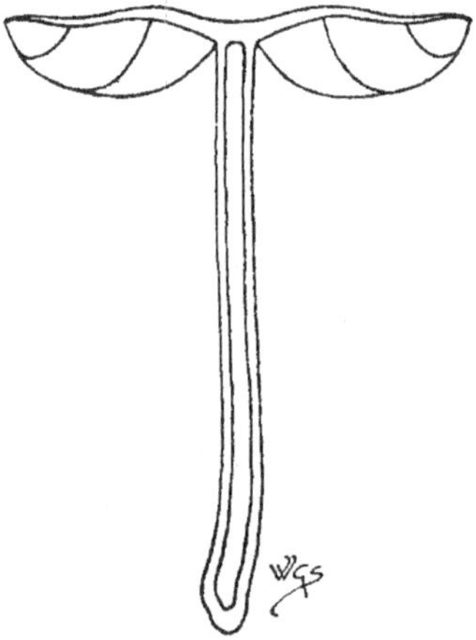

Fig. 27.—Section of *Nolanea pascua* Quél.
Natural size.

confluent with but heterogeneous from the cartilaginous stem. *Pileus* submembranaceous, more or less campanulate, subpapillate or subumbonate, not umbilicate, except **536** and **542**, striate, even or floccose, margin straight, at first adpressed to the stem, not involute. *Stem* central, simple, fistulose or stuffed, cartilaginous. *Gills* free or adfixed, not decurrent, usually whitish at first, then salmon or rose. *Spores* rose-coloured or salmon. (Fig. 27.)

The species are thin, slender, and usually brittle, some are tough. They are terrestrial except **541**, and appear in summer and autumn on grassy hills and in damp places in woods. The group generally is suspected to be poisonous.

Nolanea agrees in structure with *Mycena*, *Galera*, *Psathyra*, and *Psathyrella*. Species **529—543**

Pileus dark-coloured, hygrophanous. *Gills* grey or fuscous.	**529—533**
Gills becoming yellow or rufescent.	**534—537**
Pileus hygrophanous. *Gills* shining white then rosy.	**538, 539**
Pileus not hygrophanous. *Gills* whitish.	**540—543**

529. N. pascua Quél. (from its habitat, pastures) *a b c*.

P. flat or subumbonate, silky, grey- or pinkish-brown. *St.* striate, faintly brownish or ochreous, or almost white. *G.* adnato-adnexed, pale salmon.

Odour none. Fields, open woods; common. May–Nov. $1\frac{7}{8} \times 2\frac{3}{8} \times \frac{3}{16}$ in.

530. N. Babingtonii Sacc. (after Prof. C. C. Babington) *a*.

P. conico-campanulate, ashy-grey; mid. dark-brown-squamulose. *St.* equal, downy, dark brown. *G.* adnato-ascending, ashy-grey.

Woods; rare. Oct.–Nov. $\frac{1}{2} \times 1\frac{1}{8} \times \frac{1}{16}$ in.

531. N. mammosa Quél. (from the shape of the pileus; *mamma*, a breast) *a b c*.

P. umber or smoky. *St.* smoky white-brown. *G.* adnexed, grey then salmon.

Lawns, woods. Feb. $2\frac{5}{8} \times 3\frac{1}{2} \times \frac{3}{16}$ in. Often grows in company with **1238**.

531a. N. papillata Bres. (from the papillate pileus).

P. submembranaceous, subcampanulate, convex, expanded, striate, fuscous-bay becoming cinnamon. *St.* glabrous, shining, obsoletely white-mealy at apex, white-tomentose at base. *G.* sinuato-adnate, somewhat crowded, livid white becoming fuscous flesh-colour. *Flesh* colour as P. but paler.

Odour pleasant, sometimes absent. In grassy or woody places. Sept. $1\frac{1}{4} \times 2\frac{1}{4} \times \frac{1}{10}$ in.

532. N. juncea Quél. (from its usual habitat, rushy places; *juncus*, a rush) *a b*.

P. hemispherical, papillate or subumbilicate, smoky or umber. *St.* smooth, reddish-brown. *G.* adnato-ascending, grey- or brownish-salmon.

Odour none. Woods, commons. Oct. $1\frac{1}{4} \times 2\frac{3}{4} \times \frac{1}{8}$ in.

533. **N. fulvostrigosa** Sacc. (from the tawny strigose stem).

P. hemispherical, subwrinkled, grey. *St.* furfuraceo-squamulose, reddish. *G.* adnate, grey.

Woods. Sept. $\frac{3}{4} \times \frac{1}{2} \times \frac{1}{16}$ in.

534. **N. pisciodora** Gill. (from its fishy odour; *piscis*, a fish, *odor*, a smell) *a b c.*

P. campanulate, velvety, umbonate, tawny-cinnamon, salmon-brownish or dark brown. *St.* rarely with a rudimentary A., chestnut-brown or black. *G.* adnexed, toothed, pale yellowish-tawny, then flesh-colour or dull salmon. *Flesh* warm brown, dark at base of stem.

Odour strong of rotten fish, varying to cucumber. Woods, amongst decayed leaves, on sticks; frequent. Oct. $1\frac{3}{8} \times 2\frac{1}{2} \times \frac{1}{8}$ in. This may be only a form of **709.** See **239, 535** and **538.** Spores may possibly vary white, salmon and brown.

535. **N. nigripes** Gill. (from the black stem; *niger*, black, *pes*, a foot) *a b.*

P. hemispherical, floccose, fuscous or umber. *St.* equal, twisted, somewhat wavy, hollow. *G.* adnexed, yellow-flesh-colour or salmon.

Odour as in **534.** Swamps. Nov. $1\frac{3}{4} \times 3\frac{7}{8} \times \frac{1}{8}$ in. Perhaps a form of **534.**

536. **N. rufocarnea** Sacc. (from its colour; *rufus*, reddish, *caro*, flesh) *a b c.*

P. hemispherical, subumbilicate, red-brown. *St.* minutely fibrillose, pale rufous. *G.* adnate, rose-colour.

Taste somewhat bitter. Grassy heaths; frequent. Sept. $1\frac{1}{2} \times 2 \times \frac{1}{8}$ in.

537. **N. icterina** Quél. (from its colour; Gr. *ikteros*, jaundice) *a b.*

P. hemispherical, sometimes papillate, slightly silky, greenish-sulphur or olive-brassy. *St.* stuffed, colour as P. or full ochreous. *G.* adnexed, adnate or decurrent, salmon-white to salmon, saffron in decay. *Flesh* sulphur in P., white with a buff bark in St.

Woods, gardens. Oct. $1\frac{1}{4} \times 1\frac{1}{8} \times \frac{1}{8}$ in.

538. **N. picea** Gill. (from the more or less pitch-black colour; *piceus*, pitch-black) *a b.*

P. campanulate, papillate, smooth, at first umber; marg. lighter, shaded slate. *St.* pruinose, colour as P., or dull salmon, blotched and shaded slate. *G.* ascending, emarginate, flesh-colour.

Odour of decaying fish and cucumber. Grassy places, gardens. $1\frac{3}{4} \times 2\frac{1}{2} \times \frac{1}{8}$ in. Perhaps the same as **534, 535** and **709.**

539. **N. infula** Gill. (perhaps from the lustrous gills, at first dazzling white then rose; *infula*, a fillet or ornament) *a b.*

P. plane, papillate, fawn or smoky when damp, ochreous-tan when dry. *St.* smooth, colour as P. or grey-brown. *G.* varied in attachment.

Odour none. Lawns, woods, burnt ground. Sept.–Oct. $1 \times 2\frac{1}{2} \times \frac{1}{8}$ in.

540. N. verecunda Gill. (from its appearance ; *verecundus*, modest) *a*.
P. convex, umbonate, watery-reddish. *St*. faintly ochreous. *G*. adnate, salmon-white. *Spores* reddish.

Densely gregarious or in troops. Amongst short grass. Sept. $\frac{3}{4} \times \frac{7}{8} \times \frac{1}{16}$ in.

541. N. cœlestina Gill. (from the colour of the pileus, sometimes azure ; *cœlum*, the sky) *a*.
P. convex, umbonate, pale slate. *St*. colour as P. *G*. adnate, salmon-grey or hoary-white. *Spores* faintly rosy.

Woods, pine ; on wood, rotten oak. Oct. $1\frac{1}{2} \times 2\frac{1}{4} \times \frac{1}{8}$ in.

541a. N. exilis Quél. (*exilis*, small).
P. membranous, conical, expanded, glabrous, striate, livid grey to lilac ; mid. papillose, darker. *St*. hollow, glabrous, tough, apex naked, greyish-green. *G*. adnexed, subcrowded, whitish with a tinge of pink.

Amongst grass, leaves, etc., under beeches, pines ; in swamps. $\frac{3}{8} \times 3 \times \frac{1}{24}$ in. Allied to **526**, but more slender and delicate.

542. N. RUBIDA Sacc. (from its frequent reddish tinge) *a b c*.
P. convex, subumbilicate, greyish or faintly ruddy. *St*. attenuate below, colour as P or paler. *G*. adnate, rose.

Odour of new meal. Amongst grass in a conservatory. Mar.–Nov. $\frac{1}{2} \times \frac{1}{2} \times \frac{1}{8}$ in.

543. N. RHODOSPORA Sacc. & D. Sacc. (from the rose-coloured spores) *a b c*.
P. convex, sooty-fibrillose or rufescent-pilose. *St*. subbulbous, white. *G*. sinuate or free, salmon or rose.

On earth and wooden borders in stoves. May–Sept. $1 \times 1\frac{1}{4} \times \frac{1}{16}$ in.

XIX. ECCILIA Quél.

(From the hollowed out or depressed pileus ; Gr. *ekkoiloo*, to hollow out.)

Veil represented by pruina, fibrillæ or dot-like scales. *Hymenophore* confluent with but heterogeneous from the cartilaginous stem.

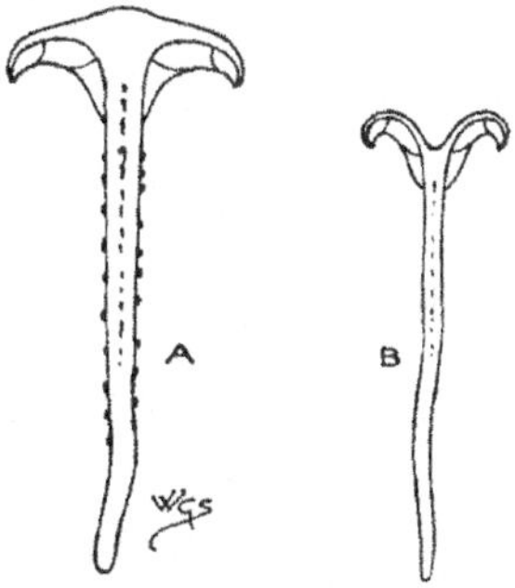

Fig. 28.—A, section of *Eccilia atropuncta* Gill. B, section of *E. acus* Sacc. Natural size.

Pileus somewhat membranous, margin at first incurved, umbilicate, except **548.** *Stem* central, simple, fistulose, stuffed in **550.** *Gills* decurrent, not separating from their attachment to the stem as in *Nolanea.* *Spores* smooth or warted, salmon or rose-colour. (Fig. 28.)

Most of the species grow on the ground, one grows on rotten wood and on walls, another amongst leaves.

Eccilia agrees in structure with *Omphalia* and *Tubaria.*

Species **544—552**

544. E. parkensis Quél. (from its being first noted in the King's park, Upsala) *a c.*

P. plano-convex, fuscous or blackish, not pale when dry; marg. substriate. *St.* attenuate downwards, not dotted, fuscous. *G.* dull flesh-colour.

Grassy places. July–Aug. $\frac{3}{4} \times 1\frac{1}{8} \times \frac{1}{16}$ in.

545. E. carneogrisea Gill. (from the colour of the pileus; *caro,* flesh, *griseus,* grey) *a b.*

P. plane; marg. striate. *St.* colour as P. *G.* distant, rosy, edge darker.

Gregarious. Amongst fir-leaves, grass. Aug.–Oct. $1\frac{1}{8} \times 1\frac{1}{2} \times \frac{1}{8}$ in.

545a. E. Smithii W. G. Sm., *Agaricus Smithii* Mass. (after Worthington George Smith) *a b.*

P. soon plane, orbicular, umbilicate, even, atomate, yellowish- or buff-white; marg. somewhat undulate, becoming salmon-striate. *St.* equal or slightly attenuate upwards, surface undulate, fibrillose, salmon, paler above, white downy below. *G.* broadly adnato-decurrent, salmon-rose. *Flesh* somewhat thick except at marg., pale salmon-white.

On the ground. Oct. $1\frac{3}{4} \times 2\frac{1}{8} \times \frac{1}{8}$ in. Differs from **512** in the plane pileus, coloured stem, different spores, etc.

546. E. griseorubella Quél. (from the reddish-grey pileus) *a.*

P. hygrophanous, plane, brown or grey-brown; marg. striate. *St.* smooth, colour as P. *G.* subdistant flesh-colour.

Gregarious. Suspected poisonous. Inodorous. On earth and wood. Autumn. $1\frac{1}{8} \times 2 \times \frac{1}{8}$ in.

547. E. atrides Quél. (*ater,* black) *a.*

P. plane, fuscous, black-streaked, becoming pale; marg. striate. *St.* pallid, black-dotted upwards. *G.* distant, slightly serrulate, pallid with a black edge.

Gregarious. Woods, moist places. Autumn. $1\frac{1}{2} \times 1\frac{3}{8} \times \frac{1}{8}$ in. Perhaps a form of **520.**

548. E. atropuncta Gill. (from the black-dotted stem; *ater,* black, *punctum,* a dot) *a b c.*

P. convex, pale ashy or dull umber; marg. lighter. *St.* umber, dark or pale, black-punctato-scaly. *G.* distant, ashy flesh-colour.

Taste disagreeable. On the ground amongst moss. Sept.–Oct. $\frac{7}{8} \times 1\frac{1}{2} \times \frac{1}{8}$ in.

549. E. nigrella Gill. (from the blackish pileus; *niger*, black) *a b*.
P. smooth, deep umber. *St.* glaucous, pale salmon-brown, not black-punctate. *G.* flesh-colour or subashy, not serrulate or with a black edge.

Sometimes cæspitose. Pastures. Autumn. $\frac{1}{2} \times 2 \times \frac{1}{16}$ in.

550. E. rhodocylix Gill. (from the resemblance of the pileus to a rose-coloured cup; Gr. *rhodon*, a rose, *kulix*, a cup) *a b*.
P. convex, fuscous, grey or tan; marg. striate when moist. *St.* stuffed, ashy, tan, or white. *G.* very distant, flesh-colour.

Rotten wood, stumps, alder, wall-tops. Sept. $1\frac{1}{8} \times 1 \times \frac{1}{16}$ in. In the young state when the gills are white this resembles **386**.

551. E. FLOSCULUS Sacc. (from its likeness to a small flower) *a b*.
P. slightly irregular, pruinoso-crystalline, pink-tan; mid. dark brown. *St.* pruinose or innately fibrillose, pinkish-tan, dark at base. *G.* slightly wavy, bright rose.

In conservatories, on the ground and on tree-fern stems. June. $\frac{7}{8} \times \frac{5}{8} \times \frac{1}{16}$ in.

552. E. ACUS Sacc. (from the needle-like stem; *acus*, a needle) *a b*.
P. deeply umbilicate, pruinose, white; marg. incurved, striate. *St.* smooth, white. *G.* thick, distant, deeply decurrent, rose.

Coco-nut fibre in conservatories. Aug. $\frac{1}{2} \times 1\frac{1}{4} \times \frac{1}{32}$ in.

XX. CLAUDOPUS Gill.

(From its dwarfed stem; *claudus*, lame, Gr. *pous*, a foot.)

Veil represented by tomentum or down. *Hymenophore* confluent and homogeneous with the fleshy stem when the latter is present. *Pileus* excentric, lateral or resupinate. *Stem*, when present, lateral,

Fig. 29.—A, sections of *Claudopus variabilis* Gill.; B, sections of *C. byssisedus* Gill. Natural size.

simple, without cartilaginous bark. *Gills* reaching to the excentric stem or axis. *Spores* smooth or warted, rubiginous. (Fig. 29.)

The species are small. They grow on wood or on the ground.

Claudopus agrees in structure and habit with *Pleurotus* and *Crepidotus*. Species **553—555**

553. C. variabilis Gill. (from its variability) *a b c*.
P. resupinate, then reflexed, white-tomentose. *St.* minute, excentric or none. *G.* salmon-pink or salmon-brownish.

Sticks, stumps, leaves, oak, beech, maple, bramble, apple, sedge, moss, living chickweed; common. May–Mar. Diam. $1\frac{1}{2}$ in. Sometimes sterile with white gills. Sometimes $\frac{1}{16}$ in. or less in diam.

554. **C. depluens** Gill. (from its watery substance; *depluo*, to rain) *a b c*.

P. resupinate, then reflexed, hygrophanous, rufescent-hoary. *St.* small, lateral, rarely central, or none, white villous. *G.* adnato-decurrent, rufescent.

On the ground, amongst moss, sawdust, wood-ashes, sometimes in stoves; rare. Oct. Diam. 1 in.

555. **C. byssisedus** Gill. (from the fibrils on the stem; *byssus*, fine linen thread, *sedeo*, to sit) *a b c*.

P. resupinate, then horizontal and reniform, villous, pale grey. *St.* small, white-villous. *G.* adnato-decurrent, salmon.

On the ground, rotten wood, beech; rare. Sept.–Oct. Diam. ⅜ in.

SERIES III. DERMINI Fr.

(From the membrane of the pileus; Gr. *derma*, a skin.)

Spores various shades of reddish-brown, brown, red, yellowish-brown or dull yellow. Genera XXI—XXXI.

XXI. **ACETABULARIA** Mass.

(From the cup-like volva; *acetabulum*, a vinegar-cup.)

Veil universal, forming a volva. *Hymenophore* distinct from the fleshy stem. *Stem* central, simple, volvate. *Gills* adnexed. *Spores* pallid-tawny or brownish.

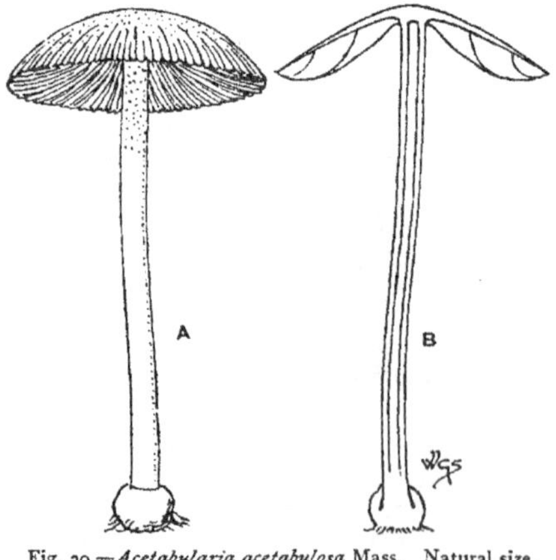

Fig. 30.—*Acetabularia acetabulosa* Mass. Natural size. A, perfect plant; B, plant in section.

Acetabularia agrees in structure with *Amanitopsis*, *Volvaria* and *Chitonia*.

Saccardo (*Sylloge Fungorum*, vol. v. p. 761) makes Berkeley's *Acetabularia* analogous with Gillet's *Locellina*. The two genera are quite distinct; *Locellina* is not represented in Britain.

Locellina.	*Acetabularia.*
Pileus smooth, glabrous.	*Pileus* sulcate, white-furfuraceous.
Gills broadly adnate.	*Gills* adnexo-free.
Stem fibrilloso-annulate.	*Stem* simple.

556. A. acetabulosa Mass. (from the cup-like volva; *acetabulum*, a vinegar-cup) *a b*.

P. convex, salmon; mid. sienna-reddish, white-furfuraceous; marg. striate, denticulate, splitting. *St.* hollow, whitish, white-mealy above. *G.* adnexed, subdistant, glandular, pale brownish-salmon with a lighter edge.

By the Thames at Millbank. May. $1\frac{3}{8} \times 2 \times \frac{1}{8}$ in. The glands on the gills, as illustrated by Sowerby, may be cystidia.

XXII. TOGARIA W. G. Sm.

(From the annulus, like a cloak, well seen in the young state of **557**; *toga*, a Roman garment.)

Veil universal, manifest in the membranous annulus and in the

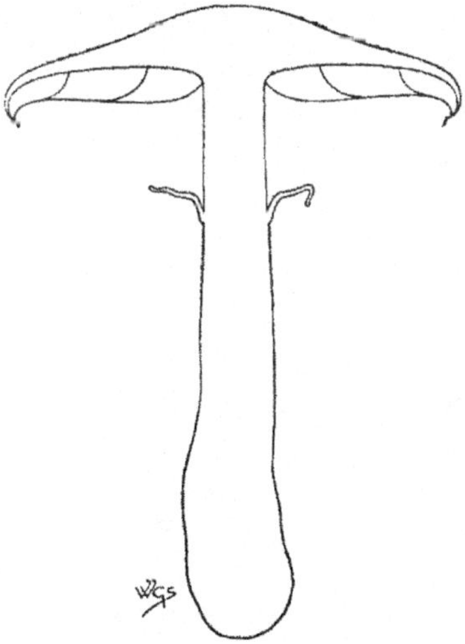

Fig. 31.—Section of *Togaria aurea* W. G. Sm.
One-quarter natural size.

squamules, fibrillæ and flock of the pileus; the squamules, etc. being sometimes appendiculate at the margin. *Hymenophore* not truly distinct from the fleshy stem, except in **564**, although in the majority of cases it is nearly so. *Pileus* fleshy. *Stem* central, fleshy, annulate. *Gills* adnato-decurrent, adnate, adnexed, or rounded-free. *Spores* ferruginous. (Fig. 31.)

All the species grow on the ground.

Togaria agrees in nearly all points of structure and habit with *Lepiota*, *Annularia* and *Psalliota*.

In **564** the gills are normally rounded-free, and in several other species the gills are frequently adnexed. In *Lepiota* one species—**47**—has adnate gills and several others have approximate gills; in *Psalliota* several species have approximate gills. Species **557—566**

a. Euderminæ. Spores ferruginous. **557—564**

b. Phæotæ. Spores fuscous-ferruginous. **565, 566**

a. Euderminæ.

557. T. aurea W. G. Sm., *Agaricus aureus* Mattusch. (from its golden colour; *aurum*, gold) *a b c.*

P. convex, velvety, squamulose; marg. partially appendiculate with V. *St.* solid, striate and scurfy below A. *A.* superior, large. *G.* adnato-decurrent to adnexo-free, ochreous then sienna. *Flesh* whitish-sulphur above, golden-sienna below.

Subcæspitose. Plantations, cedar; rare. Aug.–Nov. 10 × 11¾ × 1¾ in. Var. *Vahlii* W. G. Sm. (*Agaricus Vahlii* Schum.). *P.* even, smooth. Var. *herefordensis* W. G. Sm. (*Agaricus aureus* Mattusch. var. *herefordensis* Renny.) *St.* granulate, tuberculate.

558. T. caperata W. G. Sm., *Agaricus caperatus* Pers. (from the wrinkled pileus; *capero*, to wrinkle) *a b c.*

P. convex, white-floccose on a deep yellow or buff-tan ground; marg. deeply wrinkled or lobed. *St.* solid, white, shaded dull salmon; white-squamulose above A. *A.* median, somewhat large. *G.* adnate or adnexed, dull ochreous or clay-cinnamon, sometimes colour of iron-oxide.

Solitary. Woods. Aug.–Dec. 4½ × 4½ × ⅝ in. Sometimes there is an obscure volva. Pale forms resemble **789.**

559. T. terrigena W. G. Sm., *Agaricus terrigenus* Fr. (from its habitat; *terra*, the earth, *gigno*, to bear) *a.*

P. convex, silky-fibrillose, yellowish; marg. partially appendiculate with V. *St.* colour as P., clad with ferruginous-yellow warts. *A.* apical. *G.* adnate with a decurrent tooth, yellowish then brownish.

Woods, damp places, old earthy stumps. Aug.–Nov. 2⅜ × 2⅝ × 5/16 in. Perhaps **589** is a form of this. Small forms must not be confounded with **1085.**

560. T. erebia W. G. Sm., *Agaricus erebius* Fr. (Gr. *erebos*, dark) *a b c.*

P. convex, at first viscid, dark ferruginous brown or umber, becoming paler; marg. often streaky-wrinkled. *St.* hollow, smooth, pale salmon-brownish, darker at base and white above, white squamulose below. *A.* apical, membranous, white. *G.* adnate, sometimes with a decurrent tooth, somewhat pale livid-brown. *Flesh* pale brownish.

Gregarious. Taste fungoid, insipid; odour slight or obsolete. Woods, shrubberies, grassy places, under birches. Aug.–Oct. $2 \times 2\frac{3}{4} \times \frac{1}{4}$ in. Not unlike some forms of **792.** Must not be confounded with **69a.**

561. T. ombrophila W. G. Sm., *Agaricus ombrophilus* Fr. (Gr. *ombros*, rain, *philos*, loving) *a b.*

P. convex, gibbous, smooth, clay-colour, sometimes shaded ferruginous; marg. with traces of the white V. *St.* hollow, paler than P. or almost white. *A.* distant, white. *G.* adnate with a tooth or adfixed, pallid brownish, then ferruginous-brown.

Grassy places after heavy rain. Oct. $2\frac{1}{8} \times 2\frac{1}{2} \times \frac{5}{16}$ in. Var. *brunneola* W. G. Sm. (*Agaricus ombrophilus* var. *brunneolus* Fr.). *P.* brown. $1\frac{1}{2} \times 1\frac{1}{2}$ in.

562. T. molliscorium W. G. Sm., *Agaricus molliscorium* Cooke & Mass. (from the skin of the pileus; *molliscorium*, with a soft, leather-like skin) *a b.*

P. convex, smooth, polished, dry, subumbonate, tawny yellow; mid. ochreous or sienna. *St.* hollow, equal, pale ochreous, silky fibrillose, broken up into minute sq. above A. *A.* deciduous, brownish. *G.* adnate or uncinate, crowded, ferruginous.

Taste and odour none. June. $2\frac{5}{8} \times 2\frac{3}{4} \times \frac{1}{4}$ in.

563. T. togularis W. G. Sm., *Agaricus togularis* Bull. (from the annulus, like a little cloak, *togula*) *a b.*

P. convex, pallid ochreous; mid. sienna or umber. *St.* hollow, whitish, yellowish or brownish. *G.* adnate or adnexed, yellow then pallid ferruginous. *Flesh* colour of exterior.

In troops. Grassy woodlands, gardens, wood-yards, amongst chips; rare. May–Nov. $2\frac{1}{4} \times 3\frac{1}{8} \times \frac{3}{16}$ in.

564. T. blattaria W. G. Sm., *Agaricus blattarius* Fr. (from the colour of the pileus; *blatta*, a cockroach) *a.*

P. flat, subumbonate, ferruginous; mid. darker. *St.* fistulose, whitish, brownish at base. *A.* white. *G.* rounded-free or slightly adnexed, watery cinnamon.

Gardens and cultivated places. Autumn. $1\frac{1}{8} \times 1\frac{1}{2} \times \frac{1}{8}$ in. Like an annulate *Galera.*

b. Phæotæ.

565. T. dura W. G. Sm., *Agaricus durus* Bolt. (*durus*, hard) *a b c.*

P. convexo-plane, kid-leather-like, becoming cracked, tawny, pale ochreous or almost white; marg. partially appendiculate with V. *St.* paler than P., sometimes with root-like fibres at base. *A.* ample, fugitive. *G.* adnate with a decurrent tooth, livid brown to deep ferruginous.

Possibly poisonous. Fields, gardens; uncommon. May–Oct. $2\frac{3}{4} \times 3\frac{1}{4} \times \frac{3}{8}$ in. Sometimes gathered for **792.**

566. T. præcox W. G. Sm., *Agaricus præcox* Pers. (*præcox*, early) *a b c*.

P. convex, obtuse, smooth, whitish or whitish-ochre. *St.* hollow above, mealy-floccose, white, sometimes with root-like fibres at base. *A.* ample, fugitive, white. *G.* rounded-adnexed or sub-adnate, at length fuscous.

Said to be edible. Taste rank; odour strong, not unpleasant. Gardens, pastures, road-sides; common. May–Oct. 3 × 4 × ⅜ in. Sometimes wholly whitish. May not be distinct from last. Sometimes gathered for **792.**

566a. T. sphaleromorpha W. G. Sm., *Agaricus sphaleromorphus* Bull. (from the difficulty of distinguishing this from the two last species; Gr. *sphaleros*, delusive, *morphe*, form) *a b*.

P. globose, convex, plane, obtuse, even, smooth, expanded, white or pale buff to ochre. *St.* stuffed, then partially hollow, in-crassate and villose below, when young clavate, silky, pale buff to yellowish. *A.* entire, median to distant, ample, membranous. *G.* arcuato- or adnato- to sinuato-decurrent, yellowish-white, becoming ochre, ferruginous-tan or brownish.

Leaf-soil. Oct. 4⅛ × 3⅜ × ½ in. Must not be confounded with **565** and **566.**

XXIII. PLUTEOLUS Gill.

(Diminutive of *Pluteus*, Genus XIV.)

Veil indicated by gluten, fibrillæ and meal. *Hymenophore* distinct from the fleshy stem. *Pileus* conical or campanulate, then expanded,

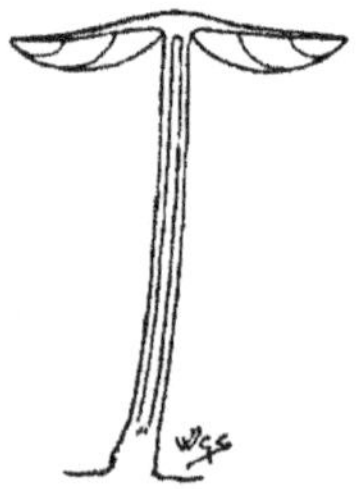

Fig. 32.—Section of *Pluteolus reticulatus* Gill. One-half natural size.

slightly fleshy, viscid, margin striate, at first straight and adpressed to the stem. *Stem* fistulose, simple. *Spores* ferruginous or saffron. *Gills* rounded-free. (Fig 32.)

The species grow on dead wood, rotten sticks, etc.

Pluteolus agrees in structure with *Hiatula*, *Pluteus*, and *Pilosace*.

Species **567, 568**

567. P. reticulatus Gill. (from the network of veins on the pileus; *rete*, a net) *a b*.

P. expanded, subumbonate, violaceous, fuscous or greyish; mid. faintly ochreous. *St.* fibrillose, whitish. *G.* dull cinnamon. *Spores* dark ferruginous.

Dead wood; rare. Sept. $1\frac{7}{8} \times 2\frac{1}{4} \times \frac{1}{8}$ in.

568. P. aleuriatus Karst. (from the flour-dusted appearance of the stem; Gr. *aleuron*, flour) *a*.

P. convexo-plane, bluish-grey, rose-colour or tan; mid. darker. *G.* saffron-ochraceous.

Rotten sticks. Oct. $1 \times 1 \times \frac{1}{16}$ in.

XXIV. PHOLIOTA Quél.

(From the scaly pileus; Gr. *pholis*, a scale.)

Veil partial in the form of an annulus, or universal as scales and gluten on the pileus and stem. *Hymenophore* confluent and homogeneous with the fleshy stem. *Pileus* fleshy, viscid, dry, naked or

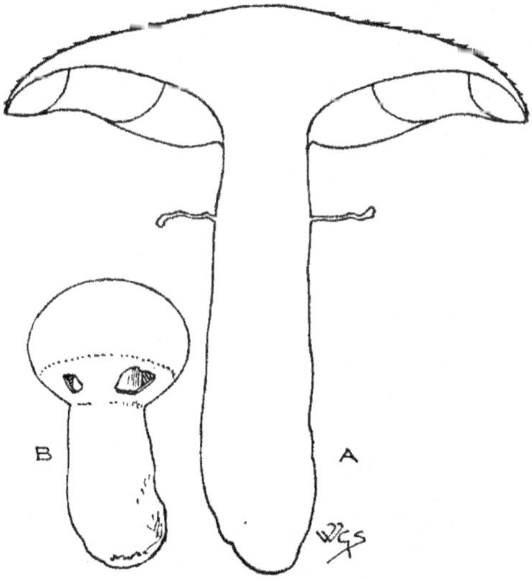

Fig. 33.—*Pholiota spectabilis* Gill. One-half natural size. A, plant in section; B, young state.

scaly. *Stem* central, annulate. *Annulus* persistent, membranous, spreading, not arachnoid or fibrillose. *Gills* usually adnate, with or without a decurrent tooth, variously attached, at first pale or tawny, then usually becoming dark fuscous, but not the bright ferruginous colour of *Cortinarius*. *Spores* sepia-brown, bright yellowish-brown, or light red. (Fig. 33.)

All the species except **596** and **597** grow on trees, stumps and branches.

Pholiota agrees in structure and habit with *Armillaria* and *Stropharia*.

Pholiota must not be confounded with *Cortinarius*, where all the species are terrestrial, have an arachnoid veil and spores the colour of iron-oxide. Species **569—597**

TRUNCIGENÆ. Mostly cæspitose, growing on trunks and stumps.

a. *Ægeritinæ*. *Pileus* naked, not scaly, but here and there rimoso-rivulose. *Gills* pallid, then rufescent, becoming fuscous. **569—574**

b. *Squamosæ*. *Pileus* scaly, not hygrophanous. *Gills* changing colour.

Gills pallid, then fuscous, olivaceous, clay-coloured, not purely ferruginous. **575—579**

Gills yellow, then purely ferruginous or tawny. **580—586**

c. *Hygrophanæ*. *Gills* cinnamon, not at first light yellow. Hygrophanous. **587—595**

MUSCIGENÆ. Hygrophanous, resembling annulate *Galeræ*; mostly growing on moss. **596, 597**

TRUNCIGENÆ.

a. *Ægeritinæ*.

569. P. radicosa Quél. (from the rooting stem) *a b c*.

P. convexo-plane, viscid, pale tan-brownish, clouded darker. *St*. spongy-solid, enlarged below and fusiform rooting, whitish-tan. *A*. distant, fugitive. *G*. adnato-toothed or rounded.

Solitary or gregarious, rarely cæspitose. Taste pleasant or obsolete; odour strong, like prussic-acid or sweetish, not unpleasant. Woods, about stumps, on wood; frequent. Aug.–Oct. $4\frac{1}{2} \times 10\frac{3}{4} \times \frac{5}{8}$ in. P. sometimes 6 in. in diam.

570. P. pudica Gill. (from its modest colour; *pudicus*, modest) *a b c*.

P. convex, moist, hygrophanous, rivuloso-wrinkled, tan; mid. sienna. *St*. solid, firm, rigid, paler than P. *A*. persistent. *G*. adnato-rounded, colour as St. *Flesh* pallid white.

Single or cæspitose. Odour strong, penetrating, disagreeable. Stumps, elder, elm, rarely on lawns, downs. May–Nov. $2\frac{5}{8} \times 2\frac{1}{4} \times \frac{3}{8}$ in. Sometimes mistaken for **792**.

571. P. leochroma Sacc. (from its lion-tawny colour; Gr. *leon*, a lion, *chroma*, colour) *a b*.

P. convex, then depressed; mid. yellowish-white or pale sienna; marg. pale. *St*. solid, whitish above, brown below. *A*. persistent. *G.* adnato-rounded, dark cinnamon. *Flesh* pale livid-brownish, darker in St.

Cæspitose. Said to be edible. Stumps, elm, fir, larch. July–Sept. $3\frac{1}{2} \times 4\frac{1}{4} \times \frac{1}{2}$ in. Perhaps a form of **570.**

572. P. capistrata Sacc. (from a fancied resemblance in the annulus to a halter, *capistrum*) *a b.*

P. flat, subviscid, tan; mid. pale sienna; marg. rivuloso-wrinkled. *St.* stuffed, squamulose, tan-white. *A.* large, inrolled, persistent. *G.* decurrent or rounded, thin, tan-white.

Cæspitose. Said to be edible. Taste varying pleasant to unpleasant. Stumps, elm. Aug.–Nov. $4 \times 4 \times \frac{5}{8}$ in. Intermediate forms occur between this and **570** and **571.**

573. P. ægerita Quél. (from its frequent habitat, poplar trunks; Gr. *aigeiros*, the black poplar) *a b.*

P. convex, rivuloso-wrinkled, pale buff to bright full tawny. *St.* stuffed, sometimes substriate from the decurrent base of G., tan-white, white-mealy above. *A.* superior, swollen. *G.* serrulate, adnate with a tooth, fuscous.

Cæspitose. Odour pleasant. Stumps, *Populus fastigiata*, willow, ash, elm. July–Nov. $4 \times 5\frac{1}{4} \times \frac{1}{2}$ in. The "Pioppini" of Italy.

574. P. luxurians Gill. (from its frequent luxuriant growth) *a b.*

P. flat, silky, squamulose, yellowish-white then rufous. *St.* stuffed, subsquamulose, colour as P. *A.* apical, fugacious. *G.* adnate with a tooth, fuscous.

Cæspitose. Stumps, elm, poplar, oak, ash. July. $3 \times 2\frac{3}{4} \times \frac{5}{8}$ in.

b. Squamosæ.

575. P. comosa Quél. (from the hairy scales of the pileus; *coma*, a hair) *a b.*

P. convex, obtuse, viscid, tawny; sc. paler; marg. appendiculate with V. *St.* solid, bulbous, fibrillose, brassy-white. *A.* superior, subpersistent. *G.* adnate or adnato-decurrent, fuscous.

Subcæspitose. Odour faint. Stumps, beech, poplar, birch; rare. Oct.–Nov. $7 \times 4\frac{1}{2} \times 1\frac{1}{4}$ in.

576. P. heteroclita Quél. (from the curved stem; Gr. *heteros*, other, *klino*, to lean) *a b c.*

P. convex, obtuse, viscid when old and wet, whitish, yellowish or brownish; marg. appendiculate with V. *St.* solid, bulbous, fibrillose, whitish-brown at base. *G.* adnate or rounded-adnexed.

Solitary. Odour strong and pungent of horse-radish. Trunks, birch, poplar; small examples on chips. Sept.–Nov. $5\frac{1}{2} \times 3\frac{3}{4} \times 1$ in.

577. P. aurivella Quél. (from a fancied resemblance in the pileus to a golden fleece; *aurum*, gold, *vellus*, fleece) *a b c.*

P. convex, subumbonate, sulphur or deep ferruginous yellow. *St.* stuffed, fibrillose, colour as P. *A.* superior. *G.* adnate with a tooth or sinuato-adnexed, brown.

Cæspitose or solitary. Odour faint. Trunks, ash, willow, beech; uncommon. Sept.–Nov. $4\frac{5}{8} \times 4 \times \frac{1}{2}$ in.

578. P. squarrosa Quél. (from the rough-scaly pileus; *squarrosus*, rough) *a b c*.

P. convex, gibbous, dry, ochreous-sienna; marg. appendiculate with V. *St*. stuffed, attenuate downwards, brown-squarrose below A. *G*. adnate with a decurrent tooth, olivaceous-ferruginous. *Flesh* light yellow.

Cæspitose, sometimes in fascicles of 50 or more. Sometimes eaten. Odour strong and nauseous of putrid cheese or none. On or near stumps, beech, oak, ash, laburnum, acacia, apple, plum; common. July–Dec. $3\frac{7}{8} \times 4\frac{1}{2} \times \frac{5}{8}$ in. Sometimes twice the size of type. Var. *verruculosa* Gill. *P*. glutinous when wet; maple. Var. *reflexa* Gill. *P*. umbonate. Var. *Mülleri* Gill. small, wholly sulphur.

579. P. subsquarrosa Quél. (from the pileus, less scaly than **578**) *a c*.

P. plane, viscid, brown; marg. appendiculate with V. *St*. sub-squarrose below A., yellow-ferruginous. *G*. adnate with a decurrent tooth, or almost emarginato-free, pale brown. *Flesh* pale ochreous-brownish.

Subcæspitose. Odour almost obsolete. Trunks, fir, and on the ground. Sept. $2 \times 2\frac{5}{8} \times \frac{5}{16}$ in.

579a. P. grandis Rea (*grandis*, large) *b*.

P. convex, then expanded, broadly gibbous, dry, fulvous-tawny, covered with small innate deeply tawny sq. at the circ., revolute at mid. *St*. solid, fusiform, slightly squamulose and tawny below the distant, almost fugacious A., deeply striate above. *G*. broad, attenuate in front, deeply sinuato-decurrent behind, somewhat crowded, pallid, then fuscous. *Flesh* light yellow, ferruginous in St.

Cæspitose. Taste and odour very pleasant. Trunks, ash, oak. Oct. $9 \times 8\frac{1}{4} \times 2\frac{1}{4}$ in.

580. P. spectabilis Gill. (*spectabilis*, notable) *a b c*.

P. convex, obtuse, dry, rich sienna-yellow. *St*. solid, squamulose or smooth, ventricose, rich yellow. *G*. adnate with a tooth or subdecurrent. *Flesh* sulphur.

Cæspitose. Taste, bitter-aromatic. Stumps, oak, beech, hornbeam, hawthorn; frequent. Aug.–Dec. $5\frac{5}{8} \times 5\frac{5}{8} \times 1\frac{1}{8}$ in. Stem sometimes 9 in. Resembles **1049** in appearance.

581. P. adiposa Quél. (from the pileus, as if covered with fat) *a b c*.

P. plane, subdepressed, very glutinous, sulphur; sc. floccose, ferruginous; mid. deep sienna-yellow; marg. appendiculate with V. *St*. squarrose, viscid, colour as P. *A*. small, floccose. *G*. adnato-rounded.

Cæspitose. Trunks, beech, ash, plum, silver-fir, birch; frequent. July–Oct. $7 \times 5 \times \frac{3}{4}$ in.

581a. P. lucifera Quél. (from the flame-like colours; *lux*, light, *fero*, to carry) *a*.

P. fleshy, convex, almost plane becoming umbonate, viscid, covered at first with darker adpressed sc. which soon disappear, yellow, or orange with sulphur marg. *St*. stuffed to hollow, sometimes solid, equal, somewhat squamulose, becoming smooth, whitish above, brown below without and within. *A.* distant. *G.* emarginato-adfixed, narrow, crenulate, yellow-white, luminous yellow, then olive or ferruginous.

Trunks, branches, straw, almost terrestrial. Odour strong. Autumn. $2\frac{1}{8} \times 2 \times \frac{3}{8}$ in.

582. **P. flammans** Karst. (from its fiery colours) *a b c.*

P. convex, subumbonate, yellow or orange; sc. sulphur. *St.* squarrose, sulphur or yellow. *A.* superior. *G.* sulphur, then brown. *Flesh* sulphur.

Gregarious or in clusters. Odour none. Woods, pine, pine-leaves, dead pine-branches. July–Oct. $2 \times 2\frac{7}{8} \times \frac{5}{16}$ in. Mycelium often deep orange.

583. **P. Junonia** Karst. (from its beauty, belonging to Juno) *a b.*

P. convex-plane, subumbonate, yellow to deep sienna-orange. *St.* even, solid, yellow above, sienna below. *A.* distant. *G.* adnate.

Usually solitary. Trunks. Oct. $3\frac{3}{8} \times 3\frac{1}{8} \times \frac{3}{8}$ in. Must not be confounded with **695.**

584. **P. tuberculosa** Gill. (from the bulbous base of the stem) *a b.*

P. convexo-plane, obtuse, tawny or pale sulphur; mid. brown. *St.* hollow, subbulbous, attenuato-rooting, fibrilloso-scaly, light yellow or ochre. *A.* fugitive. *G.* adnate or emarginate, serrulate.

Solitary or cæspitose. Trunks, birch; sawdust. Oct. $2 \times 1\frac{5}{8} \times \frac{1}{4}$ in.

585. **P. curvipes** Quél. (from the frequently curved stem; *curva*, a curve, *pes*, a foot) *a b c.*

P. expanded, subumbonate, tawny or orange; marg. at first appendiculate with white V. *St.* fistulose, fibrillose, light yellow. *A.* floccose, fugitive. *G.* adnate with a white edge. *Flesh* light yellow.

Gregarious, subcæspitose. Taste somewhat pleasant-insipid; not strongly scented. Trunks, sawdust, deal boards in greenhouses. Oct. $2\frac{1}{8} \times 1\frac{5}{8} \times \frac{1}{4}$ in.

586. **P. cruentata** Sacc. (from the blood-coloured scales of the pileus; *cruentatus*, blood-stained) *a b.*

P. convex, obtuse, yellow; sc. crimson-brown. *St.* solid, attenuato-rooting, subsquamulose below A., colour as P. *A.* fibrillose. *G.* adnate or emarginate, brown, somewhat olive. *Flesh* pale yellow.

Taste insipid. Trunks, oak; burnt ground. Aug.–Sept. $1\frac{3}{8} \times 1\frac{1}{2} \times \frac{1}{4}$ in.

c. Hygrophanæ.

587. **P. Paxillus** Gill. (from its resemblance to **1181,** *Paxillus involutus* Fr.) *a.*

P. expanded, subumbonate, ochreous-buff; mid. darker. *St.* solid, smooth, colour as P., brown at base. *A.* narrow, spreading. *G.* subdecurrent, crowded.

Trunks. Autumn. $6 \times 6\frac{3}{8} \times \frac{3}{8}$ in. Sometimes wholly cinnamon.

588. **P. dissimulans** Sacc. (from its appearing like, but differing from, **560**; *dissimulo*, to dissemble) *a c*.

P. convex, subumbonate, buff-brown, becoming whitish. *St.* fistulose with transverse dissepiments, attenuate upwards, white. *A.* fragile. *G.* adnexed or adnate with a tooth, clay-brown.

Sticks, hawthorn, sloe. Oct.–Nov. 1⅜ × 1⅞ × $\frac{3}{16}$ in.

589. **P. Cookei** Sacc. (after Dr. M. C. Cooke) *a b c*.

P. convex, obtuse, viscid, dull yellowish; sc. adpressed, faintly brownish. *St.* solid, whitish above, fibrillose and yellowish-brown at base. *A.* narrow. *G.* adnate.

Damp places, amongst grass, stumps. Sept.–Oct. 2½ × 2¼ × ¼ in. Perhaps the same as **559**.

590. **P. sublutea** Karst. (*subluteus*, yellowish) *a c*.

P. flat, subumbonate, squamulose, dull brownish-yellow; sc. darker. *St.* stuffed, smooth, attenuate upwards, paler than P. *A.* narrow, spreading. *G.* decurrent, crowded.

Subcæspitose. Grassy ground. Oct. 2½ × 4¼ × ¼ in.

591. **P. confragosa** Karst. (*confragosus*, rough) *a b*.

P. convexo-plane, obtuse, densely furfuraceous, red, cinnamon-rufous or umber, tawny when dry; flocci fugitive, white. *St.* hollow, shining pale warm-umber; flocci fugitive. *G.* adnate or subdecurrent.

Somewhat cæspitose. Fallen elm. Oct. 1⅞ × 2 × $\frac{3}{16}$ in. Resembles **917**.

592. **P. mutabilis** Quél. (*mutabilis*, changeable) *a b c*.

P. convex, obtusely umbonate, cinnamon, pale when dry, or mid. deep cinnamon, with marg. straw-colour; sometimes appendiculate with V. *St.* pale above, cinnamon below, subsquarrose below A. *A.* narrow, spreading, fugitive. *G.* adnato-decurrent, crowded.

Sometimes eaten, taste of gingerbread. Stumps, lime, ash, sawdust. April–Dec. 2 × 3¾ × ¼ in.

593. **P. marginata** Quél. (from the striate margin of the pileus) *a b c*.

P. expanded, subumbonate, honey-colour, tan when dry. *St.* fistulose, fibrillose, paler than P. *A.* spreading, fugitive. *G.* adnate or somewhat rounded.

Solitary, gregarious, cæspitose. Woods, fir, beech, pine, chips; frequent. Sept.–Dec. 1⅛ × 2 × ⅛ in.

594. **P. mustelina** Quél. (from its colour; *mustella*, a weasel) *a*.

P. convex, dry, yellow or reddish. *St.* fistulose, whitish. *A.* superior, reflexed. *G.* adnate.

Solitary, subgregarious. Woods, pine, stumps. Sept. ⅝ × ¾ × $\frac{1}{16}$ in.

595. **P. unicolor** Gill. (from its being wholly of one colour) *a c*. Bay-brown then ochreous.

P. convex, subumbonate. *St.* fistulose, somewhat smooth. *G.* adnexed, adnate or subdecurrent.

Subgregarious, subcæspitose. Trunks, branches, larch. Sept.–Nov. 1½ × 2⅛ × ½ in.

MUSCIGENÆ.

596. P. pumila Gill. (*pumilus*, diminutive) *a b c*.

P. hemispherical, smooth, ochreous; marg. substriate. *St*. fistulose, smooth, colour as P. or paler; base often darker. *A*. floccose. *G*. adnate with a tooth, subdistant, pallid-brownish.

Solitary, gregarious. Moss, walls, earth, sticks, sawdust, firs; common. Aug.–Jan. $\frac{3}{4} \times 2\frac{3}{8} \times \frac{1}{8}$ in.

597. P. mycenoides Quél. (from its resemblance in stature to a *Mycena*) *a b c*.

P. hemispherical, ferruginous, tawny pale or medium; marg. striate. *St*. fistulose, smooth, colour as P. *A*. persistent, white. *G*. adnate with a decurrent tooth.

Amongst moss in damp places, lawns. Sept.–Oct. $\frac{1}{2} \times 1\frac{7}{8} \times \frac{1}{32}$ in. Often accompanied by **283**.

XXV. INOCYBE Quél.

(From the fibrous pileus; Gr. *is*, *inos*, a fibre, *kube*, a head.)

Veil concrete with the cuticle of the pileus, fibrillose or somewhat arachnoid, often appendiculate at the margin of the pileus and sometimes forming an indefinite annulus. *Hymenophore* confluent

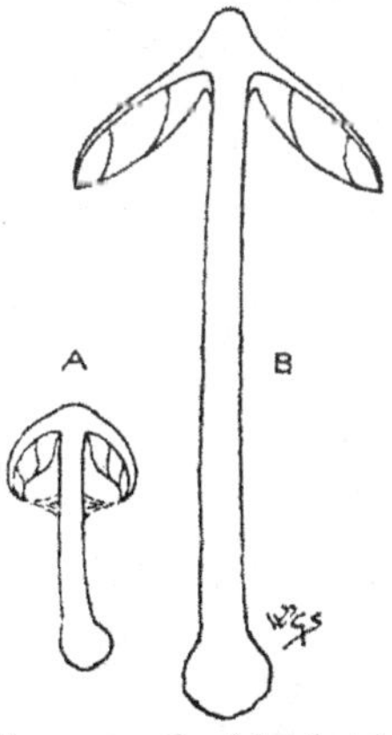

Fig. 34.—A, section of *Inocybe scabra* Gill.; B, ditto *I. fastigiata* Quél. One-half natural size.

and homogeneous with the fleshy stem. *Pileus* squarrose, fibrillose or velvety; in a few species becoming smooth or viscid. *Stem* central, scaly, fibrillose or smooth, simple or imperfectly squarroso-annulate.

Gills usually sinuate, sometimes adnexed, adnate or decurrent, adhering to the hymenophore, except **602** and **621**, at first pale, changing colour, but not to fuscous-ferruginous. *Spores* often angular; pale dull tan, brownish or brown. (Fig. 34.)

All the species, except **644**, grow on the ground; they are somewhat small; none are edible, and the majority have a somewhat nauseous odour.

Inocybe as defined by Fries (*Hymenomycetes Europæi*, p. 226) has no genus corresponding wholly with it in structure amongst the white-, pink-, purple- or black-spored groups. Certain of the fibrous-pileated species of *Tricholoma*, *Entoloma* and *Stropharia* might, however, be separated as analogues. *Inocybe* was formerly included in the next genus, *Hebeloma*. Species **598—646**

Mr. George Massee, in his monograph of the genus (Annals of Botany, XVIII. (1904), p. 459), has rearranged the species in two groups, viz.: I. Spores rough, and II. Spores smooth; these groups he has again divided into—A. Cystidia present, and B. Cystidia absent, with minor distinctions as to a whitish or coloured stem and brownish or olive-tinted gills.

a. Squarrosæ. *Pileus* squarrose, somewhat fuscous. *Stem* scaly, smooth above, same colour as pileus, solid except **204**. **598—607**

b. Laceræ. *Pileus* torn into scales or fibres, not split into cracks. *Stem* fibrillose, paler than the pileus. **608—619**

c. Rimosæ. *Pileus* longitudinally fibrous, soon cracked, partially adpressed-scaly. *Stem* fibrillose, whitish, slightly tinged with the colour of the pileus. Many become yellow when old. **620—634**

d. Velutinæ. *Pileus* not cracked, the cuticle fibrilloso-woven, becoming even or adpressedly-scaly. *Stem* shining, smooth, whitish, mealy above. **635—642**

e. Viscidæ. *Pileus* viscid, even. **643—646**

a. Squarrosæ.

598. I. hystrix Karst. (from the scaliness; *hystrix*, a porcupine) *a b c.*

P. plane, subgibbous, mouse-colour to brown. *St.* attenuate downwards; sc. fuscous, brownish or mouse-colour. *G.* adnate, fuscous.

Woods, beech. Oct. $3 \times 2\frac{1}{8} \times \frac{3}{8}$ in. Resembling a small **578** in appearance.

599. I. relicina Quél. (from the recurved floccose scales of the pileus; *reclinus*, curled backwards) *a c.*

P. expanded, obtuse, fuliginous or rufescent livid-brown. *St.* slightly attenuate downwards, floccoso-scaly, colour as P. *G.* adnexed, olivaceous.

Odour almost obsolete. Woods, fir, marshy places, amongst *Sphagnum*. July–Oct. $2\frac{7}{8} \times 3\frac{1}{4} \times \frac{5}{16}$ in.

599a. I. mimica Mass. (from its resemblance to **26** as figured in Cooke's Illustrations, p. 941).

P. campanulate, obtusely umbonate, fibrillose, yellow-brown, covered with large adpressed fibrous slightly darker sc. *St.* solid, equal, fibrillose, paler than P. *G.* deeply sinuate, broad, yellow-brown.

Woods. Autumn. $3 \times 3 \times \frac{3}{8}$ in.

600. I. calamistrata Gill. (from the curled scales of the pileus; *calamister*, a curling iron) *a b c.*

P. campanulate, fuscous-scaly on a tan ground. *St.* squarrose, somewhat bulbous, at first pale, then as P., tinged with dull blue at base. *G.* adnexed, ferruginous with a white edge. *Flesh* becoming red when cut.

Sometimes subcæspitose. Odour strong, not unpleasant. Woods, pine, by paths and ditch-sides; rare. Aug.–Oct. $2 \times 2\frac{1}{4} \times \frac{3}{16}$ in.

601. I. hirsuta Quél. (from the bristly pileus; *hirsutus*, bristly) *a b.*

P. campanulate, acutely umbonate; mid. fuscous or greenish; marg. pale tan. *St.* fibrillose, pale tan above, greenish at base. *G.* adnexo-ascending, umber with a white edge. *Flesh* faintly rose when cut.

Odour fainter than **600.** Woods, beech, damp places, pastures; rare. Oct. $1\frac{1}{8} \times 2\frac{1}{2} \times \frac{1}{16}$ in.

602. I. lanuginosa Quél. (from the woolly pileus; *lanuginosus*, woolly) *a b c.*

P. hemispherico-convex, obtuse, umber then yellow. *St.* squamuloso-fibrillose, whitish-fuscous. *G.* separating-free, denticulate, pallid clay-colour.

Odour faint or none. Woods, mixed, near paths, sandy places, near firs. July–Nov. $1\frac{3}{8} \times 1\frac{5}{8} \times \frac{1}{8}$ in.

602a. I. calospora Quél. in Bres. (from the beautiful spores; Gr. *kalos*, beautiful) *a.*

P. convex or campanulate, then expanded and umbonate, fibrillose with darker sq. at mid., yellowish-brown or tawny-grey; edge paler. *St.* pale then reddish, or as P. *G.* sinuato-free, tawny-ochre or brownish.

Woods, shady places. Autumn. $1 \times 2\frac{1}{2} \times \frac{1}{8}$ in.

603. I. dulcamara Quél. (from the taste—sweet, becoming bitter; *dulcis*, sweet, *amarus*, bitter) *a c.*

P. campanulato-convex, umbonate, olivaceo-fuscous. *St.* squamuloso-fibrillose below, paler than P. *G.* adnexed, olivaceous.

Gregarious. Woods, plantations, fir, mixed. Aug.–Oct. $1\frac{3}{8} \times 1\frac{7}{8} \times \frac{1}{8}$ in.

604. I. plumosa Quél. (from the downy tufts on the pileus; *plumosus*, feathery) *a b.*

P. hemispherico-convex with squarrose flocci in erect fascicles, mouse-colour or shaded rufous. *St.* stuffed or hollow, floccoso-scaly, paler than P. or smoky. *G.* subadnate, pallid mouse-colour.

Odour faint, not unpleasant. Woods, pine, moist places; rare. Aug. $1\frac{3}{8} \times 2\frac{5}{8} \times \frac{1}{8}$ in. The colours are not unlike those of Series II., *Hyporhodii.*

605. I. cincinnata Quél. (from the curved scales of the pileus; *cincinnus*, a curl) *a.*

P. convexo-plane, usually subumbonate, villoso-scaly, fuscous. *St.* fibrilloso-scaly, fuscous-violaceous or lavender above, brownish below. *G.* adnexed, fuscous-violaceous or lavender.

Subgregarious. Odour none. Woods, beech. Autumn. $1 \times 2 \times \frac{1}{8}$ in. Compare **618** and **641a.**

606. I. hæmacta Sacc. (from the change of colour to blood-red when broken; Gr. *haimaktos*, mingled with blood) *a b.*

P. plano-convex, floccoso-fibrillose, umber; mid. dark; marg. pallid. *St.* scaly-fibrillose, whitish above, dull verdigris below inside and out. *G.* adnate, clay-umber. *Flesh* crimson in P., greenish-blue below.

Subcæspitose. Lawns. Aug.–Oct. $1\frac{3}{4} \times 2 \times \frac{1}{4}$ in.

607. I. fasciata Sacc. (from its growing in bundles; *fasciatus*, bound together in bundles) *a b.*

P. convex, obscurely acuto-umbonate, minutely brown-squarroso-scaly on a pale ground. *St.* solid, attenuate below, pallid above, vinous within and without at base. *G.* adnato-sinuate, crowded, thin, soft, pallid.

Densely cæspitose. Taste and odour none. Grassy places. $3 \times 3\frac{1}{4} \times \frac{5}{16}$ in.

b. Laceræ.

608. I. pyriodora Quél. (from its odour of pears; *pyrus*, a pear, *odor*, a smell) *a b c.*

P. convex, subumbonate, fuscous to tan-ochreous. *St.* solid, whitish. *G.* adnato-emarginate, fuscous, often olive-shaded, edge whitish. *Flesh* becoming pale reddish.

Taste none; odour strong, sweet, pleasant of pears, decaying pears, violets, cinnamon, of *Muscari racemosum.* Woods, gardens. May–Dec. $2\frac{7}{8} \times 3\frac{3}{4} \times \frac{5}{16}$ in. Sometimes *P.* and *St.* whitish-tan, *G.* tan-rufescent.

609. I. incarnata Bres. (from the flesh-coloured pileus; *incarnatus*) *a b.*

P. expanded, broadly umbonate, sometimes tinged with pale orange-scarlet, crimson or purple; marg. appendiculate with V. *St.* solid, base enlarged, colour as P. *G.* sinuate, whitish to olive-brownish, often spotted orange, scarlet or crimson, or wholly rufescent, edge crenulate, white. *Flesh* pale orange or pale purple-hyacinth.

Odour strong of meal or pears. Woods, pine, etc. June–Oct. $3 \times 3\frac{1}{4} \times \frac{1}{2}$ in. Red when bruised or broken. Closely allied to **608.**

609a. I. adequata Sacc. (from its claim to equal specific rank with **610**).

P. campanulate, then expanded, squarrose and fibrous-scaly, yellow and red-brown or pale, somewhat vinous-umber with darker umber sc. *St.* whitish- or reddish-fibrous, or as P. *G.* adnexed, thick, somewhat distant, pale greyish-white or yellowish-umber, becoming greyish-brown, always with a white edge.

Odour none. Woods. Summer and autumn. $3\frac{1}{2} \times 4\frac{1}{4} \times \frac{9}{16}$ in. Allied to **610**.

610. I. scabra Gill. (*scaber*, rough) *a b c.*

P. convex, subgibbous, pale fuliginous, tan or umber. *St.* solid, whitish. *G.* adnexed, often separating from stem, dull whitish-brown. *Flesh* white.

Suspected poisonous. Woods, plantations, fir, mixed; frequent. June–Oct. $1\frac{3}{4} \times 1\frac{3}{4} \times \frac{1}{4}$ in. Var. *firma* Mass. fuscous-tan with fuscous scales.

611. I. maritima Karst. (from its being first found on the sea-shore) *a b c.*

P. hygrophanous, expanded, obtuse or umbonate, fuscous or mouse-colour, hoary when dry. *St.* solid, paler than P. *G.* adnexo-rounded, or somewhat sinuate, grey, then ferruginous.

Sometimes cæspitose. Odour faint. Woods, pine, sandy places, damp sand by sea-shores. Sept.–Oct. $1\frac{1}{4} \times 1\frac{5}{8} \times \frac{3}{16}$ in.

612. I. lacera Quél. (from the scaly-torn pileus; *lacerus*, torn) *a c.*

P. expanded, subumbonate, mouse-colour, pale and yellow when old. *St.* stuffed, paler than P., white and mealy. *G.* attenuato-adnexed, mouse-colour. *Flesh* reddish.

Commonly gregarious. Woods, pine, mixed. July–Oct. $1\frac{1}{4} \times 1\frac{3}{4} \times \frac{3}{16}$ in.

613. I. flocculosa Sacc. (from the small flocci on the pileus) *a b c.*

P. convex, umbonate, rufescent-sienna. *St.* paler than P., solid, stuffed or hollow. *G.* adnato-ascending, colour as St., then chocolate with a white edge.

Odour of new meal, but nauseous. Amongst grass, woods. Sept.–Oct. $1 \times 2\frac{1}{4} \times \frac{1}{8}$ in. Compare **602** and **612**.

614. I. Bongardii Karst. (after H. G. Bongard, a Russian botanist) *a b c*

P. campanulate, obtuse, greatly to slightly scaly, often cracking between the sc. and showing the flesh, fuscous, pale when dry. *St.* solid, subbulbous, lighter than P., sometimes dark below. *G.* adnexed, ventricose, reddish, cinnamon or umber, edge white. *Flesh* reddish.

Usually solitary. Odour pleasant of ripe pears or bergamot. Sandy ground, sand-hills, pine-woods, pastures. May–Oct. $2\frac{1}{8} \times 4 \times \frac{3}{8}$ in.

615. I. mutica Karst. (from the at first obtuse pileus; *muticus*, curtailed) *a b.*

P. obtuso-convex, plane, then subdepressed, whitish, tan or straw. *St.* hollow, attenuate downwards, colour as P. *G.* adnate, white or tan, then subfuscous.

Woods, shady places, roadsides. Sept.–Oct. $2\frac{1}{4} \times 2 \times \frac{1}{4}$ in.

616. **I. carpta** Quél. (from the torn woolly pileus; *carpo*, to tear) *a b c.*

P. flat or depressed, dark fuscous or umber. *St.* hollow, attenuate downwards, paler than P. *G.* adfixed to adnate, ventricose, brown-fuscous with a pale edge.

Taste bitterish-sweet. Woods. Aug. $1\frac{1}{8} \times 1\frac{3}{4} \times \frac{1}{8}$ in.

617. **I. deglubens** Gill. (from the fibrils of the pileus which peel off; *deglubens*, peeling off) *a b.*

P. expanded, obtusely umbonate, date-brown-rufescent, becoming yellowish. *St.* solid, colour as P. above, paler and yellowish below. *G.* adnate, ventricose, grey then cinnamon.

Woods, pine, moist places; rare. Aug.–Sept. $1\frac{5}{8} \times 2\frac{1}{4} \times \frac{1}{8}$ in.

617a. **I. cervicolor** Quél. (from its dark tawny colour, as of a deer; *cervus*, a stag) *a.*

P. campanulate, covered with brown recurved fibrils. *St.* firm, whitish, fibrillose, with brown recurved filaments throughout its length. *G.* emarginate, ventricose, distant, rusty brown; marg. denticulate, whitish. *Flesh* white, tinged purplish when cut.

Odour strong, unpleasant. Amongst grass, woods. Autumn. $\frac{7}{8} \times 3 \times \frac{1}{8}$ in. Compare **614**.

618. **I. obscura** Gill. (from the dark scales of the pileus; *obscurus*, dark) *a b c.*

P. expanded, umbonate, purplish-brown or shaded lavender or faint crimson. *St.* stuffed, lilac or purplish above, yellow-brownish below, lilac within. *G.* sinuato-adnexed, olivaceous, then brown.

Subcæspitose. Odour strong, somewhat bitter. Amongst pines, damp places, woods; rare. July–Nov. $2\frac{1}{2} \times 1\frac{3}{4} \times \frac{1}{4}$ in. Var. *rufa* Sacc., *P.* brownish-rufous, *G.* violet.

619. **I. Rennyi** Sacc. (after James Renny, mycologist) *a b.*

P. hemispherical, fawn-colour; mid. brown. *St.* equal, whitish at top and bottom, yellow-brown in mid. *G.* adnexo-ascending with a tooth, whitish-brown then cinnamon.

Woods, fir. Nov. $1\frac{1}{8} \times 2\frac{5}{8} \times \frac{1}{8}$ in. The var. *major* Mass. is illustrated in the Public Gallery, the type is little more than half this size.

c. Rimosæ.

620. **I. schista** Sacc. (from the clefts in the pileus; Gr. *schistos*, a cleft), *a b.*

P. campanulate, obtuse, shining bay-brown; mid. sienna. *St.* solid, twisted, splitting, paler than P. *G.* adnato-ascending with a tooth, separating, sienna-cinnamon; edge serrate, white.

Taste pleasant, somewhat insipid. Lawns. May. $2\frac{3}{4} \times 3 \times \frac{1}{2}$ in.

621. I. fibrosa Gill. (from the fibrous pileus) *a b c.*

P. campanulato-expanded, umbonate; mid. foxy-red, lighter at marg. *St.* solid, splitting, pale ochreous. *G.* adnexo-sinuate, separating, foxy.

Odour fetid. Woods, fir. May–Sept. $5\frac{7}{8} \times 4\frac{1}{4} \times \frac{7}{8}$ in.

621a. I. proximella Karst. (from its superficial resemblance to **628**) *a.*

P. conico-convex, then expanded and umbonate, even, then longitudinally fibrosely cracked, pallid; mid. and um. rusty-brown or bay. *St.* stuffed, sometimes wavy, subfibrillose, pallid. *G.* adnate, crowded, broad, ventricose, pallid, tan, then brown. *Flesh* white.

Woods. Sept. $1\frac{3}{8} \times 2\frac{3}{4} \times \frac{1}{4}$ in. Distinguished by its ventricose gills.

622. I. perlata Sacc. (from the pileus; *perlatus*, very broad) *a.*

P. expanded, broadly umbonate, straw-ochreous; mid. black-fuscous; marg. pale. *St.* solid, straw-pallid, black-fuscous at base. *G.* adnexed, subrounded or slightly sinuate, umber.

Under trees, hornbeam. Aug. $4\frac{3}{4} \times 2\frac{7}{8} \times \frac{1}{2}$ in.

623. I. phæocephala Sacc. (from the dusky pileus; Gr. *phaios*, dusky, *kephale*, a head) *a.*

P. conico-campanulate, umbonate, bay-brown, fuliginous or rufous. *St.* solid, subbulbous, pale brown above, white below. *G.* adnexo-free, olive-brown.

Woods, pine, mixed. Oct. $3\frac{3}{4} \times 4\frac{3}{4} \times \frac{3}{8}$ in. Perhaps not belonging to *Inocybe.*

624. I. fastigiata Quél. (from the pileus, pointed like a gable, *fastigium*) *a b c.*

P. conico campanulate, acutely umbonate, dull yellow-brownish. *St.* solid, paler than P. *G.* adnexo-free, crowded, olive-brown.

Gregarious, sometimes cæspitose. Woods, grassy places; uncommon. June–Oct. $3\frac{5}{8} \times 5 \times \frac{5}{16}$ in.

625. I. hiulca Gill. (from the furrow-like cracks of the pileus; Gr. *hio*, to gape) *a c.*

P. expanded, umbonate, fuscous, olivaceous or ochreous. *St.* stuffed, whitish. *G.* adnexo-free, subdistant, olive-brown.

Woods, pine; uncommon. Aug.–Nov. $2\frac{1}{4} \times 2\frac{3}{4} \times \frac{5}{16}$ in. Young examples sometimes wholly white. Rufescent when bruised.

626. I. Curreyi Sacc. (after Frederick Currey, mycologist) *a b.*

P. expanded, dull yellowish-brown. *St.* solid, pale rufescent-brown. *G.* free, olive-brownish. *Flesh* pale olive-brown.

Odour none. Woods. Aug. $2\frac{7}{8} \times 3\frac{3}{8} \times \frac{3}{8}$ in.

627. I. rimosa Quél. (from the cracked pileus; *rima*, a crack) *a b c.*

P. campanulate, subumbonate or obtuse, rufescent- or olive-brown; mid. darker; cracks yellowish. *St.* solid, slightly marginato-bulbous, pallid above, yellowish and rufescent below. *G.* attenuato-adnexed, olivaceous-brown or fuscous.

Subgregarious. Odour earthy. Woods and open places amongst trees. June–Oct. $2\frac{7}{8} \times 3\frac{1}{4} \times \frac{5}{16}$ in.

627a. I. subrimosa Sacc. (from the slightly cracked pileus; *rima*, a crack) *a*.

P. conico-campanulate, then expanded, prominently umbonate, even, smooth, longitudinally fibrillose, at length cracked, pale ochre, brown or pallid ferrugineo-ochreous; mid. brown. *St*. solid, equal, shining, base somewhat marginately bulbous, white-pruinose. *G*. attenuate behind, adnexed or free, slightly ventricose, whitish clay-colour, then ferruginous.

Taste insipid; odour none. Grassy places, gardens. Sept. $1\frac{1}{2} \times 1\frac{3}{4} \times \frac{1}{8}$ in.

628. I. asterospora Quél. (from the star-shaped spores; *aster*, a star) *a b*.

P. expanded, umbonate, shining, sometimes viscid, sooty-brown, brown-striate. *St*. marginato-bulbous, whitish, often reddish at base. *G*. emarginato-adnexed to adnate, cinnamon.

Taste none; odour mouldy or none. Woods and open places amongst trees. Aug.–Oct. $2\frac{7}{8} \times 3\frac{3}{8} \times \frac{5}{16}$ in.

628a. I. brunnea Quél. (from its brown colour) *a*.

P. campanulate, umbonate, fibrilloso-silky, then cracked, chestnut-brown. *St*. stuffed, thickened below, fibrilloso-striate, light brown, apex white-pruinose. *G*. emarginate, broadest in front, ivory then umber, edge white, indented.

Amongst grass, near pines. Oct. $1\frac{3}{8} \times 1\frac{7}{8} \times \frac{1}{8}$ in.

629. I. eutheles Quél. (from the well-formed umbo; Gr. *eu*, well, *thele*, a teat) *a b c*.

P. expanded, umbonate, fawn-colour, with darker striæ. *St*. solid, subbulbous, whitish above, brownish below. *G*. adnate, sub-attenuate, tan- or olive-brownish.

Taste not unpleasant; odour of new meal, but disagreeable, or none. Amongst fir-leaves, woods. Aug.–Oct. $2\frac{1}{2} \times 3\frac{1}{2} \times \frac{1}{4}$ in.

630. I. margarispora Sacc. (from the pearl-like spores; Gr. *margarites*, a pearl) *a*.

P. expanded, broadly umbonate, fawn-colour or olive tan-brown. *St*. solid, whitish or shaded tan-olive, darker below. *G*. ascending, adnexo-adnate, faintly rufescent-pallid.

On the ground. Oct. $2 \times 3\frac{3}{8} \times \frac{1}{4}$ in.

630a. I. prætervisa Quél. in Bres. (from the fact that it had been previously confused with **624**; *præter*, before, *video*, to see) *a*.

P. conico-campanulate, then expanded, umbonate, fibrilloso-virgate, slightly viscid, fawn. *St*. solid, bulbous, pubescent, white then pale straw. *G*. attenuate behind, adnexed, white, then fawn with a white edge. *Flesh* white.

Taste mild; odour somewhat earthy. Woods, pine. Aug. $1\frac{3}{4} \times 2\frac{1}{4} \times \frac{1}{4}$ in. Distinguished by its viscid pileus. Related to **624** and **627**, but distinct in its white or straw-coloured stem.

630b. I. Bucknallii Mass. (after John Bucknall, mycologist) *a*.

P. campanulato-convex, fibrillose; mid. subsquamose, brownish. *St*. fibrillose, brownish. *G*. adnexed, thick, subdistant, rusty-brown, edge minutely fimbriate.

Under bushes. Autumn. $\frac{5}{8} \times 1\frac{1}{2} \times \frac{1}{16}$ in. Distinguished by the large size of the basidia, spores and periphyses.

630c. I. duriuscula Rea (from its firm substance) *a*.

P. fleshy, campanulate, expanso-gibbous, floccose, soon longitudinally fibrillose, tawny-ochreous; mid. paler; marg. at length revolute. *St*. solid, firm, apex and base slightly enlarged, striate, white. *G*. sinuato-adnate, decurrent as ribs on apex of St., somewhat crowded, white, then brown. *Flesh* white.

Open places in woods. Sept. $2\frac{3}{8} - 2\frac{3}{4} \times 3\frac{1}{4} \times \frac{5}{8}$ in.

631. I. destricta Quél. (from the fibres of the pileus stripping off; *destringo*, to strip off) *a b c*.

P. flat, umbonate, whitish to vinous or rich sienna-brown, fibres stripping off and often showing white cracks. *St*. solid, sometimes subbulbous, light or dark rufescent or whitish above or below. *G*. uncinato-adnate, grey-brown or olive-shaded. *Flesh* rufescent.

Odour unpleasant. Woods, pine, oak. July–Oct. $2\frac{1}{2} \times 2\frac{1}{8} \times \frac{1}{4}$ in. Sometimes 4 inches or more high.

632. I. perbrevis Gill. (from the comparatively short stem; *perbrevis*, very short) *a b*.

P. expanded, undulate, umbonate, dry, silky, fibrillose, tawny or rufous, becoming yellowish. *St*. stuffed, attenuate downwards, colour as P. *G*. uncinato-adnexed or sinuate, subdistant, tawny to cinnamon.

Woods, shady places. Autumn. $2 \times 1\frac{1}{4} \times \frac{3}{16}$ in.

633. I. descissa Quél. (from the split pileus; *scissus*, split) *a b c*.

P. campanulate, umbonate, whitish to pallid fuscous or ochreous-sienna. *St*. hollow or solid, white or pallid, darker below. *G*. adnexo-free, fuscous.

Suspected poisonous. Woods, pine; rare. Sept.–Nov. $1\frac{7}{8} \times 1\frac{3}{4} \times \frac{1}{8}$ in. Var. *auricoma* Gill. smaller, *P*. golden-fibrillose. Somewhat resembling 640, but differing in colour of P. and absence of odour.

633a. I. Godeyi Gill. (after Dr. Godey, French botanist) *a*.

P. campanulate, obtusely umbonate, silky, fibrillose, rimose, whitish, then tinged rosy, sometimes rosy and ochreous. *St*. equal, subbulbous, colour of P., apex white-pruinose. *G*. adnexo-free, narrowed behind, whitish, then dusky cinnamon with an olive tinge, edge minutely flocculose, white.

Autumn. $1\frac{3}{4} \times 2\frac{3}{4} \times \frac{1}{4}$ in. Rosy-red and ochreous-rosy when bruised.

634. I. Trinii (after Carl Bernard Trinius, Russian botanist) *a b*.

P. convex, obtuse, whitish-rufescent, tawny or pale brown-reddish. *St*. stuffed, colour as P., or paler. *G*. rounded-adnexed, ventricose, cinnamon with a white slightly jagged edge. *Flesh* pale salmon.

Odour strong of clove-pinks, evanescent. Grassy places, woods. Aug.–Sept. $\frac{3}{4} \times 1\frac{7}{8} \times \frac{1}{16}$ in. Sometimes $1\frac{1}{2} \times 1\frac{1}{2}$ in.

d. Velutinæ.

635. I. sambucina Quél. (from its being first noted under elder, *sambucus*) *a.*

P. convex, obtuse, often repand, whitish. *St.* solid, striate, colour as P. *G.* adnexed, colour as P. or pale tan-brownish.

Solitary. Odour strong, disagreeable. Woods, pine, grassy places. $2\frac{1}{8} \times 1\frac{5}{8} \times \frac{1}{2}$ in. Compare **638**, which has narrower gills.

636. I. cæsariata Karst. (from the woven-hairy pileus; *cæsariatus*, covered with hair) *a b c.*

P. expanded, subumbonate, dull tawny. *St.* solid, pale brownish-ochreous. *G.* adnexo-adnate, dull whitish to brown.

Gregarious, subcæspitose. Odour disagreeable. Under beeches. Sept.–Oct. $2\frac{3}{8} \times 1\frac{7}{8} \times \frac{5}{16}$ in.

637. I. lucifuga Gill. (from its habit, apparently shunning light; *lux*, light, *fugio*, to flee) *a b c.*

P. plane, subumbonate, brownish-olivaceous or fawn. *St.* solid, paler than P., or pallid. *G.* adnexed, yellowish or olivaceous, then dark olive.

Odour strong, pleasant or unpleasant, somewhat of radish. Woods, pine; frequent. Sept.–Oct. $1\frac{3}{8} \times 2\frac{3}{4} \times \frac{1}{8}$ in.

638. I. sindonia Karst. (from the woven surface of the pileus; Gr., *sindon*, muslin) *a b c.*

P. convex, gibbous or subumbonate, whitish or faintly ochreous; marg. when young fibrillose with V. *St.* hollow, whitish. *G.* attenuato-adnexed, whitish-fuscous.

Mixed woods, damp shady places; rare. Sept.–Oct. $2\frac{1}{8} \times 2\frac{1}{2} \times \frac{3}{16}$ in. Resembling **640**, but differing in the hollow stem and absence of odour.

639. I. Clarkii Sacc. (after J. Aubrey Clark) *a.*

P. campanulate, obtuse, whitish. *St.* stuffed to solid, flocculose, colour as P. *G.* ascending, adnexed, whitish-fuscous with a white edge.

On the ground, shady places in woods. Oct. $1 \times 1\frac{3}{8} \times \frac{3}{16}$ in.

639a. I. corydalina Quél. (from its odour of *Corydalis cava*) *a.*

P. campanulate, then expanded, fibrillose, white, the prominent um. glaucous-green. *St.* fragile, pruinose, white, then fuscous. *G.* adnato-emarginate, narrow, brown, with a white edge. *Flesh* white, sometimes tinged violet.

Under oaks. Sept. $2 \times 2\frac{5}{8} \times \frac{3}{8}$ in.

640. I. geophylla Quél. (from the earthy colour of the gills; Gr. *ge*, the earth, *phullon*, a leaf) *a b c.*

P. expanded, umbonate, never truly squamulose, white, lilac or pale or dark bright purplish-blue; mid. sometimes brown;

or whole P. brown. *St.* stuffed, sometimes with slight trace of A., white, or as P. *G.* adnexo-free, separating from hymenophore, umber.

Odour usually faint, sometimes very strong and unpleasant. Woods, under trees, hedges. Autumn. $\frac{5}{8} \times 2 \times \frac{1}{16}$ in. Var. *lateritia* W. G. Sm., *Agaricus geophyllus* Sowerb., var. *lateritius* Stev. *P.* red, varying yellow or white, sometimes white-scaly. There is a large form, twice the size of type or larger.

641. I. scabella Quél. (from the roughish pileus; *scaber*, rough) *a b.*

P. expanded, umbonate, scaly-torn, fuscous. *St.* white, or as P. *G.* adnato-adnexed, yellow-brown or fuscous.

Gregarious. Odour none. Woods, amongst short grass. Sept.–Nov. $1\frac{1}{2} \times 1\frac{3}{4} \times \frac{1}{8}$ in.

641a. I. fulvella Bres. (*fulvus*, yellowish-brown).

P. subhygrophanous, conico-campanulate, then expanded and umbonate, floccosely silky, yellowish, then brownish-olive; mid. tawny. *St.* stuffed, narrowed downwards, glabrous, lilac, then rufescent, apex white-pruinose. *G.* subdistant, ventricose, rounded behind and nearly free, pale lilac, then ochraceous-cinnamon; edge fimbriate. *Flesh* yellow, rufescent-lilac at apex of stem.

Shady places. Autumn. $\frac{1}{2} \times 1\frac{1}{8} \times \frac{1}{16}$ in. Allied to **641.**

642. I. violaceofusca Sacc. (from the dusky-violet colour of the stem and gills) *a.*

P. expanded, obtusely subumbonate, squamose, umber; marg. fimbriate with V. *St.* solid, pale violaceous above, pallid below. *G.* adnate or emarginate, pale dull violaceous, then umber.

Subcæspitose. Amongst grass in open places. Autumn. $2\frac{1}{8} \times 1\frac{3}{4} \times \frac{1}{4}$ in.

e. Viscidæ.

643. I. trechispora Karst. (from the rough spores; Gr. *trachus*, rough, *spora*, a seed) *a b c.*

P. expanded, umbonate, whitish with mid. tawny. *St.* stuffed, white with a mass of white mycelium at the base. *G.* emarginate, or attenuato-adnexed, pinkish-grey, then brown.

Woods, damp places, amongst ferns. July–Oct. $1\frac{1}{4} \times 2\frac{5}{8} \times \frac{1}{8}$ in. Sometimes fatal to pheasants.

644. I. vatricosa Karst. (from the stem, twisted below; *vatricosus*, with bad feet) *a c.*

P. plane, broadly subumbonate, whitish or shaded light-yellowish. *St.* hollow. colour as P., white-pulverulent. *G.* adnexed, broadly emarginate, ochreous-fuscous.

Odour none. Woods, pine, bare places, on dead stumps, twigs, chips. Sept. $1\frac{3}{4} \times 2\frac{5}{8} \times \frac{3}{16}$ in.

645. I. Whitei Sacc. (after Dr. F. Buchanan White, botanist) *a c.*

P. hemispherico-campanulate, obtuse, tawny; marg. whitish. *St.* solid, white, becoming tawny. *G.* adnexed, tawny.

Woods, pine. Oct. $1\frac{1}{8} \times 1\frac{3}{4} \times \frac{1}{8}$ in. Allied to **640.**

646. I. Tricholoma Sacc. (from its resemblance to species of *Tricholoma*) *a b c.*

P. convexo-plane, subdepressed, ivory-whitish, shaded pale buff; marg. fringed with V. *St.* stuffed, fibroso-scaly above, colour as P., or faintly shaded salmon. *G.* adnato-decurrent, crowded, pale clay-fuscous or faintly olive-shaded.

Woods, mixed. Nov. $1\frac{1}{2} \times 1\frac{1}{2} \times \frac{3}{16}$ in.

XXVI. HEBELOMA Quél.

(From the fringe-like veil, seen in certain young examples; Gr. *hebe*, youth, *loma*, a fringe.)

Veil partial—in **648, 654a** and **655** forming an imperfect annulus—or universal, indicated by gluten, squamules, fibrillæ or silkiness. *Hymenophore* confluent and homogeneous with the fleshy stem. *Pileus* fleshy, cuticle continuous, smooth, damp, subviscid or

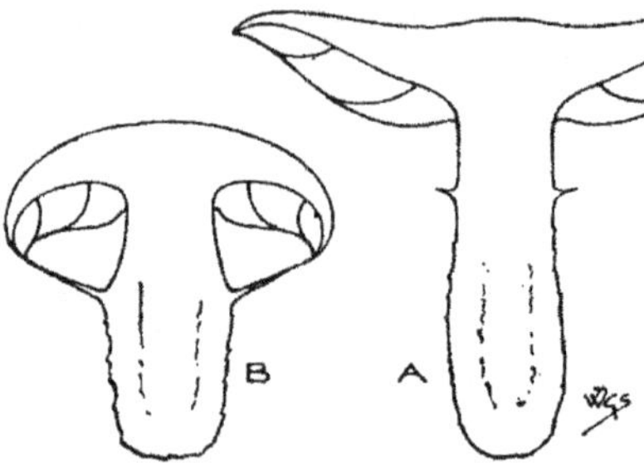

Fig. 35.—A, section of *Hebeloma fastibile* Quél.; B, ditto half-grown example showing veil. One-half natural size.

glutinous, margin at first incurved, sometimes silky from veil. *Stem* central, simple, fibrous-coated, submealy at the apex, simple or rarely imperfectly annulate. *Gills* sinuato-adnate or adnexed, at first pallid-whitish or clay-colour, then darker. *Spores* dull tan or somewhat clay-colour or brown. (Fig. 35.)

All the species, except **649**, grow on the ground; they are gregarious and strongly scented; some are poisonous, none are edible; they appear early in autumn and last till late in the season.

Hebeloma corresponds with the non-fibrillose species of *Tricholoma*, *Entoloma* and *Hypholoma*. Species **647—669**

a. Indusiatæ. *Pileus* furnished with a cortina from the veil, by which the margin is often superficially silky. *Stem* usually white-pruinate or white-mealy above or minutely white-scaly. **647—657**

b. Denudatæ. *Pileus* smooth; cortina obsolete. **658—667** *c. Pusillæ.* *Pileus* scarcely an inch broad; stature that of *Naucoria*, but without the cartilaginous bark. **668, 669**

a. Indusiatæ.

647. H. mussivum Sacc. (from its uncertain position amongst the *Indusiatæ; mussivus*, undecided) *a.*

P. plane, obtuse or subumbonate, pale primrose-yellowish; mid. sienna. *St.* stuffed or hollow above, fibrillose, light yellow. *G.* emarginate or somewhat rounded adnate, light yellow then subferruginous. *Flesh* becoming yellow.

Odour faint, not unpleasant. Woods, pine. Oct. $4\frac{3}{4} \times 5\frac{1}{4} \times 1$ in.

647a. H. sinuosum Quél. (from the frequently irregular pileus) *a b.*

P. fleshy, convex, expanded, obtuse, even, almost glabrous, viscid, then dry, yellowish-red to pale buff, sometimes whitish; marg. wavy. *St.* stuffed to hollow, equal, fibrilloso-striate, at first flocculoso-squamulose above, soft, biscuit. *G.* slightly adnexed, broadly emarginate to emarginato-free, dry, crowded, dull white to brownish-salmon and ferruginous. *Flesh* white.

Odour faint, pleasant. Woods. Aug. $3\frac{5}{8} \times 5 \times \frac{7}{8}$ in.

648. H. fastibile Quél. (from its disagreeable odour) *a b c.*

P. convexo-plane, viscid in wet weather, pale yellowish-tan; mid. darker, sometimes sienna. *St.* solid, fibrillose, scaly or white silky, sometimes with an imperfect silky-floccose A. *G.* emarginate, subdistant, dull whitish-clay-colour or tan-brownish.

Subcæspitose, often in large dense rings. Poisonous. Taste like radishes, but disagreeable and bitter; odour disagreeable, fetid. Woods, mixed pastures, near trees and bushes, sometimes on mushroom-beds; common. July–Oct. $4 \times 2\frac{3}{8} \times \frac{3}{4}$ in. Often mistaken for the mushroom and sold in markets with mushrooms. Often slug-eaten. Var. *elegans* Mass., *P.* purple-brown.

649. H. glutinosum Sacc. (from the glutinous pileus) *a b c.*

P. plane, obtuse, yellow-white or pale buff with mid. rich sienna-buff, white-squamulose. *St.* stuffed, attenuate upwards, white, dull faint ochreous below, clad with fugitive white sc. *G.* sinuato-adnate, light yellowish then clay-cinnamon. *Flesh* pale reddish in stem.

Taste and odour mild, not unpleasant of radish. Woods, amongst leaves and branches, oak, beech; frequent. Sept.–Dec. $3\frac{1}{2} \times 3\frac{3}{8} \times \frac{1}{2}$ in. Sometimes one-third the size of type. Compare **678.**

650. H. testaceum Quél. (from the colour of the pileus; *testa*, a brick) *a b c.*

P. convex, obtuse or subumbonate, crimson-sienna through red and yellowish to tan. *St.* hollow above, fibrillose, colour as P. or paler. *G.* attenuato-sinuate or almost free, clay-colour then subferruginous or crimson-sienna.

Taste disagreeable, sickly-nauseous; odour faint of radish or none. Woods, fir, pastures; frequent. Sept.–Nov. $2\frac{3}{4} \times 3\frac{1}{4} \times \frac{5}{16}$ in.

651. H. firmum Gill. (from its firm substance) *a*.

P. flat, subumbonate, subsquamulose, yellowish-brown or reddish; mid. darker. *St.* solid, brownish, darker below, white-squamulose. *G.* adnexed or adnate, rounded, ferruginous, edge white, serrulate.

Odour faint. Woods, fir. Feb.–Oct. $2\frac{5}{8} \times 3 \times \frac{5}{8}$ in.

652. H. claviceps Quél. (from its resemblance, when young, to a nail; *clavus*, a nail, *caput*, the head) *a*.

P. hemispherical, gibbous-expanded, viscid, whitish or faintly yellowish; mid. sienna; marg. fibrillose with V. *St.* stuffed, whitish, fuscous below. *G.* emarginate, pallid brown.

Woods, mixed; uncommon. Sept.–Oct. $1\frac{3}{4} \times 3\frac{1}{8} \times \frac{1}{4}$ in.

653. H. punctatum Quél. (from the viscid-dotted pileus; *punctus*, a point) *a*.

P. plano-gibbous, pale sienna-brown, whitish at marg. *St.* hollow, silky-fibrillose, faintly brownish. *G.* sinuato-adnate, brownish.

Gregarious, in troops. Odour faint, not unpleasant. Woods, pine; uncommon. Sept. $2 \times 3\frac{1}{4} \times \frac{1}{4}$ in.

654. H. versipelle Gill. (from its changeable appearance; *verto*, to turn, *pellis*, skin) *a b c*.

P. flat, silky-agglutinate, sienna-reddish or dull tan. *St.* hollow, tan-white, fuscous below. *G.* rounded, serrulate, whitish-flesh-colour.

Subcæspitose. Odour faint, not unpleasant, but sometimes strong, fetid-oily or like bugs. Grassy places, amongst fir-leaves. Aug. $2\frac{1}{8} \times 3 \times \frac{3}{16}$ in.

654a. H. strophosum Sacc. (from a fancied resemblance in the annulus to a sword-belt, Gr. *strophos*) *b*.

P. convex, plane, slightly subumbonate, somewhat viscid, bay or vinous-sienna, clouded deep claret-brown; marg. white-silky. *St.* hollow, equal, silky, whitish, shaded brown or red-salmon within and without near base. *A.* superior, silky-white. *G.* adnexo-free, crowded, ventricose, cream to dull flesh-colour and dull cinnamon. *Flesh* white.

On the ground. Oct. $1\frac{1}{8} \times 1\frac{3}{4} \times \frac{3}{16}$ in.

655. H. mesophæum Quél. (from the dusky centre of the pileus; Gr. *mesos*, the middle, *phaios*, dusky) *a b c*.

P. plane, subgibbous, pale yellowish or pale pinkish-brown. *St.* fistulose, fibrillose, whitish or faintly ochreous, ferruginous below, sometimes with an imperfect fugitive A. *G.* rounded or emarginate, pale clay-ferruginous.

In troops, gregarious. Odour faint. Woods, pine; uncommon. Sept.–Oct. $1\frac{3}{4} \times 2\frac{3}{4} \times \frac{3}{16}$ in.

656. **H. subcollariatum** Sacc. (from the gills which form an imperfect collar at the apex of the stem) *a b*.

P. convex, subviscid, pallid; mid. fuscous; marg. at first white and scaly. *St*. stuffed or fistulose, brown at base. *G*. sinuato-adnate, clay-colour with white edge.

Subcæspitose. Taste and odour strong and rank, like **1527**. On the ground. May–Oct. $1\frac{1}{2} \times 2\frac{1}{8} \times \frac{3}{16}$ in.

657. **H. senescens** Sacc. (from the hoary margin of the pileus and the white apex of the stem; *senseco*, to grow old) *a*.

P. obtuse or subdepressed, ochreous-tawny. *St*. solid, squamulose, white above, fuscous below. *G*. rounded-adnexed, crowded, cinnamon.

Odour acrid. Amongst firs. Sept. $4 \times 3\frac{3}{8} \times \frac{3}{4}$ in.

b. Denudatæ.

658. **H. sinapizans** Gill. (from its odour; *sinapi*, mustard) *a b*.

P. expanded, undulate, subdepressed, subviscid, pale clay-colour; mid. yellowish. *St*. somewhat solid, whitish. *G*. emarginate, crowded, unspotted, not distilling drops, clay-cinnamon.

Solitary. Odour strong, usually of radish. Under trees, fields, woods; uncommon. Sept.–Oct. $5\frac{3}{8} \times 2\frac{7}{8} \times 1\frac{1}{2}$ in. *P*. sometimes 9 in. in diam. Compare with **473**.

659. **H. crustuliniforme** Quél. (from the shape of the pileus; *crustulum*, a small pie) *a b c*.

P. expanded, obtuse or subumbonate, subviscid, whitish; mid. pale reddish-ochreous. *St*. whitish. *G*. adnexed, crowded, distilling drops, clay-colour or brown.

Sometimes forming large rings. Poisonous. Odour very strong of radish, but fetid or like laurel flowers. Woods, mixed, lawns, pastures, on heavy soil; common. Aug.–Nov. $4\frac{1}{4} \times 2\frac{1}{4} \times 1$ in. Variable in size. Often mistaken for the horse-mushroom, **789**.

659a. **H. subsaponaceum** Karst. (from the somewhat soapy odour; *sapo*, soap).

P. fleshy, obtuse, pallid, darker when dry. *St*. equal, somewhat wavy, adpressedly fibrillose, apex somewhat mealy. *G*. adnate.

Taste bitter; odour strong of soap. Woods, mixed, fir. Aug. $1\frac{1}{4} \times 1 \times \frac{1}{8}$ in.

660. **H. elatum** Gill. (*elatus*, tall) *a b*.

P. expanded, subviscid, tan; mid. faintly ochreous, slightly reddish or shaded brown. *St*. stuffed, twisted, whitish or faintly ochreous. *G*. adnato-rounded with a tooth, pale cinnamon, becoming cinnamon-salmon, sometimes with a white edge.

Odour very strong of radish. Woods, pine, commons, under oaks. Sept.–Nov. $2 \times 4\frac{1}{4} \times \frac{1}{2}$ in.

661. H. longicaudum Quél. (from the long stem; *longus*, long, *cauda*, a tail) *a b*.

P. expanded, sometimes umbonate, viscid, clay-colour; mid. darker. *St.* stuffed, subbulbous, fibrous-elastic, white above, tawny below within and without. *G.* emarginate, crowded, serrulate, cinnamon.

Odour faint, not unpleasant. Woods. Sept.–Nov. $5 \times 7\frac{3}{4} \times \frac{5}{8}$ in. Var. *radicatum* Sacc. *St.* fusiform-rooting.

662. H. lugens Gill. (from its sombre colour; *lugeo*, to mourn) *a*.

P. convexo-plane, subviscid, brown or yellowish. *St.* solid, shining, fibrilloso-striate, whitish. *G.* attenuato-adnexed or almost free, ferruginous.

Odour strong, not radish-like. Under beeches. $2\frac{3}{4} \times 2\frac{1}{2} \times \frac{3}{8}$ in.

663. H. truncatum Sacc. (from the short stem; *truncus*, maimed) *a b*.

P. obtuse, repand, irregular, ochreous-rufous or tan-ochreous, faintly shaded rose. *St.* solid, pruinate, white. *G.* emarginate or emarginato-free, watery-ferruginous.

Subcæspitose. Odour faint, not unpleasant. Grassy places in woods. Sept. $3\frac{1}{4} \times 2\frac{1}{8} \times \frac{3}{4}$ in.

664. H. nudipes Karst. (from the naked stem; *nudus*, naked, *pes*, a foot) *a b*.

P. expanded, subumbonate, viscid, pale clay or ochreous-white; marg. thin, exceeding G. *St.* solid, smooth, whitish. *G.* emarginate, sienna-tan.

Odour faint, not unpleasant, somewhat of meal. Woods. Oct. $2\frac{3}{8} \times 3 \times \frac{1}{4}$ in.

665. H. ischnostylum Sacc. (from the thin stem; Gr. *ischnos*, thin, *stulos*, a pillar) *a b*.

P. expanded, broadly subumbonate, subviscid, whitish; mid. pallid. *St.* naked, colour as P. *G.* adnate, clay-colour.

Odour none, or faint of meadow-sweet. Fields, plantations, woods, amongst grass. Sept.–Oct. $1\frac{3}{4} \times 2 \times \frac{1}{8}$ in.

666. H. nauseosum Sacc. (*nauseosus*, rank-smelling) *a b*.

P. convex, gibbous, subviscid, ochreous-white; mid. sienna-tan. *St.* solid, silky-fibrillose, white-mealy above, base at length black. *G.* sinuate, broad, subdistant, salmon then ferruginous, sometimes with a purplish shade.

Odour very strong and offensive. Woods, mixed. Oct. $2\frac{1}{4} \times 3\frac{1}{4} \times \frac{1}{2}$ in.

667. H. capniocephalum Gill. (from the sooty-coloured pileus; Gr. *kapnos*, smoke, *kephale*, the head) *a c*.

P. convexo-plane, viscid, pale yellowish; mid. sienna or sooty; marg. at length black. *St.* stuffed, attenuate below, rufescent, fibrilloso-striate on a whitish ground. *G.* emarginate, subcrowded, brown-ferruginous.

Woods, mixed, elm-stumps. Oct.–Nov. $3\frac{1}{8} \times 2\frac{1}{4} \times \frac{3}{8}$ in.

c. Pusillæ.

668. H. magnimamma Karst. (from the large umbo; *magnus*, large, *mamma*, breast) *a.*

P. plane, ochreous-red, then pale yellowish. *St.* stuffed or fistulose, smooth, naked, colour as P. *G.* adnate, subsinuate, ferruginous.

Amongst grass, under apple-trees. Sept. $1 \times 1\frac{1}{4} \times \frac{1}{8}$ in.

669. H. petiginosum Quél. (from the scurfy-hoary pileus; *petigo*, scab) *a b c.*

P. convex, subumbonate, hoary-silky, brown or shaded ochreous, rufescent, slate or purplish. *St.* stuffed, rufescent, paler above, white-pulverulent. *G.* slightly adnexed, or free, crowded, olivaceous-brown or shaded rufescent.

Woods, beech. Oct. $1 \times 1\frac{7}{8} \times \frac{1}{8}$ in.

XXVII. FLAMMULA Quél.

(From the frequent flame-like colours; *flamma*, a flame.)

Veil fibrillose, fugitive or obsolete. *Hymenophore* confluent and homogeneous with the fleshy stem. *Pileus* fleshy, margin at first involute. *Stem* central, subannulate or simple, fleshy-fibrous, not

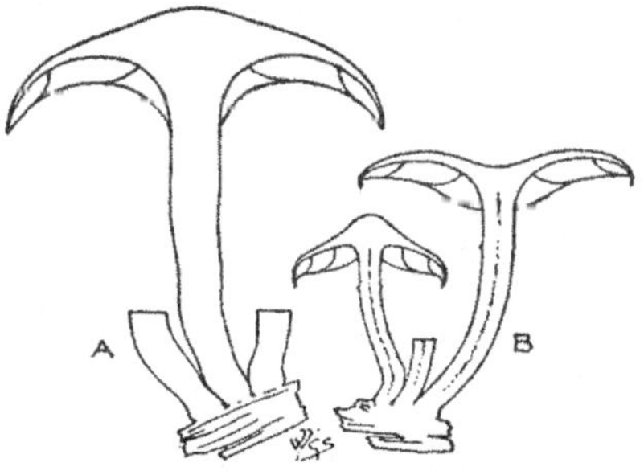

Fig. 36.—A, section of *Flammula gymnopodia* Quél.; B, ditto *F. flavida* Quél. One-third natural size.

mealy above. *Gills* decurrent or adnate, without a sinus, commonly entire and of one colour, at first whitish, clay-colour or yellowish, then coloured by the spores. *Spores* mostly pure ferruginous, sometimes fuscous-ferruginous or tawny-ochraceous. (Fig. 36.)

The species usually grow on wood, some grow on the ground, many are bright yellow, orange or orange-brown in colour. Some agree in structure with *Clitocybe* and *Clitopilus*, those growing on wood approach *Pholiota*, and agree with *Armillaria* and in part with *Hebeloma* and *Stropharia*. Species **670—702** *a. Gymnotæ.* *Pileus* dry, most frequently squamulose. *Veil* none, except **673** and **676.** *Spores* ferruginous, fuscous-ferruginous in **675.** **670—677**

b. Lubricæ. *Pileus* covered with a continuous, subseparable, smooth, viscid pellicle. *Veil* fibrillose. *Gills* usually not truly sinuate. *Spores* ferruginous, fuscous-ferruginous in **685,** not tawny. Gregarious, terrestrial, rarely growing on wood. Allies of *Hebeloma.* **678—685**

c. Udæ. *Pileus* with a continuous, non-separable cuticle, slightly viscid in rainy weather. *Veil* appendiculate at the margin of the pileus, or forming an imperfect fibrillose annulus. *Spores* not tawny or ochraceous. Cæspitose. Growing on wood. Allies of *Pholiota.* **686—694**

d. Sapineæ. *Pileus* scarcely pelliculose, flesh scissile or torn above into scales, not viscid. *Veil* fibrilloso-adpressed to the stem, not appendiculate at the margin of the pileus. *Gills* light yellow or yellow, then tawny. Subcæspitose. On pines, or terrestrial amongst pine-branches. **695—698**

e. Sericellæ. *Pileus* dry, or at first viscid, slightly silky. *Veil* fibrillose. **699—702**

a. Gymnotæ.

670. F. gymnopodia Quél. (from the naked stem; Gr. *gumnos*, naked, *pous*, a foot) *a.*

P. convex, gibbous, squamulose, ochreous-sienna; mid. darker. *St.* solid, paler than P. *G.* deeply decurrent, crowded, colour as P., cinnamon or orange.

Cæspitose. Pine sawdust, on the ground. Autumn. $4\frac{3}{4} \times 3 \times \frac{1}{2}$ in.

671. F. Aldridgei Mass. (after Miss Emily Aldridge) *a.*

P. infundibuliform, velvety, rich orange; mid. sienna. *St.* hollow, colour as P., black below. *G.* deeply decurrent, rich orange.

Gregarious. Woods, amongst moss. Sept. $2\frac{1}{4} \times 2\frac{3}{4} \times \frac{1}{4}$ in.

672. F. vinosa Gill. (from its colour; *vinum*, wine) *a c.*

P. infundibuliform, flocculose, umber, pink-shaded. *St.* solid, attenuate upwards, flocculose, paler than P. *G.* decurrent, colour as St., faintly pinkish at base.

On the ground; rare. Oct. $2\frac{3}{4} \times 1\frac{1}{2} \times \frac{5}{16}$ in.

673. F. PURPURATA Sacc. (from its purple colour) *a b.*

P. expanded, subumbonate, minutely flocculose, purple to brownish-purple. *St.* solid, pallid above, purplish-brown below. *A.* superior, red-fibrillose. *G.* adnate, lemon-yellow then bright ferruginous, purple-brown where touched.

Taste very bitter. Tree-fern stems. May. $2 \times 1\frac{3}{8} \times \frac{3}{16}$ in.

674. F. floccifera Sacc. (from the white fibrils on the pileus; *floccus*, a flock of wool, *fero*, to carry) *a*.

P. expanded, tawny. *St.* hollow, attenuate downwards, white-silky, pale ochreous below. *G.* adnate, wrinkled, ferruginous with a white edge.

Cæspitose. Lime-stumps. Oct. $2 \times 1\frac{1}{2} \times \frac{1}{4}$ in.

675. F. decipiens Sacc. (from its likeness to **685**; *decipio*, to deceive) *a b c*.

P. convex, subumbonate, sienna; mid. almost white. *St.* stuffed, attenuate downwards, striate, tawny. *G.* decurrent, crowded, sienna. *Flesh* yellowish.

Subfasciculate. Burnt gorse-stumps; in company with **685**. June. $1\frac{1}{4} \times 1\frac{3}{4} \times \frac{1}{4}$ in.

676. F. clitopila Sacc. (from the depressed pileus; Gr. *klitos*, a declivity) *a b*.

P. expanded, dry, pale purplish-brown, becoming whitish. *St.* hollow, subventricose, attenuate downwards, slightly fibrillose towards base, at first white then paler than P. *A.* imperfect, fibrillose. *G.* sinuato-adnexed, crowded, colour as St.

Amongst firs. Nov. $2\frac{1}{2} \times 2\frac{1}{4} \times \frac{3}{8}$ in.

677. F. nitens Sacc. (from the shining pileus; *nitens*, shining) *a b*.

P. expanded, obtuse, dry, silky, deep purple-brown or deep bright chestnut. *St.* solid, fibrillose, reddish-salmon. *G.* adnate, crowded, clay-umber. *Flesh* salmon-white.

Cæspitose. Taste and odour none. On the ground. Aug.–Sept. $1\frac{3}{4} \times 2\frac{1}{2} \times \frac{3}{8}$ in.

b. Lubricæ.

678. F. lenta Gill. (*lentus*, tough) *a b c*.

P. plane, obtuse, whitish, clay-colour to lemon-white. *V.* white, stretching from marg. to St. in infancy. *St.* substuffed, paler than P., pale rufous below, white-scaly. *G.* adnate, sinuate or decurrent, ferruginous. *Flesh* white to pale sulphur, brownish at base of St.

Sometimes cæspitose. Woods, mixed, beech, oak, fir, on wood, leaves, on the ground, meadows. Sept.–Dec. $2\frac{1}{2} \times 2\frac{1}{2} \times \frac{1}{4}$ in. See **649**.

679. F. lubrica Quél. (from the slimy pileus; *lubricus*, slimy) *a*.

P. flat, obtuse or depressed; mid. deep sienna; marg. orange-yellow. *St.* solid, laxly fibrillose, white above, fuscous below. *G.* adnate or with a decurrent tooth, clay-colour.

Almost scentless. On or near trunks. Autumn. $2\frac{1}{2} \times 2\frac{3}{4} \times \frac{5}{16}$ in.

680. F. lupina Karst. (from the colour of the pileus; *lupus*, a wolf) *a c*.

P. obtuse, then plano-depressed, tawny. *St.* stuffed, whitish above, ferruginous and adpressed-fibrillose below. *G.* adnato-decurrent, light yellowish. *Flesh* white.

Taste bitter; odour mild to strong and pungent. Pastures. Aug.–Oct. $1\frac{5}{8} \times 2\frac{1}{4} \times \frac{1}{4}$ in.

681. F. mixta Karst. (from its position amongst its allies; *mixtus*, mixed) *a b c.*

P. plane, subdepressed, tan or shaded-ochreous; mid. slightly sienna. *St.* white-ochreous above, laxly fibrillose and rufous below, minutely fibrilloso-annulate. *G.* adnate or adnato-decurrent, separating from stem, clay-colour.

Subcæspitose. Taste disagreeable; odour slight or none. Woods, pine, mixed; rare. Aug.–Nov. $2\frac{1}{8} \times 3 \times \frac{3}{16}$ in.

682. F. juncina Sacc. (from its habitat; *juncus*, a rush) *a b c.*

P. convex, sulphur-ochreous; mid. brown. *St.* stuffed, attenuate downwards, subfibrillose, colour as P. *G.* adnate, crowded, very thin, red-brown. *Flesh* sulphur-whitish or livid-brownish.

Taste nauseous, disagreeable, somewhat bitter. Dead bulrushes. Nov. $1\frac{5}{8} \times 3\frac{3}{4} \times \frac{3}{16}$ in. A close ally of **681.**

683. F. gummosa Quél. (from the viscid-gummy pileus) *a b c.*

P. flat, obtuse or depressed, light yellowish or olive-shaded; marg. appendiculate with V. *St.* silky-fibrillose, paler than P., ferruginous below. *G.* adnate, crowded, cinnamon.

Cæspitose. Suspected poisonous. Taste not bitter; odour none. On and about stumps. Oct.–Dec. $2\frac{1}{2} \times 2\frac{7}{8} \times \frac{5}{16}$ in.

684. F. spumosa Karst. (from the frothy-viscid pileus; *spuma*, froth) *a b c.*

P. flat, obtuse or depressed, light yellowish or olive-shaded; mid. sienna; or whitish with mid. sulphur and yellow; marg. appendiculate with V. *St.* silky-fibrillose, paler than P., ferruginous below within and without. *A.* slight, fibrillose, evanescent. *G.* adnate, crowded, pale olive-brown to cinnamon.

Solitary, gregarious or subcæspitose, sometimes in rings. Taste not bitter; odour none. Woods, fir, stumps, sawdust. Sept.–Dec. $2\frac{5}{8} \times 2\frac{7}{8} \times \frac{5}{16}$ in.

685. F. carbonaria Quél. (from its habitat; *carbo*, charcoal) *a b c.*

P. plane, sometimes depressed, tawny. *St.* fibrilloso-squamulose, brownish-white above, brown below. *G.* adnate, fuscous, clay-colour. *Flesh* brownish-white. *Spores* fuscous-ferruginous.

Single or subcæspitose. Taste sweetish. Burnt wood, earth and leaves in and near woods; frequent. Sept.–Dec. $2\frac{1}{2} \times 2\frac{3}{4} \times \frac{1}{4}$ in. Sometimes 4×4 in.

c. Udæ.

686. F. filia Mass. (*filia*, a daughter; derivation obscure) *a b.*

P. expanded, subumbonate, buff; mid. sienna; marg. white-zoned with V. *St.* stuffed, fibrilloso-striate, rooting, partially annulate, whitish above, buff or salmon-buff below. *G.* sinuato-decurrent, tan-buff or olive-shaded.

Odour none. Woods, logs, on the ground. Oct. $4 \times 6 \times 2$ in.

687. F. fusus Gill. (from the spindle-shaped stem; *fusus*, a spindle) *a*.
P. plane, depressed, whitish-brown; mid. sienna. *St*. stuffed, whitish above, pale brown below. *G*. subdecurrent, ferruginous.

Solitary, gregarious, sometimes cæspitose. Taste mild. Woods, stumps, on the ground. Sept.–Nov. $4 \times 2\frac{1}{4} \times \frac{5}{8}$ in. Var. *superba* Mass., bright deep orange; *G*. bright yellow.

688. F. astragalina Quél. (from its colour; Gr. *astragalinos*, a goldfinch) *a b*.
P. expanded, blood-saffron; mid. almost vermilion; marg. yellow, with fibrillose sulphur-white V. when young. *St*. hollow, saffron; base darker. *G*. sinuato-adnate, white, sulphur, then yellowish-brown or olive-shaded. *Flesh* saffron.

Taste and odour bitter, nauseous, as in **69**. Stumps, pine, fir. Aug.–Oct. $1\frac{7}{8} \times 3\frac{1}{2} \times \frac{1}{4}$ in. Sometimes turns black when cut or bruised.

688a. F. rubicundula Sacc. (from the vinous stains; *rubicundulus*, somewhat reddish) *a b*.
P. fleshy, convex, plane, viscid and innato-fibrillose, then smooth, yellow, becoming vinous-shaded; marg. paler. *St*. solid, whitish then red-shaded, white-mealy above. *G*. sinuato-adnate to adnato-decurrent, crowded, light ochre then ferruginous; edges reddening with age. *Flesh* bright yellow, then lighter.

Taste acrid; odour none. Woods, under scrub-oak. July–Sept. $3 \times 4\frac{1}{4} \times \frac{9}{16}$ in. Allied to **688**.

689. F. alnicola Quél. (from its frequent habitat; *alnus*, alder) *a b c*.
P. flat, sulphur-yellow; mid. sienna; marg. clouded sulphur-greenish. *St*. attenuato-rooting, sulphur-yellow above, sienna below, sometimes with an imperfect fibrillose A. *G*. adnate with a small tooth, bright ferruginous. *Flesh* sulphur-yellow. *Spores* umber.

Usually fasciculate. Taste and odour strong, dry-insipid, bitter or acrid. On or near stumps, alder, willow, oak, thorn, birch, sallow; rare. Sept.–Oct. $2\frac{1}{4} \times 4\frac{1}{4} \times \frac{1}{4}$ in.

690. F. flavida Quél. (*flavus*, light yellow) *a b c*.
P. expanded; mid. darker yellow. *St*. colour as P., ferruginous at base. *A*. slight, fibrillose, fugitive. *G*. adnate, tawny-ferruginous. *Flesh* light-yellowish.

Cæspitose, sometimes in large clusters. Tasteless or insipid. Woods, stumps, fir, lime, on the ground; frequent. Aug.–Nov. $3\frac{1}{4} \times 3\frac{1}{2} \times \frac{5}{16}$ in.

691. F. inaurata Sacc. (from the colour; *inauro*, to gild) *a b c*.
P. expanded, sulphur-ochreous; marg. appendiculate with V. *St*. lemon-white, ferruginous within at base. *V*. slight, fibrillose, fugitive. *G*. adnate with a tooth, pale yellowish clay-colour, then vinous-clay.

Single or cæspitose. Tasteless or insipid. Willows. Nov. $1\frac{1}{4} \times 1\frac{3}{8} \times \frac{3}{16}$ in.

692. F. conissans Gill. (from its abundant dust-like spores; Gr. *konis*, dust) *a b c*.

P. flat, subdepressed, light yellowish-tan or pale bronzy-ochre. *St.* attenuate downwards, lighter than P., ferruginous below, base white-villous. *G.* adnate with a tooth, crowded, fuscous-ferruginous. *Flesh* white.

Densely cæspitose. Taste and odour none or acid. Woods, willows. Oct.–Nov. $2\frac{1}{4} \times 4 \times \frac{1}{4}$ in. Often passed over as **826**.

693. F. inopus Karst. (from its fibrillose stem; Gr. *is*, *inos*, a fibre, *pous*, a foot) *a b c*.

P. expanded, obtuse, pale yellowish-tan or sulphur; mid. sienna. *St.* fistulose, attenuato-rooting, pallid above, sienna-brown below within and without, sometimes with a fibrillose A. when young. *G.* adnate, yellowish-white, pale olive-brown or sulphur-greenish, sometimes fuscous. *Flesh* paler than P.

Fasciculate. Woods, stumps, fir, larch, rooting amongst pine-leaves, on the ground; rare. Vernal and Aug.–Dec. $3 \times 8\frac{1}{2} \times \frac{5}{16}$ in. Must not be confounded with **824**.

694. F. apicrea Gill. (from its taste; Gr. *apikros*, not bitter) *a*.

P. expanded, pale ochreous-tan; mid. sienna. *St.* hollow, colour as P., fibrillose, ferruginous below. *G.* adnate or adnato-decurrent, ferruginous.

Subcæspitose. Odour mild. Woods, pine, trunks. Sept.–Oct. $4\frac{1}{2} \times 3\frac{1}{4} \times \frac{7}{16}$ in.

d. Sapineæ.

695. F. hybrida Gill. (from its hybrid characters) *a b c*.

P. flat, obtuse or subdepressed, tawny-buff, orange or sienna. *St.* attenuate upwards, tan-brownish, brownish below. *G.* adnate, tawny or rufous, unspotted. *Flesh* yellowish or brownish.

Subcæspitose or in troops. Fir, stumps, branches, under beeches, on saw-dust and wood; rare. Aug.–Dec. $2\frac{1}{8} \times 2\frac{1}{8} \times \frac{1}{4}$ in.

696. F. sapinea Quél. (from its growing on pine-wood; *sapinus*, a pine) *a b c*.

P. flat, golden-tawny or orange-vermilion; mid. darker; marg. appendiculate with white V. in young examples. *St.* sulcate, orange or fuscous-white above, white or brown below. *A.* fibrillose, fugitive. *G.* adnate, colour as P. or tawny-cinnamon.

Single or cæspitose. Taste unpleasant, bitter; odour strong, not unpleasant. Fir-stumps, branches, sawdust, wood, in and near woods, sometimes on the ground, under beeches and larches. Aug.–Jan. $2\frac{5}{8} \times 3\frac{1}{2} \times \frac{1}{4}$ in. Var. *terrestris* Sacc., fusiform-rooting.

697. F. liquiritiæ Quél. (from its taste of liquorice) *a c*.

P. flat, subumbonate, bay-brown or orange-tawny; mid. darker. *St.* hollow, attenuate upwards, striate, tawny or ferruginous. *G.* adnate or sinuate, golden or tawny.

Gregarious or cæspitose. Taste sweet; odour acid. Firs. Oct. $1\frac{7}{8} \times 2\frac{7}{8} \times \frac{3}{16}$ in.

698. F. picrea (Gr. *pikros*, bitter) *a b c*.

P. convex, obtuse, rufous; traces of white V. at marg. in young examples. *St.* fistulose, umber, darker at base; when young wholly white-pulverulent. *G.* adnato-ascending or decurrent, ferruginous.

Cæspitose. Pine-stumps, old deal boards; rare. Oct.–Dec. 2¼ × 4⅜ × ¼ in.

e. Sericellæ.

699. F. FILICEA Sacc. (from its habitat, tree-fern stems; *filix*, a fern) *a b c*.

P. flat, subumbonate, golden-yellow, appendiculate with V. *St.* stuffed or hollow, fibrillose, colour as P., ferruginous below. *V.* annulate, fibrillose, fugacious, reddish. *G.* adnate, tawny.

Tree-ferns in greenhouses. Spring and summer. 1⅞ × 1¾ × ⅛ in.

700. F. ochrochlora Karst. (from the colour of the pileus; Gr. *ochros*, yellow, *chloros*, green) *a b*.

P. plane, subumbonate, subviscid, pale dull olive-ochre, minutely white-squamulose. *St.* fistulose, paler than P., ferruginous at base, white-squamose. *V.* slight, seen in young examples. *G.* adnate, olivaceous, then olive-brown.

Cæspitose. Old trunks, furze-roots, hazel-sticks. Aug.–Nov. 2⅛ × 2¾ × ¼ in.

701. F. helomorpha Quél. (from its resemblance to a nail; Gr. *helos*, a nail, *morphe*, form) *a c*.

P. convex, gibbous or obtusely umbonate, viscid, white or clay-white. *St.* solid, smooth, tan-white. *G.* plano-decurrent, clay-white. *Spores* pale brown.

Scattered. Wood, fir. Nov. ¾ × 1⅝ × ⅛ in.

702. F. scamba Sacc. (from the frequently curved stem; Gr. *skambos*, bow-legged) *a b*.

P. plane, subdepressed, viscid in wet weather, whitish or faintly ochreous. *St.* stuffed, flocculose or mealy, colour as P.; base sometimes faintly rufescent. *G.* adnate, light clay-yellowish. *Flesh* brownish.

Gregarious. Suspected poisonous. Odour strong like **1527**. Woods, pine, rotten wood, larch, on the ground; uncommon. Sept.–Oct. 1⅜ × 1⅜ × ⅛ in.

XXVIII. **NAUCORIA** Quél.

(From the almost obsolete veil; *naucum*, a trifle.)

Veil squamulose, fugacious or obsolete. *Hymenophore* confluent with, but heterogeneous from the cartilaginous stem. *Pileus* more or less fleshy, convex, conical or convexo-plane, becoming flat, naked, flocculose, squamulose, silky or atomate; margin at first incurved, not usually striate. *Stem* normally central, cartilaginous, fistulose or spongy-stuffed, simple or imperfectly annulate. *Gills* free or adnate, not decurrent. *Spores* ferruginous. (Fig. 37.)

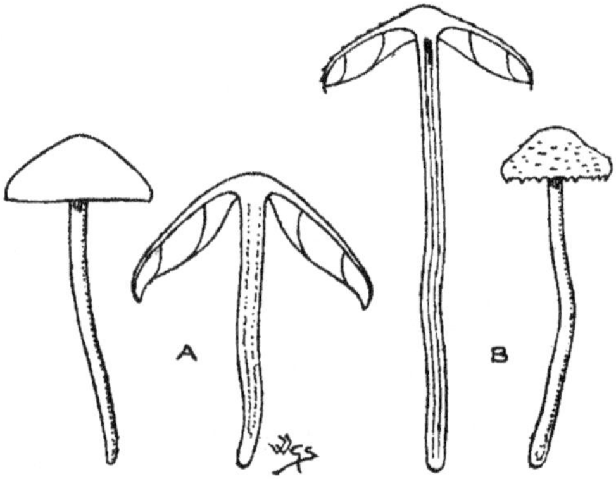

Fig. 37.—A, *Naucoria Cucumis* Gill.; B, *N. porriginosa* Karst. Entire and in section; one-half natural size.

The species grow on the ground or are epiphytal, sometimes rooting; they are mostly small in size and brown in colour.

Naucoria corresponds in structure with *Collybia*, *Leptonia*, and partially with *Psilocybe* and *Panæolus*. Species **703—746**

a. *Gymnotæ*. *Pileus* smooth. *Veil* obsolete, except **711** and **717**. *Spores* ferruginous, not becoming fuscous-ferruginous, except **712**.

Gills usually free or slightly adnexed. **703—716**
Pileus convexo-plane. *Gills* usually adnate. **717—721**
Pileus campanulate, then expanded, except **722** and **723**. *Gills* usually adnate. **722—728**

b. *Phæotæ*. *Pileus* naked. *Veil* potential, but rarely manifest; when visible, in the form of a fugacious cortina. *Gills* and *spores* fuscous-ferruginous.

Growing in fields and on plains, except **730**. **729—734**
Growing in moist uncultivated woody places. **735—738**

c. *Lepidotæ*. *Pileus* flocculose or squamulose. *Veil* manifest. *Spores* ferruginous, except **744**.

Pileus with separating, superficial squamules. **739, 740**
Pileus with innate squamules. **741—744**
Pileus without scales, silky or atomate. **745, 746**

a. Gymnotæ.

703. N. lugubris Sacc. (from its sombre colours; *lugeo*, to mourn) *a*.
P. expanded, gibbous or subumbonate, sometimes glutinous, ochreous-tan to ferruginous. *St.* fusiform-rooting, whitish above, ferruginous below. *G.* ferruginous.

Often cæspitose. Mountainous fir-woods. Autumn. $2 \times 3\frac{7}{8} \times \frac{3}{8}$ in. Resembling **229**.

704. N. festiva Bres. (*festivus*, handsome) *a b*.
P. expanded, umbonate, glutinous, brownish-olive, whitish when dry. *St.* attenuate downwards, buff-white. *G.* ventricose, crowded, salmon-ferruginous.

Woods, amongst dead leaves and grass. Sept. $1\frac{5}{8} \times 4 \times \frac{3}{16}$ in.

705. N. obtusa Sacc. (from the obtuse pileus) *a b*.
P. campanulate, sienna-reddish, becoming pale; marg. substriate. *St.* paler than P. *V.* obsolete. *G.* adnato-ascending, edge serrulate, colour as P. or St.

On the ground. Autumn. $1\frac{3}{8} \times 2\frac{1}{2} \times \frac{3}{16}$ in.

706. N. subglobosa Sacc. (from the shape of the pileus) *a c*.
P. subviscid, yellowish or pale brassy-yellow. *St.* striate, paler than P. *G.* ascending, ventricose, ochreous flesh-colour. *Spores* pale, pallid.

On the ground. Autumn. $1 \times 2\frac{1}{8} \times \frac{1}{8}$ in.

707. N. hamadryas Sacc. (from the habitat; Gr. *hamadruas*, a wood-nymph) *a b*.
P. hygrophanous, expanded, gibbous or umbonate, bay-brown-ferruginous, becoming buff or ochreous-tan. *St.* shining silky-fibrillose, whitish-tan, with slight A. low down in some young examples. *G.* sinuato-adnexed, crowded, edge eroded, ferruginous.

Woods, on the ground. Nov. $2\frac{1}{8} \times 3\frac{1}{4} \times \frac{1}{4}$ in.

708. N. cidaris Sacc. (from a fancied resemblance in the pileus to a tiara, Gr. *kidaris*) *a*.
P. campanulate, clay-cinnamon to tan. *St.* attenuate downwards, colour as P. above, deep sienna-blackish below. *G.* adnexo-ascending, honey-colour.

Odour none. Woods, pine. Nov. $2\frac{1}{8} \times 3\frac{1}{8} \times \frac{3}{16}$ in.

709. N. Cucumis Gill. (from its frequent odour of cucumber, *Cucumis*). *a b c*.
P. campanulate, deep chestnut-umber; marg. tan-buff; or wholly tan-buff. *St.* attenuate downwards, bay or rich purple-brown. *G.* adnexo-ascending, or strongly sinuate, ventricose, tan-buff. *Flesh* purple-brown.

Odour strong, rank and penetrating of stale or rotten fish, sometimes of cucumber or a combination of fish and cucumber as in the odour of the smelt. Woods, fir, grassy places, gardens, sawdust, fragments of wood; frequent. Sept.–Nov. $2\frac{1}{4} \times 2\frac{3}{8} \times \frac{3}{16}$ in. See **534**.

710. N. ECHINOSPORA Sacc. (from the spinulose spores; *echinus*, a hedgehog) *a b c*.

P. flat, subumbonate, moist hygrophanous, slightly furfuraceous, buff, then pale; marg. substriate. *St.* brownish-salmon, paler above, rufescent below, white-flocculose. *G.* sinuate, subdistant, ochre, olive-shaded.

Greenhouses. Aug. $\frac{1}{2} \times \frac{7}{8} \times \frac{1}{16}$ in.

711. N. anguinea Sacc. (from the stem, spotted like a snake, *anguis*) *a b*.

P. expanded, gibbous, ochreous to tan; marg. white-zoned with patches of V. *St.* attenuate upwards, tan above, bay-brown below, white-fibrillose with V. *G.* adnexo-ascending, pale ferruginous. *Flesh* brownish.

Odour none. On the ground. May–Nov. $2\frac{1}{2} \times 3\frac{3}{4} \times \frac{3}{16}$ in.

712. N. centuncula Gill. (from the change of colour in the pileus; *cento*, patchwork) *a*.

P. plane, often excentric, fuscous-olive to light ochreous-greenish or ashy. *St.* attenuate upwards, ashy-light-yellowish. *G.* adnate, thick, colour as St. *Spores* ochraceous.

Gregarious or cæspitose. Rotten wood, beech; rare. Oct. $1 \times 1\frac{3}{8} \times \frac{1}{8}$ in.

713. N. horizontalis Quél. (from the partially horizontal stem) *a b c*. Watery-cinnamon or rufescent.

P. convexo-plane, obtuse, subexcentric, splitting. *St.* curved. *G.* rounded-free, broad.

In troops. Branches, logs, felled trees, elm; rare. Dec. $\frac{3}{4} \times \frac{1}{2} \times \frac{1}{8}$ in.

714. N. rimulincola Sacc. (from its growing in small cracks of bark; *rimula*, a small crack, *incolo*, to inhabit) *a c*.

P. hemispherical, umbilicate, plicate, subexcentric, deep cinnamon. *St.* curved, colour as P. *G.* broad, crenulate, whitish-cinnamon. *Spores* cinnamon.

Twigs, branches, elm, pear. Oct.–Dec. $\frac{1}{4} \times \frac{1}{2} \times \frac{1}{16}$ in.

715. N. semiflexa Sacc. (from the slightly curved stem; *semi*, half, *flecto*, to bend) *a*.

P. convex, subexcentric, chestnut. *St.* tan. *G.* adnexed, distant, tawny.

On the ground, on branches. Oct. $\frac{1}{2} \times \frac{1}{4} \times \frac{1}{16}$ in.

716. N. rubricata Sacc. (from the white pileus tinged with red; *rubrico*, to colour red) *a*.

P. expanded. *St.* tan. *G.* adnexed, whitish to brownish. *Spores* pale.

Twigs, bramble. Autumn. $\frac{1}{4} \times \frac{1}{2} \times \frac{1}{32}$ in.

717. N. abstrusa Sacc. (from its secluded habitat; *abstrusus*, hidden) *a*.

P. viscid, ferruginous clay-colour; mid. brown. *St.* lustrous, colour as P., darker at base. *G.* crowded, cinnamon. *Flesh* pallid ferruginous.

On the ground, woods, sawdust. Oct. $1\frac{1}{4} \times 2\frac{1}{4} \times \frac{1}{8}$ in.

718. N. innocua Sacc. (*innocuus*, harmless) *a*.

P. striate, pale ochreous-rufous to rufous, pale when dry. *St.* tan-flesh-colour, white-fibrillose, base woolly. *G.* light yellow-ochreous. *Spores* rubiginous.

Damp places. Autumn. $1\frac{1}{8} \times 1\frac{1}{8} \times \frac{1}{8}$ in.

719. N. cerodes Quél. (from its waxy appearance; Gr. *keros*, wax) *a c*.

P. depressed, striate, watery-cinnamon to tan. *St.* pallid above, brown below. *G.* adnate or adnexed, cinnamon.

Gregarious. Woods, amongst moss, on the ground, on burnt earth. May–Sept. $1\frac{3}{8} \times 1\frac{3}{8} \times \frac{1}{8}$ in.

720. N. melinoides Quél. (from its resemblance to honey in colour; Gr. *mel*, honey, *eidos*, appearance) *a b c*.

P. obtuse, striate, shining, ochreous to sienna-ochreous. *St.* paler than P., whitish at top and bottom. *G.* crowded, somewhat tawny.

Lawns, pastures, roadsides. June–Nov. $1\frac{1}{8} \times 1\frac{3}{8} \times \frac{1}{16}$ in.

721. N. pusiola Gill. (from its small size; *pusus*, a little boy) *a*.

P. subviscid, shining, yellow, tawny or lemon. *St.* subviscid, lemon. *G.* crowded, watery-cinnamon.

On the ground, amongst moss, grass. Autumn. $\frac{1}{2} \times 1\frac{1}{2} \times \frac{1}{32}$ in.

722. N. nucea Sacc. (from the shape and colour; *nux*, a nut) *a*.

P. globose, lobed, umbilicate, pale chestnut; marg. strongly incurved. *St.* bulbous, silky-fibrillose, then even, white. *G.* adnexo-free, often waved, cinnamon.

Woods, pine, amongst heath and furze; rare. Autumn. $\frac{3}{4} \times 3\frac{1}{2} \times \frac{1}{16}$ in.

723. N. glandiformis Sacc. (from the shape of the pileus; *glans*, an acorn) *a b*.

P. at length hemispherical, obtuse or slightly depressed, lobed, tan, sienna-shaded; marg. strongly incurved. *St.* equal, striate, twisted, splitting, pale brownish-tan. *G.* adnexo-ascending, broad, crowded, serrate, whitish-umber.

On the ground. Oct. $1\frac{1}{8} \times 3\frac{1}{8} \times \frac{1}{8}$ in.

724. N. scolecina Quél. (from the flexuous stem; Gr. *skolex*, a worm) *a b*.

P. sometimes umbonate, bay-brown-ferruginous; paler at the striate marg. *St.* whitish above, fuscous below, at first wholly white-mealy. *G.* subdistant, ferruginous.

Dead leaves, moist ground under alders. Sept. $\frac{3}{4} \times 1\frac{3}{4} \times \frac{1}{16}$ in.

725. N. striæpes Sacc. (from the striate stem; *pes*, a foot) *a b*.

P. obtuse, sometimes rugulose or lacunose, ochreous. *St.* brittle, white. *G.* adnexo-ascending, serrate, crowded, tawny-ferruginous.

Gregarious or cæspitose. Lawns, amongst grass on soil-heaps. Nov. $1\frac{5}{8} \times 3\frac{3}{8} \times \frac{3}{16}$. Perhaps a var. of 723.

726. N. sideroides Gill. (from the colour of the stem, like oxide of iron; Gr. *sideros*, iron, *eidos*, appearance) *a c*.

P. subumbonate, subviscid, pale yellowish to cinnamon; marg. slightly striate. *St*. slightly attenuate downwards, white above, becoming yellowish or ferruginous. *G*. with a tooth, somewhat cinnamon.

Stumps, trunks, chips, ash; rare. Oct.–Nov. $1\frac{1}{4} \times 2\frac{1}{2} \times \frac{1}{16}$ in.

727. N. badipes Sacc. (from the brown stem; *badius*, bay-brown, *pes*, a foot) *a b*.

P. umbonate, tan to cinnamon; mid. darker; marg. striate. *St*. ferruginous, fuscous-blackish below, base white-floccose. *G*. ventricose, pale yellowish-ferruginous.

Under larches. Oct.–Nov. $1\frac{1}{8} \times 3 \times \frac{1}{16}$ in.

728. N. nasuta Sacc. (from the shape of the umbo; *nasutus*, with a conspicuous nose) *a b*.

P. acutely umbonate, ochreous to citron- or greenish-ochreous; marg. striate; mid. orange or sienna. *St*. equal, flexuous, smooth, even, yellow, ferruginous below. *G*. broadly adnate, then rounded adnate, subcrowded, pallid, then brownish. *Flesh* thin, citron-lemon, ferruginous in St.

Gregarious. Swampy places under trees. Autumn. $1\frac{1}{4} \times 1\frac{7}{8} \times \frac{1}{16}$ in. Massee regards the British form as a distinct species, *N. hydrophila*.

b. Phæotæ.

729. N. vervacti Quél. (from its habitat; *vervactum*, fallow ground) *a c*.

P. plane, subviscid, light yellow; mid. darker. *St*. smooth, whitish. *G*. adnate, toothed, ferruginous-fuscous. *Flesh* white.

Pastures, gardens; uncommon. May–Nov. $1\frac{1}{8} \times 1\frac{3}{8} \times \frac{1}{8}$ in.

730. N. triscopus Sacc. (from the hair-like stem; Gr. *thrix*, a hair, *pous*, a foot) *a*.

P. convexo-plane, umbonate, bay-brown to ochreous. *St*. flexuous, ferruginous. *G*. adnate or adnexed, subcrowded, dark ferruginous.

Copses, rotten wood, old wood in cellars. Autumn. $\frac{1}{2} \times 1\frac{1}{8} \times \frac{1}{32}$ in.

731. N. pediades Quél. (from its frequent habitat; Gr. *pedion*, a plain) *a b c*.

P. plane, obtuse, rimoso-rivulose, yellow or somewhat buff. *St*. with a small basal bulb, colour as P., paler above and below. *G*. adnexed or adnate, subdistant, dull cinnamon. *Flesh* whitish.

Pastures, lawns, roadsides, etc.; frequent. July–Nov. $1\frac{1}{2} \times 2\frac{5}{8} \times \frac{1}{16}$ in.

732. N. arvalis Quél. (from its habitat; *arvum*, a cultivated field) *a b c*.

P. plane, subviscid, subrugose, tan-buff. *St*. enlarged downwards, rooting, or branched-rooting, colour as P., furnished with an arachnoid A., when young. *G*. adnexo-sinuate, subdistant, warm-umber.

Taste disagreeable. On the ground, sea-sands. July. $1\frac{1}{2} \times 1\frac{5}{8} \times \frac{1}{8}$ in. Rooting base $1\frac{1}{2}$ in.

733. **N. semiorbicularis** Quél. (from the hemispherical pileus; *semi*, half, *orbicularis*, round) *a b c.*

P. smooth, viscid, corrugate when dry, tawny-ferruginous to tan. *St.* pallid-ferruginous or tan, rarely with trace of A. *G.* adnate, subrounded or sinuate, ferruginous.

Pastures, roadsides, etc., examples in Brit. Mus. from interior of skull of a stranded whale; common. June–Nov. $1\frac{1}{8} \times 2\frac{3}{4} \times \frac{1}{8}$ in.

734. **N. tabacina** Gill. (from the colour of the pileus as of tobacco, *tabacum*) *a.* Bay-brown-fuscous.

P. plane, subobtuse, hygrophanous, traces of V. at marg. *St.* lighter above, darker below. *G.* adnate or adnexed.

Subcæspitose. Waysides, amongst short grass. Autumn. $\frac{3}{4} \times 1\frac{1}{4} \times \frac{1}{16}$ in.

735. **N. tenax** Gill. (*tenax*, tough) *a b c.*

P. expanded, obtusely umbonate, viscid, cinnamon or shaded olive, fuscous or pale fuscous-slate. *St.* adpressedly fibrillose, light-yellowish to slate-white. *G.* adnate, subcrowded, whitish-fuscous.

Fields, amongst sticks. Oct. $1 \times 2\frac{1}{4} \times \frac{1}{16}$ in.

736. **N. Myosotis** Quél. (from its growing in wet places with forget-me-not, *Myosotis*) *a b.*

P. expanded, umbonate, viscid, olivaceous or olive-brownish, becoming light; marg. striate, cortinate with white V. *St.* often flexuous, fibrillose, white above, brownish below, sometimes with a fugitive A. *G.* adnate with a tooth, subdistant, brown-ferruginous, with a serrate white edge.

Bogs with *Myosotis* and *Potentilla Comarum.* July–Aug. $2\frac{1}{4} \times 5\frac{5}{8} \times \frac{1}{4}$ in. Very variable.

737. **N. temulenta** Quél. (from the bibulous substance and flexuous stem; *temulentus*, drunken) *a b.*

P. expanded, subumbonate, moist, ferruginous or reddish, ochreous or whitish when dry; marg. striate. *St.* pale dull orange or reddish, white at base. *G.* adnate, subdistant, attenuate in front, reddish.

Woods, open moist places. Aug.–Sept. $1\frac{1}{2} \times 2\frac{1}{2} \times \frac{1}{8}$ in.

738. **N. latissima** Sacc. (from the very broad gills; *latissimus*, very broad) *a b c.*

P. hemispherical, umber, orange-sienna-shaded, or orange-yellow; mid. flat. *St.* greatly attenuate downwards, tan with a brownish base. *G.* rounded adnate, separating, greyish then brown.

Amongst grass. Sept. $1\frac{1}{8} \times 2\frac{1}{8} \times \frac{3}{16}$ in. Pileus sometimes $1\frac{3}{4}$ in. in diam. The mycelium forms an earthy ball at base of stem.

c. Lepidotæ.

739. **N. porriginosa** Karst. (from the scurfy pileus; *porriginosus*, full of scurf) *a b.*

P. expanded, obtuse or subumbonate, dull yellow, clad with saffron flocci; marg. appendiculate with V. *St.* silky, then smooth, whitish above and below; mid. ochreous. *G.* adnate, cinnamon. *Flesh* lemon-white.

Amongst twigs and rubbish. Oct. $2 \times 4 \times \frac{3}{16}$ in.

740. **N. sobria** Gill. (from its not being hygrophanous; *sobrius*, sober) *a c.*

P. convex, obtuse or subumbonate, honey-colour. *V.* appendiculate at marg., fugacious. *St.* tan above, ferruginous below, sometimes white spotted with V. *G.* adnate, paler than P.

Woods, mixed, on the ground. July–Oct. $\frac{7}{8} \times 1\frac{3}{8} \times \frac{1}{8}$ in. Var. *dispersa* Sacc. is half the size of type.

741. **N. erinacea** Gill. (from the bristly pileus; *erinaceus*, a hedgehog) *a b c.*

P. expanded, depressed or umbilicate, umber or sienna. *St.* shaggy, deep sienna, lighter above and below. *G.* adnate with a tooth, clay-umber.

Solitary. Taste insipid. Dead branches; rare. Jan.–Dec. $1 \times \frac{1}{2} \times \frac{1}{8}$ in. Often difficult to see, being exactly the colour of the bark-matrix to which it is frequently adpressed.

742. **N. siparia** Gill. (from the covering formed by the veil; *siparium*, a small curtain) *a.* Rufous-ferruginous, upper part of St. and G. somewhat paler.

P. convex, densely villoso-squamulose. *V.* appendiculate at marg. *St.* villoso-downy, except upper part. *G.* adnate.

Wood, earth, dead fern-stems, caddis-worm cases, twigs; rare. July–Oct. $\frac{1}{2} \times 1\frac{1}{8} \times \frac{1}{16}$ in. Often differs greatly in length of stem from **741**.

743. **N. conspersa** Quél. (from the besprinkled scales of the pileus; *conspersus*, besprinkled) *a b c.*

P. convex, hygrophanous, bay-brown or crimson-shaded. *St.* fibrillose, cinnamon or ochreous. *G.* adnate, crowded, dark cinnamon.

Gregarious. On the ground or amongst dead leaves and *Sphagnum* in woods; uncommon. Sept.–Oct. $\frac{7}{8} \times \frac{1}{2} \times \frac{1}{16}$ in.

744. **N. escharoides** Quél. (from the scurfy pileus; Gr. *eschara*, a scar, *eidos*, appearance) *a b.*

P. campanulate, obtuse, umbonate or umbilicate, tan or whitish; mid. ochreous or fuscous, with an evanescent white V. *St.* adpressedly fibrillose, then smooth, pallid or pale ochre to fuscous. *G.* adnate with a tooth or emarginato-free, somewhat cinnamon, with a pale edge.

Gregarious, subcæspitose. Bare damp ground, under alders. Aug.–Oct. $\frac{7}{8} \times 1\frac{1}{2} \times \frac{1}{16}$ in.

745. **N. carpophila** Quél. (from its growing on fruit, beech-mast, etc.; Gr. *karpos*, fruit, *philos*, loving) *a.*

P. convex, obtuse, furfuraceous, atomate, tan-whitish. *St.* furfuraceous, then naked, pallid. *G.* adnexo-rounded or adnate, ochreous.

Beech-mast, leaves, etc.; rare. Sept.–Oct. $\frac{3}{16} \times \frac{3}{4} \times \frac{1}{32}$ in.

746. N. graminicola Gill. (from its habitat; *gramen*, grass, *colo*, to inhabit) *a c.*

P. convex, sometimes papillate, shaggy-tomentose, fuscous to fawn-ochraceous. *St.* hirsute, fuscous. *G.* slightly adnexed, pallid ochreous.

Decaying rushes, straw, grass, twigs. Oct. $\frac{3}{8} \times 1\frac{1}{2} \times \frac{1}{32}$ in. The stem is rarely branched.

XXIX. GALERA Quél.

(From the shape of the pileus; *galerus*, a cap.)

Veil fibrillose or obsolete. *Hymenophore* confluent with, but heterogeneous from the cartilaginous stem. *Pileus* more or less membranous, conical or semi-ovate, sometimes expanding; margin striate, at first straight and adpressed to the stem. *Stem* central,

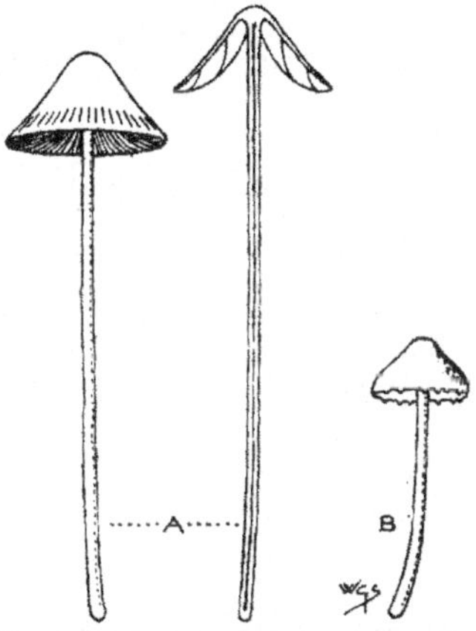

Fig. 38.—A, *Galera tenera* Quél., entire and in section. B, *G. ravida* Quél. One-half natural size.

somewhat cartilaginous and, except **752**, fistulose, simple or imperfectly annulate. *Gills* adnate or with a decurrent tooth, or adnexed and becoming almost free. *Spores* elliptical, smooth, ochreous-ferruginous or tawny-ochreous. (Fig. 38.)

The species are usually small, slender, tender and fragile; they grow on the ground or amongst moss. Mostly autumnal.

Galera agrees in structure with *Mycena*, *Nolanea*, *Psathyra* and *Psathyrella*. Species **746a—761** *a. Conocephalæ.* *Veil* obsolete. *Pileus* usually conico-campanulate, hygrophanous, somewhat even, when dry dotted with soft particles. *Stem* tense and straight. *Gills* ascending, inserted at the top of the cone, somewhat crowded. **746a—754**

b. Bryogenæ. *Veil* cortinate, fugacious. *Pileus* membranous, campanulate, striate, smooth, hygrophanous, even when dry, opaque, very slightly silky. *Stem* thin, lax, flexile, slender. *Gills* broadly and planely adnate, broad, somewhat denticulate. Habitat amongst moss except **758.** **755—759**

c. Eriodermeæ. *Veil* manifest, superficial, separating, at first chiefly round the margin, silky and squamulose. *Pileus* somewhat membranous. **760, 761**

a. Conocephalæ.

746a. G. apala Quél. (from its habitat, rich pastures, not marshes as in some allied plants; *a* without, *palus*, a marsh) *a c.*

P. obtusely campanulate, submembranous, smooth, hygrophanous, somewhat pale warm brown, or pallid-livid to white. *St.* hollow, attenuate upwards, fragile, velvety, white. *G.* almost free, somewhat narrow, ventricose, pale whitish-ochre.

In rich grassy places. Sept. $7\frac{1}{8} \times 2\frac{1}{4} \times \frac{3}{16}$ in. The form *Sphærobasis* has a smooth stem and bulbous base.

747. G. lateritia Quél. (from its reddish-brick-colour; *later*, a brick) *a b c.*

P. conico-cylindrical, membranous, pale yellowish to ochreous-brown; marg. striate when moist. *St.* shining-whitish or paler than P., white-pruinose. *G.* adnexo-ascending, cinnamon.

Rich pastures, grassy places; rare. June–Oct. $1 \times 4\frac{1}{2} \times \frac{1}{8}$ in.

748. G. tenera Quél. (*tener*, tender) *a b c.*

P. submembranous, deep buff or ferruginous to whitish; marg. striate when moist. *St.* shining, striate above, colour as P. *G.* adnato-ascending, cinnamon.

Pastures, gardens, flower-pots, roadsides, grassy places in woods; common. April–Dec. $1\frac{1}{4} \times 4\frac{1}{2} \times \frac{1}{8}$ in. Var. *pilosella* Karst. *P.* pubescent; rotten wood.

749. G. siliginea Quél. (from the colour, like bread; *siligineus*, wheaten) *a b.*

P. campanulate, smooth, membranous, whitish to sienna-bay; when dry whitish and atomate-sparkling; mid. sometimes brown. *St.* subpruinose, tan-whitish, lilac or purplish, darker below. *G.* adnato-ascending, subdistant, ochreous then bright cinnamon.

Taste none; odour strong. Amongst short grass, road scrapings. Oct. $1 \times 2\frac{1}{2} \times \frac{1}{16}$ in. *G. campanulata* Mass. is a form of this.

750. G. ovalis Gill. (from the semi-oval pileus) *a b c*.

P. ferruginous to dull yellowish. *St*. colour as P., at first imperfectly annulate. *G*. adnexo-ascending or free, ferruginous.

Amongst grass, on dung; uncommon. Sept.–Oct. $1\frac{5}{8} \times 4\frac{1}{4} \times \frac{1}{8}$ in.

751. G. antipus Quél. (from the sinuous, rooting base differing from the straight stem; Gr. *anti*, opposite, *pous*, a foot) *a b c*.

P. expanded, hygrophanous, atomate, deep ochreous to whitish; marg. paler. *St*. attenuate below, the rooting base long, tortuous, tan-ochreous, sometimes with an imperfect A. *G*. adnexo-ascending, cinnamon. *Spores* rubiginous.

Single. Gardens, dung, newly made lawn; rare. Mar.–Aug. $1\frac{3}{8} \times 2\frac{1}{4} \times \frac{1}{16}$ in. Rooting base 2 in.

752. G. CONFERTA Gill. (from the crowded habit; *confertus*, crowded) *a*.

P. somewhat acutely pointed, white; mid. yellowish-brown. *St*. solid, shining white. *V*. downy, fugitive. *G*. ascending, adnexed or free, crowded, pale brownish. *Spores* fuscous-ferruginous.

Cæspitose. Amongst tan in hothouses. Nov. Withering and becoming like soft paper in decay. Resembles 41 in habit.

752a. G. SPICULA Quél. (from the atomate pileus; *spiculum*, a little hard point) *b*.

P. membranous, conico-campanulate, then expanded, smooth, hygrophanous, striate when moist, flocculose and atomate when dry, brown-ochre. *St*. hollow, thickened below, densely white-floccose. *G*. adnate, ventricose, ochreous then cinnamon.

Coco-nut fibre. Nov. $\frac{3}{4} \times 1 \times \frac{1}{16}$ in.

753. G. spartea Quél. (from the rigid grass-like stem; *spartum*, esparto-grass) *a b*.

P. expanded, ferruginous-cinnamon, becoming paler; marg. striate when moist. *St*. smooth, whitish. *G*. adnate, darker than P.

Gregarious. Amongst short grass and moss. Sept. $\frac{1}{2} \times 2\frac{7}{8} \times \frac{1}{32}$ in. Sometimes wholly whitish-orange.

754. G. pygmæoaffinis Quél. (from its resembling *Naucoria pygmæa*, but differing in the elongate stem) *a b*.

P. flat, reticulato-wrinkled fuscous, honey-colour or tan. *St*. shining-white. *G*. adnexo-free, ferruginous-ochreous.

Tree-shaded walks, amongst grass, in a cucumber house, base of trunks. Nov. $1\frac{3}{8} \times 2\frac{1}{2} \times \frac{1}{16}$ in.

b. Bryogenæ.

755. G. vittæformis Quél. (from a fancied resemblance in the striate margin to a head-band or chaplet, *vitta*) *a*.

P. brown; mid. sometimes paler. *St*. bright rusty-brownish. *V*. almost obsolete. *G*. adnato-ascending, colour as St.

Amongst moss. Nov. $\frac{3}{4} \times 2\frac{3}{4} \times \frac{1}{16}$ in.

756. G. rubiginosa Gill. (from the dark ferruginous stem; *rubigo*, rust) *a c*.

P. wholly striate, cinnamon or honey-colour to tan; mid. darker. *G.* adnato-ascending, distant, ochreous.

Amongst moss. Sept.–Oct. $1\frac{1}{4} \times 2\frac{3}{4} \times \frac{1}{8}$ in.

757. G. hypnorum Quél. (from its growing on moss; *Hypnum*, a genus of mosses) *a b c*.

P. campanulato-expanded, obtuse or umbonate, pale ochre or pale sienna to tan; mid. darker. *St.* sienna below, paler and white-pruinose above. *G.* ventricose, distant, cinnamon-tawny.

Odour very strong, alkaline. Amongst mosses and hepatics in woods. May–Nov. $1 \times 4 \times \frac{1}{16}$ in. Var. *bryorum* Gill. Larger than type. Var. *sphagnorum* Quél. On *Sphagnum* in swamps.

758. G. mniophila Gill. (from its growing on moss; Gr. *mnion*, moss, *philos*, loving) *a*.

P. subpapillate, fuscous-light-yellowish to clay-colour. *St.* fibrillose, mealy above, floccose below, yellow. *G.* adnato-ascending, fuscous-clay-colour.

On moss. Sept.–Oct. $\frac{5}{8} \times 3\frac{1}{4} \times \frac{1}{16}$ in.

759. G. minuta Quél. (from its very small size) *a*.

P. tan-ochreous. *St.* tawny; base forming a small white disc. *G.* triangular, adnate, clay-colour, edge whitish.

Decayed wood, amongst moss. Sept.–Oct. $\frac{1}{8} \times \frac{5}{8} \times \frac{1}{32}$ in.

c. Eriodermeæ.

760. G. ravida Quél. (*ravus*, grey) *a*.

P. tan-greyish or tan-umber; marg. appendiculate with white V. *St.* pallid, white-silky. *G.* adnexed, ventricose, ochreous-salmon.

Gregarious. On the ground, amongst chips and rotten wood. Autumn. $1\frac{1}{4} \times 2\frac{1}{8} \times \frac{1}{8}$ in.

761. G. mycenopsis Quél. (from its resemblance to a *Mycena*; Gr. *opsis*, resemblance) *a*.

P. expanded, umbonate, pale honey-colour; marg. appendiculate with white V. *St.* white-yellowish, white silky. *G.* adnexed, ventricose, yellowish.

On *Sphagnum*, marshy ground. Aug.–Oct. $1\frac{1}{4} \times 4\frac{3}{8} \times \frac{1}{8}$ in. Somewhat like 350.

XXX. TUBARIA Gill.

(From the trumpet-shape of typical species; *tuba*, a trumpet.)

Veil universal, floccose at margin of pileus. *Hymenophore* confluent with, but heterogeneous from the cartilaginous stem. *Pileus* submembranaceous, sometimes umbilicate or depressed, margin—except

770—striate. *Stem* central, subcartilaginous, fistulose, simple or imperfectly annulate. *Gills* subdecurrent, broadest behind and

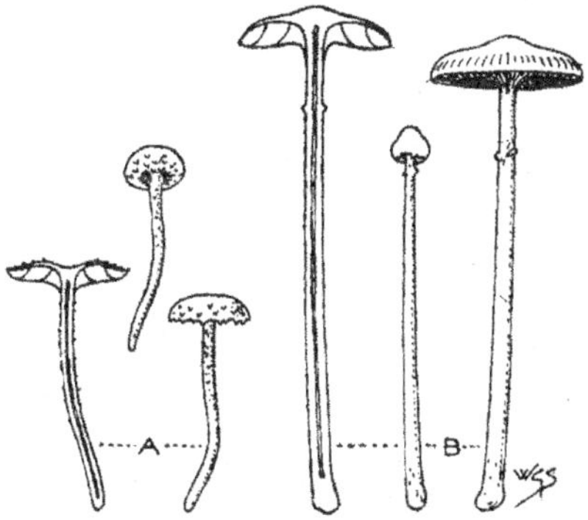

Fig. 39.—A, *Tubaria furfuracea* Gill.; B, *T. paludosa* Karst. Entire and in section. One-half natural size.

somewhat triangular. *Spores* ferruginous or fuscous ferruginous. (Fig. 39.)

The species are all small; some grow on the ground, others on mosses, twigs, chips, leaves, etc.

Tubaria agrees in its chief structural characters with *Omphalia* and *Eccilia*. Species **762—771**

a. Genuinæ. *Spores* ferruginous. **762—768**

b. Phæotæ. *Spores* fuscous ferruginous. **769—771**

a. Genuinæ.

762. T. cupularis Gill. (from the shape of the pileus; *cupularis*, cup-shaped) *a.*

P. plano-depressed, hygrophanous, smooth, rufescent to light yellowish. *St.* whitish or faintly ochreous. *G.* serrulate, reddish.

Mountainous heaths, amongst grass. Aug.–Sept. $1\frac{1}{8} \times 2 \times \frac{1}{16}$ in.

763. T. furfuracea Gill. (from the scurfy pileus; *furfur*, bran) *a b c.*

P. flat, umbilicate, cinnamon to tan-whitish; marg. substriate when moist, appendiculate with white-squamulose V. *St.* floccose, colour as P., villous at base. *G.* subdecurrent, bright cinnamon or clay.

Gregarious. Tasteless or pleasant. Thatch, chips, twigs, wood, shavings, Ivy, humus. Jan.–Dec. $1\frac{5}{8} \times 1\frac{3}{4} \times \frac{1}{8}$ in. Polymorphous. Var. *trigonophylla* Sacc. Smaller than type; *G.* triangular. Var. *heterosticha* Karst. *P.* depressed, umbonate.

764. **T. paludosa** Karst. (from its habitat; *palus*, a marsh) *a b*.
P. expanded, papillate, sometimes striate, pale yellowish-fuscous to whitish. *St.* colour as P., flocculose above, white-villous below. *G.* adnato-decurrent, paler than P.

Amongst *Sphagnum*. May–Sept. $1\frac{1}{4} \times 3\frac{3}{4} \times \frac{1}{8}$ in.

765. **T. stagnina** Gill. (from its habitat; *stagnum*, a swamp) *a b*.
P. expanded, brown, or somewhat pale and ochreous; marg. white-scaly with V. *St.* attenuate below, rubiginous. *G.* adnato-decurrent, broad, ferruginous.

Amongst *Sphagnum* in bogs. Sept. $1 \times 4 \times \frac{1}{16}$ in.

766. **T. pellucida** Gill. (from the translucent pileus; *pellucidus*, transparent) *a b*.
P. campanulate, umbonate, cinnamon to ochreous-tan; marg. striate, silky-squamulose with V. *St.* shining, paler than P., pruinose above. *G.* adnato-decurrent, subtriangular, colour as P., faintly sienna-tinted.

Roadsides, amongst leaves. Autumn. $\frac{7}{8} \times 1\frac{3}{8} \times \frac{1}{16}$ in.

767. **T. embolus** Karst. (from the wedge-shaped gills; Gr. *embolos*, a wedge) *a b c*.
P. expanded, obtuse, pale yellowish-tawny to ochreous-tan. *St.* attenuate downwards, smooth, naked, paler than P. *G.* thick, very distant, separating from hymenophore, bright tawny-cinnamon.

Damp heathy ground, amongst moss and grass; rare. June–Nov. $\frac{7}{8} \times 2 \times \frac{1}{16}$ in. Stem sometimes drawn up abnormally amongst grass.

768. **T. autochthona** Sacc. (from its terrestrial habit; Gr. *autos*, self, *chthon*, the earth) *a b c*.
P. flat, ochreous-white; marg. flocculose with V. *St.* paler than P., base white. *G.* adnato-toothed, honey-colour.

On the ground, amongst grass. July–Oct. $1 \times 1\frac{1}{8} \times \frac{1}{16}$ in.

b. Phæotæ.

769. **T. muscorum** Gill. (from its growing on moss, *muscus*) *a b*.
P. expanded, subdepressed, tawny-yellow; mid. sienna. *St.* colour as P. *G.* horizontal, adnate with a decurrent tooth, paler than P.

Amongst mosses on heaths, on tree-trunks. July. $\frac{7}{8} \times 2\frac{1}{8} \times \frac{1}{16}$ in.

770. **T. crobulus** Karst. (from a fancied resemblance in the pileus to a knot of hair on the crown of the head, Gr. *krobulos*) *a*.
P. expanded, obtuse, sienna- to whitish-tan, white-floccose with V., then naked. *St.* white-floccose, sometimes subannulate. *G.* adnato-decurrent, crowded, fuscous-ferruginous.

Amongst sticks, chips. Sept.–Oct. $1 \times 1\frac{1}{4} \times \frac{1}{8}$ in.

771. **T. inquilina** Gill. (from its growing on other substances; *inquilinus*, a lodger) *a b c*.
P. plane, depressed, broadly umbonate, viscid when moist, yellowish-sienna, becoming pale; marg. striate, sienna. *St.*

attenuate downwards, paler than P., white-fibrillose, base white-floccose. *G.* adnato-decurrent, subdistant, umber.

Gregarious. Grass, ferns, stalks, chips; common. May–Oct. $\frac{7}{8} \times 1\frac{1}{8} \times \frac{1}{16}$ in. Must not be confounded with **858** or **860**. Var. *ecbola* Karst. Stem rooting; roots of grasses.

XXXI. CREPIDOTUS Quél.

(From the shape of some of the species; Gr. *krepis*, a slipper.)

Veil obsolete. *Hymenophore* confluent and homogeneous with the fleshy stem. *Pileus* excentric, lateral or resupinate, commonly thin and with soft flesh. *Stem* lateral, simple or absent. *Gills*

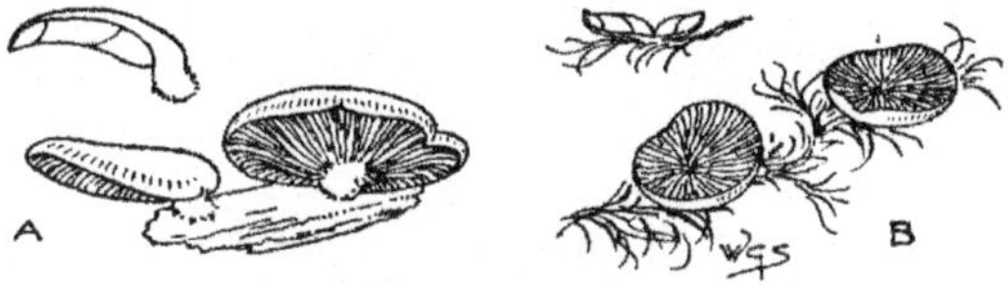

Fig. 40.—A, *Crepidotus applanatus* Karst. One-half natural size. B, *C. epibryus* Quél. Twice natural size. Entire and in section.

more or less decurrent, reaching the stem or axis, or radiating from an excentric point. *Spores* pale ferruginous or cinnamon. (Fig. 40.)

The species, except **784**, grow on wood, trunks, twigs, grass, moss, etc., they are mostly small in size.

Crepidotus agrees in structure with *Pleurotus* and *Claudopus*.

Species **772—785**

772. C. palmatus Gill. (from a fancied resemblance in the pileus to the palm of the hand) *a c.*

P. fleshy, convex, expanded, smooth, ferruginous to warm brownish-buff. *St.* solid, excentric or lateral, curved, smooth, whitish. *G.* rounded, subdistant, brownish-buff.

Trunks. Oct. $5 \times 1\frac{3}{4} \times \frac{1}{2}$ in.

773. C. alveolus Quél. (from a fancied resemblance, when inverted, to a little trough; *alveus*, a trough) *a c.*

P. obovate, dimidiate, ochreous-fuscous. *St.* a villous base. *G.* subdecurrent on base, clay-fuscous.

Stumps, oak; rare. Aug.–Nov. Diam. 2 in.

774. C. mollis Quél. (*mollis*, soft) *a b c.*

P. obovate, dimidiate, pale yellowish-tan; marg. at first incurved, then undulate. *St.* a strigose base. *G.* rounded on base, sometimes broad, cinnamon.

Stumps, poplar, willow, pear, sticks, logs, sawdust, squared timber, humus; common. May–Dec. Diam. $2\frac{3}{4}$ in.

775. C. applanatus Karst. (from the plane pileus) *a c*.

P. reniform, watery, pale cinnamon to whitish; marg. substriate when moist. *St*. when present white-tomentose. *G*. subrounded, watery-cinnamon or ochreous.

Decaying wood. Oct. $1\frac{1}{4} \times \frac{1}{4} \times \frac{1}{4}$ in. Much like **1444** in appearance.

776. C. calolepis Karst. (from the beautiful scales of the pileus; Gr. *kalos*, beautiful, *lepis*, a scale) *a b*.

P. reniform, pale dull-ochreous, rufescent-scaly. *St*. a small villous base. *G*. rounded on base, pale brown then fuscous-ferruginous.

Dead wood, poplar. Oct. Diam. $1\frac{3}{8}$ in., sometimes $2\frac{3}{4}$ in.

777. C. haustellaris Karst. (from its watery substance; *haustellaris*, a drawer of water) *a*.

P. obovate or reniform, pale yellowish-tan, then paler. *St*. excentric, slightly attenuate upwards, white. *G*. rounded, adnexed, fuscous-cinnamon.

Gregarious. Dead trunks, poplar; rare. $1 \times \frac{3}{8} \times \frac{1}{8}$ in.

778. C. Rubi (from its habitat; *Rubus*, bramble) *a c*.

P. obovate or reniform, crystalline-mealy, yellow to livid-grey, then pallid. *St*. excentric, whitish, strigose at base. *G*. subdecurrent, brown. *Spores* brown.

Gregarious; uncommon. July–Oct. $\frac{5}{16} \times \frac{1}{8} \times \frac{1}{16}$ in.

779. C. chimonophilus Sacc. (from its time of growth; Gr. *cheimon*, winter, *philos*, loving) *a*.

P. circular, convex, white, villous-downy. *St*. excentric, almost obsolete, white. *G*. subdecurrent on base, few, distant, pale yellow-brown. *Spores* pale yellow-brown.

Dead leaves, *Pyrus torminalis*, logs. Dec. Diam. $\frac{1}{4}$ in.

780. C. epibryus Quél. (from its frequent habitat, moss; Gr. *epi*, upon, *bruon*, moss) *a c*.

P. resupinate, at first cup-like, sessile, membranous, white. *G*. rounded to a central or excentric axis, pale yellowish.

Mosses, grass, leaves, holly, *Vaccinium*. Oct. Diam. $\frac{3}{16}$ in.

781. C. Phillipsii Sacc. (after William Phillips, mycologist) *a b c*.

P. ovate, expanded, smooth, pale umber; marg. striate. *St*. curved, colour as P. *G*. rounded-adnate, paler than P.

Dead grass. May–Oct. $\frac{1}{4} \times \frac{1}{8} \times \frac{1}{32}$ in.

782. C. proboscidens Gill. (from the trunk-like stem; Gr. *proboskis*, a snout) *a b*.

P. obovate, pendulous, scabrous-spotted, pale sepia; marg. incurved. *St*. attenuate to base, inserted on and darker than P., shaded umber-sienna. *G*. inferior, adnate, rounded to an excentric axis, flesh-colour.

Pendulous from a wooden ceiling. Dec. $\frac{5}{8} \times \frac{5}{8} \times \frac{1}{16}$ in.

783. C. pezizoides Gill. (from its resemblance to certain cup-fungi of the genus *Peziza*) *a*.

P. resupinate, saucer-shaped, then reflexed, sessile, mealy, whitish. *G*. radiating from a central axis, tawny.

Gregarious. Rotten branches; rare. Diam. $\frac{1}{4}$ in.

784. C. epigæus Sacc. (from its growing on the ground; Gr. *epigaios*, upon the earth) *a*.

P. subreniform, becoming horizontal, fragile, reddish-grey. *St*. a villous lateral base. *G*. rounded on base, watery, rufescent.

On marlstone clay. Nov. Diam. $1\frac{1}{4}$ in.

785. C. Ralfsii Sacc. (after J. Ralfs) *a b c*.

P. reniform, horizontal, slightly hispid-furfuraceous, yellow. *St*. almost obsolete. *G*. rounded to a lateral base, clay-colour, edge whitish.

Decaying wood, dead branches. March. Diam. $\frac{1}{2}$ in.

785a. C. putrigenus Sacc. (from its habitat, rotten logs; *puter*, putrid, *gigno*, to bear) *a c*.

P. membranous, dimidiate or reniform, pubescent, whitish; marg. substriate, base villose. *G*. broad, irregular, unequal, whitish, then coloured by the ferruginous-brown spores.

Gregarious, or imbricate, in close clusters, closely sessile. Damp or putrid logs, beech. Sept. P. $\frac{3}{4} \times \frac{3}{8}$ in.

Series IV. PRATELLÆ Fr.

(From the habitat; *pratum*, a meadow.)

Spores typically black-purple or fuscous-purple, often fuscous, brown-black or dark brown. Genera XXXII—XXXVIII.

XXXII. CHITONIA Karst.

(From the veil which in infancy covers the plant like a cloak; Gr. *chiton*, a tunic.)

Veil universal, separate from the pileus and remaining as a volva at the base of the stem. *Hymenophore* distinct from the fleshy

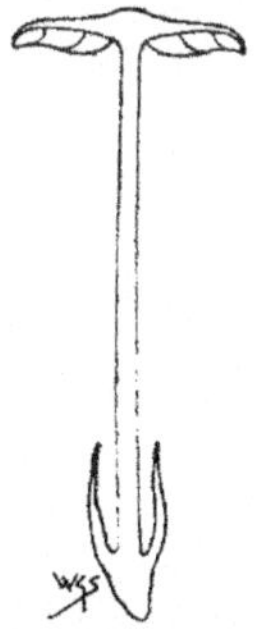

Fig. 41.—Section of *Chitonia rubriceps* Sacc.
One-half natural size.

stem. *Pileus* subfleshy. *Stem* central, simple, hollow, volvate. *Gills* free. *Spores* fuscous-purple. (Fig. 41.)

Represented only by one introduced species.

Chitonia agrees in structure with *Amanitopsis*, *Volvaria* and *Acetabularia*. Species **786**

786. C. RUBRICEPS Sacc. (from the red pileus; *ruber*, red, *caput*, the head) *a*.

P. expanded, umbonate, smooth, even, brick-red. *St*. rooting, dull pale reddish-ochreous, paler below. *Vo*. large, sheathing, saccate, torn at edge. *G*. subcrowded, purplish-brown.

Soil in an aroid house. Dec. $1\frac{3}{8} \times 3\frac{1}{2} \times \frac{1}{8}$ in.

XXXIII. PSALLIOTA Quél.

(From the annulus; Gr. *psalion*, a ring.)

Veil universal, concrete with the cuticle of the pileus and forming an annulus on the stem. *Hymenophore* distinct from the fleshy stem. *Pileus* fleshy. *Stem* central, fibrous-fleshy, annulate. *Gills*

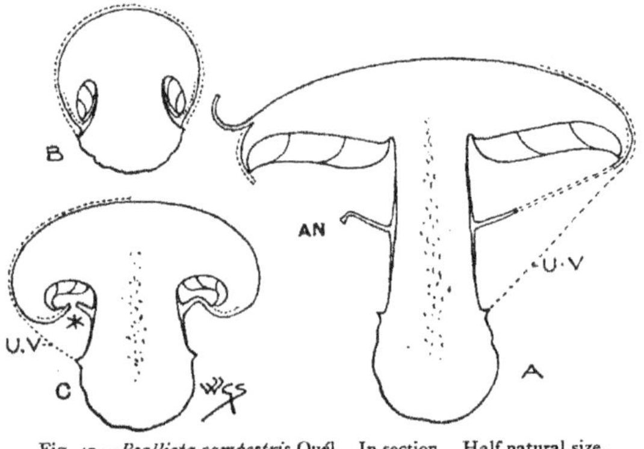

Fig. 42.—*Psalliota campestris* Quél. In section. Half natural size. A, full grown example; AN, annulus; UV, universal veil; B, young example, showing universal veil; C, half-grown example, showing rupture of veil at *.

free, at first whitish, then frequently salmon-colour, at length deep purple-brown. *Spores* dark brownish-purple, brown or reddish-purple. (Fig. 42.)

All the species grow on the ground and are late æstival or autumnal, rarely vernal; they mostly grow in rich pastures and on manured ground, sometimes in woods; a non-British species, *Agaricus geniculatus* Briz., grows on wood.

Several species are valued for their esculent qualities.

Psalliota agrees in structure with *Lepiota*, *Annularia* and *Togaria*. Species **787—800**

a. *Edules*. Larger, fleshy. Edible. **787—794**

b. *Minores*. Usually smaller, thinly fleshy. Not used as food. **795—800** *a. Edules.*

787. P. augusta Quél. (from its appearance; *augustus*, noble) *a b.*
P. expanded, fibrilloso-squamose, rich ochreous-brownish; marg. almost white, sometimes appendiculate with V. *St.* solid, enlarged below, tan, faintly sienna below. *G.* crowded, faintly brownish.

An inferior esculent. Orchards, woods, under spruce-firs, on ant-hills. Aug.–Sept. 7 × 6½ × 1¼ in. Might when young be mistaken for a *Lepiota*.

788. P. Elvensis Henn. (from its place of growth when first named, near the River Elwy, N. Wales, 1863) *a b.*
P. expanded, fibrilloso-scaly, warm brown with darker scales. *St.* solid, equal or attenuate downwards, fibrillose, paler than P. *A.* large. *G.* brownish flesh-colour. *Flesh* changing to red or pale rose when cut.

Sometimes cæspitose. Under oaks, beech, fir. Aug.–Sept. 6½ × 5 × 1½ in. A drawing in Brit. Mus. is dated 1847.

789. P. arvensis Quél. (from its habitat; *arvum*, a cultivated field) *a b c.*
P. at length flat, floccose, silky or squamulose, whitish; mid. faintly yellowish; marg. ragged with V. *St.* hollow, attenuate upwards, even, smooth, white. *A.* large. *G.* ventricose, at length dark purple-brown. *Flesh* white, changing to pale yellowish or brownish.

Spring to winter. 7¾ × 4½ × 1¼ in. The cultivated mushrooms of gardeners belong to this species. There are several named varieties. 1. *purpurascens* W. G. Sm., *Agaricus arvensis* Schaeff., var. *purpurascens* Cooke, *P.* purple. 2. *albosquamosa* W. G. Sm., *P.* white-scaly. 3. *vaporaria* W. G. Sm., *A. vaporarius* Otto, *P.* even, brown-pilose. 4. *hortensis* W. G. Sm., *A. campestris* L., var. *hortensis* Cooke, *P.* brownish fibrillose or squamulose. 5. *Buchananii* W. G. Sm., *A. campestris* L., var. *Buchananii* Berk., *P.* smooth as if whitewashed, *A.* slight, fugacious. 6. *cryptarum* W. G. Sm., *A. cryptarum* Letell, the name applied to the form grown in caves, grey-white or brownish-white.

789a. P. perrara Magnus (*perrarus*, very rare) *a b.*
P. fleshy, globoso-hemispherical, then expanded, scaly, whitish-ochre; sc. concentric, floccose, brownish-yellow to sienna. *St.* stuffed, then hollow, equal or attenuate upwards and sub-marginately-bulbous, smooth above, squamose below, white to pale yellowish or salmon-shaded; sc. salmon-brown. *A.* large, superior, membranous, floccose, persistent. *G.* free to remote, crowded, whitish, then salmon to full brownish-slate or fuscous. *Flesh* white, becoming yellowish, faint reddish-ochre or salmon when broken.

Taste somewhat sweet; odour somewhat of anise. Damp ground, under oaks, near conifers. Aug. 5½ × 4 × ⅝ in. Analogous with **21**, **33** and **792a.**

789b. P. xanthoderma W. G. Sm., *Agaricus xanthodermus* Genev. (from the yellow-stained outer surface and flesh; Gr. *xanthos*, yellow, *derma*, skin) *a.*

P. fleshy, convex, even, greyish-white, becoming yellowish when touched. *St.* stuffed to hollow, white, yellow at base. *A.* membranous. *G.* sinuato-free, pale rose, becoming purple-brown. *Flesh* becoming yellow.

Pastures. Autumn. 3½ × 4 × ¾ in.

790. P. cretacea Quél. (from its chalky-white colour; *creta*, chalk) *a b c.*

P. convexo-flattened, obtuse; marg. even, rivulose or squamulose. *St.* hollow, attenuate upwards, smooth. *G.* becoming fuscous. *Flesh* unchanging white.

Pastures, stoves; rare. Feb.–Oct. 3¾ × 4 × ⅝ in. Resembling **40.**

791. P. pratensis Quél. (from its habitat; *pratum*, a meadow) *a c.*

P. expanded, obtuse, floccoso-farinose, then smooth, whitish, then greyish. *St.* hollow, colour as P. *G.* arcuate, brown.

Taste and odour pleasant. Pastures, woods. Oct. 3½ × 3 × ⅝ in.

792. P. campestris Quél. (from its habitat; *campus*, a field or plain) *a b c.*

P. expanded, silky-even, sometimes squamulose, whitish. *St.* stuffed, even or squamulose, whitish. *G.* crowded, bright salmon, then umber-fuscous. *Flesh* white.

Odour pleasant. May–Dec. 5 × 3¾ × 1 in. Variable in size and colour. The common edible mushroom. Sometimes the flesh becomes faintly reddish or brownish when cut or broken. Sometimes the odour is strong and somewhat unpleasant. The following varieties have been described:—1. *alba* W. G. Sm., *Agaricus campestris* L. var. *albus* Viv., *P.* silky-white; *St.* short. 2. *praticola* Magnus, *P.* rufous-squamulose; flesh rufescent when cut. 3. *rufescens* W. G. Sm., *A. campestris* L., var. *rufescens* Fr., *P.* rufous-squamulose, *St.* long. 4. *umbrina* Magnus, *P.* at length even; umber. 5. *fulvastera* W. G. Sm., *A. campestris* L., var. *fulvaster* Viv., *P.* ochreous-tawny; *St.* solid. 6. *costata* W. G. Sm., *A. campestris* L., var. *costatus* Viv., *P.* sulcate. 7. *silvicola* Magnus, *P.* smooth, shining; *St.* elongate, somewhat bulbous, hollow; *A.* large. 8. *elongata* W. G. Sm., *A. campestris* L., var. *elongatus* Berk., *P.* at length even, shining-white; marg. permanently appendiculate with veil; *St.* elongate, bulbous. 9. *exannulata* W. G. Sm., *A. campestris* L., var *exannulatus* Cooke, *A.* fugitive or obsolete. The gills of cultivated mushrooms are frequently thickened or obliterated by disease; this is caused by a white parasitic fungus, *Mycogone perniciosa.* Diseased examples are not fit for food. Sound examples of true mushrooms sometimes cause diarrhœa and sweating.

792a. P. villatica Magnus (from its frequent habitat, near houses in the country; *villa*, a country house) *a b.*

P. convex then expanded, silky, soon squamose, ground pale yellowish or brownish straw-colour then vinous-brownish; sc. vinous-brown. *St.* solid, somewhat ventricose, paler than P., clouded pale vinous-brown. *A.* superior, large, tomentose beneath. *G.* free, very crowded, not at first bright salmon, becoming vinous-brown. *Flesh* white, pale reddish-yellow when broken.

Edible. Taste of almonds, odour somewhat rank. Rich pastures, lawns, manured fields, in cellars, on mushroom-beds in and out of doors; rare. 1ft. 2 in. × 7½ × 3½ in. This is the "gigantic mushroom" of newspapers; it attains a diam. of 1 ft. 5 in. or more and a weight of 4½ lbs. By some authors considered a var. of **792** or **789**, but probably distinct. Badham, "Esculent Funguses of England," p. 100, mentions an example with a circumference of 43 ins., and weight of 5 lbs. 6 ozs.

793. P. silvatica Quél. (from the habitat; *silva*, a wood) *a b*.

P. at length flat, subumbonate, floccoso-squamulose, pale rusty-brown or rufescent to pale ashy-brown. *St.* fibrillose below A., much paler than P. or white. *G.* thin, arid, reddish to umber-fuscous.

Edible, with a delicate flavour. Woods, under cedars; frequent. July–Sept. 3⅝ × 3¼ × ⅝ in. Usually thinner than its allies. *St.* varying long and short.

794. P. hæmorrhoidaria Magnus (from the flesh changing colour to blood-red when bruised; Gr. *haimorrhoides*, veins which discharge blood) *a b*.

P. at length expanded, obtuse, adpressed-scaly, rufous-fuscous. *St.* hollow, fibrillose, slightly attenuate upwards, whitish. *G.* crowded, bright salmon to umber purple.

Suspected sometimes poisonous. Woods, parks. Aug.–Nov. 4⅝ × 4⅜ × ⅝ in.

794a. P. peronata W. G. Sm., *Agaricus peronatus* Mass. (from the scaly stem; *perônatus*, sheathed) *a*.

P. hemispherical then expanded, pale dull ochreous, densely scaly; sc. small, silky, brown. *St.* hollow, equal, marginately bulbous, smooth above A., white-scaly below. *A.* large, spreading. *G.* very distant, crowded, pink then pale purple-brown. *Flesh* white, brownish in St. when cut.

Edible. Pine-woods. Oct. 3½ × 6¼ × ½ in. Attaining a size of 4½ × 8 in. Allied to **787** (but differing in the hollow, peronate stem) and **794**.

b. Minores.

795. P. comtula Quél. (from the beautifully coloured gills; *comptus*, gaily adorned) *a b*.

P. flat, adpressedly fibrilloso-scaly, yellowish-white; marg. appendiculate with V. *St.* hollow, colour as P. or paler. *A.* fugacious. *G.* crowded, fuscous-flesh-colour. *Flesh* colour as P.

Edible. Taste and smell strong of anise. Woods. May–Oct. 1½ × 1⅝ × $\frac{3}{16}$ in. Might be mistaken for a small **792**.

796. P. subgibbosa Henn. (from the somewhat obtuse pileus; *subgibbosus*, somewhat gibbous) *a c*.

P. plane, smooth, sienna-tan; marg. silky-fibrillose. *St.* hollow, tan-white. *A.* fugacious. *G.* ashy-brown.

Woods. Sept. 4 × 3¼ × ½ in. Might be mistaken for **792**.

797. P. sagata Henn. (from the pelliculose pileus, clothed with a cloak ; *saga*, a mantle) *a b*.

P. plane, smooth, bright yellowish-brown. *St*. hollow, colour as P., ochreous or reddish. *A*. distant, white. *G*. pale purple-brown or brown-salmon.

Taste and odour of **792**. Under beeches. Oct. $2\frac{3}{8} \times 2\frac{5}{8} \times \frac{1}{4}$ in.

798. P. rusiophylla W. G. Sm., *Agaricus rusiophyllus* Lasch. (from the rose-coloured or reddish gills ; *ruseus*, reddish, Gr. *phullon*, a leaf) *a b*.

P. somewhat fleshy, convex then plane, subumbonate, fibrillose, pale flesh-colour or pale purple, becoming pallid. *St*. imperfectly hollow, attenuate upwards, silky fibrillose, white, shaded faint ochre. *A*. persistent, drooping. *G*. free, crowded, thin, dry, rosy, then pale purple, pinkish-sienna and brownish. *Flesh* whitish, unchangeable.

On the ground. Autumn. $2\frac{1}{8} \times 2\frac{1}{8} \times \frac{1}{4}$ in.

799. P. dulcidula Henn. (from its taste ; *dulcis*, sweet) *a*.

P. plane, subgibbous, glabrous, dry, whitish to pale ochreous. *St*. hollow, sometimes subbulbous, colour as P. *A*. persistent. *G*. crowded, slate-colour.

Odour mild. Heaths. Oct. $2\frac{1}{2} \times 2 \times \frac{1}{4}$ in. Might be mistaken for **792**.

800. P. echinata Quél. (from the rough pileus ; *echinus*, a hedgehog) *a b c*.

P. expanded, obtuse, floccoso-pulverulent, dark brown ; mid. darker ; marg. appendiculate with V. *St*. fistulose, usually attenuate downwards, floccose, light or dark dull-brownish, crimson, purple or grey-brown. *G*. crowded, rose, cinnamon-crimson, crimson or purple. *Flesh* brownish or reddish. *Spores* very pale whitish-dull-purple.

Gregarious or cæspitose. Odour strong and disagreeable, mephitic or like cucumber or like **35**. Woods, beech, greenhouses. May–Nov. $1\frac{7}{8} \times 3 \times \frac{1}{4}$ in. Springing from a compact ball of earth or mycelium. Becomes jet-black when dry. The vesiculose meal on the pileus is transparent yellow. Sometimes placed in *Inocybe*, but retained here on account of the colour of the spores when fresh.

XXXIV. **PILOSACE** Quél.

(From the pelliculose pileus ; Gr. *pilos*, felt, *sakos*, a garment.)

Veil obsolete. *Hymenophore* distinct from the fleshy stem. *Pileus* fleshy. *Stem* central, simple, without cartilaginous bark. *Gills* free or remote. *Spores* purple-brown. (Fig. 43.)

The only British species grows on the ground.

Pilosace agrees in structure with the fleshy species of *Hiatula*, *Pluteus* and *Pluteolus*. Species **801**

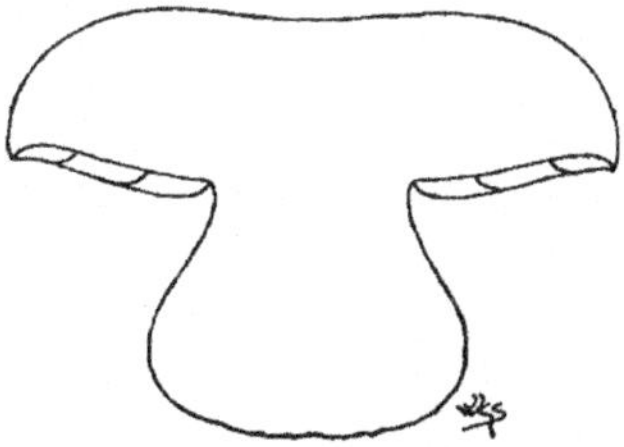

Fig. 43.—*Pilosace algeriensis* Quél. In section. Half natural size.

801. P. algeriensis Quél. (from its being first found in Algeria) *a b.* *P.* plane, gibbous or subumbonate, smooth, white, becoming brownish above when fully developed. *St.* solid, silky, white. *G.* narrow, salmon to rose, becoming dark purple-madder.

Edible. On the ground. Aug. $3\frac{3}{4} \times 1\frac{1}{2} \times 1\frac{1}{4}$ in. Intermediate between *Psalliota* and *Stropharia* with the aspect of an exannulate **789.**

XXXV. STROPHARIA Quél.

(From a fancied resemblance in the annulus to a sword-belt, Gr. *strophos.*)

Veil annular on stem. *Hymenophore* confluent and homogeneous with the fleshy stem. *Pileus* with or without a pellicle, even, scaly

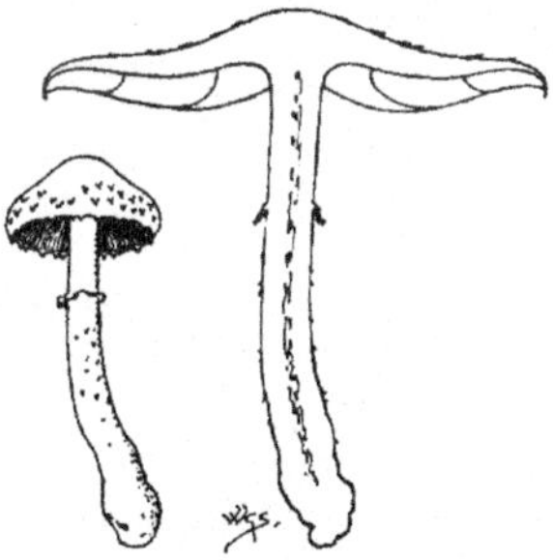

Fig. 44.—*Stropharia æruginosa* Quél.; entire and in section. One-third natural size.

and viscid, or innato-fibrillose and dry. *Stem* central, fleshy, annulate. *Gills* more or less adnate, at first whitish, or pale clay-brownish. (Fig. 44.)

Most of the species are epiphytal, some grow on the ground. *Stropharia* agrees in structure with *Armillaria* and *Pholiota*.

Species **802—820**

a. *Viscipelles*. *Pileus* with an even or scaly, usually viscid pellicle.

Not usually growing on dung. **802—810**

Growing on dung; annulus often imperfect. **811—814**

b. *Spintrigeræ*. *Pileus* without a pellicle, innato-fibrillose, not viscid. **815—820**

a. *Viscipelles*.

802. S. æruginosa Quél. (from its verdigris colour; *ærugo*, verdigris) *a b c*.

P. expanded, subumbonate, brownish-yellow, covered with transparent blue gluten, clad with white scales; marg. appendiculate with V. *St*. hollow, white scaly and viscid below A., blue or verdigris, often furnished with stringy root-like growths. *G*. purple.

Sometimes cæspitose. Odour strong of rancid ointment. Woods, meadows, damp thatch, wood, stumps, hazel, amongst stones; common. July–Nov. $3\frac{1}{2} \times 4\frac{1}{4} \times \frac{1}{2}$ in.

802a. S. squamulosa Mass. (from the small scales of the pileus; *squama*, a scale)

P. soon plane, dry, silky, deep verdigris-green with darker, innate, floccose squamules; mid. yellow-orange; marg. white with veil. *St*. at length hollow, paler than P., white at base. *A*. fragmentary in the form of large hard white sq. at apex. *G*. crowded, brown, with a vinous tinge. *Flesh* white in P., pale verdigris in St.

Amongst stones, Ripon. Sept. 1895. $2\frac{3}{8} \times 2\frac{7}{8} \times \frac{5}{8}$ in.

803. S. albocyanea Quél. (from its colours; *albus*, white, *cyaneus*, blue) *a b c*.

P. plane or subumbonate, bright blue or verdigris from the gluten, becoming dull yellowish or white. *St*. hollow, not viscid, white. *A*. fibrillose, purple. *G*. fuscous-salmon. *Flesh* white, pale verdigris-green or pale blue.

Sometimes cæspitose. Pastures, woods, sometimes on dung; uncommon. Sept.–Nov. $1\frac{1}{2} \times 3\frac{3}{4} \times \frac{1}{8}$ in.

804. S. inuncta Quél. (from the besmeared pileus; *inungo*, to besmear) *a b c*.

P. convex, subumbonate, light yellowish or brownish, slate-purple from the gluten. *St*. fistulose, not viscid, white. *A*. fibrillose. *G*. adnate, dull fuscous-purple. *Flesh* white.

Gregarious, sometimes cæspitose. Taste and odour like radishes, but disagreeable. Rich pastures, graveyards, lawns, stumps, sticks. Sept.–Nov. $1\frac{3}{4} \times 3\frac{3}{8} \times \frac{3}{16}$ in. Form *lundensis* Sacc., *P*. convex, expanded; *St*. stuffed. Form *upsaliensis* Sacc., *P*. convex, plane, umbonate; *St*. hollow. Var. *pallida* W. G. Sm., *Agaricus inunctus* Fr., var. *pallidus* B. & Br., pale.

805. S. coronilla Quél. (from a fancied resemblance to a crown in the whitish-floccose margin of the pileus; *corona*, a crown) *a b*.

P. convexo-expanded, obtuse, light yellow; mid. sub- or wholly-vinous. *St*. stuffed, fibrillose, white. *G*. crowded, vinous-brownish.

Waysides, parks. May–Oct. $1\frac{1}{4} \times \frac{7}{8} \times \frac{1}{8}$ in. Perhaps the same as **807.**

806. S. obturata Quél. (from the stuffed stem; *obturo*, to stuff) *a b c*.

P. very fleshy, convex, obtuse, often rimoso-squamulose, pale ochreous; mid. shaded sienna. *St*. not scaly, white, with a membranous collapsing A. *G*. crowded, dull purplish-brown.

On the ground, amongst grass; rare. Sept.–Oct. $1\frac{1}{2} \times 1\frac{5}{8} \times \frac{3}{16}$ in. Might be mistaken for **792.**

807. S. melasperma Quél. (from the blackish spores; Gr. *melas*, black, *sperma*, a seed) *a c*.

P. expanded, biscuit-colour or pale buff, or white with a yellowish marg. *St*. hollow, white. *G*. rounded, ventricose, purplish, then deep purple-brown. *Flesh* white, yellowish in St.

Meadows and borders of thickets; rare. Aug.–Oct. $2\frac{3}{8} \times 2 \times \frac{3}{16}$ in. Might be mistaken for **792.**

808. S. squamosa Quél. (from the scaly pileus; *squama*, a scale) *a b c*.

P. expanded, obtuse, ochreous or brownish; mid. darker, clad with small whitish or brown scales. *St*. hollow, viscid-scaly below distant A., pallid above, sienna below. *G*. rounded-adnate, blackish, with a whitish edge.

Woods, fields, amongst chips, sometimes from interior of beech-nuts. Sept.–Nov. $2\frac{1}{4} \times 5\frac{5}{8} \times \frac{3}{16}$ in. Var. *thrausta* Mass. *P*. soon glabrous, more slender than type. Var. *aurantiaca* Mass. *P*. orange or brick-red.

809. S. Percevalii Sacc. (after Cecil H. Spencer Percival) *a b c*.

P. expanded, subumbonate, warm brown-ochreous; marg. appendiculate with floccose white V. *St*. hollow, attenuate upwards, fusiform-rooting, pallid above, pale brownish and squarrose below distant A. *G*. distant, pallid umber.

Sawdust, rotten wood. Nov. $2\frac{5}{8} \times 3\frac{5}{8} \times \frac{3}{8}$ in. Possibly the same as *S. depilata* Karst., but that has a solid stem. Might be mistaken for a *Flammula*. Allied to **808.**

810. S. Worthingtonii Sacc. (after Worthington G. Smith) *a b*.

P. hemispherico-conical, smooth, golden-yellow. *St*. fistulose, cobalt, or turquoise-blue. *A*. imperfect. *G*. adnato-ascending, pallid brown. *Flesh* pale sulphur-yellow in P., pale bright blue in St.

Pastures, woods. Nov.–Dec. $\frac{7}{8} \times 2\frac{3}{4} \times \frac{1}{16}$ in.

811. S. luteonitens Quél. (from the shining yellow pileus; *luteus*, yellow, *nitens*, shining) *a*.

P. expanded, umbonate, sometimes whitish, with a few small sienna scales. *St*. hollow, even, pruinose above distant A., colour as P., sienna at base. *G.* adnexed, ventricose, pale purplish-brown.

Dung, sawdust, earth. Sept. $2\frac{3}{4} \times 2\frac{1}{4} \times \frac{1}{4}$ in. Not unlike **818**.

812. **S. merdaria** Quél. (from its habitat; *merda*, dung) *a b c.*

P. convexo-plane, smooth, hygrophanous, yellow or brownish then straw-colour; marg. appendiculate with white V. *St.* stuffed, floccose, dry, straw-white or brownish, base white-villous. *G.* adnato-rounded or subdecurrent, purplish-fuscous.

Gregarious, subcæspitose. Dry dung, road scrapings. Aug.–Oct. $1\frac{1}{2} \times 1\frac{1}{2} \times \frac{1}{8}$ in. **817** is perhaps form *major* Fr. of this species. Easily overlooked, colour of pileus exactly that of dry horse-dung.

813. **S. stercoraria** Quél. (from the habitat; *stercus*, dung) *a b c.*

P. hemispherical, smooth, yellow; mid. darker; marg. substriate. *St.* stuffed, yellowish. *A.* viscid. *G.* very broad, subdistant, purple-slate.

Dry dung in pastures. May–Nov. $1\frac{1}{2} \times 3\frac{1}{4} \times \frac{1}{8}$ in.

814. **S. semiglobata** Quél. (from the half-globose pileus) *a b c.*

P. yellowish. *St.* fistulose, smooth, colour as P., thickly covered with transparent gluten below the glutinous A. *G.* very broad, pale purplish, clouded purple-black.

Gregarious. Dung, rabbit, horse, in pastures; common. April–Nov. $1\frac{1}{4} \times 5\frac{1}{2} \times \frac{1}{8}$ in. Usually smaller than **813**, but sometimes twice the size of type.

b. Spintrigeræ.

815. **S. caput-medusæ** Karst. (from a fancied resemblance in the pileus to the head of Medusa; *caput*, a head) *a b c.*

P. expanded, obtusely umbonate, scaly-squarrose, soon naked, pale white umber; mid. umber; sc. sepia. *St.* hollow, white above, squarrose and white umber below superior A., if present. *G.* ventricose, at length separating free, pallid umber or olive-shaded, spotted ferruginous, edges white.

Subcæspitose. Woods, pine, about fir-stumps, base of elm; rare. Sept.–Oct. $3 \times 3\frac{1}{8} \times \frac{1}{2}$ in. Becoming darker when dry.

816. **S. scrobinacea** Sacc. (from the appearance of the pileus as if sprinkled with sawdust, *scrobis*) *a b.*

P. expanded, obtuse, tan; mid. ochreous-tan, or grey or brown shaded, clad with minute, fugitive sepia squamules. *St.* hollow, fibrillose, whitish, squamulose below the fugitive A. *G.* crowded, purple.

Subcæspitose. On or near stumps, ash; rare. Aug.–Oct. $1\frac{7}{8} \times 3\frac{3}{8} \times \frac{1}{4}$ in. Thinner and more graceful than **815**.

817. **S. ventricosa** Mass. (from the ventricose gills) *a.*

P. expanded, broadly gibbous, even, smooth, subviscid in wet weather, tawny-ochreous, becoming paler. *St.* hollow, ventricose below with a rooting base, minutely squamulose, pale yellow. *A.* large, persistent. *G.* slightly adnexed, ventricose, sub-crowded, purplish-brown.

Cæspitose. On the ground. $1\frac{1}{2} \times 3 \times \frac{1}{4}$ in. Perhaps only a form of **812** form *major* Fr.

818. S. Jerdonii Sacc. (after Archibald Jerdon) *a*.

P. expanded, obtusely umbonate, minutely rivulose, pale ochreous; mid. slightly darker; sc. white. *St.* hollow, attenuate upwards, white above A., subscaly, tan-ochreous and brownish below. *G.* transversely striate, purple-brownish.

Stumps, fir; rare. Sept.–Nov. 2 × 2½ × ¼ in.

819. S. spintrigera Sacc. (from the annulus; *spinther*, a bracelet, *gero*, to carry) *a*.

P. flat, obtuse, even, smooth, faintly ochreous or brownish. *St.* hollow, attenuate upwards, floccoso-fibrillose below A., white. *A.* distant, fugacious. *G.* crowded, fuscous-brown.

Cæspitose. Stumps, beech. Sept. 3¼ × 3⅛ × 3/16 in.

820. S. hypsipus Karst. (from the long stem; Gr. *hupsipous*, high-footed) *a b*.

P. plane, subobtuse, even, smooth, hygrophanous, pallid, pale brownish, or shaded dull flesh-colour; marg. striate. *St.* hollow, smooth, white. *A.* distant. *G.* subdistant, brown or dull purplish-slate.

Swampy places, amongst grass, sometimes on moss and sticks. Oct. 3½ × 6½ × 5/16 in.

XXXVI. HYPHOLOMA Quél.

(From the web- or fringe-like veil; Gr. *huphe*, a web, *loma*, a fringe.)

Veil fringe-like, adhering to the margin of the pileus, sometimes forming a fibrillose annulus. *Hymenophore* confluent and

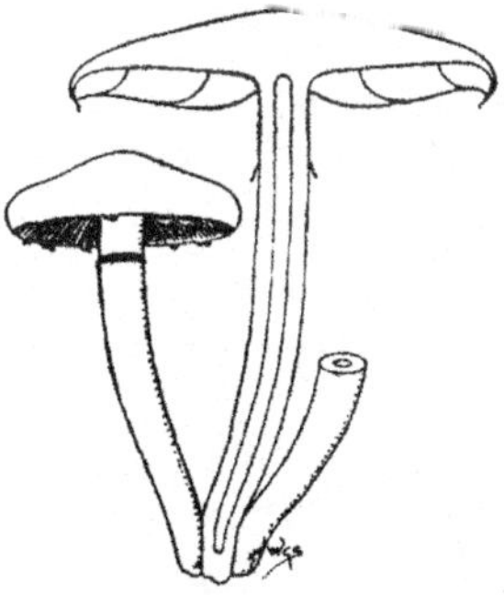

Fig. 45.—*Hypholoma sublateritium* Quél.
Entire and in section.
One-third natural size.

homogeneous with the fleshy stem. *Pileus* more or less fleshy, the margin at first incurved. *Stem* central, simple, or slightly fibrilloso-annulate. *Gills* adnate or sinuate. *Spores* slate-purple. (Fig. 45.)

The species grow on wood and have a cæspitose habit. None are edible, some are nauseous-bitter.

Hypholoma agrees in structure with *Tricholoma*, *Entoloma* and *Hebeloma*. Species **821—844**

a. Fasciculares. *Pileus* tough, smooth, dry, except **821** and **828**, brightly coloured, not hygrophanous. *Gills* adnate. **821—828**

b. Viscidæ. *Pileus* naked, viscid. **829, 830**

c. Velutinæ. *Pileus* silky with innate fibrils or streaked. **831—834**

d. Flocculosæ. *Pileus* with floccose superficial separating scales. **835, 836**

e. Appendiculatæ. *Pileus* smooth, hygrophanous. **837—844**

a. Fasciculares.

821. H. silaceum Quél. (from its colour; *sil*, yellow-ochre) *a.*

P. subexpanded, obtuse, viscid, brown-ochreous; marg. cracked, whitish. *St.* bulbous, fibrilloso-striate, shining, colour as P. *G.* purplish-brown or grey-olivaceous.

Solitary to cæspitose. Taste mild; odour mealy. Old pastures, under firs. Aug.-Nov. 2 × 3½ × ¼ in. Not unlike some forms of **822**; intermediate forms occur.

822. H. sublateritium Quél. (from the colour of the pileus; *sub*, somewhat, *later*, a brick) *a b c.*

P. convexo-plane, obtuse, tawny brick-red; marg. paler, appendiculate with V., sometimes with sc. *St.* stuffed, attenuate downwards, scaly-fibrillose, pale yellow above, ferruginous below. *G.* sinuate, arcuate or decurrent, dull yellow-greenish to pale slate-purple. *Flesh* sulphur-white.

Cæspitose to solitary, sometimes in rings near trees. Probably poisonous. Taste usually bitter-nauseous. Forests, hedgerows, stumps, rails, old pine wood, oak; common. April-Dec. 3½ × 4¾ × ½ in. Polymorphic, *P.* sometimes 7 in. in diam. Var. *Schaefferi* Sacc. *G.* deeply decurrent. Var. *squamosum* Sacc. *P.* brown-squamulose.

823. H. capnoides Quél. (from the smoky-tinted gills; Gr. *kapnos*, smoke, *eidos*, appearance) *a b c.*

P. subexpanded, obtuse, shining, pale ochreous with mid. sienna-red, or wholly orange. *St.* hollow, silky, even, elastic, pale ochreous, or whitish above and brownish below. *G.* sub-crowded, arid, grey to fuscous-purple.

Taste and odour mild. Woods, pine, pine-stumps; uncommon. April-Dec. 2⅝ × 5 × ⅜ in.

824. H. epixanthum Quél. (from the yellowish-brown colour; Gr. *epixanthos*, yellowish-brown) *a b c.*

P. expanded, obtuse, silky, apt to split; marg. lobed, undulate. *St.* hollow, floccoso-fibrillose, reddish-sienna, darker below.

G. crowded, light yellow, sometimes cinereous. *Flesh* white, becoming light yellow, sienna in St.

Taste acrid ; odour strong, acid. Stumps, fir, fallen firs, on the ground ; frequent. Aug.–Dec. 3¾ × 6¼ × ⅜ in. Sometimes wholly bright yellow ; often sterile. One form of this is *Clitocybe Sadleri* Sacc.

825. **H. elæodes** Gill. (from the olive colour of the gills ; Gr. *elaios*, olive, *eidos*, appearance) *a b c*.

P. plane, subumbonate, tan or buff. *St*. fibrillose, colour as P., becoming ferruginous. *G*. crowded, thin, white to olivaceous. *Flesh* pale buff.

Taste bitter ; odour sour. Stumps, on the ground. Oct. 2⅝ × 2⅛ × ¼ in. Sometimes wholly ivory-whitish. Sometimes regarded as a form of **826**.

826. **H. fasciculare** Quél. (from the fasciculate habit) *a b c*.

P. flat, subobtuse, sulphur-yellow ; mid. sienna. *St*. hollow, fibrillose, sulphur above, sienna below or wholly sulphur-yellow, often fibrilloso-annulate. *G*. crowded, sulphur to green. *Flesh* sulphur.

Gregarious, densely cæspitose. Probably poisonous. Taste and odour acrid, nauseous, bitter, sickening. Stumps, on the ground ; very common. April–Dec. 2⅜ × 4⅝ × ¼ in. There are many varieties. Recorded from Glyncorrwg collieries with stems 4 feet long.

827. **H. instratum** Mass. (from the naked, uncovered pileus ; *instratus*, uncovered) *a*.

P. convex, broadly umbonate, rugose, rich dark-brown. *St*. hollow, smooth and white above, fibrillose and pale brown below. *G*. adnato-ascending, brown to purple-brown, edge paler. *Flesh* pale sienna-brownish.

Stumps. Autumn. 1⅜ × 2⅝ × $\frac{5}{16}$ in.

828. **H. dispersum** Quél. (from the scattered habit of growth ; *dispergo*, to scatter) *b*.

P. expanded, viscid, obtuse, tawny-yellowish to purplish-umber. *St*. fistulose, sometimes rooting, silky-fibrillose, faintly ferruginous above, rich fuscous-brown below. *G*. distant, thick, brownish flesh-colour or shaded olive, with a white edge. *Flesh* pale yellow-tawny.

Woods, pine, dry to marshy places, on the ground, on sawdust. April–Nov. 1⅝ × 4¼ × ⅛ in. Variable in colour and habit.

b. Viscidæ.

829. **H. incomptum** Mass. (from its simple, unadorned appearance ; *incomptus*, rough, without ornament) *a*.

P. expanded, broadly gibbous, even, deep bay-brown, or orange-tawny. *St*. hollow, equal, pale above, dark ferruginous below, brown-fibrillose and yellow tomentose. *G*. adnato-rounded, crowded, deep olive, purple-clouded.

Stumps. Autumn. 4 × 3½ × ½ in.

830. H. œdipus Sacc. (from the swollen base of the stem; *oidipous*, with a swollen foot) *a*.

P. hemispherical; mid. umber; marg. pallid, appendiculate with V. *St.* solid, semibulbous, fibrillose below the distant A., whitish, shaded pale sienna. *G.* adnate, umber.

Solitary or in groups of two or three. Sticks and dead leaves. Feb. $\frac{7}{8} \times 1\frac{1}{2} \times \frac{1}{8}$ in.

c. Velutinæ.

831. H. pseudostorea Rea (from its being at first considered the same as *H. Storea*) *a b*.

P. convex, obtuse or subumbonate, then expanded, fleshy, at first purplish-brown with mid. darker, soon breaking up into large fibrillose purplish-brown scales, silky-white to pale fawn beneath; marg. appendiculate and entire when young. *St.* hollow, equal or attenuate downwards, at first even, then silky-brown-fibrillose, white above, pale yellowish below. *G.* adnate, at first white, then rose, becoming brown to purple-brown, not exuding drops of water.

Usually cæspitose, sometimes solitary. Taste insipid, somewhat disagreeable; odour strong, disagreeable. Woods, plantations, under larches. Sept.–Nov. $2\frac{7}{8} \times 4\frac{1}{2} \times \frac{1}{2}$ in. This is the plant referred to *Agaricus Storea* Fr. by Berkeley, and afterwards named *A. Storea*, var. *cæspitosa* by Cooke. Sometimes confounded with **832**.

832. H. lacrymabundum Quél. (from the tear-like drops on the gills; *lachrymabundus*, tearful) *a b c*.

P. expanded, obtuse, innato-pilose, whitish, then fuscous; sc. darker; marg. appendiculate with V. *St.* hollow, fibrilloso-scaly, fuscous-white. *G.* adnato-rounded, whitish to white-purple-brownish, studded with drops of water.

Cæspitose. Stumps or on the ground, yards. Sept.–Dec. $3 \times 3\frac{1}{2} \times \frac{5}{16}$ in. Often confounded with **831** and **834**. Used for inferior ketchup. Sometimes sold for the mushroom.

833. H. pyrotrichum Quél. (from the fiery-red fibrils of the pileus; Gr. *pur*, fire, *thrix*, a hair) *a*.

P. expanded, obtuse, pale bright vermilion-ochre; marg. appendiculate with V. *St.* hollow, fibrillose, whitish to pale orange, sometimes imperfectly annulate with V. *G.* adnate, dull purplish-brown, studded with drops of water.

Cæspitose. Woods, pine, beech, about roots. Aug.–Oct. $3 \times 3\frac{1}{4} \times \frac{5}{16}$ in. Var. *egregium* Mass., with a densely shaggy stem.

834. H. velutinum Quél. (from the velvety pileus of young examples; *vellus*, fleece) *a b c*.

P. expanded, obtusely umbonate, tawny, umber or sienna; marg. appendiculate with V. *St.* hollow, fibrillose, whitish above, sienna or umber below. *G.* adfixed, brown-fuscous, dark-spotted, studded with drops of water.

Subcæspitose. Taste insipid. Stumps; common. July–Oct. $4 \times 5 \times \frac{1}{2}$ in. Variable in stature; pileus sometimes 6 in. in diam. Often confused with

832. Intermediate forms occur between this and **832** and **833.** Sometimes sold for the mushroom. Used for inferior ketchup. Var. *leiocephalum* Mass. *P.* mid. rugged, much smaller than type.

d. Flocculosæ.

835. H. cascum Quél. (from its grey and white colours; *cascus*, old) *a b.* Fragile.

P. hemispherical, obtuse, dull greyish-ochre or ochre; marg. paler, appendiculate with white V. *St.* hollow, fibrillose, white. *G.* ascending, adnexo-rounded, arid, at first pale dove-colour then blackish-brown, edge whitish.

Gregarious or cæspitose. Pastures. July. $1\frac{1}{2} \times 3\frac{1}{4} \times \frac{3}{16}$ in.

836. H. punctulatum Mass. (from the squamuloso-punctate pileus) *a.*

P. convex, pallid, shaded yellowish, brownish or white sulphur; sc. dot-like, pale. *St.* stuffed, punctiform-squamulose from base to A., colour as P.; sc. faintly rufescent. *G.* sinuato-adnate, subcrowded, pale umber to whitish-sienna. *Flesh* colour as P. *Spores* brown.

Densely cæspitose. Chips, sawdust, rotten twigs. $1\frac{3}{8} \times 1\frac{3}{8} \times \frac{1}{4}$ in. Berkeley, Cooke and Massee assign this plant to the present position. Kalchbrenner places it in *Pholiota* near **578.** Fries includes it in *Stropharia* after **816.**

e. Appendiculatæ.

837. H. LANARIPES Sacc. (from the woolly stem; *lana*, wool, *pes*, a foot) *a b c.*

P. expanded, obtuse, thin, tough, apt to split, somewhat hairy-scaly, atomate, rugose, tawny or brownish; marg. appendiculate with white V. *St.* hollow, fibrous, striate above, subfibrillose or tomentoso-squarrose below. *G.* adnato-ascending, at first white, then pale-brown, at length purplish-brown.

Subcæspitose. Soil in conservatories. June–July. $3\frac{1}{4} \times 4\frac{1}{4} \times \frac{1}{4}$ in.

838. H. Candolleanum Quél. (after Alphonse de Candolle) *a b c.*

P. flat, obtuse, depressed, smooth, apt to split, thin, brown, becoming white; mid. ochreous. *St.* hollow, striate above, white. *G.* adnexo- or adnato-rounded, ventricose, serrulate, fuscous-cinnamon. *Flesh* white.

Gregarious, cæspitose. Taste slightly styptic, disagreeable or none; odour strong. Stumps, elm, on the ground; frequent. April–Nov. $3 \times 3\frac{1}{2} \times \frac{1}{4}$ in. Firmer than **839.**

839. H. appendiculatum Quél. (from the appendiculate veil at the margin of the pileus) *a b c.*

P. hygrophanous, expanded, rugose, thin, apt to split, atomate, faintly ochreous-white; mid. ochreous. *St.* fistulose, pruinate above, white. *G.* adnato-rounded, purplish flesh-colour or brownish.

Cæspitose. Stumps, sticks, under elm, on the ground; common. June–Oct. $3 \times 3\frac{3}{4} \times \frac{1}{8}$ in. Thinner and more fragile than **838.**

840. H. felinum Sacc. (from the tawny pileus, coloured like members of the cat genus, *Felis*) *a b*.

P. expanded, smooth, ochreous, biscuit-colour when dry. *St.* fistulose, shining, striate above, white-floccose at base. *V.* annulate. *G.* dove-colour, shaded brown.

Gregarious, subcæspitose. Parks, roadsides, etc., on the ground. Sept. 1 × 1½ × ⅛ in.

841. H. leucotephrum Sacc. (from the colour of the gills; Gr. *leukos*, white, *tephros*, ash-coloured) *a b c*.

P. convexo-expanded, wrinkled, whitish or faint-salmon or ochreous-shaded. *St.* hollow, striate above, silky-fibrillose below, whitish. *A.* large. *G.* adnato-ascending, deep slate-black, edge whitish.

Cæspitose. Base of ash-trees. Sept.–Oct. 2 × 4¾ × ¼ in.

842. H. egenulum Sacc. (from its poor appearance; *egenulus*, poor) *a*.

P. hemispherical, broadly umbonate, smooth, white; marg. finely striate. *St.* fistulose, minutely adpressedly scaly, white, faintly sienna-ochreous below. *G.* adnate with a tooth, ventricose, purple-umber, edge white.

Solitary. On the ground amongst grass. May. 1⅜ × 1¾ × ⅛ in.

843. H. hydrophilum Quél. (from its hygrophanous habit; Gr. *hudor*, water, *philos*, loving) *a b*.

P. flat, brittle, smooth, rugose, deep ferruginous-sienna or tawny-ochreous, paler when dry. *St.* hollow, fibrilloso-mealy, shining, whitish above, faintly sienna below. *G.* adnexed or adnato-rounded, crowded, exuding drops of water, brown or sienna.

Cæspitose. About and on trunks and stumps. Oct.–Nov. 2⅝ × 4¼ × ¼ in. Pileus sometimes 5 in. in diam. Sometimes placed in *Bolbitius*.

844. H. pilulæforme Gill. (from the appearance of the pileus, like a little ball, *pilula*) *a*.

P. even, smooth, fuscous, ochreous when dry. *St.* fistulose, white. *V.* annulate in young examples. *G.* adnexo-ascending, fuscous.

Cæspitoso-crowded. Mossy trunks; rare. Autumn. ¾ × 1⅛ × $\frac{1}{16}$ in.

XXXVII. PSILOCYBE Quél.

(From the naked pileus, owing to the usually obsolete veil; Gr. *psilos*, naked, *kube*, a head.)

Veil very fugacious, usually obsolete. *Hymenophore* confluent with, but heterogeneous from the cartilaginous stem. *Pileus* more or less fleshy, smooth; the margin at first incurved. *Stem* central, fistulose or stuffed, subcartilaginous, rigid or tough, often rooting, simple. *Gills* becoming fuscous, brownish or purplish. (Fig. 46.)

Nearly all the species grow on the ground and are gregarious and sometimes cæspitose, a few grow on dung; they are inodorous and none are edible.

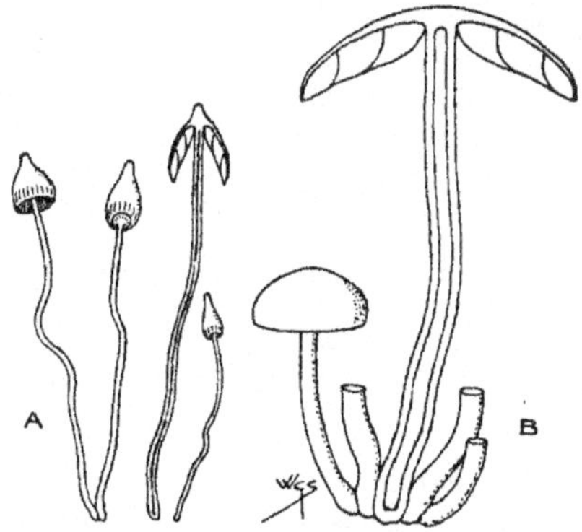

Fig. 46.—A, *Psilocybe semilanceata* Quél.; B, *S. spadicea* Quél. Entire and in section. One-half natural size.

Psilocybe agrees in structure with *Collybia*, *Leptonia*, *Naucoria* and *Panæolus*. Species **845—871**

a. Tenaces. *Veil* rarely conspicuous. *Pileus* pelliculose, slightly viscid in wet weather, bright in colour, becoming pale. *Stem* with a thick tough bark, flexile, usually coloured.

Gills ventricose, not decurrent. **845—854**
Gills plane, very broad behind, subdecurrent **855—860**
Gills ascending sublinear. **861, 862**

b. Rigidæ. *Veil* obsolete. *Pileus* scarcely pelliculose, flesh usually scissile, hygrophanous. *Stem* rigid. *Gills* adnexed, rarely adnate. **863—871**

a. Tenaces.

845. P. sarcocephala Gill. (from the fleshy pileus; Gr. *sarx*, flesh, *kephale*, the head) *a b.*

P. expanded, obtuse, dry, pale tawny, salmon or pale ferruginous. *St.* whitish to rusty. *G.* adnato-rounded, salmon, then sooty. *Flesh* white, pale salmon in stem.

Gregarious, subcæspitose. Woods, grassy places, usually about or near stumps, elm. Sept.–Nov. $4 \times 5\frac{3}{4} \times \frac{3}{4}$ in. Resembles **230.**

845a. P. helvola Mass. (from the yellowish colour; *helvolus*, pale yellow) *a.*

P. campanulate or conical, convex, then expanded, obtuse, even, glabrous, tawny-ochreous; mid. darker; cuticle often cracked into patches near mid. *St.* equal, hollow, fibrillose, often twisted, white or tinged ochreous, base thickened, white-downy. *G.* adnate, then separating, broad, subdistant, purplish-umber, clouded, edge white.

On the ground amongst grass. Sept. $1\frac{1}{2} \times 2\frac{3}{4} \times \frac{3}{16}$ in.

846. P. ericæa Quél. (from the habitat; *erica*, heath) *a b*.

P. flat, obtuse or subumbonate, tawny-ferruginous or umber, ochre when dry. *St.* smooth, subsilky, tan above, brownish below, base white-villous. *G.* adnate, subdistant, blackish with a light edge.

Gregarious. Woods, pastures; rare. May–Oct. $1\frac{1}{4} \times 4\frac{1}{2} \times \frac{1}{8}$ in. A smaller form occurs in wet places.

847. P. subericæa Sacc. (from its partial resemblance to **846**) *a b*.

P. plane, obtuse or subumbonate, tawny or whitish-yellow, paler when dry. *St.* pallid or pale yellowish. *G.* adnato-sinuate, blackish.

Taste acid. Fields, under firs. June–Aug. $2\frac{1}{8} \times 2\frac{3}{8} \times \frac{3}{16}$ in.

848. P. uda Gill. (from the moist pileus; *udus*, moist) *a b c*.

P. flat, obtuse, smooth or wrinkled, brick-tawny or yellowish. *St.* fibrillose, tan above, pale tawny-ferruginous below. *G.* adnato-sinuate, broad, fuscous-purple.

Scattered. Swampy places amongst *Sphagnum*. Oct.–Nov. $1\frac{1}{2} \times 3\frac{3}{4} \times \frac{1}{8}$ in. Gills sometimes remaining pallid and sterile.

849. P. canofaciens Mass. (from the hoary appearance; *canofaciens*, making hoary) *a b*.

P. expanded, obtuse, umber; mid. subareolate, sienna; marg. darker with feeble traces of a fibrillose or arachnoid V. *St.* equal, hollow, brittle, umber-greyish, sooty below, white scaly. *G.* adnato-ascending, broad, livid-brownish, purple clouded.

Cæspitose. Rotten straw and rubbish. May. $1\frac{5}{8} \times 2\frac{7}{8} \times \frac{3}{16}$ in.

850. P. areolata Sacc. (from the cracked-areolate pileus) *a b c*. Not brittle.

P. subexpanded, minutely fibrillose, white, then ochreous or fuscous. *St.* fibrillose, striate, tan and pulverulent above, sienna below. *V.* fibrillose, evanescent. *G.* adnato-rounded, dull purple with a white edge.

Densely cæspitose. Gardens, on wood. May–Oct. $1\frac{3}{4} \times 2\frac{1}{2} \times \frac{1}{4}$ in.

851. P. virescens Mass. (from the green tints; *virescens*, green) *a b*.

P. expanded, at first minutely silky, cracking, bright dark-brown, interstices of cracks clear pale verdigris-green, then yellowish. *St.* downy, apex striate, pale green above, ferruginous below. *G.* adnexed, subcrowded, smoky-purple with paler edge.

Solitary. Taste and odour none. Rotten stumps and chips. Autumn. $\frac{1}{2} \times 2\frac{1}{2} \times \frac{1}{4}$ in.

852. P. agraria Karst. (from the habitat; *ager*, a field) *a*.

P. flat, obtuse or umbonate, white or ashy. *St.* smooth, white. *G.* adnate, ashy to fuscous. *Flesh* white.

Near stumps. Sept.–Oct. $1\frac{3}{4} \times 2\frac{3}{8} \times \frac{3}{16}$ in.

853. P. chondroderma Sacc. (from the parchment-like pellicle of the pileus; Gr. *chondros*, cartilage, *derma*, skin) *a*.

P. conico-campanulate, smooth, cracked, dark date-brown; marg. jagged-appendiculate with V. *St.* fibrillose, pale sienna, darker below. *G.* ascending, adfixed, purplish-brown, edge white.

Woods, fir. Sept. $1\frac{1}{4} \times 1\frac{1}{2} \times \frac{3}{16}$ in. Stains paper etc., yellow.

854. P. scrobicola Sacc. (from its habitat; *scrobis*, sawdust) *a*.

P. convex, umbilicato-depressed, whitish; marg. undulate. *St.* attenuate downwards, fibrillose, colour as P. *G.* arcuato-adnate, brown, with a red tinge.

Pine sawdust, branches. Nov. $1\frac{3}{4} \times 1\frac{3}{8} \times \frac{3}{16}$ in.

855. P. ammophila Gill. (from the habitat; Gr. *ammos*, sand, *philos*, loving) *a b*.

P. expanded, subumbonate, sienna-buff; marg. striate. *St.* base clavate, apex striate, white to pale buff, half buried in sand. *G.* adnate or with a decurrent tooth, pale purplish, becoming slate-pulverulent with spores.

Sea-shores, amongst sand. Aug.–Sept. $1\frac{3}{8} \times 2 \times \frac{1}{8}$ in.

856. P. coprophila Quél. (from the habitat; Gr. *kopros*, dung, *philos*, loving) *a c*.

P. expanded, obtuse or subumbonate, downy, white, then rufescent. *St.* shaggy-flocculose, whitish. *G.* adnate, very broad, crowded, fuscous.

Amongst grass, rabbits'-dung, on dung-heaps; rare. Oct.–Jan. $1\frac{3}{8} \times 1\frac{3}{4} \times \frac{1}{8}$ in. Sometimes infesting mushroom-beds.

857. P. bullacea Quél. (from a fancied resemblance in the pileus, when young, to a knob, *bulla*) *a b c*.

P. expanded, subumbonate, dark tawny, tan when dry; marg. appendiculate with V. *St.* subfibrillose, tawny or yellow, base fuscous-ferruginous, fibrilloso-annulate when young. *G.* adnate with a small sinus, very broad, crowded, fuscous-ferruginous.

Gregarious. Dung, rich soil, amongst *Polytrichum*. April – Sept. $1\frac{1}{8} \times 1 \times \frac{1}{8}$ in.

858. P. physaloides Quél. (from a fancied resemblance in the pileus to a bladder; Gr. *phusalis*, a bladder, *eidos*, appearance) *a b c*.

P. flat, umbonate, fuscous-purple. *St.* adpressedly fibrillose, date-brown below, paler above. *G.* adnato-decurrent, crowded, ventricose, pallid ferruginous-fuscous.

Rich grassy places, woods, on moss, horse-dung; old walls; uncommon. Feb.–Sept. $1\frac{3}{8} \times 1\frac{1}{8} \times \frac{1}{16}$ in. Must not be confounded with 771.

859. P. atrorufa Quél. (from the blackish-red pileus; *ater*, black, *rufus*, red) *a*.

P. expanded, obtuse, black-rufous or purple-fuscous; mid. darker; marg. striate. *St.* at first subfibrillose, subpruinate above, colour as P., base darker. *G.* adnate with a small tooth, umber to purple-umber.

Woods, mixed. Nov. $1\frac{1}{4} \times 2\frac{1}{8} \times \frac{1}{16}$ in.

860. P. nuciseda Mass. (from the frequent habitat, nuts; *nux*, a nut, *sedeo*, to sit) *a.*

P. expanded, subumbonate, subsilky when dry, light yellow or shaded sienna; mid. sienna. *St.* slightly attenuate downwards, fuscous, or white below. *G.* adnate, broad, ventricose, fuscous to umber-black.

Beech and hazel nuts, chips, wood. $1\frac{1}{8} \times 1\frac{1}{8} \times \frac{1}{16}$ in. Habit of **771.**

861. P. compta Sacc. (from the adorned pileus, sprinkled with minute shining spots; *comptus*, ornamented) *a.*

P. expanded, obtuse, pallid to ochreous; marg. striate, subcrenulate. *St.* flexuous, silky-shining, tan, faintly rufous below. *G.* adnato-ascending or adnexed, distant, rosy-umber.

Woods, amongst grass; rare. Sept.–Oct. $1\frac{1}{8} \times 2\frac{1}{4} \times \frac{1}{16}$ in.

862. P. semilanceata Quél. (from a fancied resemblance in the pileus, as seen in vertical section to a lance-head; *semi*, half, *lancea*, a spear) *a b c.*

P. acutely conical, whitish-ochre; mid. pale; marg. slightly striate, slate, sometimes fuscous-shaded. *St.* flexuous, smooth, fibrilloso-annulate when young, colour as P., base pale sienna. *G.* adnexo-ascending, crowded, purplish to purple-black.

Gregarious. Reputed poisonous. Pastures, roadsides; common. Stem sometimes 6 in. high. Aug.–Nov. $\frac{1}{2} \times \frac{3}{8} \times \frac{1}{16}$ in. Var. *cærulescens* Sacc. *P.* subobtuse. *St.* more or less distinctly tinged with blue.

b. Rigidæ.

863. P. canobrunnea Quél. (from the greyish-brown pileus; *canus*, hoary, *brunneus*, brown) *a b.*

P. plane, subviscid when moist, flesh-brown, pale clay when dry. *St.* rooting, squamulose, whitish. *G.* adnexed, crowded, broad, ventricose, brownish-purple.

Solitary or laxly gregarious. On the ground in sunny places, on scorched ground. Autumn. $2 \times 1\frac{3}{8} \times \frac{3}{16}$ in.

864. P. spadicea Quél. (from the colour of the pileus; *spadiceus*, date-brown) *a b c.*

P. plane, usually cracked, dark to ochreous-brown, becoming paler in patches. *St.* smooth, whitish, base pale sienna. *G.* adnexo-rounded, adnate or adnate with a tooth, crowded, brown.

Laxly clustered. Stumps, earth, manure-heaps, amongst short grass; common. Feb.–Nov. $2\frac{3}{8} \times 4 \times \frac{1}{4}$ in. Var. *hygrophila* Karst. larger than type. Var. *polycephala* Gill. densely crowded.

865. **P. squalens** Karst. (from its dirty colours; *squalens*, dirty) *a*.
P. flat, pale livid-ferruginous, pale dull clay-colour when dry. *St*. striate above, colour as P. *G*. adnato-decurrent, crowded, cinnamon-umber.

Solitary or cæspitose. On and near trunks; rare. $1\frac{3}{4} \times 2\frac{1}{4} \times \frac{1}{8}$ in.

866. **P. cernua** Quél. (from the frequently drooping pileus; *cernuo*, to fall head foremost) *a b c*.
P. expanded, slightly wrinkled when dry, whitish. *St*. smooth, flexuous, colour as P. *G*. adnato-ascending, cinereous-blackish or slate-sprinkled.

Solitary or gregarious, usually cæspitose. Dead wood, chips, leaves, lawns; uncommon. Aug.–Dec. $1\frac{1}{2} \times 3\frac{1}{2} \times \frac{1}{8}$ in.

867. **P. hebes** Karst. (from the obtuse pileus; *hebes*, blunt) *a b*.
P. becoming flat, striate when moist, lurid, obscure olive, pale when dry. *St*. naked, whitish. *G*. adnexo-rounded, broad, crowded, brownish or brown.

Usually solitary. Amongst grass and leaves, near chestnut-trees, trunks. Nov. $1\frac{1}{2} \times 1\frac{1}{2} \times \frac{1}{8}$ in.

868. **P. fœnisecii** Quél. (from its habitat, meadows; *fœnisicia*, hay-harvest) *a b c*.
P. campanulate, obtuse, pale ferruginous-fuscous to deep umber, becoming paler in patches. *St*. pulverulent, then naked, rufescent, sienna-tan or flesh-colour. *G*. adnato-ascending, ventricose, separating from stem, umber to black-purple.

Gregarious. Fields, lawns, roadsides, manure, horse-dung; common. Feb.–Dec. $\frac{3}{4} \times 2\frac{3}{4} \times \frac{1}{16}$ in.

869. **P. callosa** Quél. (from the tough stem; *callosus*, having a hard skin) *a b c*.
P. conico-campanulate, obtuse, pale buff. *St*. smooth, paler than P. *G*. adnato-ascending, ventricose, sooty or purple-black.

Scattered or in small groups. Fields, lawns, roadsides. Aug.–Oct. $\frac{5}{16} \times 2\frac{3}{8} \times \frac{1}{32}$ in.

870. **P. clivensis** Sacc. (from the place where it was first observed, King's Cliffe, Northamptonshire) *a b*.
P. expanded, shining-atomate, striate, pallid brown or ochre to white. *St*. silky, whitish. *G*. adnate, widely emarginate, ventricose, umber, edge white.

Amongst grass, on downs. June–Oct. $1\frac{1}{4} \times 1\frac{3}{4} \times \frac{1}{8}$ in.

871. **P. catervata** Mass. (from its fasciculate habit; *caterva*, a troop) *a*.
P. campanulate, obtuse, satiny, brittle, white. *St*. white. *G*. slightly adnexed, somewhat broad, crowded, purple-shaded or cinnamon, edge white.

On the ground. Oct. $\frac{5}{8} \times 2\frac{5}{8} \times \frac{1}{16}$ in.

XXXVIII. **PSATHYRA** Quél.

(From the friable substance; Gr. *psathuros*, friable.)

Veil when present universal, floccoso-fibrillose, not annulate, sometimes obsolete. *Hymenophore* confluent with, but heterogeneous from the cartilaginous stem. *Pileus* conical or campanulate, then expanded, submembranous, margin at first straight and adpressed to

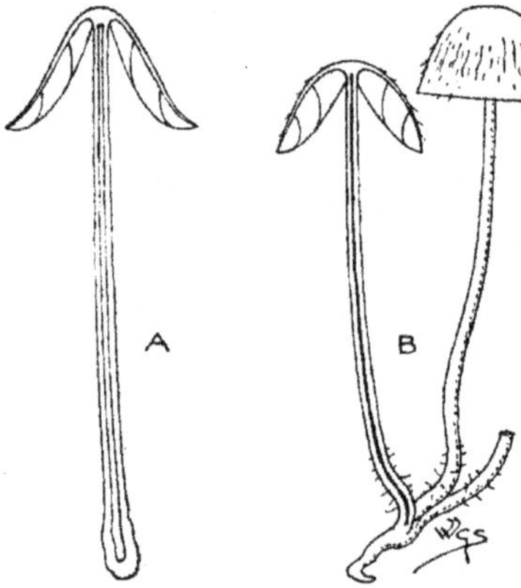

Fig. 47.—A, section of *Psathyra conopilea* Quél.; B, *P. bifrons* Sacc. entire and in section. One-half natural size.

the stem, except **884**. *Stem* central, somewhat cartilaginous, fistulose, lustrous, fragile, simple. *Gills* at first pale, becoming purple or fuscous. *Spores* purple, brownish-purple, dark purple or dark brown. (Fig. 47.)

The species grow on the ground, or on rotten wood, they are slender, fragile, hygrophanous and with fugitive colouring.

Psathyra agrees in structure with *Mycena*, *Nolanea*, *Galera* and *Psathyrella*. Species **872—893**

a. *Conopileæ*. *Veil* obsolete. *Pileus* conico-campanulate. *Stem* tense and straight, fistulose. *Gills* usually adnexo-ascending, often free. On the ground in grassy or damp places. **872—877**

b. *Obtusatæ*. *Veil* obsolete. *Pileus* campanulato-convex, becoming flat; smooth or atomate. *Gills* plano- or arcuato-adfixed. **878—880**

c. *Fibrillosæ*. *Pileus*—except **884** and **888**—and stem at first fibrillose or floccose from the veil. *Stem* fistulose. **881—893**

a. Conopileæ.

872. P. elata Mass. (from its long stem; *elatus*, tall) *a b*.

P. even, glabrous, atomate, dark brown, pale ochreous when dry. *St*. hollow, silky-shining, white. *G*. broadly adnate, crowded, purplish-brown.

Amongst grass, on hedge-banks. Aug. $2 \times 7\frac{5}{8} \times \frac{1}{4}$ in.

873. P. conopilea Quél. (from the conical pileus) *a b c*.

P. even, smooth, whitish, greyish-brown or umber with marg. tan. *St*. attenuate upwards, whitish, silvery-lustrous. *G*. slightly adnexed, subventricose, dark slate or fuscous-purple.

Grassy places, gardens, ditch-bottoms, under beeches; rare. Sept.–Nov. $1\frac{5}{8} \times 4\frac{3}{8} \times \frac{1}{8}$ in. Stem sometimes $6\frac{1}{2}$ in. long. Var. *superba* W. G. Sm., *Agaricus superbus* Jungh. larger. *P*. chestnut-brown to buff.

874. P. mastigera Sacc. (from the breast-like shape; Gr. *mastos*, the breast, *gero*, I bear) *a b*.

P. acutely umbonate, dark chestnut-brown, umber-tan when dry. *St*. fibrillose, furfuraceous or smooth, white. *G*. narrow, umber. *Flesh* pale brown.

Amongst grass, roadsides. July–Nov. $\frac{7}{8} \times 3\frac{5}{8} \times \frac{1}{8}$ in.

875. P. glareosa Sacc. (from the habitat, gravelly soil; *glareosus*, gravelly) *a*.

P. obtuse or umbonate, minutely tomentose, atomate, striate, grey; mid. pale chestnut. *St*. brown, white fibrillose. *G*. broadly adnate, umber. *Flesh* brown.

Gravelly soil after wet weather. June. $\frac{3}{4} \times 2\frac{1}{4} \times \frac{1}{16}$ in.

876. P. corrugis Quél. (from the pileus, at length wrinkled; *corrugis*, having folds) *a b c*.

P. subumbonate or obtuse, smooth, substriate, whitish, pale rose, ochreous-tan or pale umber. *St*. smooth, colour as P. *G*. adnato-ascending or somewhat sinuate, pale, then purple-black or dark slate.

Pastures, grassy places, open places in woods, amongst grass and sticks, old manure patches; frequent. April–Jan. $1\frac{3}{4} \times 3\frac{1}{8} \times \frac{3}{16}$ in. Stem sometimes 5 in. long. Var. *vinosa* Mass. *P*. rosy-vinous half-way up. Var. *gracilis* Mass. more slender than type.

876a. P. pellosperma W. G. Sm., *Agaricus pellospermus* Cooke (from the dark coloured spores; Gr. *pellos*, dark coloured, *sperma*, a seed) *a*.

P. subcampanulate or subovate, even, then striate, sometimes rugose, white or ochreous-white, becoming fuliginous with age. *St*. fistulose, naked, nearly equal, white or as P. *G*. broad, free, much narrowed at the tips, cinereous, then fuliginous, at length black. *Flesh* thin, white.

On the ground. Autumn. $1\frac{3}{8} \times 4\frac{7}{8} \times \frac{1}{8}$ in.

877. P. gyroflexa Quél. (from the weak flexuous stem; *gyrus*, a circle, *flexus*, bent) *a b*.

P. atomate, striate, whitish or pale grey; mid. pale buff or rufescent. *St.* white, silky-shining. *G.* adnato-ascending, ventricose, greyish-purple to slate.

Scattered or subcæspitose. Margins of woods, amongst rubbish. Aug. $1\frac{1}{8} \times 2 \times \frac{1}{16}$ in.

b. Obtusatæ.

878. P. spadiceogrisea Quél. (from the brown-grey pileus; *spadix*, bay-brown, *griseus*, grey) *a b c*. Watery, fragile.

P. even, pellucid-striate to mid., date-brown, greyish and buff when dry. *St.* subpulverulent, striate above, whitish. *G.* adnexed, crowded, narrow, umber-fuscous or brown.

Solitary or gregarious, subcæspitose. On trunks, old boards, chips, on the ground. Aug.–Nov. $1\frac{5}{8} \times 3\frac{1}{8} \times \frac{1}{8}$ in.

879. P. obtusata Gill. (from the obtuse pileus) *a b c*.

P. even, brown-fuscous or pale umber, paler and striate at marg. *St.* fibrilloso-silky, not pulverulent, even above, whitish. *G.* adnate, broad, subdistant, ashy-brown or umber.

Single or cæspitose. Woods, on the ground, on wood, trunks, oak; rare. June–Dec. $1\frac{5}{8} \times 2\frac{7}{8} \times \frac{1}{8}$ in. Stem sometimes rooting.

880. P. neglecta Mass. (from its having been overlooked) *a b*.

P. almost plane, even, glabrous, atomate, pale rufous, almost white when dry; mid. pale clay. *St.* somewhat wavy, pellucid, pruinose above, white, tinged pale rufous below. *G.* slightly attached, ventricose, crowded, pink then purple-brown.

On the ground, gardens. Oct.–Nov. $\frac{5}{8} \times 1 \times \frac{1}{32}$ in.

c. Fibrillosæ.

881. P. frustulenta Sacc. (from its fragility, breaking at a touch; *frustulentus*, full of pieces) *a*.

P. hygrophanous, glabrous, substriate, sienna-tan, tan when dry. *St.* subwavy, whitish. *G.* adnato-ascending, crowded, brownish.

Gravel in damp places. Sept. $1\frac{1}{4} \times 2\frac{1}{4} \times \frac{1}{16}$ in.

882. P. bifrons Sacc. (from the two-coloured pileus; *bifrons*, with two faces) *a b*.

P. obtuse, subwrinkled, ochreous-brown tinged with red, whitish when dry, evanescent white-fibrillose. *St.* subrooting, minutely satiny, fibrillose below, white. *G.* adnate, broad, ventricose, pink-cinereous; edge white.

Sometimes two or three connate. Woods, amongst sticks, twigs, chips; rare. Aug.–Sept. $1\frac{1}{8} \times 3\frac{7}{8} \times \frac{1}{16}$ in. Var. *semitincta* Sacc. *P.* vinous-purple; mid. pale ochre.

883. P. semivestita Sacc. (from the pileus white-fibrillose halfway to the middle; *semi*, half, *vestio*, to clothe) *a b*.

P. obtuse, chestnut-brown, tan when dry. *St.* fibrilloso-silky, whitish. *G.* adnato-ascending, grey, becoming umber.

Solitary and large, or cæspitose and smaller. Rich pastures, waysides, woods. Oct. Solitary 2¼ × 4⅝ × ⅛ in. Cæspitose ⅞ × 3¼ × $\frac{1}{16}$ in.

884. P. Loscosii Sacc. (after Francisco Loscos) *a b*.

P. flat, radiato-pitted, sulcato-wrinkled, moist, cracking, atomate, tan-umber. *St.* striate, twisted, subfibrillose, elastic, tan, white-pruinose above, rufescent below. *G.* adnate, ventricose, thin, purple-umber with pale edge. *Flesh* sienna.

Cæspitose. Taste and odour slight, fungoid. Gardens, on mushroom-bed. Nov. 2 × 7¼ × $\frac{3}{16}$ in.

885. P. fatua Quél. (from its resemblance to its allies; *fatuus*, foolish) *a*.

P. rugose, at length smooth, dingy-ochreous, then whitish; mid. sienna. *St.* striate and mealy above, becoming smooth, white. *G.* adnato-ascending, crowded, brown.

Gardens. Autumn. 1⅛ × 3⅝ × ⅛ in.

886. P. fibrillosa Quél. (from the fibrillose covering) *a b*.

P. conical then flat, striate, becoming smooth, livid or whitish, the white fibrils becoming black with age or disappearing. *St.* white squamulose on pale steel-grey ground, or spreading white-fibrillose. *G.* adnate, pale slate then purple-black.

Solitary. Woods, damp places, rotten branches, sticks and leaves; rare. Oct. 2⅛ × 2⅛ × $\frac{3}{16}$ in.

887. P. Gordonii Gill. (after the Marchioness of Huntly) *a*.

P. sulcato-striate, pale cinereous to white. *St.* transversely undulate, white-pruinose above, then shining, smooth, white. *G.* ascending, narrowly adnate, cinereous.

Densely cæspitose. Odour faint to strong and nauseous. Stumps, amongst chips. Oct. 1⅝ × 2¼ × ⅛ in.

888. P. helobia Sacc. (from its habitat; Gr. *helos*, a marsh, *bios*, life) *a c*.

P. flat, subumbonate, radiato-rugose; mid. with concentric elevated ridges, sooty-brown, becoming ochreous-salmon or reddish. *St.* reddish-umber, salmon when dry. *G.* adnate, sooty-brown. *Flesh* pale pink.

Woods. Sept. 2¼ × 5¾ × $\frac{3}{16}$ in. Allied to **884**.

889. P. pennata Quél. (from the feathery scales; *penna*, a feather) *a b c*.

P. livid, whitish or pale brown; mid. sienna; marg. appendiculate with large white sc., then naked. *St.* villous, white pulverulent above, white. *G.* adnexo- or adnato-ascending, ventricose, fuscous-blackish.

Gregarious. Gardens, scorched ground, earth, burnt sawdust; rare. June–Oct. 1⅝ × 2⅝ × ⅛ in.

890. P. gossypina Gill. (from its cotton-like fibrils; *gossypium*, cotton) *a b c*.

P. expanded, ochreous-clay; mid. pale sienna; marg. striate. *St.* whitish-tomentose. *G.* adnexed, ventricose, white, then fuscous-black.

Subcæspitose. Woods, twigs, on the ground. May–Oct. $2\frac{1}{2} \times 2\frac{5}{8} \times \frac{3}{16}$ in.

891. P. noli-tangere Quél. (from its fragility, breaking at a touch; *noli tangere*, do not touch) *a b*.

P. expanded, obtuse, hygrophanous, pale umber when large, dark fuscous when small, becoming pale; marg. striate, appendiculate with sc. *St.* subnaked, even, grey-white above, fuscous or slate below. *G.* adnexo- or adnato-rounded, fuscous. *Flesh* pale brownish.

Gregarious. Oak chips, amongst moss, damp sandy ground. Sept.–Dec. $1 \times 1\frac{3}{8} \times \frac{1}{16}$ in.

892. P. microrrhiza Sacc. (from the small rooting stem; Gr. *mikros*, small, *rhiza*, a root) *a b c*.

P. ochreous or rufous, at first yellow-pilose, then pale, atomate. *St.* attenuato-rooting, whitish-scaly. *G.* adnexo-ascending, black-brown.

Gregarious. Gardens, on the ground. Sept. $1\frac{3}{8} \times 4 \times \frac{1}{16}$ in.

893. P. urticæcola Sacc. (from its habitat, nettle-roots; *urtica*, a nettle, *colo*, to inhabit).

P. white-flocculent; marg. striate. *St.* white-flocculent. *G.* adnexo-ascending, rich chocolate.

At the base of old nettle-stems. Aug. $\frac{5}{16} \times 1 \times \frac{1}{16}$ in.

Series V. COPRINARII Fr.

(From the habitat; Gr. *kopros*, dung.)

Spores black. *Gills* from the first separate, not cohering and becoming fluid as in *Coprinus*.

Intermediate between the preceding series and *Coprinus*.

Genera XXXIX–XLI.

XXXIX. **ANELLARIA** Karst.

(From the annulus; *anellus*, a little ring.)

Veil appendiculate at the margin of the pileus. *Hymenophore* confluent with, but heterogeneous from the cartilaginous stem. *Pileus* subfleshy, conico-campanulate, without striæ, shining when dry, not expanding, the margin exceeding the gills. *Stem* central, lustrous, somewhat firm, fistulose, annulate. *Gills* ascending, black or variegated slate-blackish, not deliquescent. (Fig. 48.)

The species usually grow in rich grassy places or on dung, from summer to late autumn, often near towns and villages.

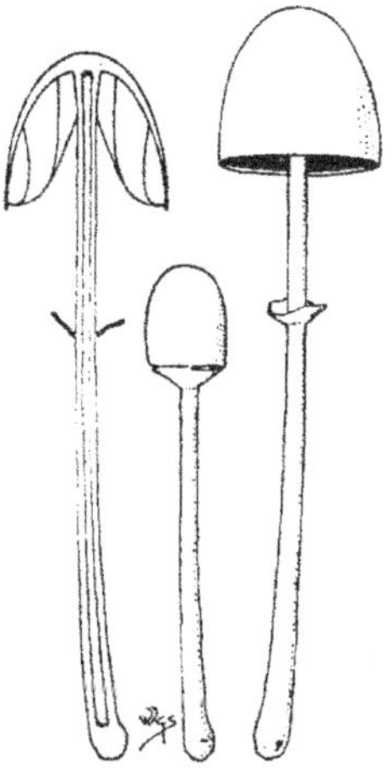

Fig. 48.—*Anellaria separata* Karst.
Entire and in section.
One-third natural size.

The nearest structural allies are found in *Amanita*, and the non-British genera *Metraria* and *Locellina*. Species **894—896**

894. A. separata Karst. (from its appearance, distinct from species placed near it) *a b c*.

P. whitish to tan-ochreous; mid. sienna-tinted. *St.* attenuate upwards, colour as P. *A.* distant, persistent, white. *G.* adnato-ascending, broad, dark purplish-slate.

In pastures on dung; common. April–Dec. $2\frac{1}{2} \times 6\frac{3}{4} \times \frac{1}{4}$ in. Stem sometimes 8 in.

895. A. scitula Sacc. (*scitulus*, handsome) *a b*.

P. dull pale ochre. *St.* white, base peronate, ending as a persistent A. below mid. *G.* adnexo-ascending, ashy-grey to black. *Flesh* white.

On cow-dung, on soil in a flower-pot, Scarborough, 1885. $\frac{7}{16} \times 1\frac{1}{4} \times \frac{1}{16}$ in.

896. A. fimiputris Karst. (from its habitat, rotten dung ; *fimus*, dung, *putris*, rotten) *a b c*.

P. innato-silky, pale umber to dark lead-colour; marg. appendiculate with white V. *St.* colour as P., umber below. *A.* superior, small, imperfect. *G.* adfixo-ascending.

Solitary, rarely cæspitose. Taste insipid. Fields, gardens, horse-dung ; common. April–Dec. $1\frac{1}{2} \times 5\frac{3}{8} \times \frac{1}{8}$ in. Examples with a wrinkled pileus must not be confounded with **900**.

XL. PANÆOLUS Quél.

(From the variegated gills; Gr. *panaiolos*, all variegated.)

Veil appendiculate at the margin of the pileus. *Hymenophore* confluent with, but heterogeneous from the cartilaginous stem. *Pileus* subfleshy, conical without striæ, not expanding, the margin exceeding the gills, viscid when moist, lustrous when dry. *Stem* central, lustrous, somewhat firm, fistulose except **905—8**, simple.

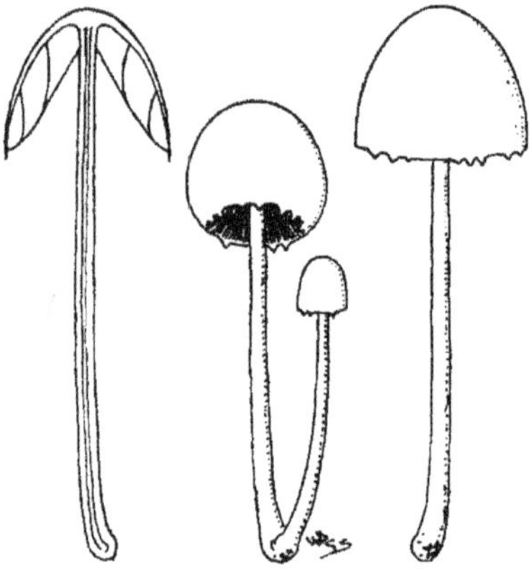

Fig. 49.—*Panæolus campanulatus* Quél.
One-half natural size. Entire and in section.

Gills ascending, black or variegated slate-blackish, not deliquescent. (Fig. 49.)

The species usually grow in rich grassy places or on dung from summer to late autumn, often near towns and villages.

The nearest structural allies are found in *Collybia*, *Leptonia*, *Naucoria* and *Psilocybe*. Species **897—908**

Pileus conico-campanulate with a gelatinous-liquescent cuticle, viscid, shining when dry. *Stem* fistulose, minutely so in **898**. **897—899**

Pileus hemispherical to conico-campanulate, moist or damp, opaque, bibulous, somewhat flocculose when dry. *Stem* fistulose. **900, 901**

Pileus hemispherical to conico-campanulate, dry, smooth, slightly shining, not zoned. *Stem* fistulose. **902—904**

Pileus conical to campanulate, dry, smooth, zoned round the margin. **905—908**

897. P. leucophanes Sacc. (from its white appearance; Gr. *leukos*, white, *phaino*, to appear) *a b*.

P. expanded, obtuse, innato-silky, white or faintly ochreous; marg. appendiculate with V. *St*. mealy, fibrillose, white. *G*. adnato-ascending, black with white edge.

Pastures, grassy places, manure-heaps. Aug.–Nov. $1\frac{1}{2} \times 3\frac{3}{8} \times \frac{1}{8}$ in.

898. P. egregius Sacc. (*egregius*, distinguished) *a b*.

P. slightly agglutinato-downy, watery, virgate when dry, bright orange-brown; mid. darker. *St*. solid, attenuate upwards, fibrous, fibrillose, splitting longitudinally, warm umber, base white-cottony. *G*. adnexo-ascending, ventricose, crowded, brownish- or purple-black. *Flesh* ochreous.

Solitary. Odour none. On the ground, on horse-dung in fields. Jan. $2\frac{1}{4} \times 4\frac{3}{4} \times \frac{3}{8}$ in.

899. P. phalænarum Quél. (from the variegated gills, like the wings of certain moths; Gr. *phallaina*, a moth) *a b c*.

P. obtuse, clay-white; marg. appendiculate with fugacious V. *St*. subnaked, tan-rufescent. *G*. adnexo-ascending, broad, ventricose, ashy-black.

Rich pastures, on dung. Sept.–Oct. $1\frac{3}{4} \times 4\frac{7}{8} \times \frac{1}{8}$ in.

900. P. retirugis Gill. (from the net-like wrinkles on the pileus; *rete*, a net, *ruga*, a wrinkle) *a b c*.

P. obtuse, atomate, fleshy-tan; mid. ochreous-tan or dark umber with paler marg., appendiculate with V. *St*. pruinose, flesh-colour, then purplish, rufescent or cinereous. *G*. adfixo-ascending, cinereous-blackish.

Sometimes fasciculate. Dung, cow, manure-heaps, woods; uncommon. Feb.–Nov. $1\frac{1}{2} \times 4\frac{1}{4} \times \frac{1}{8}$ in

901. P. sphinctrinus Quél. (from the pileus, contracted, never expanded; Gr. *sphigkter*, a muscle naturally contracted) *a b c*.

P. opaque, silky or fibrillose when dry, fuliginous or greyish-shaded, livid when dry; marg. appendiculate with V. *St*. smooth, pruinose above, fuliginous-grey. *G*. adnato-ascending, crowded, cinereous-blackish.

Pastures, on dung and on garden soil. April–Nov. $1 \times 3\frac{1}{4} \times \frac{1}{16}$ in.

902. P. papilionaceus Quél. (from the speckled pileus, as in some butterflies; *papilio*, a butterfly) *a b c*.

P. rimoso-scaly when dry, dull tan whitish or greyish; mid. faintly ochreous. *St*. white pulverulent above, whitish or faintly ochreous. *G*. adnato- or adnexo-ascending, broad, ventricose, blackish.

Pastures, lawns, manured ground, dung; common. July–Nov. $1\frac{3}{4} \times 4\frac{1}{4} \times \frac{1}{8}$ in.

903. P. campanulatus Quél. (from the bell-shaped pileus; *campanula*, a little bell) *a b c*.

P. sometimes umbonate, not cracked, fuliginous-fuscous, or somewhat umber-slate with umber mid., sienna and umber or grey when dry. *St*. smooth, pruinose and grey above, rufescent or purplish below. *V.* fugacious or obsolete. *G.* adnato-ascending, crowded, grey and black, edge whitish.

Pastures, rich ground, horse-dung, amongst grass; common. June–Nov. $1\frac{3}{8} \times 4\frac{1}{4} \times \frac{1}{8}$ in.

904. P. caliginosus Gill. (from its gloomy colour; *caligo*, gloom) *a b c*.

P. obtuse, even, smooth, ochreous-brown or grey. *St.* naked, rufescent-brown, lighter above. *V.* feeble or obsolete. *G.* adnato- or adnexo-ascending, ventricose, fuliginous-black.

Rich pastures, lawns. Oct.–Nov. $\frac{9}{16} \times 2\frac{3}{4} \times \frac{1}{16}$ in.

905. P. subbalteatus Sacc. (from the somewhat zoned margin of the pileus; *balteus*, a band) *a*.

P. obtuse or subumbonate, hygrophanous, irregular, subwrinkled, dull fawn or pale ochreous; mid. pale sienna. *St.* splitting longitudinally, white-fibrillose, red-brown and ochreous above, pale sienna below. *G.* adnato-ascending, ventricose, brownish; edge toothed, white.

Cæspitose. Borders of fields, on the ground. Sept.–Oct. $\frac{3}{4} \times 3\frac{3}{4} \times \frac{1}{16}$ in.

906. P. acuminatus Quél. (from the sharp apex of the pileus; *acuminatus*, sharp pointed) *a b*.

P. even, shining, flesh-tan; mid. faint sienna. *St.* pruinose, tan above, fuscous below. *G.* adnexo-ascending, crowded, grey and black.

Dung, roadsides. Oct.–Nov. $\frac{7}{8} \times 1\frac{1}{2} \times \frac{1}{8}$ in.

907. P. fimicola Quél. (from the usual habitat; *fimus*, dung, *colo*, to inhabit) *a b c*.

P. obtuse, smooth, opaque, fuliginous-grey, clay-hoary when dry; marg. with a fuscous zone. *St.* slightly silky-striate, white-pruinose above, dingy below. *G.* adnato-ascending, ventricose, grey and fuliginous.

Dung, rich pastures; rare. Oct. $\frac{7}{8} \times 3\frac{1}{4} \times \frac{1}{16}$ in.

908. P. cinctulus Sacc. (from the zone round the margin of the pileus; *cinctus*, a girdle) *a*.

P. even, smooth, reddish-cinnamon; marg. zoned, dark brown. *St.* dull fuscous. *G.* ascending, adnexo-free, dusky- or olivaceus-black or slate.

Dunghills after rain. June–July. $2\frac{5}{8} \times 4\frac{1}{4} \times \frac{1}{8}$ in.

XLI. **PSATHYRELLA** Quél.

(Diminutive of *Psathyra*.)

Veil almost obsolete. *Hymenophore* confluent with, but heterogeneous from the cartilaginous stem. *Pileus* membranous, striate; margin not exceeding the gills, straight and at first adpressed to the stem. *Stem* central, fistulose, simple. *Gills* adnate or free, not variegated or becoming fuscous or purple, not deliquescent. *Spores* black or nearly so. (Fig. 50.)

The species grow on rich ground, in gardens, by hedges and stumps, sometimes on wood.

Psathyrella agrees in structure with *Mycena*, *Nolanea*, *Galera* and *Psathyra*. Species **909—920**

Pileus conical to campanulate. *Stem* tense, straight, fistulose. **909—914**

Pileus hemispherical to campanulate or conical. *Stem* flexuous, pruinate at the apex, fistulose. **915—920**

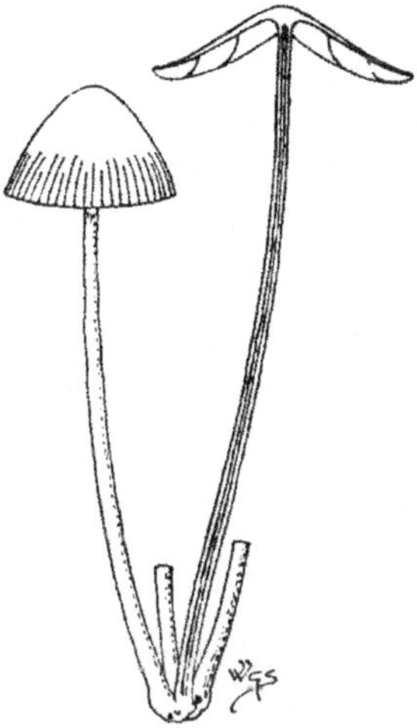

Fig. 50.—*Psathyrella subatrata* Gill.
One-half natural size.
Entire and in section.

909. P. subatrata Gill. (from the blackish pileus; *ater*, black) *a b*.

P. obtuse, subexpanded, umber-rufescent, pallid-rufescent when dry. *St.* smooth, whitish. *G.* adnexo-ascending, crowded, fuliginous, blackish-umber.

Gregarious. Rich ground, grassy places, woods. Sept. $2\frac{1}{8} \times 5\frac{3}{8} \times \frac{1}{8}$ in.

910. P. gracilis Quél. (*gracilis*, slender) *a b c*.

P. obtuse, smooth, dark- or pale-fuliginous or livid with pale shades of rose, tan when dry. *St.* smooth, whitish. *G.* adnate, cinereous-blackish; edge rose.

Gregarious, waysides, hedgerows, etc.; common. July–Oct. $1\frac{7}{8} \times 4 \times \frac{1}{8}$ in.

911. P. hiascens Quél. (from the pileus opening in furrows; *hiasco*, to gape) *a*.

P. expanded, obtuse, thin, cracked-revolute, white or warm tan, becoming pale tan-yellow; mid. pale sienna. *St.* smooth, whitish. *G.* adnate, distant, black.

Grassy places, hedgerows, damp woods, rubbish-heaps; rare. Nov. $1\frac{5}{8} \times 3\frac{7}{8} \times \frac{1}{8}$ in.

912. P. arata Sacc. (from the furrowed pileus; *aro*, to plough) *a*.

P. attenuato-conical, deeply sulcate, sienna-brown. *St.* smooth, whitish or faint sienna-white. *G.* ascending free, purplish-black.

Hedgerows. Autumn. $1\frac{7}{8} \times 5\frac{1}{4} \times \frac{1}{8}$ in.

913. P. trepida Gill. (from its trembling habit; *trepidus*, trembling) *a b c*.

P. obtuse, smooth, fuliginous; mid. date-brown. *St.* naked, tan-whitish. *G.* adnate, crowded, fuliginous-black.

Moist ground, bases of trunks, gardens. Oct. $\frac{7}{8} \times 3\frac{1}{8} \times \frac{1}{16}$ in.

914. P. hydrophora Quél. (from its hygrophanous substance; Gr. *hudor*, water, *phero*, to carry) *a b c*.

P. expanded, obtuse, then cracked-revolute; mid. rufous; marg. paler. *St.* white, studded with drops of moisture. *G.* adnate, crowded, black.

Gardens. Autumn. $1\frac{3}{4} \times 3\frac{1}{2} \times \frac{1}{8}$ in.

915. P. caudata Quél. (from the tail-like base of the stem; *cauda*, a tail) *a b c*.

P. gibbous then flat, smooth, sometimes areolato-cracked, atomate, date-brown or umber. *St.* with rooting base, undulate on surface, white. *G.* adnate, ventricose, ashy-black or dull purplish-brown.

Wooden pavement, rotten wood, rich soil, gardens, amongst dead leaves. Sept.–Dec. $2\frac{3}{4} \times 4\frac{5}{8} \times \frac{1}{8}$ in.

916. P. prona Gill. (from its frequent prostrate state; *pronus*, prostrate) *a b c*.

P. subexpanded, obtuse, subsilky, fuliginous, atomate and hoary when dry. *St.* smooth, whitish. *G.* adnate or adnexed, livid fuliginous, black-dotted.

Grassy places; rare. May–Oct. $\frac{5}{8} \times 1\frac{7}{8} \times \frac{1}{16}$ in. Var. *Smithii* Mass. *P.* hemispherical $\frac{1}{8}$ in. in diam.

917. P. empyreumatica Sacc. (from its odour of burning; Gr. *empuros*, scorched) *a b*.

P. hygrophanous, atomate, rufous, becoming pale. *St.* silky-furfuraceous, pale sienna. *G.* adnate with a tooth, distant, veined, rufous to brown-purple; edge pallid. *Flesh* colour as St.

Wooden pavement. Oct. $1\frac{5}{8} \times 2\frac{1}{2} \times \frac{1}{8}$ in. Resembles **591**.

918. P. atomata Quél. (from the atomate pileus) *a b c*.

P. obtuse, hygrophanous, livid, tan or pale flesh-colour, sometimes reddish. *St.* white. *G.* adnate, cinereous-blackish or purple-brown.

Solitary or gregarious. Woods, hedgerows, waysides, amongst leaves; common. June–Jan. $1 \times 2\frac{5}{8} \times \frac{1}{16}$ in.

919. P. crenata Gill. (from the scalloped margin of the pileus; *crenatus*, scalloped) *a b*.

P. hygrophanous, atomate, subochreous, rufescent or flesh-colour, then pale; mid. sometimes sienna. *St.* attenuate below, whitish. *G*. adnato-ascending, dull brownish-purple to blackish.

Amongst grass, amongst beech leaves with **1973**. Nov. $1\frac{1}{8} \times 2\frac{1}{8} \times \frac{1}{8}$ in.

920. P. disseminata Quél. (from the scattered habit; *dissemino*, to scatter) *a b c*.

P. scurfy, then smooth, whitish or pale umber, then purple-grey and ashy; mid. grey-umber. *St*. subscurfy, then smooth, whitish. *G*. adnato-ascending, blackish.

Densely crowded, gregarious, cæspitose. About decayed stumps, poplar, willow, piles, in hedgerows, on brick walls, heaps of weeds and rubbish, etc. April–Nov. $\frac{1}{2} \times 2\frac{1}{2} \times \frac{1}{16}$ in.

XLII. **COPRINUS** Pers.

(From the habitat; Gr. *kopros*, dung.)

Veil usually universal, floccose, furfuraceous or obsolete. *Hymenophore* distinct from the stem. *Pileus* at first cylindrical or ovate,

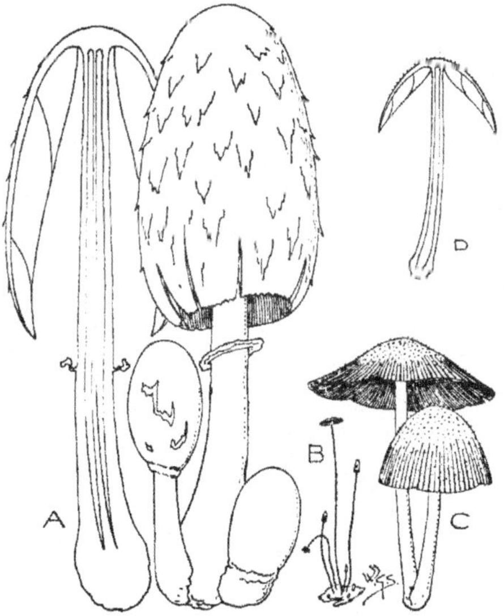

Fig. 51.—A, *Coprinus comatus* Fr., entire and in section; B, *C. radiatus* Fr.; C, *C. domesticus* Fr.; D, ditto in section. One-third natural size.

usually floccose or scurfy; margin at first straight and adpressed to the stem; flesh thin to membranous. *Stem* central, hollow, usually attenuate upwards, annulate, volvate or simple, confluent with or distinct from the hymenophore. *Gills* membranous, linear or ventricose, at first white and closely pressed together, cohering, scissile, at length melting into black fluid. *Trama* obsolete. *Spores* oval, even, black or dark brown. (Fig. 51.)

The species are of rapid growth, often ephemeral. They grow on rich and highly manured places in fields, parks, gardens and near human habitations, often on dung, sometimes on decaying wood, rotting straw, etc. Some are edible. Species **921—963**

Mr. George Massee, in a revision of the genus (Ann. Bot. x. (1896), p. 123), establishes new sectional characters, and the sequence of the species, as here given, is somewhat altered.

PELLICULOSÆ. *Pileus* with a pellicle not opening into furrows over the gills, but becoming torn and revolute.

a. Comatæ. *Pileus* slightly fleshy, at first ovate, the cuticle torn into large or small scales from the concrete veil; margin striate or sulcate. *Stem* annulate, volvate or both. *Gills* free or remote. **921—925**

b. Atramentariæ. *Pileus* slightly fleshy, at first ovate, dotted or spotted with innate scurfy squamules from the veil. *Stem* imperfectly ridged-annulate at the base. *Gills* free. **926—928**

c. Picaceæ. *Pileus* submembranous, at first continuous, then broken up into usually large flocculose superficial scales —warts in **932.** *Stem* simple. *Gills* free. **929—932**

d. Tomentosæ. *Pileus* membranous, at first evanescent-floccose or laxly villous from the veil, margin striate. *Stem* simple. *Gills*, except **936**, free. **933—937**

e. Micaceæ. *Pileus* membranous, covered with fine, evanescent scurf and minute glittering particles from the veil, margin striate or sulcate. *Stem* simple, or with a slight basal annular ridge. *Gills* reaching the stem, adnexed in **938.** **938—941**

f. Glabratæ. *Pileus* membranous, smooth; margin striate; veil obsolete. *Stem* simple, smooth. **942—946**

VELIFORMES. *Pileus* very thin, without a pellicle, opening into furrows over the gills and becoming plicato-sulcate. *Veil* present. *Stem* thin, fistulose. *Gills* melting into thin, black lines. **947—963**

g. Cyclodeæ. *Pileus* membranous, striate. *Stem* annulate or volvate. *Gills* free. **947, 948**

h. Lanatulæ. *Pileus* submembranous to membranous, clothed with superficial, separating floccules. *Stem* simple. *Gills* free or remote. **949—955** *i. Furfurellæ.* *Pileus* micaceous or furfuraceous, striate or sulcate. *Stem* simple. *Gills* adnate, adnexed, free or remote. **956—961**

j. Hemerobiæ. *Pileus* smooth. *Stem* simple. **962**

k. Basipedes. *Pileus* flocculose. *Stem* dilated, disc-like at the base. **963**

PELLICULOSÆ.

a. Comatæ.

921. C. comatus Pers. (from the shaggy pileus; *coma*, a hair) *a b.*

P. cylindrical, then expanded, white; sc. large. *St.* fibrillose, white. *A.* ragged, movable, evanescent. *Vo.* adnate, almost obsolete. *G.* white, through reddish and purple to black.

Often cæspitose. Edible when young, used for inferior ketchup, sometimes candied with sugar, sold in American markets, at Boston, as the "English Mushroom." Gardens, parks, fields and roadsides, often near human habitations, on the ground, on rotten leaves; frequent. April–Nov. $4\frac{1}{2} \times 8\frac{3}{4} \times \frac{3}{4}$ in.

922. C. ovatus Fr. (from the egg-shaped pileus; *ovum*, an egg) *a.*

P. at length expanded, densely imbricato-scaly, white; mid. pale buff. *St.* flocculose, rooting, scaly. *A.* ragged, movable, evanescent. *Vo.* adnate, almost obsolete. *G.* white to umber-blackish.

Usually solitary. Rich ground, pastures. $3 \times 5\frac{1}{2} \times \frac{5}{8}$ in. Intermediate forms occur between this and **921.**

923. C. umbrinus Cooke & Mass. (from the colour of the pileus) *a.*

P. campanulato-plane then revolute, dark umber; sc. white. *St.* subrooting, bulbous from the Vo., lustrous dark umber. *G.* black, edge white. *Flesh* pale umber.

Manured ground. Aug. $2 \times 6\frac{1}{2} \times \frac{3}{16}$ in. Perhaps a form of *C. stenocoleus* Lindbl.

924. C. sterquilinus Fr. (from the habitat; *sterquilinium*, a dunghill) *a.*

P. campanulato-expanded, whitish; marg. pale umber; sc. small. *St.* whitish. *A.* inferior, sometimes forming an adnate Vo. *G.* slate-purple.

Dung, cow, manured grounds, dung-heaps; rare. July–Aug. $2 \times 4\frac{3}{4} \times \frac{5}{16}$ in.

925. C. oblectus Fr. (from its growing on—delighting in—dung; *oblecto*, to delight) *a.*

P. expanded, minutely scaly, then silky-smooth, white then sienna; mid. ochreous; marg. dull purplish. *St.* silky-white. *Vo.* large, free, spreading. *G.* black.

Newly-made dunghills, manure-heaps, street sweepings. Nov. $1\frac{1}{2} \times 4\frac{1}{2} \times \frac{1}{4}$ in. A close ally of **924.** Mr. C. Crossland, who has found this plant at Halifax, says that the pileus is covered with white silky scales and always passes from white through bright flesh-colour to black, and the annulus sometimes projects nearly $\frac{1}{4}$ inch from the stem. In **924** the gills are at length ventricose, but in this plant, as described by Bolton, they are linear and slightly arcuate and pass from white through purple to black. In **924** the stem turns black when bruised, but there is no change of colour in the Halifax plant.

925a. C. squamosus Morg. (from the scaly pileus and stem; *squama*, a scale) *a*.

P. submembranous, ovoid then expanded, cinereous, covered with persistent reddish-brown sc., becoming split and revolute. *St.* hollow, almost equal, smooth and white above, scaly as P. below. *A.* median, small. *G.* free, ventricose, white, then reddish-brown and purple-brown, becoming black.

Cæspitose. Base of trunks, elm. Sept. $3\frac{1}{8} \times 4\frac{1}{4} \times \frac{3}{8}$ in.

b. Atramentariæ.

926. C. atramentarius Fr. (from the ink-like fluid into which the gills deliquesce; *atramentum*, ink) *a b c*.

P. campanulato-expanded, ribbed, sulcate, slate-white, then slate-purplish. *St.* fibrillose, white, pale umber at base. *A.* forming a fugacious ridge near base of St. *G.* grey-umber to umber-purple and black.

Cæspitose. Said to be edible, used for inferior ketchup, preferred by some to **921**. Rich soil by grassy roadsides, near stumps, usually near human habitations, amongst rubbish, has appeared in abundance on burnt coffee, rich soil, does not grow on dung; common. June–Nov. $3\frac{1}{2} \times 6\frac{3}{4} \times \frac{3}{8}$ in. A group of this fungus lifted a large mass of asphalt paving in Hampton Road in 1889; a similar occurrence took place at Dunstable in 1899.

927. C. soboliferus Fr. (from its cæspitose habitat; *soboles*, an offshoot, *fero*, to bear) *a b*.

P. irregularly conical, abruptly truncate, then expanded, ribbed, grey; mid. pale sienna-brownish. *St.* furrowed, bluntly rooting, whitish. *A.* forming a fugacious ridge near base of St. *G.* dark brown-purplish or blackish.

Cæspitose. Base of rotten wooden palings, fences, stumps, elm. Spring and Autumn. $2\frac{1}{8} \times 8 \times \frac{3}{4}$ in. The mycelium produces two crops annually. Intermediate forms between this and **926** occur. Sometimes both forms grow in one group.

928. C. fuscescens Fr. (from the dusky colour of the pileus; *fuscus*, dark, swarthy) *a b*.

P. campanulate, then revolute, somewhat ribbed, tan; mid. brownish; marg. purplish. *St.* whitish. *A.* forming a fugacious ridge near base of St. *G.* purple-umber to black.

Cæspitose. Stumps, rotten wood, wood in cellars, elm, in woods; uncommon. May–Dec. $2\frac{3}{8} \times 4\frac{1}{2} \times \frac{1}{4}$ in. Var. *rimososquamosus* Cooke. *P.* cracked into angular squamose patches.

c. Picaceæ.

929. C. picaceus Fr. (from the white patches on the black ground of the pileus; *pica*, a magpie) *a b c*.

P. campanulato-expanded, very smooth, then revolute. *St.* smooth, somewhat bulbous, biscuit, furfuraceous below. *G.* ventricose, grey, then black.

Somewhat cæspitose. Considered poisonous. Odour often fetid. Grassy places, roadsides, borders of woods, beech, rarely on rotten wood; rare. Sept.–Dec. $4 \times 11 \times \frac{1}{2}$ in.

930. C. aphthosus Fr. (from the pileus, like a "thrush"-diseased tongue; Gr. *aphthai*, the "thrush") *a b.*

P. campanulate, pale dull ochreous; sc. darker, grey-brown or dull sienna. *St.* fibrillose, glossy white. *G.* adnate, linear, black.

Subcæspitose. Hollow trees, wood in cellars, kitchens; rare. Autumn. $1\frac{3}{4} \times 3\frac{1}{2} \times \frac{3}{16}$ in. Var. *Boltonii* Mass. *P.* repand, olivaceous.

931. C. flocculosus Fr. (from the white-floccose scales of the pileus) *a.*

P. campanulate, pale purplish-grey-tan; mid. brownish; marg. striate. *St.* whitish. *G.* slate-purple to fuscous-black.

Usually solitary, sometimes cæspitose. Pastures, gardens; rare. June–Sept. $4 \times 5 \times \frac{3}{8}$ in.

932. C. similis B. & Br. (from its resemblance to **930**; *similis*, like) *a.*

P. campanulate, biscuit-white; sc. wart-like, acute, brownish at apex; marg. striate. *St.* white. *G.* adnate, sublinear, brownish near edge.

Trunks of dead trees. Sept. $1\frac{1}{2} \times 3 \times \frac{1}{8}$ in.

d. Tomentosæ.

933. C. exstinctorius Fr. (from the pileus, shaped like an extinguisher, *exstinctor*) *a b c.*

P. campanulate, tan- or ochreous-white; mid. livid; marg. splitting. *St.* subrooting, becoming smooth, whitish. *G.* lanceolate, fuscous-blackish to black.

On the ground, sandy places, about trunks, ash. Sept. $3 \times 6 \times \frac{1}{4}$ in.

934. C. fimetarius Fr. (from the habitat; *fimetum*, a dunghill) *a b c.*

P. campanulato-expanded, then cracked and revolute, white-floccose, becoming grey-slate and black. *St.* subrooting, white villous. *G.* narrow-lanceolate, black.

Frequently clustered, sometimes solitary. Dung, horse, rich soil; common. April–Oct. $2\frac{5}{8} \times 8\frac{1}{4} \times \frac{3}{16}$ in. Var. *pullatus* Fr. *P.* soon naked and black. Dunghills, rich meadows. Var. *macrorhizus* Fr. *St.* bulbous, rooting. Var. *cinereus* Fr. *P.* ashy-grey.

935. C. tomentosus Fr. (from the downy pileus; *tomentum*, down) *a b.*

P. campanulato-expanded, flocculose, grey-white. *St.* white-villous. *G.* narrow-lanceolate, fuscous-blackish, edge at first white.

Subcæspitose. Dung, rich pastures, gardens. Sept. $2\frac{1}{4} \times 3 \times \frac{3}{16}$ in.

936. C. niveus Fr. (from the snow-white pileus; *nix*, *nivis*, snow) *a b c.*

P. at length flat, revolute, mealy-floccose or squamulose. *St.* white-downy, sometimes imperfectly annulate. *G.* adnexed or adnate, blackish to black.

Cæspitose. Dung, horse, deer, sometimes invading mushroom-beds. June–Dec. $2\frac{1}{4} \times 4\frac{1}{2} \times \frac{3}{16}$ in. There is a var. with white gills and (when present) colourless spores.

936a. C. roseotinctus Rea (from the rose-coloured meal on the pileus and stem) *a*.

P. membranous, cylindrical, then expanded and revolute, umbonate, fuscous, densely powdered with deep rose-coloured meal; marg. sulcate. *St.* hollow, semibulbous below, when young powdered as P., less so at maturity, base white-floccose. *G.* adnexed, at first white, becoming black.

On the ground. Aug. $1\frac{1}{8} \times 2\frac{3}{8} \times \frac{1}{10}$ in. Allied to **936** and **946a.**

937. C. cothurnatus Godey in Gill. (from the stem, sheathed with woolly squamules at the base, as if buskined; *cothurnus*, a buskin) *a b*.

P. conico-expanded, revolute, reddish-white or pale buff. *St.* white. *G.* narrow lanceolate, salmon to blackish.

Dung, cow. Autumn. $1\frac{1}{4} \times 1\frac{7}{8} \times \frac{1}{8}$ in.

e. Micaceæ.

938. C. micaceus Fr. (from the minute shining particles on the pileus; *mica*, a shining grain) *a b c*.

P. campanulato-expanded, lobed, splitting, yellow-ferruginous; mid. darker; marg. plicate, becoming purple-brown. *St.* white-silky or fibrillose, with an annular ridge near base, white or pale ochreous-whitish. *G.* adnexed, purple-brown to dark slate.

Densely clustered. Amongst grass, at base of stumps and posts, gardens; common. Jan.–Nov. $2\frac{7}{8} \times 7 \times \frac{3}{16}$ in. Two to three crops annually. Pileus sometimes white.

938a. C. truncorum Fr. (from its habitat, tree-trunks) *a*.

P. at first globose, then campanulate, densely covered with micaceous meal, soon naked, then tawny ochreous; marg. striate, not becoming sulcate. *St.* glabrous, hollow, white. *G.* free, rosy, then black.

Rotten willows, etc. $1\frac{1}{4} \times 2\frac{3}{8} \times \frac{1}{8}$ in. Distinguished from **938** by the P. being at first globose, with marg. striate, not plicate, and with free rosy gills.

939. C. aratus B. & Br. (from a fancied resemblance in the sulcate pileus to furrows; *aro* to plough) *a b*.

P. campanulate then revolute, umber. *St.* smooth or silky, with a slight annular ridge near base, white, pale sienna-tan within. *G.* lanceolate, dark brown to purple-brown and black.

Solitary. Hollow trees, rich ground, dunghills. May–Aug. $2\frac{3}{4} \times 6\frac{1}{2} \times \frac{1}{4}$ in.

940. C. radians Fr. (from the radiating mycelium at the base of the stem) *a b c*.

P. campanulato-expanded, then revolute, tawny; mid. sienna. *St.* at first with an annular ridge, then smooth, white or faintly ochreous. *G.* lanceolate, reddish-purplish-brown to violaceous-black.

Subcæspitose. Damp plaster walls, sculleries, cracks in ceilings; uncommon. Jan.–July. $1\frac{3}{4} \times 4\frac{1}{2} \times \frac{1}{8}$ in. The usually white or olive-ochre mycelium is sometimes black, resembling radiating threads of black silk.

941. C. papillatus Fr. (from the minute umbo on the pileus; *papilla*, a teat) *a b*.

P. campanulato-flat, then revolute; um. prominent, ochreous-umber; mid. darker. *St*. smooth, flocculose below, whitish. *G*. ventricose, blackish.

On the ground, amongst grass, dung. June. $\frac{1}{8} \times 1\frac{1}{4} \times \frac{1}{16}$ in. Less fugacious than other *Coprini*. Var. *oxygena* Fr., whitish or greyish. *P*. and *St*. sparingly flocculose.

f. Glabratæ.

942. C. alternatus Fr. (from the striæ of the pileus, alternately broad and narrow) *a c*.

P. hemispherical, whitish; mid. umber, sometimes pale. *St*. whitish. *G*. adnate, narrow, purple-black to black.

Subcæspitose. On the ground. $1\frac{3}{4} \times 4\frac{1}{4} \times \frac{3}{16}$ in.

943. C. deliquescens Fr. (from the dissolving gills; *deliquesco*, to dissolve) *a c*.

P. campanulato-expanded, grey or livid; marg. granuloso-punctate, faintly ochreous. *St*. shining white. *G*. free, slate to blackish.

Subcæspitose. Stumps, amongst fallen leaves, thatch; frequent. Aug.–Sept. $2\frac{3}{4} \times 4 \times \frac{3}{16}$ in.

944. C. tardus Karst. (from its sometimes late appearance; *tardus*, slow, late) *a b*.

P. campanulato-expanded, viscid, bright brown to pale ochreous-tan. *St*. thinly pruinose, shining white, minutely striate above, woolly below. *G*. adnexed, brown to shining black.

Densely cæspitose. Odour none. On the ground. Oct. $2\frac{1}{8} \times 6\frac{1}{4} \times \frac{3}{16}$ in. Brittle, hygrophanous.

945. C. digitalis Fr. (from the finger-like young state; *digitus*, a finger) *a*.

P. conico-hemispherical, faint ochre; mid. sienna, then yellowish-grey or olive tinted. *St*. faint ochreous-white. *G*. adnexed, blackish.

Tufted. On the ground, woody places. Autumn. $1\frac{1}{2} \times 4\frac{1}{4} \times \frac{3}{16}$ in.

946. C. congregatus Fr. (from the densely cæspitose habit), *a b c*.

P. campanulato-expanded, viscid, ochreous; mid. sienna. *St*. faintly ochreous-white. *G*. adnexo-free, linear, black.

On the ground by roadsides, hothouses. Oct.–Nov. $1 \times 1\frac{3}{8} \times \frac{1}{16}$ in.

VELIFORMES.

g. Cyclodeæ.

946a. C. dilectus Fr. (from its beauty; *dilectus*, beloved) *a b*.

P. cylindrical, then campanulate, obtuse, floccose or mealy, rosy-white, then pale, tawny; marg. finely striate, at length splitting and becoming revolute and naked. *St*. hollow, whitish, sprinkled rosy-pulverulent. *Vo.* reduced to whitish spreading sq. *G.* free, sublanceolate, reddish-brown, then black.

Cæspitose. On the ground, on dead and on partially buried hawthorn sticks. Nov. $\frac{7}{8} \times 2\frac{3}{4} \times \frac{1}{10}$ in. Allied to **936a.** Differing from **925** in very rudimentary volva and generally smaller size.

947. C. Hendersonii Fr. (after J. L. Henderson) *a b c.*

P. campanulato-expanded, minutely granular, tan; mid. sienna; marg. grey. *St.* tan. *A.* inferior, erect. *G.* subdistant, black, edge white.

Hotbeds, in fields on dung, horse, cow; rare. Feb.–Sept. $\frac{1}{2} \times 1 \times \frac{1}{32}$ in.

947a. C. Bresadolæ Schulz. (after J. Bresadola).

P. subcylindrical, membranous, greyish-white, faintly brownish at mid. *St.* hollow, attenuate upwards, glabrous, white. *A.* deciduous. *G.* narrow, crowded, black, with a white edge.

Gregarious. Odour none. On wood, squared wood, near woods. $\frac{5}{8} \times 4\frac{3}{4} \times \frac{1}{8}$ in. Not unlike a small **926**, but whiter in colour.

948. C. volvaceominimus Crossl. (from its volva and small size) *a.*

P. campanulate, ashy-grey, white-squamulose; mid. darker. *St.* hyaline. *G.* adnexo-free, linear, blackish-purple.

Decaying straw on dunghills. $\frac{3}{16} \times \frac{7}{8} \times \frac{1}{32}$ in.

h. Lanatulæ.

949. C. lagopus Fr. (from the woolly stem, like a hare's foot, Gr. *lagopous*) *a b.*

P. campanulato-expanded, whitish or greyish; mid. sienna. *St.* densely white-woolly. *G.* remote, linear, black.

Sometimes cæspitose. Inodorous. Rich soil, dung, rotten wood, shady places in woods, rooting in rotten sawdust; rare. July–Oct. $2\frac{1}{4} \times 5\frac{1}{2} \times \frac{1}{4}$ in.

950. C. narcoticus Fr. (from the opium-like odour) *a b.*

P. campanulato-expanded, then revolute, villous-scaly, then naked, white or greyish; mid. pale sienna. *St.* villous, then smooth, faintly ochreous-white. *G.* lanceolate, blackish.

Odour very powerful, like opium, but fetid, or narcotic-alkaline, like **1211.** Manure-heaps. Oct. $1 \times 1\frac{3}{4} \times \frac{1}{16}$ in.

951. C. tuberosus Quél. (from the tuber at the base of the stem) *a.*

P. campanulato-expanded, then revolute, striate, white-pulverulent. *St.* attenuate above and below, villous-white, springing from a comparatively large sclerotium. *G.* linear, violet-black.

Dung, cow, decaying vegetable matter. Oct.–Dec. $\frac{3}{8} \times 2\frac{3}{4} \times \frac{1}{16}$ in.

952. C. macrocephalus Berk. (from the long-headed young state; Gr. *makros*, long, *kephale*, the head) *a.*

P. campanulate, slate; mid. brown, white-scaly. *St.* downy, sprinkled with fine spreading fibres, dull white. *G.* linear, black.

Subcæspitose. Putrid dung; rare. March–Sept. $1 \times 2 \times \frac{1}{16}$ in.

953. C. nycthemerus Fr. (from its length of life, a night and day, Gr. *nuchthemeron*) *a b*.

P. at length flat, then revolute, furfuraceo-flocculose, then naked, grey; mid. sienna. *St*. flaccid, smooth, white. *G*. linear, blackish.

Subcæspitose. Dung, manured ground, decayed straw, tan; rare. July–Oct. $1\frac{1}{8} \times 3 \times \frac{1}{32}$ in. Sometimes very small.

953a. C. cordisporus Gibbs (from the heart-shaped spores; *cor*, the heart, Gr. *sporos*, a seed) *a*.

P. cylindrico-ovate, then expanded and upturned, very thin, plicato-sulcate; marg. crenate, whitish or pallid ochraceous, furfuraceo-papillose and tawny at mid. *St*. filiform, glabrous, except at the densely strigoso-squamulose base, whitish hyaline. *G*. free, somewhat narrow. *Spores* heart-shaped, compressed.

Dung, horse, sheep, rabbit. April. $\frac{5}{16} \times 1 \times \frac{1}{32}$ in.

954. C. radiatus Pers. (from the radiating striæ or furrows of the pileus) *a b c*.

P. at length flat, downy, then naked, whitish, soon translucent; mid. pinkish-sienna. *St*. minutely pruinate or smooth, hyaline. *G*. few, pallid blackish

In troops. In fields on dung, horse, cow; common. May–Nov. $\frac{3}{8} \times 3 \times \frac{1}{64}$ in. Often minute.

954a. C. Gibbsii Mass. & Crossl. (after Thomas Gibbs of Sheffield, the first observer) *a*.

P. hemispherical, then expanded, striate, glabrous, minutely atomate, pale ochreous or very pale buff; mid. darker. *St*. very slender, glabrous, pellucid shining white, attached by a few white strands of mycelium. *G*. adnate, 5–7 in number.

Dung, sheep, horse. Nov. $\frac{1}{4}$ in. high. Probably the smallest agaric known; differs from **954** in its smaller size, glabrous pileus and stem, and in its circular, compressed spores.

955. C. Spraguei Berk. & Curt. (after Charles J. Sprague) *a*.

P. campanulato-expanded, tomentose, grey; mid. sienna. *St*. pale cinnamon. *G*. few, narrow, grey-black.

Gardens. July. $\frac{3}{4} \times 2\frac{1}{8} \times \frac{1}{16}$ in.

i. Furfurellæ.

956. C. domesticus Fr. (from its usual habitat; *domus*, a house) *a b c*.

P. campanulato-expanded, furfuraceous, whitish, greyish or brownish; mid. sienna. *St*. adpressedly silky, white. *G*. adnexed, linear, reddish to fuscous-blackish.

Often cæspitose. In and near human habitations, greenhouses, gardens, cellars, cupboards, ceilings, under floors, on burnt ground, matting, damp carpets, rubbish, etc.; frequent. April–Dec. $3 \times 3\frac{1}{2} \times \frac{1}{4}$ in. Sometimes very light in colour.

957. C. stercorarius Fr. (from the habitat; *stercus*, dung) *a b c.*

P. campanulato-expanded, then revolute, clad with dense white micaceous meal; mid. pale sienna; marg. faint greyish or brownish. *St.* pruinate, white. *G.* adnexed, narrow, black.

Rich soil, dung, roadsides. $1\frac{1}{2} \times 4\frac{1}{4} \times \frac{1}{8}$ in. Brefeld illustrates this plant as growing from a sclerotium. Must not be confused with **936**, which is usually much smaller and with a white floccose pileus.

957a. C. stellaris Quél. (from the pileus splitting in a star-like manner) *a.*

P. ovoid then campanulate, striate, afterwards split in star-shape and crowned with points formed of minute diaphanous vescicles, white, becoming grey. *St.* filiform, fistulose, hyaline, velvety with long silky hairs, white. *G.* narrow, adnexed, attenuate, greyish then sooty-brown.

Dung, fox, man. Spring and Summer. $\frac{1}{10} \times \frac{5}{8} \times \frac{1}{40}$ in.

958. C. ephemerus Fr. (from its length of life; Gr. *ephemeros*, lasting for a day) *a b c.*

P. campanulato-expanded, delicately scurfy, white; mid. prominent, faint sienna; marg. greyish. *St.* smooth, whitish. *G.* adnexed, linear, fuscous then black.

Dunghills, manured ground, roadsides; common. May–Oct. $1\frac{1}{8} \times 2\frac{3}{4} \times \frac{1}{16}$ in. Must not be confounded with **960**, which has a depressed disc to the pileus.

958a. C. velox Godey in Gill. (from its rapid growth; *velox*, swift) *a.*

P. obovate, striate then plicato-sulcate, scurfy between the furrows and on the mid., grey. *St.* white-floccose. *G.* narrow, grey then black.

Dung, ox, horse. $\frac{3}{8} \times 2\frac{3}{8} \times \frac{1}{20}$ in. Must not be mistaken for **954a.**

958b. C. aquatilis Peck (from its habitat, watery places; *aqua*, water) *a.*

P. ovate, then campanulate, coarsely sulcate to mid., at first somewhat furfuraceous, when young pale orange, then red-brown; mid. darker. *St.* somewhat swollen at base, hollow, smooth below, pulverulent above, dull white. *G.* adnexed, grey then black. *Flesh* in P. pale orange.

Decaying twigs, moss, *Luzula sylvatica*, etc., partly submerged in shaded swamps. June. $1 \times 2\frac{1}{8} \times \frac{1}{12}$ in.

959. C. sociatus Fr. (from its habit, growing in groups; *socio*, to associate) *a.*

P. at length expanded, slightly scurfy, umbilicate, fuscous; mid. brown, becoming pale. *St.* white. *G.* adnexed, ashy-black.

Damp, loose soil, gardens, walls. Oct. $1\frac{1}{2} \times 2$ in.

960. C. plicatilis Fr. (from the folds or furrows of the pileus; *plico*, to fold) *a b c.*

P. expanded, grey and livid-umber; mid. depressed, sienna. *St.* smooth, pallid to ochreous. *G.* remote, linear, slate.

Pastures, gardens, roadsides, under trees, woods; common. April–Oct. $1\frac{5}{8} \times 4 \times \frac{1}{16}$ in. Sometimes very small.

961. C. filiformis B. & Br. (from the thread-like stem; *filum*, a thread) *a*.

P. campanulato-expanded, grey, white-mealy; mid. sienna. *St*. pallid or white. *G*. linear, adnate, blackish.

On the ground in woods. Sept. $\frac{1}{8} \times \frac{5}{16} \times \frac{1}{32}$ in.

j. Hemerobiæ.

962. C. hemerobius Fr. (from the length of its life—one day; Gr. *hemera*, a day, *bios*, life) *a*.

P. campanulato-expanded, umber. *St*. smooth, pallid. *G*. adnate, linear, blackish.

Damp, shady, rich grassy places, roadsides, etc.; rare. Sept.–Oct. $1\frac{5}{8} \times 5 \times \frac{1}{8}$ in.

k. Basipedes.

963. C. PLATYPUS Berk. in Cooke (from the basal disc; Gr. *platus*, broad, *pous*, a foot) *a*.

P. campanulato-expanded, white, then ochreous; mid. umber. *St*. even, whitish. *G*. free, narrow, distant, white.

Palm-stems in conservatories. $\frac{3}{16} \times \frac{5}{16} \times \frac{1}{32}$ in.

XLIII. BOLBITIUS Fr.

(From a frequent habitat of the species; Gr. *bolbiton*, cow-dung.)

Veil rudimentary, indicated by squamules on the pileus or stem, or obsolete. *Hymenophore* distinct from the stem. *Pileus* moist,

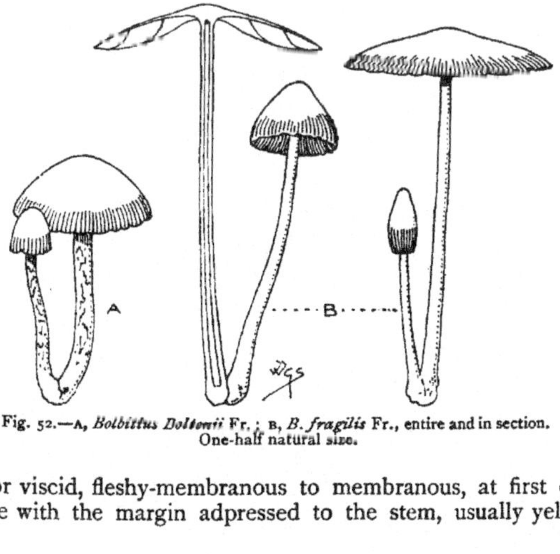

Fig. 52.—A, *Bolbitius Boltonii* Fr.; B, *B. fragilis* Fr., entire and in section. One-half natural size.

mucid or viscid, fleshy-membranous to membranous, at first conical or ovate with the margin adpressed to the stem, usually yellow or salmon; margin striate or sulcate. *Stem* central, simple, equal or attenuate upwards, fistulose. *Gills* adnexed or free, membranous, becoming moist, not deliquescing. *Spores* somewhat ferruginous or salmon. (Fig. 52.)

A group of species resembling the *Coprini* in habit and short life, but not deliquescing. They chiefly grow on dung or manured ground, often near and in roads, frequently near towns, villages and human habitations. None are edible. Species **964—973**

964. B. grandiusculus Cooke & Mass. (from its well-grown appearance; *grandiusculus*, well-grown) *a b*.

P. campanulato-expanded, papillose, ochreous-tan; mid. brownish-orange; marg. whitish-salmon. *St.* white. *G.* linear, adnexed or free, orange or rusty-ochre.

Amongst grass. Autumn. 2 × 4¼ × ⅛ in.

965. B. flavidus Mass. (*flavus*, light yellow) *a b*.

P. campanulato-expanded, glutinous, olivaceous-ochre; mid. somewhat elevated, sienna; marg. striate. *St.* pale yellow within and without, fibrillose below, faintly olivaceous. *V.* fugacious, white. *G.* adnexo-free, white, yellow then dusky brown.

Dunghills after rain, rotten cloth in wood. June–Aug. 2½ × 2½ × 3/16 in. "Whole plant dissolves in a brown jelly," Bolton.

966. B. Boltonii Fr. (after James Bolton of Halifax) *a b*.

P. flat, sulcate, viscid, sulphur-white, then dull salmon-whitish; mid. somewhat depressed. *St.* floccose, sulphur-white. *V.* fugacious. *G.* subadnate, white to yellow, pale dull salmon or livid brown.

Heaps of leaves and dung. May–Sept. 1⅝ × 3 × ⅛ in.

967. B. vitellinus Fr. (from the colour of the pileus, like yoke of egg, *vitellus*) *a*.

P. campanulate, lobed, sulcate, viscid. *St.* white squamulose, faint sulphur at base. *G.* slightly adnexed or free, ochreous-tan.

Rich pastures, dung, horse. Autumn. 1½ × 3⅜ × 3/16 in.

968. B. fragilis Fr. (from its fragility) *a b c*.

P. flat, viscid, sulphur-white, straw or yellow. *St.* at first covered with fine matted down, colour as P. *G.* attenuato-adnexed, yellow, faintly olivaceous then pale cinnamon.

Road-drift and dung, horse, roadsides; common. April–Nov. 2¼×4¼×⅛ in.

969. B. titubans Fr. (from its tremulous habit; *titubans*, shaking) *a b c*.

P. flat, subdepressed, viscid, faint salmon-purplish; mid. faintly ochreous. *St.* even, shining-yellowish, sometimes white, squamulose at base. *G.* adnexed, narrow, colour as P.

Odour unpleasant. Grassy richly-manured soil, mushroom-beds, by rotten stumps; common. May–Oct. 2 × 3¾ × 1/16 in.

970. B. apicalis W. G. Sm. in Cooke (from the well defined apex of the pileus) *a b*.

P. conical, plicate to apical disc, tan-sienna or tan-umber; mid. abruptly ochreous. *St.* striate, minutely pruinose, salmon-whitish. *G.* free, ventricose, sienna.

Pastures. June. $\frac{1}{2} \times 1\frac{1}{2} \times \frac{1}{16}$ in.

971. B. NIVEUS Mass. (from the white pileus; *niveus*, snowy) *a b*.

P. campanulato-expanded, striate halfway, subumbonate, slightly viscid; mid. cracked-areolate, faint ochre. *St.* clavato-bulbous, striate above, mealy-pruinose, white. *G.* free to adnate, whitish to rufescent salmon.

On earth, in palmhouse at Kew. $1\frac{1}{4} \times 4\frac{1}{4} \times \frac{1}{8}$ in. Very delicate and fragile.

972. B. tener Berk. (from its tender substance; *tener*, tender) *a b*.

P. smooth, even, white; mid. faintly ochreous; marg. faint salmon, sometimes wholly salmon. *St.* bulbous, white. *G.* adnexed, salmon.

Rich grassy ground; uncommon. May–Sept. $1 \times 3\frac{3}{4} \times \frac{1}{16}$ in. Sometimes 8 ins. high.

973. B. RIVULOSUS B. & Br. (from the rivulose pileus) *a*.

P. campanulate, dull tan or clay. *St.* white. *G.* free, narrow, cinnamon.

Earth in orchid-house. July. $1\frac{1}{4} \times 3 \times \frac{1}{8}$ in.

XLIV. CORTINARIUS Fr.

(From the *cortina* or *veil.*)

Veil single, partially double or double. When single there is only an arachnoid cortina extending from the margin of the pileus to the stem, on the latter of which it often forms a more or less perfect fibrillose zone. When partially double the pileus and stem are more or less fibrous and the arachnoid cortina distinct. When double the whole plant is enveloped in infancy in a veil of gluten from which the arachnoid cortina is distinct. *Hymenophore* continuous with the stem. *Pileus* more or less fleshy. *Stem* central, more or less viscid, annulate from the veil or fibrillose-zoned from the cortina; annulus and zone often evanescent. *Gills* usually adnate, but sometimes adnexed, free or sinuate with a decurrent tooth, membranous, persistent, arid, at first often purple or blue changing to cinnamon—from the spores. *Spores* the colour of iron-rust, rusty-ochre, somewhat ochreous or pale or dark cinnamon.

The species are variable in size and changeable in colour, purple and blue being common in infancy; they present a different appearance in different stages of growth and according as they are moist or dry; young as well as mature examples must therefore be studied. All are terrestrial except **1140**, which sometimes grows on wood.

Some species of *Pholiota* and *Flammula* somewhat resemble the

Cortinarii, but these usually grow upon stumps and are structurally distinct. Species **974—1167**

Subgenus 1. **Phlegmacium.** *Veil* partial. *Pileus* viscid. *Stem* dry. **974—1021**

Subgenus 2. **Myxacium.** *Veil* universal. *Pileus* glutinous. *Stem* glutinous. **1022—1034**

Subgenus 3. **Inoloma.** *Veil* single. *Pileus* fleshy, silky or fibrillose, then smooth. *Stem* somewhat bulbous. Not hygrophanous. **1035—1057**

Subgenus 4. **Dermocybe.** *Veil* single, fibrillose. *Pileus* thinly fleshy, innato-silky, then smooth, dry. *Stem* equal. Not hygrophanous. **1058—1085**

Subgenus 5. **Telamonia.** *Veil* universal, somewhat double. *Pileus* thin, moist, smooth or sprinkled with whitish fibres. *Stem* annulate or peronate with scales. Hygrophanous. **1086—1128**

Subgenus 6. **Hydrocybe.** *Veil* thin, fibrillose. *Pileus* smooth or superficially white-fibrillose, not viscid, changing colour when dry. *Stem* somewhat rigid, sometimes irregularly annulato-zoned. **1129—1167**

Subgenus 1. **Phlegmacium.**

(From the clammy pileus ; Gr. *phlegma*, shiny moisture.)

Veil partial, arachnoid, dry. *Pileus* equally fleshy, with a con-

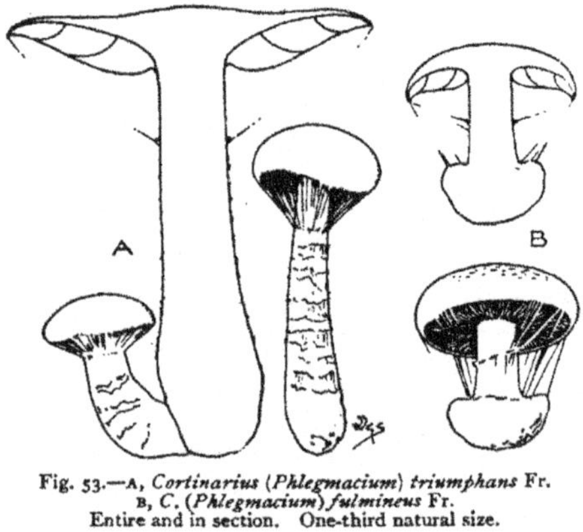

Fig. 53.—A, *Cortinarius (Phlegmacium) triumphans* Fr. B, *C. (Phlegmacium) fulmineus* Fr. Entire and in section. One-third natural size.

tinuous viscid pellicle. *Stem* firm, dry, usually solid, more or less zoned with the cortina. (Fig. 53).

a. Cliduchiæ. Partial veil depending in the form of a ring from the apex of the nearly equal or clavate stem.

Gills pallid then clay-colour. **974—980**
Gills violaceous or purplish then cinnamon. **981—985**
Gills yellow, cinnamon, ferruginous, not at first whitish or violaceous. **986, 987**
Gills olivaceous. **988—990**

b. Scauræ. *Stem* fleshy-fibrous, marginato-bulbous, rounded bulbous in **988** and **1007**. *Cortina* commonly inferior, arising from the margin of the bulb. *Gills* more or less sinuate.

Gills whitish, then clay-colour or pale cinnamon. **991—994**
Gills violaceous, dark blue, purplish, then pale cinnamon. **995—998**
Gills ferruginous, tawny or yellow. **999—1005**
Gills olivaceous. **1006—1009**

c. Elasticæ. *Veil* simple, thin, fugacious, median or inferior. *Pileus* usually somewhat thin, often hygrophanous. *Stem* commonly thin, not usually bulbous, generally attenuate upwards, rigid-elastic, externally subcartilaginous, lustrous, usually becoming hollow.

Gills whitish, then clay-colour or dull cinnamon. **1010—1014**
Gills violaceous, purplish or flesh-coloured. **1015—1017**
Gills usually ochre, tawny or ferruginous. **1018—1021**

a. Cliduchiæ.

974. C. triumphans Fr. (from the appearance of the stem, like an enwreathed triumphal column) *a b c.*

P. convexo-plane, obtuse, yellow; marg. fringed with Co. *St.* striate, yellow, with transverse evanescent bands of sienna rings and an imperfect Z. *G.* emarginate, crowded. *Flesh* white or sulphur.

Grassy ground, woods, under birch; rare. Sept.-Oct. 5⅛ × 5⅛ × ⅞ in.

975. C. claricolor Fr. (from the clear colour; *clarus*, clear) *a b c.*

P. convexo-plane or depressed, often broken into sc.; marg. fringed with Co. *St.* white-scaly. *G.* emarginate or adnate, sometimes almost free, crowded.

Mixed woods, under birch, amongst heather. Aug.-Nov. 4⅛ × 4⅛ × ⅞ in.

976. C. turmalis Fr. (from growing in troops, *turma*) *a b c.*

P. convexo-plane, yellow tan; marg. fringed with Co. *St.* evanescent, white or faintly ochreous-woolly, zoned. *G.* variously attached, crowded, serrate.

Densely cæspitose. Mixed woods, chiefly beech, larch. Sept.-Oct. 3¼ × 4¼ × ¾ in.

977. C. crassus Fr. (from the thick stem; *crassus*, thick) *a b*.

P. expanded, plane or slightly depressed, subfibrillose, dull yellow or tan-brown. *St.* fibrillose, mealy above, whitish, with an imperfect Z. *G.* rounded, crowded.

Moist woods. Aug.–Oct. 6 × 4⅜ × 1⅛ in.

978. C. balteatus Fr. (from the band or zone round the stem from the veil; *balteatus*, belted) *a b*.

P. convexo-expanded, gibbous, innato-floccose, umber; marg. lilac. *St.* tomentose, palest lilac. *G.* rounded or variously attached, crowded.

Chiefly pine-woods, under birch, dry places. Sept.–Oct. 4⅛ × 2⅛ × ⅞ in.

979. C. sebaceus Fr. (from its colour and clamminess, tallow-like; *sebum*, tallow) *a*.

P. campanulato-convex, then repand, smooth; marg. fringed with Co. *St.* subfibrillose, paler than P. with an imperfect Z. *G.* emarginate, veined, not crowded.

Taste mild. Woods, mixed, pine. Sept.–Nov. 4⅜ × 5¼ × ⅞ in.

980. C. lustratus Fr. (from the absence of colour; *lustratus*, purified, whitish) *a*.

P. convexo-expanded, smooth; marg. fringed with Co. *St.* zoned. *G.* crowded.

Sunny places, amongst heath, grass. Autumn. 2¼ × 1⅝ × ½ in.

981. C. varius Fr. (from its changeable characters) *a b c*.

P. convexo-expanded, obtuse, yellow-tawny or umber; marg. fringed with Co. *St.* adpressedly flocculose, whitish,. zoned. *G.* emarginate, subcrowded.

Taste somewhat nauseous. Woods, fir; uncommon. Sept.–Nov. 4⅛ × 3⅜ × ⅞ in. Must not be confounded with **1015**.

982. C. cyanopus Fr. (from the dark to pale purple stem; Gr. *kuanos*, dark blue, *pous*, a foot) *a b c*.

P. convexo-plane, obtuse, date-brown-livid, then tan. *St.* naked, colour as P. *G.* adnate or emarginate, edge unequal. *Flesh* sometimes changing to violet.

Woods, under oak; frequent. Aug.–Nov. 4 × 3⅜ × ½ in.

983. C. variicolor Fr. (from its variable colour) *a b*.

P. convexo-expanded, reddish-bay; marg. violet, fringed with Co. *St.* solid, grey-bluish and whitish, transversely 3–5 zoned with Co. *G.* emarginato-decurrent, crowded.

Woods, pine. Sept.–Nov. 5¼ × 6⅛ × 1¼ in. Var. *nemorensis* Fr. smaller. *P.* rivulose, yellow-bay. *St.* hollow. Woods, beech.

984. C. largus Fr. (from its large size) *a b*.

P. convexo-plane, obtuse, tawny brick-red, sometimes deep bright purple when young; marg. pale purple, fringed with Co. *St.* fibrillose, pale purple above, tan below, sometimes becoming red when bruised. *G.* emarginate or adnate, crowded.

Somewhat cæspitose. Woods, pine. Sept.–Oct. 2⅝ × 4⅜ × ⅝ in.

985. C. Riederi Fr. (after M. Rieder of St. Petersburg) *a b*.

P. campanulato-expanded, obtuse, ochreous; marg. fringed with Co. *St.* lilac-silky and tan-fibrillose. *G.* adnate or rounded, eroded, somewhat thick.

Woods, pine. Sept.–Oct. $2\frac{1}{4} \times 4\frac{5}{8} \times \frac{3}{8}$ in.

986. C. saginus Fr. (from its great size; *sagina*, corpulence) *a c*.

P. convexo-expanded, obtuse, repand, yellow; mid. darker; marg. paler. *Co.* fugacious. *St.* somewhat bulbous, fibrillose, light yellowish. *G.* decurrent, eroded.

Gregarious, somewhat cæspitose. Woods, larch. Oct. $5\frac{5}{8} \times 4 \times 1\frac{3}{8}$ in.

987. C. russus Fr. (*russus*, red) *a b*.

P. campanulato-expanded, obtuse; marg. fibrillose. *St.* not bulbous, whitish. *G.* adnate or subadnate, crowded, veined.

Taste nauseous-bitter. Woods, moist places; uncommon. Sept.–Oct. $4\frac{5}{8} \times 3 \times \frac{1}{2}$ in.

988. C. infractus Fr. (from the broken and bent margin of the pileus; *infractus*, broken, incurved) *a c*.

P. expanded, slightly depressed, virgate, olive then yellowish; marg. at first broken, then revolute and wavy. *St.* adpressedly fibrillose, dull olive becoming pale. *Co.* fugacious. *G.* adnate, broad, crowded.

Taste nauseous-bitter. Woods, beech. Oct. $3 \times 1\frac{1}{4} \times \frac{7}{8}$ in. Becoming olive-blackish in drying.

989. C. anfractus Fr. (from the margin of the pileus; *anfractus*, recurved) *a b*.

P. convexo-expanded, then repand, dark olivaceous; marg. thinner than **988**, broken, usually sinuous and very irregular. *St.* fibrillose, violaceous above, dull olivaceous below. *Co.* superior. *G.* emarginate, distant. *Flesh* reddish when cut.

Woods; rare. Sept.–Oct. $3\frac{3}{4} \times 3 \times \frac{1}{2}$ in.

990. C. Berkeleyi Cooke (after the Rev. M. J. Berkeley) *a*.

P. convexo-expanded, obtuse or subumbonate, silky, bright ochreous-brown, white-patched with Co.; marg. plicato-sulcate tan-brown, lilac or violaceous, appendiculate with Co. *St.* fibrillose, whitish. *G.* emarginate, arcuate.

Woods. Autumn. $7\frac{1}{8} \times 4\frac{3}{4} \times 1\frac{1}{2}$ in. In infancy the whole plant is covered with a whitish veil.

b. Scauræ.

991. C. multiformis Fr. (from its varied shapes) *a b c*.

P. convexo-plane, slightly depressed, light yellow, clay or tawny; marg. fringed with Co. *St.* white or pale yellow. *G.* emarginate or free, serrulate, crowded, thin.

Taste and odour usually mild, odour sometimes of *Stephanotis*. Woods, beech; uncommon. Aug.–Nov. $3\frac{1}{8} \times 2\frac{5}{8} \times \frac{5}{8}$ in. Var. *flavescens* Cooke. *P.* tawny-yellow. *G.* yellow. *Flesh* tinged yellow.

992. C. napus Fr. (from the turnip-shaped base of the stem; *napus*, a turnip) *a*.

P. convexo-plane, obtuse, fuliginous or date-brown-tawny; marg. fringed with Co. *St.* white, becoming yellowish below. *G.* emarginate or broadly adnate, subdistant, crisped.

Woods, pine. Oct. $2\frac{5}{8} \times 1\frac{1}{2} \times \frac{5}{8}$ in.

993. C. allutus Fr. (from its viscidity; *allutus*, bathed) *a b*.

P. convexo-plane, reddish-russet or orange-yellow. *St.* viscid, white above, reddish-striate below. *G.* adnate, subcrowded, thin, crenulate.

Woods, pine. Oct. $1\frac{1}{2} \times 1\frac{5}{8} \times \frac{3}{16}$ in.

994. C. talus Fr. (from the base of the stem, like swollen ankles; *talus*, an ankle bone) *a b*.

C. convexo-plane, dull ochre; marg. faintly olivaceous. *St.* smooth, tan or dull ochreous tan. *G.* adnato-emarginate. *Flesh* pallid whitish.

Woods. Sept.–Nov. $3\frac{5}{8} \times 2\frac{7}{8} \times \frac{1}{2}$ in.

995. C. glaucopus Fr. (from the pale blue stem; Gr. *glaukos*, pale blue, *pous*, a foot) *a b*.

P. convexo-expanded, unequal, dull ochre, commonly floccoso-scaly with a raised dark zone at marg. *St.* fibrilloso-woolly and steel-blue above, base ochreous. *G.* rounded emarginate, crowded.

In troops. Woods, pine, beech; frequent. Aug.–Oct. $3\frac{1}{4} \times 2\frac{1}{2} \times \frac{3}{4}$ in. Becomes bright violet when cut or bruised.

996. C. calochrous Fr. (from its beautiful colours; Gr. *kalos*, beautiful, *chros*, colour) *a b c*.

P. convexo-expanded, guttate, tawny; marg. yellow. *St.* fibrillose, tan, yellow at base. *G.* emarginate, crowded, serrate.

Taste acrid; odour fetid. Woods, oak, beech, pastures; frequent. Sept.–Nov. $2\frac{5}{8} \times 2\frac{1}{4} \times \frac{3}{4}$ in.

997. C. cærulescens Fr. (from the azure-clouded pileus; *cæruleus*, azure) *a b c*.

P. convexo-plane, dull yellow, clouded pale purple, sometimes dull or bright purple or violet. *St.* at first fibrillose, faint or deep purple, bright or dull. *G.* adnate, thin, crowded. *Flesh* pale violaceous.

Taste insipid; odour almost obsolete. Mixed woods, mossy places; frequent. Aug.–Nov. $2\frac{7}{8} \times 2\frac{5}{8} \times \frac{5}{8}$ in. Not changing colour when broken.

998. C. purpurascens Fr. (from the gills becoming purple when bruised) *a b c*.

P. convexo-plane, then repand, brown clouded with purple at the sometimes zoned marg. *St.* solid, fibrillose, clouded pale brown and blue. *G.* emarginate, crowded.

Single or cæspitose. Mixed woods, beech, pine; common. Sept.–Nov. $4\frac{1}{4} \times 3 \times 1\frac{1}{8}$ in. The purple hues have sometimes a tendency to crimson.

Sometimes the whole plant is purple and not unlike 139. Var. *subpurpurascens* Fr., thinner than type, slightly virgate. *P.* tawny smoke-colour. *St.* hollow, pale bluish or whitish.

999. **C. dibaphus** Fr. (from the two colours; Gr. *dibaphos*, twice dipped or dyed) *a b c.*

P. convexo-plane, or slightly depressed, yellow or sienna with purple or lilac marg. *St.* deep yellow to straw. *G.* sinuato-adnate, somewhat crowded, straw to purple-ferruginous. *Flesh* yellow.

Taste and odour none. Woods, beech, oak. Sept.–Oct. 4 × 3⅝ × ⅞ in. Wholly deep claret when dry. Var. *xanthophyllus* Cooke, the young state. *P.* for a long time yellow.

1000. **C. turbinatus** Fr. (from the shape of the base of the stem; *turbo*, a spinning-top) *a b.*

P. convexo-depressed, dull yellowish; marg. somewhat olive-brownish. *St.* stuffed to hollow, faintly dull yellowish. *G.* attenuato-adnate or decurrent, thin, crowded.

Woods, beech; uncommon. Sept. 5⅞ × 2⅞ × ⅝ in.

1001. **C. corrosus** Fr. (*corrosus*, gnawed to pieces, in reference to the late condition of the pileus) *a.*

P. convexo-depressed, subfloccose, ferruginous to pale tan. *St.* almost hollow, tan, white above. *G.* emarginate, narrow.

Woods, pine. Sept. 3 × 1⅛ × ⅝ in.

1002. **C. fulgens** Fr. (from its shining colour; *fulgeo*, to shine) *a b c.*

P. convexo-plane, silky-fibrillose, brilliant yellow. *St.* fibrilloso-woolly, colour as P. *G.* emarginate. *Flesh* yellow.

Woods, pine; rare. Sept. 3⅝ × 2⅛ × ⅞ in.

1003. **C. fulmineus** Fr. (from the shining pileus; *fulmen*, lightning) *a.*

P. convexo expanded, agglutinato-scaly, hard, bright yellow to tawny. *St.* yellow. *G.* rounded.

Shady woods, oak. Oct. 2⅛ × 1⅞ × ½ in.

1004. **C. orichalceus** Fr. (from the colour, like yellow copper ore, *orichalchum*) *a b.*

P. convexo-depressed, sienna-crimson; mid. spotted-scaly; marg. dull yellow or faintly olive. *St.* fibrillose, sulphur to dull yellow. *G.* rounded-adnate.

Under trees, woods, pine. Sept.–Oct. 3 × 3⅝ × ⅝ in.

1005. **C. testaceus** Cooke (from the colour; *testaceus*, brick-red) *a b.*

P. convex, flat, umbonate, vinous brick-red. *St.* fibroso-striate below, white above, rufous below. *G.* adnato-emarginate.

Woods, amongst leaves. Oct.–Nov. 3⅞ × 3⅝ × ½ in.

1006. **C. prasinus** Fr. (from the colour; Gr. *prason*, a leek) *a c.*

P. convexo plane, obtuse, spotted, verdigris or tawny-fuliginous. *St.* greenish-pallid. *G.* emarginate, subdistant, somewhat crisped.

Commonly sporadic. Edible. Taste mild. Woods, beech. Sept.–Oct. 3⅛ × 3⅞ × ⅝ in.

1007. C. atrovirens Kalch. (from the pileus, often dark green, *atrovirens*) *a b*.

P. convex, slightly depressed, green, sometimes olive-umber or bluish-green. *St*. pale-green, ferruginous below. *G*. adnate. *Flesh* tinged with green. *Myc*. sulphur or tawny.

Taste bitter; odour none. Woods, pine. Oct.–Nov. $3\frac{3}{4} \times 2\frac{3}{4} \times \frac{3}{4}$ in.

1008. C. scaurus Fr. (from the stem, as if with swollen ankles, *scaurus*) *a b*.

P. convex, plane, depressed, spotted, tawny-fuliginous to vinous; mid. darker. *St*. fibrilloso-striate, bluish-grey or pale olive, becoming white. *G*. adnate or adnexo-rounded, crowded.

Woods; rare. Sept.–Nov. $3\frac{1}{4} \times 3 \times \frac{3}{8}$ in.

1009. C. herpeticus Fr. (from the lengthening stem; Gr. *herpetikos*, creeping) *a b*.

P. convex, obtuse, spotted, olive or yellow-olive, then tan. *St*. stuffed, fibrillose, greyish above, yellowish or tan below. *G*. adnate.

Woods. Sept. $2\frac{1}{4} \times 2\frac{1}{8} \times \frac{3}{8}$ in.

c. Elasticæ.

1010. C. cumatilis Fr. (from the colour of the pileus; *cumatilis*, sea-coloured, blue) *a b c*.

P. convex, slate-grey or purple-grey; marg. fringed with Co. *St*. subbulbous, sometimes slightly rooting, white. *G*. attenuato-adnexed or adnate, serrate.

Sometimes subcæspitose. Woods. Sept.–Oct. $3\frac{5}{8} \times 4\frac{1}{4} \times \frac{3}{4}$ in.

1011. C. serarius Fr. (from the whey-like moisture of the pileus, like serum) *a*.

P. convexo-plane, gibbous, ochreous, buff-whitish or white. *St*. fibrillose, subrooting, ochre-whitish or white. *G*. broadly adnate, crowded, at first pale olive, then sienna-olive with ochre. *Flesh* pale yellowish.

Woods. Autumn. $3\frac{1}{2} \times 3\frac{1}{4} \times \frac{5}{8}$ in.

1012. C. emollitus Fr. (from its soft substance; *emollitus*, soft) *a b*.

P. convex, obtuse, subrepand, fibrilloso-virgate, tawny ochreous-yellow. *St*. fibrillose, white, then tinged yellow. *G*. emarginate, broad, subdistant.

Often cæspitose. Taste acrid. Woods, beech, grassy places. Sept.–Oct. $2\frac{1}{4} \times 1\frac{1}{4} \times \frac{1}{2}$ in.

1013. C. crystallinus Fr. (from the shining white pileus; Gr. *krustallinos*, crystalline) *a b*.

P. expanded, obtuse, at first watery-pallid, or ivory. *St*. fibrillose, whitish-straw, zoned. *G*. emarginate.

Taste acrid. Woods, beech, fir, oak. Oct. $1\frac{3}{4} \times 1\frac{7}{8} \times \frac{1}{4}$ in.

1014. **C. decoloratus** Fr. (from the change of colour in the bibulous pileus, drying lighter in patches) *a b c*.

P. campanulato-expanded, obtuse, clay-colour or tan, corrugate when old. *St.* whitish. *G.* emarginate, adnate, or decurrent.

Taste acrid. Woods, birch, pine, beech. Aug.–Oct. Protean, varying large and small. Large $3\frac{1}{4} \times 3\frac{1}{2} \times \frac{3}{4}$ in. Small $\frac{3}{4} - 1\frac{5}{8} \times 1\frac{3}{4} \times \frac{1}{4}$ in.

1015. **C. decolorans** Fr. (from the gills, soon changing colour from purple to cinnamon, but unchanging when bruised) *a b*.

P. convexo-flat, yellow; marg. fringed with Co. *St.* shining-white, zoned. *G.* emarginate or sinuato-adnexed, crowded, thin. *Flesh* white.

Woods, fir. Sept.–Oct. $2\frac{5}{8} \times 3\frac{1}{4} \times \frac{3}{8}$ in. Must not be confounded with **981**.

1016. **C. porphyropus** Fr. (from the purple stem; Gr. *porphureos*, purple, *pous*, a foot) *a b c*.

P. convexo-plane, innately streaked, livid brownish-clay. *St.* purple, becoming whitish, but again lilac or purple if bruised. *G.* adnato-emarginate or rounded, at length cinnamon, but again purple if bruised. *Flesh* deep purple.

Taste insipid; odour in drying sometimes of mice. Woods. Sept.–Oct. $2\frac{1}{8} \times 2\frac{1}{8} \times \frac{1}{4}$ in.

1017. **C. croceocæruleus** Fr. (from the saffron-gills and blue-purple pileus; *croceus*, saffron, *cæruleus*, azure) *a b*.

P. gibbous, plane or depressed. *St.* naked, whitish or pale purple, zoned. *G.* emarginate. *Flesh* pallid.

Odour disagreeable. Woods, beech. Oct. $1\frac{1}{2} \times 1\frac{5}{8} \times \frac{3}{16}$ in.

1018. **C. coruscans** Fr. (from the glittering pileus; *coruscans*, glittering) *a*.

P. obtuse, flat, ochreous, often spotted tawny. *St.* solid fibrilloso-striate, white. *G.* plano-decurrent.

Taste and odour obsolete. Copses, beech. $5\frac{1}{2} \times 4\frac{1}{2} \times \frac{1}{2}$ in.

1019. **C. papulosus** Fr. (from the granular patches on the pileus; *papulosus*, full of pimples) *a*.

P. convexo-plane, honey-tan; mid. glabrous, ferruginous-patchy when dry. *St.* solid, fibrillose, white. *G.* adnato-decurrent.

Woods, pine. Oct.–Nov. $2\frac{7}{8} \times 3\frac{3}{8} \times \frac{1}{2}$ in.

1020. **C. vespertinus** Fr. (from the dull colours, as of evening, *vesper*) *a b*.

P. convexo-plane, pale umber; marg. fringed with Co. *St.* solid, fibrillose, white. *G.* emarginate, broad, crowded, umber, shaded olive.

Woods, oak. Oct. $2\frac{5}{8} \times 4 \times \frac{1}{4}$ in.

1021. **C. olivascens** Fr. (from the olive-greenish pileus) *a c*.

P. convexo-plane, glabrous; marg. striate, fringed with Co. *St.* tan-white. *G.* emarginate or adfixed, olive-brown. *Flesh* yellowish near surface of P. and St.

Taste acrid. Damp places amongst pines. Sept. $2\frac{1}{4} \times 3\frac{1}{4} \times \frac{5}{16}$ in.

SUBGENUS 2. **MYXACIUM.**

(From the glutinous pileus and stem ; Gr. *muxa*, mucus.)

Veil universal, glutinous. *Pileus* fleshy, glutinous, somewhat thin. *Stem* scarcely bulbous, viscid, more or less zoned. *Gills* adnate or decurrent. (Fig. 54.)

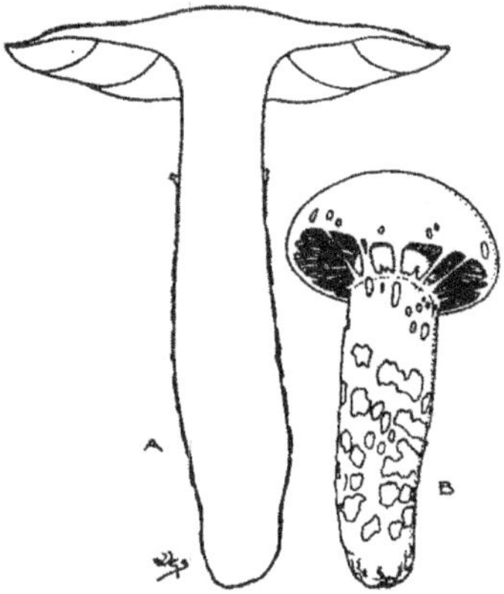

Fig. 54.—*Cortinarius* (*Myxacium*) *collinitus* Fr.
A, section of mature example ; B, young state.
One-third natural size.

a. Collinitæ. *Stem* floccoso-peronate, the flocci at first covered with gluten. **1022—1027**

b. Delibutæ. *Veil* entirely viscid. *Stem* not floccoso-peronate, viscid, with a varnished appearance when dry. **1028—1034**

Gills whitish, then clay-colour. **1028**

Gills at first violaceous, dark blue or reddish. **1029—1031**

Gills at first ochraceous or cinnamon. **1032—1034**

a. Collinitæ.

1022. C. arvinaceus Fr. (from the viscid greasy pileus and stem ; *arvina*, grease) *a b*

P. flat, slightly depressed, repand, orange-tawny ; marg. striate. *St.* solid, equal, silky-viscid, white. *Co.* fugitive, forming an imperfect Z. *G.* adnato-decurrent, subdistant, crenulate, straw or ochreous-tan to bright ochreous.

Woods, beech. Oct. $3\frac{1}{2} \times 5\frac{1}{2} \times \frac{1}{2}$ in.

1023. C. collinitus Fr. (from the pileus and stem, besmeared with gluten ; *collino*, to besmear) *a b c.*

P. convexo-flat, obtuse, orange-tawny or umber, white-patched with Co. ; marg. not striate. *St.* solid, equal or attenuate upwards, floccoso-glutinous and concentrico-scaly with the fugitive viscid-floccose Co., tan-ochreous. *G.* adnate, sub-crowded, whitish-bluish-grey to cinnamon.

Woods, amongst heather; common. July–Nov. $4\frac{5}{8} \times 5\frac{1}{2} \times 1$ in. Var. *mucosus* Fr., firmer, stem even, silky. Woods, pine.

1024. C. mucifluus Fr. (from the abundant flowing gluten; *mucifluus*, flowing with mucus) *a b.*

P. campanulato-expanded, obtuse, reflexed, livid-clay or shaded olivaceous, white-patched with Co.; marg. striate. *St.* attenuate downwards, white, with steel-blue gluten forming zones. *G.* adnate, clay-colour then watery-cinnamon.

On the ground. Aug.–Oct. $3\frac{1}{2} \times 5 \times \frac{5}{8}$ in.

1025. C. elatior Fr. (from its tall growth; *elatus*, tall) *a b c.*

P. convexo-flat, subreflexed, plicato-wrinkled, light yellow, dull ochreous, deep orange, whitish, brown, grey or blackish; whitish, violaceous or olivaceous at marg. *St.* attenuate downwards, longitudinally fibrous, viscid-scaly, whitish, greyish or pale-purplish with a fugitive viscid Co. *G.* adnate, veined, brown-cinnamon.

Taste mild. Mixed woods, pine, beech; common. Aug.–Nov. $4\frac{1}{4} \times 6\frac{5}{8} \times 1\frac{1}{8}$ in.

1026. C. grallipes Fr. (from the long, stilt-like stem; *grallæ*, stilts, *pes*, a foot) *a b.*

P. flat, umbonate, repand, hygrophanous, ferruginous to ochreous-tan. *St.* stuffed to hollow, equal, fibroso-striate, yellowish-tawny to ochreous, with a fugacious, arachnoid Co. *G.* adnate with a decurrent tooth, crowded, clay to cinnamon.

Cæspitose. Mixed woods, grassy places near trees, poplar, oak. Oct. $3\frac{1}{8} \times 4\frac{5}{8} \times \frac{3}{16}$ in.

1027. C. livido-ochraceus Berk. (from the livid ochre colour of the pileus) *a b.*

P. convexo-flat, white, with faint ochre, pale dull-lavender, or faint brown mid. or wholly brownish; lavender-purplish when young. *St.* violet, steel-blue or white above, brownish below, with a fugitive Co. which forms an imperfect ochre Z. *G.* rounded-adnexed, subdistant, ochre to cinnamon; edge pale.

Odour none. Woods; uncommon. Sept.–Oct. $2\frac{1}{8} \times 2\frac{1}{4} \times \frac{1}{4}$ in.

b. Delibutæ.

1028. C. nitidus Fr. (from its shining, beautiful appearance; *nitidus*, bright) *a.*

P. convex, plane, obtuse then depressed, honey-colour, sometimes white. *St.* clavate, pallid-white or yellowish-tan. *G.* decurrent, crowded, serrate.

Subcæspitose. Woods, beech. Autumn. $3\frac{1}{2} \times 3 \times \frac{1}{2}$ in.

1029. C. salor Fr. (from its greyish colour, like the sea, *salum*) *a b.*

P. convex, obtuse or broadly umbonate; mid. grey; marg. violaceous or steel-grey. *St.* attenuate upwards, white, when young covered with bluish gluten. *Co.* forming an imperfect Z. *G.* emarginate or adnate, distant, grey or cinnamon.

Woods. Oct. $3 \times 3\frac{3}{4} \times \frac{5}{8}$ in.

1030. C. delibutus Fr. (from the besmeared gluten; *delibuo*, to besmear) *a b.*

P. convexo-flat or subdepressed, light yellow. *St.* stuffed or hollow, attenuate upwards, white, sometimes pale steel-blue above. *Co.* forming an imperfect Z. *G.* adnate, slightly emarginate or subdecurrent, subdistant, serrulate.

Taste watery to slightly pungent. On the ground, damp grassy places. Sept.–Oct. $2\frac{1}{4} \times 3\frac{3}{4} \times \frac{5}{16}$ in.

1031. C. illibatus Fr. (from the gills, the margin entire, not serrulate, as in the last; *illibatus*, entire) *a.*

P. convexo-plane, obtuse or subumbonate, yellow; mid. darker. *St.* stuffed or hollow, slightly attenuate above, white, often red-dotted upwards. *G.* adnato-decurrent, crowded.

Woods, pine. Autumn. $1\frac{7}{8} \times 2\frac{7}{8} \times \frac{3}{16}$ in. Smaller and more slender than **1030**, with no trace of violet colour.

1032. C. stillatitius Fr. (from the dripping gluten; *stillo*, to drip) *a b.*

P. convexo-plane, subumbonate, hygrophanous; mid. tan-grey; marg. steel-grey and appendiculate with gluten. *St.* hollow, attenuate upwards, steel-grey with gluten. *Co.* forming an imperfect Z. *G.* emarginate, subdistant.

Odour none. Mossy places, pastures, amongst dead leaves; uncommon. Oct. $2\frac{1}{2} \times 3\frac{1}{4} \times \frac{3}{8}$ in.

1033. C. vibratilis Fr. (from a fancied tremulous habit; *vibratilis*, quivering) *a b.*

P. plano-gibbous, hygrophanous, shining golden-yellow. *St.* stuffed, attenuate upwards, whitish. *Co.* forming a glutinous fugitive Z. *G.* emarginato-decurrent, thin, crowded.

Taste and odour disagreeable. Woods. Sept. $1\frac{5}{8} \times 1\frac{3}{4} \times \frac{1}{4}$ in.

1034. C. pluvius (from its viscidity in rainy weather; *pluo*, to rain) *a*

P. convexo-gibbous, hygrophanous, pale yellow-tawny; marg. striate. *St.* stuffed to hollow, nearly equal, naked, obsoletely viscid, white. *G.* adnexed, crowded.

Taste watery, then acrid-pungent. Woods, pine. Sept.–Nov. $1\frac{1}{4} \times 2 \times \frac{1}{8}$ in.

SUBGENUS 3. **INOLOMA.**

(From the fibrillose or fringed veil; Gr. *is*, *inos*, a fibre, *loma*, a fringe.)

Veil single, fibrillose. *Pileus* equally fleshy, dry, subcompact, at first silky with innate scales or fibrils, becoming smooth, flesh continuous, not hygrophanous. *Stem* fleshy, somewhat bulbous. (Fig. 55.)

Handsome and easily distinguished. Species **1035—1057**

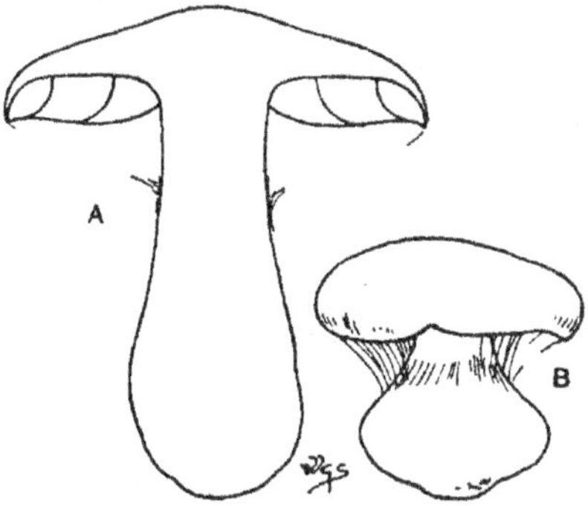

Fig. 55.—*Cortinarius* (*Inoloma*) *violaceus* Fr.
One-half natural size.
A, section of mature example; B, young state.

Gills at first white or pallid.	**1035—1038**
Gills, *Veil* and *Stem* violaceous.	**1039—1045**
Gills or *Veil* cinnamon, red or ochreous.	**1046—1052**
Gills or *Veil* dark, fuscous, olivaceous.	**1053—1057**

1035. C. argutus Fr. (from the graceful habit; *argutus*, graceful) *a b*.
P. expanded, gibbous then subdepressed, ochreous. *St.* solid, ventricose, fibroso-squamose, yellowish-white. *G.* adnate, rounded, subdistant, white clay-colour.

Woods, amongst leaves, oak. Oct. $4\frac{1}{2} \times 3\frac{1}{2} \times \frac{3}{8}$ in.

1036. C. argentatus Fr. (from the silvery white pileus and stem; *argentum*, silver) *a b c*.
P. convexo-plane, broadly gibbous; marg. pale lilac then dun. *St.* solid, slightly attenuate upwards, yellowish at base. *G.* emarginate, crowded, becoming watery-cinnamon or sienna.

Taste insipid; odour weak to strong, sweet, not fetid. Woods. Sept.–Oct. $5 \times 3\frac{3}{4} \times \frac{3}{4}$ in. Var. *pinetorum* Cooke, smaller and more graceful, odour faint. Woods, pine.

1037. C. opimus Fr. (from its obese appearance; *opimus*, fruitful) *a*.
P. convexo-plane, hard, dry, wavy-deformed, rimoso-rivulose, white, covered with short ochreous down. *St.* solid, hard, subturbinate, white-fibrillose with Co., base rooting. *G.* emarginate, subcrowded, whitish then tan.

Solitary. Woods. Autumn. $4\frac{1}{2} \times 1\frac{1}{2} \times 1\frac{1}{4}$ in.

1038. C. turgidus Fr. (from the stem, frequently swollen or bulbous; *turgidus*, swollen) *a.*

P. convexo-obtuse, flat, clay-white; marg. white-silky. *St.* solid, bulbous, rimoso-striate, surface often undulate, colour as P., with an imperfect Z. *G.* emarginate, narrow, crowded, whitish to tan.

Odour strong. Woods, beech. Autumn. 3 × 2½ × ¾ in.

1039. C. violaceus Fr. (from the violet colour) *a b c.*

P. convex, obtuse, flat, villous-scaly, dark purple, smeared copper-brown. *St.* solid, bulbous, zoned. *G.* adnate, distant, dark violaceous, clouded cinnamon. *Flesh* dark blue or dark purple.

Edible. Taste like that of **792.** Woods and open ground. Aug.–Nov. 4 × 4 × ⅞ in. Often confounded with **138** and **139.**

1040. C. cyanites Fr. (from the dark blue colour; Gr. *kuanites*, dark blue) *a b.*

P. convex, flat, not villous-scaly, at first dark blue, then pale turquoise; marg. fringed with Co. *St.* solid, bulbous, colour as P., zoned. *G.* rounded, crowded, turquoise to grey-bluish. *Flesh* colour as P., becoming reddish, with a red juice.

Woods, mixed. Oct. 5 × 6 × ¾ in. Var. *major* Fr. *P.* and *St.* compact; *G.* subdistant, cinereous, dark blue; flesh turning red more slowly than type. When bruised the blue colour changes to salmon.

1041. C. muricinus Fr. (from the purple colour, as of the mollusc *Murex*) *a.*

P. convex, obtuse, becoming smooth, violaceous to reddish-liver; marg. whitish, fibrillose. *St.* bulbous, villous, zoned, not juicy. *G.* emarginate, subcrowded, colour as St.

Odour strong. Woods, larch, fir. Oct. 3⅜ × 2½ × ⅝ in.

1042. C. alboviolaceus Fr. (from the violaceous-white pileus; *albus*, white, *violaceus*, violet) *a b.*

P. convexo-expanded, broadly umbonate or gibbous, dry. *St.* solid, clavato-bulbous, white villous, zoned. *G.* adnate, subdistant, cinereous-violaceous to cinereous-cinnamon. *Flesh* palest azure-blue.

Taste insipid; odour none. Woods, beech, shady places. Sept.–Oct. 3¼ × 4 × ½ in.

1043. C. malachius Fr. (from the pale purple colour, as of the mallow) *a.*

P. convex, obtuse, white fibrillose becoming smooth, lilac, brownish or reddish when dry. *St.* bulbous, bluish-lilac, then whitish, zoned. *G.* emarginate, crowded, purple to watery ferruginous. *Flesh* whitish.

Odour none. Woods, fir. Sept. 2⅝ × 2½ × ⅝ in. Must not be confounded with **1133.**

1044. C. camphoratus Fr. (from its powerful odour; *camphoratus*, strong scented) *a c*.

P. convex, obtuse then flat, lilac, whitish, yellowish or tan-livid; marg. fringed with Co. *St.* clavato-bulbous, pale violet or lilac, zoned. *G.* emarginate or adnato-decurrent, grey-blue to purplish. *Flesh* pale bluish.

Odour strong, fetid, penetrating, like fenugreek or curry-powder. Woods, fir. Autumn. 4 × 4½ × ⅝ in.

1045. C. hircinus Fr. (from its strong odour; *hircus*, a he-goat) *a*.

P. convex, obtuse, violet; mid. becoming ferruginous; marg. fringed with Co. *St.* bulbous, becoming pallid, yellowish at base, zoned. *G.* emarginate, subdistant, azure blue or violet to cinnamon. *Flesh* dingy, becoming yellowish.

Fetid. Woods, fir. Autumn. 2⅞ × 2⅜ × ⅝ in.

1046. C. traganus Fr. (from its strong odour; *tragus*, a he-goat) *a b c*.

P. convex, obtuse, flat, lilac-purplish, becoming yellowish. *St.* bulbous, spongy-solid, violaceous to whitish, zoned, sub-annulate. *G.* emarginate, broad, distant, thick, saffron-ochreous to cinnamon. *Flesh* yellowish.

Woods, pine, amongst ling. Taste strong, pleasant to unpleasant; odour fetid. Aug.–Sept. 4¼ × 3½ × ⅞ in. Var. *finitimus* Fr., odour pleasant, peculiar, like fermenting gum.

1047. C. suillus Fr. (from its odour of the pig; *suillus*, relating to swine) *a*.

P. convex, obtuse, pallid brick-red or ochreous-cinnamon; marg. silky or floccoso-squamulose. *St.* clavate or equal, spongy, pallid, apex shaded light violet, darker when bruised, zoned. *G.* adnate, subdistant, fragile, connected by veins, cinnamon.

Woods, fir. Autumn. 2¾ × 2¾ × ¾ in.

1048. C. tophaceus Fr. (from the colour; *tophus*, tufa) *a b*.

P. hemispherico-convex, very obtuse, glittering golden-tawny, with small sienna sq. *St.* solid, subbulbous, villoso-squamulose, tawny, zoned. *G.* emarginate, distant, tawny, cinnamon.

Subcæspitose or solitary. Odour sometimes strong. Woods, beech. Aug. 2⅝ × 4 × ½ in. Var. *redimitus* Fr. *St.* not bulbous or villoso-scaly.

1049. C. callisteus Fr. (Gr. *kallistos*, most beautiful) *a b c*.

P. convexo expanded, obtuse, yellow-tawny to yellow-ferruginous. *St.* clavato-bulbous, solid, fibrillose, yellow, rhubarb-colour at base, imperfectly zoned. *G.* adnate, subdistant, yellow-tawny.

Odour pungent or none. Woods, oak, pine; rare. July–Oct. 2⅞ × 4¼ × ½ in. Resembles **580** in appearance.

1050. C. Bulliardii Fr. (after Pierre Bulliard, French mycologist) *a b c*.
P. convex, flat, crimson bay-brown or umber. *St*. solid, bulbous, fibrillose, blood-red below, pale above, irregularly zoned. *G*. adnato-subdecurrent, purplish to ferruginous, edge whitish. *Flesh* pale purplish-rose, flecked pale blue.

Odour strong. Woods, mixed, beech. Sept.–Oct. $2\frac{1}{2} \times 2 \times \frac{1}{2}$ in. Sometimes approaching **1041**, at other times **1064**.

1051. C. vinosus Cooke (from its colour ; *vinosus*, like red wine) *a*.
P. flat, subdepressed, shining. *St*. solid, marginato-bulbous, pale violet above, reddish below, zoned. *G*. adnato-subdecurrent, scarcely crowded, ferruginous-cinnamon.

Under trees. Oct. $2\frac{3}{8} \times 2 \times \frac{1}{2}$ in.

1052. C.. bolaris Fr. (from its being marked with red ; *bolare*, to mark) *a b*.
P. convexo-plane, obtuse, light yellow-red, covered with sienna-red sc. *St*. stuffed to hollow, apex white, elsewhere covered with sienna-red sc., often in irregular zones. *G*. decurrent, adnate or emarginate, crowded, pale to dark cinnamon.

Solitary or subcæspitose. Woods, beech ; uncommon. Aug.–Nov. $2\frac{1}{8} \times 2\frac{1}{2} \times \frac{1}{4}$ in. When young P. and St. wholly sienna-crimson. Very dark when dry.

1053. C. pholideus Fr. (from the scaly pileus and stem ; Gr. *pholis*, a scale) *a b c*.
P. convexo-plane, sometimes subumbonate, fawn-colour, covered with smaller umber and sepia sc. *St*. solid, pale white-purplish above, fawn below, clad with dark umber sc., often in zones, zoned cortinate above. *G*. adnexo-emarginate, violaceous or pale slate to cinnamon.

Woods. Sept.–Oct. $3 \times 3\frac{1}{4} \times \frac{1}{2}$ in.

1054. C. sublanatus Fr. (from the somewhat woolly pileus ; *lana*, wool) *a b c*.
P. campanulate, expanded, umbonate, fawn, becoming ferruginous, covered with dark sq. *St*. solid, clavato-bulbous, apex faint azure-purplish, colour as P. below, covered with dark sc., sometimes zoned. *G*. adnate, scarcely crowded, olivaceous-yellowish to cinnamon.

Odour of radish. Woods, larch, fir ; rare. Oct. $3\frac{3}{4} \times 3\frac{1}{2} \times \frac{1}{2}$ in. Intermediate between **1053** and **1084**.

1055. C. phrygianus Fr. (from the colour, like gold embroidery ; *phrygio*, an embroiderer in gold) *a*.
P. convex, obtuse, honey-colour, covered with small black sc. *St*. subbulbous, paler than P., black-scaly below. *G*. rounded-adnate, subcrowded, dull yellow.

Odour of radish. Shady moist places, under beeches. Autumn. $2\frac{3}{8} \times 2\frac{1}{4} \times \frac{1}{2}$ in. Habit like that of **69**.

1056. C. arenatus Fr. (from the granulose pileus, as if sand-covered; *arena*, sand) *a b c*.

P. convexo-expanded, sometimes umbonate, pale yellowish-fuscous; mid. umber, minutely squamulose. *St.* usually attenuate upwards, pallid above, fuscous and dark squamulose below. *G.* adnexo-emarginate, yellowish-cinnamon or salmon.

Woods; uncommon. Aug.–Oct. $2\frac{7}{8} \times 2\frac{3}{4} \times \frac{3}{8}$ in.

1057. C. penicillatus Fr. (from the pencilled pileus) *a b*.

P. campanulato-convex, subumbonate, ferruginous-fuscous, densely floccoso-scaly. *St.* stuffed, attenuate upwards, clothed with ferruginous-fuscous, concentric sc. *G.* rounded-adnate, dark brown. *Flesh* ferruginous-fuscous.

Woods, pine. Sept.–Oct. $1\frac{1}{4} \times 2\frac{1}{4} \times \frac{3}{16}$ in.

Subgenus 4. **Dermocybe.**

(From the thinly fleshy pileus; Gr. *derma*, skin, *kube*, a head.)

Veil single, fibrillose, forming a zone in **1064.** *Pileus* thin, equally fleshy, at first silky, with innate villous down, becoming smooth when old, dry, not viscid or hygrophanous. *Stem* hollow or stuffed, except **1058,** often equal or attenuate, rigid, elastic or brittle. *Gills* changeable in colour. *Flesh* when moist watery or coloured. (Fig. 56.)

Polymorphous species defined with difficulty owing to the changeable colour of the gills. Species **1058—1085**

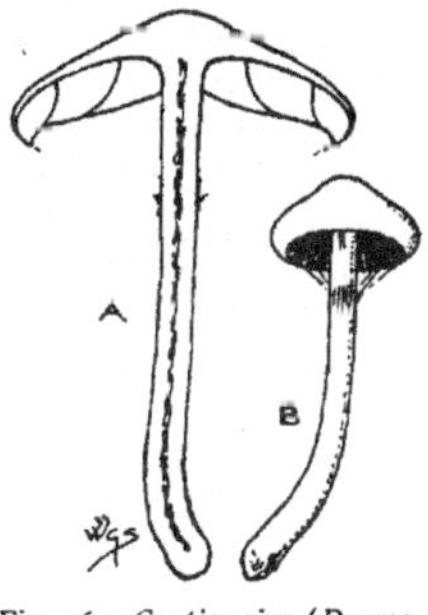

Fig. 56.—*Cortinarius* (*Dermocybe*) *cinnamomeus* Fr. A, section of mature example; B, young state. One-half natural size.

Gills at first whitish or pallid. **1058—1063**

Gills at first violaceous, becoming purple. **1064—1070**

Stem and fibrillose *Cortina* coloured. *Gills* bright cinnamon, red or yellow. **1071—1080**

Olivaceous. *Pileus* not torn into scales. *Veil* dull pallid or fuscous. **1081—1085**

1058. C. ochroleucus Fr. (from the yellowish-white pileus; Gr. *ochros*, yellowish, *leukos*, white) *a b*.

P. convex, gibbous. *St.* solid, ventricose or subbulbous, naked, fibrillose above with Co., colour as P. *G.* adnexo- or adnato-rounded, clay to ochreous or cinnamon.

Taste none or bitterish, not unpleasant; odour none or sweetish. Woods, mixed; rare. Sept.–Nov. $2\frac{3}{8} \times 1\frac{3}{4} \times \frac{3}{8}$ in.

1059. C. decumbens Fr. (from the usually decumbent stem) *a*.
P. convex, gibbous or subumbonate, shining white or yellowish. *St*. stuffed to hollow, subclavate, smooth, white, with a fugacious Co. *G*. adnexed or adnato-rounded, white to clay and ochreous-cinnamon.

Woods and grassy places. Autumn. $1\frac{3}{8} \times 1\frac{1}{4} \times \frac{3}{16}$ in.

1060. C. riculatus Fr. (from the cortina or veil; *rica*, a veil) *a*.
P. convexo-plane, gibbous or umbonate, atomate, buff, brown ochre or whitish; marg. adpressedly silky-floccose from Co. *St*. hollow, subclavate, smooth, naked, buff-white, with a slight and imperfect pallid Z. *G*. adnato-rounded, subcrowded, colour as P., tan to watery-ferruginous. *Flesh* palest yellow.

Woods, pine. Autumn. $2\frac{1}{2} \times 2\frac{1}{4} \times \frac{3}{8}$ in.

1061. C. tabularis Fr. (from the flat pileus; *tabula*, a flat board) *a c*.
P. convex, broadly gibbous, then plane, clay or fuscous-clay, becoming pale; marg. fringed with Co. *St*. stuffed to hollow, equal or subclavate, white, zoned. *G*. emarginate, crowded, white to clay.

Woods; common. Sept.–Oct. $2\frac{3}{4} \times 2\frac{1}{4} \times \frac{1}{4}$ in.

1062. C. camurus Fr. (from the frequently crooked stem; *camurus*, crooked) *a b c*.
P. campanulato-convex, obtuse or umbonate, pallid fuscous-hoary becoming pale pallid-yellowish; mid. darker; marg. fringed with Co. *St*. subhollow, twisted, equal, white. *G*. adnato-rounded or emarginate, grey-clay to subfuscous-watery-cinnamon.

Cæspitose. Taste not remarkable; odour somewhat rancid. Woods, mixed, oak. Sept.–Oct. $2\frac{1}{4} \times 3\frac{3}{4} \times \frac{5}{16}$ in. Very fragile.

1063. C. diabolicus Fr. (from its doubtful position in *Dermocybe*) *a c*.
P. hemispherical, obtuse, flat, often unequal and cracked, fuscous, crusted with grey, then smooth and fuscous-yellowish. *St*. attenuate downwards, white, pale bluish-grey above. *G*. adnato-emarginate, pale bluish-grey to whitish and clay.

Odour none. Woods, mixed, beech; uncommon. Aug.–Oct. $1\frac{3}{8} \times 2 \times \frac{3}{16}$ in.

1064. C. caninus Fr. (from its commonness; *canis*, a dog) *a b c*.
P. convexo-plane, tan-sienna to fuscous-rufescent or tawny. *St*. stuffed to hollow, slightly attenuate upwards, whitish, violaceous above, zoned. *G*. emarginate, subdistant, bluish-grey or purplish to cinnamon.

Edible. Mixed woods; common. Sept.–Nov. $3\frac{5}{8} \times 3\frac{7}{8} \times \frac{5}{8}$ in.

1065. C. myrtillinus Fr. (from its chestnut-brown colour; *myrteus*, chestnut-brown) *a b*.
P. convex, gibbous, plane, fuliginous, violet or dusky-brown.

St. clavato-bulbous, stuffed, whitish or pale violet. *Co.* obsolete. *G.* adnate, amethyst-azure-blue, or pale violet to dusky, not purple. *Flesh* watery fuscous or whitish.

Mixed woods, near trunks, beech. Sept.–Oct. $2\frac{7}{8} \times 2\frac{1}{2} \times \frac{3}{8}$ in. Greatly resembles **138** and **139**, but these have no arachnoid veil, etc.

1066. C. azureus Fr. (from the sky-blue pileus; *azureus*, sky-blue) *a b*.

P. expanded, obtuse, atomate, lilac, then fuscous and pallid. *St.* stuffed, slightly attenuate upwards, smooth, striate, bright azure then whitish, zoned. *G.* adnato-emarginate, bright bluish-violet. *Flesh* white in P., blue in St.

Solitary. Woods, beech, amongst moss and leaves. Autumn. $1\frac{3}{4} \times 2\frac{1}{8} \times \frac{5}{16}$ in.

1067. C. albocyaneus Fr. (from the white pileus and blue gills; *albus*, white, *cyaneus*, sea-blue) *a b*.

P. convex, plane, obtuse or umbonate, becoming yellowish, fringed with Co. *St.* subclavate, stuffed, whitish, zoned. *G.* emarginate, crowded, bluish-purple to subochreous.

Sometimes cæspitose. Woods, beech. Sept. $2\frac{5}{8} \times 4\frac{1}{4} \times \frac{5}{16}$.

1068. C. anomalus Fr. (from the colour of all its parts being very changeable; *anomalus*, irregular) *a b c*.

P. convex, expanded, gibbous or subumbonate, fuliginous or olive-shaded to rufescent, then yellowish with ochreous-sienna mid. *St.* stuffed to hollow, attenuate upwards, fibrillose or scaly, whitish, then palest yellow, zoned. *G.* adnato-emarginate, crowded, thin, violaceous to cinnamon.

Gregarious. Odour none. Woods; common. Aug.–Oct. $2\frac{3}{8} \times 2\frac{1}{2} \times \frac{5}{16}$ in.

1069. C. spilomeus Fr. (from the spotted pileus and stem; Gr. *spilos*, a spot) *a*.

P. convex, expanded, gibbous, rufescent to clay-colour, spotted with reddish-tawny sc. *St.* hollow, almost equal, white-lilac above, pale yellowish below, with many zones of reddish tawny sc. *G.* adnato- or adnexo-emarginate, crowded, thin, bluish-grey or violaceous to watery cinnamon.

Subcæspitose. Woods; uncommon. Oct. $2 \times 3\frac{3}{4} \times \frac{3}{8}$ in.

1070. C. lepidopus Cooke (from the fibrillose zones round the stem; Gr. *lepis*, a scale, *pous*, a foot) *a b*.

P. convex, flat, slightly depressed, shining umber; mid. becoming rufescent; marg. steel-grey or pale violet. *St.* clavate, becoming hollow, violet or steel-grey above, whitish below, with concentric biscuit-coloured zones. *G.* adnato-rounded, thin, violet to cinnamon, or olive-shaded.

Gregarious or fasciculate. Woods, amongst heath. Sept.–Oct. $2\frac{3}{8} \times 4 \times \frac{1}{4}$ in. Perhaps a form of **1064**.

1071. C. miltinus Fr. (from the colour, like red lead, Gr. *miltos*) *a b*.

P. expanded, gibbous or subumbonate, bay-brown-cinnamon to brick-red. *St*. hollow, attenuate upwards, cinnamon, reddish or crimson, salmon above and below, zoned red with Co. *G*. adnate, crowded, reddish-cinnamon to ferruginous. *Flesh* rose in St., pale brown in P.

Odour none. Woods, mixed, mossy places. Oct. $1\frac{5}{8} \times 1\frac{1}{2} \times \frac{3}{16}$ in.

1072. C. cinnabarinus Fr. (from the vermilion colour; *cinnabaris*, vermilion) *a b c*.

P. convex, obtuse or subumbonate, vivid scarlet-red. *St*. equal, solid, colour as P. *G*. adnate or subdecurrent, blood-red, and then blood-sienna. *Flesh* orange-vermilion in P., vermilion in St.

Taste mushroom-like; odour of radish. Woods, beech. Sept.–Oct. $2 \times 2\frac{3}{8} \times \frac{1}{4}$ in. Wholly dark-brown when dry.

1073. C. sanguineus Fr. (from the wholly blood-red colour; *sanguis*, blood) *a b c*.

P. convex, obtuse, plane. *St*. equal, stuffed to hollow. *G*. adnate.

Odour of radish. Woods, pine, oak. Sept.–Nov. $2\frac{1}{8} \times 2\frac{1}{8} \times \frac{1}{4}$ in. Giving out a blood-red juice when pressed. Sometimes very small.

1074. C. anthracinus Fr. (from the colour of the stem and gills, like burning coal, Gr. *anthrax*) *a b c*.

P. convex, umbonate, dark chestnut. *St*. hollow, equal, deep blood-red, fuscous below. *G*. rounded adnate, crowded, red to fiery-red, blood-red when bruised.

Woods, amongst moss. Aug.–Sept. $1\frac{1}{4} \times 1\frac{3}{4} \times \frac{1}{8}$ in.

1075. C. cinnamomeus Fr. (from the cinnamon colour) *a b c*.

P. expanded, obtuse or subumbonate, somewhat pale or bright cinnamon. *St*. stuffed to hollow, zoned. *G*. adnate, crowded, blood-red, red-cinnamon, saffron, golden, or light yellow. *Flesh* yellowish.

Tasteless. Woods, mixed, fir; common. Aug.–Feb. Variable in size. Large $2\frac{7}{8} \times 3\frac{1}{2} \times \frac{5}{16}$ in. Small $\frac{3}{4} \times 1\frac{3}{8} \times \frac{1}{8}$ in. Var. *semisanguineus* Fr. *G*. blood-red. Var. *croceus* Fr., smaller than type. *G*. yellowish or olive. Intermediate forms occur between the type and its varieties.

1076. C. croceoconus Fr. (from the saffron colour of the conical pileus; *croceus*, saffron, *conus*, a cone) *a*.

P. conical, campanulate, acutely umbonate, tawny-cinnamon to yellowish-tawny. *St*. stuffed to hollow, yellow, zoned. *G*. adnexo-rounded, crowded, yellow to cinnamon. *Flesh* yellow.

Woods, pine, mossy places. Autumn. $1 \times 1\frac{1}{8} \times \frac{1}{16}$ in.

1077. C. uliginosus Berk. (from its habitat; *uligo*, marshy ground) *a b c*.

P. campanulato-conical, expanded, strongly and somewhat acutely umbonate, bright brown-red. *St.* stuffed to hollow, paler than P., zoned. *G.* adnate with a tooth, distant, yellow to olive and cinnamon.

Woods, boggy places amongst *Sphagnum*. Autumn. $1\frac{1}{2} \times 3 \times \frac{3}{16}$ in.

1078. C. orellanus Fr. (derivation obscure) *a b c*.

P. convex, obtusely umbonate, orange-tawny. *St.* solid, subequal, striato-fibrillose, tawny, orange-striate. *G.* adfixed or adnate, subdistant, orange to tawny-cinnamon. *Flesh* as P., reddening.

Taste mushroom-like. Woods, mixed. Aug.–Oct. $2\frac{1}{8} \times 1\frac{3}{4} \times \frac{3}{4}$.

1079. C. malicorius Fr. (from the colour, like the fruit of the pomegranate; *malicorium*, the rind of the pomegranate) *a*.

P. convexo-plane, obtuse, golden-tawny, or golden to fuscous or olive. *St.* hollow, golden. *G.* adnexo-rounded, crowded, golden-tawny. *Flesh* yellow to greenish-olive.

Taste and odour pleasant. Woods, pine. Autumn. $1\frac{5}{8} \times 2\frac{7}{8} \times \frac{1}{4}$ in.

1080. C. infucatus Fr. (from the bright colour; *fuco*, to paint) *a*.

P. convex, obtuse, bright yellow to pale sulphur; marg. fringed with Co. *St.* solid, clavate, fibrillose, pallid light yellow, zoned. *G.* adnate, crowded, tawny to cinnamon. *Flesh* whitish, light yellow or pale olivaceous.

Woods. Sept.–Oct. $2\frac{1}{2} \times 4\frac{5}{8} \times \frac{5}{16}$ in.

1081. C. cotoneus Fr. (from the colour, like the olive, Gr. *kotinos*) *a b*.

P. campanulate, expanded, obtuse, subrepand, velvety, olivaceous. *St.* solid, attenuate upwards, pallid olivaceous, with one or more fuscous zones. *G.* adnexo or adnato-rounded, subcrowded, olivaceous to cinnamon.

Odour of radish. Woods, mixed, oak. Autumn. $3\frac{1}{2} \times 5\frac{1}{2} \times \frac{1}{2}$ in.

1082. C. subnotatus Fr. (from its characters, hardly distinct; *subnotatus*, hardly marked) *a b*.

P. convex, flat, hoary-floccose, olive or shaded yellowish to dark fuscous. *St.* stuffed, conical to equal, naked above, yellow fibrillose below, zoned. *G.* adnate, slightly emarginate, subdistant, yellowish to olivaceous-cinnamon.

Odour strong. Under trees, beech, fir. Nov. $3\frac{1}{2} \times 3\frac{3}{4} \times \frac{3}{8}$ in.

1083. C. raphanoides Fr. (from the strong odour of radish; Gr. *raphanos*, radish, *eidos*, resemblance) *a b*.

P. campanulate, expanded, umbonate, silky-fibrillose, fuscous-olivaceous to tawny. *St.* stuffed, slightly attenuate upwards, paler than P., yellowish above. *G.* adnate, scarcely crowded, somewhat olivaceous or orange-sienna to cinnamon ferruginous, edge paler. *Flesh* pale yellow or sulphur-white.

Taste acrid-bitter. Woods, mixed, beech, fir. Sept.–Oct. $2 \times 2\frac{1}{2} \times \frac{3}{16}$ in.

1084. C. valgus Fr. (from the frequently twisted stem; *valgus*, bow-legged) *a*.

P. convex, gibbous, olivaceous to brick-red. *St*. somewhat hollow, clavato-bulbous, whitish, apex pale violaceous, zoned. *G*. adfixed or rounded-adnate, yellowish to brick-red and cinnamon.

Odour none. Woods, pine, amongst moss. Oct. $3 \times 3\frac{3}{4} \times \frac{5}{8}$ in.

1085. C. venetus Fr. (from the sea-green pileus and stem; *venetus*, sea-coloured) *a b*.

P. hemispherico-obtuse, convex, umbonate, becoming greenish-yellow. *St*. stuffed or hollow above, equal or subclavate, colour as P. *G*. adnate, scarcely crowded, olivaceous to orange-cinnamon. *Flesh* pallid yellowish.

Gregarious. Woods. Aug.–Oct. $1\frac{3}{8} \times 1\frac{7}{8} \times \frac{5}{16}$ in.

SUBGENUS 5. TELAMONIA.

(From the whitish, lint-like fibres of the veil; Gr. *telamon*, lint.)

Veil universal, somewhat double. *Pileus* moist, hygrophanous, at first smooth or sprinkled with the superficial, arachnoid, whitish

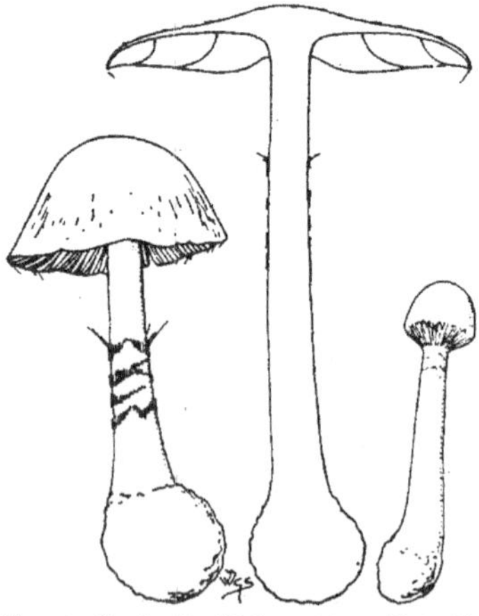

Fig. 57.—*Cortinarius (Telamonia) armillatus* Fr. Young examples and section of a mature specimen. One-third natural size.

fibres of the veil, thin or moderately compact at the middle, thin towards margin, splitting. *Stem* annulate from the inferior veil, or peronate with scales, somewhat cortinate at the apex. *Flesh* thin throughout or becoming so abruptly near the margin, scissile. (Fig. 57.)

The species are mostly large and handsome. Species **1086—1128**

a. Platyphyllæ. Gills broad.

Stem and *Cortina* white or whitish. **1086—1092**

Stem violaceous. *Cortina* usually white-violaceous. Universal *Veil* white. *Gills* violaceous. **1093—1098**

Stem and *Veil* reddish or yellow, except **1104.** *Gills* tawny or cinnamon, neither violaceous nor becoming brown. **1099—1107**

Stem becoming fuscous. *Veil* fuscous or dull. *Gills* dark coloured. **1108—1114**

b. Leptophyllæ. Pileus thin. *Stem* externally more rigid, subcartilaginous, stuffed or hollow, often attenuate downwards. *Gills* more or less crowded, thin, narrow.

Stem whitish, pallid, not floccoso-scaly. **1115, 1116**

Stem inclining to violet. **1117—1119**

Pileus and *Stem* tawny, ferruginous. **1120—1122**

Pileus fuscous. *Stem* floccoso-scaly, fuscous. **1123—1128**

a. Platyphyllæ.

1086. C. macropus Fr. (from the long stem; Gr. *makros*, long, *pous*, a foot) *a c.*

P. convex, obtuse, flat, brick-red to ferruginous, hoary with small sc. *St.* solid, equal, fibrillose, dull whitish, then as P., zoned. *G.* adnexed or adnate, distant, pallid to watery cinnamon.

Woods, moist places. Oct. $3\frac{1}{2} \times 6\frac{1}{2} \times \frac{5}{8}$ in.

1087. C. laniger Fr. (from the woolly pileus; *lana*, wool) *a.*

P. hemispherical, expanded, obtuse, bright tawny, clad with white sc. *St.* solid, equal, white, annulate or bi-annulate. *G.* rounded-adnexed, tawny-saffron.

Odour strong. Woods, larch, pine, beech, mossy places. Sept. $3\frac{1}{2} \times 6 \times \frac{5}{8}$ in.

1088. C. bivelus Fr. (from the double veil; *bis*, double, *velum*, a veil) *a c.*

P. convex, obtuse, plane, tawny-ferruginous; mid. spotted or darker. *St.* solid, clavato-bulbous, dull whitish, squarrose below. *G.* adnexo-rounded, bright tawny-cinnamon.

Taste mild; odour strong. Woods. Sept.–Oct. $3\frac{3}{8} \times 3 \times \frac{1}{2}$ in.

1089. C. bulbosus Fr. (from the bulbous stem) *a b.*

P. campanulato-expanded, gibbous, olivaceous, or date-brown to reddish-fuscous. *St.* solid, paler than P., yellow below, fugaciously zoned white. *G.* adnate, subdistant, cinnamon, never violet.

Odour of radish. Woods, mossy places; rare. Sept.–Oct. $3\frac{1}{4} \times 3\frac{3}{4} \times \frac{1}{2}$ in.

1090. **C. urbicus** Fr. (from its habitat, near trees and human habitations; *urbs*, a city) *a b*.

P. hemispherical, convexo-plane, clay-whitish; marg. crenato-torn. *St.* solid, equal, white zoned. *G.* emarginate, crowded, watery-ferruginous.

Grassy places. Sept.–Oct. $2\frac{1}{4} \times 2\frac{1}{4} \times \frac{3}{8}$ in.

1091. **C. licinipes** Fr. (from the woolly stem; *licinium*, lint, *pes*, a foot) *a b*.

P. campanulate, flat, obtusely umbonate, pale yellow to pallid tan. *St.* stuffed to hollow, equal, whitish, fugitive floccose, villous below, annulate. *G.* adnate, subcrowded, watery cinnamon.

Odour none. Woods, fir, damp places. Oct. $2\frac{1}{4} \times 2 \times \frac{3}{16}$ in. Var. *robustior* Cooke is larger and stouter than type.

1092. **C. microcyclus** Fr. (from the veil, forming a small circle; Gr. *mikros*, small, *kuklos*, a circle) *a b*.

P. campanulato-expanded, reddish-fuscous; mid. darker, paler when dry. *St.* stuffed, attenuate upwards, whitish, zoned white. *G.* adnate, distant, lilac to cinnamon.

Under trees, pines. Oct. $1\frac{3}{8} \times 2 \times \frac{3}{16}$ in.

1093. **C. torvus** Fr. (from its habit, growing in wild places as contrasted with **1090**; *torvus*, savage) *a b c*.

P. convex, obtuse or subumbonate, brick-red, brown, coppery or pale brown, shaded purplish. *St.* solid, clavate or equal, pale violaceous above, whitish and floccoso-scaly below. *A.* white. *G.* adnate, thick, distant, becoming purplish-umber and dark cinnamon. *Flesh* faintly dull purplish, often much eaten by larvæ.

Taste insipid, not unpleasant. Odour disagreeable, like **648**. Woods, mixed, beech; frequent. Aug.–Oct. $4\frac{3}{4} \times 5\frac{1}{4} \times \frac{3}{4}$ in. Intermediate forms occur between this and **1094**.

1094. **C. impennis** Fr. (from its smoothness as contrasted with **1095**; *impennis*, without feathers) *a b*.

P. hemispherico-convex, glabrous, brown to reddish; marg. silky and pale violet when young. *St.* solid to hollow, equal or subclavate, fibrillose, pallid, with a white Z. *G.* adnate or emarginate, very broad, cinnamon, shaded flesh-violet, becoming watery ferruginous. Substance soft and watery.

Taste none. Woods, pine, amongst dead leaves. Sept.–Oct. $4\frac{3}{8} \times 5\frac{1}{2} \times \frac{7}{8}$ in. Var. *lucorum* Fr. *St.* sometimes tinged with violet above. *G.* emarginate, cinnamon slightly shaded violet. Massee (*Eur. Fung. Fl.* 192) regards this as a species.

1095. **C. plumiger** Fr. (from the scales on the pileus and stem; *plumiger*, bearing feathers) *a*.

P. convexo-plane, umbonate, fuscous, olive-shaded to brick-tan, white floccoso-scaly. *St.* solid, clavate, floccoso-scaly below, whitish. *A.* white. *G.* adnate, scarcely crowded, becoming pale cinnamon.

Woods, mixed. Sept.–Oct. $2\frac{1}{4} \times 2\frac{1}{2} \times \frac{5}{16}$ in.

1096. C. scutulatus Fr. (from the stem, chequered with the veil; *scutula*, lozenge-shaped) *a b*.

P. hemispherico-expanded, squamuloso-rivulose or lacunoso-wrinkled, purple-umber or brick-fuliginous; marg. white silky. *St.* solid, subclavate, deep violaceous to fuscous, annulate and 4- to 7-zoned with white V. *G.* adnate, becoming cinnamon. *Flesh* violaceous.

Odour strong of radish. Woods, moist places. Aug.–Oct. $2\frac{3}{8} \times 4\frac{3}{8} \times \frac{3}{8}$ in.

1097. C. evernius Fr. (from the stem growing tall; *evernius*, sprouting) *a b c*.

P. conico-campanulate, expanded, obtuse or subumbonate, purple-bay-brown to pale pink-brown or umber with purplish marg., hoary when old. *St.* stuffed, slightly attenuate upwards, squamulose, violaceous or paler than P., obsoletely multi-zoned. *G.* adnate, distant, becoming cinnamon. *Flesh* yellowish-white in P., purplish or greyish in St.

Woods, pine, mixed, damp places; frequent. Sept.–Dec. $2\frac{1}{4} \times 5\frac{1}{2} \times \frac{1}{2}$ in.

1098. C. quadricolor Fr. (from the four colours, yellow, violaceous, cinnamon, and white) *a b*.

P. campanulate, expanded, umbonate, yellow to tawny. *St.* stuffed to hollow, fibrillose, violaceous-whitish. *A.* fugitive, white. *G.* adnate, distant, sometimes violet, becoming cinnamon, edge white.

Woods, beech. Oct. $1\frac{7}{8} \times 3 \times \frac{1}{4}$ in.

1099. C. armillatus Fr. (from the rings or zones round the stem; *armilla*, a ring) *a b c*.

P. campanulate, flat, dry, innately fibrillose or squamulose, brick-red or sienna, often cinnabar-zoned or spotted at marg. *St.* solid, bulbous, fibrillose below, 1- to 6-zoned vermilion. *G.* rounded or sinuato-adnate, becoming dark ferruginous.

Taste agreeable or insipid; odour of radish or of 1394, or of cold cooked potatoes. Woods, mixed, under hazel; frequent. Aug.–Oct. $4\frac{1}{2} \times 7 \times \frac{1}{2}$ in. The vermilion zones are permanent after drying.

1100. C. hæmatochelis Fr. (from the blood-red zone of the stem; Gr. *haima*, blood, *chelus*, the chest) *a b c*.

P. convex, gibbous, flat, pallid fuscous-brick-colour or claret-sepia. *St.* solid, clavate, zoned rufous. *G.* rounded, adnate, crowded, pallid cinnamon.

Woods, beech. Aug.–Oct. $5 \times 4\frac{1}{2} \times \frac{1}{2}$ in. Colour of bright red zone permanent after drying.

1100a. C. paragaudis Fr. (from the fibrillose veil; *paragauda*, a border or trimming of a garment) *a*.

P. somewhat fleshy, campanulato-expanded, umbonate, becoming somewhat fibrillose, brown, tawny, bay or tawny-tan, sometimes orange. *St.* hollow, twisted, soft, pallid reddish, clad with reddish adpressed fibrillose sc. *G.* adnexed, ventricose subdistant, opaque cinnamon.

Damp places under pines. Autumn. 2½ × 6¼ × ⅜ in. Var. *præstigiosus* Fr. *P.* submembranous, striate, silky-fibrous towards edge, dark bay, then lighter. *St.* red-squamulose. *G.* distant, tawny-cinnamon. Under pines.

1101. **C. croceofulvus** Fr. (from the colour of the pileus; *croceus* saffron, *fulvus*, tawny) *a*.

P. campanulate, obtuse, flat, bright tawny-orange. *St.* hollow, equal, striate, yellow, rufous-tinged. *Z.* a bright red line. *G.* adnato-sinuate, thick, subdistant, becoming ferruginous.

Woods, fir. Sept. 2¾ × 2⅛ × ⅜ in.

1102. **C. limonius** Fr. (from the lemon-coloured pileus; *limoneus*, lemon-coloured) *a*.

P. convexo-plane, obtuse, tawny lemon-yellow, sometimes orange. *St.* solid, equal, orange-yellow, light yellow floccoso-scaly. *G.* adnate, distant, at first light yellow. *Flesh* colour as P.

Odour none. Woods, pine; rare. 3¼ × 3½ × ½ in.

1103. **C. helvolus** Fr. (from the pileus, becoming pale yellow, *helvolus*) *a b*.

P. convexo-plane, obtuse or subumbonate, at first dark tawny-cinnamon. *St.* solid, equal, fibrillose, paler than P. with a median ferruginous A. *G.* emarginate, distant, thick, somewhat veined.

Woods, mixed, wooded pastures. Sept.–Oct. 2½ × 3½ × $\frac{5}{16}$ in.

1104. **C. hinnuleus** Fr. (from its fawn colour, *hinnuleus*) *a b c*.

P. campanulato-expanded, umbonate, often pierced-dotted, tawny-cinnamon; marg. silky. *St.* stuffed, slightly attenuate downwards, dull tawny; *Z.* broad, fugitive, white. *G.* emarginato-adnexed or adnate, distant, thin, veined.

Tasteless. Woods, mixed; common. Aug.–Dec. 3¼ × 4 × ⅜ in. Variable in size.

1105. **C. gentilis** Fr. (from its relationship to its neighbours; *gentilis*, of the same race) *a b c*.

P. campanulate, expanded, obtuse, orange, tawny-cinnamon or yellowish. *St.* stuffed to hollow, equal, colour as P., with 3–5 yellow or sulphur zones. *G.* adnate, very distant, thick. *Flesh* yellowish or pale brownish.

Gregarious. Woods, chiefly pine. Aug.–Oct. 1⅝ × 6 × ¼ in.

1106. **C. helvelloides** Fr. (from a fancied resemblance to certain species of *Helvella*, a Discomycetous fungus) *a b c*.

P. convex, flat, obtuse or umbonate, hygrophanous, ferruginous to tawny; marg. striate. *St.* hollow, equal, flexuous, subferruginous, white-silky above. *Z.* broad, yellowish. *G.* adnate, very thick, very distant, violaceous umber to cinnamon and ferruginous.

Woods, mixed, beech, among rotten leaves; uncommon. Aug.–Oct. 1 × 4 × ⅛ in.

1107. C. rubellus Cooke (*rubellus*, ruddy) *a*.

P. campanulate, expanded, umbonate, rufous-orange; mid. darker. *St*. solid, equal or subclavate, paler than P., 3- to 6-zoned ferruginous. *G*. adnato-sinuate, scarcely crowded, pale then bright ferruginous-red.

Swampy places. Autumn. 2¾ × 3¼ × ½ in. With the appearance of an abnormal **1099** or **1111**.

1108. C. bovinus Fr. (from its stout habitat; *bovinus*, like a bull) *a b*.

P. convex, obtuse, pallid to watery-cinnamon and tawny. *St*. solid, spongy-bulbous, brownish-cinnamon, whitish above the brown Z. *G*. adfixed or adnate, subdistant, cinnamon, becoming dark.

Woods, mixed, pine. Sept. 3 × 2½ × 1 in.

1109. C. nitrosus Cooke (from the nitrous odour) *a b*.

P. convexo-expanded, obtuse, flat, becoming scaly, fawn to tawny; mid. brownish and darker. *St*. solid, subequal, ochreous, paler than P., marked below with squamose zones. *G*. emarginate, subdistant, violet to watery-cinnamon. *Flesh* pale brown.

Woods, mixed. Autumn. 2½ × 2 × ⅜ in.

1110. C. brunneus Fr. (from its brownish colour; *brunneus*, brown) *a b*.

P. campanulate, flat, subumbonate, umber to dull reddish-tan. *St*. stuffed, subequal, brownish, white-striate. *A*. dull brownish-white. *G*. rounded-adnate, thick, distant, purplish to brownish-cinnamon.

Woods, pine, moist places, amongst grass. Sept.–Oct. 3⅝ × 3⅝ × 1 in.

1111. C. brunneofulvus Fr. (from its colours; *brunneus*, brown, *fulvus*, tawny) *a b*.

P. convexo-flat, subumbonate, fibrilloso-virgate, tawny-cinnamon; marg. at first white fibrillose. *St*. stuffed to solid, subequal, fibrilloso-striate, tawny, patched cinnamon, with a fugitive dull-white Z. *G*. adnate, ventricose, subdistant, tawny-cinnamon.

Woods, pine, oak, amongst moss. Sept. 1⅜ × 2¼ × 5/16 in.

1112. C. injucundus Fr. (from its unpleasant odour; *injucundus*, unpleasant) *a b*.

P. convex, plane, obtuse, fuscous-cinnamon. *St*. solid, subclavate, colour as P., then tawny-yellow, faintly lilac above Z. *G*. emarginate, lilac-tan, then clay.

Odour musty. Woods, fir. Nov. 5 × 4½ × ¾ in.

1113. C. glandicolor Fr. (from its colour as of the acorn, *glans*) *a*.

P. campanulate, expanded, umbonate, becoming dull pale-ochreous; marg. striate. *St*. hollow, subfibrillose with a fugacious white Z. *G*. adnexo-adnate.

Woods, pine, *Sphagnum* bogs, mossy places. Autumn. 1⅜ × 3 × 3/16 in. Var. *curtus* Fr. *St*. 1 in., flexuous, peronate with white Co., cingulate; um. becoming blackish.

1114. **C. punctatus** Fr. (from the minute dots or punctures on the pileus; *punctatus*, dotted) *a*.

P. campanulate, flat, subumbonate, tan or somewhat cinnamon. *St*. fistulose, fibrilloso-striate, yellow-fuscous, with a fugacious brownish Z. *G*. adnato-rounded, very distant, brown-cinnamon.

Woods, beech, fir. Autumn. 1¾ × 2⅝ × ⅛ in.

b. Leptophyllæ.

1115. **C. triformis** Fr. (from its assuming three forms) *a*, the type form, *Schæfferi*.

P. flat, gibbous or subumbonate, fawn to yellowish and whitish-grey. *St*. stuffed, subbulbous, pallid or faintly pinkish-ochre. *A*. distant, white. *G*. emarginato-adnate, ventricose, watery-ochreous to pale cinnamon.

Woods, beech. Oct. 3⅜ × 3½ × ⅝ in. 1. *Schæfferi*, as above. 2. *Melleo-pallens*. *P*. hemispherical, honey-colour, then pale tan. *G*. at first yellowish. *St*. conically attenuate, pallid yellow. Woods, pine. 3. *Fuscopallens*. *P*. thin, fuscous, becoming pale. *G*. at first watery-white. Woods, pine.

1116. **C. biformis** Fr. (from its resembling two subgenera; a *Telamonia* with the appearance of a *Hydrocybe*) *a*.

P. campanulate, expanded, umbonate, dark ferruginous-brown to pale bay. *St*. stuffed, attenuate downwards, fibrilloso-striate, paler than pileus. *A*. fugitive, white. *G*. adnato- or emarginato-rounded, greyish to watery-cinnamon or whitish-umber.

Woods, mixed. Oct. 2¼ × 2⅝ × ¼ in. With the habit of 1140.

1117. **C. periscelis** Fr. (from the garter-like annulus; Gr. *periskelis*, a garter) *a c*.

P. campanulate, convex, umbonate, lilac, greyish or brownish, white-silky. *St*. stuffed, colour as P., or pale steel-grey, fibrillose, multi-zoned fuscous. *G*. adnato-ascending, pallid or salmon-tan to dark ferruginous.

Bogs and under beeches. Autumn. 2 × 3½ × 3/16 in.

1118. **C. flexipes** Fr. (from the flexuous stem; *flexus*, bent, *pes*, a foot) *a b*.

P. campanulate, expanded, acutely umbonate, dark-brown-fuscous, shaded violaceous to pale yellow, hoary-fibrillose. *St*. stuffed, pallid violaceous or steel-grey, floccoso-scaly below the white Z. *G*. adnate, slightly rounded, purple or umber-violaceous to cinnamon, edge whitish.

Woods, larch, fir; frequent. Sept.–Oct. 1⅝ × 3 × ⅛ in.

1119. **C. flabellus** Fr. (from a fancied resemblance to a small fan, *flabellus*) *a b*.

P. conical, expanded, umbonate, olivaceo-fuscous to yellow-umber, then tan, at first white scaly, then incised and torn-fibrillose. *St*. stuffed, floccoso-scaly, pallid, apex violet.

Z. or *A.* white or none. *G.* adnate, dark violaceous to cinnamon and olive.

In troops, gregarious. Odour strong, somewhat of radish. Woods, beech, moist places. $\frac{3}{4} \times 2\frac{3}{4} \times \frac{1}{8}$ in.

1120. C. psammocephalus Fr. (from the furfuraceo-squamulose pileus, as if sand-covered; Gr. *psammos*, sand, *kephale*, the head) *a*. Tawny-cinnamon.

P. convex, plane, subumbonate, at length revolute. *St.* stuffed to hollow, slightly attenuate upwards, squamulose below A. *G.* sinuato-adnate, becoming umber-cinnamon.

Odour none. Woods, mixed, fir, charcoal heaps; frequent. Sept.–Oct. $2\frac{1}{4} \times 1\frac{5}{8} \times \frac{3}{16}$ in.

1121. C. incisus Fr. (from the cracked, incised pileus) *a b*.

P. convex, expanded, broadly umbonate, naked, then fibrilloso-scaly, brown of various shades. *St.* stuffed, fibrous, fibrillose, tan to ochreous and brown. *A.* white or obsolete. *G.* adnate, subdistant, salmon to deep sienna.

In troops or subcæspitose. Odour slight. Woods, pine, naked ground, dry places, dry turf bogs. Sept. $1\frac{1}{2} \times 1\frac{3}{8} \times \frac{1}{8}$ in.

1122. C. iliopodius Fr. (from the colour of the stem, as if dirty; Gr. *ilus*, mud, *pous*, a foot) *a b c*.

P. expanded, umbonate, opaque cinnamon to tan. *St.* stuffed, equal, fuscous, pale saffron-cinnamon within. *A.* median white. *G.* adnate, subdistant, thin, salmon to cinnamon.

Taste somewhat of radish. Woods, mixed, fir, beech; common. July–Dec. $1\frac{3}{8} \times 2\frac{3}{4} \times \frac{3}{16}$ in.

1123. C. hemitrichus Fr. (from the partly hairy pileus and stem; Gr. *hemi*, half, *thrix*, a hair) *a b*.

P. convex, expanded, umbonate, grey or dark fuscous to fuscous-tan, erect white-fibrillose, especially round marg.; um. dark. *St.* hollow, equal, white flocculose below the median white A. *G.* adnate, very crowded, clay to cinnamon or sienna. *Flesh* sienna.

Woods, mixed, mossy places, boggy ground, amongst fallen leaves; uncommon. April–Oct. $1\frac{7}{8} \times 2\frac{3}{4} \times \frac{1}{4}$ in.

1124. C. stemmatus Fr. (from the silky marginal band of the pileus, like a wreath, Gr. *stemma*) *a*.

P. convex, expanded, subumbonate, dark to pale brown. *St.* stuffed to hollow, equal or slightly attenuate downwards, ferruginous-brown within and without, with grey-white zones. *G.* adnate, crowded, opaque date-brown.

Woods, moist places. Autumn. $1\frac{1}{2} \times 3\frac{3}{8} \times \frac{3}{16}$ in.

1125. C. rigidus Fr. (*rigidus*, firm) *a b*.

P. conico-campanulate, umbonate, cinnamon-bay to yellowish-tan; marg. pellucidly striate. *St.* stuffed to hollow, equal, colour as P. variegated with tan-white and umber zones. *G.* adnate, ventricose, subcrowded, salmon to cinnamon.

Odour strong. Woods, damp places. Nov. $1\frac{3}{8} \times 2\frac{3}{4} \times \frac{3}{16}$ in.

1126. **C. paleaceus** Fr. (from the chaffy white scales of the pileus; *palea*, chaff) *a b c*.

P. convex, expanded, acutely umbonate, fuscous to greyish; mid. umber; marg. paler or whitish. *St.* fistulose, floccoso-scaly, fuscous within and without. *A.* and 1–7 zones whitish. *G.* adnate, crowded, grey pallid to cinnamon.

Odour weak. Woods, beech, moist places. Sept.–Nov. $1\frac{1}{2} \times 2\frac{3}{4} \times \frac{1}{8}$ in.

1127. **C. iris** Mass. (from the rainbow-like colours; *iris*, the rainbow) *a*.

P. conical, expanded, umbonate, pale ochreous-brown, densely white-fibrillose; marg. silky, splitting. *St.* stuffed to solid, clavate, densely fibrillose, violet, then pale above the fibrillose, bright-brown A., orange-brown at base within and without. *G.* sinuato-adnexed, subcrowded, dull ochreous to orange-brown.

Solitary, or in clusters of two to four. Woods. Oct. $1\frac{1}{4} \times 3 \times \frac{1}{8}$ in.

1128. **C. Cookei** Quél. (after Dr. M. C. Cooke) *a b*.

P. conical, hemispherical, umbonate, tawny-yellow; marg. woolly, white. *St.* stuffed, pale yellowish with 4–5 floccose yellow zones. *G.* adnate, violet to reddish and rust-colour.

Woods, fir. $\frac{1}{2} \times 1\frac{5}{8} \times \frac{1}{16}$ in.

Subgenus 6. **Hydrocybe.**

(From the moist or hygrophanous pileus; Gr. *hudor*, water, *kube*, a head.)

Veil thin, fibrillose, rarely collapsing and forming an irregular zone on the stem. *Pileus* smooth or only covered with white

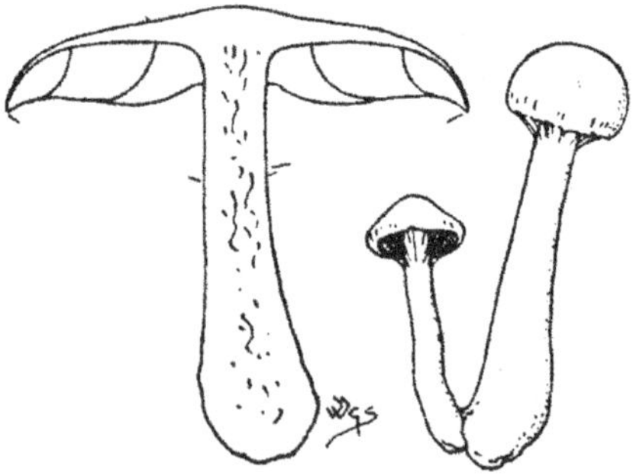

Fig. 58.—*Cortinarius (Hydrocybe) saturninus* Fr., and section. One-half natural size.

superficial fibrils, not viscid, but moist when fresh, changing colour when dry; flesh very thin, scissile, rarely more compact in the middle.

Stem slightly rigid, subcartilaginous externally, not sheathed, annulate or scaly. (Fig. 58.) Species **1129—1167**

a. Firmiores. Pileus subfleshy, convex, or campanulato-convex, then expanded, obtuse or at length gibbous, margin at first incurved. *Stem* for the most part attenuate upwards.

Stem and cortina white. **1129—1137**
Stem and gills inclining to violet. **1138—1141**
Stem and somewhat obsolete veil yellow or rufous. **1142—1146**
Stem inclining to fuscous. *Cortina* pallid, dull or white. *Gills* dark. **1147—1150**

b. Tenuiores. Pileus somewhat membranous, conical, then expanded, umbonate; umbo acute or rarely obtuse and vanishing; margin at first straight. *Stem* usually almost equal, or attenuate downwards.

Stem white. **1151—1156**
Stem inclining to violet or reddish. **1157—1160**
Stem yellowish, commonly becoming pale. **1161—1163**
Stem inclining to fuscous. **1164—1167**

a. Firmiores.

1129. C. firmus Fr. (from the compact substance; *firmus*, firm) *a.*
P. ferruginous to ochreous; marg. tan. *St.* solid, clavate, fibrilloso-striate. *Z.* ferruginous. *G.* adfixed or emarginate, pale umber to cinnamon.

Woods, mixed, grassy places. Autumn. 5 × 3½ × 1½ in.

1130. C. subferrugineus Fr. (from the colour of the pileus, somewhat rust-colour, *ferrugo*), *a c.*
P. umbonate, hygrophanous, dry, ferruginous to watery-cinnamon tawny or tan; *V.* at marg. only. *St.* solid, attenuate upwards, adpressedly fibrillose, pallid, saffron-yellow at base. *G.* emarginate or adnate, subcrowded, pallid to dark ferruginous.

Taste and odour unpleasant. Woods, oak, amongst rotting pine-leaves. Sept.–Oct. 2⅛ × 4⅜ × ⅜ in.

1131. C. armeniacus Fr. (from the colour of the pileus, like the apricot, *Pyrus armeniaca*) *a b c.*
P. obtuse or subumbonate, tawny-cinnamon to ochreous; marg. substriate. *St.* stuffed, attenuate upwards. *Z.* white. *G.* adnate, pallid to clay-cinnamon.

Woods, pine; uncommon. Aug.–Oct. 2¼ × 2⅜ × ⅜ in. Var. *fulsarius* Fr. *P.* yellow, white when dry.

1132. C. damascenus Fr. (from the colour of the pileus, like the damson) *a.*
P. rivulose when dry, bay-cinnamon or deep umber with a crimson shade to brick-red. *St.* solid, equal. *Z.* whitish. *G.* rounded-adnate, cinnamon or paler than P.

Subcæspitose. Taste acrid. Under trees, grassy places, pastures. Sept. 3¼ × 3½ × ½ in.

1133. C. privignus Fr. (from its being fragile in a non-fragile group; *privignus*, a stepson) *a b*.

P. obtusely umbonate, flat, reflexed, hygrophanous, fuscous to tan. *St.* stuffed to hollow, equal, silky with V. and Z. *G.* rounded-adnate, not crowded, watery- then opaque-cinnamon.

Taste scarcely acrid. Odour strong, unpleasant. Woods, pine, oak, damp places. Oct. $2\frac{1}{4} \times 2 \times \frac{1}{4}$ in.

1134. C. duracinus Fr. (from its firmness; *durus*, hard, *acinus*, a berry) *a b*.

P. obtusely umbonate, pale brick-colour or sienna to tan, or mid. sienna and marg. tan; marg. clad with V. *St.* solid, ventricose, attenuato-rooting, imperfectly zoned. *G.* adnate or emarginate, ventricose, pale cinnamon or whitish-sienna.

Woods, old pastures, near stumps. Aug.–Oct. $2\frac{1}{2} \times 3 \times \frac{1}{2}$ in.

1135. C. illuminus Fr. (dark, as contrasted with the allied non-British *C. candelaris*; *illuminus*, dark) *a b*.

P. gibbous, or subumbonate, pale brick-red, or orange with mid. orange-sienna to tan. *St.* stuffed, equal or slightly attenuate upwards, silky-fibrillose, pallid or tan-ochre, or white above and ochre to sienna below. *Z.* ochre. *G.* adnate, subcrowded, subdistant, pallid to cinnamon or whitish-sienna.

Woods, pine. Sept. $1\frac{1}{2} \times 2\frac{5}{8} \times \frac{3}{16}$ in.

1136. C. tortuosus Fr. (from the somewhat twisted stem; *tortuosus*, twisted) *a b*.

P. gibbous, ferruginous-bay to brick-red, or warm-tan with mid. and marg. somewhat dark crimson-umber, purple or turkey-red when bruised. *St.* stuffed, equal, silvery, spotted with white imperfect zones. *G.* adnate, crowded, tawny, purple or turkey-red when bruised.

Woods, pine, damp places. Autumn. $2\frac{1}{2} \times 4\frac{1}{2} \times \frac{3}{8}$ in.

1137. C. dilutus Fr. (from the watery colour, *dilutus*) *a*.

P. obtuse or subumbonate, opaque bay-brown or pale brick-red to tan; marg. white-silky and fringed with V. *St.* stuffed, equal, silky-white. *G.* deeply emarginate, ventricose, crowded, pale cinnamon or brownish-ochre.

Odour none. Woods; rare. Nov. $1\frac{3}{4} \times 2\frac{3}{8} \times \frac{3}{16}$ in.

1138. C. saturninus Fr. (from its colour; *saturninus*, gloomy) *a b*.

P. obtuse or umbonate, dark bay-brown or umber, soon pale; marg. silky and appendiculate with V. *St.* stuffed to hollow, attenuate upwards, deep violet or purple to pale dove-grey, partially white-zoned. *G.* rounded-adfixed or emarginate, crowded, thin, pale purple-grey to purplish and ferruginous.

Sometimes subcæspitose. Odour faint. Grassy places. Oct.–Nov. $3\frac{1}{2} \times 2\frac{1}{4} \times \frac{1}{2}$ in.

1139. **C. imbutus** Fr. (from the moist, stained flesh; *imbuo*, to soak) *a*.

P. flat, obtuse, sienna-brown to pale yellowish; marg. white hoary-fibrillose. *St.* solid, equal, whitish, pale violaceous above. *G.* rounded-adnate, subdistant, dark bluish-grey to watery-cinnamon. *Flesh* dingy pale violaceous at top of St.

Subcæspitose. Woods. Sept. $3\frac{5}{8} \times 2 \times \frac{1}{2}$ in.

1139a. **C. sciophyllus** Fr. (from the dusky gills; Gr. *skia*, shade, *phullon*, a leaf) *a*.

P. thin, convexo-expanded, obtuse, blue-purple-slate, paler towards marg.; mid. fleshy; edge narrow, silky white; *Co.* white. *St.* solid, slightly attenuate upwards, pale lavender to violet, gradually reddish towards the white base, clad with irregular white Z. *G.* adnate, crowded, narrow, purple-brownish or dusky umber. *Flesh* purple-slate above, salmon-buff and reddish below.

Single or cæspitose. Under beeches. Autumn. $1\frac{3}{4} \times 2\frac{5}{8} \times \frac{1}{2}$ in.

1140. **C. castaneus** Fr. (from its colour; *castanea*, a chestnut) *a b c*.

P. obtuse or subumbonate, shining fuscous- or purplish-chestnut; mid. darker. *St.* stuffed, fibrillose with V., pale purplish above, purple-brownish below. *Z.* white or ochre. *G.* adnate or emarginate, thin, violaceous to ferruginous, edge whitish.

Gregarious, sometimes cæspitose. Edible. Taste like that of **1394** or **792**; odour none or fungoid. Woods, pastures, gardens, on the ground, rarely on wood; common. June–Nov. $1\frac{5}{8} \times 2\frac{5}{8} \times \frac{3}{16}$ in.

1141. **C. bicolor** Cooke (from its two prevailing colours, purple and tan) *a b*.

P. broadly and acutely umbonate, whitish or lilac shaded. *V.* fugacious, white. *St.* solid, equal or attenuate downwards, pallid violet or steel-grey to whitish, zoned. *G.* adnate with a tooth, subcrowded, purplish-violet to cinnamon. *Flesh* whitish to bright lavender, purple at base of St.

Woods, mixed, on the ground. Aug.–Oct. $2\frac{1}{8} \times 2\frac{1}{2} \times \frac{3}{8}$ in.

1142. **C. balaustinus** Fr. (from the colour of the pileus, like the flower of the pomegranate, Gr. *balaustion*) *a*.

P. convex, flat, virgate, innato-fibrillose, shaded sienna-crimson or vermilion. *St.* solid, fibrilloso-striate, pallid to tawny-ferruginous within and without, zoned red-sienna. *G.* adnate, subcrowded, ferruginous-red or paler than P.

On the ground, woods, mixed, beech. Oct. $2\frac{1}{4} \times 2\frac{3}{4} \times \frac{5}{16}$ in.

1143. **C. colus** Fr. (from the fibrillose stem; *colus*, a distaff) *a b*.

P. convex, obtuse or obtusely umbonate, brown-rufescent, deep bay or umber, becoming paler. *St.* stuffed, attenuate upwards, paler than P., base encircled by blood-red or fiery-saffron mycelium. *G.* ascending-adnate, subcrowded, pale lilac or brownish to dark cinnamon.

Woods, pine. Oct. $2\frac{3}{8} \times 3\frac{3}{4} \times \frac{3}{16}$ in.

1144. **C. isabellinus** Fr. (from the dirty-yellow colour, like unwashed linen; from Isabel of Austria, daughter of Philip II., who vowed not to change her linen till Ostend was taken) *a b*.

P. subumbonate, at first honey-colour then paler. *St.* stuffed, equal, striate, colour as P., or tan, imperfectly zoned. *G.* rounded-adnate, subdistant, yellow to cinnamon.

Woods, pine. Autumn. $1\frac{3}{4} \times 3\frac{1}{8} \times \frac{3}{8}$ in.

1145. **C. renidens** Fr. (from the glistening pileus; *renidens*, shining) *a*.

P. expanded, flat, at first broadly and acutely umbonate, hygrophanous, ferruginous-tawny to ochreous. *St.* stuffed, at first bulbous, then equal, fibrillose, colour as P., imperfectly zoned yellow. *G.* rounded-adnate, subcrowded, pallid cinnamon to tawny.

Cæspitose. Woods, shady places. Sept. $3 \times 3 \times \frac{3}{8}$ in.

1146. **C. angulosus** Fr. (from the small, pointed scales of the pileus; *angulosus*, full of corners) *a b*.

P. convex, expanded, shining tawny-yellow, variegated with small sienna or ochreous sc. *V.* fugacious, tawny. *St.* equal, striate, tan-orange, zoned tawny-sienna. *G.* adnate, thick, subdistant, orange-sienna.

Woods, fir. Aug.–Sept. $2\frac{1}{2} \times 2\frac{1}{4} \times \frac{1}{3}$ in. Var. *gracilescens* Mass. *P.* somewhat fleshy. *St.* hollow, base somewhat attenuate, tortuous.

1147. **C. uraceus** Fr. (from its mouse-colour; Gr. *hurax*, a mouse) *a b*.

P. umbonate or obtuse, subfibrillose, brown, olive, red or buff. *St.* stuffed or hollow, equal, pale or yellowish above, fuscous or blackish below, imperfectly zoned. *G.* adnate or sinuate, distant, cinnamon.

Taste and odour not unpleasant. Woods, pine. Nov. $1\frac{7}{8} \times 3\frac{1}{4} \times \frac{3}{16}$ in.

1148. **C. jubarinus** Fr. (from the shining pileus; *jubar*, radiance) *a b c*.

P. umbonate, often repand, rich tawny-cinnamon; marg. white-silky with V. *St.* stuffed or hollow, fibrilloso-striate, pale tawny within and without. *G.* ascending-adnate, ventricose, subdistant, tawny-cinnamon.

Woods, pine, grassy places, on pine-leaves. May. $2 \times 2\frac{1}{8} \times \frac{3}{16}$ in. *C. irregularis* Fr. has been published as British by Massee and Crossland (*Fung. Fl. Yorks.* p. 97) following Bolton (*Hist. Brit. Fung.* t. 13), but this figure and description differ entirely from Fries, *Hym. Eur.* p. 394, and *Hym. Suec.* ii. p. 106. Thus—

In Fries—*P.* dry, glabrous. *St.* brick-red. *G.* decurrent, ferruginous. Hab. amongst pine-leaves.

In Bolton—*P.* covered with glutinous matter. *St.* pale grey. *G.* adnexed or free, pale dusky flesh-colour. Hab. pastures and meadows. Bolton's plant is probably an *Entoloma* with viscid P.

1149. **C. pateriformis** Fr. (from the shape of the pileus when depressed, like a *patera* or libation saucer) *a b*.

P. plane to depressed, shining dark chestnut, sprinkled with fugacious white fibrils. *St.* stuffed or hollow, equal, fibrillose, white to dusky, zoned sienna. *G.* emarginato-adnate, ventricose, brick-red to vinous-ochre.

Woods, damp grassy places. Sept.–Oct. $2\frac{5}{8} \times 3 \times \frac{5}{16}$ in.

1150. C. unimodus Britz. (from its uniform colour; *unimodus*, of one kind) *a b*.

P. umbonate, brown, shaded rufous, vinous or sienna. *St.* stuffed, equal, paler than P., sometimes pale purple above, imperfectly 2–5-zoned. *G.* emarginato-adnate, distant, brown.

Grassy places. Sept.–Oct. $2 \times 3\frac{1}{2} \times \frac{1}{4}$ in.

b. Tenuiores.

1151. C. dolabratus Fr. (from the shape of the stem and gills as seen in section; *dolabra*, a pickaxe) *a b*.

P. obtusely umbonate, smooth, brick-red to tan; marg. silky. *St.* stuffed or hollow, smooth, white, sometimes faintly ochreous above. *G.* rounded-adnate, distant, tawny-cinnamon or orange.

Odour strong, pungent-nauseous, stinking. Woods, pine, damp places amongst *Vaccinium*, *Sphagnum*, etc. Sept.–Nov. $3\frac{1}{2} \times 3\frac{1}{4} \times \frac{1}{2}$ in.

1152. C. rigens Fr. (from the rigid stem) *a b*.

P. obtusely umbonate, smooth, tan-clay to tan-whitish. *St.* stuffed, cartilaginous externally, zoned. *G.* sinuato-adnate, watery-clay to cinnamon or sienna.

Woods, pine, mixed; uncommon. June–Oct. $1\frac{3}{4} \times 2\frac{1}{2} \times \frac{5}{16}$ in.

1153. C. Krombholzii Fr. (after J. V. Krombholz) *a b*.

P. obtusely or acutely umbonate, smooth, yellowish-brown to ochreous; mid. always dark; marg. appendiculate with V. *St.* fistulose, naked, whitish. *V.* membranous, fugitive, whitish. *G.* rounded-adnexed, dull purplish-brown, edge yellowish.

Often cæspitose. Amongst moss. $2\frac{5}{8} \times 4\frac{1}{2} \times \frac{3}{16}$ in. Much like a *Hypholoma*.

1154. C. Reedii Berk. (after Miss Reed, sister of Mrs. Hussey) *a b*.

P. conical, expanded, strongly umbonate, brown; mid. areolate. *St.* solid, bulbous, shaded pale umber. *G.* ascending, attenuato-free, ventricose, white to cinnamon.

Taste and odour none. Amongst moss and beech-mast. May. $1\frac{1}{4} \times 1\frac{1}{4} \times \frac{3}{16}$ in.

1155. C. leucopus Fr. (from the white stem; Gr. *leukos*, white, *pous*, a foot) *a b c*.

P. conical, expanded, umbonate, smooth, moist, yellowish or somewhat buff to tan. *St.* hollow, equal, attenuate upwards, shining, zoned. *G.* adnexed, crowded, ventricose, pallid to cinnamon.

Woods, fir, mixed; frequent. Sept.–Nov. $1\frac{1}{4} \times 1\frac{1}{2} \times \frac{3}{16}$ in.

1156. C. scandens Fr. (from the long wavy stem; *scando*, to climb) *a b*.

P. conico-campanulate, obtusely or acutely umbonate, moist, smooth, tawny-ferruginous to honey-colour and tan; marg.

striate; *Co.* white. *St.* fistulose, flexuous, rooting, greatly attenuate downwards, yellowish to whitish. *G.* ascending-adnate, tan to yellow and tawny-cinnamon.

Woods, pine, moist places. Oct.–Nov. $1\frac{1}{8} \times 2\frac{1}{2} \times \frac{3}{16}$ in.

1157. C. erythrinus Fr. (Gr. *eruthros*, red) *a*.

P. umbonate, smooth, bay-brown-rufous, or vinous-sienna to tawny; *Co.* white. *St.* stuffed to hollow, equal, smooth, striate, white, pale purple or violaceous. *G.* adnexed or somewhat adnate, pallid to pale cinnamon or sienna.

Woods, amongst leaves in damp places. Oct. $1\frac{1}{8} \times 1\frac{7}{8} \times \frac{1}{16}$ in. Var. *argyropus* Fr. more slender than type. *St.* silvery, white-mealy above.

1158. C. decipiens Fr. (from its resemblance to several other species; *decipio*, to deceive) *a b c*.

P. acutely or obtusely umbonate, smooth, bay-brown to shining brick-red; mid. darker. *St.* solid, stuffed or hollow, fibrillose, pale rufescent, clouded brick-red. *Z.* imperfect, fugacious, white. *G.* adnate, sometimes with a tooth, crowded, thin, sienna-umber or sienna-tan.

Woods, mixed; frequent. Sept.–Oct. $2\frac{1}{2} \times 3\frac{1}{2} \times \frac{1}{4}$ in. Var. *insignis* Fr. *P.* paler; *St.* flexuous, glabrous; *G.* less crowded.

1159. C. germanus Fr. (from its close relationship to adjoining species; *germanus*, born of the same parents) *a b*.

P. obtusely or acutely umbonate, silky, hygrophanous, fuscous to clay or tan-umber; mid. sepia; marg. whitish, or wholly tan. *St.* fistulose, equal, silvery-pale or somewhat lilac. *Z.* fugacious, ochre. *G.* adnate, slightly rounded, subdistant, watery-cinnamon.

Odour powerful. Woods, pine, beech. Oct. $1 \times 2\frac{5}{8} \times \frac{1}{8}$ in.

1160. C. ianthipes Fr. (from the violet stem; Gr. *ianthios*, violet-coloured, *pous*, a foot) *a b c*.

P. obtusely umbonate, fibrillose, brown; marg. yellowish. *St.* stuffed, smooth, equal, violaceous, base rufescent, imperfectly zoned. *G.* adnexed or adnate, subcrowded, brown olivaceous, then dark umber-sienna or deep red-brown.

Woods. Sept. $\frac{5}{8} \times 1\frac{1}{4} \times \frac{1}{16}$ in.

1161. C. detonsus Fr. (from the smooth pileus and stem; *detondeo*, to shear) *a c*.

P. subumbonate, slightly silky to somewhat smooth, brick-red, umber or yellowish. *St.* fistulose to stuffed, yellowish-pallid, dull purple at mid. *G.* adnate, subdistant, yellowish brick-colour.

Woods, pine, amongst moss in damp places. Sept. $1\frac{3}{8} \times 1\frac{5}{8} \times \frac{1}{8}$ in.

1162. C. obtusus Fr. (from the obtuse umbo which at length disappears) *a*.

P. bay-brown-ferruginous, cinnamon, or orange-sienna to pale ochreous or tan-whitish; marg. striate. *St.* stuffed to hollow attenuate downwards, lighter than P.; *Co.* whitish. *G.* adnate, ventricose, orange-sienna; edge white.

Gregarious. Odour strong. Woods, chiefly pine; requent. April–Oct. $1\frac{1}{2} \times 2\frac{1}{2} \times \frac{1}{4}$ in.

1163. **C. acutus** Fr. (from the acute umbo) *a b c.*

P. honey-colour or pale yellow to whitish; marg. white fibrillose. *St.* fistulose, equal, white-fibrillose, colour as P., zoned. *G.* adnate, crowded, thin, white-ochre, ochreous-cinnamon or watery-brown.

Scattered, in troops, or cæspitose. Odour at length nitrous. Woods, fir, mixed, moist places; frequent. Sept.–Nov. $1\frac{1}{4} \times 2\frac{1}{2} \times \frac{1}{8}$ in.

1164. **C. Junghuhnii** Fr. (after Francis Junghuhn) *a.*

P. papillate, cinnamon or dull crimson-orange to tawny; marg. striate. *St.* stuffed to hollow, reddish, whitish below, adpressedly fuscous-fibrillose. *G.* adnate, subdistant, thin, reddish-ochre or olive shaded. *Flesh* reddish.

Odour none. Woods, mossy places. Aug. $1\frac{1}{2} \times 2\frac{1}{4} \times \frac{3}{16}$ in.

1165. **C. depressus** Fr. (from the pileus, at length depressed) *a.*

P. smooth, vinous- or ochreous-umber; marg. at first silky then striate. *St.* hollow, attenuate downwards, reddish or yellow-brownish. *G.* adnate, ventricose, saffron-yellow, reddish or reddish-brown-ochre.

Odour faint of stale fish or cucumber, as in **709**, which resembles this in habit. Woods, moist places. Autumn. $1\frac{1}{2} \times 1\frac{3}{4} \times \frac{1}{4}$ in.

1166. **C. milvinus** Fr. (from the coloration, like a kite's back; *milvus*, a kite) *a b.*

P. obtusely umbonate, olivaceous-fawn or olive to tan; marg. striate, white squamulose. *St.* stuffed, equal, pale fuscous, with white silky spots, or whitish above, tan-brown below; *Co.* white. *G.* adnate, subdistant, ochreous olive-tan to olive.

Gregarious. Odour strong. Woods. Oct. $1\frac{3}{8} \times 2\frac{5}{8} \times \frac{1}{4}$ in.

1167. **C. fasciatus** Fr. (from the fibrous-splitting stem, as if in bundles, *fasciæ*) *a b.*

P. acutely umbonate, smooth or subsilky, brick colour or umber; mid. sepia. *St.* fistulose, equal, splitting longitudinally into fibres, pallid to cinnamon-fuscous, or variegated with bands of white, tan and brown. *G.* adnate, subventricose, thin, distant, cinnamon.

Woods, pine, damp places, pastures under pines. Sept.–Oct. $1\frac{5}{8} \times 2\frac{5}{8} \times \frac{1}{8}$ in.

XLV. GOMPHIDIUS Fr.

(From the shape, like a large bolt or nail, Gr. *gomphos.*)

Veil universal, viscoso-floccose. *Pileus* at length turbinate, glutinous when moist. *Stem* central, continuous and homogeneous with the hymenophore, imperfectly floccoso-annulate or cortinate;

annulus fugitive. *Gills* decurrent, distant, soft and somewhat branched, edge acute, composed of a mucilaginous membrane which often admits of being detached from the hymenophore and stretched

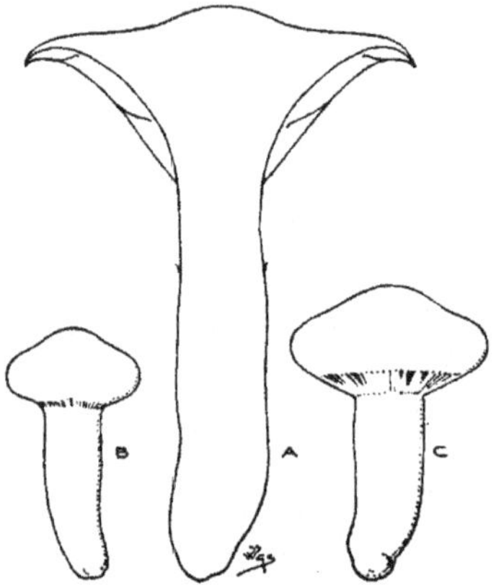

Fig. 59.—*Gomphidius viscidus* Fr.
A, section; B, young state; C, half-grown example showing floccose veil. One-third natural size.

out into a continuous membrane, not deliquescent. *Spores* large, fusiform, black or nearly so. (Fig. 59.)

The species grow on the ground, chiefly in pine woods, solitary, subpersistent.

The species are most nearly allied to *Hygrophorus*, but they differ greatly in the annulate stem and the blackish spores. There is also an affinity with *Cortinarius*, as shown by the cortina, but the spores differ greatly in shape and colour. Species **1168—1171**

1168. G. glutinosus Fr. (from the glutinous pileus) *a b c.*

P. becoming flat, smooth, purple-umber or umber-crimson, at first paler. *St.* solid, viscid and zoned with the fugitive V., whitish above, yellowish or brownish below; *Co.* white. *G.* whitish or olive-whitish, then cinereous. *Flesh* whitish, greyish or pale rose.

Said to be edible. Eaten at Hereford Fungus Forays, but not much approved, 1870. Taste watery, mouldy; odour faint. Woods, chiefly pine, larch; common. July–Nov. $4\frac{1}{4} \times 3\frac{3}{4} \times \frac{5}{8}$ in. Must not be confounded with **155.** Var. *roseus* Fr. *P.* rose-colour, sometimes dark crimson-red. *St.* white above, pale rose below.

1169. G. viscidus Fr. (from the somewhat viscid pileus) *a b c.*

P. becoming flat, rich apricot, shaded vermilion, sometimes brick-red, often mottled bright purple. *St.* solid, orange, pale purplish above. *G.* olive, then crimson-purple, purplish or whitish-purple. *Flesh* yellowish, often with purple stains, rhubarb-colour at base.

Said to be edible; odour not unpleasant. Woods, chiefly pine; frequent. July–Oct. $4\frac{5}{8} \times 4\frac{1}{2} \times 1\frac{1}{8}$ in. Var. *testaceus* Fr. *P.* brick-red. Woods, beech. Every intermediate form occurs between this and **1168**, including the varieties *roseus* and *testaceus*.

1170. G. maculatus Fr. (from the spotted pileus and stem; *maculatus*, spotted) *a.*

P. becoming flat and depressed, white to brownish-salmon with large umber or blackish spots; marg. striate. *St.* solid, slightly attenuate downwards, yellow or yellowish below, white, grey or purplish in middle, whitish above. *G.* pale slate, shaded umber. *Flesh* reddish.

Woods, fir. Autumn. $2\frac{1}{4} \times 2\frac{1}{2} \times \frac{1}{2}$ in. Var. *Cookei* Mass. *St.* whitish above, blackish at base.

1171. G. gracilis B. & Br. (*gracilis*, slender) *a b c.*

P. becoming flat and depressed, pale vinous-brown covered with dingy-fuliginous gluten, at length black-spotted and black-bordered with the drying gluten. *St.* solid, slightly attenuate downwards, white-scaly above, yellow and virgate below, clouded vinous-tan. *G.* slate-white, white umber or greenish-white.

Woods, fir; frequent. July–Oct. $1\frac{3}{4} \times 2\frac{1}{4} \times \frac{1}{4}$ in. Berkeley's illustration, *Outlines*, t. 12, fig. 7, does not agree with his description. Every intermediate form occurs between this and **1170**.

XLVI. PAXILLUS Fr.

(From the form, like a small stake or peg, *paxillus*.)

Veil obsolete. *Pileus* fleshy, margin at first involute. *Stem* central, excentric or obsolete, when present fleshy, continuous and

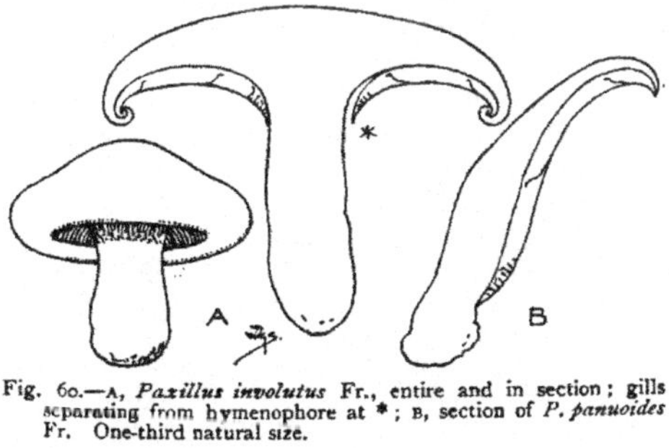

Fig. 60.—A, *Paxillus involutus* Fr., entire and in section; gills separating from hymenophore at *; B, section of *P. panuoides* Fr. One-third natural size.

homogeneous with the hymenophore. *Gills* decurrent, often anastomising behind and forming spurious pores as in *Boletus*, readily separating from the hymenophore owing to the—except in **1185**—absence of a trama, usually separated from the flesh of the pileus by a thin differentiated horny or cartilaginous layer. *Spores* whitish to ferruginous. (Fig. 60.)

Fleshy, putrescent fungi, some growing on the ground, others on tree-trunks or sawdust. Some of the members of the first subgenus are allied to *Tricholoma* and *Clitocybe* amongst the white-spored Agarics; those of the second to *Flammula*, amongst the brown-spored Agarics.

The gills of **1181** are sometimes wholly porous, in **1181a** always wholly porous, as in *Boletus*.

Most of the species are said to be edible. Species **1172**—**1186**

SUBGENUS 1. **LEPISTA.** *Spores* dull white—reddish- or fuscous-pallid in **1174**—faintly ferruginous in **1176.** **1172**—**1179**

SUBGENUS 2. **TAPINIA.** *Pileus* usually more or less excentric, sometimes lateral and stemless, at length depressed. *Spores* ferruginous. **1180**—**1186**

SUBGENUS 1. LEPISTA.

(From the type, **1174** *P. lepista.*)

1172. P. giganteus Fr. (from its great size) *a b c.* Tan white.

P. depressed or infundibuliform, not umbonate; marg. sulcate, very involute. *St.* solid, smooth, stained faintly vinous. *G.* deeply decurrent, very crowded, inclined to branch and anastomose.

Odour strong, like **1394**. Woods and pastures; uncommon. Aug.–Nov. 1–1¼ × 4½ × ⅝ in. Very different from **190**, with which it is often confounded.

1173. P. Alexandri Fr. (from P. Alexander, who gathered the first examples described) *a b.*

P. plano-depressed, dry, fawn to umber-whitish; marg. striate, involute. *St.* colour as P., or faintly ochreous. *G.* sub-decurrent, crowded, yellowish or brownish.

Amongst moss, on the ground. Sept.–Oct. 3¼ × 1⅝ × ⅝ in.

1174. P. lepista Fr. (from the shape of the pileus when young; *lepista*, a pan) *a b c.*

P. plano-depressed, dry to subviscid; marg. involute, undulato-flexuous, rimuloso-squamulose, not striate, whitish or yellowish-white. *St.* solid, whitish, sometimes inclining to fuscous below. *G.* decurrent, very crowded, whitish or pale yellowish-white.

Solitary, sporadic, changeable in habit. Odour of rancid meal, or strong and acid. Woods, oak, fir, moist places. Oct.–Nov. 5⅝ × 3¼ × 1 in.

1175. P. extenuatus Fr. (from the thin margin of the pileus; *extenuatus*, made thin) *a b.*

P. campanulato-convex, broadly umbonate or obtuse, tan, clay or yellowish; marg. involute, smooth, even or striate. *St.*

solid, smooth, somewhat tuberous-rooting, colour as P. *G.* deeply decurrent, very crowded, white to mouse-colour.

Woods, fir, grassy places. Oct. 4 × 2⅛ × ⅝ in.

1176. P. panæolus Fr. (from the change of colour of the flesh to black; Gr. *panaiolos*, all variegated) *a b c.*

P. convexo-plane, depressed, smooth, moist, whitish; marg. subinvolute. *St.* stuffed, attenuate upwards, striato-fibrillose, rufescent or as P., ochreous below. *G.* subdecurrent, crowded, becoming watery ferruginous or cinnamon.

Subgregarious. Woods, fir, mixed. Oct.–Nov. 1¼ × 1½ × ⅛ in. Var. *spilomæolus* Fr. *P.* spotted with dots.

1177. P. orcelloides Cooke & Mass. (from its resemblance to **504**, *Clitopilus orcella*) *a b.*

P. convexo-plane, silky, white, then clouded greyish or livid; marg. involute. *St.* solid, attenuate downwards, elastic, silky-fibrillose, livid-ochreous. *G.* adnato-decurrent, crowded, whitish to salmon-brown.

Woods, amongst grass. Sept.–Nov. 2⅛ × 1¼ × 3/16 in.

1178. P. lividus Cooke (from its livid colour) *a.*

P. convex, then depressed, dull whitish or livid-ochreous; marg. subinvolute. *St.* stuffed to hollow, attenuate downwards, fibrillose, whitish. *G.* subdecurrent, somewhat crowded, white.

In small clusters. Odour pleasant. Woods, fields. Oct. 2 × 2⅞ × ⅜ in. Allied to **1179.**

1179. P. revolutus Cooke (from the, at length, slightly revolute margin of the pileus) *a.*

P. campanulato-obtuse, pale ochreous; mid. darker; marg. sometimes pale lavender. *St.* solid, attenuate downwards, whitish or pale lavender. *G.* arcuato-decurrent, pallid to clay.

Odour of meal. Fields. Oct. 1⅝ × 1⅜ × ⅜ in.

SUBGENUS 2. TAPINIA.

(From the pileus, at length depressed; Gr. *tapeinoo*, to depress.)

1180. P. paradoxus Berk. ex Cooke (from its being a new, unexpected species; *paradoxus*, strange, unexpected) *a b c.*

P. convex to plane, densely tomentose, deep rufous-umber to yellowish-brown; marg. subinvolute. *St.* solid, attenuate downwards, sometimes rooting, yellow or reddish, or crimson to rose above, and yellow to brownish below. *G.* decurrent, distant, veined, yellow to deep yellow, reddish when bruised.

Woods, on the ground, on earth inside hollow trees, oak. Aug.–Nov. 4 × 1½ × ⅝ in. With the general appearance of a dark pileated **1472.**

1181. P. involutus Fr. (from the involute margin of the pileus) *a b c.*

P. convexo-plane, then depressed, ochreous-ferruginous; marg. very involute. *St.* solid, slightly enlarged downwards, dull yellowish, dull purplish below. *G.* decurrent, branched, sometimes forming elongate Po. near the St., pallid, pale or umber-yellow to ferruginous, changing to vinous when bruised.

Said to be edible and highly esteemed in Belgium and Russia; tasteless when raw. Eaten at Hereford Fungus Forays, but not much approved, 1870. Woods, by and on stumps; very common. June–Nov. 7 × 2⅜ × 1⅛ in. A yellow parasitic mould, *Hypomyces chrysospermus*, frequently grows on the gills. Var. *excentricus* Mass. *St.* excentric. Var. *subinvolutus* W. G. Sm. (*Agaricus subinvolutus* Batsch) margin of *P.* less involute than type.

1181a. P. porosus Berk. (from the porous hymenium) *a b.*

P. fleshy, viscid when moist, reddish-claret or olive-brown; marg. thin, even, not involute. *St.* excentric, tough, equal or attenuate downwards, claret-brown, darker below, somewhat reticulate above with the Po. *G.* decurrent, shallow, poriform, yellow to sulphur-green, changing to pale blue and then brownish when bruised, dull green when old. *Po.* round to elongate, irregular, large to small. *Flesh* dull pale vinous-brown, mottled and streaked, darker in stem, changing colour as St.

Tasteless. Odour very strong, unpleasant. Moist woods, under firs. Autumn. 4⅜ × 3⅝ × ⅜ in. Not unlike **1181**, to which it is closely allied, but the margin is never involute and the pores are different in colour. Small forms somewhat resemble **1468.**

1182. P. leptopus Fr. (from the thin stem; Gr. *leptos*, thin, *pous*, a foot) *a b.*

P. depressed, excentric, downy, then torn-villous-scaly, fuscous-yellowish to yellow-olive-umber; marg. subinvolute. *St.* solid, attenuate downwards, yellowish. *G.* deeply decurrent, sometimes poriform at base, yellowish to full yellow or shaded olive. *Flesh* straw-white to yellow.

On the ground, about stumps, fir. Aug.–Sept. 3½ × ⅝ × ⅜ in.

1183. P. atrotomentosus Fr. (from the black downy pileus; *ater*, black, *tomentum*, down) *a b c.*

P. convexo-plane, then depressed, deep opaque umber or deep orange-umber, somewhat orange-sienna at involute marg. *St.* solid, equal, somewhat rooting, excentric, rarely lateral, blackish-umber or intense crimson-umber, pale yellowish at apex. *G.* subdecurrent, branching and somewhat porous at base, yellowish to deep yellow. *Flesh* white or sulphur-white in P., rose-purple in St., mottled dull purplish.

Taste insipid; odour when cut rank and strong like **2086.** Stumps, pine, on the ground. Aug.–Nov. 5⅜ × 2¼ × 1⅜ in.

1184. P. crassus Fr. (from the stem, thick above; *crassus*, thick) *a.*

P. plane, even, ferruginous; marg. subinvolute. *St.* stuffed, attenuate downwards, excentric, colour as P. *G.* subdecurrent, subdistant, cinnamon.

On the ground, woods. Nov. 3⅛ × 1 × ½ in.

1185. P. panuoides Fr. (from its shape, as in genus *Panus;* Gr. *eidos*, appearance) *a b c.* Dull, opaque, yellow.

P. conchato-dimidiate, imbricate; marg. subinvolute. *St.* obsolete. *G.* decurrent to base, crowded, branched and anastomosing, trama well developed, tan-sienna or somewhat pale yellow-ochre, sometimes studded with drops.

Beech, fir, hawthorn, sawdust, in cellars, on wood; uncommon. July–Nov. Diam. 4 in. *P.* sometimes whitish-ochre with reddish marg. *G.* sienna. The resupinate and cup-shaped form is *Gomphus pezizoides* Pers.

1186. P. Fagi B. & Br. (from its habitat, stumps of beech, *fagus*) *a.*

P. dimidiate, crisped, becoming revolute, pallid. *St.* obsolete. *G.* decurrent to base, crisped, orange.

Gregarious. Autumn. Diam. 3½ in.

XLVII. HYGROPHORUS Fr.

(From the water-bearing character; Gr. *hugros*, moist, *phero*, to bear.)

Veil universal and viscid or obsolete. *Pileus* more or less fleshy, glutinous, viscid or watery; margin slightly incurved. *Stem* central, continuous and homogenous with the hymenophore, usually exannulate. *Gills* furnished with a subfloccose trama, usually distant and

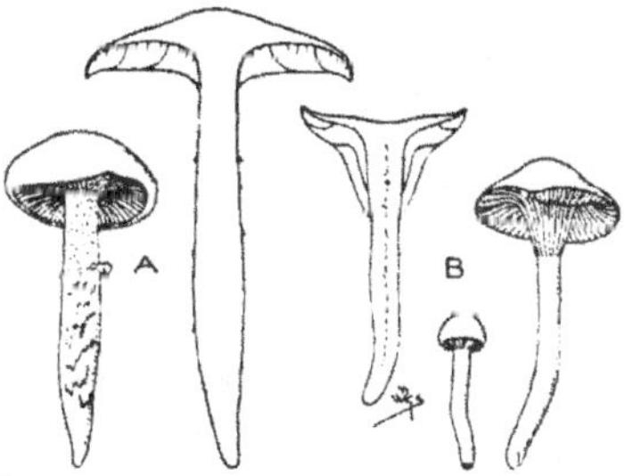

Fig. 61.—A, *Hygrophorus olivaceoalbus* Fr., entire and in section; B, *H. virgineus* Fr., entire and in section. One-third natural size.

thick, but acute at the edge, not membranous, often branched and veined, the hymenium waxy and at length removable from the trama, which is of the same substance as the pileus. *Spores* subglobose or elliptical, smooth, white, pale umber in **1208.** (Fig. 61.)

The species are putrescent, often brilliant in colour as scarlet, orange, sulphur, green and shining-white; they all, except **1215**, grow on the ground, and often appear after the first frosts of late autumn. They differ from *Cortinarius*, *Gomphidius* and *Paxillus* in the colour of the spores, from *Lactarius* in being non-milky, from *Russula* in not being vesciculoso-rigid, and from *Cantharellus* in the sharp-edged gills. Many have a pleasant taste and are edible.

Typical species of the genus are best seen in Subgenus 3, *Hygrocybe*. Species **1187—1228**

SUBGENUS 1. **LIMACIUM.** *Pileus* glutinous. *Universal veil* viscid, with occasionally a floccose partial veil which is annular or marginal. *Stem* clothed with minute scales or rough with dots above. *Gills* adnato-decurrent.

White or yellowish-white. **1187—1190**
Reddish. **1191—1194**
Tawny or light yellow. **1195—1198**
Olivaceous umber. **1199—1202**
Fuscous-cinereous or livid. **1203—1206**

SUBGENUS 2. **CAMAROPHYLLUS.** *Veil* none. *Pileus* firm, opaque, moist in rainy weather, not viscid, except **1216, 1219, 1220** and **1224**, where the pileus is somewhat viscid. *Stem* even, smooth or fibrillose, not rough with dots. *Gills* distant, arcuate.

Gills deeply and at length obconically decurrent. **1207—1217**
Gills ventricose, sinuato-arcuate or plano-adnate. **1218—1224**

SUBGENUS 3. **HYGROCYBE.** *Veil* none. *Pileus* viscid when moist, shining when dry, rarely floccoso-scaly. *Stem* hollow, except **1225, 1228**, stuffed in **1232**, soft, not punctate. *Gills* soft. Whole fungus thin and watery-succulent, fragile, mostly bright coloured and lustrous.

Gills decurrent. **1225—1234**
Gills adnexed, adnate in **1241** and **1243**, broadly emarginate in **1244**, somewhat separating. **1235—1246**

SUBGENUS 1. LIMACIUM.

(From the pileus and stem glutinous with the veil; *limax*, a slug.)

1187. H. chrysodon Fr. (from the golden, tooth-like scales; Gr. *chrusos*, gold, *odous*, a tooth) *a b c*. White or sulphur-white. *P.* convexo-plane or flat, yellow squamulose at mid. and marg. *St.* stuffed, equal, colour as P., minutely yellow squamulose above, sometimes imperfectly floccoso-annulate. *G.* decurrent, distant, white with a faintly yellowish edge.

Woods, lime, beech, under trees; rare. Aug.–Nov. $2\frac{3}{4} \times 3\frac{1}{2} \times \frac{1}{2}$ in. Var. *leucodon* Stev., with shining-white pubescence.

1188. H. eburneus Fr. (from its ivory-white colour; *ebur*, ivory) *a b c*. *P.* convexo-plane, flat, repand; marg. pubescent, soon naked. *St.* stuffed to hollow, equal or nearly so. *G.* adnato-decurrent, distant.

Said to be edible. Odour mild, sweetish, not unpleasant or none. Woods, grassy roadsides, pastures; frequent. Aug.–Oct. $2\frac{3}{8} \times 3\frac{1}{2} \times \frac{1}{2}$ in. Not unlike **115**.

1189. H. Cossus Fr. (from its odour of the goat-moth, *Cossus ligniperda*) *a b c.* Faintly yellowish-white.

P. convexo-plane, obtuse; mid. slightly ochreous; marg. naked. *St.* stuffed, slightly attenuate downwards. *G.* adnato-decurrent, distant.

Odour powerful, persistent. Woods, oak; frequent. Sept.–Nov. 2¾ × 3¾ × ⅜ in.

1189a. H. melizeus Fr. (a fanciful name to express its beauty, like the melody of a song; Gr. *melizo*, to sing) *a.* Straw colour.

P. fleshy, convexo-expanded, depressed, even, viscid; marg. thin, at first involute. *St.* stuffed, attenuate downwards, white punctato-squamulose above. *G.* decurrent, thin, distant.

Odour pleasant. Woods. Nov. 3½ × 4½ × ⅜ in.

1189b. H. discoxanthus Rea (from the yellow middle of the pileus; Gr. *diskos*, a disc, *xanthos*, yellow) *a.*

P. convex, then expanded and depressed, viscid, white then yellowish, deeper coloured at mid.; marg. becoming brownish with age. *St.* solid, attenuate downwards, often curved, viscid, white, soon becoming reddish-brown. *G.* somewhat distant, decurrent, veined at base, exceeding marg. of P., white, then yellowish to reddish-brown. *Flesh* white, becoming reddish in St.

Odour pleasant of anise. Amongst grass. Oct. 2 × 1⅜ × ¼ in.

1190. H. penarius Fr. (from its edible qualities; *penus*, provisions) *a b.* Tan-whitish.

P. convexo-expanded, umbonate obtuse, then flat, sometimes clouded fawn-vinous; mid. palest sienna or ochre. *St.* solid, attenuate downwards, fusiform. *G.* adnato-decurrent, distant, thick, veined.

Taste sweet; odour pleasant. Woods, mixed. Oct. 3¾ × 3½ × ⅝ in.

1191. H. pulverulentus B. & Br. (from the powdered stem; *pulvis*, dust) *a b.*

P. pulvinate, white; marg. tomentose, sometimes rose- or pale rufous-mealy. *St.* stuffed, attenuate downwards, fusiform, wholly powdered with rose, rose-sienna or pale rufous meal. *G.* decurrent, thick, whitish, edge obtuse.

Amongst pine-leaves. Nov.–Dec. ¾ × 1⅛ × ⅛ in.

1192. H. erubescens Fr. (from becoming red, *erubesco*) *a b c.*

P. gibbous, convexo-flat, adpressedly dotted-squamulose, white or pale buff, becoming red, rose-salmon or vinous-crimson-brown; mid. sienna. *St.* solid, equal or attenuate downwards, red-fibrillose, red or pale dull brownish-purple, pale above. *G.* decurrent, distant, narrow, white, red-spotted or rose, or sienna-shaded. *Flesh* white or pale rose.

In troops, sometimes in large circles. Woods, pine. Oct. 4 × 3½ × ⅜ in. Perhaps allied to **88.**

1193. H. pudorinus Fr. (from its flesh-colour ; *pudorinus*, blushing) *a b.*

P. convex, flat, salmon-rose ; mid. darker. *St.* solid, attenuate upwards or subventricose, white or salmon-rose above, paler or faintly ochreous below. *G.* decurrent, thick, distant, white or pale flesh-colour, edge sometimes salmon.

Odour none. Woods, fir. Oct. $4 \times 3 \times \frac{5}{8}$ in.

1194. H. glutinifer Fr. (from its gluten-bearing habit ; *fero*, to carry) *a b c.*

P. convex, obtuse, rufescent, mid. wrinkled-dotted, buff, deep buff or umber ; marg. whitish. *St.* stuffed, ventricose downwards, white above, gradually sienna-umber below. *G.* decurrent, white or faintest olive-white. *Flesh* white.

Gregarious. Taste mild. Woods, grassy places. Sept.–Oct. $4\frac{3}{4} \times 4 \times \frac{5}{8}$ in. Must not be confused with **1201.**

1195. H. arbustivus Fr. (from its habitat, plantations, *arbustum*) *a b.*

P. convexo-plane, subrepand, innato-fibrillose, smooth, buff-tan, fawn-tan or sienna ; marg. paler. *St.* solid to stuffed, equal, white above, tan at base. *G.* adnate, subdecurrent, distant, white.

Edible. Taste and odour pleasant. Woods, birch, fir. Sept.–Dec. $2\frac{1}{4} \times 2\frac{1}{8} \times \frac{1}{4}$ in.

1196. H. aureus Arrh. ex Fr. (from the golden-yellow pileus ; *aureus*, golden) *a.*

P. convex, flat, even. *St.* stuffed, equal, pale sulphur above, yellow below, sometimes annulate with the glutinous reddish V. *G.* decurrent, distant, thin, whitish to deep sulphur.

Woods. Nov. $1\frac{1}{2} \times 2 \times \frac{1}{4}$ in.

1197. H. discoideus Fr. (from the disc-shaped pileus) *a b c.*

P. convex, obtuse, flat, even, smooth, ochreous-white ; mid. pale reddish-brown, sometimes dark. *St.* stuffed, slightly attenuate upwards, whitish, faintly ochreous or brownish below. *G.* decurrent, becoming distant, faint yellowish-tan.

Gregarious, forming rings, solitary. Woods, pine, grassy places. Oct.–Nov. $2\frac{1}{2} \times 2\frac{3}{4} \times \frac{1}{4}$ in.

1198. H. aromaticus Berk. (from its odour of cinnamon) *a b.*

P. convexo-plane, flat, cuticle netted-reticulate and separable, cinnamon. *St.* stuffed to hollow, rosy-cinnamon. *G.* rounded-adnate to decurrent, separating from hymenophore, salmon.

Taste acrid ; odour spicy. Autumn. $3\frac{3}{8} \times 2\frac{5}{8} \times \frac{3}{8}$ in.

1199. H. limacinus Fr. (from its slime ; *limax*, a slug) *a b.*

P. convexo-plane, obtuse, umber, fuliginous or olive ; marg. paler. *St.* solid, equal, fibrilloso-striate, white or sulphur-green-white, somewhat sienna below. *G.* adnato-decurrent, subdistant, thin, white or pale olive-cinereous.

Sometimes cæspitose. Woods, amongst damp leaves. Oct.–Nov. $3\frac{3}{4} \times 3 \times \frac{1}{2}$ in.

1200. **H. olivaceoalbus** Fr. (from the olivaceous pileus and white gills) *a b c.*

P. conical, convex, subumbonate, even, smooth. *St.* solid, attenuate below, white above, spotted with milky drops, spotted squamulose below with fuscous V., and more or less annulate. *G.* adnato-decurrent, distant, veined, white or sulphur-greenish-white.

Taste insipid; odour none. Woods, woodland pastures, chiefly under pine, oak. Aug.–Nov. $3\frac{1}{4} \times 4\frac{1}{2} \times \frac{1}{2}$ in.

1201. **H. hypothejus** Fr. (from the sulphur-colour under the gluten; Gr. *hupo*, under, *theion*, sulphur) *a b c.*

P. convex, flat, depressed, gluten leaf-green or olivaceous, umber or deep madder-brown. *St.* stuffed to hollow, equal or slightly attenuate downwards, at first yellow-greenish, then yellow, sometimes flesh-colour, fugaciously subannulate. *G.* decurrent, distant, white to yellow, sometimes flesh-colour. *Flesh* pale sulphur-yellow.

Woods, pine, larch, amongst heather, etc.; frequent. Sept.–Dec. $3 \times 3 \times \frac{1}{2}$ in. Must not be confused with **1194.**

1202. **H. ceracinus** Berk. (from its odour, like cherry-laurel; *Prunus Laurocerasus*) *a b c.*

P. convexo-expanded, obtuse or subumbonate, umber to grey; marg. minutely tomentose. *St.* solid, equal or attenuate downwards, white. *G.* adnato-decurrent, very distant, branched, white.

Somewhat gregarious. Woods, fir. Sept.–Oct. $2\frac{5}{8} \times 2\frac{3}{4} \times \frac{3}{8}$ in.

1203. **H. fuscoalbus** Fr. (from the brown pileus and white gills) *a b.*

P. convexo-plane, obtuse, becoming cinereous; marg. white-floccose. *St.* solid, equal, white above, tan below, sometimes floccoso-annulate. *G.* decurrent, somewhat thick.

Woods, mixed, amongst moss. Sept. $1\frac{3}{8} \times 2\frac{1}{4} \times \frac{1}{4}$ in.

1204. **H. agathosmus** Fr. (from the pleasant odour; Gr. *agathos*, good, *osme*, scent) *a b c.*

P. convexo-plane, gibbous, livid grey or dull pale yellowish-olive; mid. usually darker. *St.* solid to hollow, equal, somewhat fibrilloso-striate, white above, as P. below. *G.* decurrent, distant, white.

Odour sweet of anise, or of fruit with spirit. Woods, fir. Sept.–Nov. $3\frac{1}{4} \times 2\frac{1}{8} \times \frac{3}{8}$ in.

1205. **H. mesotephrus** B. & Br. (from the brown-grey middle of the pileus; Gr. *mesos*, middle, *tephros*, ash-coloured) *a b c.*

P. hemispherical to flat, white; mid. pale sienna or buff; marg. striate. *St.* stuffed, attenuate downwards, white, faint sienna or buff below. *G.* decurrent, subdistant, white.

Woods, damp places; rare. Oct. $1\frac{3}{4} \times 2\frac{1}{8} \times \frac{1}{4}$ in.

1206. **H. lividoalbus** Fr. (from the livid pileus and white gills) *a b c*.
P. flat, buff-white or white, with white-umber marg. *St*. stuffed, equal, somewhat even, glabrous, white. *G*. decurrent, sub-distant, white.

Woods. Oct.–Nov. $2\frac{3}{4} \times 3\frac{3}{4} \times \frac{1}{4}$ in.

SUBGENUS 2. CAMAROPHYLLUS.

(From the arcuate gills; Gr. *kamera*, a vault, *phullon*, a leaf.)

1207. **H. caprinus** Fr. (from the fibrillose stem, as if covered with goat's hair; *caper*, a goat) *a c*.
P. conical, umbonate, convex, then flat or depressed, streaked-fibrillose, blackish, cinereous-fuliginous or umber. *St*. solid, or hollow above, equal, pale umber- or grey-white. *G*. very distant, white, then glaucous.

Woods, pastures, under firs. Oct. $3\frac{1}{4} \times 4 \times \frac{1}{2}$ in.

1208. **H. leporinus** Fr. (from the colour, like a hare, *lepus*) *a c*.
P. convex, gibbous, fibrilloso-floccose, ochre-sienna or umber. *St*. solid, rigid, subfusiform, fibrillose, pallid above, brownish-tawny below. *G*. pale yellowish. *Flesh* brownish above and below.

Taste and odour none. Woods, downs; rare. Sept.–Oct. $2\frac{3}{4} \times 2\frac{1}{2} \times \frac{3}{8}$ in.

1209. **H. nemoreus** Fr. (from the habitat; *nemus*, a grove) *a b*.
P. convexo-expanded, gibbous, then depressed, orange, or yellowish-white-sienna. *St*. stuffed, squamulose, fibroso-striate, slightly attenuate downwards, pale biscuit-colour. *G*. thick, whitish to orange.

Edible. Taste mild. Woods, pastures. Oct.–Dec. $2\frac{3}{8} \times 1\frac{7}{8} \times \frac{1}{2}$ in.

1210. **H. pratensis** Fr. (from the habitat; *pratum*, a meadow) *a b c*. Pale buff or yellowish.
P. convex, obtuse or umbonate, then repand, even, smooth. *St*. stuffed, attenuate downwards. *G*. very distant, thick, veined.

Edible, but with little flavour. Pastures, roadsides, downs; common. Aug.–Nov. $3\frac{3}{4} \times 2\frac{3}{8} \times \frac{5}{8}$ in. Sometimes wholly white. Var. *umbrinus* W. G. Sm. *P*. olive-slate or whitish-umber. Var. *cinereus* Fr. *P*. thin, cinereous, margin striate. Var. *pallidus* Cooke. *P*. infundibuliform, pallid. Var. *Meisneriensis* Fr. *P*. skin easily separable. Tasteless or slightly insipid.

1210a. **H. Karstenii** Sacc. & Cub. (after P. A. Karsten) *a*.
P. fleshy, convexo-plane, compact at mid., glabrous, even, whitish. *St*. solid, attenuate downwards, wavy, white. *G*. deeply decurrent, distant, yellow.

Woods. Nov. $3\frac{1}{2} \times 5\frac{3}{4} \times \frac{3}{4}$ in. Distinguished from **1210** by its chalk-white downward tapering stem.

1211. **H. fœtens** Phill. (from its nauseous-fetid odour) *a b c*.
P. convex, plane, smooth, then squamulose, sepia. *St*. stuffed, attenuate downwards, paler than P. *G*. somewhat thick, vinous-brown-tan, somewhat slate, glaucous-pruinose.

Odour cadaverous, ordure-like, fetid-nauseous. Grassy places. Oct.–Nov. $1 \times 1\frac{1}{4} \times \frac{1}{8}$ in.

1212. **H. virgineus** Fr. (from the pure white colour) *a b c.*

P. convex, obtuse, flat, depressed, at length cracked and floccose. *St.* stuffed, attenuate downwards. *G.* very decurrent, somewhat thick.

Edible. Pastures, woods, roadsides, downs. Aug.–Dec. $1\frac{3}{4} \times 1\frac{3}{4} \times \frac{1}{4}$ in. Very variable in size. Sometimes clouded with beautiful shades of clear lilac, caused by the growth of the mould named *Verticillium Marquandii.* Must not be confounded with the white form of **1210** or with **1214.** Var. *roseipes* Mass. ex Cooke. *St.* rose-colour or salmon within and without, sometimes strigose at base.

1213. **H. ventricosus** B. & Br. (from the ventricose stem) *a b.* White.

P. convex, unequal, lobed. *St.* solid or partially hollow. *G.* narrow, forked.

Amongst grass. Sept. $1\frac{1}{4} \times 2\frac{1}{2} \times \frac{1}{2}$ in. Often tinged with red from the growth of a parasitic *Fusisporium*, which may be the cause of the hypertrophy. Greatly resembles one form of **123.**

1214. **H. niveus** Fr. (from its snow-white colour; *niveus*, snow-white) *a b c.*

P. convex, flat, umbilicate; marg. striate. *St.* fistulose, equal or slightly attenuate downwards. *G.* thin.

Pastures, roadsides, parks, downs, etc.; common. Sept.–Nov. $\frac{7}{8} \times 2 \times \frac{1}{16}$ in. Much like **207** in appearance. Sometimes minute. Sometimes stained lilac, as in **1212.**

1215. **H. Wynniæ** B. & Br. (after Mrs. Lloyd Wynne) *a b c.* Semi-transparent lemon, faintly greenish, sometimes tan or white.

P. convex, umbilicate, or subinfundibuliform; marg. striate. *St.* stuffed, equal, or attenuate downwards. *G.* narrow, thin.

On chips, twigs, old stumps, etc. Oct. $1 \times 1\frac{3}{8} \times \frac{1}{8}$ in. The habitat points to *Mycena* or *Omphalia.*

1216. **H. russocoriaceus** Berk. & Mill. (from the odour of Russian leather; *coriaceus*, leathery) *a b c.* White.

P. convex, expanded, obtuse, slightly viscid. *St.* solid, attenuate downwards, smooth. *G.* thick, distant, few.

Odour varying of *Potentilla atrosanguinea.* Pastures; rare. Sept.–Dec. $1 \times 1\frac{1}{8} \times \frac{3}{16}$ in. Odour sometimes persistent in dried examples for more than twenty years. Must not be confounded with **207.**

1217. **H. micaceus** B. & Br. (from the shining mica-like particles on the pileus) *a c.*

P. hemispherical, flat, slightly depressed, wrinkled, light yellow to cinereous. *St.* solid, equal, colour as P., then brown below. *G.* pallid umber to clay-sienna.

Clayey soil. Oct. $\frac{5}{8} \times \frac{5}{8} \times \frac{1}{16}$ in. Whole plant turns brown in drying.

1218. **H. fornicatus** Fr. (from the pileus, at first convex; *fornicatus*, arched) *a b c.*

P. becoming expanded and obsoletely broadly umbonate, viscid, white or pale livid; mid. brown. *St.* solid, hollow or stuffed, equal, smooth or subscaly, whitish to greyish. *G.* adnexed or sinuato-adnexed, ventricose, distant, white.

Pastures. Oct.–Nov. 3 × 2⅝ × ⅜ in. Var. *clivalis* Fr., white. *P.* thin, fragile; marg. incurved, striate, not viscid. *St.* short, fragile, attenuate downwards. *G.* rather thick.

1219. **H. distans** Berk. (from the very distant gills) *a b.*

P. convex, plane, subdepressed or umbilicate, viscid, white or faintly brownish. *St.* solid, attenuate downwards, white, cinereous or pale ochre-tan below. *G.* plano-decurrent, few.

Sometimes fragrant of essence of almonds, pleasant, sweet. Woods; rare. Oct. 2 × 1½ × ¼ in.

1220. **H. Clarkii** B. & Br. (after J. Aubrey Clark) *a b.*

P. obtuse, convexo-plane, viscid, opaque umber or livid-cinereous; marg. even, white. *St.* solid, stuffed or hollow, equal or attenuate downwards, viscid and white-squamulose above, pale umber-scaly below. *G.* adnate, distant, broad, thick, veined, ivory-white.

Woods, pastures. Sept.–Oct. 3¾ × 2⅝ × 1⅛ in. *H. latitabundus* Britz. is this species, as is Kalchbr. *Hymen. Hung.* t. 24, f. 1, incorrectly named *H. limacinus* Fr.

1221. **H. metapodius** Fr. (from the stem, sometimes thick above and attenuate below; Gr. *meta*, upside down, *pous*, a foot) *a b c.*

P. convex, plane, obtuse, cinereous-fuscous or pale rose-shaded sepia. *St.* stuffed, unequal, smooth, cinereous. *G.* adnate, broadly emarginate, distant, veined, grey-white. *Flesh* pallid-grey, reddish when broken.

Sometimes in clusters and deformed. Taste mild, insipid to sweet; odour of new meal. Old mossy pastures. Oct.–Jan. 2¼ × 1⅜ × ⅝ in. At maturity becoming black or deep sepia within and without. Analogous with 1306 and 1307.

1222. **H. ovinus** Fr. (from the woolly pileus; *ovis*, a sheep) *a b c.*

P. campanulate, expanded, subumbonate, revolute, incised, subviscid, then obscurely squamulose, dull-sienna or yellow; mid. sienna-slate; marg. tan. *St.* somewhat stuffed, almost equal, smooth, tan-grey. *G.* emarginate or arcuato-adnate, veined, whitish-grey then rufescent.

Odour of new meal. Woods, pastures, mossy places; uncommon. Sept.–Dec. 2¼ × 2⅜ × ½ in. Sometimes changing to black.

1223. **H. subradiatus** Fr. (from the radiato-striate pileus) *a b.*

P. convex, expanded, subumbonate, pale yellowish; mid. sienna and umber; marg. sienna, sometimes lilac. *St.* stuffed to hollow, equal, smooth, often twisted, whitish or pale lilac above, tan-brownish below. *G.* decurrent, somewhat thin, distant, veined, white to pale yellowish.

Old pastures, heathy ground. Sept.–Oct. 3½ × 2 × ½ in. Var. *lacmus* Fr. wholly or almost wholly shining-lavender.

1224. **H. irrigatus** Fr. (from its watery texture; *irrigo*, to wet) *a.*

P. convex, obtuse, flat, smooth, viscid, pale fawn; mid. brown;

marg. striate. *St.* hollow, equal, smooth, very viscid, colour as P., but with a pale yellowish tinge. *G.* adnate with a tooth, subdistant, somewhat thick, veined, white, grey, or tan.

Woods, fir, grassy places. Oct.–Nov. 2 × 2⅝ × ¼ in.

SUBGENUS 3. **HYGROCYBE.**

(From the moist pileus; Gr. *hugros*, moist, *kube*, a head.)

1225. H. Colemannianus Blox. ex Berk. (after W. H. Coleman) *a b c.*

P. convexo-expanded, umbonate, dark reddish-umber; marg. striate, paler. *St.* solid, nearly equal, somewhat silky, salmon-whitish, shaded pale brownish. *G.* distant, veined, paler than P. to whitish.

Grassy places, lawns. Oct.–Nov. 1⅛ × 1¾ × ¼ in.

1226. H. sciophanus Fr. (from its delicacy as compared with **1210**; Gr. *skia*, a shadow, *phaino*, to appear) *a.*

P. hemispherical, expanded, obtuse, deep tawny, brown, red or vermilion to white; marg. striate, paler. *St.* equal, smooth, tawny-yellowish to white. *G.* adnate to plano-decurrent, ventricose, distant, colour as P., or inclining to rose or yellow.

Densely gregarious. Grassy and mossy places. Autumn. 1¼ × 1⅞ × ⅛ in.

1227. H. lætus Fr. (from the bright colour; *lætus*, pleasing) *a b c.*

P. convex, flat, slightly depressed, slippery-smooth, deep yellow, tawny, vermilion-orange or salmon; marg. striate. *St.* fistulose, equal, slippery-smooth, tawny or yellow. *G.* plano-decurrent, distant, thin, veined, flesh-colour, whitish, fuliginous or shaded brownish. *Flesh* in P., yellow and orange, in St., tan-yellow.

In troops. Pastures, downs; frequent. Oct.–Nov. 1¼ × 2¼ × ⅛ in.

1228. H. Houghtonii B. & Br. (after the Rev. Wm. Houghton) *a b c.*

P. convex, flat, umbilicate, glutinous, rugose to mid., flesh-colour with sienna marg., pale vermilion shaded darker or wholly cinnabar, sometimes very pale. *St.* fistulose, equal, very glutinous, pale salmon above, pale or dark yellow below, sometimes wholly cinnamon. *G.* plano-decurrent, thick, pale yellowish or salmon. *Flesh* in P. flesh-colour or orange, in St. salmon above, tan below.

Tasteless; odour strong, foxy. Amongst grass. Oct. 1⅜ × 2⅞ × ⅛ in.

1229. H. vitellinus Fr. (from the egg-yellow colour, within and without; *vitellus*, yolk of egg) *a b.*

P. convexo-expanded, smooth, varying lemon-yellow with a suggestion of olive, becoming white; marg. plicato-striate. *St.* equal, even, smooth, becoming white. *G.* decurrent, subdistant.

Woods, pastures. Oct.–Dec. 1 × 2⅛ × ⅛ in.

1230. H. ceraceus Fr. (from the colour, like yellow wax; *cera*, wax) *a b c.*

P. convexo-plane; marg. striate, not becoming white; mid. sometimes vermilion. *St.* equal to unequal, even, smooth. *G.* adnato-decurrent.

In troops. Woods, pastures; common. Sept.–Dec. 1⅜ × 3¼ × ⅛ in.

1231. **H. coccineus** Fr. (from its scarlet colour, *coccineus*) *a b c.*
P. hemispherical, convex, plane, smooth, not floccoso-scaly, crimson-scarlet or yellow to orange, clouded and blotted carmine, becoming whitish or yellowish. *St.* even, not slippery-smooth, unequal, scarlet above, yellow below, sometimes rose or purplish, becoming whitish or yellowish. *G.* adnate, sometimes with a decurrent tooth, distant, veined, yellow below, purplish above, edge glaucous. *Flesh* crimson, orange or sulphur, external layer of St. orange-vermilion.

Pastures; very common. June–Nov. $2\frac{3}{4} \times 2\frac{3}{8} \times \frac{1}{2}$ in.

1232. **H. miniatus** Fr. (from its colour; *minium*, vermilion) *a b c.*
P. convex, umbilicate, even, smooth then squamulose, becoming rose or orange. *St.* stuffed, slightly attenuate downwards, smooth, shining. *G.* adnate or slightly decurrent, distant, rather thick, yellow or vermilion-shaded. *Flesh* of P. and external layer of St. scarlet.

Open and boggy places, woods, moist banks, pastures. June–Oct. $1 \times 2\frac{1}{4} \times \frac{1}{8}$ in. Sometimes very small. Does not adhere to paper in drying.

1233. **H. turundus** Fr. (from the flocculose pileus; *turunda*, lint) *a b c.*
P. convex, plane, depressed or umbilicate, yellow to orange; mid. vermilion; sq. cinereous-fuscous; marg. sometimes crenate. *St.* equal, even, lustrous, tawny-reddish or orange above, sulphur below, base white. *G.* slightly to deeply decurrent, white to orange.

Bogs amongst *Hydrocotyle vulgaris*, peaty soil, grassy places. July–Oct. $1\frac{5}{8} \times 2\frac{1}{2} \times \frac{3}{16}$ in. Adheres to paper in drying. Var. *mollis* B. & Br. *P.* golden-yellow-squamulose.

1234. **H. mucronellus** Fr. (from the shape of the pileus, acutely conical when young; *mucro*, a sharp point) *a.*
P. becoming campanulate and repand, umbonate, scarlet, bright red, orange or yellow, then hoary; marg. striate when moist. *St.* equal, even, smooth, yellow or scarlet, white below. *G.* subdecurrent, yellow, orange above.

Fields. Dec. $\frac{3}{4} \times 1\frac{1}{8} \times \frac{1}{16}$ in.

1235. **H. puniceus** Fr. (from the colour; *puniceus*, blood-red) *a b c.*
P. convex, obtuse, expanded, subumbonate, scarlet-crimson, sometimes clouded orange, sometimes orange-brown. *St.* equal or ventricose, striate; colour as P., white, yellowish or sulphur at base. *G.* ventricose, thick, distant, white-yellow, often reddish above. *Flesh* deep orange and sulphur in P., sulphur in St.

Pastures, downs, mossy places, sometimes woods; frequent. June–Nov. $4\frac{1}{2} \times 5 \times \frac{7}{8}$ in.

1236. **H. obrusseus** Fr. (from the golden colours; *obrussa*, the assaying of gold by fire) *a b c.*
P. campanulate, expanded, obtuse or umbonate, golden sulphur-yellow or golden to orange; mid. somewhat vermilion. *St.*

compressed, unequal, hollow, smooth, even, light or dark sulphur, base tawny or sulphur-white. *G.* ventricose, distant, white to sulphur and yellow-orange.

Woods, grassy places. Aug.–Oct. $2\frac{1}{2} \times 4\frac{1}{4} \times \frac{3}{8}$ in.

1237. **H. intermedius** Pass. (from its resemblance to adjoining species) *a.*

P. convex, obtuse, flat, fibrilloso-silky, golden-yellow becoming cinereous; mid. orange. *St.* compressed, fibrilloso-striate, sulphur to orange. *G.* ventricose, distant, whitish then yellowish.

Odour of meal. Roadsides, grassy places, damp ground. Autumn. $2\frac{3}{8} \times 2\frac{5}{8} \times \frac{3}{8}$ in. Like **1236**, but becoming grey.

1238. **H. conicus** Fr. (from the conical pileus) *a b c.*

P. often acute, then expanded and cracked, smooth, blood-scarlet, orange, sulphur-greenish, sulphur or smoky light yellow. *St.* equal, fibroso-striate, twisted, sulphur or yellow, reddish at base. *G.* ventricose, thin, subcrowded, white, yellowish or sulphur.

Pastures, downs, roadsides, etc.; common. July–Nov. $2\frac{7}{8} \times 3\frac{1}{2} \times \frac{3}{8}$ in. Usually, but not invariably, becoming wholly jet black at maturity.

1239. **H. calyptræformis** B. & Br. (from the pointed hood-shaped pileus; *calyptra*, a hood) *a b c.*

P. acutely conical, expanded, repand, splitting, minutely innato-fibrillose, salmon-rose, purplish-rose or white; mid. ochreous or salmon. *St.* somewhat ventricose, smooth, substriate, white. *G.* ventricose, rose to pallid, or as P.

Said to be edible. Pastures, lawns, etc.; uncommon. July – Oct. $3 \times 4\frac{1}{4} \times \frac{3}{8}$ in. Var. *niveus* Cooke, wholly white.

1240. **H. chlorophanus** Fr. (from the pale yellow colour; Gr. *chloros*, greenish-yellow, *phaino*, to appear) *a b c.*

P. hemispherical, expanded, obtuse, plane, smooth, cracked, viscid, sulphur, yellow, orange or scarlet; mid. sometimes vermilion; marg. striate. *St.* equal, viscid when moist, rich light yellow. *G.* emarginato-adnexed, ventricose, thin, distant, yellow. *Flesh* yellowish.

Woods, grassy and mossy places; common. July–Oct. $1\frac{7}{8} \times 3\frac{1}{4} \times \frac{1}{4}$ in.

1241. **H. psittacinus** Fr. (from the yellow or red and green colours; *psittacus*, the ringed green parrot) *a b c.*

P. campanulate, expanded, umbonate, yellow-reddish or whitish, smeared with green gluten. *St.* equal, even, colour as P. *G.* adnate or with a decurrent tooth, ventricose, thick, subdistant, yellow, green above.

Pastures, downs, roadsides, etc.; common. Aug.–Dec. $2\frac{1}{4} \times 2\frac{3}{8} \times \frac{1}{4}$ in. When young wholly green from the gluten.

1242. H. spadiceus Fr. (from the brown pileus; *spadiceus*, date-brown) *a.*

P. conical, subacute, campanulate, fibrillose, virgate, covered with olive-bay gluten. *St.* equal, dry, fibrillose, tawny or yellow, white below. *G.* ventricose, distant, lemon or clear yellow.

Mossy and grassy places, on the ground. July. $2\frac{3}{8} \times 2\frac{1}{2} \times \frac{3}{8}$ in. Not turning black.

1243. H. unguinosus Fr. (from the glutinous pileus and stem; *unguinosus*, oily) *a b.*

P. campanulato-convex, obtuse, sepia, umber, fuliginous or yellowish-brown. *St.* equal or slightly attenuate above and below, colour as P. *G.* adnate with a tooth, ventricose, white.

Taste and odour none. Woods, pastures, moist places; frequent. Aug.–Oct. $1\frac{7}{8} \times 3\frac{3}{8} \times \frac{1}{4}$ in.

1244. H. nitratus Fr. (from the nitrous odour) *a b c.*

P. convex, obtuse or depressed, at first viscid, then flocculoso-squamulose and rimosely incised, fuscous-cinereous, blotted, becoming pale, or deep umber with whitish marg. *St.* equal, smooth, lustrous, grey-whitish. *G.* broadly emarginate, distant, veined, whitish to glaucous.

Odour very strong and disagreeable, like compounds of nitrogen and oxygen. Woods, pastures; uncommon. Aug.–Nov. $1\frac{5}{8} \times 2\frac{3}{4} \times \frac{1}{4}$ in. Must not be confounded with **111.** There is a form larger than type. Var. *glauconitens* Fr., rigid. *P.* dark-olive or sooty, becoming pale. *G.* becoming glaucous. Persoon and others, including myself, regard the variety as a distinct species. Berkeley, I believe, looked upon it as a *Tricholoma* near **111.**

XLVIII. **LACTARIUS** Fr.

(From the milky juice; *lac*, milk.)

Hymenophore confluent and homogeneous with the stem. *Veil* more or less obsolete, present in the ragged-appendiculate or pubescent margin of the pileus in some species and in the gluten of others. *Pileus* fleshy, somewhat rigid, the texture floccose or vesciculose, not fibrous, often zoned, margin at first involute, milky. *Stem* usually central, not corticate, exannulate, milky. *Gills* adnato-decurrent, adnate in **1256, 1264, 1270, 1290** and **1302,** often branched, unequal, membranous-waxy, subrigid, edge acute, trama vesiculose, milky. *Spores* subglobose, minutely echinulate, white, rarely yellowish. (Fig. 62.)

Every part of the plant contains numerous anastomosing lactiferous cells filled with densely granular *latex* or milk which is usually white, but in some instances the colour changes on exposure to the air; in a few cases the milk is coloured before its escape from the cells; it varies in taste from mild to intensely acrid, and is of a resinous nature.

The species, except **1305**, grow on the ground. The nearest allies

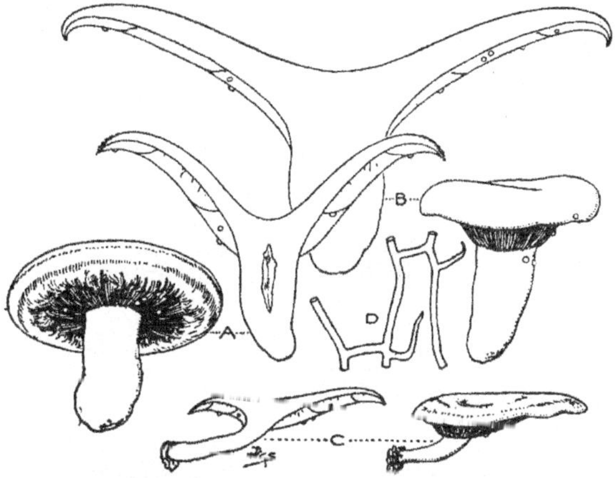

Fig. 62.—A, *Lactarius torminosus* Fr.; B, *L. piperatus* Fr.; C, *L. obliquus* Fr.: entire and in section. One-third natural size. D, Lactiferous tubes of *L. subdulcis* Fr. × 100.

are found in *Russula*, which has no latex. Some are edible, others are poisonous.

The *Lactarii* of France and Italy are popularly termed "cows" or "goats"; the acrid species are called "*peperone*" and "*poivre blanc*."

The sclerotia and perfect plants of **248** sometimes grow from the gills of *Lactarii*. *Sphæronema vitreum* (*Sphæropsideæ*) frequently grows on the the *Lactarii* and *Russulæ*. The mould *Penicillium macrosporum* occurs on *Lactarius*. Species **1245—1305**

PIPERITES. *Stem* central. *Gills* not changing colour, not pruinose. Milk white, usually acrid.

a. Tricholomoideæ. *Pileus* viscid when moist; margin at first involute and tomentose. **1245—1253**

b. Limacinæ. *Pileus* viscid when moist, pelliculose, margin usually naked. **1254—1261**

c. Piperatæ. *Pileus* without a pellicle, dry, usually not shining. **1262—1274**

DAPETES. *Stem* central. *Gills* naked. Milk from the first deeply coloured. **1275, 1276**

RUSSULARES. *Stem* central. *Gills* pallid, then changing colour to darker, glancing when turned to the light, at length white pruinose. Milk at first white, mild or from mild turning acrid.

Pileus at first viscid. **1277—1283**

Pileus not lustrous, squamulose, villous or pruinose. **1284—1294**

Pileus lustrous, smooth. **1295—1304**

PLEUROPODÆ. *Stem* excentric or lateral. Growing on trunks. **1305**

PIPERITES.

a. Tricholomoideæ.

1245. L. scrobiculatus Fr. (from the pitted stem; *scrobis*, a trench) *a b c.*

P. convex, then infundibuliform, villous-downy; marg. at first appendiculate, yellowish or buff. *St.* stuffed to hollow, equal, light yellowish or white above, tan below. *G.* thin, crowded, whitish to yellowish. *Milk* white, then sulphur-yellowish.

Taste very acrid; odour not unpleasant. Woods, on the ground, damp places. Sept.–Oct. 9 × 2½ × 1½ in. Pileus sometimes 12 in. in diam.

1246. L. intermedius B. & Br. (from its position between **1245, 1247** and **1248**) *a b.*

P. convex, then infundibuliform, smooth, ochreous or brownish-yellow, somewhat zoned; marg. at first ragged-tomentose. *St.* solid to hollow, equal, pitted as in **1245,** yellowish. *G.* whitish. *Flesh* white then pale sulphur. *Milk* white, then yellow-white.

Somewhat acrid. Woods. 3⅜ × 1½ × ¾ in. Perhaps the same as **1248.**

1247. L. torminosus Fr. (from its painful acridity; *tormina*, gripes) *a b c.*

P. convex, then infundibuliform, becoming even, somewhat zoned, pallid flesh-colour; marg. ragged-fibrillose, white. *St.* stuffed to hollow, equal or attenuate downwards, even or obsoletely pitted, white, palest rose above, ochre or tan below. *G.* very thin, crowded, paler than P. or tan. *Milk* white.

Eaten with oil and vinegar in Russia after being preserved in salt. Very acrid; odour none. Mixed woods, on the ground, fields, amongst grass, heather, etc.; common. Aug.–Nov. 5½ × 1¾ × ⅝ in. Often confounded with **1276.** *Hypomyces torminosus*, a parasitic white mould, changing to yellow and dark-brown is frequent on the gills.

1248. L. cilicioides Fr. (from the tomentose pileus; Gr. *kilikion*, goat's-hair cloth, *eidos*, appearance) *a b c.*

P. convex, then infundibuliform, wholly tomentose, flesh-colour to yellowish or fuscous; marg. fibrillose-woolly. *St.* solid to hollow, equal or attenuate downwards, even, pallid. *G.* crowded, branched, white to yellowish. *Flesh* yellowish-white. *Milk* white to yellowish-white.

Woods, pine, on the ground; frequent. Aug.–Nov. 6½ × 2½ × 1⅜ in. Intermediate between **1247** and **1250.**

1249. L. lateritioroseus Karst. (from the colour; *lateritius*, brick-red, *roseus*, rosy) *a*.

P. convex, then subinfundibuliform or umbilicate, wavy, usually zoneless, sometimes flesh-colour, becoming pale or lavender-salmon; mid. granuloso-squamulose. *St*. stuffed to hollow, unequal, slightly flocculose, colour as P. or paler. *G*. sub-distant, furcate, veined, colour as St., becoming yellowish. *Milk* white.

Acrid. Woods. Autumn. 3¼ × 2 × ⅜ in.

1250. L. turpis Fr. (from its ugly appearance; *turpis*, base) *a b c*.

P. convex, then depressed, rigid, usually zoneless, umber or olivaceous. *St*. solid to hollow, equal or attenuate downwards (sometimes upwards), pallid or dark olivaceous or shaded pale slate. *G*. thin, crowded, furcate, white straw-colour or tan. *Flesh* white or shaded pale slate. *Milk* white.

Poisonous. Gregarious. Woods, fir, mixed, roadsides, etc.; common. Sept.–Nov. 8 × 2½ × 1⅛ in. Habit of **1181**. Var. *plumbeus* Cooke. *P*. blackish-brown.

1251. L. controversus Fr. (from the pileus, at first convex then infundibuliform) *a b c*.

P. convex, then infundibuliform, rigid, flocculose to smooth, white, spotted and zoned blood-red, varying flesh-colour. *St*. solid, equal, sometimes excentric, viscid, white or spotted as P. *G*. thin, very crowded, branched, pallid, white-flesh-colour. *Milk* white.

Taste very acrid and pungent; odour weak, pleasant. Woods, pastures, under poplars, chestnuts; uncommon. Aug.–Nov. 8 × 3¼ × 1⅞ in. Said by Dr. Badham to be eaten at Lucca under the name of "Lucchese Goat."

1252. L. pubescens Fr. (from the pubescent margin of the pileus) *a b*.

P. convex, then broadly infundibuliform, zoneless, even, dry, shining, whitish to flesh-colour, rose, ochreous or faint brown. *St*. stuffed, attenuate downwards, even, colour as P. *G*. crowded, pale flesh-colour or yellowish to salmon. *Milk* white.

Taste very acrid; odour none or pungent. Pastures, grassy places, amongst pebbles. Sept.–Oct. 4 × 1⅝ × ⅜ in.

1253. L. aspideus Fr. (from the shield-like shape of the pileus; *aspis*, a shield) *a b*.

P. gibbous-convex then depressed, straw-colour, pale yellowish, or faintly lilac-shaded; marg. with a tomentose, deciduous, white zone. *St*. solid to stuffed, equal or attenuate downwards, whitish. *G*. somewhat thick, whitish, faintly shaded lilac. *Flesh* colour as G. *Milk* white then lilac.

Taste sweet. Damp pastures, moist places. Sept. 2¾ × 1¼ × ⅜ in.

b. Limacinæ.

1254. L. insulsus Fr. (from its indifferent resemblance to **1276**; *insulsus*, insipid) *a b c.*

P. umbilicate then deeply infundibuliform, rigid, smooth, yellow or orange; mid. sienna-yellow, usually zoned, spotty-pale-sienna. *St.* solid to hollow, whitish or faint yellowish. *G.* decurrent, very crowded, forked at base, colour as St. *Milk* white.

Taste acrid. Mixed woods and their borders, pastures; common. Aug.–Oct. 3⅝ × 1½ × ⅜ in. Pileus sometimes 7 in. in diam.

1255. L. zonarius Fr. (from the zoned pileus) *a b c.*

P. convexo-plane, depressed, even, smooth, 4- to 7–zoned near the minutely ruguloso-flocculose marg., pallid orange or pale yellowish; mid. sometimes umber. *St.* solid, smooth, dry, white then yellowish. *G.* thin, subcrowded, branched, whitish to dull yellowish, dingy when bruised. *Milk* white.

Taste very acrid, strong, disagreeable. Woods and their borders, grassy places; rare. Aug.–Oct. 4⅛ × 1⅝ × ⅞ in.

1256. L. utilis Fr. (from its edible qualities; *utilis*, useful) *a.*

P. convexo-plane, obtuse, then infundibuliform, even, smooth, tan, pale dull ochre or livid. *St.* solid to hollow, even, colour as P. *G.* adnate, crowded, pallid. *Milk* white.

Highly esteemed as an esculent in Russia. Taste mild to slightly acrid. Woods, on the ground. Autumn. 5⅜ × 2 × 1¼ in. Sometimes olive shaded.

1257. L. blennius Fr. (from the slimy pileus; Gr. *blennos*, mucus) *a b c.*

P. plano-depressed, pallid olivaceous, æruginous-grey or indigo-sepia, shining with gluten; mid. dark; marg. at first slightly downy. *St.* stuffed to hollow, subequal or attenuate downwards, even, viscid, pallid above, pallid grey or pale slate. *G.* thin, crowded, white, cinerous when bruised. *Milk* white.

Poisonous. Taste very acrid. Woods, beech, fields near woods, on the ground, rarely on trunks; common. July–Nov. 3⅝ × 1¼ × ⅝ in.

1258. L. hysginus Fr. (from the colour of the pileus; Gr. *husginon*, a crimson dye) *a b c.*

P. umbilicate, depressed, plane, even, red flesh-colour, reddish-brown or dark brownish sienna-salmon, somewhat zoned pale umber. *St.* solid to hollow, attenuate downwards, smooth, warm sienna-yellowish, somewhat spotted. *G.* thin, crowded, white to primrose or yellow-ochreous. *Milk* white.

Suspected poisonous. Taste acrid. Woods, under beech, amongst grass; rare. Aug.–Nov. 2⅞ × 2 × ⅝ in.

1259. L. trivialis Fr. (from its commonness; *trivialis*, common) *a b c.*

P. convex, sometimes umbonate, then depressed and infundibuliform, even, smooth, zoneless, tan-umber-lavender, dull orange-buff or dark to pale lurid. *St.* hollow, even, smooth, slippery, palest tan-grey, white, yellowish or paler than P. *G.* somewhat thin, crowded, whitish or pale yellowish. *Milk* white.

Taste very acrid. Woods, chiefly pine, moist mossy places; uncommon. Aug.–Oct. 6 × 3 × ⅝ in.

1260. **L. circellatus** Fr. (from the zones on the pileus; *circellus*, a small ring) *a b c.*

P. convexo-plane then depressed, umbilicate, even, smooth, rufous, fuscous or umber becoming pale, zoned grey or claret-brown. *St.* solid, equal, or attenuate downwards, even, smooth, pale yellowish, shaded faintly brownish or steel-whitish. *G.* thin, crowded, forked, whitish to pale yellowish. *Milk* white.

Taste stinging-acrid; odour not unpleasant. Woods, amongst grass; rare. Sept.–Oct. 4 × 2¼ × ⅜ in. Must not be confounded with **1262** or **1263**.

1261. **L. uvidus** Fr. (from the moist, clammy pileus; *uvidus*, moist) *a b c.*

P. convex, plane, then depressed, even, smooth, usually zoneless, pale lilac-umber on a yellowish ground, blotted lilac. *St.* hollow, smooth, viscid, whitish or light-yellowish, sometimes spotted yellow-ferruginous or lilac. *G.* thin, crowded, unequal, somewhat veined, whitish or yellowish, lilac when bruised. *Milk* white, becoming bright lilac.

Taste insipid to acrid; odour weak, nauseous-pungent, somewhat aromatic. Woods, moist places; uncommon. Aug.–Nov. 3⅞ × 2⅞ × ½ in.

c. Piperatæ.

1262. **L. flexuosus** (from the frequently undulate margin of the pileus; *flexus*, a bending) *a b c.*

P. hard, plane, depressed, subinfundibuliform, smooth to rivuloso-scaly, dull, zoned or zoneless, lead-grey, violet-grey, violet-brown or purplish-umber. *St.* solid, equal or attenuate downwards, whitish or greyish, base yellowish. *G.* thick, distant, branched, light yellowish to whitish flesh-colour. *Milk* white.

Said to be edible. Taste when raw very acrid. Woods, pine. July–Oct. 5 × 3⅝ × 1⅛ in. Var. *roseozonatus* Fr. *P.* rosy flesh-colour with darker z.

1263. **L. pyrogalus** Fr. (from the very acrid milk; Gr. *pur*, fire, *gala*, milk) *a b c.*

P. convex, plane, depressed, smooth, cinereous-grey, dull yellow or umber-tan, zoned sienna or madder-brown. *St.* stuffed to hollow, usually attenuate downwards, dull pallid-whitish or paler than P. and G. *G.* thin, subdistant, light wax-yellow, tan-ochre or shaded reddish-salmon or straw-colour. *Milk* white.

Poisonous. Odour sometimes of apple. Woods, fir, pastures; frequent. Aug.–Nov. 3½ × 1⅝ × ½ in. Sometimes very small.

1264. **L. squalidus** Fr. (from the dingy colouring; *squalidus*, dirty) *a b.*

P. convex, umbilicate, depressed, smooth, tan-olive-green with darker mid. or yellowish-olive-umber. *St.* solid, equal, smooth, whitish or pale brownish. *G.* adnate, narrow, yellowish or shaded faint olive or tan. *Milk* white.

Taste mild to sweet. Woods, beech, moist mossy places. Oct. $3\frac{3}{4} \times 3\frac{1}{4} \times \frac{1}{2}$ in.

1265. L. capsicum Schulz. ex Fr. (from the peppery taste; *capsicum*, red-pepper) *a*.

P. convex, obtuse, pulvinate, dark chestnut or deep sienna; mid. paler. *St.* solid, equal, whitish with tawny or rufescent fibres, paler above. *G.* subcrowded, tawny, somewhat orange or ochreous-sienna. *Flesh* yellowish, becoming pale sienna when broken. *Milk* white.

Woods, under birches, grassy places. Sept. $2\frac{1}{4} \times 1\frac{3}{4} \times \frac{5}{8}$ in.

1265a. L. glaucescens Crossl. (*glaucescens*, becoming sea-green) *a b c*.

P. fleshy, rigid, convex, umbilicato-depressed, smooth, glabrous, dry, zoneless, white then cream-colour, slightly spotted ochreous-white; marg. involute. *St.* solid, attenuate downwards, smooth, compact, colour as P. *G.* adfixed, densely crowded, somewhat forked, narrow, colour as P. *Milk* copious, white, then pale glaucous-green.

Taste acrid. Woods. Aug.–Sept. $2\frac{1}{2} \times 1\frac{7}{8} \times \frac{5}{8}$ in.

1266. L. chrysorrheus Fr. (from the gold-coloured milk; Gr. *chrusos*, gold, *rheo*, to flow) *a b c*.

P. umbilicate then infundibuliform, pale yellow-pink, brownish-salmon, apricot or whitish, zoned reddish-brown. *St.* stuffed to hollow, equal, even, white or shaded yellowish-salmon. *G.* very thin, crowded, slightly furcate, pallid-yellowish, salmon or tan-straw colour. *Flesh* white, bright sulphur when broken. *Milk* white.

Very acrid. Woods, beech, mixed; frequent. Aug.–Oct. $2\frac{7}{8} \times 2\frac{1}{8} \times \frac{3}{8}$ in.

1267. L. acris Fr. (from the acrid taste) *a b*.

P. convex, plane, sometimes subumbonate, at length depressed, scarcely zoned, umber or deep umber. *St.* stuffed to hollow, slightly attenuate downwards, tan, shaded umber. *G.* adnato-rounded to subdecurrent, thin, pallid to yellow flesh-colour or pale ochre, sparingly-spotted sienna-pink. *Milk* white, then reddish.

Odour strong, stinking. Woods; rare. Aug.–Nov. $4 \times 2 \times \frac{1}{2}$ in. Often grows in company with **1289**.

1267a. L. violascens Fr. (from the change of colour in the flesh and milk to pale violet) *a c*.

P. fleshy, depressed, glabrous, ashy; z. smoky. *St.* solid, sometimes becoming hollow, equal, ashy- or greyish-white. *G.* somewhat crowded, white. *Milk* white, then violaceous or pale lavender. *Flesh* changing as milk.

Poisonous. Taste sweet. Woods; rare. Autumn. $3\frac{3}{4} \times 2\frac{3}{4} \times \frac{3}{4}$ in.

1268. **L. umbrinus** Fr. (from the colour of the pileus) *a b c.*
P. convexo-plane, depressed, rivuloso-floccose, umber to deep sepia or dark rufous-brown, zoneless or obscurely zoned darker. *St.* solid, white to greyish, or as P. *G.* crowded, pallid-yellowish, grey-spotted. *Milk* white, becoming greyish.

Acrid. Pine woods, damp places. Autumn. $3\frac{1}{8} \times 1\frac{1}{2} \times \frac{3}{4}$ in.

1269. **L. plumbeus** Fr. (from the colour; *plumbeus*, leaden) *a b c.*
P. convex, then depressed and infundibuliform, greyish-brown, sooty-sepia with a shade of olive or tan-olive; mid. darker. *St.* solid, equal, paler than P. or tan-ochre. *G.* crowded, yellowish grey-white. *Milk* white.

Woods. Nov. $9\frac{1}{2} \times 3\frac{1}{4} \times 1\frac{1}{8}$ in.

1270. **L. pergamenus** Fr. (from its parchment colour; *pergamena*, parchment) *a b.* White or pale yellowish.
P. convex, plano-depressed, smooth, slightly wrinkled, zoneless. *St.* stuffed, smooth, equal or slightly attenuate downwards. *G.* adnate, very crowded, becoming straw-colour. *Milk* white.

Taste very acrid. Woods. Aug.–Oct. $3\frac{1}{2} \times 3\frac{1}{4} \times \frac{5}{8}$ in.

1271. **L. piperatus** Fr. (from the peppery taste; *piper*, pepper) *a b c.* White.
P. convex, then depressed and infundibuliform, rigid, even, smooth, zoneless, often spotted faint umber or grey. *St.* solid, attenuate downwards, subpruinose. *G.* crowded, branched, becoming spotted dull, pale brownish-yellow. *Milk* copious, white.

Said to be edible; poisonous when raw. Odour none. Mixed woods; common. July–Oct. $7\frac{1}{4} \times 3 \times 1\frac{3}{4}$ in. Sometimes clouded yellowish within and without.

1272. **L. vellereus** Fr. (from the pubescent pileus; *vellus*, fleece) *a b c.* White.
P. convex to depressed, dry, zoneless, rigid, becoming pale yellowish or spotted pale brownish or yellowish. *St.* solid, equal or attenuate downwards, thinly pubescent, pale sulphur within at base. *G.* somewhat thick, subdistant, branched. *Flesh* white, changing to pale sulphur. *Milk* sparing, sometimes absent, white, sometimes pale sulphur, changing to white.

Poisonous. Taste bitter-acrid. Woods, beech; common. Aug.–Dec. $8 \times 6 \times 2\frac{1}{4}$ in.

1273. **L. scoticus** B. & Br. (found first in Scotland) *a.*
P. convex to slightly depressed, pubescent, pale yellowish or faint brownish. *St.* solid, equal, pale flesh-colour. *G.* thin, broad, scarcely branched, pallid-white. *Milk* white.

Taste very acrid; odour pungent. Amongst moss. Autumn. $2\frac{1}{8} \times 1\frac{1}{8} \times \frac{3}{8}$ in.

1274. L. involutus Sopp. ex Cooke (from the involute margin of the pileus) *a b*. White or faintly ochreous.

P. convex to slightly depressed, smooth, even; marg. silky. *St.* solid, equal. *G.* very crowded, narrow, sometimes furcate. *Milk* white.

Taste very acrid. Woods, on the ground. Autumn. $1\frac{3}{4} \times 1\frac{1}{8} \times \frac{1}{4}$ in. Very firm and rigid as in **1272**.

DAPETES.

1275. L. deliciosus Fr. (from its flavour when cooked) *a b c*.

P. convex to infundibuliform, smooth, subviscid, dull orange-yellow, zoned orange-sienna, sometimes clouded green. *St.* stuffed to hollow, equal, colour as P., usually spotted orange-reddish. *G.* crowded, saffron, green when bruised. *Flesh* orange, becoming green when cut. *Milk* red-saffron, then green.

Gregarious. Edible. Sometimes acrid when raw; odour aromatic. Woods, pine. July–Nov. $5 \times 2\frac{1}{4} \times \frac{7}{8}$ in. This and **1295** are the "vegetable sheeps-kidneys" of French cooks. The gills are often attacked by a reddish parasitic fungus, *Hypomyces lateritius*. *Epicoccum micropus* grows upon this species when decaying. The deep orange milk of this plant must not be confounded with the sulphur-coloured milk of **1280**.

1276. L. sanguifluus Fr. (from the blood-coloured milk; *sanguis*, blood, *fluo*, to flow) *a b*.

P. convexo-depressed, smooth, sometimes zoned, blood-red. *St.* narrowed upwards, smooth, pallid. *G.* thin, white. *Milk* blood-red, changing to green.

Taste mild. Oct. $2\frac{1}{2} \times 2 \times \frac{3}{4}$ in. Closely allied to **1275**, but known by its blood-red, not orange, milk.

RUSSULARES.

1277. L. pallidus Fr. (from its pale colour) *a b c*. Clay flesh-colour or dull buff-whitish.

P. convex to depressed, smooth, zoneless. *St.* stuffed to hollow, equal, even, smooth. *G.* somewhat thin, crowded, slightly furcate. *Milk* white.

Taste somewhat mild, almost sweet. Woods, beech, mixed; uncommon. Sept.–Oct. $4 \times 2\frac{1}{2} \times \frac{7}{8}$ in.

1278. L. quietus Fr. (from its mild taste; *quietus*, mild) *a b c*.

P. convex to depressed, rich sienna, clouded, splashed and somewhat zoned darker. *St.* stuffed, smooth, rufescent or as P. *G.* somewhat furcate at base, whitish to yellowish sienna. *Flesh* white to rufescent. *Milk* white.

Taste sweet, somewhat of walnuts; odour oily, bug-like. Woods, moist; common. July–Nov. $4 \times 2\frac{3}{4} \times \frac{5}{8}$ in. Not unlike **202**.

1279. L. aurantiacus Fr. (from the bright orange-coloured pileus and stem; *aurantiacus*, orange-coloured) *a b*.

P. plano-depressed, even, zoneless. *St.* stuffed, smooth. *G.* crowded, very decurrent, paler than P. *Milk* white.

Taste slowly acrid. Woods, on the ground amongst moss. Sept. $2\frac{1}{2} \times 2 \times \frac{3}{8}$ in.

1280. **L. theiogalus** Fr. (from the sulphur-coloured milk; Gr. *theion*, brimstone, *gala*, milk) *a b c.*

P. convex to infundibuliform, sometimes papillate, even, smooth, somewhat rufous-tawny with a shade of liver; marg. crenulate. *St.* stuffed to hollow, equal, or nearly so, even, rufous-salmon within and without. *G.* adnato-decurrent, thin, somewhat distant, faintly ochreous to deep yellow or rufescent. *Milk* white, slowly becoming sulphur.

Taste mild, then slowly acrid; odour none. Woods, mixed, pine, oak, amongst heather, sometimes on dead root-stocks of *Pteris aquilina*; frequent. Aug.–Nov. $2\frac{5}{8} \times 2\frac{3}{8} \times \frac{3}{8}$ in. The yellow milk of this species must not be confounded with the deep orange milk of **1275**.

1281. **L. cremor** Fr. (from the thick milk; *cremor*, thick juice) *a b.* Thin, fragile.

P. convex to plano-depressed, minutely punctulate, tawny, zoneless; mid. darker; marg. striate. *St.* hollow, equal, colour as P. *G.* adnate, subdistant, pallid, or as P. *Milk* somewhat transparent white.

Taste mild or slowly acrid. Woods, larch, beech, etc. Sept. $4\frac{1}{4} \times 1\frac{1}{2} \times \frac{5}{8}$ in. Var. *pauper* Karst., smaller. P. 3 in. in diam.

1282. **L. vietus** Fr. (from its appearance; *vietus*, shrivelled) *a b c.*

P. convex, then depressed, somewhat papillate, even, smooth, flesh-colour or livid-grey, becoming whitish-grey; mid. pale umber. *St.* stuffed to hollow, equal or attenuate downwards, pale rose-brown or as P. *G.* somewhat crowded, white to yellowish. *Milk* white, then pale lavender-grey.

Taste mild, becoming slowly acrid; odour faint but pungent. Woods, mixed, damp, grassy places, on the ground. Sept.–Oct. $3\frac{1}{4} \times 1\frac{3}{8} \times \frac{3}{8}$ in.

1283. **L. cyathula** Fr. (from the cup-shaped pileus; *cyathula*, a little cup) *a b.*

P. convexo-plane, depressed, sometimes papillose, rimoso-rivulose, flesh-colour, pale yellowish or ivory, lavender-shaded, zoned, yellowish-brown or zoneless. *St.* stuffed, equal, even, smooth, whitish or yellowish, lavender-shaded. *G.* very crowded, thin, whitish, then yellowish-salmon. *Milk* white.

In troops. Taste acrid; odour at first none, then strong of bugs. Woods, birch, damp ground. Aug.–Oct. $2\frac{1}{8} \times 2\frac{1}{4} \times \frac{1}{4}$ in.

1284. **L. rufus** Fr. (from the reddish colour) *a b c.*

P. plane to infundibuliform, usually umbonate, deep crimson-sienna, sometimes paler, zoneless or with one zone; marg. sometimes sulcate. *St.* stuffed, equal, rufescent, paler than P. *G.* ochreous to rufescent. *Milk* white.

Poisonous. Taste very acrid-bitter; odour none. Woods, pine, etc., dry places. June–Nov. $3\frac{7}{8} \times 2\frac{1}{2} \times \frac{1}{2}$ in.

1285. **L. helvus** Fr. (from the colour; *helvus*, pale yellowish) *a b c.*

P. convex to slightly depressed, sometimes subumbonate, granuloso-floccose, sometimes sienna-biscuit, sometimes faintly 1- to 2-zoned. *St.* stuffed to hollow, equal, pale buff.

G. crowded, whitish, pale yellowish or pale salmon. *Milk* watery-white.

Taste mild, sweet then slightly acrid; odour faint, sweetish. Woods, fir, mossy places, swampy ground. Aug.–Oct. 3¼ × 2¼ × ⅝ in.

1286. L. tomentosus Cooke (from the woolly tomentose pileus) *a b.*

P. at first umbonate, then infundibuliform, dull salmon to deep sienna. *St.* stuffed to hollow, equal, naked, smooth, pallid or salmon above, ochreous below red-streaked at mid. *G.* yellow flesh-colour or ochre, or white-ochre near edge and red above. *Milk* whitish.

Taste mild to slightly acrid; odour pleasant. On the ground, damp places. 4⅛ × 1¾ × ½ in. Sometimes confused with **1285.**

1287. L. mammosus Fr. (from the papillate pileus; *mamma*, a breast) *a b.*

P. convex to depressed, at first acutely umbonate, dry, grey-fuscous or dull pale indigo-greenish, zoneless or faintly zoned; marg. white pubescent. *St.* stuffed to hollow, equal, pubescent, faint yellowish or brownish. *G.* crowded, whitish or pale yellowish to pallid ferruginous. *Milk* white.

Taste slowly acrid; odour none. Woods, pine, on the ground. Oct. 3½ × 2½ × ⅜ in. Var. *monstrosus* Cooke. Larger than type.

1288. L. glyciosmus Fr. (from the sweet smell; Gr. *glukus*, sweet, *osme*, smell) *a b c.*

P. convex to slightly depressed, acutely papillate, yellowish, shaded over with pale lavender-purple, zoned pale purplish. *St.* stuffed to hollow, pubescent, whitish or light yellowish, tawny when bruised. *G.* crowded, thin, white to straw-ochreous. *Milk* somewhat scanty, white, unchanging or changing to pale straw-green.

Taste mild, then slowly acrid; odour strong, pungent-sweet, fragrant-aromatic. Woods mixed, chiefly pine, on the ground. Sept.–Nov. 2⅞ × 2½ × ⅜ in.

1289. L. fuliginosus Fr. (from the smoky colour; *fuligo*, soot) *a b c.*

P. convex to subinfundibuliform, dry, full velvety ochreous- or smoky-brown; marg. paler. *St.* stuffed, equal or attenuate downwards, white, tan or pale tan-umber. *G.* somewhat distant, white, then pale ochreous, pale indian-red when bruised. *Milk* white, sometimes watery, changing to rose and saffron. *Flesh* changing to vinous when cut.

Probably poisonous. Taste mild, becoming very acrid-nauseous; odour faint nauseous-pungent. Shady woods; frequent. Aug.–Oct. 4½ × 2⅜ × 1 in. Often growing in company with **1267.**

1290. L. picinus Fr. (from the black pileus; *picinus*, pitch-black) *a b.*

P. convex, obtuse or subumbonate, minutely velvety, umber to blackish-umber. *St.* stuffed, equal, smooth, paler than P. or almost white, brownish below. *G.* adnate, thin, very crowded, pale ochreous to rufous tan. *Milk* white.

Taste acrid. Woods, pine, dried turfy swamps. Aug.–Oct. 2½ × 2 × ⅜ in.

1291. **L. lignyotus** Fr. (from the woody substance; *lignum*, wood, Gr. *ous*, an ear) *a*.

P. convex to subinfundibuliform, umbonate, plicato-rugulose, velvety, deep umber-purplish or opaque sooty-umber. *St.* solid to stuffed, equal, striate above, colour as P., yellowish or whitish at base. *G.* adnexed to decurrent, somewhat crowded, whitish or yellowish, reddish when bruised. *Flesh* white, changing to ochreous, saffron and ferruginous. *Milk* white, scanty, changing through reddish to saffron.

Poisonous. Taste mild, sweet, pleasant; odour slight. Amongst moss, in damp places under firs, beech. Oct. $3\frac{1}{4} \times 3 \times \frac{1}{2}$ in.

1292. **L. lilacinus** Fr. (from the lilac pileus) *a b*. Fragile.

P. convex to depressed, papillate, floccose, rosy-lilac, zoneless; marg. tan. *St.* stuffed to hollow, pale yellowish-brown or dull rose, lilac-shaded, or as P., white mealy. *G.* subdistant, pallid flesh-colour or yellowish-white. *Milk* white, copious.

Taste acrid. Woods, damp places. Aug.–Oct. $2\frac{1}{4} \times 1\frac{1}{8} \times \frac{1}{4}$ in.

1293. **L. retisporus** Mass. (from the reticulate spores) *a b*.

P. convex, then plane and depressed, minutely velvety, dark smoky-brown. *St.* equal, solid, even, glabrous, paler than P. or pale sepia or greyish. *G.* subdistant, deeply sinuate, slightly adnexed, pale ochreous with darker spots. *Flesh* dull yellowish or whitish-salmon, becoming reddish when cut. *Milk* sweet, white, changing to brown and becoming thick and tenacious.

Under beeches, on the ground. Sept. $3\frac{3}{4} \times 2\frac{1}{4} \times \frac{5}{8}$ in.

1294. **L. spinosulus** Quél. (from the spines on the pileus; *spinosulus*, full of little spines) *a*.

P. convex to depressed, acutely umbonate, clad with minute, erect spines, brick-red to rosy-lilac, often zoned and spotted. *St.* stuffed to hollow, rugose, granulate, shining, colour as P. *G.* thin, yellowish flesh-colour to yellowish. *Milk* white.

Taste acrid. On the ground. Sept. $1\frac{1}{2} \times 1\frac{1}{2} \times \frac{5}{16}$ in.

1295. **L. volemus** Fr. (from its resemblance to a certain large pear—the "Red Warden" or baking-pear—*volema-pira*) *a b c*.

P. plano-convex to subdepressed, compact, bright rufous-tawny or golden, becoming pale; mid. deep-sienna, zoneless. *St.* solid to stuffed, equal, pruinose, colour as P. *G.* thin, crowded, white to yellowish or yellow, orange-umber when wounded. *Milk* white, sometimes changing to yellow.

Edible. Taste sweet, pleasant when raw. Woods, mixed. Oct. $4\frac{7}{8} \times 4 \times 1\frac{1}{2}$ in. The "Golden Brötling" of Germany and Hungary. With **1275** the "vegetable sheeps-kidneys" of French cooks.

1296. **L. ichoratus** Fr. (from the watery milk; Gr. *ichor*, lymph) *a b*.

P. plane to depressed, dry, tawny-reddish; mid. brown-sienna; marg. sienna-yellow, sometimes zoned. *St.* solid, equal, colour as P. *G.* scarcely crowded, white to tan-yellow or ochreous. *Milk* white, unchangeable.

Taste sweet; odour strong. Woods. Oct. $2\frac{1}{2} \times 2\frac{5}{8} \times \frac{1}{2}$ in.

1297. **L. serifluus** Fr. (from the watery milk; *serum*, whey, *fluo*, to flow) *a b c.*

P. plane to depressed, dry, brown-tawny or orange-reddish, sometimes crimson-purplish, zoneless or with one faint zone. *St.* solid, yellowish, red-salmon or paler than P. *G.* crowded, colour as St., yellowish or ochre-brownish. *Milk* whitish, especially thin in examples from wet places.

Taste almost mild; odour disagreeable, bug-like. Woods, mixed, damp places; frequent. July–Nov. $3 \times 2\frac{1}{4} \times \frac{1}{2}$ in. Resembling **193.**

1298. **L. mitissimus** Fr. (from its mild taste; *mitissimus*, very mild) *a b c.*

P. convex to depressed, at first papillate, even, bright golden-tawny or bright orange, zoneless or faintly zoned. *St.* stuffed to hollow, even, smooth, colour as P. *G.* thin, paler than P., often rufous-spotted. *Milk* copious, white.

Taste mild to somewhat bitterish or slightly acrid-disagreeable. Woods, mixed, pine, hedge-banks, etc.; common. Aug.–Nov. $2 \times 2\frac{3}{4} \times \frac{3}{16}$ in. Like a small **1295.**

1299. **L. subdulcis** Fr. (from the somewhat sweet taste; *dulcis*, sweet) *a b c.*

P. convex to depressed, at first papillate, even, dry, rufescent or orange-sienna, usually zoneless. *St.* stuffed to hollow, equal, subpruinate, colour as P. *G.* crowded, colour as P., or paler. *Flesh* rufescent. *Milk* white.

Taste mild or sweet, then slightly bitter-acrid; odour none or like cedar-wood. Woods, mixed, fir. July–Nov. $3 \times 2 \times \frac{1}{2}$ in. Resembling **201.** Form *concavus* Fr. *P.* concavo-depressed; colours of **1284.** Form *sphagneti* Fr. Marg. of P. reflexed, crenate, shining as if varnished, red-bay.

1300. **L. camphoratus** Fr. (from the strong odour, sometimes of camphor) *a b c.*

P. convex to depressed, dry, brownish-red to sienna-orange, paler when young, somewhat zoned or zoneless. *St.* stuffed, equal, colour as P., paler above. *G.* yellow-reddish, paler than P., or salmon. *Flesh* somewhat ochre or sienna. *Milk* watery, white.

Taste mild; odour sometimes of melilot, permanent in dried examples. Woods, chiefly pine; frequent. Aug.–Nov. $2\frac{5}{8} \times 2\frac{1}{4} \times \frac{1}{4}$ in. Resembling **194.** Var. *Terrei* Cooke. P. $\frac{1}{2}$ in. in diam. corrugate, depressed, bay-brown. *St.* colour as P., clad with orange-down.

1301. **L. cimicarius** Mass. (from the odour; *cimex*, a bug) *a.*

P. convex to subinfundibuliform, deep sienna, shaded deep umber. *St.* stuffed to hollow, lighter than P., paler above. *G.* dusky ochre, sienna or rufous. *Flesh* grey to yellowish. *Milk* watery, white.

Taste somewhat acrid; odour when fresh, strong, heavy, oily, bug-like, fugitive. Woods, on the ground. Autumn. $2\frac{1}{2} \times 1\frac{5}{8} \times \frac{1}{4}$ in.

1302. **L. subumbonatus** Lindgr. (from the somewhat umbonate pileus) *a b*.

P. convex to depressed, acutely papillate, rugose, punctate, dark cinnamon, zoneless. *St*. stuffed, equal or attenuate downwards, rufescent, crimson-umber below. *G*. adnate, flesh-colour or reddish. *Flesh* pale brown. *Milk* watery, white.

Taste mild; odour none when fresh, fetid in drying. Woods, on the ground. Sept. $1\frac{3}{8} \times 1 \times \frac{3}{16}$ in.

1303. **L. obnubilis** Fr. (from the clouded pileus; *obnubilo*, to overcloud) *a b*.

P. convex to depressed, at first papillate, brown-fuliginous or umber, zoneless; marg. somewhat striate. *St*. stuffed to hollow, equal, paler than P. *G*. somewhat crowded, yellowish. *Milk* white.

Taste mild, somewhat sweet. Woods, damp places under alders, pastures. Sept.–Nov. $2\frac{1}{8} \times 1\frac{7}{8} \times \frac{1}{4}$ in. Var. *crenatus* Mass. *P*. coarsely and regularly sulcate. Usually larger than type. Fir woods.

1304. **L. minimus** W. G. Sm. (from its very small size; *minimus*, very small) *a b c*. Pallid clay-colour.

P. pulvinate, obtuse, or subumbonate. *St*. usually excentric. *G*. furcate, moderately distant, slightly darker than P. *Milk* white.

Taste mild. Woods, pastures. Oct. $\frac{5}{16} \times \frac{3}{16} \times \frac{1}{16}$ in.

PLEUROPODÆ.

1305. **L. obliquus** Fr. (from the oblique stem) *a*. White, changing to yellowish.

P. plano-depressed, lobate, silky, zoned grey. *St*. stuffed to hollow. *G*. crowded, white. *Milk* white.

Odour strong. Trunks, beech, burnt stumps, banks. Oct. $2\frac{7}{8} \times 1 \times \frac{1}{4}$ in. Pileus sometimes 6 in. or more in diam.

XLIX. **RUSSULA** Fr.

(From the frequently reddish colour of the pileus; *russulus*, reddish.)

Veil obsolete. *Hymenophore* confluent and homogeneous with the stem. *Pileus* fleshy, rigid, regular, at length depressed. *Stem* central, stout, rigid, without cartilaginous bark, shining, exannulate. *Gills* adnate, rigid, fragile, often equal in length or furcate, edge thin, acute, sometimes exuding watery drops, especially in rainy weather, but not milky as in *Lactarius*; *trama*, vesicular. *Spores* minutely verruculose or echinulate, white or pale yellowish. (Fig. 63.)

Terrestrial fleshy rigid and brittle putrescent fungi appearing in late summer and autumn. A few species are reputed edible, but most are noxious, and there is some danger of confusing the esculent with the poisonous. Odour none or unpleasant.

Russula agrees with *Lactarius* in size and some other characters, but differs in the absence of milk.

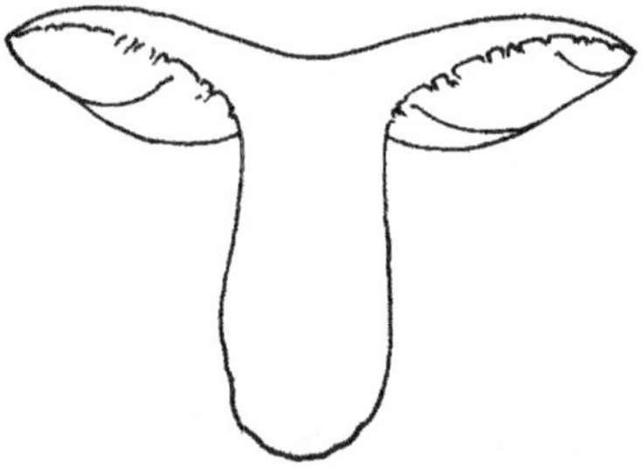

Fig. 63.—Section of *Russula lepida* Fr.
One-half natural size.

The sclerotia and perfect plants of **248** sometimes grow from the gills of **1306** and **1307.**

Sphæronema vitreum frequently grows on species of *Russula* and *Lactarius*. Species **1306—1367**

a. Compactæ. Pileus fleshy, margin bent inwards, without striæ, without a distinct viscid pellicle, colour only changing with age and the state of the atmosphere. *Stem* solid, fleshy. *Flesh* compact, firm. *Gills* unequal. **1306—1313**

b. Furcatæ. Pileus compact, firm, covered with a thin closely adnate pellicle, which at length disappears, margin abruptly thin, at first inflexed, acute, even. *Stem* at first compact, then spongy-soft within. *Gills* somewhat forked, with a few shorter ones intermixed, commonly attenuate at both ends, thin, and normally narrow. **1314—1324**

c. Rigidæ. Pileus without a viscid pellicle, dry, rigid, the cuticle breaking up into flocci or granules, margin straight, never involute or striate. *Stem* solid, at first hard, then softer or spongy. *Flesh* thick, compact, firm, ending short of the margin of the pileus. *Gills*, a few dimidiate, others divided, rigid, dilated in front, running out with a broad, rounded front, whence the margin of the pileus becomes obtuse. Handsome, somewhat rare. **1325—1334**

d. Heterophyllæ. Pileus fleshy, firm, with a thin, at first inflexed margin, then, except **1340,** expanded and striate, covered with a thin adnate pellicle. *Stem* solid, stout, spongy within, sometimes becoming hollow. *Gills* consisting of many shorter, mixed with longer ones, others furcate. **1335—1346**

e. Fragiles. Pileus more or less fleshy, covered with a pellicle which is always continuous, and in wet weather viscid and somewhat separable, margin membranous, at first convergent and not involute, in full grown plants commonly sulcate and tubercular, not striate in **1348, 1356** and **1363.** *Stem* spongy, at length wholly soft and hollow. *Gills* almost all equal, simple, broadening in front, free. *Flesh* commonly floccose, lax, friable. Several doubtful forms occur; **1357** is specially variable in colour.

Gills and *spores* white. **1347—1355**
Gills and *spores* white, then bright lemon-yellow. **1356—1361**
Gills and *spores* ochreous. **1362—1369**

a. Compactæ.

1306. R. nigricans Fr. (from its changing to jet-black in decay; *nigricans*, becoming black) *a b c.*

P. convex, depressed; streaky umber-tan, sometimes olive-shaded; mid. almost black. *St.* attenuate downwards, sienna-tan, steel-greyish or faint purplish. *G.* rounded-adnexed, very thick, distant, pale yellowish. *Flesh* white, then reddish or pale purplish, at length black.

Odour none. Woods, mixed, etc.; common. June–Nov. 5½ × 2½ × 2 in. Becoming wholly black and in this character analogous with **1221.** **1389** grows on decaying examples of this species.

1307. R. adusta Fr. (from its scorched appearance; *adustus*, scorched) *a b c.*

P. convex to depressed or subinfundibuliform, smooth, whitish to pallid yellowish, at length clouded olive-sepia. *St.* equal, colour as young P. *G.* adnate to decurrent, thin, crowded, white then dull. *Flesh* unchangeable or becoming grey or pale sooty.

Taste mild to acrid. Woods; frequent to rare, local. Aug.–Oct. 3½ × 2¼ × 1⅛ in. Not becoming wholly black. **1390** grows upon this species and on **1341.** Var. *cærulescens* Fr. becomes blue when broken or cut.

1308. R. albonigra Fr. (from the colours; *albus*, white, *niger*, black) *a b.* Very rigid.

P. convex to infundibuliform, viscid, whitish; marg. smoky. *St.* dusky, or white above, pale grey-ochreous towards base, spotted-sooty. *G.* decurrent, crowded, thick, unequal, connected by veins, dimidiate, dusky whitish or yellowish. *Flesh* white, then black or sooty, not red.

Taste somewhat acrid and unpleasant to mild; odour none. Grassy places under trees. Aug. 3⅜ × 1¾ × 1⅜ in.

1309. R. densifolia Cooke (from the crowded gills, as compared with **1306**; *densus*, thick, *folium*, a leaf) *a b.*

P. convex to depressed, smooth, whitish, becoming streaky-fuliginous, grey or brownish; mid. black. *St.* smooth, whitish to grey, then blackish. *G.* adnato-decurrent, unequal, thin, white to rosy, red when bruised. *Flesh* white, reddish when bruised, becoming black.

Taste mild to slowly acrid. Woods, under trees. Aug.–Sept. $4\frac{3}{8} \times 2\frac{1}{4} \times \frac{7}{8}$ in.

1310. **R. semicrema** Fr. (from the white stem becoming cinereous and black when broken; *semi*, half, *cremo*, to burn) *a*. White or faintly yellowish.

P. convex to depressed, even. *St*. equal or slightly attenuate upwards. *G*. decurrent, crowded, thin. *Flesh* in P. white.

Taste mild; odour none. Woods, mixed, amongst leaves. Aug.–Sept. $4\frac{1}{2} \times 2\frac{3}{4} \times \frac{7}{8}$ in.

1311. **R. delica** Fr. (from its resembling a milkless **1272** in appearance; *delicus*, weaned) *a b c*. White or whitish.

P. convex to depressed, even, smooth. *St*. even, smooth. *G*. decurrent, thin, distant, unequal, exuding small watery drops in wet weather.

Woods, mixed, usually pine; uncommon. Sept.–Oct. $5 \times 2 \times 1\frac{3}{8}$ in.

1311a. **R. chloroides** Bres. (from the greenish-yellow colour of the gills; Gr. *chloros*, greenish-yellow, *eidos*, appearance) *a b c*.

P. fleshy, plane to infundibuliform; marg. incurved, pubescent, whitish, whitish-buff or livid. *St*. solid, equal or attenuate downwards, thinly pubescent, whitish. *G*. decurrent, somewhat crowded, furcate, veined, verdigris-white, greenish-white, or turquoise-white, shaded emerald-green-white. *Flesh* thick, whitish to snow-white.

Acrid. Woods, pine. Aug.–Oct. $4 \times 2\frac{1}{2} \times 1$ in.

1312. **R. elephantina** Fr. (from Fries thinking, in error, that Bolton's obese *Agaricus elephantinus* might be one form of this comparatively slender species) *a b*.

P. convexo-umbilicate, fuscous-tan; marg. undulate, paler, dark wash-leather colour when dry. *St*. hard, even, smooth, paler than P. *G*. adnate or sinuato-adnate, then subdecurrent, arcuate, somewhat crowded to distant, thin, obtuse, divided behind, unequal in front, spotted pale yellowish when touched, dull pale salmon when dry.

Odour none. Woods. Sept. $3 \times 2\frac{1}{2} \times 1$ in. Habit of **1341**, but inodorous. *A. elephantinus* Bolton (*Hist. Fung.* t. 28) is undoubtedly **1306**, as is shown by the remote gills and the change of colour to black.

1313. **R. mustelina** Fr. (from the colour of the pileus; *mustela*, a weasel) *a b*.

P. convex to depressed, opaque yellow-ochre or sienna with yellow to full burnt sienna. *St*. white, whitish, or faintly shaded yellowish, brownish or salmon. *G*. adnexed, adnate or rounded, crowded, veined above, somewhat branched.

Woods. Oct. $3 \times 2\frac{1}{8} \times \frac{3}{4}$ in.

b. Furcatæ.

1314. **R. olivascens** Fr. (from the pileus, more or less olive) *a b.*

P. gibbous to slightly depressed, olive, ochreous-olive or olive-brownish; mid. yellow to olive-brown. *St.* equal, white or shaded pale yellowish or olive-tan. *G.* adnate, crowded, almost equal, colour as St.

Taste mild, becoming slightly acrid, gills acrid. Woods, bushy shady places. Aug. 5 × 2⅜ × 1 in.

1315. **R. furcata** Fr. (from the forked gills; *furcatus*, forked) *a b c.*

P. gibbous to depressed, even, smooth, not breaking into warts, æruginous-greenish, umber-greenish, olive-tan, dark indigo-olive, slate, or dull reddish-purple, sometimes clouded dull rose. *St.* equal, white. *G.* adnato-decurrent, rather thick, somewhat distant to crowded, white. *Flesh* white, often pale rose under olive pellicle of pileus.

Taste mild to more or less bitter. Woods, grassy places under trees, meadows; common. May–Oct. 5½ × 2⅝ × 1⅛ in. Var. *pictipes* Cooke. *St.* rosy above, greenish below. Var. *ochroviridis* Cooke. *P.* marg. ochreous, mid. olivaceous or fuliginous. *Flesh* fuliginous when cut.

1316. **R. sanguinea** Fr. (from the blood coloured pileus; *sanguis*, blood) *a b c.*

P. convex to depressed, shining, even, mid. thick; marg. becoming pale, sometimes white-clouded. *St.* equal, slightly striate, white or reddish. *G.* decurrent, sometimes adnate, crowded and veined, white. *Flesh* white, carmine under pellicle of pileus.

Poisonous. Taste acrid, peppery. Woods, chiefly fir, grassy places; uncommon. Aug.–Sept. 4 × 2¼ × 1 in. Must not be confounded with 1329, which differs in the solid firm flesh and the form of the gills.

1317. **R. rosacea** Fr. (from the rose-coloured pileus) *a b.*

P. convex to flat, even, pellicle viscid and separable in wet weather, white, salmon, pale rose to red-rose and darker at the sometimes faintly-ochreous mid., sometimes white, red-spotted. *St.* even, smooth, white to reddish. *G.* adnate, thin, crowded to distant, white. *Flesh* white, rose-colour under pellicle of P.

Taste mild, becoming slowly acrid, gills acrid. Woods, mixed grassy places; frequent. Aug.–Oct. 3⅜ × 2⅛ × ⅞ in.

1318. **R. maculata** Quél. (from the spotted pileus and stem; *maculatus*, spotted) *a b.*

P. hemispherical to depressed, viscid, reddish flesh-colour to pallid, then discoloured and spotted purple or brown; marg. striate, often dark. *St.* reticulato-striate, white or rosy then spotted ochre, not becoming grey. *G.* adnate, pallid-sulphur to somewhat peach-colour.

Taste somewhat acrid. Sept. 3¼ × 1½ × ⅝ in.

1319. R. sardonia Fr. (from the acrid taste like that of a certain plant, *Herba Sardoa, Ranunculus sceleratus*) *a b c.*

P. convex to plane or slightly depressed, viscid in wet weather, reddish-purple or pale lemon with mid. orange-yellow, often partially white; marg. sometimes striate and somewhat tuberculose. *St.* equal, even, white, yellowish or reddish. *G.* adnate, crowded, white, spotted yellowish from exuded drops in wet weather. *Flesh* white.

Woods, chiefly fir, larch, near paths; uncommon. Sept.–Nov. $4\frac{1}{2} \times 2\frac{1}{2} \times 1$ in.

1320. R. depallens Fr. (from the pileus becoming pale; *depalleo*, to be pale) *a b.*

P. convex to plane or depressed, often undulate or irregular, reddish, red-crimson, fuscous, purplish-rose or dark purple-crimson, the hues becoming pale to yellowish or white, the slightly striate marg. sometimes cobalt-purple. *St.* equal, white to pale greyish. *G.* adnexed, scarcely crowded, white to pale yellowish.

Edible. Taste mild to bitter or acrid; odour none. Woods, beech, pastures, etc.; uncommon. June–Oct. $4 \times 1\frac{7}{8} \times \frac{3}{4}$ in.

1321. R. purpurea Gill. (from the purple pileus) *a b.*

P. hemispherical to depressed, rugoso-plicate, deep purple; mid. darker; marg. even, then faintly striate. *St.* striate, white, or white above, rosy in middle, yellowish below. *G.* rounded, white to yellowish. *Flesh* yellowish, red beneath the pellicle.

Taste mild. Under larches, oak, amongst grass. Aug.–Oct. $3\frac{1}{4} \times 1\frac{7}{8} \times \frac{5}{8}$ in.

1322. R. cœrulea Fr. (from the partially azure-blue colour; *cœruleus*, azure) *a b.*

P. convex or umbonate to slightly depressed, shining, deep red-purple or bluish-purple; marg. even, somewhat cobalt, sometimes with a white line round edge; mid. bluish-purple or buff. *St.* white, sometimes rose at base. *G.* adnate, white to yellowish. *Flesh* white, deep purple under pellicle.

Taste mild. Woods, pine, etc., grassy places. Aug.–Oct. $3\frac{3}{8} \times 2\frac{3}{4} \times \frac{1}{2}$ in.

1323. R. drimeia Cooke (from the pungent taste; Gr. *drimus*, pungent) *a b.*

P. hemispherical to depressed, scarcely viscid, bright deep-purple or deep brownish-red-purple; marg. even. *St.* equal, tinged with purple. *G.* adnexed, pale sulphur to pale yellow.

Intensely acrid. Under larches. Sept.–Nov. $3\frac{7}{8} \times 2\frac{5}{8} \times \frac{3}{4}$ in. Colour and habit of **1345**, differing in yellow gills and intense acridity.

1324. R. virginea Cooke & Mass. (from its wholly white colour) *a b.*

P. convex to depressed, smooth, even, viscid when moist, lustrous when dry; marg. even. *St.* solid, slightly attenuate upwards, finely rugulose. *G.* decurrent, crowded, furcate, narrow.

Taste mild. Terrestrial. Under trees, beech. Oct. $2 \times 2\frac{1}{4} \times \frac{1}{2}$ in. Must not be confounded with other *Russulæ* bleached white with age.

c. Rigidæ.

1325. R. lactea Fr. (from its milk-white colour; *lac*, milk) *a b*.

P. convex, sometimes faintly spotted cream-buff; mid. biscuit-buff. *St.* nearly equal, even, white, sometimes rose at base. *G.* free to adnate, thick, distant, furcate at base and apex, ivory.

Said to be edible, but suspicious. Taste mild; odour of meal. Woods, mixed, fir; uncommon. Aug.–Oct. 3¼ × 2¼ × ¾ in. Var. *incarnata* Cooke. *P.* shaded with pink-buff or dull purple-rose, with lighter and a few pale sienna spots.

1326. R. virescens Fr. (from the frequently green pileus; *vireo*, to be green) *a b c*.

P. pulvinate to depressed, innato-flocculose, without a pellicle, surface soon broken up into patches or warts, verdigris-green, varying olive, dull ochre with mid. verdigris, reddish-purple and other shades. *St.* somewhat rivulose, often clouded pale verdigris. *G.* free, white to biscuit. *Flesh* white, verdigris under pellicle.

Edible. Taste mild, like hazel-nuts when raw, sometimes slowly and slightly acrid; odour of **1394.** Woods; frequent. July–Oct. 4⅛ × 2¾ × 1 in. The "Champignon des Dames" of French authors.

1327. R. cutefracta Cooke (from the broken cuticle); *cutefractus*, with the skin broken) *a b*.

P. pulvinate to depressed, even, cuticle cracking as in **1326**, surface green, purple, bluish-purple, dull red, claret-red, etc., purple beneath cuticle. *St.* equal, smooth, white, often clouded rose-purplish. *G.* adnexed or nearly free, somewhat crowded, white or pale yellowish.

Taste mild. Woods. Sept. 3¼ × 1¾ × ¾ in. Probably a form of **1326**; intermediate forms occur.

1328. R. lepida Fr. (from its appearance; *lepidus*, elegant) *a b c*.

P. convex to slightly depressed, often rimoso-squamulose, reddish-rose with deep red-rose mid. or sienna-ochreous with deep red mid., which at length becomes whitish. *St.* even, equal, white to rose. *G.* adnato-rounded, somewhat thick, somewhat crowded, white or pale yellowish, often rose at edge.

Edible. Taste mild. Woods, mixed, amongst grass; frequent. Aug.–Oct. 4⅛ × 2⅛ × 1 in. Becomes green in cooking.

1329. R. rubra Fr. (from the red pileus; *ruber*, red) *a b c*.

P. hemispherical to plane or slightly depressed, dry, becoming lustrous, often rivuloso-rimose when old, intense crimson; mid. almost black or deep purple, becoming pale. *St.* even, equal, or slightly attenuate upwards, white, sometimes clouded rose or purplish. *G.* adnate, somewhat crowded, whitish to yellowish, often red at edge. *Flesh* reddish under cuticle.

Poisonous. Flesh mild, gills very acrid. Woods, mixed, grassy places; frequent. Aug.–Nov. 4¼ × 4½ × 1 in. Must not be confounded with **1316.**

1330. R. Linnæi Fr. (in honour of Linnæus) *a*.

P. convex to plano-depressed, even, smooth, dry, dark purple or blood-red to light rose. *St.* sometimes hollow, subventricose, deep blood-red to rose. *G.* adnato-decurrent, rather thick, not crowded, white to pale yellowish.

Taste mild, sweet. Woods. Oct. $5 \times 3\frac{3}{4} \times 1\frac{1}{4}$ in. Sometimes white.

1331. R. xerampelina Fr. (from the colour of the pileus, like dying vine-leaves; Gr. *xerampelinai*, of the colour of dry vine-leaves) *a b c*.

P. convex to depressed, dry, even, at length finely granulate, rosy-purplish or dark rosy-red, clouded buff; mid. becoming yellowish-white. *St.* sometimes hollow, equal or slightly attenuate upwards, white or reddish. *G.* adnexed, somewhat crowded, whitish to yellowish. *Flesh* whitish.

Taste mild; odour strong of crab. Woods, fir, grassy places. July–Oct. $4\frac{3}{8} \times 3\frac{1}{2} \times \frac{7}{8}$ in.

1332. R. olivacea Fr. (from the colour of the pileus, at length olive) *a b*.

P. hemispherical to slightly depressed, slightly silky-squamulose, dull purple to olivaceous or fuscous-olivaceous, not lustrous. *St.* equal or ventricose, pallid, rose-colour or clouded pale purple or orange-crimson, crimson and ochre at base. *G* adnexed, yellow.

Edible. Taste mild. Woods, pine, grassy places. Aug. $4 \times 3 \times \frac{7}{8}$ in.

1333. R. serotina Quél. (from its sometimes late appearance; *serotinus*, backward) *a b*.

P. hemispherical to flat, pruinose, sooty-purplish or olive, or clouded-slate and dull olive; marg. lilac or purplish-reddish. *St.* equal or attenuate downwards, wrinkled, mealy, white. *G.* adnexed, crowded, yellowish-white.

Taste acrid. Under beeches. Aug.–Sept. $1\frac{1}{2} \times 1 \times \frac{3}{16}$ in.

1334. R. Duportii Phill. (after the Rev. Canon J. M. Du Port) *a b*.

P. convex to depressed, smooth; mid. rufous, dull purplish, brown, or flesh-red; marg. becoming indistinctly striate and tuberculate, dull bluish or greenish. *St.* minutely striate, glabrous, white or dull sienna-white. *G.* rounded, broad, distant, white or faintly yellowish. *Flesh* reddish-brown when cut.

Taste mild, pleasant; odour of crab. Woods. Sept. $2\frac{5}{8} \times 2\frac{1}{2} \times \frac{5}{8}$ in.

d. Heterophyllæ.

1335. R. vesca Fr. (from its edible qualities; *vescus*, eatable) *a b c*.

P. convex to depressed, viscid, slightly wrinkled, rose-purple, red-flesh-colour, liver, tan-buff umber or ochre, or clouded purple and buff; mid. usually darker, not becoming pale.

St. wrinkled, equal or attenuate downwards, white, sometimes shaded pale rose, purple or ochre. *G.* adnate, crowded, white or yellowish.

Edible. Taste mild, pleasant to unpleasant; odour of hog's lard or somewhat like crab, becoming like lobster or crab when cooked. Woods, mixed; frequent. Aug.–Oct. $4\frac{3}{8} \times 2\frac{1}{2} \times \frac{7}{8}$ in.

1336. R. lilacea Quél. (from its lilac colours) *a.*

P. convex to depressed, viscid, violet, purple or rosy-purple; marg. at length almost white. *St.* attenuate downwards, pruinose above, white, sometimes rosy at base. *G.* adnate, distant, white. *Flesh* violet under pellicle.

Taste mild. Woods, moist places. Aug. $3\frac{1}{2} \times 2\frac{3}{8} \times \frac{3}{4}$ in.

1337. R. azurea Bres. (from the azure-blue pileus) *a b.*

P. convex to depressed, minutely granulose, dry, bright blue, deep purple, dark indigo or greenish-slate; mid. sometimes ochre or brown; marg. scarcely striate, sometimes lilac, becoming dilute cobalt and indigo. *St.* rugulose, equal, ventricose or subclavate, white or clouded pale rose. *G.* adnexed, crowded, white.

Edible. Taste mild, sometimes bitter. Woods, fir, grassy places under trees. Aug.–Sept. $3\frac{3}{4} \times 1\frac{3}{4} \times \frac{3}{4}$ in.

1338. R. cyanoxantha Fr. (from the colours; Gr. *kuanos*, blue, *xanthos*, yellow) *a b.*

P. hemispherical to depressed, viscid, lilac, purplish, olivaceous-green, green-slate, purplish-rufous, pale rose, purplish-slate or buff; mid. becoming pale or yellowish. *St.* equal, smooth, shining, hollow when old, white. *G.* rounded, somewhat crowded, white. *Flesh* white, reddish or purplish beneath pellicle.

Said to be edible. Taste mild, pleasant. Moist places in woods, mixed; common. June–Oct. $3\frac{5}{8} \times 2\frac{1}{2} \times \frac{3}{4}$ in.

1338a. R. mitis W. G. Sm. (from the mild taste; *mitis*, mild) *b.*

P. convex, then plano-expanded and depressed, viscid, then smooth, with a separable pellicle, striate and in old examples, tuberculate; mid. pale yellowish, surrounded by purplish-mouse-colour or pale rose and purple; marg. yellowish or biscuit. *St.* stuffed, then hollow, slightly attenuate downwards, rugulose, white. *G.* adnate, branched, veined above, white, then faint straw-colour. *Flesh* white, tinging to rusty, pale ochre beneath pellicle.

Taste mild. Woods. Aug. $3 \times 2 \times \frac{3}{4}$ in.

1339. R. heterophylla Fr. (from the gills, of different lengths; Gr. *heteros*, one of two, *phullon*, a leaf) *a b c.*

P. hemispherical to depressed, even, shining, usually apple-green or indigo-apple-green; mid. darker or pale apple-green, clouded brownish or sienna, never reddish or purple.

St. equal, even, white. *G.* thin, attenuate behind and reaching the stem, crowded, white.

Edible. Taste mild, sweet, nutty, odour of hog's lard, but like lobster or crab when cooked. Woods, etc.; common. July–Oct. 4 × 2¼ × 1 in. Var. *galochroa* Fr. *P.* white to whitish-apple-green.

1340. R. consobrina Fr. (related to **1341**; *consobrinus*, a cousin) *a b.*

P. campanulate to depressed, viscid, dark cinereous to fuscous-olivaceous, tan-umber, tan-sepia, sooty-sepia or yellowish-umber; mid. darker; marg. even. *St.* even, smooth, shining-white, then cinereous. *G.* adnate, crowded, shining-white. *Flesh* white, ashy under pellicle.

Taste of flesh acrid, gills intensely acrid; not fetid as in **1341.** Mixed woods, pine, beech. July–Oct. 3⅝ × 1⅝ × ⅝ in. Var. *sororia* Fr., margin striate. Var. *intermedia* Cooke. *P.* becoming striate. *St.* usually attenuate downwards. *G.* dull white.

1341. R. fœtens Fr. (from the fetid odour) *a b c.*

P. bullate to expanded, sometimes depressed, viscid in wet weather, sienna- or brown-yellow; marg. striato-ribbed, at length tuberculate. *St.* becoming hollow, whitish or yellowish. *G.* adnexed, crowded, exuding watery drops, whitish.

Said to be poisonous. Taste acrid, rarely mild; odour fetid-burnt, empyreumatic, very strong in mature examples in wet weather, sometimes slight. Woods; very common. July–Oct. 4½ × 3¼ × 1⅛ in. **1390** grows on this species.

1342. R. subfœtens W. G. Sm. (from the somewhat fetid odour) *a b.*

P. convex to plane and subdepressed, somewhat viscid, sienna-yellowish or buff, paler at the striato-tuberculate marg. *St.* equal, even, whitish or faintly buff. *G.* thick, branched, distant, yellowish-white.

Taste slightly acrid and bitter; odour somewhat disagreeable. Grassy places. Aug.–Sept. 3¼ × 2 × ⅝ in. Whole plant tough, elastic.

1343. R. fellea Fr. (from the bitter-acrid taste; *felleus*, full of gall) *a b c.*

P. hemispherical to convex, smooth, lustrous, straw-colour to ochre or buff; mid. brownish. *St.* equal, even, white to straw or pale yellowish, or shaded pale buff. *G.* adnate, more or less crowded, thin, exuding drops in wet weather, pale yellowish or faint buff.

Poisonous. Taste intensely bitter and acrid; odour none. Woods, beech, open places; common. Aug.–Nov. 4½ × 2½ × ⅝ in.

1344. R. elegans Bres. (*elegans*, pretty) *a b.*

P. hemispherical to depressed, glutinous and densely granulate, bright rosy flesh-colour, soon ochreous at marg. or pale crimson; mid. purple-crimson or rose-buff, spotted crimson or purplish-crimson; marg. tuberculose. *St.* equal, becoming hollow, subrugulose, white, ochre-buff below. *G.* adnexed or rounded-adnate, crowded, rarely furcate, whitish, ochre above, becoming orange-ochre with age. *Flesh* very firm, white, ochre-buff with age.

Woods, moist. Sept. 3⅝ × 2 × ⅝ in.

1345. R. Queletii Fr. ex Quél. (after Dr. Lucien Quelet) *a b c.*

P. convex to infundibuliform, even, viscid, intense blackish-crimson or purple-crimson, sometimes paler at marg. or wholly intense purple, blue-purple or slate. *St.* equal or attenuate downwards, mealy, dark or pale-purple to faint rose, sometimes paler above. *G.* adnexed, exuding drops, white to biscuit, at length bluish-ashy or cinereous spotted. *Flesh* white, colour as P. under pellicle.

Taste acrid. Woods, fir ; common. Aug.–Nov. $2\frac{3}{4} \times 2\frac{1}{2} \times \frac{1}{2}$ in.

1346. R. expallens Gill. (from its losing colour ; *expallens*, becoming pale) *a b.*

P. convex to depressed, viscid, deep bright crimson-purple or deep claret-purple ; mid. darker, sometimes almost black, losing colour and becoming livid-cream, with mid. dull pale purplish. *St.* equal, mealy, rose-purple, paler or somewhat buff at base. *G.* adnexed, pallid yellowish or somewhat lemon. *Flesh* purple, pale yellowish or white.

Taste intensely acrid. Under trees. Sept.–Oct. $3\frac{3}{4} \times 3\frac{1}{4} \times \frac{3}{4}$ in.

e. Fragiles.

1347. R. emetica Fr. (from its acting as an emetic) *a b.*

P. convex to plane or slightly depressed, lustrous, blood-red to rose, sometimes bleaching to almost white or purplish, tawny when old. *St.* equal, even, longitudinally rugulose, white or reddish. *G.* adnexo-free, somewhat distant. *Flesh* white, rose-colour under the separable pellicle.

Poisonous. Taste very acrid. Woods and open damp places ; common. July–Dec. $4\frac{1}{4} \times 3\frac{1}{4} \times \frac{7}{8}$ in. Var. *Clusii* Fr. *Flesh* white, turning yellowish. Var. *fallax* Fr. is a form of **1354.**

1347a. R. luteotacta Rea (from the change of colour to light yellow when bruised ; *luteus*, yellow, *tactus*, a touch) *b.*

P. hemispherical, then plane and subumbonate, ruguloso-rivulose, carmine-scarlet, becoming whitish at the striate marg. *St.* equal, or slightly attenuate downwards, rugulose, white, shaded pale lilac, yellowish when bruised. *G.* furcate, adnate, very light straw-colour, yellow when bruised. *Flesh* white, colour as P. under pellicle, tinging yellowish when bruised or slug-eaten.

Taste of flesh acrid, of gills very acrid. On the ground, woody places. Aug. $1\frac{5}{8} \times 1\frac{7}{8} \times \frac{3}{8}$ in.

1348. R. fingibilis Britz. (from its seeming distinctness ; *fingibilis*, imaginary) *a.*

P. convex to depressed, viscid, yellow, with orange or brownish mid. ; marg. not striate. *St.* equal, white or faintly yellowish. *G.* somewhat crowded, thin. *Flesh* white.

Taste mild ; odour none. Amongst grass under trees. July. $2\frac{1}{2} \times 1\frac{7}{8} \times \frac{1}{2}$ in.

1349. R. pectinata Fr. (from the sulcate margin of the pileus, like the teeth of a comb; *pecten*, a comb) *a c*.

P. convex to flat or slightly depressed, rigid, viscid, yellowish-umber or dull ochre, with a dark or purplish mid., becoming pale; marg. tuberculoso-sulcate. *St.* equal, spongy-stuffed, rigid, slightly striate, white or faintly greyish. *G.* attenuato-free, subcrowded.

Suspected poisonous. Taste somewhat acrid; odour weak, nauseous, somewhat like **1341**. Woods, mixed. Aug. $3 \times 1\frac{5}{8} \times \frac{5}{8}$ in.

1350. R. ochroleuca Fr. (from the usually yellowish-white pileus; Gr. *ochros*, pale yellow, *leukos*, white) *a b c*.

P. convex to slightly depressed, lustrous, deep or light yellow, becoming pale, never reddish; marg. striate when old. *St.* equal, slightly reticulato-wrinkled, white to lemon-white, becoming pale steel-grey. *G.* rounded, white or pale yellowish.

Taste of flesh mild, gills acrid; odour none, or pleasant. Woods, chiefly fir; frequent. July-Nov. $3\frac{5}{8} \times 2\frac{1}{4} \times \frac{3}{4}$ in. Var. *claroflava* Cooke. *P.* deep chrome-yellow. *G.* white, becoming pale lemon then somewhat ochraceous.

1351. R. granulosa Cooke (from the, at length, minutely granular pileus and stem) *a b*.

P. convex to subinfundibuliform, pale or ochreous yellow; mid. darker; marg. paler. *St.* equal, white above, as P. or somewhat sienna below. *G.* adnexed or adnate, rarely furcate, somewhat crowded.

Taste of flesh mild, gills acrid; odour sour. Under trees. Aug.–Sept. $3\frac{3}{4} \times 2\frac{1}{2} \times \frac{3}{4}$ in.

1352. R. æruginea Fr. (from the verdigris-green pileus; *ærugineus*, like verdigris) *a b*.

P. convex to slightly depressed, even, somewhat dry, sometimes olive or umber-yellowish, blotted with olive; mid. darker. *St.* equal, even, smooth, white. *G.* attenuate behind and slightly adnexed, somewhat distant.

Taste mild, pleasant. Woods. Aug. $4\frac{3}{4} \times 1\frac{7}{8} \times 1\frac{1}{8}$ in.

1353. R. citrina Gill. (from the citron-coloured pileus) *a b*.

P. convex to depressed, smooth, sulphur-lemon with mid. brassy-yellow, sometimes faintly greenish with mid. becoming pale ochreous. *St.* equal, striate, white to faintest lemon. *G.* attenuate and furcate behind, adnexed. *Flesh* white, brassy-yellow under pellicle.

Taste sweet to acrid; odour almost obsolete. Woods, mixed. Aug.–Oct. $3\frac{1}{4} \times 1\frac{3}{4} \times \frac{5}{8}$ in.

1354. R. fragilis Fr. (from the brittle substance) *a b c*.

P. convex and sometimes umbonate to depressed, crimson, purple, rose-flesh-colour to buff or white, often clouded, changing colour; mid. darker or fuscous, not red under the pellicle. *St.* equal, slightly striate, white. *G.* adnexed, thin, crowded, ventricose.

Taste very acrid. Moist places in woods; common. Aug.–Oct. $2\frac{5}{8} \times 1\frac{7}{8} \times \frac{3}{8}$ in. Var. *nivea* Cooke. *P.* white. Var. *violacea* Cooke. *P.* violet. Var. *fallax* Mass. *P.* dirty reddish, or variously coloured as mid. pale greyish-slate and marg. rose, or mid. brown and marg. rose and white. Included under **1347** by Fries.

1355. R. punctata Gill. (from the dotted pileus; *punctatus*, dotted) *a*.

P. convex to depressed, rosy or purple, punctate with dark rufous pointed tubercles, becoming pale; mid. darker; marg. sometimes cream. *St.* attenuate downwards, colour as P., base whitish, non-punctate. *G.* adnexed, white to yellowish, edge often reddish.

Taste mild, sweet. Woods, grassy places. July. $2\frac{1}{2} \times 1\frac{1}{2} \times \frac{3}{8}$ in. Var. *leucopus* Cooke. *St.* white.

1356. R. veternosa Fr. (from its uncertain affinities; *veternosus*, languid) *a b*.

P. hemispherical to slightly depressed, deep to pale rose or purplish; mid. usually yellowish or whitish; marg. plane. *St.* equal, even, white. *G.* adnexed to adnate, crowded, white then straw-colour.

Suspected poisonous. Taste acrid. Woods, amongst grass under trees. July–Sept. $2\frac{3}{4} \times 2\frac{3}{8} \times \frac{5}{8}$ in.

1357. R. integra Fr. (from its frequently perfect form; *integer*, entire) *a b c*.

P. convex to depressed, viscid, red, purplish-crimson or rose with marg. buff, or tan, clouded crimson, or dull yellowish, clouded pale crimson-brown, or pale yellow with ochre mid. and purple marg., or umber, slate-purple or almost black; marg. sulcate and tubercular. *St.* somewhat ventricose, white, sometimes faintly rose or pale yellowish. *G.* adnexo-free, very broad, subdistant, white, sometimes rose or pale purple with ochreous base. Pileus sometimes bleaching, almost white.

Taste mild to astringent. Woods, chiefly pine, oak; common. June–Nov. $5 \times 2\frac{3}{8} \times 1\frac{1}{4}$ in. Var. *alba* Cooke. Whole plant, except gills, cream-white.

1358. R. decolorans Fr. (from the changeable colour of the pileus; *decolorans*, changing colour) *a b*.

P. spherical to slightly depressed, orange-red becoming light yellow, sometimes olive-ochre or variegated pale yellow, orange or vermilion. *St.* rugoso-striate, white, shaded steel-grey or faintest yellow-olive. *G.* rounded-free, white then yellowish.

Taste mild. Woods, pine, mixed; uncommon. Aug.–Sept. $3\frac{1}{2} \times 3\frac{3}{4} \times \frac{3}{4}$ in.

1359. R. aurata Fr. (from the golden-yellow pileus; *aurum*, gold) *a b c*.

P. hemispherical to plane; mid. scarlet, orange or red; marg. lemon; sometimes mid. sulphur; marg. vermilion. *St.*

obsoletely striate, white to lemon. *G.* rounded-free, whitish to light yellow, bright lemon at edge. *Flesh* white, as P. under pellicle.

Said to be edible, but suspicious. Taste mild to slightly acrid; odour pleasant, sweet. Woods, oak; uncommon. June–Oct. 4¾ × 2⅛ × ⅞ in.

1360. R. Barlæ Quél. (after J. B. Barla) *a b.*

P. convex to depressed, even, peach-colour, yellow, tinged with orange-red or clouded ochre, dull crimson, reddish or copper-colour; mid. darker. *St.* equal, silky-pruinose, white to faintly yellowish or reddish. *G.* white, then pallid-ochreous, sometimes orange.

Taste mild, sweet; odour of crab, or melilot. Amongst grass under trees. July–Oct. 3⅜ × 1⅝ × ⅝ in.

1361. R. nitida Fr. (from the shining pileus; *nitidus*, shining) *a b c.*

P. convex to slightly depressed, bay-brown-purplish, purplish-crimson, crimson-brown, crimson, or purple-white; mid. darker; marg. tuberculose. *St.* equal, even, white or faintly clouded rose. *G.* adnexed, crowded, bright lemon.

Suspected poisonous. Taste mild to nauseous or somewhat fetid; odour strong, unpleasant, sometimes obsolete. Woods, fir; uncommon. July–Sept. 1 × 1½ × ⅜ in. Var. *cuprea* W. G. Sm., *Agaricus cupreus* Kromb. *P.* copper-colour. Var. *pulchralis* W. G. Sm., *Russula pulchralis* Britz. *P.* spotted red or purple at mid., ochreous at marg., or the reverse.

1362. R. alutacea Fr. (from the colour of the gills; *aluta*, tanned leather) *a b c.*

P. hemispherical to depressed, even, crimson, red, rose, purple, black-purple, black, olive, greenish, rarely æruginous, becoming pale; mid. often darker; marg. tubercular. *St.* equal or slightly clavate, white, usually clouded red, rose or purplish, sometimes deep crimson. *G.* adnexed or adnate, light yellow to bright deep ochreous. *Flesh* white to pale yellowish.

Edible. Taste mild, pleasant, acrid when old. Woods, beech, mixed; common. July–Oct. 5¼ × 3 × 1⅛ in.

1363. R. armeniaca Cooke (from the colour of the pileus and the odour; *armeniacus*, of the colour of the apricot) *a b.* Very fragile.

P. convex to depressed, smooth, even, orange, ochre, shading into sienna, sometimes lemon-colour. *St.* equal or slightly attenuate upwards, smooth, white. *G.* adnato-rounded, somewhat distant, yellowish to ochre and egg-yellow.

Taste mild; odour sweet of apricot. Woods, amongst grass under trees. Aug. 2¼ × 2¼ × 5/16 in.

1364. R. puellaris Fr. (from its small size and grace; *puellaris*, girlish) *a b.*

P. hemispherical, obtuse to depressed, purplish, purple, rose or pink-orange; mid. darker, sometimes almost black, becoming pale, not shining; marg. tuberculate. *St.* equal, fragile, white, becoming yellowish or clouded pale rose. *G.* adnate or adnexed, thin, crowded, white, then pallid yellow.

Taste mild. Woods, birch, pine, waysides, amongst *Sphagnum*. Aug.–Oct. $1\frac{3}{4} \times 2\frac{3}{8} \times \frac{5}{16}$ in. Sometimes turning pale, foxy and yellow where bruised. Var. *intensior* Cooke. *P.* deep purple; mid. nearly black. Var. *roseipes* Cooke. *P.* rose, rosy-orange, becoming blanched. *St.* white, partially rosy, not changing to sienna where bruised. All the forms grow in company.

1365. **R. ochracea** Fr. (from the ochreous pileus) *a.*

P. hemispherical to infundibuliform, shining; marg. sulcate. *St.* equal, striate, yellow, pale yellowish or white. *G.* rounded-adnexed, scarcely crowded, furcate at base, ochreous. *Flesh* white to ochreous.

Taste mild. Woods, fir, mixed, grassy places. Aug. $3\frac{1}{4} \times 1\frac{3}{4} \times \frac{3}{4}$ in.

1366. **R. lutea** Fr. (from the yellow pileus; *luteus*, yellow) *a c.*

P. convex to plano-depressed, becoming pale, sometimes white, varying orange with scarlet mid.; marg. striate. *St.* equal, even, white, shaded greyish, never reddish. *G.* adnexo-free, crowded, all equal, egg-yellow.

Taste mild; odour pleasant of apricot, or almost obsolete. Woods; uncommon. July–Oct. $2\frac{3}{4} \times 2 \times \frac{3}{8}$ in.

1367. **R. nauseosa** Fr. (from the nauseous taste and odour) *a b c.*

P. hemispherical to depressed, viscid, bright yellow or purplish to livid, becoming pale or whitish, never red or reddish, with mid. deep ochre; varying orange or yellow with mid. darker; marg. tuberculoso-striate. *St.* slightly striate, equal or sub-clavate, whitish. *G.* adnexo-free, ventricose, white or yellowish.

Taste acrid, sometimes mild; odour fetid to feeble, strong, sour, like decaying oat-cake. Woods, chiefly pine. Sept.–Oct. $2\frac{1}{4} \times 2\frac{1}{8} \times \frac{3}{8}$ in. Often resembling **1341.** Var. *flavida* Cooke. *P.* clear primrose-yellow.

1368. **R. vitellina** Fr. (from its egg-yellow colour; *vitellus*, yolk of egg) *a c.*

P. convex to plane, light yellow to pallid; marg. striato-tuberculate. *St.* equal, white. *G.* adnexo-free, equal, distant, somewhat thick, connected by veins, saffron.

Taste mild; odour strong. Woods, fir, mixed; uncommon. Aug.–Sept. $1\frac{3}{4} \times 1 \times \frac{1}{4}$ in. There is a large form with P. 2–3 in. in diam.

1369. **R. chamæleontina** Fr. (from its changing colour, chameleon-like) *a b.*

P. convex to slightly depressed, flesh-colour, changing to yellow at mid., then wholly yellow, sometimes purple, olive with slate mid., scarlet, crimson, pale rose or vermilion with yellow mid. *St.* equal, slightly striate, white. *G.* adnexed, thin, crowded, equal, somewhat forked, yellow.

Taste mild, sweet; odour none. Woods, pine, beech, downs; rare. Sept.–Oct. $1\frac{1}{2} \times 1\frac{1}{4} \times \frac{1}{4}$ in.

L. CANTHARELLUS Juss.

(From the cup-shaped pileus of some of the species;
Gr. *kantharos*, a cup.)

Veil obsolete. *Hymenophore* continuous and homogeneous with the stem, when latter is present. *Pileus* fleshy to membranous. *Stem*

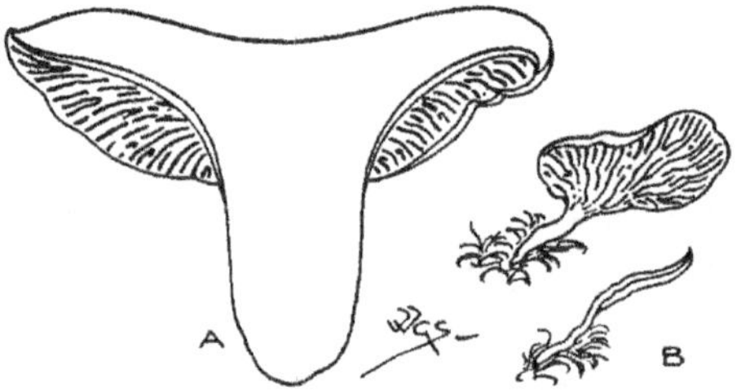

Fig. 64.—A, section of *Cantharellus cibarius* Fr. One-half natural size.
B, *C. muscigenus* Fr., entire and in section. Natural size.

central, excentric or obsolete, exannulate, without cartilaginous bark. *Gills* decurrent, thick, fleshy-waxy, fold-like, somewhat branched, edge obtuse, trama floccose. *Spores* white. (Fig. 64.)

The species grow on the ground and on mosses, rarely on wood. Some are edible, others are said to be poisonous. *Sporodinia Aspergillus* sometimes grows on *Cantharelli*. Species **1370—1388**

a. *Mesopodæ*. *Stem* central.
Pileus entire, solid. *Stem* fleshy. **1370—1376**
Pileus somewhat membranous. *Stem* tubular, shining. **1377—1384**
b. *Pleuropodæ*. Dimidiate. *Stem* lateral. **1385, 1386**
c. *Resupinatæ*. Resupinate. **1387, 1388**

a. *Mesopodæ*.

1370. C. cibarius Fr. (from its use as food; *cibaria*) *a b c*. Egg-yellow.

P. convex to depressed, turbinate, smooth; marg. often repand and irregular. *St.* solid, attenuate downwards. *G.* shallow, narrow, much swollen.

In troops, clusters or rings. Edible. Taste pleasant; odour when mature strong of apricots or ripe greengages, sometimes heavy or earthy and somewhat pungent. Woods, chiefly beech, oak, chestnut; common. July–Dec. 4 × 1¾ × ⅞ in. Sold in Italian markets under the name of *Gallinaccio*; at Cordova under the name of *Encina*; in French markets in a dried state on strings. Var. *rufipes* Gill., red at base of stem. There is a white and a yellowish-white var.

1371. C. Friesii Quél. (after Elias Fries) *a b c.*

P. convex to infundibuliform, fleshy, villous, somewhat orange. *St.* solid, attenuate downwards, yellowish-white or pale orange. *G.* narrow, yellow or orange.

Woods, on elder. Aug.–Nov. $1\frac{1}{2} \times \frac{7}{8} \times \frac{3}{16}$ in. Intermediate between **1370** and **1372.**

1372. C. aurantiacus Fr. (from its golden yellow colour; *aurantiacus*, orange-yellow) *a b c.*

P. plane to infundibuliform, somewhat thin, subtomentose; marg. undulate, at first involute, orange-ochreous; mid. sienna. *St.* stuffed to hollow, attenuate downwards, ochreous or reddish, sienna or black at base. *G.* crowded, thin, deep orange or orange-vermilion.

Gregarious. Said to be poisonous. Taste unpleasant. Woods, chiefly pine, heaths, on elder, on grass-roots. Oct.–Dec. $3\frac{1}{2} \times 2 \times \frac{1}{4}$ in. Possibly a *Clitocybe* rather than a *Cantharellus*. There is a white and a yellowish-white var. with white gills. A mould, *Dactylium dendroides*, the conidial stage of *Hypomyces rosellus*, is common on the gills of this fungus.

1372a. C. hypnorum Brond. (from its habitat, moss, *Hypnum*).

P. campanulate convex then expanded and slightly depressed; marg. incurved, minutely downy, pale primrose-yellow to pale ochre. *St.* slender, slightly flexuous, almost glabrous, yellow, sometimes darker than P. near base. *G.* thin, edge acute, somewhat crowded, branched, decurrent, yellow. *Flesh* thin, whitish.

Cudworth, Yorks. Sept. 1905. $1\frac{1}{2} \times 1$ in. Allied to **1372**, but differing in colour, *P.* less tomentose and spores smaller.

1373. C. Brownii B. & Br. (after J. Brown) *a b.* Ochreous-white.

P. convex, obtuse or umbonate, thin, obscurely silky. *St.* stuffed, equal, somewhat furfuraceous. *G.* subdecurrent, linear, narrow, subdistant, sometimes branched.

Amongst grass. Autumn. $\frac{5}{8} \times 1\frac{1}{2} \times \frac{1}{16}$ in.

1374. C. carbonarius Fr. (from its habitat, burnt wood and earth; *carbo*, charcoal) *a b c.*

P. convexo-umbilicate to infundibuliform, striato-squamulose, date-brown to dark purple-slate and black. *St.* solid, attenuate downwards, often deeply rooting, yellowish-white above, as P. below. *G.* narrow, white.

Often fasciculate. Charcoal beds, burnt ground. July–Dec. $2\frac{3}{8} \times 1 \times \frac{1}{2}$ in.; rooting stem $2\frac{1}{2}$ in. Often grows in company with **183** and **277**, to which small examples of **1374** are not unlike.

1375. C. umbonatus Pers. (from the sometimes umbonate pileus, *a b c.*

P. convex and umbonate to depressed, even, dry, floccoso-silky, cinereous-blackish, somewhat pale slate or umber-slate. *St.*

stuffed, equal or slightly attenuate upwards, cinereous. *G.* thin, crowded, white to faintly brownish.

Gregarious. Taste and odour almost obsolete. Woods, amongst mosses; rare. April–Oct. $1\frac{1}{4} \times 3 \times \frac{1}{8}$ in. Sometimes becomes reddish when bruised.

1376. **C. albidus** Fr. (from the whitish colour; *albus*, white) *a c.* Whitish to faintly yellowish.

P. subinfundibuliform, irregular, splitting, tough, somewhat faintly zoned. *St.* solid, equal or attenuate downwards, smooth. *G.* thin, subcrowded, branched.

Taste and odour heavy, earthy. Pastures, mossy places, amongst larch and fir. Sept.–Oct. $2 \times \frac{7}{8} \times \frac{1}{4}$ in. A form occurs with the gills distant and thick.

1377. **C. leucophæus** Nouel (from its white gills; Gr. *leukos*, white, *phaino*, to appear) *a c.*

P. depressed to deeply infundibuliform, smooth, tough, umber. *St.* slightly attenuate upwards, even, colour as P., darker at base. *G.* distant.

On the ground. Autumn. $1\frac{3}{8} \times 1\frac{1}{8} \times \frac{1}{8}$ in.

1378. **C. Houghtonii** Phill. ex Cooke (after the Rev. William Houghton) *a b.*

P. umbilicato-depressed, smooth, dull white, somewhat flesh-colour; marg. waved, iregular. *St.* attenuate downwards, delicately fibrillose, rooting, colour as P. *G.* narrow, scarcely forked, pallid flesh-colour to salmon.

Single or subcæspitose. On the ground. Autumn. $2\frac{1}{2} \times 3 \times \frac{1}{4}$ in.

1379. **C. tubæformis** Fr. (from the trumpet-shaped, mature pileus; *tuba*, a trumpet, *forma*, form) *a b c.*

P. umbilicato-convex to deeply infundibuliform, flocculose, umber or yellowish-umber, sometimes faintly olive-shaded, sometimes perforate to base of St.; marg. waved. *St.* compressed, lacunose, smooth, orange-yellow, paler above and below. *G.* distant, much branched, yellow-fuliginous or yellowish-umber.

Gregarious. Woods, often on fallen branches, twigs, rotten wood, and on the ground; common. Aug.–Nov. $4\frac{3}{4} \times 4\frac{1}{4} \times \frac{3}{4}$ in. Intermediate forms occur between this and **1380.** Var. *lutescens* Fr. *St.* pale yellowish.

1380. **C. infundibuliformis** Fr. (from the shape; *infundibulum*, a funnel, *forma*, form) *a b c.*

P. convex, umbilicate to infundibuliform, sometimes pervious to base of St., floccoso-wrinkled, yellow-cinereous or fuliginous, paler when dry, sometimes pale purplish-lavender; mid. deep sienna-umber; marg. undulate. *St.* even, smooth, light or bright yellow, sometimes pale purplish-lavender above. *G.* light yellowish to cinereous.

Gregarious; somewhat cæspitose. Suspected poisonous. Woods, chips, rotten wood, on the ground; common. July–Jan. $2\frac{1}{8} \times 2\frac{3}{8} \times \frac{3}{16}$ in. Intermediate forms occur between this and **1379.**

1381. **C. cinereus** Pers. (from the cinereous gills) *a b c.*
P. infundibuliform, often pervious to base of St., villoso-squamulose; marg. lobed, waved and very irregular, hoary-fuliginous or streaky umber. *St.* attenuate downwards, smooth, fuliginous-blackish or gradually dark umber to base. *G.* somewhat pale lavender.

Woods, beech, on the ground. Sept.–Nov. $2 \times 2 \times \frac{3}{8}$ in. Must not be confounded with **1782.**

1382. **C. cupulatus** Fr. (from the cup-shaped pileus; *cupa*, a cup) *a b c.*
P. when young strongly umbonate, then plane to cup-shaped, with um. sometimes at bottom of cup, smooth to flocculose, pallid fuscous, pale and slightly rufescent when dry, or pale umber or brownish-buff; marg. striate. *St.* equal, tough, smooth, pale buff or paler than P. *G.* very distant, much branched, grey, lavender or yellowish-brownish.

Waste places, on the ground; rare. Oct. $\frac{5}{8} \times \frac{3}{4} \times \frac{1}{16}$ in. Must not be confounded with **1448.**

1383. **C. Stevensonii** B. & Br. (after the Rev. John Stevenson) *a c.*
P. convexo-umbilicate, smooth, buff-white. *St.* delicately pulverulent, white to buff-white. *G.* pallid, becoming fuscous in front.

Rotten stumps and wood amongst moss Mar.–April. $\frac{3}{8} \times \frac{3}{4} \times \frac{1}{32}$ in.

1384. **C. replexus** Fr. (from the margin of the pileus, bent back, repand) *a b.*
P. membranous, convex to depressed, subumbonate, fuscous to cinereous or umber-yellowish; marg. striate. *St.* smooth, equal or slightly thickened above, white above, lavender below. *G.* adnate to decurrent, hoary-white, cinereous or pale purple.

On the ground, amongst grass, moss and leaves. Oct.–Nov. $\frac{5}{8} \times 1\frac{7}{8} \times \frac{1}{16}$ in. With the habit of a *Mycena.* Var. *devexus* Fr. *P.* hood-shaped.

b. Pleuropodæ.

1385. **C. muscigenus** Fr. (from the habitat; *muscus*, moss, *gigno*, to bear) *a b c.*
P. membranous, spathulate, smooth, undulate, fuscous, cinereous-whitish when dry, or umber-yellowish or buff. *St.* horizontal and continuous with P., colour as P., villous at base. *G.* distant, branched, colour as P.

On larger mosses, *Hypnum*, *Tortula*, sometimes on the ground; uncommon. June–Nov. $\frac{3}{4} \times \frac{1}{8} \times \frac{1}{16}$ in.

1386. **C. glaucus** Fr. (from the sea-green colour; *glaucus*, sea-green) *a b c.* Grey or indigo-whitish.
P. membranous, spathulate, silky, zoneless. *St.* pruinate. *G.* distant, dichotomous.

On the ground, sandy slopes. Autumn. $\frac{1}{2} \times \frac{3}{16} \times \frac{1}{16}$ in.

c. Resupinatæ.

1387. C. retirugus Pers. (from the reticulate gills; *rete*, a net, *ruga*, a wrinkle) *a b c.*

P. sessile, inferior, membranous, irregular, repando-lobed, cinereous-fuliginous, or pale fuliginous over biscuit *G.* superior, radiating from a central or lateral point, whitish-fuliginous or paler than P.

On sticks and mosses in bogs, *Hypnum*; uncommon. April–May. P. ¾ in. in diam. Sometimes wholly salmon or biscuit.

1388. C. lobatus Fr. (from the lobed pileus) *a b c.* Dark fuscous, pale when dry.

P. sessile, inferior, membranous, irregular. *G.* superior, radiating from a central or lateral point, somewhat distant, branched.

On mosses in bogs, *Hypnum cuspidatum*, *H. revolens*, *Sphagnum*, mossy soil; uncommon. Spring to winter. P. 1⅛ in. in diam.

LI. NYCTALIS Fr.

(From the habit, growing in dark places; Gr. *nux*, night.)

Veil obsolete or indefinite. *Hymenophore* continuous and homogeneous with the stem. *Pileus* fleshy. *Stem* central, exannulate,

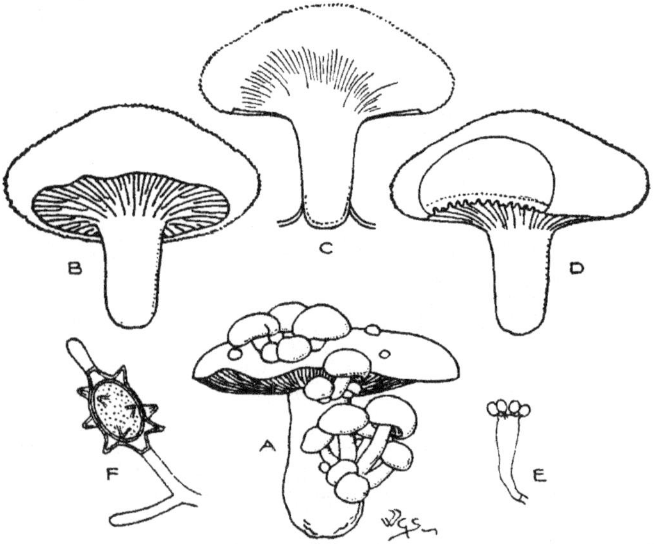

Fig. 65.—*Nyctalis asterophora* Fr. A, plants growing on *Russula adusta*. One-half natural size. B, C, entire and in section; D, with slice removed from edge, showing transverse section of vein-like gills. × 2. E, basidia and spores; F, conidium. × 400.

without cartilaginous bark. *Gills* fleshy, thick, juicy, obtuse at the edge, distant, adnate. *Spores* elliptical, subglobose, colourless. Reproduction is also effected by conidia borne on the pileus. (Fig. 65.)

Parasitic on the *Agaricaceæ*, gregarious and fasciculate.

Species **1389, 1390**

1389. N. asterophora Fr. (from the minute stellate conidia borne on the pileus; *aster*, a star, *fero*, to bear) *a b c.*

P. hemispherical, floccoso-pulverulènt, white to fawn. *St.* solid, equal, whitish to brownish. *G.* very shallow, thick, grey-white.

Gregarious or subcæspitose. Parasitic on **229, 1306, 1307** and a few other species of *Agaricaceæ*; frequent. July–Nov. $\frac{3}{4} \times \frac{3}{4} \times \frac{3}{16}$ in. The stellate conidia on the pileus have been named *Hypomyces asterophorus*.

1390. N. parasitica Fr. (from its parasitic habit) *a b c.*

P. conico-campanulate to expanded, obtuse or umbonate, pruinose, whitish fuscous to pale, or whitish-grey with mid., faint sienna. *St.* fistulose, slightly attenuate upwards, pubescent, white. *G.* somewhat broad, thick, distant, very pale rose-brown.

Gregarious or cæspitose. Parasitic on **1306, 1307, 1311** and **1341**; common. Aug.–Nov. $1 \times 1\frac{7}{8} \times \frac{1}{8}$ in. Often wholly lavender-steel colour. The pileus produces large elliptical, smooth conidia.

LII. MARASMIUS Fr.

(From the habit, drying or shrivelling, not becoming putrid; Gr. *maraino*, to wither or shrivel.)

Veil obsolete except in one non-British species—*M. subannulatus* and a slight indication in the young of **1404**. *Hymenophore* con-

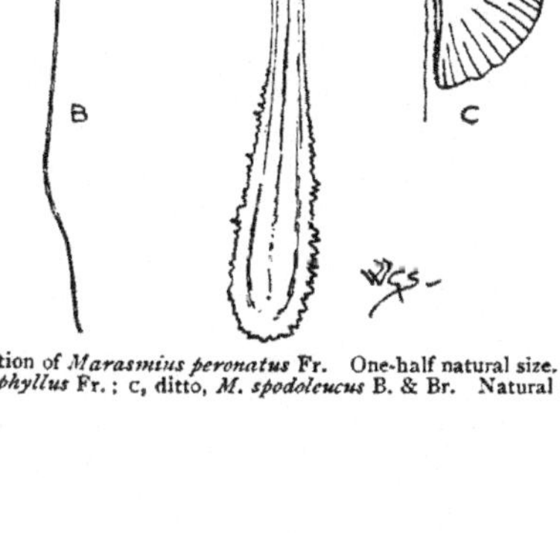

Fig. 66.—A, section of *Marasmius peronatus* Fr. One-half natural size. B, section of *M. epiphyllus* Fr.; C, ditto, *M. spodoleucus* B. & Br. Natural size.

tinuous with but heterogeneous from the stem. *Pileus* arid, pliant, more or less thin. *Stem* central, cartilaginous or horny, absent in **1431.** *Gills* usually simple, pliant, somewhat distant, edge acute, entire. *Spores* white. (Fig. 66.)

The species are commonly small, thin and epiphytal, those which appear to grow on the ground are generally on putrid leaves or roots of grasses; many have a strong odour, sometimes fetid, or of garlic, never of meal. *Marasmius* is closely allied to *Collybia*, but the true *Marasmii* shrivel in dry weather and revive in rain. Some are edible, others are esteemed for flavouring, **1391** is dangerous.

Species **1391—1431**

A. *Pileus* fleshy-pliant, at length somewhat coriaceous, sulcate or wrinkled; margin at first involute. *Stem* somewhat cartilaginous.

a. Scroteæ. *Stem* solid or stuffed, then hollow, internally fibrous, externally cartilaginous and covered with villous down which is easily removed. *Gills* separating-free.

Stem woolly or strigose at the base. **1391—1393**
Stem naked at the base, often composed of twisted fibres. **1394—1396**

b. Terginæ. *Pileus* thinner than *a* in section, hygrophanous, even or striate. *Stem* rooting, tubular, cartilaginous, not fibrous. *Gills* separating-free.

Stem woolly downwards, smooth upwards. **1397—1400**
Stem wholly pruinato-velvety. **1401—1405**

c. Calopodes. *Pileus* convexo-involute, then plane or depressed. *Stem* curt, not rooting, inserted, often with a basal floccose tubercle. *Gills* adnate or subdecurrent.

Stem smooth and shining upwards, base simple. **1406—1410**
Stem velvety or pruinate, base somewhat tubercular. **1411—1414**

B. *Pileus* somewhat membranous, campanulate then expanded; margin at first straight and adpressed to the stem. *Stem* horny, fistulose, sometimes stuffed, tough, arid. *Mycelium* rhizomorphoid, corticate.

d. Chordales. *Pileus* campanulate or convex. *Stem* rigid, rooting or adnate by a dilated disc. **1415—1417**

e. Rotulæ. *Pileus* soon becoming somewhat plane or umbilicate. *Stem* filiform, flaccid, inserted at the base. Growing on leaves.

Stem very smooth, shining. **1418—1422**
Stem velvety or pilose. **1423—1430**

C. *Pileus* sessile, resupinate. **1431**

A.

a. Scroteæ.

1391. M. urens Fr. (from its burning taste; *uro*, to burn) *a b c.*

P. convex to expanded, even or slightly squamulose or rimoso-rivulose, pale yellowish or brownish; mid. darker. *St.* solid, slightly attenuate upwards, white or yellowish, wholly white-floccose. *G.* free to remote, pale yellowish-brown to brownish.

Gregarious, somewhat cæspitose. Poisonous. Taste becoming very pungent. Woods, mixed, grassy places, on leaves, twigs, etc.; frequent. May–Sept. $2\frac{1}{4} \times 4\frac{3}{4} \times \frac{1}{8}$ in.

1392. M. peronatus Fr. (from the base of the stem, as if booted with strigose down; *pero*, a kind of boot) *a b c.*

P. hemispherical to expanded, slightly wrinkled; marg. striate, whitish-umber. *St.* stuffed to hollow, attenuate upwards, light yellow or pale brownish, densely woolly-sheathed at base with whitish or yellowish strigose down. *G.* adfixed to free, crowded, whitish, yellowish or pale rufescent.

Taste acrid. Woods, oak, amongst leaves, twigs, etc., bramble, beech-nuts; common. July–Dec. $3\frac{1}{2} \times 4\frac{1}{2} \times \frac{3}{16}$ in.

1393. M. porreus Fr. (from its odour of garlic; *porrum*, the leek) *a b c.*

P. hemispherical to flat, coriaceo-membranous, dull yellowish or pale dull brownish; mid. darker; marg. striate. *St.* stuffed to hollow, equal, pubescent, whitish or brownish above, deep purple to almost black below. *G.* adnexo-free, distant, light yellowish.

Odour strong, disagreeable, alliaceous. Woods, chiefly amongst oak-leaves, fir; frequent. Sept. Dec. $2\frac{1}{4} \times 3\frac{3}{4} \times \frac{1}{8}$ in. Dried examples more than one hundred years old retain the garlic odour.

1394. M. oreades Fr. (from its growing in fairy rings; Gr. *oreias*, a mountain nymph) *a b c.*

P. fleshy, hemispherical to plane, obtuse or umbonate, even, smooth, whitish, pale yellowish or brownish; mid. darker. *St.* solid to stuffed, equal, whitish. *G.* free, broad, distant, whitish or pale yellowish.

Gregarious. Edible. Taste pleasant; odour agreeable, stronger when dry. Pastures, lawns, roadsides; common. May–Oct. $2\frac{1}{2} \times 2\frac{5}{8} \times \frac{1}{8}$ in. A colourless ketchup can be made from this. Often pickled or dried for winter use.

1395. M. planeus Fr. (from the—at length flat—pileus; *plancus*, flat) *a c.*

P. convex to plane and depressed, even; mid. dark rosy-sienna; marg. rufescent or lightish, becoming pale. *St.* hollow, attenuate downwards, white above, yellowish to brownish below. *G.* free, distant, yellowish or brownish.

Taste mild, sweet. Woods, shady places, on the ground. Autumn. $1\frac{3}{4} \times 1\frac{3}{8} \times \frac{3}{16}$ in.

1396. M. scorteus Fr. (from its leathery substance; *scorteus*, leathery) *a c*. Whitish or white.

P. convex, somewhat obtuse. *St*. somewhat fistulose, equal, smooth, fuscous when dry. *G*. free, broad, ventricose.

Taste mild. Woods, moist places, amongst grass, under trees. Aug.–Jan. $\frac{3}{4} \times 2 \times \frac{1}{16}$ in. Habit that of a small **1394**.

b. Terginæ.

1397. M. prasiosmus Fr. (from its odour of leek or garlic; Gr. *prason*, a leek, *osme*, a scent) *a b c*.

P. campanulate to flat, obtuse, rugulose, pale buff; mid. pale sienna. *St*. fistulose, pallid above, pale rufous or fuscous below. *G*. adnexed, somewhat crowded, whitish.

Odour strong. Amongst leaves. July–Oct. $1 \times 2\frac{1}{4} \times \frac{1}{8}$ in.

1398. M. varicosus Fr. (from the blood-red juice, as if from a swollen vein, *varix*) *a c*.

P. campanulate to plane, umbonate, even, dark rose- or reddish-brown, darker when dry; marg. paler. *St*. slightly attenuate upwards, claret-brown, almost black below, filled with blood-red juice. *G*. very crowded, umber-fuscous or somewhat dark dull-purple, darker when dry.

Solitary. Odour none. Damp mossy places. Sept.–Oct. $1\frac{1}{4} \times 2 \times \frac{1}{16}$ in.

1399. M. fuscopurpureus Fr. (from the dark purplish colours) *a b c*. Juiceless.

P. hemispherical to plane, slightly umbilicate, somewhat wrinkled, dark to pale purple-brown or tan. *St*. slightly attenuate upwards, rufous or dark purple, strigoso-downy and rubiginous at base. *G*. distant, rufescent, purple-brown or slate-purple.

Solitary, cæspitose or gregarious. Odour none. Woods, on sticks, dead leaves, beech, dead grass, etc.; frequent. Aug.–Oct. $1\frac{1}{4} \times 1\frac{1}{2} \times \frac{1}{8}$ in.

1400. M. terginus Fr. (from the leathery substance; *tergum*, hide, leather) *a b c*. Juiceless.

P. campanulate to plane or slightly depressed, shining, flesh-colour to white; marg. striate. *St*. slightly attenuate upwards, shining, pallid or sulphury above, reddish below. *G*. somewhat crowded, pale yellowish, sulphury or salmon.

Solitary or in pairs. Odour none. Amongst leaves, beech, moss. Oct.–Nov. $1\frac{1}{2} \times 2\frac{1}{4} \times \frac{1}{8}$ in. This is *M. Stevensii* B. & Br.

1401. M. Wynnei B. & Br. (after Mrs. Lloyd Wynne) *a b c*.

P. convexo-plane, subumbonate, lilac-brown, purplish-lavender or purplish-salmon, clouded pale slate. *St*. attenuate downwards, colour as P., or yellowish above to brown and brown-black at base within and without. *G*. adnexed or adnate, lilac, lavender or salmon.

Gregarious or cæspitose. Odour none. Amongst leaves, twigs, etc., on the ground. Aug.–Dec. $1\frac{7}{8} \times 3\frac{3}{8} \times \frac{3}{16}$ in.

1402. **M. erythropus** Fr. (from the red stem; Gr. *eruthros*, red, *pous*, a foot) *a b c*. Juiceless.

P. campanulate to flat, obtuse, slightly wrinkled, pale buff with sienna mid., paler at marg., becoming almost white. *St.* smooth, equal, dark red. *G.* veined, not crowded, whitish to pale salmon.

Single or somewhat cæspitose. Odour none. Woods, amongst dead leaves, beech, beech-mast, near and rarely on stumps; frequent. Sept.–Oct. 2⅛ × 4⅝ × ⅛ in. Sometimes becoming purple-black in drying.

1403. **M. archyropus** Fr. (perhaps an error for *argyropus*; Gr. *arguros*, silver, *pous*, a foot, in reference to the white-pruinose stem) *a b c*.

P. hemispherical to plane or slightly depressed, smooth, tan or salmon-shaded, becoming pale. *St.* pale yellowish or pallid rufescent, white pruinose. *G.* crowded, narrow, pallid.

Cæspitose. Odour none. Amongst leaves; rare. Oct. 1¼ × 3 × $\frac{1}{16}$ in.

1404. **M. torquescens** Quél. (from the twisted stem; *torqueo*, to twist) *a b*.

P. convexo-plane, rugoso-striate, pallid-whitish; mid. tawny or fulvous. *St.* white above to brown or deepest brown below. *G.* thin, finely serrulate, distant, white or whitish.

Woods, amongst oak-leaves, twigs, etc., in moist places. Autumn. ½ × 2¼ × $\frac{1}{16}$ in.

1405. **M. impudicus** Fr. (from the odour; *impudicus*, disgusting) *a b c*.

P. convex to plane, depressed or subumbonate, striato-plicate, bay-brown-rufous or rose-umber; mid. dark umber, becoming pale. *St.* equal or attenuate downwards, velvety, rufous or violaceous-purple, white downy, white when dry. *G.* free, veined, whitish flesh-colour or salmon. *Flesh* rose-vinous.

Gregarious. Odour strong, disgusting, like burnt flannel and fetid, more powerful when cut or broken. On and about rotten pine-trunks, on pine-leaves. Sept.–Oct. 1 × 1½ × ⅛ in.

c. Calopodes.

1406. **M. scorodonius** Fr. (from the odour of garlic; Gr. *skordion*, a plant with the odour of garlic) *a b c*.

P. arid, wrinkled, crisped, rufous or yellowish-sienna; mid. vinous, becoming pale to whitish. *St.* fistulose, equal, horny, rufous, paler above, darker or blackish below. *G.* adnate, veined, whitish.

Usually gregarious. Edible, esteemed for flavouring. Odour strong. Heaths and dry pastures, on twigs and leaves; rare. Oct. 1⅝ × 2⅝ × $\frac{1}{16}$ in.

1407. **M. calopus** Fr. (from the beautiful stem; Gr. *kalos*, beautiful, *pous*, a foot) *a b c*.

P. rarely depressed, even, smooth, whitish. *St.* fistulose, slightly attenuate upwards, deep rich brown, paler above. *G.* slightly emarginate, white.

Odour none. On twigs, grass-roots, etc. Autumn. ¾ × 1⅛ × $\frac{1}{32}$ in.

1408. M. Vaillantii Fr. (after Leon Vaillant, French botanist) *a b c.*
P. tough, arid, plicato-rugose, whitish or pale yellowish. *St.* stuffed, attenuate downwards, date-brown, paler or whitish above, blackish below. *G.* adnate, subdecurrent, broad, distant, white.

Odour none. On dead wood, fallen twigs, leaves. Sept.–Oct. $\frac{7}{8} \times 1\frac{1}{4} \times \frac{1}{16}$ in.

1409. M. angulatus B. & Br. (from the angular folds of the pileus) *a b.*
P. plicato-angular, fuscous-whitish or yellowish; mid. pale brownish. *St.* slightly enlarged above and below, base hairy, rufescent-grey or white above, brown-black below. *G.* adnexed, distant, whitish.

On grass, submerged rushes, roots, on sandhills. Aug. $\frac{1}{4} \times 1 \times \frac{1}{32}$ in.

1410. M. languidus Fr. (from its weak substance; *langueo*, to be weak) *a.*
P. slightly fleshy, umbilicate, flocculose, rugoso-sulcate, whitish, light yellowish, or somewhat flesh-colour. *St.* stuffed, attenuate downwards, naked, pallid to fuscous, white villous at base. *G.* adnate to decurrent, veined, white.

Gregarious. Odour none. On dead grass, twigs. $\frac{5}{8} \times \frac{3}{4} \times \frac{1}{16}$ in.

1411. M. fœtidus Fr. (from its fetid odour) *a b c.*
P. convex to flat, umbilicate, striato-plicate, bay-brown-tawny or yellowish-sienna; mid. almost black. *St.* fistulose, attenuate downwards, sometimes with a small floccose tubercle at base, deep brown, almost black below. *G.* adnexed to subdecurrent, distant, veined, yellowish-rufescent or salmon. *Flesh* claret-brown, becoming pale.

Odour strong of burnt flannel or horse-radish, very fetid. Dead twigs, rotten branches, etc.; rare. Aug.–Jan. $1\frac{1}{8} \times 1\frac{1}{2} \times \frac{1}{16}$ in.

1411a. M. lagopinus Post ex Fr. (from the squamulose base of the stem; Gr. *lagopous*, hare's foot) *a.*
P. convex, then plane, even, glabrous, pallid. *St.* slender, equal, whitish, upper half glabrous, lower half with white spreading sq. *G.* adnate, slightly ventricose, straw-colour.

Gregarious. Dead branches. Oct. $\frac{5}{8} \times \frac{1}{2} \times \frac{1}{20}$ in.

1412. M. amadelphus Fr. (from its nearness to **1413**; Gr. *an*, together, *adelphos*, a brother) *a c.* Pale buff.
P. arid, convex to plane, umbilicate, striate; mid. darker. *St.* stuffed, equal, sometimes with an almost obsolete tubercle. *G.* adnato-decurrent, distant.

Gregarious. Odour none. Dead branches, twigs; rare. Sept.–Dec. $\frac{3}{8} \times \frac{1}{4} \times \frac{1}{32}$ in.

1413. M. ramealis Fr. (from its habitat; *ramus*, a branch) *a b c.*
P. arid, convex to plane, obtuse or depressed, white; mid. rufescent. *St.* stuffed, whitish, rufescent at base, sometimes with a small tubercle. *G.* adnate, subdistant, whitish.

Densely gregarious. Odour none. Dead twigs, branches, burnt sticks, juniper, willow, bramble; common. May–Jan. $\frac{1}{2} \times \frac{1}{2} \times \frac{1}{16}$ in.

1414. **M. candidus** Fr. (from its white colour; *candidus*, shining white) *a b c.*

P. campanulate to expanded or depressed, becoming sulcato-rugulose, wholly white or with mid. pale brown. *St.* stuffed, base floccose, white with a brownish base or wholly white. *G.* adnexed, ventricose, distant.

Gregarious, twigs, rotten stems, pine, cedar, leaves. Sept.–Nov. $\frac{3}{8} \times \frac{1}{2} \times \frac{1}{32}$ in.

B.

d. Chordales.

1415. **M. alliaceus** Fr. (from its odour of garlic, *allium*) *a b c.*

P. campanulate to expanded, subumbonate, smooth, dry, striate, white or faintly ochre to fuscous. *St.* fistulose, minutely velvety, blackish or deep dull purple, paler above. *G.* adnexed, ventricose, subdistant, whitish or fuscous-whitish.

Odour strong, persistent. Woods, leaves, rotten wood; frequent. Aug.–Oct. $1\frac{3}{8} \times 4 \times \frac{1}{16}$ in.

1416. **M. caulicinalis** Fr. (from its habitat—growing on stalks and stems; *caulis*, a stem) *a b c.*

P. striate, bright or dull yellow or yellowish-sienna. *St.* fistulose, equal, flocoso villous, sulphur above, dark sienna below. *G.* adnato-decurrent, subdistant, veined, pallid light yellow or pale sulphur.

Taste very bitter. On fir-cones, leaves, etc.; rare. Aug.–Nov. $1 \times 1\frac{7}{8} \times \frac{1}{16}$ in.

1417. **M. torquatus** Fr. (from the sometimes twisted stem; *torqueo*, to twist) *a b.*

P. membranous, campanulate, obtuse, plicato-sulcate, whitish; mid. bright sienna. *St.* hollow, smooth, shining, base dilated-tuberculose, ivory. *G.* adnate, distant, thick, connected by veins, forming a collar near St., white. *Flesh* in P. pale sienna, elsewhere white.

Stems and leaves. Autumn. $3\frac{1}{8} \times 1\frac{1}{8} \times \frac{1}{8}$ in.

e. Rotulæ.

1418. **M. rotula** Fr. (from the resemblance of the pileus—seen from below—to a little wheel; *rota*, a wheel) *a b c.*

P. arid, hemispherical, umbilicato-depressed, whitish or brownish; marg. undulato-crenulate. *St.* fistulose, horny, often branched, sometimes barren, blackish, paler above. *G.* adnate to a free collar, broad, few, distant, white.

Gregarious or cæspitose, with root-like branches. Dead twigs, stumps, old roots, branches, grass-roots, yew, elm, maple, oak; common. June–Jan. $\frac{1}{2} \times 2\frac{1}{4} \times \frac{1}{64}$ in.

1419. **M. graminum** B. & Br. (from its usual habitat, leaves and stems of grass, *gramen*) *a b c.*

P. campanulate to expanded, subumbonate, sulcate, pale, rufous-brown or salmon; mid. darker. *St.* shining, hair-like, black, whitish above. *G*. adnate to a free collar, ventricose, distant, veined, white or pale yellowish.

On grass, leaves, twigs, etc.; uncommon. July–Feb. $\frac{3}{8} \times 2\frac{1}{4} \times \frac{1}{64}$ in.

1420. **M. androsaceus** Fr. (derivation uncertain) *a b c*.

P. arid, convex to expanded, subumbilicate, smooth, striate, white pale buff or somewhat fuscous. *St*. fistulose, horny, black, paler above. *G*. adnate, without a collar, distant, whitish to salmon.

Woods, leaves, twigs, briar, beech, oak, acorn-cups, rushes, fern-stems, fir, juniper, larch; common. April–Dec. $\frac{1}{2} \times 2 \times \frac{1}{64}$ in. Sometimes growing from entangled masses of black horsehair-like mycelium. Sometimes non-pileate stems only produced.

1421. **M. splachnoides** Fr. (from the gut-like stem; Gr. *splagchna*, intestines, *eidos*, appearance) *a b c*.

P. convex to flat, umbilicate, smooth, striate, white; mid. pale yellowish. *St*. fistulose, horny, usually red, becoming fuscous or pale above and sienna below, or dark brown. *G*. subdecurrent, crowded, anastomosing.

Odour none. Amongst pine-leaves. Oct. $\frac{3}{4} \times 2 \times \frac{1}{64}$ in.

1422. **M. Curreyi** B. & Br. (after Frederick Currey) *a*.

P. convex to expanded or plane, sulcate, pallid rufous; mid. fuscous or yellowish-sienna. *St*. black, white above. *G*. attached to a free collar, few, subventricose, veined, pale yellowish.

On grass. Aug. $\frac{3}{8} \times 1\frac{1}{4} \times \frac{1}{64}$ in.

1423. **M. perforans** Fr. (from the habit, the stem perforating the leaves on which it grows) *a c*.

P. hemispherical to convexo-plane, not umbilicate, white or whitish, becoming pale rufescent. *St*. fistulose, tough, equal, bay-brown to black. *G*. adnate, subdistant, whitish.

Odour very fetid, not of garlic. Leaves, fir, larch; rare. Aug.–Oct. $\frac{3}{8} \times 2\frac{1}{4} \times \frac{1}{64}$ in.

1424. **M. insititius** Fr. (from the habit—the stem ingrafted on the host-plant; *insero*, to insert) *a b c*.

P. arid, convexo-plane, subumbilicate, plicato-sulcate, white or purplish-salmon-whitish. *St*. fistulose, horny, attenuate downwards, rufous to fuscous or sienna, paler above. *G*. adnate, distant, white or pale salmon-purplish.

Gregarious. Odour none. Leaves, oak, decayed grass, etc.; rare. Aug.–Oct. $\frac{5}{8} \times 2 \times \frac{1}{32}$ in.

1424a. **M. sclerotipes** Bres. (from the hard sclerotium at the foot of the stem; *pes*, a foot) *a*.

P. membranous, convex, then plane, umbilicate, ruguloso-striate, subflocculose; marg. reflexed, white; mid. yellow. *St*. filiform, pruinose under a lens, pallid rufescent, apex whitish, base adnate to a rufescent yellowish sclerotium. *G.* distant, adnate, white.

On the ground, swampy places. $\frac{1}{4} \times \frac{7}{8} \times \frac{1}{20}$ in. Autumn. Sclerotium $\frac{1}{8}$ in. long.

1425. **M. Hudsonii** Fr. (after William Hudson, author of "Flora Anglica") *a b c.*

P. hemispherical, usually sulcate, pale fuscous-rufescent, rose-purple to white with sienna mid. beset with erect purple hairs. *St.* horny, dark purple, paler above. *G.* adnexed, white.

Odour none. On fallen holly-leaves; frequent. Mar.–Dec. $\frac{1}{4} \times 1\frac{3}{4} \times \frac{1}{64}$ in.

1426. **M. epichloe** Fr. (from its habit, growing on grasses; Gr. *epi*, upon, *chloe*, grass) *a b.*

P. plano-convex, somewhat papillate, not striate, whitish or pale brownish-yellow; mid. darker. *St.* striate, bay-brown. *G.* adnexo-rounded, somewhat crowded, pale yellowish, faintly brown.

Base of dead grasses, spines of *Robinia.* Aug.–Oct. $\frac{3}{8} \times 1\frac{1}{2} \times \frac{1}{32}$ in.

1427. **M. actinophorus** B. & Br. (from the brownish star on the pileus; Gr. *aktis*, a ray of light, *phero*, to bear) *a b c.*

P. convex, umbilicate, bay-brown or pale sienna, clouded with an indefinite brownish star at mid. *St.* pallid or brown, paler above. *G.* few, adnato-decurrent, white or pale yellowish.

Fallen twigs, sometimes terrestrial. Aug. $\frac{1}{4} \times \frac{7}{8} \times \frac{1}{64}$ in.

1428. **M. saccharinus** Fr. (from the pileus, as if powdered with white sugar, *saccharon*) *a b c.*

P. campanulate to convexo-expanded, papillate, sulcate, white. *St.* flocculose to smooth, reddish, pale above. *G.* adnate, thick, distant, reticulato-wrinkled, whitish.

Dead twigs, leaves; rare. Aug. $\frac{3}{8} \times \frac{3}{4} \times \frac{1}{32}$ in. Pileus sometimes marked as in **1427.**

1429. **M. epiphyllus** Fr. (from the habit—growing on leaves; Gr. *epi*, upon, *phullon*, a leaf) *a b c.*

P. convex to plane, umbilicate, plicato-rugose, white. *St.* fistulose, subhorny, date-brown or sienna, white above. *G.* adnate, distant, veined and branched, white.

Gregarious. Odour none. Dead leaves, especially ivy, twigs, etc., oak, beech, poplar; common. Sept.–Dec. $\frac{3}{8} \times 2\frac{1}{8} \times \frac{1}{64}$ in. Pileus sometimes marked as **1427**, sometimes without gills, or gills replaced by wrinkles.

1430. **M. polyadelphus** Cooke (from its growing in crowded patches; Gr. *polus*, many, *adelphos*, a brother) *a.* White.

P. hemispherical, sulcate, flocculose. *St.* tough, floccose below. *G.* subdecurrent, almost fold-like.

Densely gregarious. Dead leaves, etc. Autumn. $\frac{3}{16} \times \frac{1}{2} \times \frac{1}{64}$ in.

C.

1431. **M. spodoleucus** B. & Br. (from its grey pileus and white gills; Gr. *spodos*, wood-ash, *leukos*, white) *a b c*.

P. at first flat, resupinate, then campanulate, at length pendulous, pulverulent or subfurfuraceous, cinereous or pale umber. *G*. inside bell, few, short, leaving a naked space at apex of bell, interstices even. *Flesh* umber.

Dead twigs, elm. Nov.–Mar. Diam. ¾ in. The same as *M. Broomei* Berk. Sometimes not unlike **1449**.

LIII. **LENTINUS** Fr.

(From the tough, pliant substance; *lentus*, tough or pliant.)

Veil ragged-pendulous at the margin of the pileus and more or less fugitive-annulate in the first section, in the other—except **1439**—obsolete. *Pileus* fleshy-coriaceous, pliant or hard, persistent. *Stem* hard, central, lateral or obsolete, continuous with the hymenophore when present. *Gills* concrete with the hymenophore, thin, membranous, unequal, tough, edge acute, often serrate, crenate or irregular; trama none. *Spores* white. (Fig. 67.)

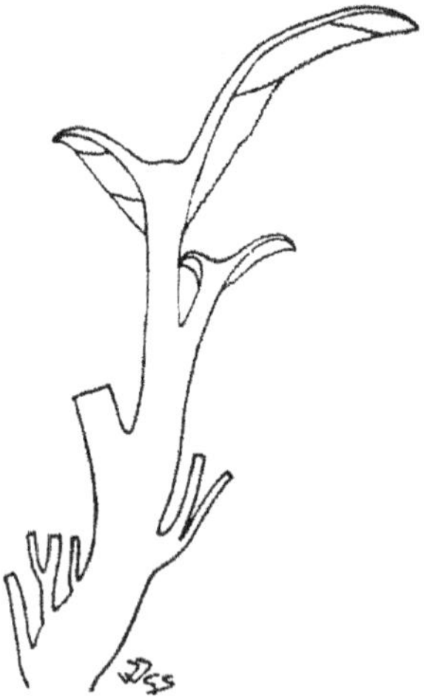

Fig. 67.—Section of *Lentinus cochleatus* Fr. One-half natural size.

Growing on wood, rarely terrestrial, polymorphous, somewhat irregular. Distinguished from *Pleurotus* by the hardness of the flesh and the serrate gills. Species **1432—1441**

MESOPODÆ. *Pileus* nearly entire, margin at first inrolled. *Stem* distinct.

a. *Lepideæ*. *Pileus* scaly. More or less veiled. **1432, 1433**

b. *Pulverulentæ*. *Pileus* villous or pulverulent. **1434—1436**

c. *Cochleatæ*. *Pileus* smooth. **1437**

PLEUROTÆ. Dimidiate, sessile or furnished with a somewhat lateral stem. **1438—1441**

MESOPODÆ.

a. *Lepideæ*.

1432. **L. tigrinus** Fr. (from the tiger-like markings of the pileus) *a b c*.

P. convex, umbilicate to infundibuliform, pale yellowish, clad with umber sc.; mid. sepia. *St*. solid, equal or attenuate downwards, central to excentric, fugitive-torn-annulate, pale yellowish. *G.* decurrent, crowded, white or pale pallid yellowish.

Often subgregarious or cæspitose. Tasteless; odour strong, acrid or acid. Stumps, oak, ash, poplar, fir; rare. April–Oct. $2\frac{3}{4} \times 1\frac{1}{8} \times \frac{1}{4}$ in. Var. *Dunalii* Fr. *P.* with adpressed separating, spot-like sc., then even.

1433. **L. lepideus** Fr. (from the scaly pileus; Gr. *lepis*, a scale) *a b c*.
P. convex to depressed, pale yellowish, clad with umber sc.; mid. darker. *St.* solid, almost woody, irregular, usually somewhat excentric and attenuate downwards, colour as P., dark sienna-umber at base, fugitive-torn-annulate. *G.* decurrent or sinuate, crowded, pale yellowish.

Solitary or subcæspitose. Odour strong, from pleasant to that of rotten cheese. On pine wood, trunks, stumps, fallen pines, elm, railway sleepers, wooden bridges, in houses, cellars, railway tunnels. Mar.–Oct. $5\frac{3}{8} \times 5\frac{1}{4} \times 1$ in. Often abnormal and *Clavaria*-like. Sometimes phosphorescent. Var. *contiguus* Fr. *P.* thinner, plano-depressed, even. *St.* slender, even.

b. Pulverulentæ.

1434. **L. leontopodius** Schulz. ex Fr. (from a fancied resemblance in the shape and colour of the pileus to a lion's foot; Gr. *leon*, a lion, *pous*, a foot) *a*.
P. expanded, depressed, boldly lobed, tan-clay colour, cinnamon or salmon-umber. *St.* woody, equal, excentric, pulverulent, pale chestnut or umber-salmon, blackish downwards. *G.* decurrent, veined, white.

Gregarious or cæspitose in small clusters. Taste pleasant, sweet. Decayed willow. Autumn. $7 \times 3 \times 1\frac{1}{2}$ in.

1435. **L. pulverulentus** Fr. (from the pulverulent pileus) *a*.
P. convex, yellow, ochreous-cinnamon or umber-yellowish, lighter at marg., white-pulverulent. *St.* solid, excentric, velvety dark umber above, brownish-yellow below. *G.* crowded, subdecurrent, white or pale yellowish.

Somewhat cæspitose, sometimes connato-branched and deeply rooting. Dead trunks, pines. Oct. $1\frac{3}{4} \times 3 \times \frac{1}{2}$ in. Var. *resinaceus* Fr. *P.* somewhat gummy villous.

1436. **L. adhærens** Fr. (from the sticky gluten of the pileus and stem) *a*.
P. convex, subumbonate, then depressed to subinfundibuliform, sulcate, umber-tan or umber-yellowish. *St.* somewhat hollow, rooting, white, then as P. *G.* sinuato-decurrent, thin, white.

Taste at length astringent. Woods, pine; rare. Autumn. $1\frac{3}{4} \times 2\frac{1}{2} \times \frac{1}{4}$ in.

1436a. **L. suffrutescens** Fr. (from the somewhat branching habit; *frutex*, a shrub) *a*.
P. tough, convex, then infundibuliform, unequal, glabrous, whitish, becoming brown stained. *St.* elongated, somewhat

branched, woody, pubescent, colour as P. *G.* crenato-torn, yellowish.

Squared wood in cellars. Feb. $3\frac{1}{4} \times 6 \times \frac{1}{2}$ in. Often abnormal in habit.

c. Cochleatæ.

1437. L. cochleatus Fr. (from the shape of the pileus, somewhat shell-like; *cochlea*, a snail-shell) *a b c.*

P. convexo-umbilicate, subinfundibuliform, at length irregularly lobed, brownish-yellow; mid. darker. *St.* usually branched, proliferous and irregularly indented, central to almost lateral, colour as P., sienna-umber at base. *G.* decurrent, crowded, lighter than P., or white flesh-colour.

Solitary to cæspitoso-proliferous. Said to be edible. Odour strong of anise, spicy, often scenting the air for long distances. Trunks, stumps and on the ground; frequent. Aug.–Oct. $2\frac{1}{2} \times 5 \times \frac{1}{4}$ in. Changes to rich brown or blackish in drying.

PLEUROTÆ.

1438. L. scoticus B. & Br. ex Fr. (from its first observed place of growth—Scotland) *a b c.*

P. convexo-umbilicate to expanded and depressed, pallid to brownish or yellowish-buff, hygrophanous; marg. lobed, sinuate or plicate. *St.* solid, darker than P., or yellowish-sienna. *G.* decurrent, subdistant, white.

Solitary or cæspitose. Odour none. Decayed stumps and rotten wood, furze, birch, spruce. Nov.–Jan. $2\frac{3}{4} \times \frac{3}{4} \times \frac{1}{4}$ in. Stem sometimes absent and pileus reniform.

1439. L. fimbriatus Curr. (from the fugitive fimbriate annulus in infancy) *a c.*

P. convex to deeply depressed, floccose-scaly, fawn-colour or umber-salmon; marg. substrigose. *St.* solid, lateral, rough with reflexed sc., brown-white. *G.* subdecurrent, pale brown or salmon-white.

Imbricate. Stumps, posts. Sept. $\frac{3}{4} \times \frac{1}{4} \times \frac{1}{8}$ in.

1440. L. vulpinus Fr. (from its strong, often foxy, odour; *vulpes*, a fox) *a b c.*

P. sessile, reniform-conchate, surface rough and corrugate with raised radiating ribs or adnate teeth, warm yellow-brownish; marg. rose flesh-brown, base darker. *G.* extended to base, crowded, brownish, pale yellow or pale yellowish-rose.

Imbricate. Odour sometimes of field-mint. Trunks, stumps; rare. Mar.–Oct. Diam. 5 in. Becomes black in drying.

1441. L. flabelliformis Fr. (from the fan-shaped pileus; *flabellum*, a small fan, *forma*, form) *a c.*

P. sessile, reniform; marg. corrugate, at length curled backwards, smooth, fawn-colour or yellowish-brown, darker at base. *St.*

rudimentary. *G.* extended to base, subdistant, somewhat thick, whitish or faintly shaded as P.

Imbricate. Old trees, stumps, trunks; rare. Feb. Diam. 2¾ in.

LIV. PANUS Fr.

(A name given to an arboreal fungus by Pliny.)

Veil obsolete. *Pileus* fibrous, fleshy-coriaceous. *Stem* lateral or none, when present confluent with the hymenophore. *Gills* entire,

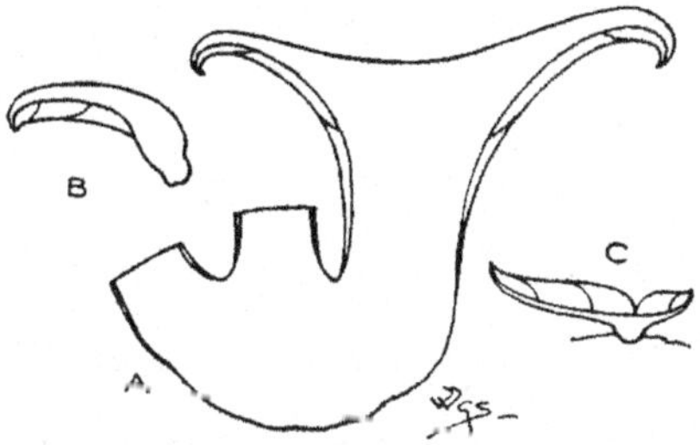

Fig. 68.—A, section of *Panus torulosus* Fr. One-half natural size. B, section of *P. stypticus* Fr.; C, ditto, *P. patellaris* Fr. Natural size.

concrete with the hymenophore, at length coriaceous, edge acute, not serrate. *Spores* white. (Fig. 68.)

Growing on stumps and lasting long. Species **1442—1447**

Pileus irregular. *Stem* excentric. **1442, 1443**

Stem lateral. **1444, 1445**

Pileus resupinate, sessile or extended behind. **1446, 1447**

1442. P. conchatus Fr. (from the shape, like certain shells; *concha*, a shell) *a b c.*

P. more or less dimidiate, deeply depressed, fleshy, pliant, thin, cinnamon or brownish-yellow, pale sienna-scaly in old examples. *St.* solid, equal or attenuate downwards, brownish-yellow. *G.* deeply decurrent, narrow, whitish or pale flesh-colour, at length yellowish.

Cæspitose, often imbricate. Trunks, stumps, beech, poplar, lime; rare. June–Oct. 4½ × ¾ × ½ in. Stem sometimes branching.

1443. P. torulosus Fr. (from the hairy down on the stem; *torulus*, a tuft of hair) *a b c.*

P. entire, plane to infundibuliform, compactly fleshy, even, smooth, brownish-yellow, buff or rufescent-livid, sometimes minutely sienna-scaly. *St.* solid, equal, oblique, grey, violaceous or buff-downy. *G.* deeply decurrent, subdistant, reddish, tan or yellowish.

Cæspitose. Old trunks, stumps, ash, birch, willow, oak; frequent. May–Aug. 3¾ × 1½ × ¾ in.

1443a. P. rudis Fr. (from its rough appearance; *rudis*, rough) *a b*.

P. becoming corky, thin, unequal, bristling with hairs, ochreous-fawn or reddish, lilac shaded; marg. incurved. *St*. short, unequal, shaggy-velvety, ochreous-fawn, hairs lilac. *G*. very decurrent, very crowded, narrow, whitish-pink, then pale ochreous-fawn, sometimes lilac shaded. *Flesh* white.

Cæspitose. Odour pleasant of anise. Trunks, beech. May. $2\frac{7}{8} \times 2\frac{1}{2} \times \frac{1}{2}$ in.

1444. P. stypticus Fr. (from its astringent taste; *stypticus*, astringent) *a b c*.

P. arid, thin, reniform, sometimes minutely white-furfuraceous scaly, tan or cinnamon, becoming pale. *St*. attenuate downwards, pruinose, paler than G. *G*. adnate, thin, crowded, salmon or ochreous-salmon to cinnamon.

Gregarious, cæspitose, usually densely imbricate. Taste hot, pungent. On decaying stumps, trunks, dead trees, etc., fir, alder, beech, oak; common. Feb.–Dec. $1\frac{1}{2} \times \frac{3}{4} \times \frac{1}{4}$ in. Becoming wholly rich orange when dry.

1445. P. farinaceus Fr. (from the scurfy pileus; *farina*, meal) *a b c*.

P. somewhat coriaceous, flexuous, reniform, cinnamon-umber with pale grey scurf. *St*. colour as P. *G*. adnexed, distinct, paler than P. or yellow ochre. *Flesh* whitish-grey.

Trunks, branches, fir. Dec. $1 \times \frac{1}{8} \times \frac{1}{16}$ in. Not unlike **1444** or **439** in appearance. Var. *albidotomentosus* Cooke & Mass. *P*. clad with short velvety-whitish down. *St*. short or absent. Often in imbricate tufts. P. $\frac{2}{3}$–$\frac{1}{2}$ in. in diam.

1446. P. patellaris Fr. (from the shape, like a small dish, *patella*) *a*.

P. sessile, resupinate, coriaceous, at first viscid, umber or dark sepia to buff; marg. white. *St*. obsolete or rudimentary. *G*. concurrent to a subcentral point, arid, yellowish to somewhat pale cinnamon.

Branches, cherry, beech. Mar. $\frac{3}{4}$ in. in diam.

1447. P. Stevensonii B. & Br. (after the Rev. John Stevenson).

P. spathulate, olivaceous light yellow. *St*. attenuate downwards, convex, slightly hispid, golden. *G*. narrow, entire.

Oak. Sept.–Oct.

LV. **XEROTUS** Fr.

(From the dry substance and ear-like shape; Gr. *xeros*, dry, *ous*, an ear.)

Veil obsolete. *Pileus* coriaceo-membranous. *Stem* central, continuous with the hymenophore, which descends into the trama. *Gills* adnato-decurrent, coriaceous, broadly fold-like, dichotomous, edge entire, obtuse. *Spores* white. (Fig. 69.)

Rigid, persistent, analogous with *Cantharellus*, but differing in its leathery substance.

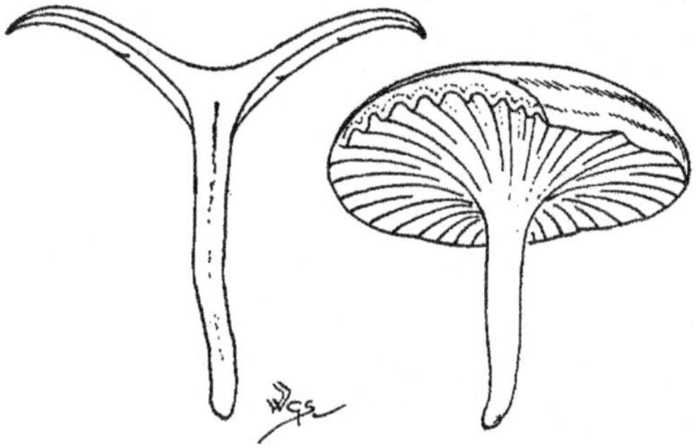

Fig. 69.—*Xerotus degener* Fr. in section and entire with portion removed to show the gills. Natural size.

1448. X. degener Fr. (from its ill-developed condition in comparison with other species of *Xerotus*, *degener*, degenerate) *a b*.

P. thin, very tough, plano-infundibuliform, somewhat zoned, striate when moist, flocculose when dry, brown-grey to grey, cinnamon, salmon or pale yellowish-brown. *St.* solid to stuffed, attenuate downwards. *G.* very distant, whitish-grey, yellowish or salmon-white.

Peaty ground, amongst short grass in sterile gravelly places. Jan. $1\frac{3}{4} \times 1\frac{1}{4} \times \frac{1}{8}$ in.

LVI. TROGIA Fr.

(After Jacob Gabriel Trog, Swedish botanist.)

Pileus membranous, soft, flaccid, arid, persistent, texture fibrillose. *Stem* lateral or none. *Gills* fold-like, venose, edge in the only

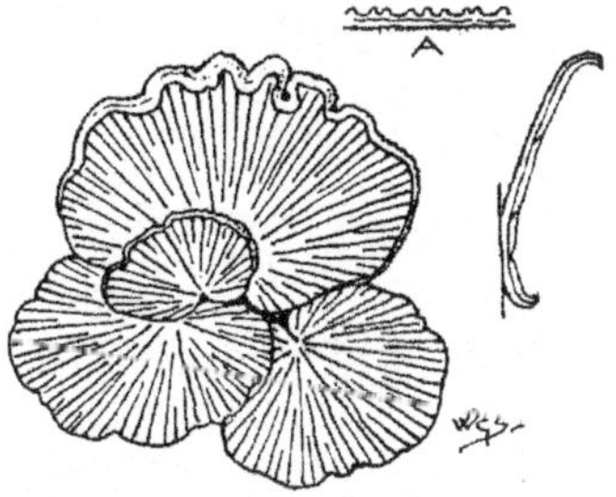

Fig. 70.—*Trogia crispa* Fr., entire and in section. Natural size. A, section across gills.

European species obtuse, crisped, not channelled as in non-European species. *Spores* white. (Fig. 70.)

Growing on wood, becoming dry and reviving with moisture.

1449. T. crispa Fr. (from the crisped gills) *a b c.*

P. cup-shaped, expanded, resupinate, light-yellowish-rufescent, zoned rufous; marg. whitish, sometimes almost white. *St.* obsolete or rudimentary. *G.* vein-like, dichotomous, narrow, white to steel-grey.

Gregarious, imbricate. Logs, branches, twigs, beech, birch; uncommon. Jan.–Dec. Diam. 1⅛ in. Sometimes almost wholly white.

LVII. SCHIZOPHYLLUM Fr.

(From the gills, split longitudinally at the edge; Gr. *schizo*, to split, *phullon*, a leaf.)

Veil obsolete. *Pileus* thin, arid. *Stem* lateral or none. *Gills* coriaceous, channelled longitudinally with edges revolute. *Spores* white. (Fig. 71.)

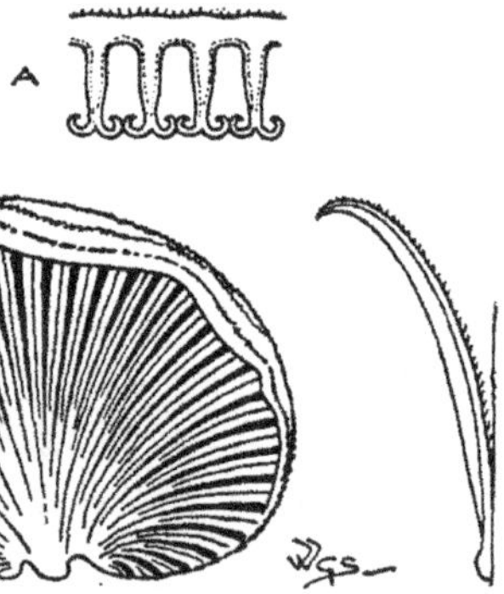

Fig. 71.—*Schizophyllum commune* Fr., entire and in section. Natural size. A, section across gills. × 10.

1450. S. commune Fr. (from its being common in certain countries; *communis*, common) *a b c.*

P. arid, at first cup-like, then expanded, reniform or excentric, resupinate, downy, white or umber-white, commonly zoned greyish. *St.* rudimentary or none. *G.* splitting, the edges revolute, fuscous-grey to purplish, cinereous or clay-white to tan-salmon.

Said to be edible. Sold in a dry state in China for soup; 800 lbs were exported from Auckland, New Zealand, to China in 1872. Dead wood, rotten trunks, logs, water-butts, beer-casks, wood in greenhouses, horse-chestnut, beech, alder, dry hay in silos; rare in Britain. May-Jan. Diam. 2 in.

LVIII. LENZITES Fr.

(After Harold Othmar Lenz, German botanist.)

Veil obsolete. *Pileus* corky or coriaceous, texture arid, floccose, dimidiate, sessile. *Stem* lateral or none. *Gills* coriaceous, firm,

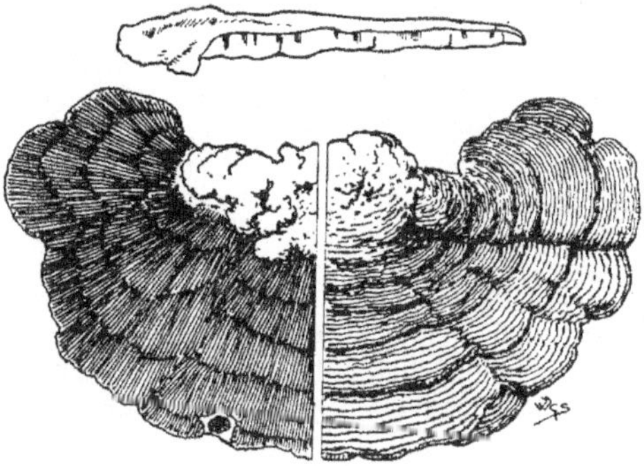

Fig. 72.—*Lenzites betulina* Fr., entire and in section. Lower surface on left, upper on right. One-third natural size.

sometimes simple and unequal, sometimes anastomosing and forming pores behind, edge obtuse to acute; trama floccose, similar to the pileus. (Fig. 72.)

Growing on wood, persistent. Allied most nearly to *Trametes* and *Dædalea* amongst the *Polyporaceæ*. Species **1451–1455**

On wood of deciduous trees. **1451, 1452**

On coniferous wood. **1453—1455**

1451. L. betulina Fr. (from its frequent habitat, birch trunks; *betula*, birch) *a b c*.

P. firm, rigid, tomentoso-woolly, flat, lobed, white, greyish, buff or umber-whitish; marg. and z. darker. *G.* reaching the base, simple, branched or anastomosing, edge acute. *Flesh* white.

Often imbricate. Stumps, logs, gate-posts, birch, oak, bird-cherry, alder; perennial, common. Jan.–Dec. Diam. 8¼ in. When young small, resupinate, with gills radiating from a central point. Often green with unicellular *Algæ*. Surface and coloration of pileus like that of **2001**.

1452. L. flaccida Fr. (from the flaccid substance) *a b c*.

P. thin, easily bent, flat, lobed, whitish to dingy, zoned same colour. *G.* reaching the base or an excentric point, broad, crowded, simple or branched towards base, not anastomosing, white to salmon. *Flesh* white.

Stumps, beech. Mar.–Jan. Diam. 4 in. Pileus often green as in **1451**.

1453. L. sæpiaria Fr. (from the habitat, fences; *sæpis*, a hedge) *a b c.*
P. thin, flat, lobed, strigoso-tomentose to squamulose, zoned; tawny-yellow to brown, black when old. *G.* reaching to and anastomosing at the base, rigid, firm, yellowish to umber. *Flesh* tawny to dark sienna.

Often imbricate. Trunks, fences, pine. Jan.–Dec. Diam. 5 in.

1454. L. abietina Fr. (from its habitat, firs, *abies*) *a b c.*
P. thin, flat, tomentose, becoming smooth, umber or yellowish-rich-brown or deep umber-sienna, zoned dark umber; marg. bright sienna-yellow, or wholly vinous. *G.* reaching the base, distant, simple, unequal, not anastomosing but torn into teeth at the base, salmon or yellowish-brown.

Imbricate. Chiefly dressed fir, deal, trunks, branches. Oct. Diam. 3¾ in. Sometimes remaining resupinate.

1455. L. heteromorpha Fr. (from its varied shapes; Gr. *heteros*, different, *morphe*, shape) *a b c.*
P. gibbous, rugose, thin to thick, yellowish- or buff-white, or colour of washleather. *G.* reaching the base, branched, crowded, often broken into teeth, sometimes effused, colour as P.

Imbricate. Fences, sticks, chips, bark, pine. Oct. Diam. 3 in., sometimes effused to 7 in. or more.

FAM. II. POLYPORACEÆ.

Hymenium inferior, in resupinate species superior, consisting of tubes with poriform orifices which are round or angular, sometimes sinuous or torn.

Fleshy, coriaceous or woody fungi, intermediate between *Agaricaceæ* and *Hydnaceæ*, connected with the former by *Trametes* and *Dædalea*, and with the latter by *Fistulina*. Species **1456—1686**

The sequence of the genera and species is almost identical with the latest views of Fries as expressed in his *Hymenomycetes Europæi*, 1874. An exception is made in the genus *Polyporus*, which is here broken up into four genera—*Polyporus*, *Fomes*, *Polystictus* and *Poria*, as given by Cooke in 1885 and 1886, in his *Præcursores ad Monographia Polypororum*, in *Grevillea*, xiii., pp. 80 and 114, xiv., pp. 17, 77 and 109, and xv., p. 19, with additions and corrections on p. 50. Cooke based his work on the *Novæ Symbolæ Mycologicæ* of Fries, published in 1851, but in this Fries only suggested that one section of *Polyporus* should be elevated to generic rank, viz. *Polystictus*. Saccardo, in his *Sylloge Fungorum*, vii., 1888, immediately adopted Cooke's views. It is in deference to the laborious work of Cooke that the old genus *Polyporus* is here broken up into four.

Key to the Genera.

Terrestrial.		
Tubes easily separable from the hymenophore. Stem central	59	Boletus.
Tubes not separable, very large with transverse dissepiments. Stem central	60	Strobilomyces.
Mostly growing on trunks, stumps or wood.		
Stipitate to sessile. Tubes small, without dissepiments. Substance fleshy, hard, tough or somewhat soft	62	Polyporus.
Sessile.		
Substance fleshy, juicy	61	Fistulina.
Substance woody	63	Fomes.
Substance coriaceous, spongy or membranous.		
Pores round	64	Polystictus.
Pores oval or elongate	66	Trametes.
Pores sinuous	67	Dædalea.
Sessile to resupinate. Substance cartilagino-soft to subgelatinous	68	Merulius.
Resupinate.		
Substance thin, woody or waxy	65	Poria.
Hymenophore covered with granules	69	Porothelium.

LIX. BOLETUS Dill.

(From the clod-like shape of the pileus; Gr. *bolos*, a clod.)

Veil universal, forming an annulus, perfect to imperfect or obsolete. *Pileus* fleshy. *Stem* central, fleshy, continuous and homo-

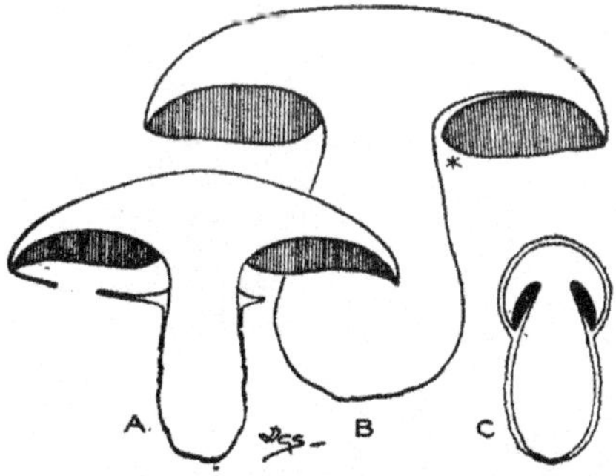

Fig. 73.—A, section of *Boletus luteus* L., showing gelatinous membranous veil. B, section of *B. edulis* Bull., showing at * tubes separating from hymenophore. C, section of *B. cyanescens* Bull., young example showing universal flocco-scaly veil. One-third natural size.

geneous with the hymenophore, annulate to simple. *Tubes* distinct from the hymenophore, connected in a stratum, the surface of which is dotted with their orifices or pores, easily separating from the hymenophore and from one another; the mouths of the tubes or pores are round or angular, in the subgenus *Gyrodon* sinuous or gyroso-plicate; trama none. *Spores* usually fusiform, rarely oval or roundish, white, rosy, yellowish or brown. (Fig. 73.)

Terrestrial, fleshy, putrescent fungi; mostly edible, a few poisonous. Species **1456—1515**

SUBGENUS I. **EUBOLETUS.** *Pores* round or angular, not sinuous or gyroso-plicate. *Trama* not descending between the tubes. **1456—1513**

EUCHROÆ. *Tubes* usually yellow, not white or grey.

a. Viscipelles. Stem solid, not bulbous or reticulate, except **1458.** *Tubes* adnate, decurrent in **1457, 1467** and **1468.** *Pores* rarely sinuate, of one colour. *Spores* usually yellowish. **1456—1468**

b. Subtomentosæ. Pileus villous when young, rarely at length becoming smooth. *Stem* not bulbous or reticulate; here and there rugose or striate. *Flesh* rarely changing colour. *Tubes* adnate, of one colour. **1469—1477**

c. Subpruinosæ. Pileus pruinose to smooth. *Stem* equal, even, not bulbous or reticulate. *Tubes* adnate, yellowish. **1478—1481**

d. Calopodes. Stem stout, at first bulbous, reticulate; reticulations usually red. **1482—1488**

e. Edules. Stem stout, bulbous, not reticulate or dotted red except **1489, 1490, 1491** and **1493.** *Tubes* somewhat free, adnate in **1494.** *Pores* not red. *Flesh* scarcely changing colour. *Taste* pleasant. **1489—1497**

f. Luridæ. Pileus pulvinate, compact then soft. *Stem* stout at first, curt and bulbous, then elongate and clavate to nearly equal, somewhat reticulate or dotted. *Tubes* rounded-free. *Pores* red to red-brown. *Flesh* juicy. Suspected poisonous. Growing chiefly in woods of deciduous trees. **1498—1500**

TEPHROLEUCÆ. *Tubes* at first white or grey.

g. Favosæ. Tubes large, angular, unequal, often shortened near the stem, adnate, not rounded-free. *Spores* brown. **1501—1504**

h. Versipelles. Tubes free, ventricose. *Pores* minute, round, equal. *Spores* ferruginous. **1505—1508**

i. Hyporhodiæ. Tubes whitish to pale salmon. *Spores* rosy to white flesh-colour. **1509, 1510**

j. Cariosæ. Stem not reticulate, stuffed to hollow. *Tubes* white to light yellowish. *Pores* minute, round. *Spores* white. **1511—1513**

SUBGENUS 2. **GYRODON.** *Tubes* short, somewhat adnate. *Pores* sinuous or gyroso-plicate. **1514, 1515**

SUBGENUS 3. **BOLETINUS.** *Trama* descending somewhat between the tubes, the hymenophore is therefore not smooth after their removal; in this respect *Boletinus* resembles *Strobilomyces*. The relationship between *Boletus* and *Boletinus* is similar to that between *Polyporus* and *Trametes*. **1515a**

SUBGENUS 1. **EUBOLETUS.**

(From being typical examples of the genus; Gr. *eu*, well.)

EUCHROÆ.

a. Viscipelles.

1456. B. luteus L. (from the yellow tubes; *luteus*, yellow) *a b c.*
P. gibbous, pulvinate to convex, deep vinous brown, covered with dark purplish gluten which is at first continuous over the tubes, forming a V. *St.* equal, dull light yellowish. *A.* large, membranous, white to dark purplish. *T.* yellow or brownish-yellow. *Po.* minute.
Edible. Woods, pine; common. July–Nov. 5 × 2¾ × ⅞ in.

1457. B. elegans Schum. (from its elegance) *a b c.*
P. convex, dark rich yellow, shaded orange. *St.* sulphur to orange above A., darker below. *A.* broad, fugacious, sulphur. *T.* decurrent, sulphur to bright yellow. *Po.* minute.
Woods, chiefly larch, oak, pine; common. June–Oct. 4⅛ × 2¾ × ⅞ in.

1458. B. flavus With. (from the yellow pileus; *flavus*, light yellow) *a b c.*
P. convex or pulvinate, full luminous yellow, covered when young with claret-brown gluten. *St.* equal, reticulate, yellow above, brownish-yellow below. *A.* membranous, fugacious, white to yellow-white. *T.* adnate, bright yellow or brownish-yellow. *Po.* larger than in **1456** and **1457.**
Woods, mixed; common. Aug.–Nov. 4¼ × 3½ × ⅞ in.

1459. B. flavidus Fr. (from the yellowish pileus; *flavidus*, light yellowish) *a b c.*
P. convex, obtuse, brownish-yellow or pale yellowish-buff. *St.* equal, whitish-yellow or straw-colour, granular above the brownish viscid A. *T.* adnate to slightly decurrent, yellow, somewhat sulphur. *Po.* angular. *Flesh* pale ochre.
Woods, pine. July–Oct. 2⅛ × 2¼ × ¼ in.

1460. B. collinitus Fr. (from the gluten-besmeared pileus; *collino*, to besmear) *a b c.*
P. pulvinate to expanded and slightly depressed, brownish-buff, sometimes pale, covered with rich vinous brown gluten. *St.* attenuate downwards, rooting, white, becoming brownish. *T.* adnate to subdecurrent, pallid to yellow. *Flesh* white.
Solitary. Edible. Woods, fir. Nov. 3⅝ × 2½ × 1¼ in. When young resembles **1456.**

1461. B. granulatus L. (from the granulate pores and stem) *a b c*.

P. convex, obtuse, luminous- or cinnamon-yellow, covered with claret-brown, vinous or rich sienna gluten. *St*. white-sulphur above, brownish below, scabrous with honey-coloured granules. *T*. pale sulphur to orange. *Po*. simple, exuding a thin gummy juice which dries in granules. *Flesh* pale sulphur.

Commonly gregarious, sometimes cæspitose. Odour rank, heavy, disagreeable, like that of **410**. Woods, fir; rare. July–Nov. $3\frac{3}{4} \times 2\frac{1}{4} \times \frac{7}{8}$ in. Sometimes springing from yellow mycelium. Much eaten by squirrels.

1462. B. tenuipes Mass. (from the slender stem; *tener*, slender, *pes*, a foot) *a b*.

P. convexo-pulvinate, smooth, becoming minutely fibrillose, slightly to very viscid, red- or rose-purplish over yellow, or red-buff, shaded Indian red. *St*. attenuate downwards, somewhat striate above, slightly to very viscid, yellow, vinous in mid., becoming rufescent, or pale yellow above to gradually vinous below. *T*. large, angular, bright yellow. *Flesh* pale vinous under pellicle, white in mid., pale buff-vinous in St.

Taste mild. Woods. Oct. $1\frac{3}{4} \times 2\frac{3}{4} \times \frac{3}{8}$ in.

1463. B. bovinus L. (from the buff-reddish colour of the pileus, as of the ox, *bos*) *a b c*.

P. convex to expanded, buff, rose-buff or orange-buff, membrane of P., involute and overlapping the tubes. *St*. equal, even, ochre-rose or pale buff, rhubarb tinted at base. *T*. decurrent, ochre, olive-ochre or rose- or yellow-buff. *Po*. compound. *Flesh* pale buff, sometimes changing to cobalt.

Gregarious, subcæspitose. Taste mild; odour pleasant. Woods, chiefly pine; uncommon. Sept.–Nov. $4\frac{1}{2} \times 2\frac{3}{4} \times \frac{1}{2}$ in.

1463a. B. pusio Howse ex B. & Br. (from its small size; *pusio*, a little boy) *a*.

P. hemispherical, pulvinate, pulverulent, reddish-buff. *St*. slightly thickened downwards, plicate, pulverulent, dull yellowish-white, white above, becoming blackish. *T*. dull sulphur-white.

1464. B. badius Fr. (from the colour; *badius*, bay-brown) *a b c*.

P. pulvinate to flat, villoso-viscid, claret-brown, umber or dark Indian-red-brown. *St*. equal or subclavate, pruinate, minutely granular, lighter than P. and streaky. *T*. adnate to slightly sinuate, whitish dull olive. *Po*. angular. *Flesh* rose under pellicle of P., cobalt over T.

Single or cæspitose. Woods, chiefly pine; frequent. Aug.–Nov. $5 \times 4 \times 1\frac{3}{8}$ in.

1465. B. paludosus Mass. (from growing in bogs, *paludes*) *a b*.

P. convex to flat, glabrous, dark burnt-sienna or vinous red-sienna; mid. darker, paler when dry. *St*. attenuate downwards, smooth, paler than P., or burnt-sienna with crimson, or dull dark sienna. *T.* adnate to subdecurrent, yellow to olive-green or whitish-green. *Po.* large, angular, compound. *Flesh* unchanging, salmon-white.

Gregarious, not cæspitose. Bogs, amongst *Sphagnum*. Sept. 4¼ × 5 × ½ in. Resembles a slender **1464**, but flesh does not change colour.

1466. B. rutilus Fr. (from the reddish pileus; *rutilus*, golden-red) *a*.
P. pulvinate, glabrous, brownish-red or umber, not dark. *St.* equal, glabrous, sulphur above, pale claret-brown below. *T.* sinuato-adnate, sulphur. *Po.* minute.

Woods, oak. Oct. 1⅝ × 1⅝ × ¼ in. Must not be confounded with **1472** or **1474**.

1467. B. aurantiporus Howse ex Cooke (from the golden-yellow pores; *aurantius*, orange) *a b*.
P. convex, expanded, sometimes umbonate, ferruginous, orange-sienna, claret-brown or buff. *St.* equal, orange above, with large purple-crimson reticulations, sulphur and pale yellow below. *T.* deeply decurrent, yellow to bright orange, becoming brownish when bruised. *Po.* broad angular. *Flesh* usually pale rose, sometimes pale yellowish, changing to pale orange-red.

Under or near trees, yews. Aug.–Oct. 3½ × 3½ × ⅝ in. Howse considered this plant to be allied to **1469**.

1468. B. piperatus Bull. (from the peppery taste; *piper*, pepper) *a b c*.
P. pulvinate to expanded, dull ochreous-orange or yellow-ochre. *St.* equal or attenuate downwards, sienna-orange. *T.* decurrent, sienna with ochre and vermilion, or rich sienna-scarlet. *Flesh* deep sulphur, orange in stem.

Sometimes cæspitose. Poisonous. Taste very acrid. Woods, mixed, under limes; frequent. Aug.–Nov. 2⅜ × 2 × $\frac{5}{16}$ in. Often small. Springing from yellow mycelium.

b. Subtomentosæ.

1469. B. variegatus Swartz (from the different colours of stem, pileus and tubes) *a b c*.
P. pulvinate, yellowish-sienna or yellowish-brown, clouded sienna or vinous; marg. extended beyond T., inrolled. *St.* equal, bright yellow to dull ochre, shaded orange, burnt-sienna or vinous. *T.* yellow, or orange- or olive-brown, not dark. *Po.* minute. *Flesh* yellowish, changing sparingly or wholly to blue, sometimes unchanging.

Tasteless; odour unpleasant. Woods, chiefly pine, heaths; frequent. July–Nov. 5⅝ × 2⅞ × 1⅛ in.

1470. B. sulphureus Fr. (from the sulphur-colour) *a b c*. Wholly sulphur-colour.
P. hemispherical to expanded-obtuse. *St.* ventricose, with rooting red base. *T.* adnato-decurrent, at length greenish. *Po.* minute, compound. *Flesh* becoming blue.

Cæspitose, sometimes in clusters of fifty or more. Tasteless. Sawdust, stumps. Sept.–Nov. 4 × 2⅛ × 1⅝ in. Sometimes changing to reddish where touched. Mycelium woolly, matting sawdust, sulphur. Pileus sometimes 9 in. in diam.

1471. B. striæpes Secr. (from the striate stem; *stria*, a line, *pes*, a foot) *a b*.

P. pulvinate to expanded, orange-buff, shaded umber; marg. paler or whitish. *St.* slightly ventricose, velvety, sulphur-white above, umber below, broadly striate with small brown dots. *T.* yellow, yellow-greenish, sulphur or apricot-sulphur. *Po.* minute. *Flesh* white, or faintly shaded pale yellowish or brownish, sparingly changing to blue.

Woods; rare. Oct. $2\frac{5}{8} \times 2\frac{1}{2} \times 1$ in.

1472. B. chrysenteron With. (from the golden-yellow flesh; Gr. *chrusos*, gold, *entera*, inward parts) *a b c*.

P. convex to plane or slightly depressed, yellowish, umber or crimson-brownish over tan, often cracked, interstices crimson-red. *St.* equal, fibroso-striate, tan-white or buff, speckled crimson or sulphur above, spotted and streaked below, base umber. *T.* sulphur to yellow-olive. *Po.* somewhat large, angular. *Flesh* yellow-white, pale rose under pellicle of P., pale blue near T.

Edible. Woods, mixed, beech, hedges, open ground; common. July–Nov. $3\frac{1}{4} \times 2\frac{5}{8} \times \frac{1}{2}$ in. Var. *nanus* Mass. *P.* olive. *St.* yellow above, crimson below. *Flesh* changing to red when cut. P. 1 in. in diam.

1473. B. sanguineus With. (from the blood-red pileus; *sanguis*, blood) *a b*.

P. pulvinate, dry, almost velvety. *St.* equal, rose above, sometimes reticulated crimson, tan or orange below, clouded crimson. *T.* adnate to subdecurrent, dull ochre to bright or olive-yellow. *Flesh* yellowish-white, changing to blue.

Woods; rare. July–Sept. $4 \times 3\frac{1}{4} \times \frac{7}{8}$ in. Pileus sometimes 9 in. in diam., sometimes very small.

1474. B. subtomentosus L. (from the somewhat tomentose pileus) *a b c*.

P. pulvinato-expanded, umber or yellowish-umber, sometimes with a red line at marg., often cracked, interstices yellow. *St.* equal, ribbed-sulcate, slightly scabrous, bright yellow, sometimes clouded crimson. *T.* rounded-adnate, bright yellow, shaded pale olive-umber. *Po.* large, angular. *Flesh* whitish-yellow, sparingly changing to blue.

Woods, mixed, oak, beech, sometimes from interior of beech-nuts; common. July–Nov. $3\frac{3}{4} \times 3\frac{1}{4} \times \frac{5}{8}$ in. Pileus sometimes $7\frac{1}{2}$ in. in diam. Var. *radicans* Mass. *P.* olive-ochre. *St.* fusiform, colour as P. *T.* colour as P. Taste insipid to bitter.

1475. B. spadiceus Fr. (from the date-brown colour of the pileus; *spadix*, a palm branch) *a b*.

P. pulvinate, rich yellowish-sienna, clouded red-brown or vinous. *St.* equal to clavate, flocculoso-furfuraceous, whitish- or pale luminous-yellow, olive-shaded above, clouded umber and vinous below. *T.* sinuato-adnate, pale yellow-olive, shaded greenish. *Po.* minute, somewhat round. *Flesh* pale sulphur or buff-white, with a faint blue line over T.

Taste insipid or pleasant, with a peculiar lasting acid flavour. Woods, mixed. July–Nov. $4\frac{5}{8} \times 2\frac{3}{4} \times 1\frac{3}{4}$ in.

1476. B. radicans Pers. (from the rooting stem; *radix*, a root) *a b.*
P. pulvinate, yellowish-umber to umber or olive, membrane of P. overlapping T. and involute. *St.* usually attenuate downwards, pale primrose to apricot-yellow, and dark ochre, pruinate, rose and sienna clouded where touched. *T.* adnate, colour as St. *Po.* unequal, large. *Flesh* yellowish-white, sometimes changing to pale blue or slate.

Taste somewhat bitter. Woods. Sept. $3\frac{1}{2} \times 2\frac{1}{4} \times \frac{5}{8}$ in. Often confused with **1472.**

1477. B. rubinus W. G. Sm. (from the ruby-coloured tubes) *a b c.*
P. pulvinate to flat, whitish orange-umber or yellow-fuscous. *St.* equal or slightly attenuate downwards, yellow clouded crimson. *T.* adnato-decurrent, carmine. *Po.* compound. *Flesh* unchanging, vivid yellow.

Taste mild. Woods, roadsides, under trees, oak, beech. Aug.–Sept. $2\frac{3}{4} \times 1\frac{1}{2} \times \frac{3}{4}$ in.

c. Subpruinosæ.

1478. B. versicolor Rostk. (from the change of colour in the flesh) *a b.*
P. pulvinate to convexo-plane, bright deep crimson-purple. *St.* slightly pruinose, smooth, colour as P. or orange-brown, lighter above and below, sometimes sulphur above. *T.* adnato-rounded, orange. *Po.* large, angular. *Flesh* whitish-yellow, changing to cobalt and vinous.

Said to be edible. Taste slightly disagreeable or none; odour strong and rank, like **2086.** Woods, beech, gardens. Aug.–Sept. $2\frac{3}{4} \times 2\frac{1}{4} \times \frac{1}{4}$ in.

1479. B. pruinatus Fr. (from the pruinose pileus) *a b.*
P. convex to flattish, rigid, yellowish vinous-umber. *St.* somewhat ventricose or clavate, even, smooth, yellow, finely lined with crimson. *T.* yellow. *Po.* minute, round. *Flesh* whitish, sparingly changing to blue.

Grassy places. Autumn. $2\frac{5}{8} \times 2 \times \frac{5}{8}$ in.

1480. B. purpurascens Rostk. (from the purple pileus and stem) *a.*
P. convex to plane, dry, glabrous, deep brownish-purple or deep purple-crimson. *St.* attenuate downwards, rooting, even, glabrous, colour as P., sometimes yellowish above and below. *T.* dull yellow. *Po.* minute. *Flesh* dull grey or yellowish, streaked or marbled.

Woods, pine. Autumn. $3\frac{1}{8} \times 2 \times \frac{3}{4}$ in.

1481. B. parasiticus Bull. (from the parasitic habit) *a b c.*
P. obtuse, plane, silky to smooth, yellowish-brown. *St.* slightly attenuate downwards, colour as P., or lighter, with pale vinous-sienna streaks. *T.* adnate to subdecurrent, colour as St.,

sienna-shaded. *Po.* compound. *Flesh* yellowish-white or pale vinous, orange-brownish at base of St.

Parasitic on **2086** and **2087.** Often overlooked, as the colour of P. agrees with that of the peridium of host. Sept.–Oct. $2\frac{1}{4} \times 2 \times \frac{1}{2}$ in.

d. Calopodes.

1482. B. appendiculatus Schæff. (from the membrane of the pileus overlapping at the margin) *a.*

P. pulvinate, expanded to plane, pubescent, crimson, bay-brown-rose, indian-red or vinous-buff. *St.* subventricose, somewhat rooting, yellowish-white, buff below, sometimes crimson in middle. *T.* adnate, sulphur-greenish, clouded pale indigo where touched. *Po.* minute, angular. *Flesh* pale sulphur, becoming pale blue.

Edible. Woods; uncommon. Oct. $4\frac{1}{8} \times 3\frac{7}{8} \times 1$ in.

1483. B. variecolor B. & Br. (from the varied colours) *a b.*

P. pulvinate, subtomentose, olivaceous; marg. involute. *St.* clavato-bulbous, pubescent and rufescent above, yellowish downwards. *T.* rounded-free, yellow. *Po.* minute. *Flesh* pale yellowish, marbled, dark purple under membrane of P.

Woods; uncommon. Aug.–Sept. $1\frac{5}{8} \times 2\frac{3}{8} \times \frac{1}{2}$ in.

1484. B. calopus Pers. (from the beautiful stem; Gr. *kalos*, beautiful, *pous*, a foot) *a b c.*

P. globose to pulvinate, subtomentose, umber, rosy-umber or olive-umber. *St.* ventricose or nearly equal, somewhat rooting, reticulate and carmine above, dark umber below, or wholly crimson or yellow above, crimson in middle, yellow and brown at base. *T.* rounded-adnate, whitish-orange. *Po.* minute, angular. *Flesh* white, changing to bright blue, or variegated blue, green, carmine and umber, crimson at base.

Taste bitter. Woods, mixed; frequent. July–Nov. $4\frac{1}{4} \times 2\frac{3}{4} \times 1\frac{1}{4}$ in.

1485. B. olivaceus Schæff. (from the olive-coloured pileus) *a b.*

P. convex, even, becoming smooth, yellowish olive-umber to olive. *St.* subclavate, salmon, sometimes white at apex, yellow-brown below, with salmon-pink or rose granular streakings or reticulations. *T.* adnate, greenish-yellow. *Po.* minute, unequal. *Flesh* salmon-umber-whitish, sometimes changing to blue.

Almost tasteless; odour somewhat strong and disagreeable. Woods, beech; uncommon. Aug.–Sept. $2\frac{3}{4} \times 2 \times \frac{3}{4}$ in.

1486. B. regius Kromb. (from the regal aspect) *a b.*

P. globose to pulvinate, minutely tomentose, bright rose, pale purple, pale claret, buff or vinous-yellow-umber. *St.* subventricose to almost equal, pale or bright yellow, dull purple, red-crimson, orange or olive at base. *T.* rounded-adnate, deep golden-yellow. *Po.* minute. *Flesh* yellow and pale salmon, sometimes changing to pale blue.

Solitary or gregarious. Edible. Open places in woods. Aug. 5×5×1¾ in.

1487. B. crassus Mass. (from the stout stem ; *crassus*, thick) *a b.*

P. convex to expanded, irregularly waved, densely tomentose, white, then pale yellow-brown or fawn colour. *St.* at first almost globose, paler than P., white-reticulate above, base almost white. *T.* free, or nearly so, white then primrose. *Flesh* white, then pale primrose or pale biscuit.

Taste sweet, nutty ; odour strong, acid. Under trees in pastures, beech. Aug. 4 × 4 × 2⅛ in. Compare **1490** and **1496**.

1488. B. pachypus Fr. (from the thick stem ; Gr. *pachus*, thick, *pous*, a foot) *a b c.*

P. pulvinate, pale warm umber, umber or yellowish-umber or biscuit-colour. *St.* equal to clavate or bulbous, pale salmon-rose above and below, rose in middle or yellow above, and umber and whitish-yellow below. *T.* rounded-adnate, bright yellow to green. *Flesh* whitish, pale vinous under membrane of pileus and at base of stem, changing sparingly and slowly to blue or unchanging.

Taste not unpleasant. Woods, chiefly pine. Sept.–Oct. 4¼ × 3¼ × 1¼ in.

e. Edules.

1489. B. candicans Fr. (from the white pileus and stem ; *candicans*, whitish, shining-white) *a b c.*

P. convex, brownish- or yellowish-white, or almost white. *St.* obese, clavate or bulbous, reticulate to smooth, ochre- or primrose-white, pale umber below. *T.* adnato- or adnexo-rounded, sulphur to primrose. *Flesh* ochre-white, changing to bright blue or indigo.

Solitary or 2–3 connate. Open places, under oaks, in company with **1490**. June–Oct. 5½ × 3½ × 1½ in. This is *B. elephantinus* Schæff.

1490. B. edulis Bull. (from its edible properties) *a b c.*

P. pulvinate to expanded and flat, pale umber, yellowish- or purplish-umber or tan, rarely sooty ; marg. usually lighter or white. *St.* obese to clavate and equal, pale umber-white, buff-white or white above and tan below, white-reticulate above. *T.* adnexo-rounded, white to greenish-yellow. *Flesh* unchanging white, sometimes salmon-white under pellicle of P.

Woods, especially beech ; common. June–Nov. 8½ × 5 × 1⅞ in. The cèpe of French markets, the *Fungo porcino*, *Porcino Bolè porcin* or *Ferrè* of Italian markets. The "suillos" of Mart. Ep. iii. 60. Sometimes attacked by a yellow parasitic mould, *Hypomyces chrysospermus*, and then not edible. Var. *lævipes* Mass. *St.* even, white. There is also a wholly white var.

1491. B. æreus Bull. (from the copper-coloured stem ; *æs*, copper) *a b c.*

P. pulvinate, smooth, umber, somewhat olive or blackish. *St.* subclavate to equal, somewhat reticulate, pale buff or yellow above, rosy-umber below. *T.* sulphur-primrose. *Po.* minute. *Flesh* pale sulphur, vinous under pellicle of P.

Woods. Autumn. 4⅝ × 4¼ × 1½ in.

1492. **B. carnosus** Rostk. (from the fleshy pileus ; *caro*, flesh) *a*.

P. pulvinate, smooth, crimson-brown. *St.* bright yellow above, orange below or warm brown, somewhat finely streaked red and crimson. *T.* ochre-greenish or dark yellow. *Po.* somewhat large, angular. *Flesh* pale ochre.

Woods. Autumn. 4 × 2½ × 1⅛ in.

1493. **B. vaccinus** Fr. (from the reddish colour of the pileus ; *vacca*, a cow) *a b*.

P. convex, expanded, minutely tomentose, rich orange-umber, brighter at marg. or deep chestnut. *St.* yellowish, reddish or brownish, not dark. *T.* adnato-rounded, white to sulphur and bright yellow. *Po.* minute. *Flesh* white to pale sienna-yellowish, vinous under pellicle of P., becoming pale blue.

Woods, beech. Autumn. 2⅝ × 2 × ¾ in.

1494. **B. Rostkovii** Fr. (after Friedrich Wilhelm Rostkovius) *a b c*.

P. convex to plano-expanded, minutely tomentose, ochre, orange, or dull olive-brown or rufous. *St.* attenuate downwards, obconic, warm umber, buff, or pale reddish-yellow. *T.* adnate, yellow, sulphur or yellow-green. *Po.* irregular, compound. *Flesh* buff- or vinous-white, becoming reddish-brown under pellicle of P., here and there blue.

Under trees, beech. Aug. 3⅝ × 1¼ × 1 in.

1495. **B. fragrans** Vitt. (from the pleasant odour) *a*.

P. pulvinate, subtomentose, yellowish rosy-umber. *St.* somewhat clavate, sometimes fusiform-rooted, yellowish, pale vinous-clouded below. *T.* cream-yellowish. *Po.* minute, round. *Flesh* buff- or yellow-whitish, changing to pale blue.

Single or cæspitose. Woods, oak. Sept.-Oct. 5⅛ × 3½ × 1⅝ in.

1496. **B. impolitus** Fr. (from the dull colour of the pileus ; *impolitus*, unpolished) *a b*.

P. pulvinate, sometimes cracking into areolæ, flocculose, dull warm- or dull yellowish-umber. *St.* clavate, pubescent, yellowish- or olive-white above, brown or olive-brown below. *T.* rounded-free, sulphur to yellow and dull olive. *Po.* minute. *Flesh* dull biscuit-colour, sparingly changing to blue.

Edible. Taste and odour pleasant. Woods, mixed ; rare. June–Sept. 5½ × 3⅝ × 1½ in. Pileus sometimes 9 in. in diam.

1497. **B. æstivalis** Fr. (from its time of growth ; *æstivalis*, pertaining to summer) *a b*.

P. convex to expanded, almost white, shaded pale buff, yellowish or brownish ; marg. often waved. *St.* clavate, even, smooth, pale yellow or pale buff above, vinous-brown below. *T.* pale to bright yellow then sulphur-greenish. *Flesh* white to pale yellowish, sometimes with a pale blue line over the T., sometimes vinous at base.

Edible. Taste and odour pleasant. Woodland pastures, roadsides, field-sides. June–July. $5\frac{1}{2} \times 3\frac{3}{4} \times 2$ in. Pileus sometimes 1 ft. 2 in. in diam.

f. Luridæ.

1498. **B. satanas** Lenz (from its fancied bad qualities) *a b c.*

P. subviscid, smooth, tan-white, shaded grey-white. *St.* carmine above, dark carmine reticulate, pale yellowish, dull brownish or crimson below. *T.* sulphur-ochre. *Po.* minute, carmine. *Flesh* whitish, clouded rose and bright blue.

Said to be poisonous, probably harmless. Taste mild. Woods, mixed, oak, woody and chalky places; rare. July–Oct. $6 \times 3\frac{1}{4} \times 2\frac{3}{8}$ in.

1499. **B. luridus** Schæff. (from the lurid colours) *a b c.*

P. subviscid, tomentose, dark umber or dark olive-brown. *St.* scabrous, dull ochreous-orange, speckled crimson. *T.* pale yellow. *Po.* round, dark dull crimson or dull dark red. *Flesh* pale ochre, changing to dark dull blue, stem streaked blue and red, crimson-streaked at base.

Said to be poisonous. Taste pleasant, sweetish; odour not unpleasant. Woods, mixed, woody places; common. June–Nov. $4\frac{1}{8} \times 4 \times 1\frac{5}{8}$ in. Var. *erythropus* Fr., smaller than type, stem within and without blood-red.

1500. **B. purpureus** Fr. (from the sometimes purple pileus, stem and flesh) *a b.*

P. pulvinate, subvelvety, dry, clouded dull claret-purple, and red or brown. *St.* stout, yellow, base deep claret-purple, sometimes wholly ochre, variegated with purple veins and dots, or red-reticulate above. *T.* adnato-rounded, yellow. *Po.* crimson. *Flesh* yellow changing to blue, deep claret-purple, or wholly crimson in St.

Taste mild. Woods, oak. June–Oct. $4\frac{3}{8} \times 3\frac{3}{4} \times 1\frac{3}{8}$ in.

TEPHROLEUCÆ.

g. Favosæ.

1501. **B. laricinus** Berk. (from the habitat, larch plantations; *larix*, larch) *a b c.*

P. convex, rarely umbonate, somewhat squamulose with V., viscid, olive-white, shaded pale olive, pale ochre, or pale slate-purple; marg. white at first, appendiculate with V. *St.* equal, annulate, cribrose above A., pitted below, whitish, pale slate-purple above, pale sepia-streaky below, or pale claret-brown above and pale buff below, or pale claret-crimson-patchy. *T.* adnato-decurrent, nearly white, pale whitish-purple. *Po.* compound. *Flesh* white, unchangeable.

Frequent. July–Sept. $3\frac{1}{4} \times 2\frac{1}{4} \times \frac{3}{8}$ in.

1502. B. rubiginosus Fr. (from the reddish-brown pileus; *rubigo*, rust) *a b*.

P. convex, pubescent to glabrous, red-brown. *St*. subclavate, wholly reticulate, whitish to pale brownish-biscuit. *T*. adnate, unequal, white. *Flesh* unchanging white.

Scattered. Under beeches. Autumn. $3\frac{1}{2} \times 2\frac{3}{8} \times 1$ in.

1503. B. viscidus L. (from its viscidity) *a b c*.

P. pulvinate, smooth, pale yellowish-brown or pale buff, clouded pale greenish or umber; marg. sometimes appendiculate with V. *St*. equal, vaguely reticulate above, somewhat annulate, pale yellowish-brown, slightly streaked, somewhat specked above. *T*. adnate, pale grey then umber and sienna, shaded umber or livid.

Woods, larch; rare. July–Aug. $5 \times 2\frac{3}{8} \times \frac{1}{2}$ in. Intermediate forms occur between this and **1501**.

1504. B. porphyrosporus Fr. (from the purplish spores; Gr. *porphureos*, purple, *sporos*, seed) *a b*.

P. convex, minutely velvety, dark olive, yellowish-umber or umber, blackish when bruised. *St*. equal, minutely punctate, brown-yellowish, shaded olive, not dark. *T*. adnate, brownish-yellow, faintly salmon or grey, brown when bruised. *Flesh* white, becoming blue near T.

Solitary or subcæspitose. Odour strong of fish. Woods, mixed, open grassy places under trees. Aug.–Sept. $5 \times 4\frac{3}{8} \times 1\frac{1}{4}$ in.

h. Versipelles.

1505. B. versipellis Fr. (from the changeable colours of the pellicle; *verto*, to turn, *pellis*, the stem) *a b c*.

P. pulvinate, dry, tomentose, becoming even, red-orange, rufous-brown or fuliginous; marg. appendiculate with the V. *St*. attenuate upwards, yellowish-white, rough with small brown scabrous sc. *T*. adnato-rounded to free or sinuate, yellowish- or pale brownish-white. *Flesh* whitish, becoming clouded pale grey.

Edible. Woods, heaths; frequent. July–Oct. $4\frac{1}{2} \times 6 \times 1\frac{1}{2}$ in. This is the "Rosetto" of Italian markets. Gathered for food by Italians in Epping Forest as the "Rossett."

1506. B. scaber Bull. (from the rough stem; *scaber*, rough) *a b c*.

P. pulvinate, smooth, viscid when moist, at length rugulose or rivulose, usually umber, but sometimes olive-umber, red, orange, fuliginous, cinereous-blackish or white. *St*. attenuate upwards, white or yellowish-white, striated and sprinkled with minute brownish or grey sc., sometimes imperfectly granuloso-annulate. *T*. adnato-rounded or sinuate, white or yellowish-white, becoming dingy brownish. *Flesh* whitish, changing to pale slate-colour.

Edible. Woods; common. June–Nov. $4\frac{1}{2} \times 7\frac{1}{2} \times 1$ in. This is the *Porcinello* or *Albarello* of Italian markets. Var. *niveus* Fr., white. Sometimes attacked by a parasitic mould, *Hypomyces chrysospermus*, and then not edible.

1506a. B. rugosus Fr. (from the wrinkled stem; *ruga*, a wrinkle) *a b*.
P. pulvinate, even, glabrous, dry; marg. naked, brown. *St.* somewhat hollow, costato-rugose, slightly scabrous, somewhat ventricose, almost equal, paler than P. *T.* free, ivory, with a faintly brownish shade. *Po.* round, minute. *Flesh* changing from white to brown or sometimes black.

Autumn. $2\frac{5}{8} \times 4\frac{3}{4} \times 1\frac{5}{8}$ in. Sowerby's t. 421 does not agree with his original drawing, and the section behind is from another drawing, not of this species.

1507. B. duriusculus Schulz. ex Fr. (from the harsh and rough stem; *duriusculus*, somewhat harsh and rough) *a b*.
P. globose to pulvinate, smooth, viscid when moist, whitish-tawny, dull chestnut, olivaceous, yellowish-brown or vinous-umber. *St.* ventricose, attenuato-rooting, whitish or yellowish, densely brown or grey punctato-squamulose. *T.* adnexo-free, white then yellowish or yellow to olive-brownish. *Po.* bright yellow. *Flesh* white, becoming rose-colour, coppery-red, rosy- or red-slate or violet-greyish.

Edible. Woods. Aug.–Oct. $3\frac{3}{4} \times 6\frac{1}{2} \times 1\frac{1}{8}$ in.

1508. B. cruentus Vent. (from the change of colour in the flesh to blood-red; *cruor*, gore) *a b*.
P. convex to plane, smooth to rugulose, brown-yellowish or olive, salmon-red or vinous where bruised. *St.* clavate, attenuate to a rooting base, longitudinally scabrous, yellowish-white or yellow below and orange above. *T.* adnate, pale yellow. *Flesh* pale yellowish or pale sulphur changing to rose or vinous, darker at base of St.

Odour fetid. Under beeches. Aug. $3\frac{1}{4} \times 3\frac{7}{8} \times 1\frac{1}{8}$ in.

i. Hyporhodiæ.

1509. B. felleus Bull. (from the bitter taste; *fel*, gall) *a b c*.
P. pulvinate to expanded or slightly depressed, at first slightly mealy, then smooth, even, pale umber, clouded darker. *St.* clavate, brown-reticulate above, paler than P. *T.* long, adnate to adnexo-rounded. *Po.* angular. *Flesh* white, more or less changing to salmon.

Said to be poisonous. Woods, fir, beech. July–Dec. $4\frac{3}{8} \times 4 \times 1$ in.

1510. B. alutarius Fr. (from the colour of the pileus; *aluta*, tanned leather) *a b*.
P. pulvinate to expanded, velvety then glabrous, umber-yellowish, not dark, clouded slightly darker. *St.* clavate, somewhat even, rugulose above, colour as P. *T.* short, sinuato-adnate, clouded brownish. *Po.* round. *Flesh* yellowish-white, almost unchangeable.

Taste watery, mild. Woodland pastures; rare. Oct. $2\frac{3}{8} \times 3\frac{3}{4} \times \frac{5}{8}$ in.

j. Cariosæ.

1511. B. cyanescens Bull. (from the change of colour in the flesh to blue; *cyaneus*, dark blue) *a b c*.

P. convex, floccoso-scaly, brownish-ochre. *St.* equal to clavate, spongy-cavernous, dry, cracking, slightly annulate when young, squamose, corticate, colour as P. *T.* adnato- or adnexo-rounded, yellowish-white or faintly-shaded olive. *Flesh* yellowish- or brownish-white, changing to blue.

Taste none. Woods, mixed; rare. July–Sept. $3\frac{1}{4} \times 3\frac{1}{2} \times 1\frac{1}{4}$ in. Whole plant dry like **1516**, drying in decay, not decomposing. Wholly veiled in infancy by a thick floccose coat. Changes to sulphur-yellow when dry.

1512. B. castaneus Bull. (from the colour of the pileus; *castanea*, the chestnut) *a b c*.

P. convex, expanded, velvety, sometimes becoming repand, yellowish-white or vinous-cream, then cinnamon. *St.* clavate, velvety, corticate, cracking transversely, stuffed to hollow, colour as P. *T.* free, whitish-sulphur to pale primrose, becoming clouded or spotted pale brownish. *Flesh* white.

Woods, pastures under trees, amongst brambles; rare. Aug.–Oct. $3\frac{1}{2} \times 2\frac{3}{8} \times \frac{3}{4}$ in.

1513. B. fulvidus Fr. (from the yellowish-brown or tawny pileus; *fulvus*, tawny) *a b*.

P. convex to plane, smooth, shining, rigid, yellowish- or brownish-orange. *St.* equal to subclavate, cracking, stuffed to hollow, colour as P., paler below. *T.* free, white to sulphur-yellowish. *Flesh* white to yellowish.

Taste sweet, nutty. Under trees, oak. Aug.–Sept. $2\frac{3}{4} \times 2\frac{3}{8} \times \frac{5}{8}$ in.

SUBGENUS 2. **GYRODON.**

(From the gyroso-toothed tubes; Gr. *odous*, a tooth.)

1514. B. cæspitosus Mass. (from the cæspitose habit) *a*.

P. hemispherical, tomentose, dry olivaceous-umber; marg. pale rose. *St.* solid, even, glabrous, yellow, base dull red. *T.* very short. *Po.* yellow. *Flesh* firm, yellow, changing to dark blue or bluish-black, then rufous fading to dull white, dingy-red at base.

Densely fasciculate, stems more or less connate at base. Under trees, amongst grass. Aug. $2\frac{1}{8} \times 2\frac{1}{8} \times 1$ in.

1514a. B. Sistotrema Fr. (from the resemblance in the tubes to those of *Sistotrema*) *a*.

P. fleshy at mid., thin at marg. where it is flattened above and below, glabrous, dry, rufous- or yellowish-brown. *St.* equal or slightly enlarged below, even, pallid rufous or as T., tinged pale brown. *T.* adnate, short, unequal, yellow-tawny or pale yellowish-sulphur. *Po.* becoming gyroso-plicate. *Flesh* white clouded pale sulphur.

Dry woods. Autumn. $2\frac{5}{8} \times 2\frac{3}{4} \times \frac{3}{8}$ in. Var. *brachyporus* W. G. Sm. (*Boletus brachyporus* Rostk.) *P.* white. *St.* white, pale sulphur below. *T.* pale sulphur. *Flesh* sulphur-white. Changing to pale green where touched.

1515. **B. McWeeneyi** W. G. Sm. (after Dr. Edmunds McWeeney), *Gyrodon rubellum* McWeeney, *a.*

P. convex, even, smooth, dry, red; mid. purplish; marg. yellowish or clouded and streaked bright crimson on a bright yellow ground. *St.* equal, smooth, even, solid, bright yellow. *T.* pale yellow. *Po.* bright yellow. *Flesh* unchanging bright yellow.

On the ground amongst moss. Oct. $\frac{3}{4} \times \frac{3}{4} \times \frac{3}{16}$ in.

SUBGENUS 3. **BOLETINUS.**

(Diminutive of *Boletus.*)

1515a. **B. cavipes** Opat. (from the hollow stem; *cavus*, hollow, *pes*, a foot) *a.*

P. convex, umbonate, dry, floccoso-scaly, dull yellow, yellow or pale buff. *St.* hollow, floccose and pale buff below white A., white above. *T.* decurrent, somewhat large, compound, pale sulphur. *Flesh* white, clouded pale buff.

Autumn. $3\frac{1}{8} \times 3\frac{3}{8} \times \frac{5}{8}$ in.

LX. **STROBILOMYCES** Berk.

(From a fancied resemblance in the pileus to a pine-cone; Gr. *strobilos*, a pine cone, *mukes*, a fungus.)

Veil white, floccose, appendiculate at the margin of the pileus and forming a fugitive annulus on the stem. *Pileus* fleshy, tough,

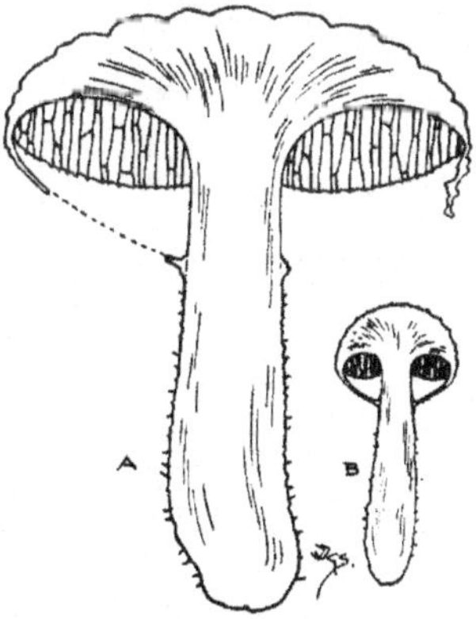

Fig. 74.—*Strobilomyces strobilaceus* Berk.
A, section, showing ruptured veil; B, ditto, young example showing veil intact.
One-third natural size.

broken into scales, hymenophore continuous and homogeneous with the stem. *Stem* solid. *Tubes* very large, anastomosing. (Fig. 74.)

1516. S. strobilaceus Berk. (from the top of the pileus, clad with pine-cone-like scales; Gr. *strobilos*, a pine cone) *a b c.*

P. globose to pulvinate, broken into large thick scales; sc. dark umber on a pale greyish ground. *St.* clavate, sulcato-reticulate above, coarsely fibrillose, white above, vinous-brown below. *T.* adnate, with transverse dissepiments, white, shaded dull brownish. *Flesh* sienna-brown.

Woods, under fir, oak, lime, amongst bracken, open grassy places. Aug.–Oct. 6 × 7⅜ × 1¼ in.

LXI. FISTULINA Bull.

(From the pipe-like character of the tubes; *fistula*, a pipe.)

Veil obsolete. *Pileus* fleshy, juicy. *Stem* obsolete. *Tubes* cylindrical, free from each other as are the spines in *Hydnaceæ*, but adnate with the substance of the pileus. (Fig. 75.)

Growing on wood. Intermediate between *Polyporaceæ* and *Hydnaceæ*.

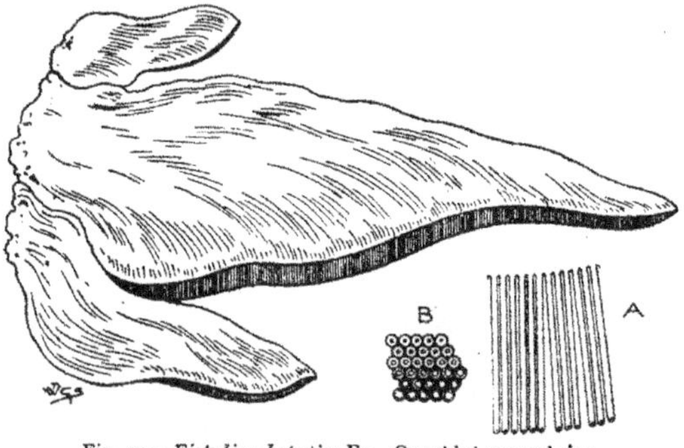

Fig. 75.—*Fistulina hepatica* Fr. One-third natural size. A, tubes. × 3. B, orifices of ditto. × 6.

1517. F. hepatica Fr. (from the liver-like appearance; Gr. *hepatikos*, of the liver) *a b c.* Imbricate.

P. roundish, dimidiate or subspathulate, rough with minute points, dull pale purplish-red, red, blood-red, liver-colour or chocolate, becoming blackish. *T.* ochre-whitish, then yellowish-red. *Po.* distinct. *Flesh* thick, juicy, red-streaked like beet-root, distilling a red juice.

Edible when quite ripe. Taste somewhat acid. Trunks, oak, ash, walnut, willow, beech, chestnut, hornbeam, elm. Aug.-Oct. Diam. 1 ft. 4 in. Sometimes attaining a great size with a weight as much as 19 lbs.; unlike many *Polyporaceæ* it perishes in about three weeks after its first appearance. The "liver" or "beef-steak" fungus. Said by Hartig to be the cause of a deep red-brown decomposition in oak. *Ceriomyces* (*Ptychogaster*) *hepaticus* is said by Saccardo to be an imperfect state of this.

LXII. **POLYPORUS** Mich.

(From the many pores or mouths of the tubes of the hymenium; Gr. *polus*, many, *poros*, a pore.)

Pileus woody, corky or somewhat soft, not zoned, flesh thick. *Stem* long, short or none, when present simple. *Hymenophore* descending and forming a trama between the tubes. *Tubes* connate

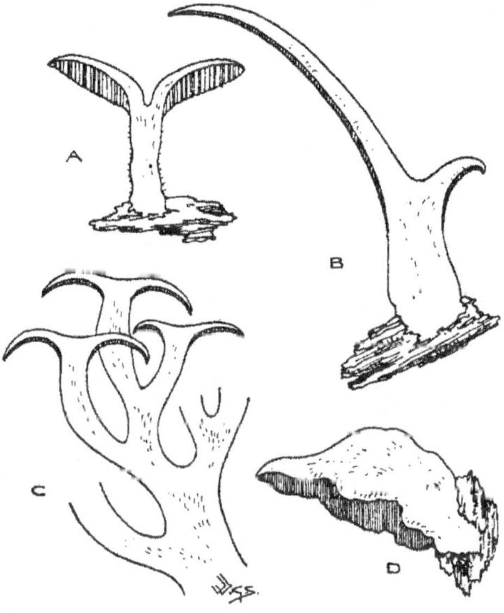

Fig. 76.—Sections of A, *Polyporus lentus* Berk.; B, *P. Rostkovii* Fr.; C, *P. umbellatus* Fr.; D, *P. cæsius* Fr. One-half natural size.

with the substance of the pileus and not separable from it or from each other, extended with round, angular or torn pores, not normally stratose. (Fig. 76.)

Persistent fungi, mostly growing on wood. Odour, if any, slightly acid. A few are said to be edible. Species **1517a—1575**

MESOPODÆ. *Pileus* entire or excentric. *Stem* distinct, vertical, central, simple, self-coloured, not blackish at the base.

a. Carnosæ. Pileus fragile or pliant, dull in colour, scaly or floccose, not coriaceous, zoneless. *Spores* white, growing on the ground. Autumnal. Edible. **1517a, 1518** *b. Lentæ.* *Pileus* fleshy-pliant, then coriaceous, zoneless. *Spores* white. Growing on trunks. Persistent.

Pileus lustrous, scaly or villous. **1519, 1520**

Pileus even, smooth. **1521, 1522**

c. Spongiosæ. *Pileus* spongy-soft, hygrophanous, tomentose, then corky or coriaceous, regular, irregular, lobed or dimidiate. *Stem* short, irregular. *Pores* irregular, changing colour. *Spores* white. Mostly terrestrial, lasting through the winter in a dead state, not perennial. **1523, 1524**

d. Subcoriaceæ. *Pileus* at first indurated, arid; when full grown corky or coriaceous, tomentose, velvety. *Stem* definite, rarely lateral. *Pores* at first sprinkled over with a white bloom. Substance ferruginous. Terrestrial or on trunks; more regular, rigid and persistent than *Spongiosæ.* Not reviving. **1524a—1526**

PLEUROPODÆ. *Pileus* pliant or corky, horizontal, not circular. *Stem* simple, lateral or excentric, black at the base. Growing on wood.

a. Lentæ. *Pileus* fleshy-pliant, zoneless. *Stem* excentric, blackish at the base. *Tubes* short. Substance somewhat fibrous, pallid. **1527—1534**

b. Suberosolignosæ. *Pileus* thick, hard zoneless. *Stem* stout, vertical, of the same colour at the base. *Tubes* elongated. **1535**

MERISMATÆ. With many branching stems proceeding from a common trunk, each as a rule bearing a pileus or pileolus. The largest of fungi are found in this group.

a. Carnosæ. Tufts central, stipitate, springing from a common, more or less concrete base. *Pileus* fleshy firm floccoso-fibrous, zoneless, not coriaceo-indurate. *Tubes* separating from the pileus. *Spores* white. Partly terrestrial. Autumnal. Edible. **1536—1539**

b. Lentæ. From the tough pileus. Tufts lateral, substipitate, in many imbricate layers. *Pileus* fleshy-pliant, then subcoriaceous, fragile when old, more or less zoned, fibrous within. *Stems* more or less connate, or growing from a common tuberous base. Growing at the base of trunks. *Spores* white. Autumnal, not lasting till the following spring. Not edible. **1540—1542**

c. Caseosæ. Tufts sessile on the stems of trees, commonly dimidiate, but in horizontal positions, central and expanded on all sides. Springing from a tubercular base into numerous pileoli. *Pileus* cheesy, at first soft and juicy, then arid, without a pellicle, zoneless. *Tubes* separating. Growing in spring and summer, soon decaying. Acid. **1543—1546** *d. Suberosæ.* Tufts somewhat sessile, lateral or central. *Pilei* or *Pileoli* corky or coriaceous, persistent, tough, free at the base, not effused. Substance floccose, somewhat soft, suitable for tinder. *Tubes* adnate. On trunks close to the ground. Not edible. **1547, 1548**

APODÆ. *Pileus* sessile, adnate by a thickened, dilated base, dimidiate or entire, attached behind, commonly by an umbo; more rarely attenuate and sessile. On wood. Polymorphous. **1549—1575**

a. Anodermeæ. *Pileus* somewhat fleshy, without a cuticle, the surface broken into flocci or fibres, zoneless, the substance transversely zoned within and more or less fibrous. Annual, not reviving.

Pileus cheesy, at first watery-soft, fragile, flocculose, not bristly-hispid; when fully grown, soft to hard, but soon putrefying, not lasting through the winter. *Tubes* white.

Pores round, entire, obtuse, without small teeth. Not becoming hard. **1549—1553**

Pores elongate, flexuous, acute and torn. **1554—1559**

Pileus tough, fleshy-pliant, soft, elastic, villoso-tomentose. *Tubes* somewhat adnate, coloured. Lasting till the following spring.

Substance coloured. **1560—1562**
Substance white or slightly brownish. **1563—1568**

At first moist, spongy, when dry firm, elastic, usually bristly-hisped, internally fibrous. Autumnal, annual. The finest examples growing on living trunks. Must not be confounded with the section *Stupposæ* of *Polystictus*, where the members are dry and rigid.

Substance coloured. **1569, 1570**
Substance and spores white. **1571, 1572**

b. Placodermeæ. *Pileus* clothed with a continuous crust, zoneless or concentrically sulcate. Persistent.

Substance coloured. **1573**
Substance white, not zoned. *Spores* white. **1574, 1575**

MESOPODÆ.

a. Carnosæ.

1517a. P. flavovirens Berk. & Rav. (from the yellowish-green colour of the pileus; *flavus*, light yellow, *vireo*, to be green) *a.*

P. soft, pulvinate or depressed, irregularly lobed, subtomentose, dull yellowish-green, yellowish-olive or olive. *St.* slightly attenuate downwards, pallid, paler than P., or shaded white and yellowish. *T.* slightly to very decurrent, short, irregular, toothed, at length torn as in *Irpex*, white, yellow, then greenish. *Flesh* white, clouded pale sulphur.

Woods, mixed, under *Pinus sylvestris*. Sept. $3\frac{1}{4} \times 2\frac{3}{4} \times \frac{7}{8}$ in.

1518. P. leucomelas Fr. (from the colours; Gr. *leukos*, white, *melas*, black) *a*.

P. expanded, thick, subfragile, slightly silky-squamulose, dark olive-umber or blackish fuliginous. *St.* equal to unequal, subtomentose, yellow-umber or as P. *T.* decurrent. *Po.* large, short, unequal, white to cinereous, black when dry. *Flesh* white, changing to faintest rufous-purplish or black.

Woods, fir. Sept. $5\frac{1}{4} \times \frac{7}{8} \times \frac{7}{8}$ in.

b. Lentæ.

1519. P. lentus Berk. (from the tough-pliant substance; *lentus*, tough) *a b c*.

P. umbilicate, thin, minutely scaly, buff-white, shaded buff, ochreous or reddish-brown. *St.* equal, hispid or furfuraceous, colour as P., shaded sienna-buff. *T.* decurrent, long, large, unequal, yellowish-white. *Po.* toothed.

Dead roots and branches, furze, hawthorn; frequent. April–Sept. $2 \times 1 \times \frac{1}{4}$ in.

1520. P. brumalis Fr. (from the time of growth; *bruma*, winter) *a b c*.

P. convexo-expanded, very thin, villous to squamulose and smooth, rich cinnamon to umber or slate-umber; marg. paler. *St.* equal, minutely dark-umber-floccose on a pale brown ground, when young brown-ochre above, ochre below. *T.* decurrent, very shallow, white, becoming yellow. *Po.* toothed. *Flesh* white.

Woods, on dead branches, sticks, hazel, birch, beech, sometimes on the ground; frequent. $2 \times 2\frac{1}{2} \times \frac{3}{16}$ in.

1520a. P. arcularius Fr. (*arcularius*, a casket-maker; in reference to the fulvous pores, covered by the pileus—as if in a casket) *a*.

P. pliant, then coriaceous, somewhat umbilicate, zoneless, orange-brown, in the first year covered with fuscous sq., in the second glabrous with a yellow strigose margin. *St.* somewhat squamulose, greyish-fuscous or sooty-brown. *T.* oblong. *Po.* rhomboidal, thin, rather large, entire, whitish then fulvous. *Flesh* white.

Dead twigs. Sept. Varying from $\frac{3}{8} \times \frac{1}{2} \times \frac{1}{16}$ in. to $2\frac{7}{8} \times 1\frac{1}{2} \times \frac{1}{4}$ in. Distinguished from **1520** by the shape of the pores, the strigose margin and fuscous squamules of the pileus.

1521. P. fuscidulus Fr. (from the somewhat fuscous pileus) *a*.

P. convexo-plane, very thin, leathery, pale yellowish-brown or pale olive-brown. *St.* equal, smooth, pale yellowish-olive. *T.* adnato-decurrent, somewhat long, warm yellowish-white.

Woods, chips; rare. $2\frac{1}{4} \times 1\frac{3}{4} \times \frac{1}{4}$ in.

1522. P. leptocephalus Fr. (from the thin pileus; Gr. *leptos*, thin, *kephale*, the head) *a b c*.

P. convexo-plane, slightly depressed or subumbilicate, thin, leathery, pale yellowish-buff or shining yellowish. *St*. smooth, colour as P., sometimes black at base. *T*. adnate. *Po*. minute, colour as P. or whitish. *Flesh* white.

Woods, on trunks, dead birch; rare. Oct.–Mar. $2\frac{1}{4} \times 1\frac{3}{8} \times \frac{3}{16}$ in.

c. Spongiosæ.

1523. P. Schweinitzii Fr. (after Ludwig David von Schweinitz) *a b c*.

P. expanded, rugged-uneven, strigoso-tomentose, thick, deep orange-sienna to sienna-umber, irregularly zoned brown; mid. almost black. *St*. rugged-uneven or obsolete, sulphur, umber-clouded below, dark umber at base. *T*. decurrent, large, sulphur to green-brownish-sulphur. *Po*. torn. *Flesh* deep sienna.

Woods, about roots, fir, pine, larch. July–Oct. $6\frac{1}{4} \times 1\frac{3}{4} \times 1\frac{3}{8}$ in. Hartig states that this fungus has an effect on living wood similar to that of the "Dry-Rot" **1680**.

1524. P. rufescens Fr. (from the rufescent colour) *a b c*.

P. depressed to infundibuliform, somewhat thin, strigose, pale vinous-umber; mid. darker; marg. irregular. *St*. equal, uneven, sometimes lateral or absent, colour as P., darker and clouded below. *T*. decurrent, pale yellowish-vinous-umber. *Po*. torn. *Flesh* colour as P.

On and about stumps, ash, elm. Sept.–Jan. $5 \times 2\frac{1}{8} \times 1$ in. Sometimes wholly biscuit-buff.

d. Subcoriaceæ.

1524a. P. tomentosus Fr. (from the minutely velvety pileus; *tomentum*, wool) *a c*.

P. plane, corky, becoming hard, often lobed and irregular, tawny; marg. thin, straight, sometimes yellowish. *St*. solid, attenuate upwards, sometimes a mere base or central point, sometimes rooting, steel-grey. *T*. subdecurrent, short. *Po*. very minute, entire, pale- or greyish-yellow, at first delicately whitish-pruinose. *Flesh* pale tawny.

On the ground. $2 \times \frac{5}{8} \times \frac{1}{4}$ in. This with **1524b**, **1525** and **1526** have been placed by Cooke, Saccardo and Massee in *Polystictus*.

1524b. P. Montagnei Fr. (after J. F. C. Montagne) *a*.

P. cyathiform, irregular, zoneless, uneven, ferruginous, becoming smooth; marg. thin. *St*. unequal, pubescent, colour as P. *T*. short. *Po*. large, round, edge entire, yellowish-white, then ferruginous. *Flesh* tawny-cinnamon.

Distinguished from **1525** by the zoneless pileus and large entire pores. Autumn. $1\frac{1}{2} \times 1\frac{1}{4} \times \frac{3}{16}$ in.

1525. P. perennis Fr. (from its perennial habit) *a b c.*

P. plano-infundibuliform, velvety to smooth, thin, pale yellowish-brown, zoned darker; marg. fimbriate. *St.* uneven, attenuate upwards, velvety, vinous-umber, pale sienna below. *T.* decurrent, short, pale umber. *Po.* small, torn, at first with a white bloom. *Flesh* pale vinous-umber.

On and about stumps, beech, burnt ground; frequent. July–Jan. $2\frac{3}{8} \times 1\frac{1}{4} \times \frac{1}{4}$ in.

1526. P. cinnamomeus Sacc. (from the cinnamon colour of the pileus) *a b.*

P. plano-depressed or subinfundibuliform, velvety to smooth, thin, leathery, flaccid, bright sienna-cinnamon, zoned deep rich brown. *St.* usually attenuate downwards, base sometimes tuberous, colour as P. *T.* adnate to adnato-decurrent, short, brownish-cinnamon, fulvous when dry. *Po.* large. *Flesh* rich sienna-cinnamon.

Solitary or in rows to subcæspitose. On the ground, under trees, burnt places. Oct. $3\frac{1}{4} \times \frac{3}{4} \times \frac{1}{4}$ in. Pileus sometimes almost jet black.

PLEUROPODÆ.

a. Lentæ.

1527. P. squamosus Fr. (from the scaly pileus; *squama*, a scale) *a b c.*

P. fan-shaped, flat, pale brownish, sienna-brown-scaly. *St.* yellowish-umber, black-spotted, blackish at base. *T.* decurrent, short. *Po.* toothed, yellowish-white, like the pores of tripe. *Flesh* white.

Single to cæspitose or imbricate. Odour rank. Trunks and stumps, ash, mulberry, walnut, maple, wych elm; common. Sometimes 2 ft. 6 in. in diam. and 34 lbs. in weight. April–Nov. 1 ft. 4 in. $\times 1\frac{3}{4} \times 1\frac{3}{4}$ in. The flesh makes good razor-strops, but is often larva pierced. In the British Museum collection of drawings Bolton illustrates an antler-like form of this species, growing from a dark brown sclerotioid base.

1528. P. Michelii Fr. (after Pier' Antonio Micheli) *a.*

P. spathulate, slightly silky, yellowish-white with darker sq. *St.* equal, rough, white, becoming fuscous at base. *T.* decurrent, short. *Po.* large, entire, white.

Somewhat cæspitose. Stumps, willow. $2\frac{3}{4} \times \frac{3}{4} \times \frac{1}{2}$ in.

1529. P. melanopus Fr. (from the black stem; Gr. *melas*, black, *pous*, a foot) *a b c.*

P. subinfundibuliform, delicately flocculose to smooth, somewhat thick, pale yellowish. *St.* attenuate upwards, somewhat velvety, umber, rich brown or black. *T.* decurrent, short, white. *Po.* minute.

Somewhat cæspitose. Dead wood, roots; uncommon. Aug.–Oct. $2\frac{7}{8} \times 1\frac{1}{2} \times \frac{3}{8}$ in.

1530. P. Rostkovii Fr. (after Friedrich Wilhelm Gottlieb Rostkovius) *a b c.*

P. dimidiato-infundibuliform, somewhat thick, yellowish-sienna, deep umber at mid. and behind. *St.* equal, reticulate, deep umber to black. *T.* decurrent, short, white to dull yellowish. *Po.* large, angular, toothed.

Somewhat cæspitose. Woods, on stumps, trunks, ash; rare. July–Jan. $4\frac{7}{8} \times 1\frac{1}{4} \times \frac{3}{4}$ in.

1531. P. picipes Fr. (from the pitch-black stem; *pix*, pitch, *pes*, a foot) *a b c.*

P. somewhat depressed at mid. or behind, red-brown or chestnut. *St.* equal or attenuate downwards, velvety to naked. *T.* decurrent, somewhat short, whitish to yellowish. *Po.* round, small. *Flesh* white.

Odour somewhat sweet. Trunks, willow; frequent. July – Dec. $6\frac{5}{8} \times 1\frac{5}{8} \times \frac{3}{8}$ in. Like a large **1533**, var. *nummularius*.

1532. P. varius Fr. (from its variable size, colour and form) *a b c.*

P. depressed at mid. or behind, somewhat virgate, yellowish-brown, shaded palish vinous-umber; marg. undulate. *St.* even, smooth, cinereous blackish downwards. *T.* decurrent, short, white to yellowish. *Po.* minute. *Flesh* white.

Single to subcæspitose. Sometimes sessile. Trunks, stumps, ash, willow, elder, birch; frequent. Aug.–Feb. $3\frac{3}{4} \times 1\frac{1}{4} \times \frac{3}{8}$ in.

1533. P. elegans Fr. (from its elegance) *a b c.*

P. convex to slightly depressed, usually excentric or dimidiate, becoming woody, ochreous, orange or reddish. *St.* even, smooth, black, abruptly pallid at apex. *T.* adnate or slightly decurrent, yellowish white, brownish yellow or slightly orange. *Po.* minute.

Single to subcæspitose. Trunks, birch, willow, beech; frequent. July–Feb. $2\frac{3}{4} \times 1\frac{3}{4} \times \frac{3}{8}$ in. Var. *nummularius* Fr. Smaller than type, *P.* almost regular. 1 in. in diam.

1534. P. petaloides Fr. (from a fancied resemblance in its shape and thinness to a leaf; Gr. *petalon*, a leaf, *eidos*, appearance) *a b.*

P. spathulate, thin, rugose, zoneless, dark rich chestnut. *St.* smooth, whitish, sometimes dark brown below. *T.* decurrent, somewhat short, white. *Po.* minute.

Old stumps. Jan. $2 \times \frac{3}{4} \times \frac{3}{8}$ in.

b. Suberosolignosæ.

1535. P. lucidus Fr. (from the shining pileus and stem; *lucidus*, bright) *a b c.*

P. fan-shaped or spathulate, sulcato-rugose, corky to woody, lustrous, deep rich blood-chestnut. *St.* lateral, irregular, uneven, lustrous, deep purple-chestnut. *T.* adnate, somewhat long, rarely stratose, yellowish-white, clouded pale brownish. *Flesh* tan-brownish.

Single to cæspitose. On and about stumps, hornbeam ; frequent. Oct.–April. $7 \times 6 \times 1\frac{1}{4}$ in. Sometimes sessile. Occurs in peat-beds in the fens of the Eastern counties and in the lake-side pile-dwellings of Switzerland and Italy. The varnished appearance of the pileus is due to the presence of resin. This species has been placed in *Fomes* by Cooke and Saccardo.

MERISMATÆ.

a. Carnosæ.

1536. P. umbellatus Fr. (from the many stems and pileoli forming an umbel) *a.*

Pi. very numerous, entire, umbilicate, fibrous-fleshy, warm-umber, or yellowish-reddish- or slate-umber, rarely white. *St.* white to yellowish-white. *T.* short, subdecurrent, colour as St. *Po.* minute.

Said to be edible. Woods, near and on stumps. July. Mass $9\frac{3}{4} \times 9\frac{3}{4}$ in. Pi. $1\frac{1}{2} \times 1\frac{1}{4} \times \frac{1}{4}$ in.

1537. P. frondosus Fr. (from the branching habit ; *frons*, a leafy branch) *a c.*

Pi. numerous, dimidiate, rugose, convex to recurved, somewhat thin and tough, yellowish-umber to pale purple-slate. *St.* yellowish-white. *T.* decurrent, somewhat short, yellowish-white. *Po.* small.

Stumps, trunks, roots, oak ; rare. Oct. Mass $10\frac{1}{2} \times 9\frac{3}{4}$ in. Pi. $3\frac{3}{4} \times 1\frac{3}{4} \times \frac{3}{4}$ in.

1538. P. intybaceus Fr. (from a fancied resemblance to succory, *Cichorium Intybus*) *a b c.*

Pi. numerous, dimidiate, fragile, sepia, not dark, lighter at mid. *St.* white. *T.* decurrent, very short, pale sepia.

Edible. Taste usually pleasant, sometimes slightly astringent when raw ; odour of mice, in decay like japanner's varnish. Stumps, trunks, oak ; rare. Sept.–Dec. Mass $12\frac{1}{2} \times 7\frac{1}{2}$ in. Pi. $3 \times 1\frac{3}{4} \times \frac{1}{2}$ in. The cauliflower-like stems are the edible parts, not the pileoli. The "cock of the woods" (grey grouse-hen) of the Vosges and Ardennes.

1539. P. cristatus Fr. (from the crested appearance ; *crista*, a crest) *a c.*

Pi. numerous, dimidiate, somewhat thick, fragile, bright yellow or greenish, sometimes clouded reddish. *St.* irregular. *T.* very short, whitish. *Po.* minute, angular, torn. *Flesh* white, becoming ochre.

Woods, beech ; rare. Mass $7\frac{3}{4}$ in. in diam. Pi. $3\frac{1}{2}$ in.

b. Lentæ.

1540. P. giganteus Fr. (from its great size) *a b c.*

Pi. several, dimidiate, thick, umber ; z. darker. *St.* short, pale yellowish-white, brownish or ashy where bruised. *T.* somewhat short. *Po.* minute, colour as St. *Flesh* cheesy, vinous-white.

Taste astringent, disagreeable ; odour strong, oppressive, like rotten cheese. Stumps, elm, oak, beech, acacia. Becoming blackish in drying. July–Jan. Mass 1 ft. 3 in. Pi. 8¾ in.

1541. P. osseus Kalch. (from its colour, like bone, *os*) *a c.*
White, shaded grey or buff-white.
Pi. several, thick, convex to flat, smooth. *St.* short. *T.* decurrent, short. *Po.* minute, shallow.

Odour acid. Stumps, larch. Oct. Mass 6¾ in. Pi. 4⅜ in.

1542. P. acanthoides Fr. (from a fancied resemblance to Acanthus ; Gr. *eidos*, appearance) *a b c.*
Pi. numerous, dimidiate, thin, pliant, yellowish-sienna ; z. sienna ; marg. paler. *St.* pale brownish. *Po.* lamelloso-sinuate, toothed, colour as St. *Flesh* faintly rufous.

Trunks, roots, buried wood. Sept. Mass 8¾ in. Pi. 4½ in.

c. Caseosæ.

1543. P. sulphureus Fr. (from the sulphur-colour) *a b c.*
P. few, thick, juicy-cheesy, sulphur-yellowish, somewhat clouded ochreous-reddish or salmon. *T.* short. *Po.* minute, colour as P. *Flesh* thick, pale sulphur, somewhat vinous above.

Imbricate. Said to be edible. Taste slightly astringent, somewhat acid ; odour fragrant to disagreeable, acid, foxy, pleasant if burnt when dry. Living trees, stumps, oak, yew, chestnut, willow, cherry, alder, poplar, walnut, pear, apple, larch, *Robinia*, ash, pine. May–Oct. Mass 9½ in. P. 7½ in. Pileus covered with crystals of oxalate of lime. One of the fungi used in past times for tinder. Usually rots the middle of a tree-trunk, leaving the outside sound. Sometimes phosphorescent. *Ceriomyces* (*Ptychogaster*) *aurantiacus* is said by Saccardo to be an imperfect state of this species.

1544. P. imbricatus Fr. (from the imbricate habit) *a b c.*
P. few to many, firmer than **1543**, thin, rugose, subviscid, buff. *T.* long, pale yellowish. *Po.* lustrous. *Flesh* thin, white.

Odour strong, like seed-cake. Trunks, oak ; rare. June–Nov. Mass 11 in. P. 6 in.

1545. P. spongia Fr. (from the spongy substance) *a b.*
P. few, thick to thin, often rugose, yellowish- chestnut- or umber-sienna, darker at base ; z. indistinct ; marg. sometimes sulphur or pale dull ochre. *T.* somewhat short. *Po.* minute, torn, pale primrose, faintly reddish where bruised. *Flesh* pale opaque sienna.

Subimbricate. Trunks, pine. Oct.–Nov. Mass 11 in. P. 6 in. *Boletus Herbergii* Rostk. is a form of this.

1546. P. alligatus Fr. (from its binding or enveloping habit; *alligo*, to bind to) *a b c*.

P. many, villous, white; marg. pale umber. *T*. short. *Po*. minutely labyrinthiform, colour as P. *Flesh* biscuit-whitish.

Imbricato-connate. Roots, etc.; rare. Sept. Mass 5 in. P. 2½ in. Sometimes faintly shaded, vinous.

d. Suberosæ.

1547. P. heteroclitus Fr. (from its irregular, leaning habit; Gr. *heteros*, one of two, *klino*, to lean) *a c*.

P. few or many, thin, shaggy-villous, zoneless, yellow, sienna behind; marg. lobed. *T*. somewhat short. *Po*. pale primrose to golden-yellow. *Flesh* white.

Imbricate. Stumps, partly buried, oak; rare. Jan. Mass 7 in. P. 3½ in.

1548. P. salignus Fr. (from a frequent habitat; *salix*, willow) *a b c*.

P. few to many, thin, adpressedly villous, whitish; marg. crested. *T*. somewhat short. *Po*. intricately flexuous, whitish. *Flesh* white.

Imbricate. Odour none. Trunks, willow, ash, poplar, oak, apple; uncommon. Oct.–Feb. Mass 7½ in. P. 3¾ in. A dark line separates the tubes from the flesh of the pileus as seen in section.

APODÆ.

a. Anodermeæ.

1549. P. epileucus Fr. (from its whitish colour; Gr. *epi*, upon, *leukos*, white) *a b c*.

P. few, pulvinate, villous-rugged, thick, cheesy-soft, yellowish-white, clouded greyish, rarely faintly zoned. *T*. long, biscuit-primrose or brownish-white. *Flesh* buff-white.

Subimbricate. Stumps, trunks, elm, fir, willow, birch; rare. Sept.–Nov. Mass 8⅜ in. P. 5½ in.

1550. P. alutaceus Fr. (from its colour, like tanned leather, *aluta*) *a*. Whitish-ochre.

P. few, tough, reniform-dilated, slightly velvety; marg. acute, even. *Po*. small.

Imbricate. Decayed stumps, pine, beech. Sept.–Oct. Mass 5⅜ in. P. 3 in.

1551. P. pallescens Fr. (from the colours becoming pale; *pallesco*, to grow pale) *a b c*.

P. few, corky, even to rugose, pale buff; marg. lighter. *T*. small, pale yellowish, pale vinous towards base. *Flesh* buff-white.

Imbricate. Stumps, branches, hornbeam; rare. Mass 5⅝ in. P. 2¼ in.

1552. P. chioneus Fr. (from its snow-white colour; Gr. *chion*, snow) *a b c*. Pearly or biscuit-white.

P. few, soft, fragile, smooth, thin, zoneless or faintly zoned. *T*. long, but varying in length.

Single to 3-imbricate. Taste astringent; odour acid. Stumps, larch, fir, birch; uncommon. June–Dec. P. 2⅛ in.

1553. **P. cerebrinus** B. & Br. (from its brain-like appearance; *cerebrum*, the brain) *a b*. White.

P. pulvinate, delicately tomentose, then smooth; marg. crenate. *T.* rounded, entire, dissepiments of the large Po. thick, obtuse.

Fir. Aug. P. 1 in. Resupinate; not unlike **1657**.

1554. **P. lacteus** Fr. (from the milk-white colour; *lac*, milk) *a c*. White.

P. 1–3 connate, gibbous, thick, flesh-fibrous, pubescent, zoneless. *T.* long. *Po.* toothed, at length labyrinthiform and torn.

Imbricato-connate. Taste astringent; odour none. Dead bark, leaves, laburnum, birch, beech; rare. Sept.–Nov. P. 3 in.

1555. **P. fragilis** Fr. (from its fragile substance) *a b c*.

P. convex to plano-depressed, rugose, thick, whitish, pale foxy-red where bruised. *T.* long. *Po.* minute, intricate, colour as P.

Single to 7-imbricate. Stumps and branches, pine, covering moss on pine-bark, larch; uncommon. Sept.–Nov. P. 2¼ in.

1556. **P. mollis** Fr. (from its softness; *mollis*, soft) *a c*.

P. effuso-reflexed, rugose, fleshy fibrous, pale flesh-colour, livid or brownish-biscuit, sometimes slightly zoned darker; marg. acute, foxy where bruised. *T.* long, unequal, soft, white or salmon-white, foxy where bruised.

Single to 3-imbricate, or connate. Dead pine. Sept. P. 2½ in.

1557. **P. cæsius** Fr. (from its bluish-grey clouding; *cæsius*, bluish-grey) *a b c*. Ivory-white, becoming pale turquoise or bluish-grey where bruised.

P. thick, fleshy, lobed, unequal, silky. *T.* long. *Po.* torn into teeth.

Single to 3-imbricate or connate. Dead fir, larch, dead fir-leaves, roots, trunks, posts, sticks. Mar.–Nov. P. 4½ in.

1558. **P. trabeus** Rostk. (from its frequent habitat, the wood of ships; *trabs*, timber) *a c*. White.

P. thick, firm, effuso-reflexed, transversely elongate, zoneless, pallid. *T.* short. *Po.* minute, toothed. *Flesh* hyaline when moist.

Single. Wood, yew, pine. Oct. P. 3⅜ in. Commonly confounded with **1559**, from which it is distinguished by being more regular and brighter in colour.

1559. **P. destructor** Fr. (from its destructive effect on wood) *a b c*. Ivory-white.

P. watery-fleshy, rugose; marg. pale brownish. *Po.* toothed or torn.

Connato-imbricate. Worked wood, which it softens and destroys, stumps, branches, fir, larch, fir-leaves, willow; common. May–Dec. P. 3⅝ in. Var. ***undulatus*** Sacc. Broadly expanded, marginate, pale foxy-brown. *Ceriomyces* (*Ptychogaster*) *albus* (fig. 77) is very commonly associated with this species and **1572**. Berkeley considers *Ptychogaster* to be an imperfect conidial state of one or both of these *Polypores*, whilst others regard it as a parasite. It has also been made a species of *Polyporus*, viz. *P. Ptychogaster*

Ludw., and as such is placed after **1559** by Saccardo. The plant grows on rotten fir-trunks and on the ground, involving fir-leaves.

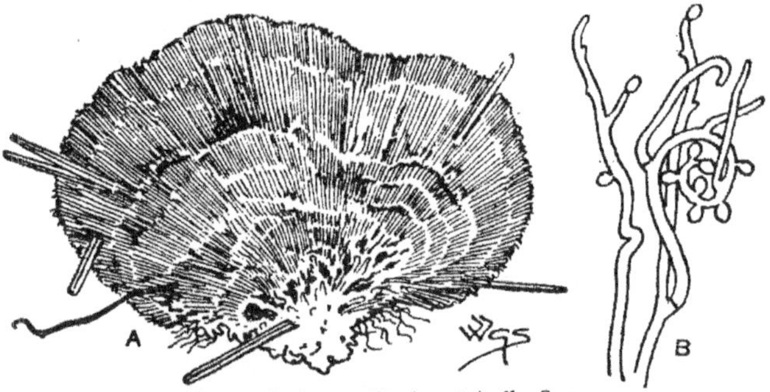

Fig. 77.—*Ceriomyces* (*Ptychogaster*) *albus* Sacc.
A, section, three-quarters natural size; B, threads and conidia. × 370.

1560. P. nidulans Fr. (from the nest-like shape; *nidus*, a nest) *a b c*. Whitish-sienna and umber, or slightly orange.

P. single, pulvinate, villous, zoneless, thick, soft. *T.* long, unequal, angular.

Single, usually in small troops. Fragrant when dry. Dead wood, branches, wych elm; uncommon. Jan.–Dec. P. 1 in. At first resupinate and from 3 to 7 in. in diam.

1561. P. rutilans Fr. (from its reddish colour; *rutilo*, to be reddish) *a b c*.

P. thin, villous to smooth, zoneless, ivory-whitish, then brownish, clouded vinous. *T.* long. *Po.* minute, equal, cinnamon or vinous. *Flesh* pale ochre, vinous towards T.

Single to 2-imbricate or connate. Odour powerful and pleasant of anise. Logs, branches, sticks, oak; uncommon. April–Dec. P. 1¾ in. At first resupinate. Perhaps this and **1560** are one species. See **1607**.

1562. P. gilvus Fr. (from its colour; *gilvus*, pale yellowish) *a b c*.

P. thick, zoneless, pale sienna-yellow or sienna-umber; marg. velvety, becoming smooth, lighter than P., or dull orange. *Po.* minute with barren patches, dark orange-brown; marg. barren. *Flesh* bright orange-brown.

Single to 3-imbricate or connate. Trunks, willow, oak; rare. Jan. P. ⅜ in. *Boletus impuber* Sow. t. 195. At first effuso-resupinate.

1563. P. fumosus (from its smoky colour; *fumosus*, smoky) *a b c*.

P. firm, zoneless, somewhat thin, silky to smooth, pale buff, pale biscuit or slate-umber, sometimes shaded vinous. *T.* whitish to ivory-umber, or smoky-livid, with a smoky tinge, darker where bruised, often with barren patches. *Po.* round, entire. *Flesh* very pale buff or brownish.

2–16 imbricate and connate. Stumps, logs, poplar, willow, birch, horse-chestnut; common. July–Feb. P. 2 in. Mr. Carleton Rea has met with a very large example on a wooden grass-roller.

1563a. P. fragrans Peck (from the fragrance).

P. fleshy, effuso-reflexed, imbricate, sometimes thickened at the base, velvety, with minute tomentum, pale reddish-grey or alutaceous; marg. thin. *T.* $\frac{1}{12}$ in. long. *Po.* minute, unequal, angular, whitish, becoming darker with age, blackish where bruised. *Flesh* slightly fibrous, zoned, concolorous.

Stumps, elm. 4 × 2 in. Often mistaken for **1563**, but distinguished by the sweet odour.

1564. P. adustus Fr. (from its scorched appearance; *aduro*, to burn) *a b c.*

P. somewhat thin, buff-ivory, with 4–6 darker z.; marg. becoming blackish. *T.* somewhat short. *Po.* minute, warm ivory-brown, becoming ragged and tinted slate-colour. *Flesh* whitish to buff-ivory.

Single to 8-imbricate. Stumps, alder, oak, willow, birch, poplar, hornbeam, dead leaves; frequent. Jan.–Dec. P. 2¾ in.

1565. P. crispus Fr. (from the crisped margin of the pileus) *a c.*

P. rugose, tuberculose behind, thin, ivory-buff, brownish behind; marg. umber-white. *T.* somewhat short. *Po.* large, unequal, labyrinthiform, slate-whitish or brown-white; marg. barren. *Flesh* buff-white.

About 4–9 connate. Old stumps; uncommon. May–Jan. Mass 3¼ in. P. 1½ in.

1566. P. amorphus Fr. (from its shapelessness; Gr. *a*, without, *morphe*, form) *a b c.*

P. thin, silky, somewhat tuberculose behind, orange to white. *T.* somewhat short. *Po.* unequal, pale orange-vermilion or pale ivory-vermilion: marg. barren.

Multi-imbricate. Stumps, leaves, fir; uncommon. Sept.–Mar. Mass 3 in. P. 1 in. Cooke refers **1622** to this, but the spores are different in shape and size. Often in part resupinate. *Ceriomyces* (*Ptychogaster*) *citrinus* is said by Saccardo to be an imperfect state of this.

1567. P. adiposus B. & Br. (from its fat-like substance; *adeps*, fat) *a c.*

P. effused, waxy-soft, shortly reflexed, thin, obscurely tomentose, bright red-brown. *T.* short. *Po.* small, salmon; marg. salmon-orange.

On the ground near stumps, on mosses, *Marchantia*; uncommon. Dec. Mass 1½ in. P. ½ in.

1568. P. Keithii B. & Br. (after the Rev. Dr. James Keith) *a b.*

P. bell-shaped, decurrent behind, rough with rigid, tooth-like processes, bright red-brown. *Po.* pallid, deep vinous when dry, dissepiments lacerate.

Single to 2-imbricate. Sticks, naked wood. P. ½ in.

1569. P. hispidus Fr. (from the hispid pileus) *a b c.*

P. thick, dimidiate, pulvinate, hispid, rich red-brown; marg. sienna-ivory. *T.* long. *Po.* minute, ivory to pale brown-ochre. *Flesh* juicy, tender, elastic, spongy, dull orange to sienna.

Single to 3-imbricate. Living trunks, pine, ash, larch, apple, mulberry, walnut, lime, oak, elm, *Euonymus;* common. May–Feb. P. 10½ in. Close to the bark of the tree the pubescence often resembles *Ozonium auricomum;* see **237**. Sometimes 18 in. in diam. Becomes dense and fibrous in age. Tubes sometimes distilling drops of water; becoming wholly black in age.

1570. P. cuticularis Fr. (from its thinness, skin-like, as compared with **1569**; *cuticula*, a thin external skin) *a b c.*

P. thin, subrugose, hairy or tomentose, obscurely zoned, yellow-buff to deep vinous-brown. *T.* somewhat long. *Po.* minute, vinous- or orange-brown, not dark. *Flesh* vinous-brown, not dark.

Single to 3-imbricate. Trunks, beech, fir. Nov.–Dec. P. 5⅝ in. Bearing the same relationship to **1569** as **1544** does to **1543**.

1571. P. spumeus Fr. (from its early condition of frothy-softness; *spuma*, froth). Ivory- or ochre-white.

P. pulvinate, gibbous, somewhat thick, rugoso-hispid to plane. *T.* somewhat short. *Po.* minute.

Single to 3-imbricate. Trunks and stumps, hornbeam, elm, apple, ash, willow, beech, birch, oak. July–Oct. P. 7½ in.

1572. P. borealis Fr. (from its being common in northern latitudes; *borealis*, northern) *a c.* White to yellowish-buff.

P. subpulvinate, hairy, somewhat thick, fibrous. *Po.* small, sinuoso-flexuous, torn; marg. barren.

Single to 3-imbricate. Stumps, pine, fir, spruce. Sept. P. 3⅜ in. Hartig states that this fungus is the cause of "white-rot" in spruce. Berkeley considers that *Ceriomyces* (*Ptychogaster*) *albus* may be an imperfect state of this species or of **1559**.

b. Placodermeæ.

1573. P. dryadeus Fr. (from its habitat, oak, Gr. *drus*) *a b c.*

P. pulvinate, thick, rugged to smooth, somewhat ferruginous, becoming fuscous. *T.* long, pale ferruginous. *Flesh* vinous-sienna, not dark, streaked and zoned darker.

Single to 2-imbricate. Taste somewhat acid or astringent. Trunks, oak, beech; frequent. July–Oct. P. 9 in. Exuding drops all over, which become black in drying.

1574. P. betulinus Fr. (from its habitat, birch, *betula*) *a b c.*

P. hoof-shaped, zoneless, smooth, thick, the vertex in the form of an um. or short stem, pellicle thin, cracking, pale umber-whitish, becoming darker. *T.* very short. *Po.* minute, white.

Trunks, branches, birch; common. May–Jan. P. 8 in. Pores sometimes stained by *Hypomyces rosellus* and *Hypocrea ochracea.* Used by rustics in past times for razor-strops. Said by Hartig to cause the "red-rot" of birch.

1574a. P. benzoinus Fr. (from its odour of benzoin) *a.*

P. woody, conchiform, but constricted at the base, fuscous-rubiginous, somewhat zoned, rugose when dry, and becoming darker. *T.* long. *Po.* at first whitish, then ferruginous. *Flesh* yellowish-white.

Subimbricate. Fallen cedars. Jan. P. 7 in. Sweet-scented, odour retained in drying. Allied to *Trametes.*

1575. **P. quercinus** Fr. (from its habitat, oak, *quercus*) *a b c*.

P. pulvinate, very thick, soft to hard, ivory- or sulphur-whitish, clouded and margined reddish. *T.* short, minute, whitish, clouded as P. *Flesh* ivory-white, zoned, orange-red and pale sulphur.

Single to 3-imbricate. Taste intensely bitter. Old oaks, oak-logs, often in company with **1517**, which it somewhat resembles. May–Dec. Mass **6** in. P. 3½ in. Often barren; sometimes 1 ft. in diam.

LXIII. **FOMES** Karst.

(From the use in kindling fire in past times; *fomes*, touch-wood, tinder).

Pileus hard and woody from the first, texture consisting of interwoven hyphæ, covered with a hard, rigid, crustaceous cuticle, zone-

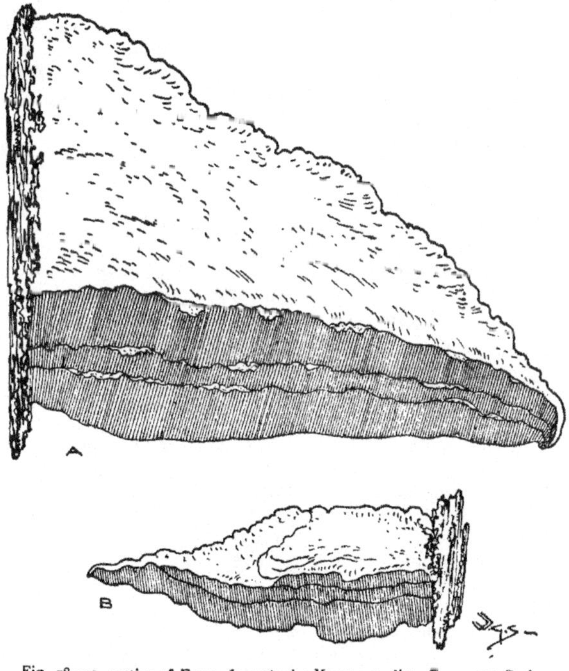

Fig. 78.—A, section of *Fomes fomentarius* Karst.; B, ditto *F. annosus* Cooke. One-half natural size.

less but often concentrically sulcate, perennial, forming successive strata. (Fig. 78.)

a. Fomentariæ. *Pileus* pulvinate, juiceless, not fleshy or spongy, floccose, covered with a hard, horny crust. *Tubes* stratose after the first year. Perennial, lasting for twenty years or more, adding new strata annually. Growing on wood.

Substance of tubes umber or fuscous. **1576**
Substance of tubes somewhat ferruginous. **1577—1585**
Substance white or pallid. **1586—1589**

b. Lignosæ. *Pileus* hard, woody, juiceless, covered with a thin, somewhat lustrous crust, smooth or becoming so, somewhat effused at the base. Flatter than *Fomentariæ* or effuso-reflexed. **1590—1594**

a. Fomentariæ.

1576. F. vegetus Karst. (from its vigorous growth, *vegetus*) *a c.*

P. concentrically and radiato-corrugate or rugose, deep olive-slate, dark brown behind; marg. sometimes brown. *T.* somewhat long, single to 5-stratose. *Po.* minute, ivory-brownish, surface brown-white; marg. barren. *Flesh* thin, floccoso-lax, yellow-brown or vinous warm-brown.

Single to 3-imbricate. Trunks, lime, elm; rare. P. 7¾ in. diam. *Polyporus Stevenii* Leveill., according to Bresadola.

1577. F. applanatus Karst. (from its flattened shape; *plano*, to level) *a b c.*

P. flattened, tuberculose, vaguely rugged-sulcate-zoned, smooth, cuticle crustaceous, fragile, somewhat shining, cinnamon or brown, becoming greyish with age. *T.* long, single to 6-stratose, subferruginous. *Po.* roundish, very minute, white, brownish when bruised. *Flesh* thick, floccoso-soft, rich brown.

Sometimes imbricate. Odour acid. Trunks, ash, beech, fir, oak, birch, poplar; uncommon. July–Mar. Diam. 1 ft. According to Bresadola *Polyporus Inzengæ* Fr. and *Polyporus roburneus* Fr. are both forms of this. Much softer than **1578.** Sometimes weighing 4–5 lbs.

1578. F. fomentarius Karst. (from its former use as tinder, *fomentum*) *a b c.*

P. hoof-shaped, more or less triangular in section, concentrically rugged-sulcato-zoned, smooth, cuticle thick, hard, yellowish-brown-reddish, fuliginous- or dingy-brown; marg. paler, at first with a white bloom. *T.* long, single to 9-stratose, ferruginous. *Po.* subangular, white, then ferruginous or ivory-brown. *Flesh* floccoso-soft, orange-brown.

Trunks, oak, birch, beech, lime, hornbeam; common. Jan.-Dec. Diam. 8 in. Formerly used as tinder in its soft natural state. Generally destroys exterior of trunk for one quarter of circumference and from base to 20 ft. up. Said by Hartig to cause "white-rot" of beech and oak. Occurs in peat-beds in the eastern counties and in the lake-side pile-dwellings of Switzerland and Italy. Formerly used as a hæmostatic. The ashes are used by the Ostiaks and Kamtschatkans as snuff.

1579. **F. nigricans** Karst. (from its black pileus : *niger*, black) *a c*.
P. hoof-shaped or pulvinate, very thick, deeply and concentrically sulcato-zoned, cuticle very hard, shining, black; marg. very obtuse, ferruginous. *T*. somewhat long, single to 13-stratose, ferruginous. *Po*. very small, obsoletely angular, dark yellowish-grey. *Flesh* very hard, yellowish-brown.

Living and dead birch, beech, hornbeam; uncommon. May–Oct. Diam. 5½ in. A var. of **1578** according to Bresadola.

1580. **F. igniarius** Karst. (from its former use as tinder; *ignis*, fire) *a b c*.
P. tuberculoso-globose, immarginate, then hoof-shaped, deeply sulcato-zoned, uneven, smooth, cuticle very hard, ochre to reddish- and blackish-brown; marg. rounded. *T*. long, single to 9-stratose. *Po*. very small, pale tawny, at length cinnamon, when old filled with white mycelium. *Flesh* very hard, zoned, ferruginous.

Trunks, ash, poplar, willow, cherry, plum, plane, apple, oak, fir, beech, alder, maple, birch; common. Mar.–Nov. Diam. 6½ in. Formerly used as a hæmostatic. The ashes are used by the Ostiaks and Kamtschatkans as snuff. Only used as tinder after baking, boiling and beating. Hartig considers this to be a parasite. Occurs in peat-beds in the eastern counties and in the lake-side pile-dwellings of Switzerland and Italy. The form on willow has been named *Polyporus pomaceus* Pers.

1581. **F. fulvus** Karst. (from the tawny pileus; *fulvus*, tawny) *a b c*.
P. ventricose, hoof-shaped, triangular in section, at first villoso-hairy, then even, rich tawny. *T*. short, ochre or brown-biscuit. *Po*. minute, colour as P. *Flesh* woody-corky, tawny.

Somewhat cæspitose and imbricate. Decayed trunks, beech, silver fir, spruce; rare. Oct.–Mar. Diam. 3⅞ in. Pores not always distinctly stratose. According to Bresadola the same as *Polyporus cinnamomeus* Trog.

1582. **F. pectinatus** Cooke (from the comb-like markings on the pileus; *pecten*, a comb) *a b c*.
P. flattened, thin, concentrically lamellato-plicate and tomentoso-scaly, whitish-brown, salmon or deep umber-sienna. *T*. short, yellow-brown; marg. barren. *Po*. very small, somewhat pale salmon-umber.

Imbricate. Trunks. Group 6⅛ in. diam. P. 4½ in. diam. A form of **1584**, according to Bresadola.

1583. **F. conchatus** Karst. (from its shape; *concha*, a shell) *a b c*.
P. flattened, thin, concentrically sulcate, somewhate corrugate, tomentose, deep yellow-brown or sienna-brown; marg. acute. *T*. short to long. *Po*. very small, yellow-orange or brownish, sometimes olive-shaded. *Flesh* thin, hard, corky, yellowish-brown.

Imbricate. Trunks, willow, beech; frequent. Feb.–Oct. Group 4½ in. diam. P. 2⅞ in. diam. Often wholly resupinate. A form of **1585** according to Bresadola.

1583a. F. Euonymi W. G. Sm., *Polyporus Euonymi* Kalchb. (from its host plant, *Euonymus europæus*) *a c*.

P. fleshy, rugose, cinnamon, darker with age, salmon at the marg., edge white. *T*. short, minute, cinnamon. *Po*. minute, salmon or whitish-cinnamon. *Flesh* cinnamon.

Mass $3\frac{3}{4} \times 3\frac{1}{4}$ in., projection 2 ins. Tubes sometimes multi-stratose, with flesh of pileus reduced to $\frac{1}{18}$ in. thick. Closely allied to *F. Loniceræ*.

1584. F. Ribis Cooke (from its habitat, currant and gooseberry bushes; *ribes*, currant) *a b c*.

P. thick, somewhat sulcato-zoned, effused amongst branches, somewhat even, indistinctly zoned, velvety, orange- or brown-sienna, becoming slate-brown behind. *T*. short, colour as flesh. *Po*. very small, yellow-sienna becoming sepia-cinnamon. *Flesh* corky-coriaceous to soft, brownish sienna-orange.

Connato-imbricate. Perennial. Common. April–July. Group $3\frac{3}{4}$ in. diam. P. $2\frac{1}{8}$ in. diam. Substance suitable for tinder. The same as **1582** and **1583a** according to Bresadola.

1585. F. salicinus Karst. (from its habitat, willow, *salix*) *a c*.

P. undulate, whitish-brown, yellowish-brown or black, 3–8 zoned, dark-umber. *T*. deep umber-sienna. *Po*. very small, ferruginous-cinnamon; marg. barren. *Flesh* woody, very hard, whitish- or yellowish-brown.

Cæspitoso-imbricate, for the most part or wholly resupinate. Common. Group $3\frac{7}{8}$ in. in diam. P. $1\frac{3}{4}$ in. diam. A form of **1583**, according to Bresadola. *Ceriomyces* (*Ptychogaster*) *spongia* is said by Saccardo to be an imperfect state of this.

1586. F. roseus Cooke (from its rose colour) *a b c*.

P. thin, somewhat flattened, even, subrugose, shining dull rose or rose-red, dark zoned; marg. dark. *T*. somewhat long; marg. barren. *Po*. minute, colour as P. *Flesh* corky-woody, floccoso-fibrous, colour as P. or paler.

Imbricate, cæspitose. Stumps, apple, dressed wood. Oct. P. $2\frac{3}{4}$ in. diam. Colour permanent in drying. The same as *F. rufopallidus* and **1592** according to Bresadola.

1587. F. ulmarius Cooke (from its habitat, elm, *ulmus*) *a b c*.

P. thick, convex, tubercular, uneven, becoming smooth, white to leather-buff, blackish in old examples; marg. white, becoming tawny. *T*. single to 6-stratose, ivory, bright salmon or buff-vinous, brown in decay. *Flesh* corky-woody, very hard, white.

Connato-imbricate. Single. Odour disagreeable. Inside old elms. Sept.–Feb. P. $6\frac{5}{8}$ in. diam.

1588. F. cytisinus Cooke (from one of its habitats, laburnum, *Cytisus*) *a b*.

P. thick, convex, smooth, whitish, biscuit, or pale brownish. *T*.

somewhat short, white-salmon, marg. barren. *Po.* biscuit or with a sulphur shade. *Flesh* thick, close, velvety-fibrous, white.

Single to imbricate. Stumps, dressed wood; rare. P. 6⅛ in. diam. Bresadola considers this and **1590** to be the same.

1589. F. connatus Cooke (from its connate habit) *a b c.*

P. somewhat thick, villous, whitish-buff, brown shaded. *T.* somewhat short, colour as P.; marg. barren. *Po.* minute, white to cream. *Flesh* corky-woody, pale buff to full buff.

Effuso-reflexed, connato-imbricate. Trunks, lime, apple, elm, maple; frequent. Oct.–Jan. P. 6⅛ in. diam. See **1631.**

b. Lignosæ.

1590. F. fraxineus Cooke (from its habitat, ash, *fraxinus*) *a b c.*

P. thick, flat, concentrically sulcato-plicate, zoneless, white or ivory to rich brown; marg. whitish. *T.* decurrent, moderate in length, white, cinnamon- or pale vinous-white. *Flesh* whitish-cinnamon.

Solitary to connate. Perennial. Odour strong, penetrating. Frequent. Jan.–Dec. P. 5½ in. diam. Bresadola considers this and **1588** to be the same.

1591. F. variegatus Cooke (from the variegated pileus, orange and bay-brown) *a b.*

P. thin, somewhat plane, even, smooth, shining tawny-orange, darker or bay-brown behind. *T.* decurrent, moderate in length, colour as P. *Po.* minute, unequal, torn, pale yellowish-white, shaded biscuit; marg. barren. *Flesh* pale buff.

Subimbricate. Trunks; rare. P. 4½ in. diam. Bresadola says this is unknown, and probably a form of *Polyporus ungulatus* Schæff.

1592. F. carneus Cooke (from its flesh-colour; *caro*, flesh) *a c.*

P. somewhat thick, convex, rugose, smooth, salmon-buff to reddish-salmon, varying deep rich brown behind, pale at marg. *T.* shallow to moderate, full opaque-salmon. *Flesh* pale bright salmon or red-salmon, varying rose-brown; marg. barren.

Connato-imbricate. Perennial. Trunks, stumps, juniper, *Picea*. Nov. P. 2⅜ in. diam. A form of **1586** according to Bresadola. Said to be the cause of "red-rot" in Red Cedar, *Juniperus virginiana*.

1592a. F. castaneus Cooke (from the colour; *castaneus*, chestnut) *a.*

P. coriaceo-woody, flattened, smooth, glabrous, zoneless, chestnut-colour. *Po.* small, yellowish. *Flesh* white.

Imbricato-connate. Black poplar. 1⅞ × 1⅜ in. ⅝ in. projection.

1593. F. annosus Cooke (from its growing in yearly strata; *annus*, a year) *a b c.*

P. convex, rugged, radiato-sulcate, sulcato-zoned, at first silky, then crusty, cinnamon-ochre when young, then deep rich vinous- or sienna-brown; marg. almost white. *T.* moderate in length, stratose; whitish-buff. *Po.* medium, salmon-white. *Flesh* colour as T.

Imbricate. About stumps, posts, roots, squared timber, pine, spruce, fir, larch, juniper, birch, beech, *Sequoia gigantea;* very common. Jan.–Dec. P. 7⅝ in. diam. Sometimes wholly resupinate. Some forms of this are *Trametes radiciperda*, said to cause "red-rot" and piping in larch, the mycelium growing from the roots to the trunk. Sometimes phosphorescent in mines and seen 20 yards off. The white mycelium so often seen running over the hymenium belongs to *Verticillium microspermum*, the conidial condition of *Hypomyces Broomeanus*.

1594. F. populinus Cooke (from its habitat, trunks of poplar, *populus*) *a b c.*

P. rigid, zoneless, rugose, villous, warm-biscuit. *T.* deeply decurrent, not stratose, shallow. *Po.* minute, colour as P. *Flesh* colour as P.

Nearly solitary to subconnate or imbricate, for the most part resupinate. Aug.–Jan. P. 2½ in. diam. *Polyporus connatus* Weinm. (not Fr.) and **1631** are forms of this according to Bresadola; see **1631**.

LXIV. **POLYSTICTUS** Fr.

(From the pores or punctures on the hymenium; Gr. *polus*, many, *stiktos*, punctured.)

Pileus coriaceous, membranaceous or somewhat spongy, cuticle thin and loosely fibrous. *Tubes* first appearing in the centre and successively developing towards the margin, at first superficial and distinct, then becoming deeper and more crowded, not stratose. (Fig. 79.) Species **1595—1608**

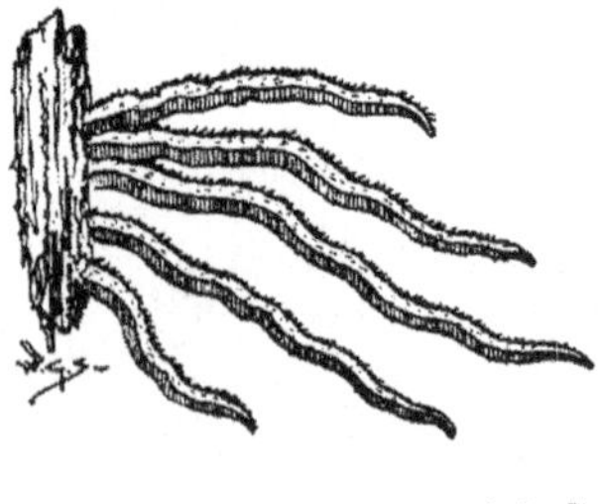

Fig. 79.—Section of *Polystictus versicolor* Fr. One-half natural size.

1595, 1595a and **1596** are placed in *Polystictus* by Cooke and Massee, in *Poria* by Saccardo.

a. Stupposæ. *Pileus* flocculose, or adpressedly villous or smooth, uneven, zoneless, texture fibrous.

Substance coloured. **1595—1597**

Substance white. **1598, 1599**

b. Coriaceæ. *Pileus* coriaceous, villous, zoned, usually in different colours. **1600—1605**

a. Stupposæ.

1595. P. radiatus Cooke (from the radiato-rugose pileus) *a b c.*

P. velvety to smooth, deep rich brown to dull orange; marg. zoned bright sienna. *T.* colour as flesh. *Po.* minute, greyish-salmon, silvery-glistening, at length ferruginous. *Flesh* rigid, corky, leathery, dull orange.

Imbricate. Not perennial. Trunks, beech, birch, oak, hazel, alder, gorse; frequent. Oct.–April. P. 4½ in. diam. Must not be confounded with **1603**.

1595a. P. nodulosus Cooke (from the nodular pileus) *a*.

P. triquetrous, nodular, corky, woody, not zoned, villous, rugose, rough, fulvous then rust-colour. *T.* somewhat long. *Po.* roundish, unequal, acutely torn, light cinnamon with a silvery sheen. *Flesh* very hard, lighter than P.

Imbricate. Dead beech. Oct. 2¼ × ⅝ in. Distinguished from **1595** by the at length confluent pores and by the absence of the brightly coloured margin of pileus.

1596. P. polymorphus Cooke (from its many forms; Gr. *polus*, many, *morphe*, form) *a*.

P. smooth, somewhat pale vinous-brown, dark shaded; marg. crisped. *T.* deeply decurrent. *Po.* somewhat large, angular, torn, colour as flesh. *Flesh* coriaceous, salmon.

Connato-imbricate. Branches, worked wood, fir. Aug. Group 2⅝ in. diam. P. 1 in. diam. Sometimes resupinate.

1597. P. cryptarum W. G. Sm., *Boletus cryptarum* Bull. (from its frequent habitat—cellars; *crypta*, a vault) *a b c*.

P. corrugate, zoneless, adpressedly silky, pale yellowish-vinous-brownish. *T.* deeply decurrent, long. *Po.* minute, colour as flesh. *Flesh* corky-cottony, paler than P.

Connato-imbricate. Fir; frequent. Group 4⅜ in. diam. Transferred to *Fomes* by Cooke and Saccardo. The same as **1608** according to Massee.

1597a. P. ravidus Cooke (from the grey or tawny pileus; *ravus*, grey) *a*.

P. coriaceo-corky, flattened, zoned, villoso-rugose, ochreous, slightly tinged brown-olive, base darker, sienna. *T.* somewhat short. *Po.* larger at mid., then smaller and absent at marg., biscuit-white. *Flesh* ivory-white.

Imbricate. Stumps of old willows. Spread 5¾ in., 1 in. thick. Colour variable, usually yellowish.

1598. P. gossypinus Cooke (from the white-cottony pileus; *gossypium*, cotton) *a c*.

P. tomentose, zoneless; marg. slightly scaly, white. *T.* long, but varying. *Po.* labyrinthine, then angular, pallid cinereous.

Somewhat connato-imbricate. Trunks, furze. Jan. Diam. 2½ in.

1599. P. fibula Fr. (from its button shape; *fibula*, a button) *a b c*.

P. velvety-hairy, zoneless, sometimes radiato-rugose, white. *T.* varying in length. *Po.* small at length torn, white, sometimes faintly sulphur.

Single to subimbricate or in troops. Wood in cellars, forests, elm, oak. Oct.–Jan. P. 2 in. diam. Sometimes shield-like, attached behind a centre of pileus, sometimes pendulous, at other times hoof-shaped and triangular in section.

b. Coriaceæ.

1600. **P. hirsutus** Fr. (from the hairy pileus; *hirsutus*, hairy) *a b c.*
P. becoming plane, depressed behind, zoned with furrows, varying white, pale slate-brown or brassy-yellow; z. yellowish-brown. *T.* long, deep, ochreous-white. *Po.* torn, whitish-salmon. *Flesh* thin, corky-coriaceous, colour as T.

Single to imbricate, effused behind. Dead trunks, branches, willow, oak, beech, poplar, plum. Oct.–Mar. P. 1⅜ in. diam. Pores much larger than in **1603.**

1601. **P. velutinus** Cooke (from the velvety pileus; *velluetum*, velvet) *a b c.*
P. plane, slightly zoned, white to ivory-ochreous; marg. white. *T.* short, white. *Po.* minute, white to yellowish; marg. barren. *Flesh* corky-coriaceous, white.

Single to connato-imbricate. Stumps, branches, etc., willow, beech, fir, birch; uncommon. Nov. Group 2¾ in. diam. P. 1¼ in. diam.

1602. **P. zonatus** Fr. (from the zoned pileus) *a b c.*
P. convex, tubercular-gibbous behind, villous; z. unicolorous to varying ivory, ochre, brown, rufous, sienna, etc., or whitish-yellow, graduated and zoned sienna. *Po.* minute, biscuit. *Flesh* corky-coriaceous, yellowish-white.

Single to subimbricate. Trunks, elm, poplar, birch. June–Oct. P. 3¼ in. diam.

1603. **P. versicolor** Fr. (from its variable colours) *a b c.*
P. even, velvety, shining pale smalt-blue, sooty-slate, or of various tints of ochre, ivory, brown or greenish. *T.* shallow. *Po.* minute, ivory; marg. barren. *Flesh* thin, rigid-coriaceous, ivory.

Imbricate. Dead wood, fences, branches, encircling twigs, back of an old brush, oak, willow, alder. April–Feb. P. 5½ in. diam. The tubes are sometimes bright orange or vermilion from the growth of a parasitic mould, *Hypomyces aurantius.* When growing on sticks, if the stick is inverted, a new hymenial surface with pores will appear on the downward pileus. Pores sometimes wholly eaten away by insects. **1774** sometimes grows on this. Var. *fuscatus* Mass. *P.* brown. *Po.* torn into teeth, yellow.

1604. **P. abietinus** Cooke (from its habitat, fir-trunks; *Abies*, fir) *a b c.*
P. villous, white, zoned dove-grey; marg. sometimes purplish. *T.* very shallow. *Po.* unequal, pale slate, becoming pale brown to vinous- or purplish-sienna. *Flesh* very thin, pale brown.

Connato-imbricate, effuso-reflexed. Fir, cedar, cypress; common. Jan.–Nov. Group 3¼ in. diam. Pileus frequently overgrown by green algæ. Sometimes wholly resupinate with a broad white woolly margin. Must not be confounded with **1610.**

1605. P. Wynnei Cooke (after Mrs. Lloyd Wynne) *a b c.*

P. membranous, marked with silky raised lines, faintly brownish, yellow-white, sometimes orange or tan. *T.* varying in length. *Po.* small, whitish.

Effuso-reflexed, almost wholly resupinate. Running over twigs, grass, etc., around stems of *Carex paniculata.* Sept.-Dec. Group 3 in. diam. Pore surface with barren places in middle and at margin.

LXV. PORIA Pers.

(From the porous hymenium.)

Resupinate, forming more or less extended patches or thin membranous expansions. *Pores* forming a continuous stratum, springing from a thin, often rudimentary subiculum. (Fig. 80.)

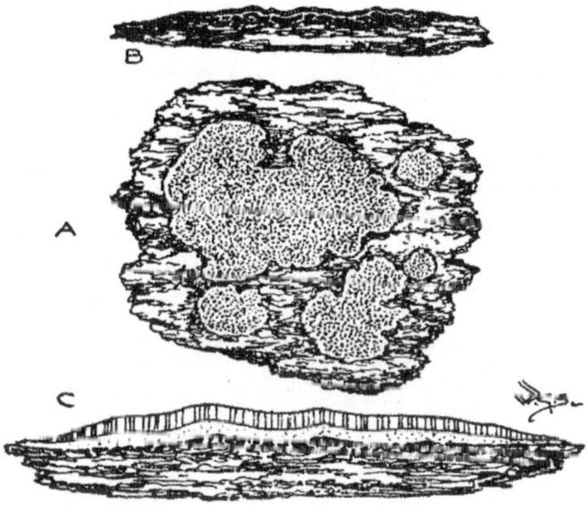

Fig. 80.—*Poria vaporaria* Cooke.
A, entire; B, in section. One-half natural size. C, section. × 4.

This is almost entirely the section of *Polyporus* named *Resupinatæ* by Fries, the sequence adopted is that of the *Hymenomycetes Europæi.*

Species **1605a—1648**

Pores ferruginous or cinnamon. **1605a—1609**
Pores purple or violaceous. **1610, 1611**
Pores red. **1612—1615**
Pores yellowish. **1616—1620**
Pores at first white, changing colour. **1621—1627**
Pores usually persistently white, minute crowded, round, equal. **1628—1635**
Pores persistently white, unequal, angular, crowded, commonly somewhat large. **1636—1643**
Pores white, superficial, distant, punctiform. **1644—1646**
Of uncertain affinity. **1647, 1648**

1605a. P. obliqua Achar. (from the oblique growth of the tubes) *a c*.
Sub. very thin; marg. barren, sometimes recurved, forming a narrow P. *T*. more or less elongate owing to oblique growth, usually minute, $\frac{1}{10}$ to $\frac{1}{5}$ in. long when not oblique, orange-brown or cinnamon, sometimes of two strata. *Flesh* colour of Po.

Trunks, logs, rotting wood. 4 × 2¼ × $\frac{1}{16}$ in. Regarded as a species of *Fomes* by Cooke and Saccardo.

1606. P. umbrina Cooke (from its brown colour; *umbrinus*, brown-coloured) *a*.
Undulato-tubercular, full to dingy umber; marg. smooth salmon- or vermilion-white, paler than pore-surface. *Sub*. thin but variable. *T*. somewhat long, varying. *Po*. minute, roundish, unequal; sienna.

Trunks. 3⅜ in. *Fomes umbrina* Saccardo.

1607. P. resupinata W. G. Sm., *Boletus resupinatus* Bolt. (from its resupinate habit) *a*.
Undulato-tubercular, spongy-coriaceous, umber, yellowish vinous-umber or ferruginous. *Sub*. thin fibrous. *T*. long, varying, sometimes stratose. *Po*. minute, pore surface nodulose to plane, umber.

Broadly effused. Trunks, branches, hazel. Feb. 3¾ in. This may be a resupinate form of **1560** or **1561**. Compare **1605a**. *Fomes resupinata* Massee.

1608. P. ferruginosa Karst. (from its rusty colour; *ferrum*, iron) *a b c*.
Undulato-tubercular, ferruginous or cinnamon; marg. barren, sometimes reflexed as a narrow P., ochre or rust colour, with white mycelium at edges of new growths. *Sub*. thin but varying. *T*. long, varying, sometimes 2-stratose or reduced to $\frac{1}{32}$ in. *Po*. medium to minute, brown. *Myc*. white to ochreous.

Imbricate, broadly effused. Trunks, branches, posts, sticks, abele, beech, oak, pine, alder, bramble, sometimes on dead leaves; common. Oct.–May. 5½ in. At first appearing as a confervoid stratum. *Fomes ferruginosa* Massee.

1609. P. contigua Karst. (from its close investment of sticks, etc.; *contiguus*, continuous) *a c*.
Somewhat plane, cinnamon or yellowish-sienna, dingy when old; marg. barren. *Sub*. thin, from ⅛ in. to a membrane. *T*. short. *Po*. somewhat large, brownish. *Myc*. ochreous.

Effused, less so than in **1608**. Trunks, branches, sticks, rotten wood, etc., pines, sweet-gale, maple, hornbeam; uncommon. 3 in. Sometimes wholly investing branches.

1610. P. violacea Cooke (from its violet colour) *a c*.
Plane, purple with marg. paler or whitish, or deep purplish to crimson-brown, with marg. light or ivory, clouded rufous.

Sub. a film. *T.* short. *Po.* round or wrinkled as in *Merulius.*

Effused, determinate. Stumps, trunks, poles, fir; frequent. Aug.–May. 4 in. Allied to *Merulius.* Must not be confounded with **1604**, in which the pores are deeper and torn and the margin reflexed.

1611. P. purpurea Cooke (from its purple colour) *a c.*
Plane, purple-lilac; marg. barren, paler. *Sub.* a film. *T.* short, *Po.* minute. *Myc.* mucedinous, flocculose, white.

Stumps, willow, beech. 3¾ in. The violet colour disappears in drying.

1612. P. incarnata Cooke (*incarnatus*, flesh-colour) *a c.*
Plano-undulate, coriaceous, marginate or immarginate, dark flesh-colour or orange, becoming dark purple-brown-crimson or purple-blackish; marg. broad, white or ochreous, sometimes vinous-clouded, sometimes shortly reflexed as a narrow P. *Sub.* a film. *Po.* short, unequal.

Effused. Fir; rare. June–Nov. 6½ in. Sometimes with barren patches. Must not be confounded with **1604.**

1613. P. rufa Cooke (from its reddish colour; *rufus*, red) *a c.*
Plano-undulate, umber, vinous-umber, blood-red-rufous or deep reddish; marg. barren, byssoid when young, greyish. *Sub.* a film. *Po.* short.

Effused. Trunks, sticks, branches, etc., broom, beech, bird-cherry. 3⅛ in.

1614. P. rhodella Cooke (from its rose-colour; Gr. *rhodon*, a rose) *a.*
Plane, soft, white, flesh-colour, vermilion-salmon or Indian-red-salmon; marg. determinate, salmon. *Sub.* a film. *Po.* short, minute, not torn.

Effused. Trunks, fir, beech. Aug. 4¾ in. Thinner than **1613.**

1615. P. micans Ehrb. (from its glittering pores; *mico*, to glitter) *a c.*
Plane, often with rain-drop-like barren depressions on surface, whitish-flesh or biscuit-colour; marg. flaxy, byssoid, white. *Sub.* a membrane. *T.* short. *Po.* minute, irregular, angular.

Effused, orbicular, confluent. Dead wood, rotten trunks, ash. Nov. 3⅜ in. Fugacious.

1616. P. nitida Pers. (from its shining pores; *nitidus*, shining) *a c.*
Plane, bright yellow to pale sienna-yellow; marg. villoso-determinate. *Sub.* a membrane. *T.* short. *Po.* round, minute.

Effused, subadnate. Dead wood; rare. Mar. 3½ in.

1617. P. Læstadii Cooke (after C. P. Læstadius) *a b.*
Plano-tubercular, brittle, separable, bright lemon-yellow, whitish-sulphur or ivory-salmon when dry; marg. not barren, edge minutely and narrowly fibrillose. *Sub.* thin, white. *T.* short, of different depths. *Po.* very minute, sinuous.

Effused, sometimes stratose as if perennial. Deal boards, sometimes in green-houses. Aug. 3½ in.

1618. P. bombycina Cooke (from the silky membranous subiculum; *bombyx*, silk) *a b c*.

Sub. very thin, translucent; marg. byssoid. *Po.* large, angular, sinuous; marg. arachnoid-velvety, pale buff or white.

Effused. Rotten wood; rare. Sept. 3 in. The grain of the wood sometimes shows through the extremely thin substance.

1619. P. aneirina Cooke (from the margin not being woolly as in **1618**; Gr. *an*, not, *eirineos*, woolly) *a c*.

Sub. thin, coriaceous, somewhat flaxy; marg. byssoid. *Po.* shallow, large, waxy, fulvous, tawny, warm biscuit-colour or bright orange-yellow; marg. flaxy, becoming waxy, white. *Myc.* buff-white.

Effused. Dead wood, branches, poplar, pine under the bark; rare. Oct. 4 in. Contracting and becoming torn in drying.

1620. P. ramentacea Cooke (from one of its habitats, chips, *ramenta*) *a b*.

Sub. thin, tomentose, white. *Po.* somewhat large, hexagonal, honey-colour; marg. almost obsolete, irregular, warm-ivory.

Somewhat orbicular. Dead branches, fir. Sept. $1\frac{1}{2} \times 1$ in. Cartilaginous and horny when dry. Barren patches occur on the hymenium.

1621. P. cincta Cooke (from the encircling, erect strigose fibres; *cingo*, to encircle) *a*. White to ivory.

Sub. membranous, sometimes barren, and with lacunæ; marg. byssoid. *Po.* extremely minute.

Old deal boards, rotten planks; rare. $3\frac{3}{4}$ in. Patches of vertical *strigæ* become confluent and form a subiculum.

1621a. P. fuscocarnea Cooke (from the brownish flesh-colour; *fuscus*, dark brown, *caro*, flesh).

Sub. thin; marg. definite. *T.* $\frac{1}{16}$ to $\frac{3}{16}$ in. long. *Po.* minute, fawn-colour when fresh or vinous-brown, graduated paler.

Rotten wood, putrid trunks. $3\frac{1}{2} \times 3$ in.

1622. P. armeniaca W. G. Sm., *Polyporus armeniacus* Berk. (from its apricot-colour; *armeniacum*, apricot) *a b*.

Sub. membranous to $\frac{1}{4}$ in. or more thick; marg. thin, well defined, minutely downy or byssoid. *Po.* shallow, to $\frac{1}{8}$ in. deep, rounded, somewhat irregular, minute, white, then bright buff, or glancing with the light, ochreous on side view, bright foxy-brown on full view.

Broadly effused, suborbicular, confluent. Bark, pine. Nov.–April. $3\frac{1}{2} \times 1\frac{1}{2}$ in. Sometimes a rugged, free pileus is formed. Cooke refers this to **1566**, but the spores are different in shape and size.

1623. P. Rennyi Cooke (after James Renny) *a b*.

Sub. thin to somewhat thick, pulvinate, at first somewhat frothy, then dry and pulverulent; marg. byssoid. *Po.* minute, few, irregular in depth, here and there elongated, white, becoming lemon or pale buff. *Myc.* gelatinous, partly byssoid.

Odour pungent. On the ground, rotten wood, running over *Marchantia*, fir. Nov. $2\frac{1}{2} \times \frac{3}{4}$ in. Pores sometimes absent or subiculum with large barren patches.

1624. **P. subfuscoflavida** Cooke (from the pale brownish-yellow colour; *subfuscus*, somewhat dusky, *flavidus*, yellow) *a c*.
Sub. thin, coriaceous, dry; marg. thin, byssoid, determinate, white. *Po.* minute, irregular, white, then yellowish-brown or glancing white and grey-brown.

Broadly effused, confluent. Dead oak wood, planks, etc. 2⅜ in.

1625. **P. viridans** Cooke (from its pale æruginous colour; *virido*, to become green) *a c*.
Sub. thin. *Po.* minute, shallow, white, ivory or pale buff, then pale verdigris; marg. pulverulento-tomentose.

Effused, crustaceo-adnate. Rotten wood; rare. Sept. 2⅜ in. Sometimes a mere thin, pale æruginous film without pores.

1626. **P. terrestris** Cooke (from its terrestrial habit) *a c*.
Sub. very thin; marg. arachnoid flaxy, fugacious with a byssoid white edge. *Po.* shallow, minute, angular, becoming smaller and scattered towards the wide, barren marg., white, becoming pale umber or faintly rufescent.

Effused. Black earth, red clay, rotten wood; uncommon. May–Nov. 4 in.

1627. **P. subgelatinosa** Cooke (from the somewhat gelatinous subiculum) *a b*.
Orbicular, forming small pulvinate masses with an obtuse raised border, at first tomentose and pallid, becoming black and subgelatinous. *Po.* somewhat large, others small, variable in size, pale grey.

On dead wood, on 1560. 1⅝ in

1628. **P. medullapanis** Cooke (from its colour like pith or bread; *medulla*, pith, *panis*, bread) *a b c*.
Somewhat undulate, firm; marg. barren, whitish, sometimes upturning and separating from matrix. *T.* somewhat long. *Po.* medium size.

Decayed branches, mulberry, oak, hornbeam, on the ground; uncommon. Sept.–Feb. 4⅛ in. Sometimes in certain positions 4–6 stratose.

1629. **P. mucida** Pers. (from its mucid substance) *a c*.
Sub. somewhat thick, soft, elastic; marg. barren, indeterminate, flaxy. *Po.* shallow, medium size, unequal, torn, white to ochre-white. *Myc.* spreading and often forming small orbicular, transparent, barren patches.

Effused in long patches. Decayed fir. 3⅛ in. Thicker than 1634.

1630. **P. vitrea** Pers. (from its shining glassy appearance as if wet; *vitrum*, glass) *a c*.
Sub. somewhat thick. *T.* somewhat long, translucent, whitish. *Po.* minute, entire, roundish, whitish, becoming buff-white or vinous-buff-whitish.

Broadly and unequally effused. Dead wood, fir, on the ground under beech; uncommon. Aug.–Mar. 3⅛ in. With the appearance of shining wax.

1631. P. obducens Cooke (from the covered-over or superimposed strata; *obduco*, to cover over) *a c*.

Sub. a hard, almost obsolete, inseparable film. *T.* somewhat long: $\frac{1}{16}$ to $\frac{1}{8}$ in. *Po.* minute, white, becoming pale dull salmon.

Effused. Rotten trunks, pine, oak, beech; rare. July–Feb. 3¼ in. Bresadola says this is a resupinate form of **1594**, Massee of **1589**. During the first year it resembles **1633**. Irregular in section, perennial by strata for six or more years.

1632. P. callosa Cooke (from the hard subiculum; *callum*, hardened skin) *a c*.

Sub. a firm, somewhat thick, leathery membrane, separable from the matrix, white, becoming umber. *T.* somewhat long. *Po.* round, equal, entire, white becoming biscuit; marg. white.

Broadly effused. Dead wood, oak, maple. April. 3½ in. Surface sometimes with small depressions as if from rain-drops. Annual, not stratose.

1633. P. vulgaris Cooke (from its commonness; *vulgaris*, common) *a c*.

Sub. a membrane, closely adnate to the matrix, almost obsolete. *Po.* somewhat shallow, white to biscuit or faintly ochreous, sometimes with a white, irregular barren marg., attached to thin, white, cloud-like growths of mycelium; marg. soon even and smooth.

Broadly effused. Dead wood, branches, pine, oak, beech; common. July–Mar. 5½ in. Patches often very large, sometimes wholly investing branches, sometimes with vacuoles. Annual not stratose. Appearance wax-like.

1634. P. mollusca Cooke (from its softness; *molluscus*, soft) *a b c*.

Sub. a membrane. *Po.* somewhat shallow, large in middle, smaller towards marg., white becoming pale biscuit or brownish-white; marg. broad, barren, white, radiate with white strands of mycelium.

Broadly effused. Dead wood, dead leaves, furze, poplar, ash; common. Jan.–Dec. 4¼ in. Sometimes proliferous or 2-stratose.

1635. P. collabefacta Cooke (from the pores which appear to arise from the collapsing substance; *collabefio*, to be brought to ruin) *a b*.

Sub. a membrane. *Po.* minute, shallow, white to ivory; marg. obtuse, barren, with a few barren places on surface.

Effused. Dead wood. Oct. 3½ in. Running over the matrix like a thin *Corticium*.

1636. P. sanguinolenta Cooke (from its changing to blood-red where bruised; *sanguis*, blood) *a b c*.

Sub. a membrane. *T.* somewhat shallow. *Po.* small, irregular, becoming torn, white to yellowish-white.

At first nodulose, soon confluent and effused. Decayed trees and branches, rails, dead wood, running over earth, clay and moss; uncommon. Aug.–Nov. 2¾ in.

1637. P. radula Pers. (from the small teeth, as of a comb, where the pores are elongated; *radula*, a scraper) *a c.*

Sub. a membrane, separable from the matrix. *T.* shallow, somewhat long in middle, sometimes 2-stratose. *Po.* tooth-like or comb-like, spinous near the sometimes barren marg., white to biscuit, not turning red.

Effused. Trunks, branches, fir, oak, hornbeam. Surface with small depressions, sometimes with barren places. Aug.–Jan. 3⅜ in.

1638. P. vaporaria Cooke (from the warm, moist air, in which it luxuriates; *vaporarium*, a stove) *a b c.*

Sub. a thin, firm, inseparable membrane. *P.* shallow, large, torn, white to ivory.

Broadly effused. Moist woods on dead branches, decaying wood, rails, pine, oak, poplar, alder, bamboo, running over dead leaves, on root wounds, barked and injured places, squared timber in buildings; very common. Jan.–Dec. 4 in. Hartig states this fungus to be as destructive as the "dry-rot" fungus **1680** on living spruce and pine. *Ceriomyces* (*Ptychogaster*) *rubescens* is said by Saccardo to be an imperfect state of this. Var. *secernibilis* Mass., separating from the matrix; white, then honey-colour; fir-leaves under moss.

1639. P. hibernica Cooke (from the place where first observed, Wicklow, Ireland) *a.*

Sub. a membrane, not separable from matrix. *Po.* small, shallow, polygonal, not fimbriato-toothed, white; marg. narrow, thin, radiato-byssoid.

At first small, orbicular, then confluent and broadly extended. Decorticated fir branches. Sept. 3¼ in.

1640. P. Gordoniensis Cooke (after the Marchioness of Huntly) *a c.*

Sub. a thin, extremely delicate film, separable from matrix. *Po.* minute, fimbriato-toothed, shining white, becoming faintly yellowish-white; marg. broad, barren, white.

Somewhat effused. Fir poles. Feb. 2⅜ in. Closely resembling **1639** in general appearance.

1641. P. Vaillantii Cooke (after Leon Vaillant, French botanist) *a b c.*

Sub. a thin film, here and there traversed by ribs. *Po.* rather large, shallow, produced here and there in clusters, white or slightly rufescent; marg. broad, barren, bordered by broad barren radiating growths ending in strands of mycelium.

Broadly effused. Dead wood. April–Oct. 6 in.

1642. P. farinella Cooke (from the pulverulent margin; *farina*, meal) *a c.*

Sub. a very delicate, thin film. *Po.* shallow, with a ragged, toothed appearance; marg. broad, barren, unwoven, edge pulverulent.

Broadly and irregularly effused. Old timber, dead wood, beech, oak, pine, cedar. Dec.–Mar. 4½ in. Fugacious to the touch.

1643. P. hymenocystis Cooke (from the cyst-like pores of the hymenium) *a b*.

Sub. very thin, arachnoid. *Po.* large, shallow, irregular, dissepiments scarious, collapsing, white, becoming pallid; marg. narrow, byssoid.

Black, rotten wood. 4 in.

1644. P. blepharistoma Cooke (from the eyelid-like appearance of the pores; Gr. *blepharon*, an eyelid, *stoma*, a mouth) *a b*.

Sub. a thin film. *Po.* small, dissepiments thin, edge finely toothed, white; marg. barren, edge minutely byssoid, somewhat mealy. *Myc.* arachnoid.

Dead wood, decorticated branches, leaf-mould. July–Nov. 4 in.

1645. P. corticola Cooke (from its growing on bark; *cortex*, bark, *colo*, to inhabit) *a*.

Sub. a film, firm. *Po.* very minute, often obsolete, or in small irregular patches, white to biscuit; marg. radiate, byssoid.

Poplar, birch, horse-chestnut, fir. 3½ in.

1646. P. reticulata Cooke (from the net-like hymenium; *rete*, a net) *a c*.

Sub. a thin film. *Po.* cup-like, somewhat large, white; marg. barren, radiato-floccose or byssoid.

Orbicular, confluent. Rotten wood. Sept.–Feb. 4 in. Very delicate, soon disappearing.

1647. P. bathypora Cooke (from the deep cup-like pores; Gr. *bathus*, deep, *poros*, a pore) *a*.

Sub. a thin film. *Po.* cup-shaped, somewhat large, dissepiments toothed, white, becoming brownish; marg. flaxy.

Effused. Dead oak, beech. 3⅛ in. Sometimes 2-stratose. Resembling the resupinate form of **1660**.

1648. P. hybrida Cooke (from the possibility of its being a monstrous form of **1559**) *a b c*.

Sub. a thick felt-like membrane, white, sometimes branched, sometimes forming small inflexed superimposed white pilei. *Po.* very minute to large, shallow, to ¼ in. deep, torn, white to biscuit; marg. irregularly radiato-branched.

On oak in ships, etc., causing "dry-rot." Jan.–Dec. 7 in. Sometimes a mere radiato-branched white membrane with a few small patches of pores or forming pendulous masses from horizontal beams, with tubes beneath.

LXVI. TRAMETES Fr.

(From the generic distinction depending on the *trama*—the substance which grows between the membranes of which the dissepiments of the pores are composed.)

Trama continuous and homogeneous with the flesh of the pileus. *Pileus*, when present, woody, corky-woody, corky or soft. *Stem* none. *Tubes* unequal in length, rarely stratose. *Pores* subround, more or less elongated radially, not labyrinthiform, entire, often unequal in depth, dissepiments somewhat thick. (Fig. 81.)

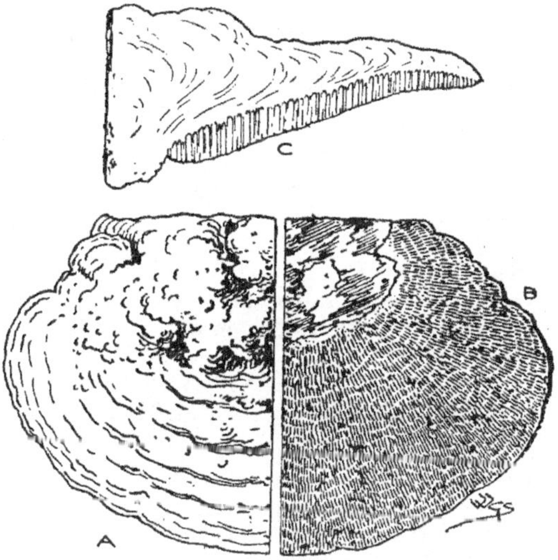

Fig. 81.—*Trametes gibbosa* Fr. One-half natural size.
A, upper surface; B, lower sufrace; C, section.

Several species are fragrant of anise, never acid. Growing on wood.

Intermediate between *Polyporus* and *Dædalea*.

Species **1649—1657a**

a. Apodæ. Without a stem.

Substance coloured.	**1649**
Substance white or whitish.	**1650—1654**
b. Resupinatæ. Resupinate.	**1655—1657a**

a. Apodæ.

1649. T. Pini Fr. (from its habitat, pine) *a b c*.

P. pulvinate, sulcato-rugged, sienna to deep sepia; marg. yellowish. *T.* somewhat long, brown-yellowish to bright ferruginous, rarely stratose. *Po.* large, irregular, oval. *Flesh* corky-woody, brownish to deep sienna.

Odour faint, pleasant. Trunks, pine, fir, larch; rare. Sept.-Oct. 5⅝ in. Often covered with algæ and lichens. Hartig states that this fungus is the cause of the diseases known as "bark-shake" and "ring-shake" or "heart-shake." The same as *Polyporus Demidoffii* Lév. according to Bresadola.

1649a. T. cinnabarina Fr. (from the red-vermilion colour of the pileus; Gr. *kinnabari*) *a*.

P. fleshy, corky, convexo-plane, pubescent to glabrous, somewhat zoned, rugulose, red-vermilion, not dark. *T.* and *Po.* somewhat crimson.

Connate. Trunks, birch, beech, etc. Oct. 3¼ × ¾ in.; projection 2 in.

1649b. T. Trogii Berk. ex Trog. (after J. G. Trog) *a*.

P. fleshy, convex, somewhat zoned, pilose, buff-white to pale buff. *T.* white-buff. *Po.* variable in size, subangular, toothed. *Flesh* white.

Connato-imbricate. Dead aspen. Oct. 5 × 2½ in.; projection 1½ in.

1650. T. gibbosa Fr. (from its thickness and convexity; *gibbus*, gibbous) *a b c*.

P. pulvinate, velvety, somewhat rugged behind, obsoletely zoned and tuberculate, whitish-ochre, greyish at the obtuse marg. *T.* long, faintly brownish-yellow. *Po.* round, oval, colour as P. *Flesh* corky, compact, yellowish-white.

Sometimes imbricate. Odour slight of anise when fresh. Stumps, posts, rails, oak, beech, willow, poplar; not common. Oct.–Feb. 6 in. Not always gibbous, sometimes thin.

1650a. T. rubescens Fr. (from the pores becoming sienna-red) *a*.

P. dimidiate, corky, thin, smooth, at length rugoso-zoned, white to rose-salmon and olive-brown, densely pulverulent. *Po.* roundish, irregular, white-salmon, clouded rose-colour.

Salix caprea. Oct. 1¾ × 1¼ in. Pores and flesh changing from white to crimson on being scratched.

1651. T. Bulliardii Fr. (after Pierre Bulliard) *a c*.

P. pulvinate, whitish, yellowish or brownish, sometimes zoned dark brown; marg. greyish or pale ochre. *T.* long, vinous-yellowish. *Po.* large, irregularly roundish or sinuous. *Flesh* faintly brownish-white, becoming fuscous.

Fragrant. Dead wood. Sept.–Oct. 5¾ in. Closely allied to *Dædalea*.

1652. T. suaveolens Fr. (from its fragrance; *suavis*, sweet) *a b c*.

P. pulvinate, villous, sometimes rugged, zoneless, whitish. *T.* long, sometimes irregular or short, whitish. *Po.* roundish, whitish, becoming fuscous. *Flesh* corky-soft, whitish or dull straw-colour.

Often imbricate. Odour of anise till dry. Trunks, willow, lime, etc.; uncommon. Oct.–Dec. 6¾ in. Often enveloping grass and twigs.

1653. T. odora Fr. (from its fragrance, *odor*) *a*.

P. pulvinate, uneven to smooth, zoneless, ivory-white or grey-whitish. *T.* shorter than in **1652**, white or pale buff-white. *Po.* minute, roundish, half the size of **1652**. *Flesh* corky-elastic, white.

Willow; rare. Jan. 3½ in.

1654. T. inodora Fr. (from its being scentless, *inodorus*) *a*.

P. expanded, thin, tomentose, smooth, zoneless, white or greyish-white. *T.* long, unchanging white. *Po.* minute, roundish. *Flesh* corky-firm, white.

Stumps, beech. 2¼ in.

b. Resupinatæ.

1655. T. mollis Fr. (from its soft substance; *mollis*, soft) *a b c*.

Sub. somewhat membranous, whitish. *T.* shallow. *Po.* large, cup-shaped, irregular, yellowish- or brownish-white, becoming fuscous, then black; marg. broadly barren, determinate, then revolute, pubescent beneath.

Often broadly effused. Dead wood, beech; common. Jan.-Dec. 4½ in. Probably lasting for two years.

1656. T. Terryi B. & Br. (after Michael Terry) *a*.

Sub. pulvinate, corky, suborbicular, 1 in. thick, whitish; marg. determinate, undulate. *T.* shallow, 4-stratose in examples first found. *Po.* irregular, with thin walls. *Flesh* dry, with 4–5 annual growths, white.

Odour none. Beech. 4 in.

1657. T. serpens Fr. (from its frequent sinuous creeping habit; *serpens*, creeping) *a b c*.

Sub. somewhat pulvinate, $\frac{1}{16}$–$\frac{1}{8}$ in. thick, arid, white. *T.* shallow to somewhat elongate. *Po.* large, somewhat round, unequal, ivory-white; marg. barren, determinate, pubescent.

At first orbicular, soon confluent. Hornbeam, privet, oak, oak chips in tan pits, squared oak timber, beech. Aug.-Jan. 5⅛ in. *Polyporus Stevensii* B. & Br. Must not be confounded with **1650.**

1657a. T. purpurascens B. & Br. (from the hymenium becoming purple).

Sub. somewhat leathery, subtomentose, chestnut. *T.* rigid, becoming purple. *Po.* small.

Dead willow. ¾ in.

LXVII. DÆDALEA Pers.

(From the Cretan labyrinth made by Dædalos—in reference to the intricate, labyrinthine pores.)

Agreeing with *Trametes*, except in the firmer and sometimes thicker dissepiments of the pores, which, when fully grown are irregularly sinuous or labyrinthine, the floccose substance of the hard pileus descending unchanged into the trama. (Fig. 82.)

Inodorous. Species **1658—1666**

Corky-woody, dimidiate, sessile. **1658—1661**

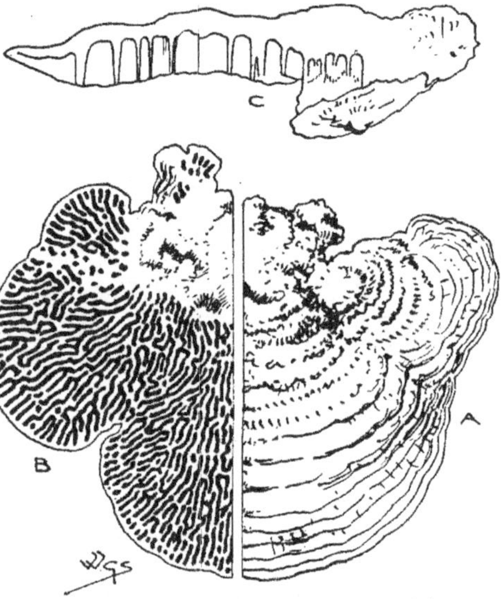

Fig. 82.—*Dædalea quercina* Pers. One-half natural size.
A, upper surface; B, lower surface; C, section.

1658. D. quercina Pers. (from its habitat—oak, *quercus*) *a b c*.
P. somewhat zoned, rugulose, uneven, somewhat smooth, buff-brown, darker behind. *T.* long. *Po.* at first simple, then labyrinthine, edge obtuse, whitish-salmon. *Flesh* pale sienna.

Single or imbricate; perennial. Sometimes on squared timber. Jan.-Dec. 7 in. Sometimes resupinate. Occurs in peat-beds in the eastern counties and in the lake-side pile-dwellings of Switzerland and Italy.

1659. D. aurea Fr. (from its golden colours; *aurum*, gold) *a c*.
P. somewhat zoned, rugulose, uneven, coarsely velvety, orange- or yellowish-brown. *T.* somewhat long. *Po.* simple then sublabyrinthine, opaque sienna-orange. *Flesh* sienna-white.

Single to imbricate. Dead oak. 2¾ in. Pores usually rounder than **1658.**

1660. D. confragosa Pers. (from the rugged pileus; *confragus*, rough) *a b*.
P. roughly rugulose and zoned, buff- or liver-brown. *T.* somewhat long. *Po.* moderately large, chiefly round and oblong, or

sublabyrinthine, pale umber, becoming darker. *Flesh* pale sienna.

Single, rarely subimbricate. Oak, willow, poplar, beech, service. Oct.–Feb. 5½ in. In drying smells of stinking fish.

1661. D. cinerea Fr. (from its ashy pileus; *cinereus*, ash-colour) *a c.*
P. undulate, zoned, tomentose, whitish-umber to greyish; z. darker. *T.* somewhat long. *Po.* round to oblong and labyrinthine, whitish to ashy or biscuit. *Flesh* white to biscuit.

Single to imbricate; perennial. Dead trunks. Jan.–Dec. 4⅞ in.

1662. D. polyzona Pers. (from the many-zoned pileus; Gr. *polus*, many) *a c.*
P. tomentose, yellowish-brown, darker or reddish behind; z. dark brown. *T.* somewhat long, sometimes in two strata. *Po.* sublabyrinthine, buff-white. *Flesh* somewhat thin, buff-white.

Imbricate. Jan.–Dec. 5½ in.

1663. D. unicolor Fr. (from its single colour; *unus*, one, *color*, colour) *a b c.*
P. somewhat rugged, villoso-strigose, zoned, yellowish-brown. *T.* moderately long. *Po.* labyrinthiform, sometimes of two strata, pale buff. *Flesh* pale buff.

Single to imbricate, sometimes resupinate and effused with small development of pileus. Stumps, rails, birch, beech, acacia, maple, oak, willow, chestnut, hornbeam; common. Jan.–Dec. 4½ in. Flesh sometimes much eaten by larvæ.

1664. D. latissima Fr. (from its broadly spreading habit; *latus*, broad) *a b c.*
Sub. from a film to ¼ in. thick, undulate. *Po.* of varying depths, small, round to sublabyrinthine, narrow, distant, colour of washleather; marg. barren.

Dead wood, branches, beech; rare. 5 in. Sometimes effused for 2 ft.

1665. D. vermicularis Pers. (from the colour, or the worm-like shape of the pores; *vermis*, a worm) *a.*
Sub. thin to ⅛ in. *Po.* shallow, partly elongated to ⅛ in. deep, sinuous, round near the barren marg., very pale salmon, or partly white and partly pale flesh-colour.

On the ground, in contact with wood and earth. Aug. 5½ in.

1666. D. ferruginea Schum. (from its colour of iron-rust; *ferrum*, iron) *a.*
P. at first white-villous, becoming yellowish-sienna or yellowish-brown; z. 2–3, umber. *T.* ⅛ in. deep, round to sublabyrinthine, tawny or yellowish-sienna; marg. barren, white-villous. *Flesh* ⅛ in., pale yellowish-sienna.

Imbricato-connate. Dead wood. 1¾ in.

LXVIII. **MERULIUS** Halle.

(From the common colours of the hymenium, yellow or orange, or partially black, the colours of the beak and plumage of the blackbird, *merula*.)

Resupinate. *Pileus*, when present, effuso-reflexed. *Stem* none. *Hymenophore* a soft, woven, mucedinous mycelium, covered with the

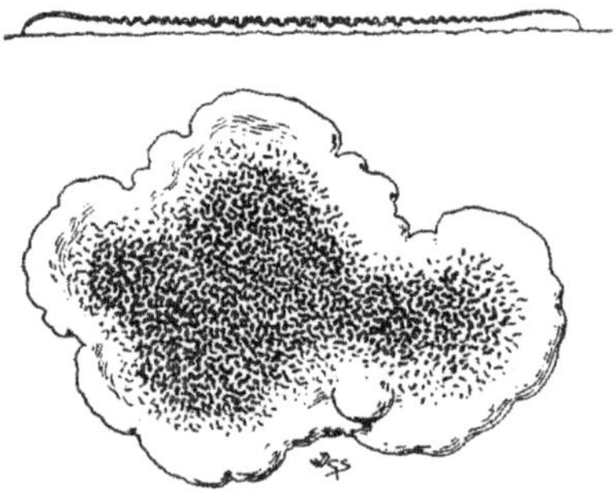

Fig. 83.—*Merulius lacrymans* Schum., entire and section above. One-third natural size.

waxy-soft, porous, gyrose or obsoletely toothed hymenium. *Spores* white or ferruginous. On wood, sawdust, leaves, mosses, etc., rarely on the ground. (Fig. 83.) Species **1666a—1681**

a. *Leptosporæ*. *Spores* white.

Pileus effuso-reflexed, circumference determinate. **1666a—1669**

Resupinato-effused, flaxy-membranous, separating, flaxy beneath and at the circumference. **1670—1674**

Crustaceo-adnate, somewhat flaxy at the circumference. **1675—1679**

b. *Coniophoræ*. *Spores* ferruginous. Hymenium at length pulverulent with spores. **1680, 1681**

a. *Leptosporæ*.

1666a. M. confluens Schwein. (from the habit, small growths becoming confluent into large patches) *a*.

Resupinate, longitudinally effused, somewhat fleshy, vinous biscuit-colour; marg. free, inflexed, subtomentose, biscuit-colour. *Po*. very small, uneven, reticulate.

When the small patches become confluent, the lines of junction become raised as irregular hexagons. Branches, alder. Aug. Small growths papillate, about 3/8 in. diam. Confluent masses about 3-4 in.

1667. M. tremellosus Schrad. (from the tremulous substance) *a b c.*
P. dentato-squarrose near marg., yellowish-white, salmon shaded. *Po.* pale ochre to yellowish-white, brownish-ochre at base; marg. barren, white. *Flesh* fleshy-tremellose, approaching *Tremella*, very pale blue-grey and pale salmon.

Connato-imbricate. Odour strong, fungoid. Decayed trees, wood, beech, birch, oak; rare. Aug.–Feb. 6 in. At first resupinate. Hymenium usually rich vinous-brown when dry.

1668. M. aurantiacus Berk. (from the colour; *aurum*, gold) *a.*
P. tomentose, sometimes zoned, dull yellowish-white, or ashy. *Po.* somewhat poriform, minute, dull orange; marg. broad, barren. *Flesh* fleshy-coriaceous.

Dead trunks, beech; rare. May. 1¾ in.

1669. M. corium Fr. (from the thin leather-like subiculum; *corium*, skin, leather) *a b c.*
P. becoming slightly reflexed, paper-like, zoned, villous-hairy, separating, white. *Po.* small, gyrose or pore-like, flesh-colour, pale tan, orange-yellow, orange or sienna-orange; marg. broad, barren, white. *Flesh* soft.

Dead wood, trunks, branches, hornbeam, oak, furze, moss, sometimes on old beams, etc., in houses, sometimes pendulous, like a small inverted cup; common. Mar.–Jan. 6¼ in.

1670. M. lœticolor B. & Br. (from the colour; *lætus*, bright) *a.*
Po. at first obsolete, then plicato-rugose, distant, bright orange; marg. broad, barren, white.

Coarse sawdust and leaves, oak, pine, mountain-ash. Nov. 3 in.

1671. M. aureus Fr. (from the colour; *aureus*, golden) *a c.*
Po. large, walls thin, plicato-porous, orange or golden-yellow; marg. broad, barren, sulphur.

Pine boards, pine-leaves and cones. Oct. 2⅛ in. Hymenium dark brown when dry.

1672. M. himantioides Fr. (from the thong-like shape of the folds of the hymenium; Gr. *himas*, a thong, *eidos*, appearance) *a b c.*
Po. porous to gyrose, folds tooth- or thong-like, yellow to dull ochre or yellow-olivaceous; marg. barren, white or sulphur.

Dead wood, fir, cherry, club-moss; rare. Sept.–Dec. 7¾ in. Hymenium olive or dark brown when dry.

1673. M. molluscus Fr. (from the soft substance; *molluscus*, soft) *a b c.*
Po. large, round, poroso-gyrose, flesh-colour, deep carnation or ochre to rich orange; marg. broad, barren, white. *Hym.* rufous or vinous-brown when dry.

Wood, branches, fir, sometimes spreading over and conjoining fallen twigs; uncommon. Sept.–Feb. 8 in.

1674. M. fugax Fr. (from its collapsing hymenium and fugitive habit; *fugio*, to flee) *a b c.* Byssoid to membranous, very thin.
Po. reticulate, feebly developed and soon collapsing, white to

yellowish-white; marg. very broad, barren, white. *Hym.* brown when dry.

Bark, pine, alder. Nov.–Jan. 3½ in. Pores sometimes not developed.

1675. M. porinoides Fr. (from the poriform folds of the hymenium) *a b c.*

Po. large, round, dissepiments feeble, pale dull yellowish or brownish-ochre; marg. broad, barren, white to yellowish-white.

Dead wood, chips, bark, fir, on the ground; rare. Sept. 4½ in.

1676. M. rufus Pers. (from the reddish hymenium; *rufus*, reddish) *a b c.* Waxy-soft.

Po. regular, round, medium to small, red-sienna, or light to dark vinous-brown; marg. obsolete, or narrow and irregular, barren, colour as Po.

Often effused. Dead wood, oak, willow; uncommon. May–Feb. 3½ in. Vinous-blackish when dry. Greatly resembling a resupinate *Polyporus*.

1677. M. serpens Tode (from its wavy, creeping habit; *serpo*, to creep) *a b c.* Inseparable, thin.

Po. wrinkled to round, angular, entire, dissepiments sometimes feeble, pale to ochreous-salmon or red; marg. broad, barren, white. *Hym.* dark vinous- or rufous-brown when dry.

Fir, lime, ash, cedar, cones; rare. Oct.–Mar. 6⅝ in.

1678. M. pallens Berk. (from its pale colour; *palleo*, to be pale) *a b c.* Inseparable, somewhat gelatinous.

Po. round, but variable, minute to somewhat large, feebly developed, pale reddish- or brownish-salmon, sometimes orange-brown; marg. indeterminate, barren. *Hym.* pale purple-umber to deep vinous-brown when dry.

Fir, oak; rare. June–Dec. 3⅜ in.

1679. M. Carmichælianus Berk. (after Capt. Dugald Carmichael) *a c.* Confluent, very thin.

Po. large to small, often hexagonal, very shallow, with feeble dissepiments, white; marg. narrow, obsoletely punctate, edge finely laciniate.

Bark. 3½ in. Brown when dry.

b. Coniophoræ.

1680. M. lacrymans Schum. (from the watery drops on the fully developed hymenium; *lachryma*, a tear) *a b c.*

Spongy-fleshy, arachnoid-velvety below, sometimes reflexed with a small extended rugose whitish more or less imperfect P. *Po.* large, round to gyroso-toothed, sienna-orange, yellowish towards the barren tomentose white or salmon marg. *Flesh* grey-whitish-flesh colour or vinous-brown. *Hym.* dark brown, sometimes almost black when dry.

Odour very strong, fungoid, penetrating. Dressed wood in buildings, spreading over brickwork, plaster, mortar, carpets, etc. Does not grow out of houses in this country. Jan.–Dec. 1 ft. 5½ in. The fungus of "dry-rot." The flesh gives a red stain to paper like **1517**. Var. *pulverulentus*

Fr., a membranous form, with a somewhat more feeble development of pores than in type. Must not be confused with **1884**, a plant which has no pores.

1681. M. squalidus Fr. (from its squalid appearance) *a b c*. Smooth, with extensive barren places.

Po. large to small, somewhat sinuous, hyaline flesh-colour; marg. broad, barren, paler, fibrillose at edge. *Hym.* dull, dark vinous-umber or purple slate-black when dry.

Hornbeam, posts, dressed wood. Feb. $3\frac{1}{8}$ in.

1681a. M. Guillemotii Boud. (after M. Guillemot) *a*.

P. dimidiate, imbricate, smooth, white, becoming greyish with age. *Po.* gyrose, subgelatinous, fulvous to brassy-orange, slightly olive, pulverulent with spores. *Flesh* white to pale lavender.

On posts, palings, willow. June–Sept. $4\frac{3}{4} \times 2\frac{3}{4}$ in.; projection $2\frac{3}{4}$ in.

LXIX. **POROTHELIUM** Fr.

(From the perforate tubercles, spines or teat-like teeth; Gr. *poros*, a pore, *thele*, a teat.)

Subiculum more or less membranous, bearing obtuse, or elongate, distinct papillæ, which are at first closed, then open in the form of pores or tubes. On dead branches, etc. (Fig. 84.)

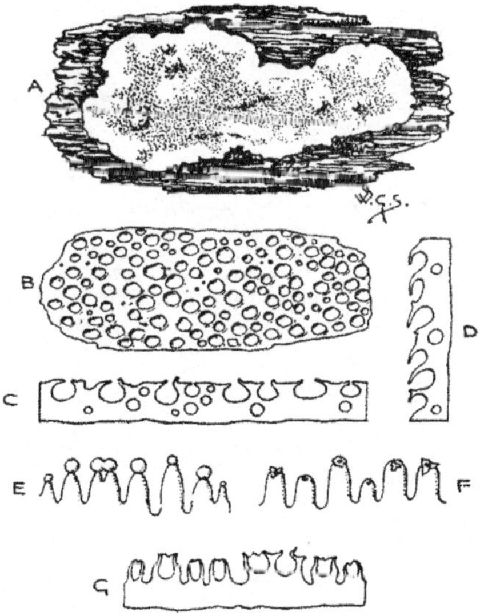

Fig. 84.—A, *Porothelium fimbriatum* Fr., one-half natural size. B, pores; C, section when grown horizontally; D, ditto when grown vertically. × 10. E, *P. Friesii* Mont., papillæ from a damp position; F, ditto from a dry position; G, section. × 10.

Closely allied to *Fistulina*, with which it agrees in the fimbriate mouths of the free tubercles or tubes. The tubes frequently distil drops of fluid in damp positions. The appearance commonly somewhat resembles effused groups of minute *Æcidia*.

Species **1682—1686**

Massee places this genus in *Hydnaceæ*.

Fries has made *Solenia* follow *Porothelium*, but in *Solenia* each cup or cylinder is a complete plant, not a mere tube. To find a place in the *Polyporaceæ* each cup should be furnished with tubes. The true place for *Solenia* is after *Cyphella* in the *Thelephoraceæ*.

1682. P. fimbriatum Fr. (from the fimbriate margin of the subiculum) *a c*.

Sub. separable. *Pa.* hemispherical, immersed, small, biscuit-colour; marg. broad, barren, fimbriate, white.

Birch, beech. 3¾ in. General appearance much like the pore surface of a minutely pored resupinate *Polyporus*. *Pa.* sometimes distant. Often with extensive barren places on subiculum.

1683. P. Friesii Mont. (after Elias Fries) *a b c*.

Sub. less fimbriate than **1682.** *Pa.* hemispherical, immersed, somewhat larger than in **1682**, varying from minute to larger, ivory or pale brown.

Pine, juniper. Sept.–Feb. 3 in.

1684. P. Stevensonii B. & Br. (after the Rev. Dr. John Stevenson) *a*.

Sub. somewhat thick, gelatinous. *Pa.* elongate, whitish or pale yellowish, each bearing a diaphanous, yellow globule; marg. rather coarsely hispid, then more or less denuded.

Pine, squared pine. 3¾ in. The mycelium imparts a sweet scent to the wood.

1685. P. Keithii B. & Br. (after the Rev. Dr. James Keith).

Sub. adnate, gelatinous in centre. *Pa.* short, collapsing, pallid; marg. very thin, at first somewhat gelatinous.

Dead fir. April.

1686. P. confusum B. & Br. (from its having been confounded with other members of the genus).

Sub. adnate, at first floccoso-pulverulent. *Pa.* very small, pallid; marg. thin, arachnoid.

Fir sticks. June–Sept.

Fam. III. **HYDNACEÆ.**

Hymenium inferior in the stipitate and dimidiate species, superior in the resupinate series, bearing from the first persistent spines, teeth, tubercles, papillæ or wart-like folds.

Basidia tetrasporous except in *Kneiffia*, where they are monosporous; *Mucronella*, which has been described as monosporous, is sometimes tetrasporous. Species **1687—1780**

Key to the Genera.

Pileate.
- Gelatinous.. 71 Tremellodon
- Fleshy. Teeth flattened, irregular .... 72 Sistotrema.

Pileate or resupinate. Fleshy. Spines acute, distinct at base 70 Hydnum.

Resupinate.
- Hymenium bearing somewhat acute teeth which spring from folds or ridges........................ 73 Irpex.
- Hymenium bearing teeth in the form of coarse deformed tubercles .... 74 Radulum.
- Hymenium covered with folds or wrinkles .... 75 Phlebia.
- Hymenium covered with granules—
  - Apices excavated .... 76 Grandinia.
  - Apices penicellate .... 77 Odontia.
- Hymenium covered with minute acute spicules .... 78 Kneiffia.
- Hymenium bearing teeth in the form of long scattered usually distant bristles. Subiculum nearly obsolete........................ 79 Mucronella.

LXX. **HYDNUM** L.

(From the Greek name for a truffle, *hydnon*, appropriated without reason by Linnæus for this genus.)

Hymenium inferior in the first four series, superior in the fifth. *Spines* awl-shaped, acute, distinct at the base. (Fig. 85.)

Species **1687—1745**

Mesopodæ. Stem central. Terrestrial, chiefly in pine woods.

a. Carnosæ. *Pileus* fleshy, somewhat fragile, stem solid. Mostly edible.

Pileus scaly or tomentose. *Spines* changing colour. **1687—1690**

Pileus even, smooth. *Spines* changing colour. **1691—1693**

Spines unchangeable, unicolorous, whitish, or pale shades of ochre, buff or brownish. **1694—1697** *b. Lignosæ.* *Pileus* corky or coriaceous.

Spines changing colour, and, as well as the spores, somewhat fuscous. **1698—1700**

Spines changing colour, and, as well as the spores, ferruginous. **1701, 1702**

Spines unchangeable, spores whitish. **1703—1706**

PLEUROPODÆ.

Stem lateral, somewhat dimidiate. **1707**

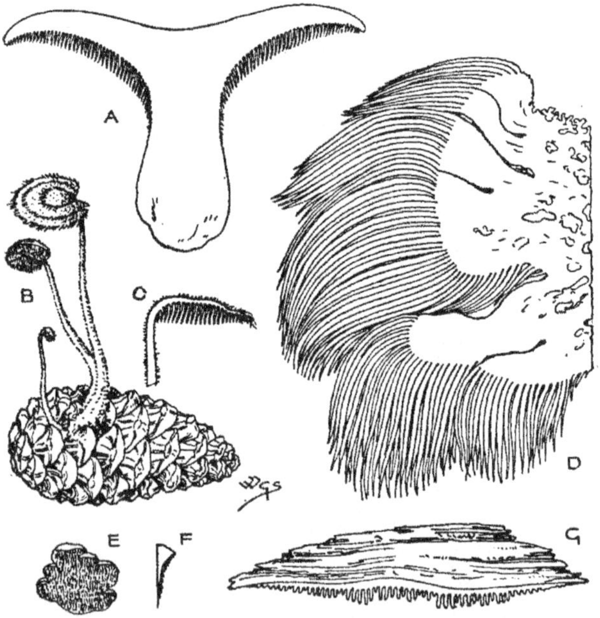

Fig. 85.—A, *Hydnum repandum* L., section, one-half natural size. B, *H. auriscalpium* L., entire, one-half natural size. C, ditto, section through pileus, natural size. D, *H. erinaceus* Bull., section, one-half natural size. E, *H. ochraceum* Pers., entire, one-half natural size. F, ditto, section, one-half natural size. G, *H. squalinum* Fr., section, one-half natural size.

MERISMATÆ. Branched.

Very much branched. **1708**

Simple, as seen without cutting, tuberculiform, immarginate, but slightly branched as seen in section. **1709, 1710**

APODÆ. Stem absent.

Pileus fleshy, soft. **1711—1713**

Cæspitose. *Pilei* imbricate, commonly pliant, tough. **1714**

Simple. *Pileus* coriaceous or corky. **1715**

RESUPINATÆ. *Pileus* none. *Spines* vertical or oblique according to situation.

Spines fuscous, ferruginous.	**1716—1723**
Spines yellowish, greenish.	**1724—1736**
Spines flesh-colour, lilac, rufescent.	**1737, 1738**
Spines white, or light yellowish when old.	**1739—1745**

MESOPODÆ.

a. Carnosæ.

1687. H. imbricatum L. (from the imbricate scales of the pileus) *a b c.*

P. at length depressed, rich, dark claret-brown, the innate sc. exposing the paler flesh. *St.* even, whitish- or yellowish-claret-brown, base as P. *Sp.* decurrent, pale brownish-biscuit. *Flesh* pale claret-brown in upper part of P., white elsewhere.

Usually single. Said to be edible. Taste when cooked like tough liver, bitter, disagreeable. Amongst firs; rare. Sept.-Nov. $9\frac{1}{2} \times 2\frac{1}{2} \times 1\frac{1}{2}$ in.

1688. H. squamosum Schæff. (from the scaly pileus; *squama*, a scale) *a.*

P. at length depressed, breaking up into rough irregular sc., brownish- or vinous-salmon; sc. brownish. *St.* equal or attenuate downwards, somewhat scaly, pale- or brownish-buff, sometimes darker below. *Sp.* decurrent, purplish-salmon. *Flesh* pale purplish-salmon, almost white.

Often connate. Amongst mosses in pine woods. Sept.-Oct. $2\frac{3}{4} \times 1\frac{1}{2} \times \frac{5}{8}$ in.

1689. H. scabrosum Fr. (from the rough, scaly pileus; *scabrosus*, rough) *a b.*

P. turbinate, then plane, usually uneven, squamulose, pale vinous-umber; mid. darker; marg. white; sc. umber. *St.* attenuate downwards, vinous, umber or cinereous, blackish at base. *Sp.* decurrent, buff-ivory or ferruginous with a grey fuscous effect. *Flesh* salmon-ivory.

Usually single. Woods, pine. Sept. $4 \times 1\frac{1}{4} \times 1$ in.

1690. H. molle Fr. (from the downy pileus; *mollis*, soft) *a c.* White. *P.* at length depressed, tomentose. *St.* almost equal. *Sp.* decurrent. *Flesh* pale, brownish-white.

Woods, pine. $4\frac{1}{2} \times 1\frac{5}{8} \times 1$ in.

1691. H. lævigatum Swartz (from the smooth pileus; *lævis*, even) *a b.*

P. at length depressed, pale yellowish-umber. *St.* equal, whitish above, colour as P. below. *Sp.* decurrent, pale brown.

Woods, pine. Aug.-Nov. $4\frac{1}{2} \times 1\frac{5}{8} \times 1$ in. Sometimes occurs three times the size of drawing. Becoming more or less black in drying.

1692. H. fragile Fr. (from its fragility) *a b*.

P. at length depressed, at first pubescent, then smooth and subsquamulose, yellowish-brown to reddish-brown-buff, zoned darker. *St*. unequal, greyish, brownish-grey or as P., sometimes vinous-brown below. *Sp*. decurrent, ivory-grey or dull pale grey-ochre. *Flesh* pale brown or pale vinous, darker at base of St.

Odour of curry powder. Woods, pine, amongst heath; rare. Sept.–Nov. $4\frac{3}{8} \times 2\frac{3}{8} \times 1$ in.

1693. H. fusipes Pers. (from the fusiform stem; *fusis*, a spindle, *pes*, a foot) *a*.

P. fibrous, tough, lobed, becoming subdepressed, cold umber to deep claret-brown; marg. paler, sometimes with a narrow white edge. *St*. colour as P., usually paler. *Sp*. decurrent, very short, salmon- or brown-whitish. *Flesh* white or shaded, pale salmon or brownish.

Woods, pine. Sept. $2\frac{1}{4} \times 1\frac{1}{4} \times 1\frac{1}{4}$ in.

1694. H. repandum L. (from the repand pileus in old examples) *a b c*.

P. somewhat smooth, rarely floccoso-pruinose, pale buff. *St*. almost equal, paler than P. *Sp*. decurrent, colour as St. *Flesh* white, becoming as P.

Subgregarious. Sometimes in rings. Edible, but requires stewing slowly for four hours. Woods; common. Aug.–Nov. $4\frac{7}{8} \times 1\frac{3}{4} \times 1$ in. Becoming wholly ochre or orange in drying.

1695. H. rufescens Pers. (from the rufescent colour) *a c*.

P. thin, somewhat lobed, pulvinate to plane, pubescent, buff, orange-brown or brown. *St*. almost equal, frequently excentric, pale yellowish, yellow- or bright-cinnamon. *Sp*. adnato-decurrent, pale ochre to bright cinnamon. *Flesh* white to pale rufescent.

Sometimes connate. Woods; frequent. Sept. $3\frac{3}{8} \times 3\frac{1}{4} \times \frac{1}{2}$ in.

1696. H. acre Quél. (from the acrid taste) *a*.

P. turbinate, flattish, pale yellowish, olive shaded; mid. darker. *St*. equal, colour as P. or darker. *Sp*. decurrent, $\frac{1}{16}$ in., colour as P. *Flesh* pale yellowish, greyish or pale brownish.

Poisonous. Taste bitter-acrid. Woods, pine, chestnut, sandy places. $5 \times 1\frac{1}{4} \times 1$ in.

1697. H. fuligineoalbum Schmidt (from the smoky-whitish stem; *fuligo*, soot, *albus*, white) *a c*.

P. convex, becoming depressed, repand, squamulose; sc. brownish on an umber-white ground. *St*. scaly, dull salmon. *Sp*. decurrent, pale vinous-brown. *Flesh* white, clouded pale rosy-brown or pale sienna.

Woods, pine, pastures near woods. $4\frac{1}{4} \times 1\frac{3}{4} \times \frac{7}{8}$ in. Becoming dull vinous-brown in drying.

b. Lignosæ.

1698. **H. compactum** Pers. (from its compact substance) *a b c*.
P. irregular, plane, tuberculose, orange-sienna or umber, sometimes ivory, shaded grey. *St.* irregular, tuberculose, orange-sienna or umber. *Sp.* decurrent, salmon-brown. *Flesh* zoned brown and bright blue, or ivory and pale blue, the blue gradually becoming indigo.

Often confluent and misshapen. Amongst firs, heather, etc. Aug.–Nov. 3⅜ × 1½ × 1¼ in. Becomes blackish in drying.

1699. **H. aurantiacum** A. & S. (from the colour; *aurum*, gold) *a b c*.
P. convex or turbinate, becoming flat, rugged, yellowish-orange or yellowish-sienna, buff when young. *St.* sienna-orange above, deep sienna below. *Sp.* decurrent, white to ivory-ochre, sometimes orange near St. *Flesh* pale dull orange.

Inodorous. Woods, pine. Sept.–Nov. 2¾ × 1¼ × ⅝ in. Becoming vinous-brown in drying.

1700. **H. ferrugineum** Fr. (from the rust-of-iron colour; *ferrum*, iron) *a c*.
P. flat or slightly depressed, rugged, sienna-umber, at first downy, exuding blood-red drops. *St.* unequal, pale ochre, umber or slate. *Sp.* decurrent, reddish-salmon. *Flesh* pale yellow, vertically brown-streaked.

Often confluent, woods, involving fir-leaves, grass, etc. Frequently connate. Woods, pine, under ling. Aug.–Oct. 4⅛ × 1⅞ × 1 in.

1700a. **H. cinereum** Bull. (from the grey pileus and spines; *cinis*, ashes).
P. corky to woody, becoming infundibuliform, covered with small elevations, villous, white to grey, varying vinous or umber. *St.* somewhat ventricose, vinous-ivory, somewhat brown below. *Sp.* decurrent, short, white to grey and pale olive-brown. *Flesh* white, vinous-buff below.

Often connate. Woods. Sept. 2¾ × 2 × ⅝ in.

1701. **H. scrobiculatum** Fr. (from the pitted pileus; *scrobis*, a trench) *a b c*.
P. rugged, tuberculose, downy; mid., tubercles and marg. deep sienna on a salmon ground. *St.* rugged, unequal, smooth, often rooting, vinous-salmon above, vinous-blackish below, or umber. *Sp.* decurrent, colour as P. or lighter. *Flesh* vinous-salmon, zoned vinous-brown.

Involving pine-leaves, etc. Often connate. Woods, fir, amongst ling. Aug.–Nov. 1⅞ × 2¼ × ⅜ in.

1702. **H. zonatum** Batsch (from the frequently zoned pileus) *a b c*.
P. depressed; mid. rugged-squarrose; marg. radiato-rugose, 1–12 zoned, z. sometimes nearly obsolete, ferruginous. *St.* unequal, rugged, floccose, colour as P. *Sp.* decurrent, small, slender, pallid to ferruginous. *Flesh* rich vinous-brown.

Commonly involving moss, leaves, etc., sometimes full of embedded fir-leaves; fragrant. Woods, fir, oak; rare. Aug.–Nov. 2¼ × 1⅜ × ⅝ in.

1702a. H. Queletii Fr. ex Quél. (after Dr. Lucien Quelet) *a*.

P. fleshy, coriaceous, plane, then depressed, zoneless, or with one zone, somewhat rugoso-radiate, vinous-sienna, darker at mid.; marg. white. *St.* colour as P., paler above. *Flesh* pale salmon-umber.

Often connate. On the ground in woods. Oct. $2 \times 1 \times \frac{3}{16}$ in.

1703. H. nigrum Fr. (from the blackish colour; *niger*, black) *a b c*.

P. turbinate, rugged-tubercular, tomentose, lavender-grey, purple or blackish-blue; z. 2–5, deep purple. *St.* unequal, lavender-umber, shaded umber, to black. *Sp.* decurrent, lavender-whitish, greyish or purple. *Flesh* slate or deep purple.

Often connate. Woods, pine. Sept.–Nov. $2\frac{3}{4} \times 1\frac{1}{2} \times \frac{3}{4}$ in.

1704. H. graveolens Fr. (from its strong odour; *gravis*, strong, *oleo*, to smell) *a b c*.

P. thin, zoneless, smooth, rugose, depressed, pale to dark umber or black-fuscous; marg. paler or whitish. *St.* equal, smooth, colour as P. *Sp.* decurrent, grey. *Flesh* umber.

Usually cæspitose and confluent, rarely solitary. Odour heavy, spicy-sweet, like curry powder. Woods, fir, oak; rare. Sept.–Nov. $1\frac{3}{4} \times 2 \times \frac{1}{8}$ in. Varied in shape and size, sometimes spathulate. Examples gathered in 1855 retained their odour in 1901. Stem sometimes very short.

1705. H. melaleucum Fr. (from the black pileus and white spines; Gr. *melas*, black, *leukos*, white) *a c*.

P. thin, plane, depressed, slate-blackish or slate-brownish or dark brown-grey; z. darker; marg. whitish; when young pale umber with white marg. *St.* smooth, ashy-grey to slate-blackish. *Sp.* adnato-decurrent, white or whitish. *Flesh* umber-white to slate.

Sometimes cæspitose. Odour none. Amongst pines. Sept. $1\frac{3}{4} \times 1\frac{1}{8} \times \frac{3}{16}$ in.

1706. H. cyathiforme Schæff. (from the pileus, sometimes cup-shaped; *cyathus*, a cup, *forma*, form) *a b c*.

P. thin, plano-infundibuliform, tomentose, somewhat rugged at mid., brown-whitish to dark claret-brown; marg. sometimes white-tomentose; z. 3–11, darker, when dry very glossy. *St.* unequal, smooth, sometimes almost obsolete, colour as P. or darker. *Sp.* adnato-decurrent, dull salmon or salmon-grey.

Sometimes connate. Woods, fir. Aug.–Nov. $2\frac{3}{4} \times 1\frac{3}{8} \times \frac{3}{8}$ in. With the exception of the spines, sometimes much like **1525**.

PLEUROPODÆ.

1707. H. auriscalpium L. (from the shape of small examples; *auriscalpium*, an ear-pick) *a b c*.

P. thin, coriaceous, reniform, hispid-hairy, brownish-salmon or brown, sometimes blackish; z. darker. *St.* rooting, hispid, often branched, colour as P. *Sp.* white, salmon or brown. *Flesh* pale umber or salmon-white.

Sometimes springing from thick rhizomorphoid growths with stems sometimes without pilei. Fir-cones, fir-leaves. May–Dec. $\frac{7}{8} \times 2 \times \frac{1}{16}$ in.

MERISMATÆ.

1708. **H. coralloides** Scop. (from its appearance, like branched white coral; Gr. *korallion*, coral, *eidos*, appearance) *a b c.*

Much and narrowly branched, wholly white, becoming faintly buff or yellowish. *Sp.* $\frac{1}{16}$ to $\frac{5}{8}$ in. long.

Trunks, beech, birch, ash, oak, etc.; rare. Oct. 12 × 7½ in. Sometimes dark brown when dry.

1709. **H. erinaceus** Bull. (from a fancied resemblance, when young, to a hedgehog, *erinaceus*) *a b c.*

Slightly and broadly branched, wholly white, becoming pale yellowish. *Sp.* 1½ to 2¾ in. long, pendulous in large tufts. *Flesh* branched, cancellate.

Taste mild. Trunks, oak, beech, hornbeam, alder, etc.; rare. Sept.–Nov. 13 × 13 in. Sometimes dark brown when dry. Sometimes very large and heavy, weighing 22 lbs. or more.

1710. **H. caputmedusæ** Pers. (from a fancied resemblance to Medusa's head; *caput*, a head) *a b c.*

Sp. in connate, tuberculiform groups, ½ to 1⅛ in. long, growing in various directions in matted locks as of hair, substipitate, wholly white or ivory, becoming grey.

Trunks; rare. Diam. 6⅝ in.

1710a. **H. Schiedermayeri** Heuf. (after K. B. Schiedermayer) *a.*

Fleshy, immarginate, ivory-white, becoming buff when old. *Sp.* ⅜ to ½ in. long. *Flesh* somewhat cancellate, pale sulphur, becoming buff below when old.

Connato-imbricate. Growing from cracks in bark of old apple trunks. In connato-imbricate groups of various sizes, each example 1½ × 1 in. Groups 6 × 3 or 4 in.

APODÆ.

1711. **H. cirrhatum** Pers. (from the pileus, clad with small curls, like little locks of hair; *cirrus*, a curl) *a b c.*

P. pulvinate, expanded, dark biscuit-colour; marg. fimbriate. *Sp.* ⅜ to ½ in. long, paler than P. *Flesh* paler than P.

Imbricate or simple. Trunks, beech. Aug. Group 7 in. Diam. 5½ in. Sometimes dark brown when dry.

1712. **H. diversidens** Fr. (from the spines, differing from the scale-like growths on the pileus; *diversus*, diverse, *dens*, a tooth) *a b c.* Wholly biscuit-colour, pale to dark.

P. somewhat lobed, thick, beset with small variously shaped sc.

Substipitate, obscurely branched. Irregularly imbricate. Trunks, hornbeam, oak, beech, birch. Oct. Diam. 5¼ in. One cause of "white-rot."

1713. **H. pulcherrimum** B. & Curt. (from its appearance; *pulcherrimus*, very beautiful) *a c.*

P. pulvinate, fleshy-fibrous, spongy-soft, dimidiate, expanded, stiffly-downy, white, shaded, pale tawny ; marg. thin, lobed. *Sp.* variable in size, 1/5 to 1/8 in., tawny.

Subimbricate. Oct. Diam. 1 3/8 in.

1714. **H. multiplex** Fr. (from the many folds of the pileus; *multus*, many, *plecto*, to twist) *a c.*

P. thin, leathery, deeply folded, spathulate, velvety, rich brown. *Sp.* 1/16 in. long, crowded, very slender, grey, shaded brownish. *Flesh* bright brown, paler than P.

Imbricate. Group 4½ in. in diam.

1715. **H. ochraceum** Pers. (from the ochreous colour) *a b c.*

Sub. thin. *P.* when present, thin, pale ochreous, zoned, usually resupinate or with marg. slightly upturned and forming more or less perfect connate pilei. *Sp.* very small, somewhat blunt, ochreous flesh-colour, orange-buff, sometimes fawn, rarely dull crimson.

Trunks, oak, branches, fir-leaves, bramble, apple, moss ; common. Oct.-April. Diam. 5 in. Readily separating from matrix.

Resupinatæ.

1716. **H. squalinum** Fr. (from a fancied resemblance in the spines to sharks' teeth ; *squalus*, a shark) *a c.*

Sub. adnate, coriaceous, pale yellowish-fawn. *Sp.* entire, stout, compressed, rounded at apex, 3/16 in. long, colour as Sub.

Trunks, beech. Diam. 2 5/8 in.

1717. **H. fuscoatrum** Fr. (from the dusky-black clouding ; *fuscus*, dusky, *ater*, black) *a c.*

Sub. a thin film, glaucous, flocculoso-pruinose, salmon-umber ; mid. darker ; or dark brown shading into sienna and brown-ochre at marg. *Sp.* short, conico-subulate, acute, colour as Sub.

Rotten wood, birch, beech, alder. Diam. 2½ in.

1718. **H. membranaceum** Bull. (from the membranous subiculum) *a b c.*

Sub. an effused waxy-membranous film, pale vinous sienna ; marg. lighter. *Sp.* awl-shaped, crowded, acute, short, darker than Sub.

Fallen sticks, pear ; frequent. Jan. Diam. 2 7/8 in.

1719. **H. molluscum** Fr. (*mollis*, soft, in reference to the thin subiculum).

Sub. membranous, dry, readily separable from the matrix, whitish. *Sp.* short, slender, reddish.

Wood. Distinguished from **1718** by the dry white subiculum, separating from the matrix.

1720. **H. Weinmannii** Fr. (after J. A. Weinmann, mycologist) *a b c.*

Sub. effused, waxy-membranous, agglutinate, fawn-grey, pale slate, or umber and white. *Sp.* minute, acute, somewhat distant, sometimes collected in small groups or fascicles, slate, somewhat umber, apex white.

Fallen branches, ash, oak. Oct.–April. Diam. 4¼ in. When dry, cracking and becoming full grey or fawn to deep brown or almost black.

1721. **H. crinale** Fr. (from the slender hair-like spines; *crinis*, a hair) *a*.

Sub. effused, villoso-interwoven, thin, sienna-ochre. *Sp.* ⅛ in. long, crowded, very slender, sienna-umber or rufous-brown.

Dead wood; rare. Diam. 3⅜ in. The hair-like spines resemble those of *Mucronella*.

1722. **H. ferruginosum** Fr. (from the rust-of-iron colour; *ferrum*, iron) *a c*.

Sub. effused, submucedinous, separable from matrix, tawny, ferruginous. *Sp.* crowded, conico-awl-shaped, acute, colour as Sub.

Decayed wood, especially under bark, rarely on the ground. July–Oct. Diam. 2 in. Sometimes mere conidial forms of other fungi, as *Isaria* amongst the Hyphomycetes, or **1780** are mistaken for this plant.

1723. **H. variecolor** Fr. (from the varying colours) *a b c*.

Sub. effused, adnate, forming a furfuraceous crust, white, becoming cream, sienna and sienna-umber in drying, sometimes resembling white miniature clouds; marg. finely white-fimbriate. *Sp.* crowded, conical, short, or minute with a ragged appearance, but here and there longer and more typically *Hydnum*-like, commonly adpressed, colour as Sub.

Dead stumps, oak; rare. Oct. Diam. 3⅞ in.

1724. **H. aureum** Fr. (from the colour; *aurum*, gold) *a*.

Sub. somewhat cartilaginous, jagged radiate; marg. irregular, narrow, byssoid. *Sp.* often developed in small patches, awl-shaped, colour as Sub.

Dead branches. Dec. Diam. 5¼ in. Appearance merulioid. Mycelium permeating wood and causing defined yellowish flesh-coloured spots.

1725. **H. denticulatum** Pers. (from the slightly toothed spines; *dens*, a tooth) *a*.

Sub. thin, somewhat fleshy, mealy, shining light-yellow-ochreous or olive-ochre, sometimes clouded pale brownish-rose. *Sp.* crowded, equal, minutely 2–3 toothed at the whitish apex.

Rotten wood. June. Diam. 2⅝ in.

1726. **H. alutaceum** Fr. (from the colour, like tanned leather, *aluta*) *a b c*.

Sub. crustose, adnate, very thin, pallid ochreous, pale brownish-yellow or biscuit. *Sp.* acute, very minute and short, here and there a little longer and typically *Hydnum*-like, colour as Sub.; marg. barren.

Dead wood, pine; rare. Nov. Diam. 4½ in. Resembling **1770**.

1727. **H. sulphureum** Schwein. (from the sulphur-colour) *a*.
Sub. effused, adnate, membranous, colour paler at marg. and gradually fading into the colours of the bark; marg. barren, byssoid. *Sp.* minute, few, subulate.

Dead wood, birch. Diam. 4 in. The barren spineless form can hardly be distinguished from **1890**.

1728. **H. sordidum** Weinm. (from the dingy colour; *sordidus*, dirty) *a*.
Sub. effused, thin, mucid, easily separating, dingy yellow; marg. gyroso-porous, sulphur-yellow. *Sp.* very crowded, somewhat gelatinous, compressed, incised, subacute, colour as Sub.

Dead wood. Nov. Diam. 3½ in. Becoming dull-reddish in drying.

1729. **H. viride** Fr. (from the colour; *viridis*, green) *a*.
Sub. effused, thin, softly velvety, somewhat pale leaf-green, becoming yellowish; marg. in places slightly upturned. *Sp.* stout, short, somewhat thick, irregular, somewhat incised, colour as Sub.

Rotten wood, alder. Diam. 3⅜ in. Sometimes effused for 8 to 10 in.

1730. **H. limonicolor** B. & Br. (from the lemon colour; *Citrus Limonum*, the lemon) *a*.
Sub. adnate, scanty. *Sp.* crowded, short, acute. *Myc.* scanty or obsolete, when present shining-white.

On stone buried amongst pine-leaves. Oct. Diam. 3 in. Distinguished from **1736** by the scanty subiculum and the absence of a persistent barren white margin.

1731. **H. pinastri** Fr. (from the habitat, pines, *pinus*) *a c*.
Sub. effused, membranous, fleecy or minutely hispid, separable, light yellow, ivory or biscuit; marg. lighter, villous beneath. *Sp.* equal, subulate, acute, in groups, oblique in different directions, pale brown.

Mossy bark, wood, pine. Diam. 4¼ in.

1732. **H. spathulatum** Fr. (from the spathulate spines) *a*.
Sub. effused, membranous, separable, yellowish; marg. fimbriate, villous beneath. *Sp.* very small, oblique, obtuse, orange.

Dead wood, *Robinia*; rare. Jan.–Dec. Diam. 3½ in. Often in cracks in bark.

1733. **H. multiforme** B. & Br. (from its many forms; *multus*, many, *forma*, form).
Sub. inseparable, at first wholly barren, then here and there fertile, ochreous-white, texture floccoso-mealy; marg. coarsely fibrillose. *Sp.* very acute, crowded, becoming pallid and fimbriate.

Dead wood. Very variable, sometimes almost tow-like with margin inflexed. Perhaps a *Corticium*, as the spines when present are barren.

1734. **H. anomalum** B. & Br. (from the anomalous spines) *a b*.
Sub. thin, gelatinous, pallid, light-yellow. *Sp.* minute, at first in the form of granules, then stipitate and obtusely divided, few, but here and there more, colour as Sub.

Rotten wood, ash. Mar. Diam. 2½ in. With immersed ovate or circular vescicles. Brown when dry. Near *Mucronella* and perhaps not a true *Hydnum*.

1735. H. melleum B. & Br. (from the honey-colour; *mel*, honey) *a c*.
Sub. thin, effused, pulverulent; marg. delicately flaxy. *Sp.* acute, minute, sometimes divided at apex with a tendency to grow in groups, partially pulverulent.

A doubtful *Hydnum*. On rails on the ground. Diam. 4¼ in.

1736. H. sepultum B. & Br. (from its habitat; *sepultus*, buried) *a*.
Sub. golden; marg. white. *Sp.* acute to obtuse, minute to medium size.

On wood, bark and stones buried amongst pine-leaves. Sept. ⅛ to ⅝ in. diam. Forming little scattered patches. Pale ochreous when dry.

1737. H. udum Fr. (from its moist substance; *udus*, moist) *a b c*.
Sub. effused, thin, subgelatinous, agglutinate to matrix, flesh-colour to watery-yellowish, sometimes bright sulphur. *Sp.* acute, 2–3 forked, small, but varying in length, crowded, arranged in groups, very small near the barren marg., colour as Sub.

Dead branches, willow, oak, alder; frequent. Sept.–April. Diam. 5½ in. Forming large patches.

1738. H. bicolor A. & S. (from the white subiculum and brown spines; *bis*, two, *color*, colour) *a c*.
Sub. widely effused, tomentose, adpressed, thin, white; marg. broadly barren. *Sp.* minute, white-villous, naked at acute apex, pale umber to rufous-brown.

Bramble, fir. March. Diam. 2 in.

1738a. H. Hollii Fr. (after F. Holl) *a*.
Sub. adnate, effused, waxy membranous, whitish-brown; marg. finely powdery. *Sp.* very minute, blunt, deformed.

On naked wood. Oct. 4 × 2½ in.

1739. H. nodulosum Fr. (from the nodulose subiculum) *a*.
Sub. broadly effused, crustaceous, nodules (less in size than a pea) sterile above, furnished with spines below, whitish to pallid yellowish. *Sp.* free and pendulous below nodules, adnate and elongate on the interstices, colour as Sub.

Trunks, stumps, fir. Oct. Diam. 4 in.

1740. H. Stevensonii B. & Br. (after the Rev. Dr. John Stevenson) *a c*.
Sub. effused, mealy beneath, here and there flaxy, white. *Sp.* minute, crowded, cylindrical, obtuse, pulverulent at apex, colour as Sub.

Dead wood, sometimes spreading and encrusting mosses, leaves, etc. March. Diam. 1½ in. Having the appearance of a resupinate *Polyporus*.

1741. H. plumosum Duby (from the plumose apex of the spines) *a c*.
Sub. very delicate, downy, white. *Sp.* very small, divided, colour as Sub.

Dead wood, pine; rare. Diam. 1⅝ in. Cracking when dry and becoming ivory-brown.

1742. H. niveum Pers. (from the snow-white colour; *nix*, *nivis*, snow) *a b c.*

Sub. effused, thin, membranous, lustrous, adnate; marg. fibrilloso-flaxy. *Sp.* at first granulose, then acutely pointed, smooth.

Dead wood, leaves, pine, birch, beech; uncommon. Nov. – April. Diam. 2½ in. Persistent. Ochreous or sulphur-white when dry.

1743. H. farinaceum Pers. (from the mealy subiculum; *farina*, meal) *a b c.*

Sub. effused, indeterminate, adnate, white; marg. subflocculose. *Sp.* thin, subdistant, acute, interstices finely farinose, colour as Sub.

Dead wood, fir, oak. Sept.–May. Diam. 5 in. Varying pale brownish, pale yellowish, or sulphur- or ochre-white.

1744. H. argutum Fr. (from the acute spines; *argutus*, acute) *a b c.*

Sub. effused in scattered patches, tomentoso-interwoven, spongy, adnate, white. *Sp.* acute, awl-shaped, unequal, slightly serrate, colour as Sub.

Wood and bark, alder, willow. Sept. Diam. 2½ in. The hymenium has a finely mossy appearance from the unequal, divergent spines. Becoming colour of coffee with milk when dry.

1745. H. stipatum Fr. (from the crowded spines; *stipatus*, crowded) *a c.*

Sub. widely effused, flocculoso-farinaceous, forming an adnate crust, white to brown-white or light yellowish. *Sp.* at first granulose then forming obtuse teeth, colour as Sub.

Rotten wood. Dec.–Jan. Diam. 3½ in. Has the appearance of a resupinate *Polyporus*.

LXXI. TREMELLODON Fr.

(From its resemblance to a *Tremella*; Gr. *odous*, a tooth.)

Gelatinous, pileate, hymenium inferior, covered with acute awl-shaped spines. *Basidia* ovate, somewhat obliquely 2–4 divided

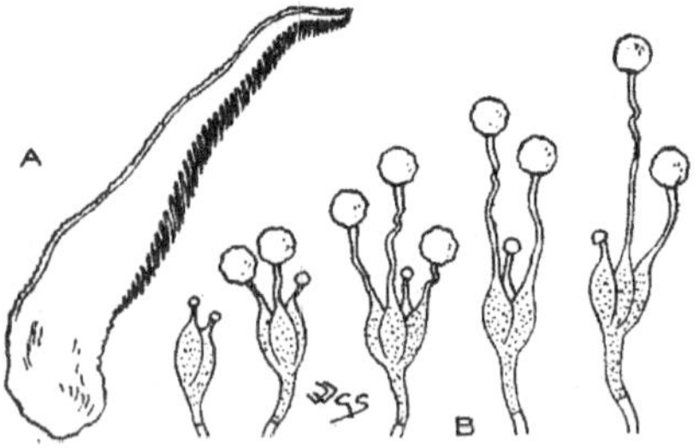

Fig. 86.—*Tremellodon gelatinosum* Fr.
A, section, one-half natural size. B, basidia, sterigmata and spores. × 500.

almost to the base, sometimes undivided for twice the length of the spore, sterigmata 1–5 lengths of spore. (Fig. 86.)

Mr. Fred. Currey (*Journ. Linn. Soc.* vol. v. p. 181, 1861) was the first to call attention to the cloven basidia in this species, and to point out the resemblance in the fruit to some of the *Tremellinaceæ*. Mr. George Massee (*British Fungus Flora*, vol. I. p. 65) has placed *Tremellodon* at the end of the *Tremellinaceæ*, but the nature of the basidium and sterigmata is an insufficient reason for the removal, and if adopted many other removals would have to be made.

1746. T. gelatinosum Fr. (from the gelatinous substance) *a b c*. Wholly translucent, grey-white.

P. fan-shaped, fleshy, dimidiate, substipitate, white-frosted with shining particles. *St.* when present, white-frosted above.

Taste pleasant. Rotten pine, sawdust; rare. Sept.–Nov. $3\frac{1}{4} \times 3\frac{1}{2}$ in.

LXXII. SISTOTREMA Pers.

(From the irregular arrangement of the spines and their interstices, as if displaced; Gr. *seistos*, shaken, *trema*, a pore.)

Pileus fleshy, irregular or dimidiate. *Hymenium* inferior. *Teeth* resembling broken, jagged gills, not radiate, readily separating from

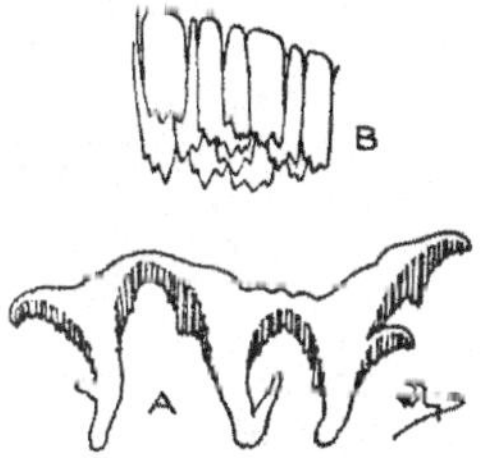

Fig. 87.—A, section of *Sistotrema confluens* Pers., one-half natural size; B, teeth of ditto, × 3.

the hymenophore. *Stem* central, excentric, flattened or almost obsolete. *Basidia* 4-spored. (Fig. 87.)

Closely allied to *Irpex*, but the latter is not fleshy and is usually sessile or resupinate and seldom terrestrial.

1747. S. confluens Pers. (from the confluent habit) *a b c*. Wholly white or whitish, becoming pale yellowish or brownish.

P. horizontal, subdepressed, villous. *St.* attenuate downwards. *Sp.* decurrent.

Gregarious. On the ground, investing sticks, leaves, moss, grass, etc., under pines; uncommon. Sept.–Nov. Group $2\frac{5}{8} \times 1\frac{5}{8}$ in.

LXXIII. **IRPEX** Fr.

(From a fancied resemblance in the fruiting surface to the teeth of a harrow, *irpex.*)

Hymenium inferior, toothed, teeth firm, somewhat coriaceous, acute—spathulate in **1753**—concrete with the pileus or subiculum,

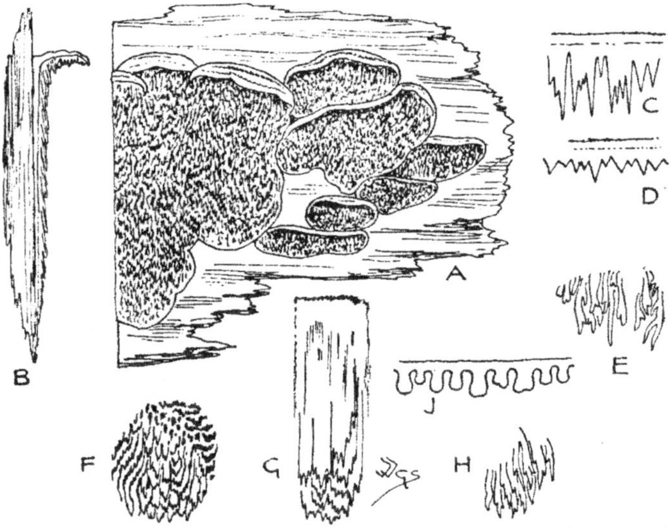

Fig. 88.—A, *Irpex fuscoviolaceus* Fr.; B, section of same; two-thirds natural size. C, D, teeth of ditto; × 3. E, *I. pendulus* Fr., teeth enlarged. F, *I. obliquus* Fr., ditto; G, ditto; × 6. H, *I. deformis* Fr., teeth enlarged. J, *I. spathulatus* Fr., teeth; × 3.

arranged in irregular rows, like broken gills or broken honeycomb, connected at the base by veins or folds. *Basidia* 4-spored. (Fig. 88.)

Generally growing on wood, somewhat sessile or resupinate, approaching *Lenzites* and *Dædalea*. Species **1748—1756**

Pendulous, with the pileus extended behind. **1748**
Effuso-reflexed, marginate. **1749—1751**
Resupinate. **1752—1756**

1748. I. pendulus Fr. (from the pendulous habit) *a c.*

P. membranous, slightly rugose or plicate, adpressedly squamuloso-pilose, light yellow to white-brown. *Te.* in rows, large, incised, white.

Sticks, pine, larch, spreading on to moss; rare. 1⅝ in.; group 2¼ in. Must not be confounded with **1757**.

1749. I. fuscoviolaceus Fr. (from the brown-violet colour of the teeth; *fuscus*, brown) *a b c*.

P. coriaceous, silky, zoned, brown-white to ochre-white, zoned brownish. *Te*. in rows in the form of much incised plates, of the purple-slate-whitish tint seen in the gills of some *Pratellæ*.

Single to imbricate. Pine, beech; rare. 2⅝ in.; group 4¼ in. Sometimes there is a considerable development of pileus.

1750. I. lacteus Fr. (from the milk-white colour; *lac*, milk) *a b c*. White.

P. coriaceous, villous, concentrically sulcate. *Te*. crowded, gyrose in rows, acute, somewhat incised. Irregularly porous at marg., becoming ragged at middle and base.

Sometimes subimbricate. Mountain-ash, birch, fir, beech. Nov. Group 3⅛ in.

1751. I. hypogæus Fück. (from its sometimes growing in and under earth; Gr. *hupo*, under, *ge*, the earth) *a b c*.

Myc. dispersed, white to yellowish. *Sub*. marg. determinate, compactly byssoid, white, then pallid brown-ochreous to dark brown; hymenium often barren or exhibiting barren patches. *Te*. irregular, somewhat labyrinthiform, lax, variable in size, straight, incised 2–6 mm. long, base usually compressed, colour as P.

Woods, pine, investing pine-leaves, twigs, grass, earth, pebbles; rare. Oct. Group 4⅜ in.

1752. I. Johnstonii Berk. (after Dr. George Johnstone) *a*.

Effused, coriaceo-membranous, separable from matrix; marg. naked, reflexed all round, white. *Te*. compressed, unequal, arranged in rows, resembling the spines of *Hydnum*, but seated upon fine folds.

Dead beech. May be a form of **1750**. 1⅛ in.; group 3¼ in.

1752a. I. candidus Weinm. (from the shining white colour; *candidus*, white) *a*. Snow white.

Broadly effused, thin, separable, arachnoid, flaxy at the marg. *Te*. subulate, or compressed, thin, toothed.

Dead wood, pine. Feb.

1753. I. spathulatus Fr. (from the shape of the teeth; *spatha*, a spatula) *a b c*.

Effused, membranous, flaxy, then smooth, white to biscuit. *Te*. entire, somewhat large, reticulato-connected with obsolete veins. Marg. sometimes obsolete, at other times broadly barren, with other barren places on hymenium. Never porous.

Larch. Nov. 4⅞ in.

1754. I. obliquus Fr. (from the oblique teeth) *a b c*.

Effused, forming an adnate crust, white, clouded and shaded pale sienna; marg. broadly barren; circ. flaxy. *Te*. extended from a base resembling small honeycomb, compressed, unequal, incised. Often with large pores near the marg.

Stumps, dead branches, fir, larch, beech, birch, palms in nurseries. Oct.–May. 4¾ in.

1755. I. deformis Fr. (from the misshapen teeth; *deformis*, misshapen) *a c.*

Effused, crustaceous, thin, white; circ. pubescent, somewhat flaxy. *Te.* awl-shaped, growing from a minutely porous base, thin, somewhat incised, unevenly and partially spread over the subiculum, matrix seen between the groups of teeth.

Wood. 2⅛ in. Allied to the *Polyporaceæ*.

1756. I. carneus Fr. (from its flesh-colour; *caro*, flesh) *a b c.*

Effuso-gelatinous, membranous, adnate, collapsing, reddish; marg. barren. *Te.* awl-shaped, entire, united at the base.

Wood and bark, beech. Autumn. 2⅛ in. Placed in *Radulum* by Fuckel. Must not be confounded with 1720.

LXXIV. RADULUM Fr.

(From the root-like appearance of the processes of the hymenium; *radix*, a root.)

Hymenium resupinate, tubercular. *Tubercles* distinct at the base, coarse, irregular in shape, commonly elongate, obtuse, mostly

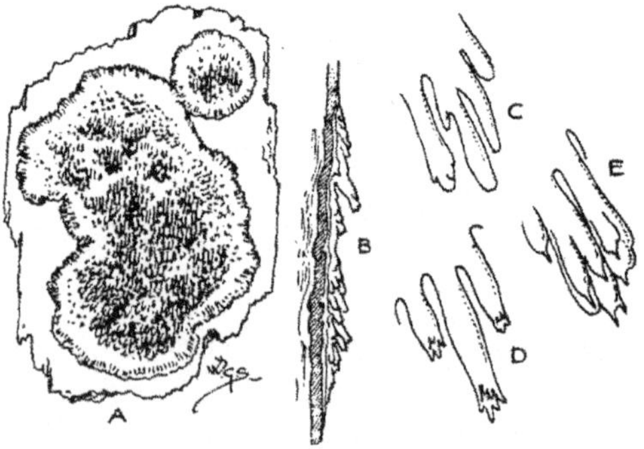

Fig. 89.—A, *Radulum quercinum* Fr., one-half natural size; B, ditto, section, natural size; C. tubercles of *R. orbiculare* Fr.; D, ditto of *R. quercinum* Fr.; E, ditto of *R. fagineum* Fr.; enlarged.

waxy, scattered or fasciculate. *Basidia* 4-spored. Growing on wood and on and under bark. (Fig. 89.) Species 1757—1764

1757. R. pendulum Fr. (from the pendulous habit) *a c.*

Longitudinally effused, thick, softly coriaceous, whitish or ivory, upper marg. shortly reflexed and rugose, lower marg. barren *Tub.* elongate, somewhat lax, irregular.

Alder, birch, pine. 5 in. Must not be confounded with 1748.

1758. **R. orbiculare** Fr. (from its frequent orbicular shape) *a b c.*
Orbicular, confluent; circ. flaxy, becoming waxy-fleshy, white to ochre-ivory; mid. faintly vinous; marg. white. *Tub.* short to long and broad to narrow, mere granules near the barren marg.

Birch, oak, larch, cherry; frequent. Oct.–Mar. 2½ in.; group 4⅝ in. Ends of tubercles usually crested.

1759. **R. quercinum** Fr. (from its habitat, oak, *quercus*) *a b c.*
Roundish, broadly confluent, innate, crustaceous, white to ochre-white. *Tub.* round, stout, obtuse, scattered or fasciculate, irregular in shape, floccoso-villous at apex; marg. barren; edge pure white.

Oak, sometimes on old squared oak timber. Nov.–Mar. 3½ in. Sometimes confounded with **1716**, **1723** and species of *Irpex.*

1760. **R. tomentosum** Fr. (from its downy appearance, especially at the margin; *tomentum*, down) *a b c.*
Effused, innate, somewhat thick, opaque or translucent white; circ. swollen or erect, tomentose. *Tub.* short, crowded, angular, obtuse, smooth, confluent, resembling white tomentum, smaller at marg.

Willow, mountain-ash, pine, sawdust. Dec. 5⅛ in.

1761. **R. deglubens** B. & Br. (from its habit of peeling from the matrix; *deglubo*, to peel off) *a.*
Orbicular, ferruginous, somewhat diaphanous. *Tub.* erect, somewhat cylindrical, irregular, scattered, interstices even, pulverulent with white spores.

Naked wood, bark, ash. Jan. ½ in.; group 5½ in. Growing in troops of small patches.

1762. **R. corallinum** B. & Br. (from its coral-like appearance) *a.*
Somewhat effused, very thin, pelliculose, shining white. *Tub.* fasciculate, fascicles ¼ in. or more across, surrounded by barren sub., divided, obtuse, coralloid, white.

On partially lichen-covered branches, oak. Sept. Fascicles 1 in.; group 6 in.

1763. **R. epileucum** B. & Br. (from the hymenium growing upon a white subiculum; Gr. *epi*, upon, *leukos*, white) *a b.*
Effused. *Hym.* waxy, edge slightly upturned, pale, honey-colour. *Sub.* forming a powdery white marg. *Tub.* scattered, distant, cylindrical, short, under a lens fimbriate at apex, deciduous exposing the sub.

Decorticated wood. Aug. 8⅛ in.

1764. **R. fagineum** Fr. (from the habitat, beech, *fagus*) *a b.*
Effused, somewhat fleshy, white or pale yellowish-brown; marg. radiate with ivory-white myc. *Tub.* enlongate, round, varied, entire, sometimes crested, white to sienna-white.

Chiefly on bark, rarely underneath, sometimes on birch. Nov.–Mar. 5⅜ in.

[*Radulum aterrimum* Fr. is not a Basidiomycete but apparently one of the *Dematieæ.* Under the microscope the appearance is that of conceptacles of *Rhizoctonia crocorum*, but it bears 1–4-septate brown spores after the manner of *Cladosporium*; the fruiting branches are compacted together and take the form of short, obtuse teeth, as in the compacted form of

Penicillium named *Coremium*. It grows on naked birch wood or bark. Sowerby named his example *Hydnum erectum*; this grew upon a *Polyporus*. There is no drawing by Sowerby in the British Museum collection. It has also been named *Sphæronema hydnoideum* Fr. and *Hydnum aterrimum* Fr.]

LXXV. PHLEBIA Fr.

(From the veined appearance of the hymenium; Gr. *phleps*, a vein.)

Resupinato-effused, amphigenous, waxy-soft or subgelatinous, from the first covered with crowded interrupted persistent veins which

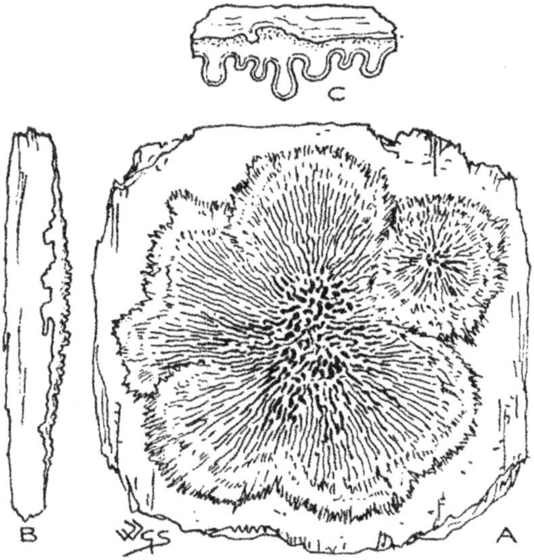

Fig. 90.—A, *Phlebia merismoides* Fr.; B, section of ditto; two-thirds natural size; C, veins enlarged.

are entire, not jagged, at the edge, rarely anastomosing and forming pores. Becoming cartilaginous or horny when dry. *Basidia* tetrasporous. (Fig. 90.)

Dead wood, stumps and branches. Species **1765—1769**

1765. P. merismoides Fr. (from the resemblance of the wrinkled hymenium to that of the non-British section *Merismatæ* of *Stereum*) *a b c*.

Whitish-orange or whitish-vermilion; mid. livid, pale purplish; marg. strigose, vermilion-white. *Ve.* simple or slightly branched, straight, crowded, never reticulate or porous.

Stumps, branches, naked wood, bark, leaves, moss, grass, oak, beech, holly plum, cherry. Aug.–Jan. 3⅝ in.

1766. P. radiata Fr. (from the radiate veins) *a b c.*

Orbicular, vermilion-orange or vermilion; mid. darker; marg. vivid orange-whitish. *Ve.* straight.

Dead wood, birch, mountain ash; rare. 2¼ in.; group 5½ in. When in perfect condition forming somewhat large, conjoined, more or less hexagonal growths as in the carapace of a tortoise. Thinner and brighter in colour than the other British species.

1767. P. contorta Fr. (from the intricate veins; *contorqueo*, to twist) *a b c.*

Yellowish, sienna or brownish or dull red flesh-colour or red-vermilion. *Ve.* conglomerate, branched, flexuous, much smaller than **1765–6**, or as in *Dædalea* on a reduced scale.

Decayed wood, furze; rare. ½ in.; mass 2⅝ in. Forming more or less conjoined circular patches.

1768. P. vaga Fr. (from the wandering appearance of the veins; *vagus*, wandering) *a b c.*

Adnate, waxy, dull yellowish, central parts sometimes purplish or grey; marg. flaxy-fibrillose and powdery. *Ve.* forming intricate wrinkles which often terminate in a granulose papilla.

Dead wood; common. Jan. 6⅜ in.

1769. P. lirellosa B. & Br. (from the veins, like small furrows; *lira*, a ridge or furrow) *a b.*

Hym. veined, interstices resembling small, round or elongate pores, umber-grey to white.

Dead branches, poplar. 3⅜ in. Purple-blackish in drying. The interstices resemble the pores of *Trametes.* Sometimes confused with *Dædalea.*

LXXVI. **GRANDINIA** Fr.

(From the granular appearance of the hymenium; *grando*, hail.)

Crustaceous, effused, thin, soft. *Hymenium* resupinate, amphigenous, waxy, not dry as in *Odontia*. *Granules* small, persistent,

B

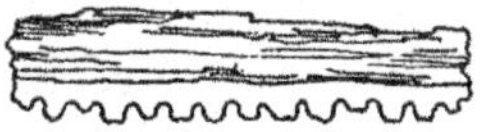

A

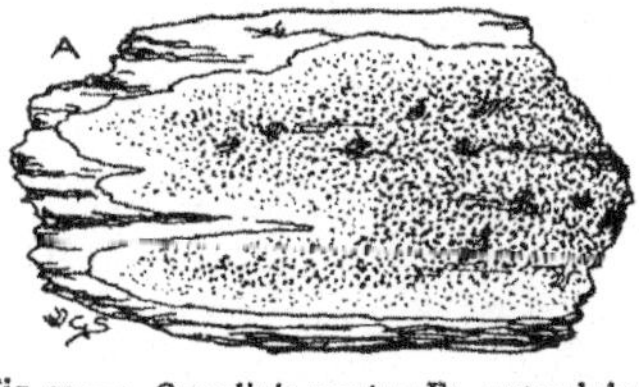

Fig. 91.—A, *Grandinia crustosa* Fr., natural size; B, section of ditto., × 10.

entire, smooth, crowded; apices not pointed as in *Hydnum*, usually indented or excavated as in *Porothelium*. *Basidia* 4-spored. (Fig. 91.)

The hymenial surface is often spuriously papillose from *Sphæriæ* on the bark beneath. Species **1770—1774**

1770. G. granulosa Fr. (from the minutely granular hymenium) *a b c.*
Agglutinate, tan to sienna-ivory; marg. determinate, ivory. *Gr.* small, crowded in groups, smaller at marg., hemispherical to somewhat elongate.

Dead branches, pine, *Myrica Gale.* Oct.–May. 5 in. Cracking when dry.

1771. G. mucida Fr. (from its waxy mucid substance) *a.*
Subinnate, somewhat gelatinous, biscuit; marg. determinate, subradiate, minutely villous, ivory-white. *Gr.* large, unequal, soft, scattered or in groups.

Rotting wood, pine, birch; rare. May–Nov. 4½ in. Cracking when dry.

1772. G. ocellata Fr. (from the minute eyelet, sometimes seen at the apex of the granules; *ocellus*, a little eye) *a b.*
Waxy, livid, rufous flesh-colour or warm brownish-salmon; marg. indeterminate, sterile, yellowish. *Gr.* crowded, subconical, obtuse, somewhat scattered or in groups.

Dead trunks. Aug.–Mar. 6½ in. Cracking when dry.

1773. G. papillosa Fr. (from the minute teat-like granules; *papilla*, a teat) *a c.*
Membranous, somewhat round, separating, white, becoming slightly sienna-ivory; marg. furfuraceous. *Gr.* minute, crowded, equal, roundish.

Sticks, pine, juniper, bird-cherry. 3 in. Cracking when dry.

1774. G. crustosa Fr. (from its crust-like appearance) *a b c.*
Adnate, irregularly effused, floccoso-mealy, white to ivory, becoming pale brownish-yellow; marg. indeterminate. *Gr.* crowded, obtuse, roundish, minute, regularly distributed, but with barren lacunæ.

Branches, willow, pine, **1603.** 6¾ in. Not cracking when dry.

LXXVII. ODONTIA Pers.

(From the minute teeth which invest the hymenium; Gr. *odous*, *odontos*, a tooth.)

Resupinate, effused, arid, not waxy as in *Grandinia*. *Hymenium* spreading from and between interwoven branched fibres, covered with minute teeth or warts which are crested or penicillate at the apex. (Fig. 92.)

1775 tends towards *Hydnum*. Species **1775—1777**

1775. O. barba-Jovis Fr. (from the minute orange fringe at the apex of each tooth, a fanciful comparison with the beard of Jupiter, *barba Jovis*) *a c*.

Flaxy membranous, whitish to yellowish-orange, faintly shaded brownish; marg. determinate, broadly barren, white. *Te.* papillose, then conical, white with an orange fimbriate crest.

Decayed wood, oak, birch. 6¼ in. The hymenium usually exhibits barren patches.

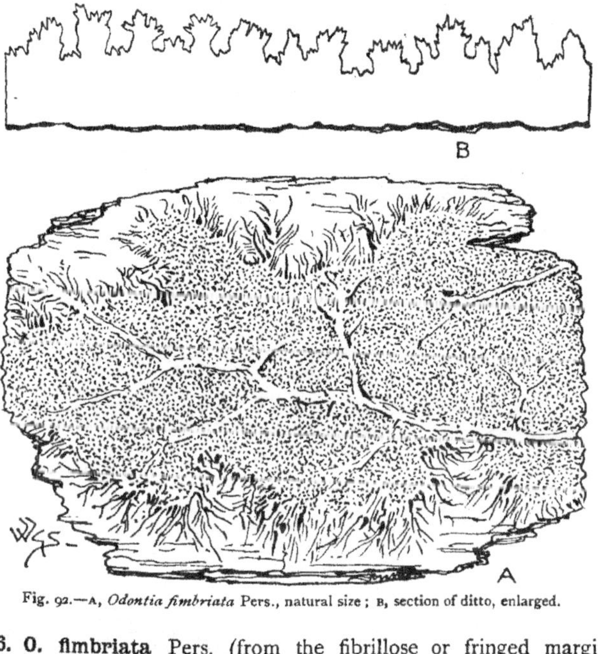

Fig. 92.—A, *Odontia fimbriata* Pers., natural size; B, section of ditto, enlarged.

1776. O. fimbriata Pers. (from the fibrillose or fringed margin; *fimbriæ*, a fringe) *a b c*.

Membranous, separating, traversed by prominent root-like branching threads continuous with the fringed marg., pallid or yellowish, suffused with dull salmon or pale fawn.

Dead wood, willow, plane, dog-wood, bark, moss. Aug.–April. 7¼ in. Extensive barren patches sometimes occur on the hymenium. Fawn-colour when dry, sometimes becoming deep rich brown.

1777. O. cristulata Fr. (from the minutely crested teeth) *a b c*.

Crustaceo-adnate, thin, not ribbed, pale warm brown or brownish-white; marg. broadly barren, here and there finely downy.

Rotten branches, birch. Feb. 5⅝ in. Cracking all over when dry as in *Corticium*.

1777a. **O. Pruni** Lasch. (from its host plant, sloe, *Prunus spinosa*) *a.* Crustaceo-adnate, effused, thin, white, becoming pallid; marg. byssoid, warts minute, rounded.

LXXVIII. KNEIFFIA Fr.

(After Friederich Gotthard Kneiff, mycologist of Baden.)

Resupinate, effused, thin. *Hymenium* strigose and roughened with rigid, scattered or fasciculate bristles which are barren outgrowths from the fertile hymenial surface. Laxly fleshy, soft, collapsing and becoming flocculose when dry. *Basidia* monosporous. (Fig. 93.)

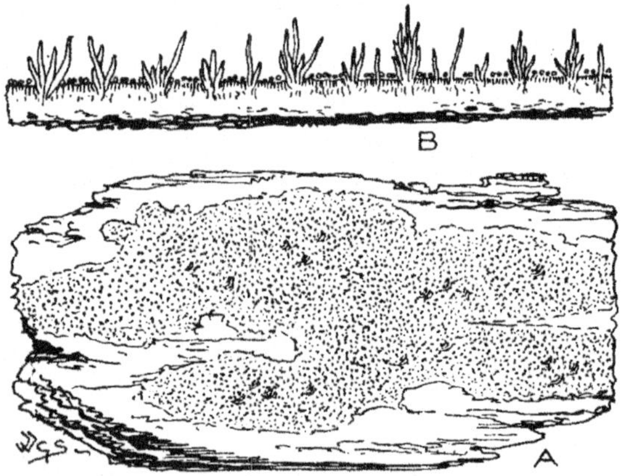

Fig. 93.—A, *Kneiffia setigera* Fr., natural size; B, section, enlarged.

The most perfect condition, where the growth is hemispherical and three inches in diameter, has not been observed in Britain.

Must not be confounded with *Peniophora*, where the basidia are 4-spored and the cystidia, which are bristle-like, are covered with minute particles of oxalate of lime. Species **1778, 1779**

1778. K. setigera Fr. (from the minute hairs or bristles borne on the hymenium; *seta*, a bristle, *gero*, to carry) *a b c.*

Sub. white, becoming ochreous-ivory, sometimes pale grey; marg. sometimes pulverulent. *Br.* minute, single or several connate, more or less scattered, hyaline; the basidia occur between the barren bristles.

Dead wood, blackberry, pine, juniper, magnolia, grass. Jan.–Dec. 4⅛ in. Sometimes the spuriously tuberculose hymenium resembles that of **1770** or **1839.**

1779. K. subgelatinosa B. & Br. (from the subgelatinous substance) *a b.*

Hyaline yellowish, ivory or pale buff, surface very finely granuloso-crested.

Stumps and cut surfaces of fir. 4 in. Sometimes accompanied by a green alga which penetrates the tissue of the fungus. Cracking when dry as in *Corticium*.

LXXIX. MUCRONELLA Fr.

(From the sharp-pointed spines; diminutive of *mucro*, a sharp point.)

Subiculum an extremely thin film, which soon becomes obsolete, spines subulate, simple, acute, glabrous, scattered or fasciculate, and more or less connate at the base. *Basidia* 1–4-sporous. (Fig. 94.)

Mucronella was at first placed under *Isaria* (*I. calva*) by Fries, but its fruit shows it to be a true Basidiomycete, and its fugitive

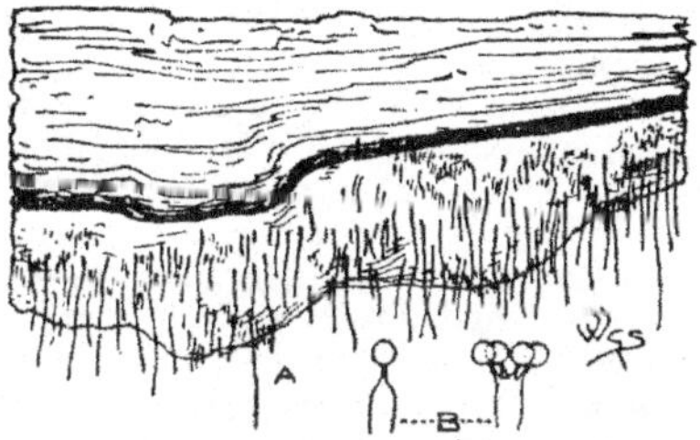

Fig. 94.—A, *Mucronella calva* Fr., one-half natural size; B, basidia and spores, × 500.

subiculum to be one of the *Hydnaceæ.* Fries seems not to have observed the feeble subiculum, sometimes distinct in *M. fascicularis.*

1780. M. calva Fr. (from the habit, a fanciful comparison with an almost bald head, bearing a few scattered short hairs; *calva*, the bald scalp) *a.*

Sub. obsolete or nearly so. *Sp.* slender, smooth, even, sharp-pointed, covered except at the barren tips with 1–4-sporous basidia, whitish, then grey. The growth of the spines is in a downward direction; dry examples revive with moisture.

Scattered or gregarious. Rotten wood, pine. Spines vary in size from a minute granule to 1 in. long.

Fam. IV. THELEPHORACEÆ

Erect and stipitate, cup-shaped, effused, with the upper portion free and reflexed or resupinate. Hymenium inferior or amphigenous, coriaceous, waxy or gelatinous, even to rugulose, ribbed or papillose, the surface glabrous, finely pulverulent or minutely velvety, with projecting setæ. Basidia clavate, 1–4-spored, normally tetrasporous. Spores usually aseptate, 3-septate in the non-British *Heterobasidium chlorascens*. The papillæ are for the most part accidental, generally arising in the resupinate species from *Sphæriæ*, growing on the bark beneath and not existing when the fungus grows on a surface free from *Sphæriæ*. True papillæ, however, sometimes occur on the pileate, terrestrial species. Some species are occasionally phosphorescent, as **1861**. In the gelatinous species the aseptate and tetrasporous basidia distinguish them from certain members of *Tremellinaceæ*, where the basidia are longitudinally and cruciately septate, or if the spores are aseptate the basidia are deeply bifurcate and bisporous. *Hypomyces rosellus* often stains the hymenium of *Thelephora*, *Corticium*, and other genera of *Thelephoraceæ* a bright purple-rose colour.

Massee has published a *Monograph of the Thelephoraceæ* in *Journ. Linn. Soc.* (Bot.) xxv. (1889) and xxvii. (1890). The sequence of the genera adopted here is that of P. A. Saccardo, *Sylloge Fungorum*, vol. vi., which is a slight modification of the order as given in Fries's *Hymenomycetes Europæi*. Species **1781—1925**

Key to the Genera.

Large.		
Erect, funnel-shaped; hymenium glabrous	80	Craterellus.
Erect to resupinate; hymenium rugulose	81	Thelephora.
Effuso-reflexed; hymenium even, glabrous	83	Stereum.
Horizontal, attached by a small base behind	82	Cladoderris.
Resupinate.		
Hymenium smooth-setulose	84	Hymenochæte.
Hymenium rough-setulose	86	Peniophora.
Hymenium glabrous	85	Corticium.
Hymenium pulverulent	87	Coniophora.
Small, cup-shaped, usually single	89	Cyphella.
Minute, cup-shaped, crowded	90	Solenia.
Parasitic on living leaves and stems	88	Exobasidium.

LXXX. **CRATERELLUS** Pers.

(From the shape, like a small crater, cup or bowl.)

Hymenium waxy-membranous, distinct, but adnate to the hymenophore, inferior, continuous, smooth, even, or rugose. *Pileus*

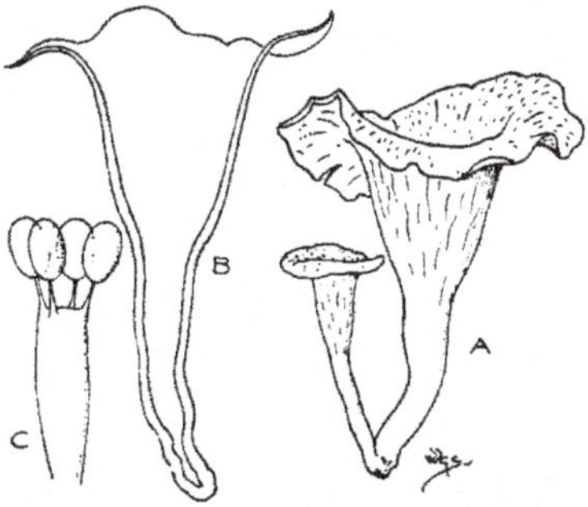

Fig. 95.—A, B, *Craterellus cornucopioides* Pers., entire and in section, one-third natural size. C, basidium and spores, × 500.

entire, stipitate, fleshy or membranous. *Spores* white. (Fig. 95.)

Usually large, erect and funnel-shaped. Growing on the ground. Autumnal. Allied to *Cantharellus*. Species **1781—1785**

Tubiform, pervious to the base of the stem. **1781, 1782**
Infundibuliform, stem stuffed. **1783, 1784**
Irregularly shaped, pileus and stem fleshy. **1785**

1781. C. lutescens Fr. (from the yellowish colour; *lutescens*, yellowish) *a*.

P. thin to submembranous, undulate, flocculose, warm vinous, orange-brown or dull ochre. *St.* smooth, ochre or salmon, or as P. *Hym.* rugoso-veined, ochre-cream, golden-sulphur or salmon, sometimes greyish. *Flesh* white-sulphur-ochre.

Sometimes cæspitose. Odour strong as of spirit. Moist woods; rare. Sept.–Oct. 2¾ × 4 × ⅜ in. Sometimes turning black in decay. Easily confounded with **1379**.

1782. C. cornucopioides Pers. (from the shape, like a cornucopia) *a b c*.

P. thin to submembranous, undulate, squamulose, umber-sepia. *St.* pale grey-slate, often blackish at base. *Hym.* even, at length slightly wrinkled, dull purplish-grey.

Solitary to cæspitose. Edible. Taste and odour agreeable. Woods, mixed, fir. Aug.–Oct. 3½ × 5½ × ⅜ in. Never insect eaten. Sometimes confounded with **1381**, in which there are thick gills. This is the Italian "Trombetta di morte." Mrs. Hussey records the sale of this species in Covent Garden market.

1783. C. sinuosus Fr. (from the undulate pileus; *sinus*, a curve) *a b c*.
P. subfleshy, floccose, pale warm umber, pale ochre-umber or fuscous-grey. *St.* pallid cinereous or umber, becoming sooty-umber. *Hym.* even, at length veined, folds broad, sometimes rather distant, pallid, pale ochre-umber, almost white or dull in hue. *Flesh* colour as St.

Odour strong, musky. Woods, fir, mixed; frequent. Aug. – Nov. $1\frac{3}{4} \times 2\frac{7}{8} \times \frac{1}{4}$ in. Small forms are often confounded with **1789.**

1784. C. crispus Fr. (from the crisped or curled pileus) *a b c*.
P. fleshy-membranous, brownish-ochre, fuliginous or fuscous. *St.* pallid, pale warm- or yellowish-umber. *Hym.* even, pallid, pale ochre-umber, almost white or pale dull ochre. *Flesh* pale ochre-umber.

Edible; odour strong, musky. Woods, mixed; frequent. Aug.–Nov. $2\frac{1}{2} \times 2 \times \frac{1}{4}$ in. Perhaps a variety of **1783.** Small forms are almost identical with **1789.** They may, with **1733,** be forms of one species.

1784a. C. pusillus Fr. (*pusillus*, very small) *a c*.
P. convex, umbilicate, thin, rugose, villous, cinereous. *St.* stuffed, somewhat compressed, grey. *Hym.* smooth or slightly wrinkled, pruinose, bluish-grey.

Single to cæspitose. Bare soil under beeches. Aug. $\frac{3}{8} \times \frac{5}{8} \times \frac{1}{8}$ in.

1785. C. clavatus Fr. (from the club shape when young; *clavis*, a club) *a c*.
P. turbinate, truncate or depressed, dull pale yellowish; mid. pale purplish-salmon. *St.* attenuate downwards, colour as P. *Hym.* even, then corrugate, dull brown-salmon, becoming lavender-white or brown-purplish. *Flesh* pale dull yellowish or pale purple-salmon.

Solitary or cæspitose, sometimes branched. Edible. Woods, beech, fir. Oct. $6\frac{1}{2} \times 3 \times 1\frac{1}{2}$ n. Sometimes strongly resembles certain of the larger *Lactarii*.

LXXXI. **THELEPHORA** Ehrh.

(From the sometimes papillose hymenium; Gr. *thele*, a teat, *phero*, to bear.)

Hymenium inferior or amphigenous, continuous with and similar to the hymenophore, even, rugulose or ribbed, ribs sometimes more or less papillose. Without the fibrillose stratum present in *Stereum*. *Basidia* 4-spored. *Spores* warted or echinulate. (Fig. 96.)

Coriaceous, dry, fibrous, destitute of a cuticle, varied in form, pileate, clavate or resupinate. Terrestrial or on stumps, branches, leaves, earth and stones; sometimes investing seedlings and living plants. Some species more or less resemble brown flowers with laciniate petals; **1794** and **1801** are fetid. The sequence of species is that of Fries, Berkeley, Stevenson and Saccardo.

Species **1786—1804**

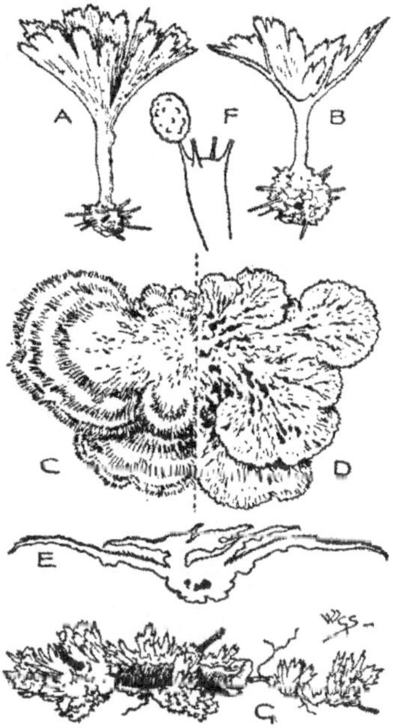

Fig. 96.—A, B, *Thelephora caryophyllea* Pers., entire and in section, one-half natural size. C, D, E, *T. laciniata* Pers., upper and under surface and section, one-half natural size; F, basidium and spore, × 750. G, *T. fastidiosa* Fr., one-half natural size.

Erect, pileus entire or branched. **1786—1794**

Pileate, dimidiate, horizontal, subsessile or effuso-reflexed. **1795—1799**

Resupinate, for the most part incrusting, of various forms; hymenium even. Soft, subgelatinous when moist and with a compact, not strigose pileus; forming effused patches or dendritic masses on twigs, grass, etc. **1800—1804**

1786. T. Sowerbei Berk. (after James Sowerby) *a b*. White, becoming yellowish.

P. infundibuliform, more or less laciniate and toothed, rough with radiating projecting processes. *St.* attenuate downwards. *Hym.* plain, setulose.

On the ground amongst moss, under beech; rare. Oct.–Nov. 2 × 2¾ in. Sometimes occurs with a tuberous base as in **1791**.

1786a. T. vitellina Plowr. (from the colour, like yolk of egg; *vitellus*, egg-yellow) *a*.

P. flabelliform, expanded upwards; marg. slightly incurved. *St*. attenuate downwards, but enlarged at base. *Hym*. rugulose, zoneless.

Gregarious. Dead fir-branches amongst moss, interior of hollow firs. Sept. 1 × 2 × ⅛ in. Differs from **1786** in its habit and pale egg-yellow colour; sometimes pale sulphur with marg. brighter and base of stem luminous yellow.

1787. T. pallida Pers. (from the pallid colour) *a c*. Pale yellowish-brown.

P. infundibuliform, strigoso-squamose, more or less laciniate and toothed. *St*. short or obsolete. *Hym*. costate, setulose.

Cæspitose. Woods, pine, beech. Nov. 1⅝ × 2½ in. Intermediate forms occur between this and **1786**.

1788. T. multizonata B. & Br. (from the many-zoned pileus; *multus*, many) *a b*.

P. infundibuliform, formed from confluent lobes and stems; marg. lobed, crenulate, bright rufous flesh-colour, zoned, darker. *St*. obsolete. *Hym*. slightly ribbed, smooth, paler than P.

Cæspitose. Woods, on the ground; rare. Oct. 6½ × 5 in. Simple fan-shaped forms occur, with the substance split.

1789. T. undulata Fr. (from the undulate pileus) *a c*. Pale yellow-brownish.

P. coriaceo-membranous, depressed, even, somewhat smooth; marg. entire. *St*. central, equal or attenuate downwards, villous. *Hym*. ribbed, somewhat setulose.

On the ground. Oct. ½ × ⅜ in. Often confounded with small forms of **1783** and **1784**.

1790. T. caryophyllea Pers. (from a fancied resemblance to a carnation, *Dianthus Caryophyllus*) *a c*. Deep claret-brown or purple-brown.

P. subcoriaceous, depressed, fibrous-torn. *St*. equal. *Hym*. smooth, somewhat even. *Flesh* very dark.

Odour none. On the ground, chiefly about the roots of fir; rare. Sept.–Nov. 2½ × ⅜ × ⅛ in.

1791. T. tuberosa Fr. (from the tuberous base) *a*. Brownish salmon-white.

P. subcoriaceous, divided to top of stem into compressed branches, infundibuliform. *St*. obscurely furrowed or lacunose, bulbous. *Hym*. smooth. *Flesh* dark brown with a whitish cortex.

Scattered. On the ground; rare. 3½ × ¼ × ⅛ in.

1792. T. anthocephala Fr. (from a fancied resemblance to a flower; Gr. *anthos*, a flower, *kephale*, the head) *a b c*.

P. coriaceous-soft, pubescent, divided into large and small divisions, apices usually fimbriate, yellowish- or claret-brown. *St.* short or obsolete, soon furcate, colour as P.

Odour none. Woods, on the ground. Aug.–Nov. 2⅝ × 3 in. Somewhat resembles a *Clavaria* or *Xylaria hypoxylon* in habit. Very variable in the form of the laciniæ.

1793. T. clavularis Fr. (from the nail-like branches; *clavula*, a small nail) *a.* Claret-brown or rufous-fuscous.

P. coriaceous-soft, much and irregularly branched, branches round, attenuate, even, smooth, apices acute, pubescent, whitish or pale yellowish. *St.* short, thick, somewhat tuberous.

Odour none. On the ground. 1⅝ × ½ × ½ in. Habit as **1792.**

1794. T. palmata Fr. (from a fancied resemblance to the palm of the hand) *a b c.*

Coriaceous-soft, much branched, pubescent, dark yellowish-brown or dark claret-brown, apices cottony-fringed, brownish-white. *St.* branched, flattened even, larger lobes wedge-shaped, the smaller sometimes in form of small infundibuliform pilei.

Odour fetid, carrion-like. Woods, fir, maple. Aug.–Nov. 3⅝ × 4⅝ in. Habit as **1792.**

1795. T. intybacea Pers. (from a fancied resemblance to the flowers of chicory, *Cichorum Intybus*) *a b c.*

P. fibrous, somewhat jagged and toothed, slightly laciniate, zoned or zoneless, rufous-ferruginous, deep sienna or deep claret-brown; marg. white. *Hym.* papillose, subfloccose, veined near marg., brownish-vinous, pale vinous at marg. *Tufts* central and obconic, lateral and subtriangular or spathulate.

Gregarious, imbricate. Pine-woods, pine-twigs, *Vaccinium Myrtillus.* Nov. 2 × ½ × ⅛ in. Intermediate forms occur between this and **1796** and **1797.**

1796. T. terrestris Pers. (from the terrestrial habit) *a b c.*

P. soft, fibrous-strigose, lobed, usually zoneless, extended into a somewhat lateral, almost obsolete stem or base, deep vinous-brown; z. when present darker; marg. pale brownish-white, often with a narrow white-pubescent edge. *Hym.* radiato-rugose, claret-brown-yellowish or salmon.

Gregarious, imbricate. Woods, fir. July–Oct. Diam. 1⅝ in.

1797. T. laciniata Pers. (from the fringed margin of the pileus; *lacinia*, a flap) *a b c.*

P. coriaceous-soft, fibrous-scaly, effuso-reflexed, somewhat zoned, deep sienna, somewhat claret; marg. fibrous-fringed, whitish. *St.* when present incrusting branches as a thick membrane, paler than P. *Hym.* veined and papillose, colour as St.

Imbricate, growing in strata, perennial. Woods, pine, larch, branches, twigs, leaves, oak, heather, *Vaccinium Myrtillus*, earth, often on compacted masses of fir-leaves. Aug.–Nov. Diam. 2⅝ in. At first forming a resupinate, yellowish or salmon-brown film.

1797a. T. atra Weinm. (from the black colour; *ater*, black) *a.* Soft. *Pi.* springing from a tuberous base, purple-black above and below, sometimes cinereous; mid. paler. *Hym.* smooth, setulose, black, white-pruinose. *St.* attenuate downwards, deformed sooty-black.

Cæspitose. Dead logs. 3 × 2¼ × ¼ in.

1798. T. biennis Fr. (from its period of growth, two years, *biennis*) *a b c.*

Coriaceous-soft, broadly incrusting, cinerous-fuscous. *P.* narrowly reflexed, tomentose, pale claret-brown; marg. fringed. *Hym.* in part resupinate, smooth, very slightly bristly, more or less plicate, the resupinate parts white-edged.

Subimbricate. Stumps, fir, on the ground amongst wood, leaves and stones; common. Aug.–Nov. Resupinate hymenium 4¼ in. Pilei 1 in.

1799. T. mollissima Pers. (from its very soft substance; *mollis*, soft) *a b c.*

P. subtomentose, white to pale or dark claret-brown with a broad white tomentose marg. *Hym.* smooth, even, claret-brown or pale snuff-colour. *Flesh* claret-brown, clouded and streaked darker to almost black.

Subimbricate. On the ground, or on twigs, leaves, etc., oak, beech; frequent. Sept.–Oct. 5¾ in. Very variable, wholly effused, palmate or like a soft stemless imbricate *Polyporus;* sometimes forming a thin faint brown-ochre membrane with a paler margin. Short stems are sometimes developed.

1800. T. cristata Fr. (from the short, fringe-like crest; *crista*, a crest) *a b c.* Pallid or ivory-colour.

Incrusting, slightly tough, becoming branched, the apices crested or fringed. *Hym.* papillose.

On the ground in woods or amongst or on moss, dead leaves and grass; frequent. Oct. 5 in.

1801. T. fastidiosa Fr. (from the disgusting odour; *fastidiosus*, disgusting) *a b c.* White.

Effused, incrusting, confluent, soft, irregular in shape, shortly branched and toothed. *Hym.* papillose, white or biscuit.

On the ground, involving twigs, rootlets, branches, leaves, etc., sloe, ivy, beech, beech-mast, oak, acorns, moss. Sept.–Nov. Group 5½ in. Odour the same as **1794.**

1802. T. sebacea Pers. (from its tallow-like appearance and consistency; *seba*, tallow) *a b c.* White or pale yellowish-white.

Effused, fleshy-waxy, becoming hard, incrusting, irregular in shape like melting and running tallow; marg. determinate. *Hym.* collapsing, flocculoso-pruinose.

Involving twigs, grass, moss, leaves, cones, clover, ivy, *Hydnum auriscalpium*, etc. Jan.–Oct. 5 in.

1803. T. crustacea Schum. (from its forming a crust) *a b c.* Fuscous-brown.

Broadly effused, somewhat fleshy, undulate and tubercular; marg. white-floccose. *Hym.* papillose, somewhat bristly.

On the ground, involving grass, leaves, etc., and on rotten sawdust, which it resembles in colour. Feb.–Nov. 3½ in.

1804. T. cæsia Pers. (from its colour; *cæsius*, bluish-grey) *a b c*. Cinereous bluish-grey.

Effused, soft, continuous, often minutely veined or folded; marg. determinate. *Hym.* even or slightly papillose, slightly bristly.

On the ground, stumps, etc., involving earth, bark, twigs, etc. Aug. 3½ in.

LXXXII. CLADODERRIS Fr.

(From the branched and coriaceous structure; Gr. *klados*, a young branch, *derris*, a leathern coat.)

Coriaceous. *Hymenium* woody, radiate or rugulose with branched ribs, or narrow greatly ramified folds, persistent, at length roughened with warts. Horizontal and attached by a small base behind. (Fig. 97.)

Typical species of *Cladoderris* resemble *Lentinus* in appearance, with a squarroso-hispid pileus of the colour and texture of leather and a veined hymenium. The species are usually stemless, but sometimes there is a pervious stem as in **1702**. The plants are not usually small although small examples occur. *Cladoderris dendritica* varies from 7 in. to ¾ in. in diam. The prevailing size is from 4 to 6 in. in diam. Some species are large, infundibuliform and almost stemless.

The species chiefly grow in Australia, Cuba and the Philippines.

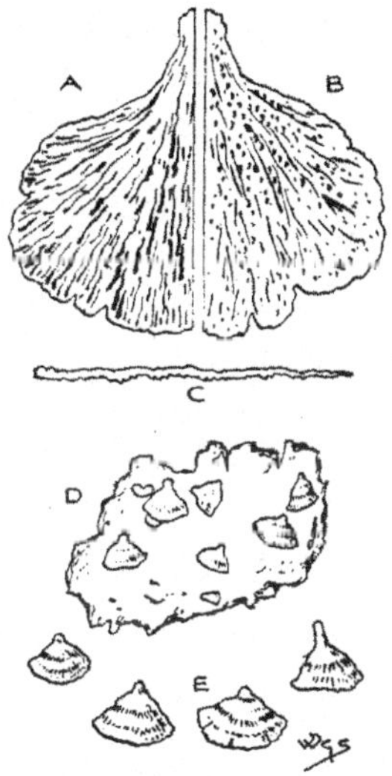

Fig. 97.—A, B, C, upper and lower surface and section of *Cladoderris australica* Berk., drawn as a type of the genus, one-half natural size. D, *C. minima* B. & Br., natural size. E, ditto, × 2.

1805. C. minima B. & Br. (from its very small size; *minimus*, the least) *a b*. White.

Flabelliform, springing from a stem-like or obsolete base, resupinate. *P.* tomentose, somewhat zoned. *Hym.* radiate on branched ribs.

Birch. Dec. Diam. $\frac{3}{16}$ in. Resembling a dwarfed **1806**.

LXXXIII. STEREUM Pers.

(From its hard substance; Gr. *stereos*, hard.)

Hymenium inferior, coriaceous, separated from the inodermous velvety-silky or fibrous-coated pileus or subiculum by an indeterminate fibrillose stratum, even, smooth, composed of basidia and hymenial cells compactly arranged side by side and of uniform height. *Basidia* 4-spored. *Spores* minute, roundish-ovate, white or coloured. (Fig. 98.)

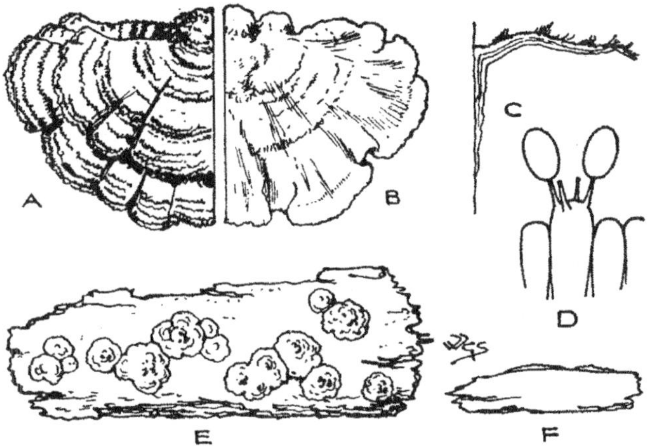

Fig. 98.—A, B, C, *Stereum hirsutum* Pers., upper and lower surface and section, two-thirds natural size. D, basidium and spores, × 660. E, F, *S. Pini* Fr., upper surface and section, two-thirds natural size.

Coriaceous or woody, mostly perennial, somewhat zoned, entire, definite in form, effuso-reflexed. Species **1806—1818**

a. Apodæ. Pileus dimidiate, sessile or at first resupinate, then effusedly reflexed, marginate. **1806—1814**

b. Resupinatæ. Agglutinate, effused; marg. not free, or scarcely so. **1815—1818**

a. Apodæ.

1806. S. hirsutum Pers. (from the hairy pileus; *hirsutus*, hairy) *a b c.* Coriaceous, rigid.

P. hispid-zoned, effused, reflexed, lobed and crisped, usually ivory-buff varying dull primrose, orange or brownish; marg. subobtuse, white or whitish. *Hym.* even, smooth, obscurely zoned, juiceless, usually warm-biscuit.

Imbricate. Stumps, oak, beech, birch, poplar, laburnum. Jan.–Dec. 4 in. Said to cause "fly"-wood and "yellow- and white-piped" wood. Pileus often green at the base from the growth of algæ. A deep rose-coloured stain is frequent on this plant caused by a parasitic mould, *Hypomyces rosellus*. Var. *subcostatum* Mass. *Hym.* costate or rugose. Var. *cristulatum* Quél. *P.* strigose, grey.

1807. **S. ochroleucum** Fr. (from the yellowish-white pileus; Gr. *ochros*, pale yellow, *leukos*, white) *a c*. Coriaceous-membranous.

P. reflexed, expanded, flaccid, silky, zoned, hoary-yellowish, shaded pale brownish. *Hym.* even, smooth, cracking, yellowish or greyish.

Wood, fir, maple, oak, hornbeam. Jan. 2¼ in.

1808. **S. purpureum** Pers. (from the purple hymenium; *purpureus*, purple) *a b c*. Coriaceous-soft.

P. effuso-reflexed, subimbricate, zoned, villoso-tomentose, pale yellowish. *Hym.* even, smooth, bright rose- or dark-purple, clouded, salmon. *Flesh* sometimes 1 in. thick, bright rose-purple, 5–7 zoned.

Dead wood, horse-chestnut, sycamore, hornbeam, fir, poplar, willow, birch, elm, twigs, grass. May–Dec. 3½ in. Said by Prof. J. Percival to be the cause of "silver-leaf" disease in plum, peach, apricot and sloe. Var. *atromarginatum* W. G. Sm. A beautiful and apparently distinct plant, illustrated with drawings by Purton and Sowerby, in the British Museum collection; it has been named *Auricularia elegans* Sow. t. 412, f. 1. Berkeley regarded the drawings as a variety of *Stereum purpureum* Pers. *P.* woolly or frosted, white, with a narrow black zone near the white margin. On a moss-covered tree.

1809. **S. sanguinolentum** Fr. (from the habit of the hymenium, bleeding when bruised; *sanguis*, blood) *a b*. Coriaceous-thin.

P. effuso-reflexed, substriate, adpressedly silky, pale brownish-yellow; marg. acute, white. *Hym.* even, smooth, pruinose when old, brown-whitish or ashy, rarely as P.

Wood, branches, twigs, pine; common. Jan.–Dec. 3½ in.

1810. **S. conchatum** Fr. (from the shell-like shape; *concha*, a shell) *a*. Coriaceous-thin.

P. conchato-reflexed, subcrisped, rugose, glabrous, obscurely zoned, dull yellowish or brown-buff-whitish. *Hym.* even, smooth, dry, brown.

Subimbricate. Fir. 1½ in.

1811. **S. spadiceum** Fr. (from the date-brown hymenium; *spadix*, date-brown) *a b*. Coriaceous.

P. effuso-reflexed, hispid-zoned, very hispid at base, pale yellowish or orange-brownish. *Hym.* even, smooth, brown-vinous over pale yellow, not dark, or claret-brown, exuding red juice when bruised.

Stumps and branches, ash, oak; common. Jan.–Dec. 4¾ in.

1812. **S. rugosum** Pers. (from the commonly wrinkled hymenium; *ruga*, a wrinkle) *a b c*. Coriaceous-rigid.

P. effuso-reflexed, shortly reflexed, obtusely marginate, pruinose, then smooth, buff, paler behind, zoned sienna. *Hym.* pruinose, yellowish-brown or cream-buff, shaded pale livid, becoming pale vinous where bruised.

Stumps, fir, hornbeam, oak, hazel; very common. Jan.–Dec. 6¼ in. *Var. Lauro-cerasi* Stev. grows on cherry laurel, *Prunus Lauro-cerasus*.

1813. S. vorticosum Fr. (from a fancied resemblance in the habit to waves in a vortex or whirlpool) *a b*. Coriaceous.

P. effuso-reflexed, imbricate, obscurely zoned, rugose, strigose, crisped, pale buff; marg. white. *Hym.* smooth, somewhat ribbed, salmon, or purplish-salmon, sometimes orange shaded.

Beech. 4¾ in. Intermediate between **1806** and **1808.**

1814. S. rufum Fr. (from the rufous pileus) *a b c*. Rigid, coriaceo-cartilaginous.

P. effuso-reflexed, rugose-rough, deep vinous-brown, with pale yellowish marg. and white edge. *Hym.* minutely cracked, grey-pruinose.

Dead pine, ash, poplar. Sept.–Mar. 3½ in. Effused and immarginate when young.

b. Resupinatæ.

1815. S. Pini Fr. (from the habitat, pine bark) *a b c*.

Coriaceous-cartilaginous, adnate as a shield, marginate, smooth beneath, at length bullate. *Hym.* noduloso-rough, purple flesh-colour, then fuscous, deep sienna or deep vinous-brown.

On Scotch fir. Nov. 3¼ in. Sometimes identical in colour with the pine bark and easily overlooked.

1816. S. frustulosum Fr. (from the habit, breaking to fragments; *frustrum*, a piece) *a c*.

Woody, tubercular, thick, crowded and subconfluent, breaking to fragments, smooth beneath, and at the obsoletely marginate circumference. *Hym.* convex, pruinose, blackish-brown, cinnamon, becoming salmon.

Wood and bark, oak, beech, branches, hard to rotten sticks. 2¼ in.; frustules ⅛ × $\frac{3}{16}$ in.

1816a. S. quercinum Potter (from the habitat, oak, *quercus*) *a*.

Coriaceous, at first filling in between crevices of bark. *Hym.* slightly papillose, young examples with a central papilla, pale buff, becoming cinnamon-whitish; marg. slightly raised from matrix, dark brown, finely zoned beneath, edge lighter. Abundant white mycelium in and under bark.

Oak bark. Jan.–Nov. 2¾ × 1⅛ in.

1817. S. acerinum Fr. (from its usual habitat, maple branches; *acer*, maple) *a b c*.

Crustaceous, adnate, effused, even, usually smooth, sometimes minutely nodulose or rugoso-rough, often transversely cracked, white, pale brownish-white or ivory.

Maple, fir, juniper, elm. Dec. 4¼ in. A slight repand hirsute pale-yellowish margin or pileus is sometimes developed.

1818. S. stratosum B. & Br. (from the stratose habit) *a*.

Stratose, the pallid strata at length separating, effused, smooth, here and there wrinkled, bright ochreous-white, becoming yellow.

3¾ in.

LXXXIV. HYMENOCHÆTE Lév.

(From the bristly hymenium ; Gr. *humen*, a membrane, *chaite*, a bristle.)

Pileus coriaceous, variable in form. *Hymenium* plane to papillose, velvety, with minute, rigid, persistent, bright brown setæ or modified

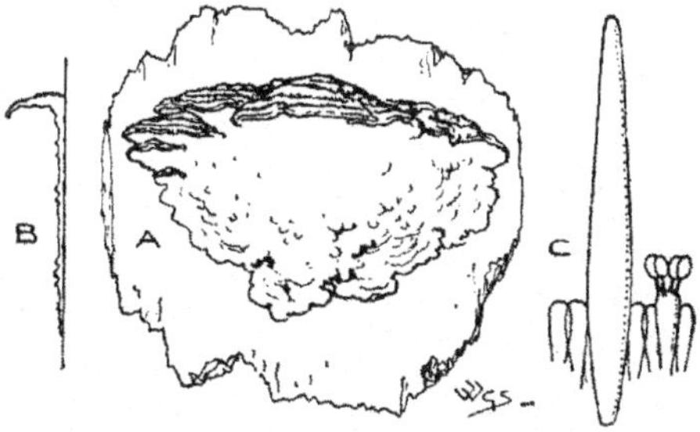

Fig. 99.—A, B, *Hymenochæte rubiginosa* Lév., entire and in section, one-half natural size. C, part of hymenium, showing cystidium, basidium and spores, × 500.

cystidia, three or four times the length of the basidia; in **1830** the setæ at first very pale. *Basidia* tetrasporous. *Spores* white or olive. (Fig. 99.)

Allied to *Peniophora*. Species **1819—1831**

a. Apodæ. No stem. *Spores* white, *cystidia* acuminate, plane. **1819, 1820**

b. Resupinatæ. With a resupinate habit, except **1827.**
Spores white, *cystidia* acuminate, plane or rough. **1821—1824**
Spores olive. **1825—1827**
Cystidia subclavate, sometimes rough. **1828—1831**

a. Apodæ.

1819. H. rubiginosa Lév. (from the iron-rust colour of the hymenium; *rubigo*, rust) *a b c.*

P. effused, reflexed, somewhat imbricate, zoned, velvety, deep orange-sienna; marg. orange. *Hym.* papillose, ferruginous or umber, sometimes deep claret-brown.

Hard wood, twigs, oak, alder, sometimes spreading to moss. Jan.-Dec. 6 in.

1820. H. Avellana Lév. ex Cooke (from the habitat, hazel, *Corylus Avellana*) *a c.*

P. effused, reflexed, zoned, villous, bay-brown; marg. obtuse, free, narrowly reflexed. *Hym.* papillose, pale ferruginous or pale brown, here and there bleeding, sometimes minutely cracked all over.

Hazel, beech, fir, etc. ½–1 in.

b. Resupinatæ.

1821. H. nigrescens Mass. (from the at length blackish hymenium; *nigrico*, to be blackish) *a b*.

P. circular, flattened; marg. free, a little reflexed. *Hym.* papilloso-rough, brownish- or greyish-umber, becoming blackish, with irregular black zones. *Flesh* indigo-umber.

Solitary, gregarious or confluent. Dry wood. May. 3½ in.

1822. H. Stevensonii B. & Br. (after the Rev. Dr. John Stevenson) *a b*.

Hym. pallid, fawn-colour, or ivory-salmon with an obtuse slightly elevated paler marg.

Yew. Sept. 2 in. This is *Stereum rufohispidum* Stev.

1823. H. leonina B. & C. (from the tawny colour; Gr. *leon*, a lion) *a c*.

Hym. somewhat papillose, ferruginous, saffron, orange or orange-sienna, sometimes dark, with a few lighter spots; marg. tomentose, lighter.

Dead wood, holly. 3¼ in.

1824. H. fuliginosa Lév. (from the smoke-coloured hymenium; *fuligo*, soot) *a c*.

Hym. somewhat papillose, dark smoky bay-brown or opaque claret-brown; marg. lighter.

Wood. 2½ in.

1825. H. corrugata Lév. (from the wrinkles or folds on the hymenium; *corrugo*, to wrinkle) *a c*.

Somewhat effused, closely adnate, soon grumous. *Hym.* papillose, red, red- pale- or vinous-brown, or deep claret-sienna, sometimes darker at marg.

Bark, alder, maple, oak, blackberry, hawthorn, holly, ling, *Vaccinium*, *Myrica*. Mar.-Dec. 5½ in. Twigs, leaves, etc., are commonly agglutinated to the hymenium. Becomes much cracked in drying.

1826. H. croceoferruginea Mass. (from the orange-brown hymenium; *crocus*, saffron, *ferrugo*, iron-rust) *a*.

Effused, broadly encrusting, very thin. *Hym.* papillose, saffron, yellow-sienna, tawny, pale sienna or yellowish-chocolate; marg. paler or yellow.

Dead dog-rose. 3½ in. Hymenium cracking.

1827. H. tabacina Lév. (from the tobacco colour) *a b c*.

Somewhat coriaceous, thin, flaccid, effused, imbricate. *P.* silky, then smooth, sienna-orange, pale or deep sienna or purplish-umber, zoned sienna, base blackish-umber; marg. reflexed, ivory. *Hym.* ferruginous, deep- or crimson-sienna, with a white, ivory, yellow or orange marg.

Somewhat uncommon. Trunks, branches, etc., willow, hazel, plum, oak, birch, lilac, blackberry, cotoneaster, rhododendron, *Andromeda*, *Vaccinium*, sometimes spreading on to moss. Oct.-Mar. 3½ in.

1828. H. crassa Cooke (from the thick substance; *crassus*, thick) *a*.

P. velvety, pallid-rufous; marg. thick, at length free. *Hym.* unequal, colour as P.

Trunks. 4½ in.

1829. H. abietina Mass. (from the habitat, trunks of fir, *abies*) *a b*.
P. effused, tomentose, sienna-brown. *Hym.* papillose, fuscous-ferruginous, dull yellowish-purple, brownish-salmon or claret-brown, clouded ivory-ochre or purple-lilac, becoming pale.

Decaying fir, spruce-fir. 2½ in. Cracking when dry.

1830. H. disciformis W. G. Sm., *Thelephora disciformis* DC. (from its frequent disc shape) *a*. White to ivory-white.
P. disciform, thinner towards the definite free naked marg. *Hym.* unequal.

Trunks, oak. 1½ in. This is *Stereum disciforme* Fr. Cooke has placed it in *Peniophora*, and has illustrated the plant with granular setæ. Saccardo follows Cooke. My plants and those in British Museum have smooth setæ.

1831. H. Boltonii Cooke (after James Bolton of Halifax) *a*.
P. villous, white to ochre or pale lavender, zoned brown or black. *Hym.* white to pale brown, becoming ferruginous-fuscous or dark red; marg. crisped, free, thin.

Bird cherry. Feb. 1⅜ in. Cracking when dry.

LXXXV. CORTICIUM Pers.

(From the usual habitat, bark, *cortex*.)

Resupinate. *Hymenium* often broadly effused, in a few species the margin is slightly upraised, amphigenous, even or tuberculose,

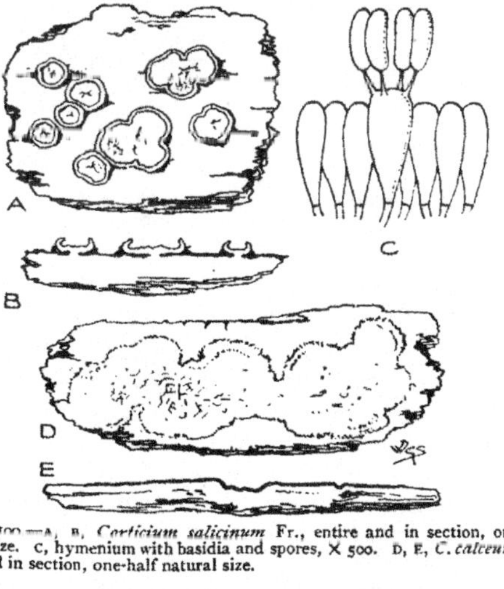

Fig. 100.—A, B, *Corticium salicinum* Fr., entire and in section, one-half natural size. C, hymenium with basidia and spores, × 500. D, E, *C. calceum* Fr., entire and in section, one-half natural size.

waxy or fleshy-soft, usually contracting and cracking when dry, smooth, arising immediately from the mycelium without an intermediate fibrillose stratum; without setæ or modified cystidia. *Spores* white. (Fig. 100.)

Usually growing on bark and wood, sometimes on sawdust, earth, fir-cones, stems of herbaceous plants, leaves, moss, etc. Some species are phosphorescent. Young examples of some species of *Corticium* may readily be confounded with *Pezizæ*. The hymenium of some species occasionally exhibits pale rose-coloured to deep crimson stains caused by the presence of a parasitic mould, *Hypomyces rosellus*. Species **1832—1866**

Marginate, margin free, determinate. **1832—1837**

Margin usually indefinite, byssoid, strigose, pulverulent or mucedinous-clouded.

Hymenium white, or ochreous. **1837a—1851**

Hymenium brightly coloured. **1852—1864**

Amphigenous, very thin, innate, throwing off the bark. **1865, 1866**

1832. C. salicinum Fr. (from the habitat, willow, *salix*) *a c*.

Coriaceous, soft, rigid when dry, small, fixed by centre. *Hym.* somewhat rugulose, becoming finely but sparingly cracked, deep claret-crimson; marg. reflexed all round, villous, white or pale rose.

Gregarious. Willow, rarely poplar. Aug.–Sept. ¼–1 in. Resembles a *Peziza*.

1833. C. amorphum Fr. (from the supposed ambiguous structure of the hymenium; Gr. *a*, without, *morphe*, form) *a b c*.

Somewhat coriaceous, small, waxy-pliant, cup-shaped then flattened. *Hym.* white- grey- or yellowish-umber; marg. white.

Gregarious. Silver fir, larch. Aug.–Nov. ⅛–⅜ in. Resembles a *Peziza*. This is *Peziza amorpha* Pers., *Thelephora amorpha* Fr. It grows in company with *Peziza calycina*, with pallid examples of which it may be easily confounded. When dry the *Corticium* and *Peziza* resemble each other closely.

1834. C. evolvens Fr. (from the habit, as of new plants emerging from the margins of older ones; *evolvo*, to unroll) *a b c*.

Soft, cup-shaped to effuso-reflexed, gregarious, becoming confluent. *Hym.* somewhat rugose or tubercular, cracking, pale sienna, clouded darker or pale yellowish-brown.

Bark, cherry, willow, poplar; frequent. Oct.–April. ¼–1¾ in.

1835. C. porosum B. & C. (from the pore-like markings on parts of the hymenium) *a*.

White, becoming pale ochre; marg. free, reflexed.

Wood. ¾–3 in. The "pores" somewhat resemble the true pores of *Porothelium*, they are of different sizes and appear as thin dried circular lines, such as are left by minute drops of moisture in drying.

1836. C. populinum Fr. (from the habitat, poplar, *populus*).
Soft, tuberculiform, soon confluent and effused, at length involute. *Hym.* uneven, ferruginous-cinerous, whitish-tomentose beneath.
Poplar.

1837. C. Lycii Cooke ex Mass. (from the usual habitat, *Lycium*) *a*.
Effused, thin. *Hym.* papillose, lilac or white-salmon; marg. free, somewhat byssoid, pallid.
Lycium, lilac. ¾–3 in.

1837a. C. seriale Fr. (from the habit, growing in long, gregarious rows; *series*, a row).
Elongato-effused, adfixed to matrix, thin. *Hym.* rimose, thin, papillose, dull ochreous, white pruinose; marg. fibrillose whitish.
Wood, pine.

1838. C. calceum Fr. (from the chalk-white colour; *calx*, lime) *a b c*.
Broadly effused, agglutinate, waxy. *Hym.* even, smooth, rigid, cracking, white or pallid, becoming pale dull salmon; marg. finely pulverulent, white.
Wood, pine, blackberry, lilac, willow, lime, hornbeam, birch, *Myrica*, leaves, earth. May–Aug. ¼–5 in. This is *C. cretaceum* Pers.

1839. C. serum Fr. (from its time of growth—winter; *serus*, late, growing in winter) *a b c*.
Broadly effused, incrusting, fleshy to thin. *Hym.* smooth to pruinose, crowded with papillæ, splitting into flocci, white; marg. fading in a mucedinous manner into the wooden matrix, sometimes forming finely pulverulent cloud-like growths.
Trunks, birch, alder, hornbeam. ½ 3½ in. This is *Hypochnus* Sacc., *Lycomyces* Karst. Often resembling spluttered drops, as from a wax candle.

1840. C. scutellare B. & C. (from the salver-like form it sometimes assumes; *scutella*, a salver) *a c*.
Broadly effused. *Hym.* cracking into minute areolæ, with intermediate finely-folded rugose patches, whitish, becoming warm biscuit-buff; marg. forming thin cloud-like growths in mature examples.
Rotting logs, furze, oak, cedar, dead herbaceous stems. 6 in.

1841. C. fœtidum B. & Br. (from its fetid odour) *a*.
Effused. *Hym.* cracking, white to ivory and ochreous, clouded pale grey and faint sienna.
Sawdust. ¼–2½ in.

1842. C. lacteum Fr. (from the milk-white colour; *lac*, milk) *a b c*.
Broadly effused, submembranous. *Hym.* waxy, cracking and parting when dry, white to pale biscuit with mid. darker, sometimes pale straw or pale orange-colour; sub. and marg. loosely fibrillose, often fringed with long branching strands of white mycelium, sometimes with pilose or farinose cloud-like growths.
Trunks and beneath bark, chestnut, beech, holly, lime; frequent. Nov.–Feb. 6 in.

1843. C. læve Pers. (from the even hymenium; *lævis*, even) *a b c.*
Effused, membranous, separating from matrix. *Hym.* even, smooth, rarely papillose, commonly with vacuoles exposing the wood, cracking, white to sienna-yellowish, flesh-colour, livid, salmon or pale red; marg. sometimes broad, edge pilose, ivory to white, sometimes with radiating strands of white mycelium.

Wood, bark, sticks, pine, poplar, birch, alder, oak; very common. Nov.–May. 8½ in. Some conditions resemble 1770.

1844. C. confluens Fr. (from the habit—forming confluent patches) *a b c.*
Agglutinate, submembranous. *Hym.* even, naked, cracking, somewhat concentrically rugose when young, hyaline becoming white, lavender-brown-white, yellowish-brown, pale brownish or pale rose-purple; marg. spreading in a dentate manner.

Bark, beech, alder, ash, willow. Mar.–Aug. ⅝–3⅝ in.

1845. C. nudum Fr. (from the hymenium becoming bare, *nudus*) *a b c.*
Waxy, agglutinate, very thin, becoming rigid. *Hym.* flesh-colour, white-salmon, umber-ivory or yellowish-buff, clad with fugacious white meal, cracking when dry; marg. determinate, smooth.

Dead wood, bramble, alder, yew, willow. Aug.–April. ¼–4½ in.

1846. C. arachnoideum Berk. (from the mycelium—like spider-web; Gr. *arachne*, a spider) *a b c.*
Very thin, effused, delicately flaxy. *Hym.* waxy, continuous or somewhat papillose, somewhat cracked when dry, white or faintly clouded sulphur, yellowish or brownish, floccose beneath; marg. white-fibrillose.

Wood, bark, oak, pine, twigs, leaves, grass, lichens, etc. Oct.–Mar. 7½ in.

1847. C. radiosum Fr. (from the radiate habit) *a b c.*
Membranous, subrotund with radiate lobes. *Hym.* smooth, glabrous, irregularly papillose, cracking when dry, primrose-white to dark ochreous or buff, adpressedly fibrillose beneath; marg. fibrilloso-radiate, often with long strands of mycelium.

Elm, birch, pine, hornbeam, naked wood. Oct.–Feb. 4½ in.

1848. C. Typhæ Fück. (from the usual habitat, bulrushes; *Typha*, a bulrush) *a b c.*
Longitudinally effused, thin, at first forming white byssoid spots, smooth then mealy, cracking when dry, ivory to tan; marg. usually ragged or ill-defined.

Dry leaves of *Typha latifolia* and species of *Carex*. 1½ in.

1849. C. Sambuci Pers. (from the habitat, elder, *Sambucus*) *a b c.*
Broadly effused, indeterminate, encrusting, somewhat innate, cracking and breaking into floccules when dry, white.

Very common. Aug.–May. 7 in. Sometimes forming hyaline-whitish spotty cloud-like growths.

1850. C. lactescens Berk. (from its habit of exuding a watery milk-like fluid; *lac*, milk) *a b c.*

Waxy-soft, agglutinate, undulate or rugose at length cracked, pale flesh-colour or pale salmon; marg. shortly byssoid.

Dead wood, ash, willow. Oct.–Feb. 6 in. Interstices of cracks silky. Superficially resembles **1893.**

1850a. C. strigosum Pers., var. **filamentosum** W. G. Sm., *Thelephora byssoides* DC. (from the thread-like habit; *filum*, a thread).

Sub. web-like, filamentous, string-like, dull yellowish, externally pulverulent.

On *Amaryllis.*

1850b. C. violeum W. G. Sm., *Hypochnus violeus* Quél. (from the colour).

Byssoid, thin, orbicular. *Hym.* velvety-floccose varying lilac and rose; marg. thin pallid.

Dead branches, pear, apple.

1851. C. lacunosum B. & Br. (from the hymenium, full of openings; *lacuno*, to hollow out) *a.*

Broadly effused, soft, pulverulent, lacunose, exposing matrix beneath, tawny, ochreous or cinnamon, clouded pale umber. *Myc.* woolly, tawny.

Wood, branches. 5 in.

1852. C. roseolum Mass. (from the rose-coloured hymenium) *a b.*

Broadly effused, indeterminate, very thin, waxy. *Hym.* smooth, continuous, not cracking, pale rose or salmon; marg. whitish, silky.

Worked wood, palings. Nov. 7 in.

1853. C. aurora B. & Br. (from the rosy colour, like the dawn, *aurora*) *a b.*

Effused, thin, agglutinate, rosy, becoming pale salmon with a lighter indeterminate marg.

Dead leaves and stems of *Carex*, *C. paniculata*. Feb. 2¼ in.

1854. C. anthochroum Fr. (from the rosy flower-like cloudings; Gr. *anthos*, a flower, *chros*, colour) *a b c.*

Broadly effused, membranous, floccoso-velvety, clouded brick-red or rose on a pale brownish ground.

Investing dry bark, twigs, etc., sycamore, birch, spreading to earth; rare. Nov.–Mar. 6¼ in. Sometimes forming small, dark vinous patches.

1855. C. molle Fr. (from its soft substance; *mollis*, soft) *a c.*

Irregularly rounded, floccoso-fleshy, loosely interwoven. *Hym.* waxy, papillose, cracked when dry, pallid, pale sienna to reddish-buff; marg. naked, paler, sometimes byssoid.

Fir. 3½ in.

1856. C. polygonium Pers. (from the hymenium, often covered with small polygons, from the confluence of small growths) *a b c.*

Adnate, determinate, soon hardened, somewhat grumous. *Hym.*

pruinate, at length more or less cracked, lavender-white or flesh-colour, red beneath.

Dead branches, hornbeam, poplar, sloe, lilac, elm, *Robinia*; frequent. Mar.–April. ⅛–3½ in. Growing in small round patches from the ostiola of *Sphæriæ*, at length confluent and forming large patches which become rugged with age.

1857. C. maculæforme Fr. (from the rose-coloured spots on the hymenium; *macula*, a spot, *forma*, form) *a*.

Orbicular, confluent, indurate, thin. *Hym.* spuriously papillose, purple with rose-coloured spots and smaller purple spots; marg. determinate, pale purple.

Dead wood. 2½ in.

1858. C. sanguineum Fr. (from its blood-red colour; *sanguis*, blood) *a b c*.

Broadly effused, indeterminate, even, smooth, loosely adherent, arachnoid beneath, slightly cracking when dry, bright deep cinnabar-crimson, clouded purple and black-crimson; marg. sometimes broadly floccose with cinnabar-white mycelium, sometimes forming cinnabar-white cloud-like growths.

Wood, squared wood, pine, larch, spreading to leaves. Oct.–Feb. 10½ in. Deeply staining wood a permanent blood-red colour. Rarely fertile.

1859. C. Carlylei Mass. (after Dr. David Carlyle of Carlisle) *a*.

Effused in elongate patches, agglutinate, between waxy and soft. *Hym.* even, smooth, dull orange; marg. white.

Oak bark. 3⅛ × ⅞ in.

1860. C. FLAVEOLUM Mass. (from the yellowish hymenium; *flavus*, light yellow) *a*.

Effused, membranous, loosely adhering. *Hym.* pallid yellowish, or very pale lemon with a determinate white marg.

Trunks of tree-ferns. 3 in.

1861. C. cœruleum Fr. (from its frequent azure-blue colour; *cœruleum*, azure-blue) *a b c*.

Broadly effused, adnate. *Hym.* waxy-soft, papillose, tomentose, slightly cracking, deep smalt or Prussian-blue, sometimes almost black or clouded deeper and lighter azure-blue; marg. determinate, bordered with a fine white line or narrowly flaxy and pale blue, sometimes upturned.

Sticks, dressed wood, dead wood, gates, rails, fir, chestnut, oak, alder; common. Dec.–April. 6 in. Sometimes phosphorescent. Sometimes black when dry.

1862. C. violaceolividum Fr. (from its violet and livid clouding) *a c*.

Somewhat effused, adnate, indurate. *Hym.* spuriously corrugate, tuberculose, mealy, cracking all over when dry, livid purple, clouded darker or indigo-white, clouded pale brown; marg. paler or yellowish.

Dead wood, willow, lilac. 4⅞ in.

1863. C. lividum Pers. (from the livid colour) *a b c.*

Effused, closely adnate, between waxy and soft. *Hym.* even, sometimes rugose or tubercular, somewhat viscid when moist, cracking when dry, variable in colour, clouded pale buff, pale liver or livid azure-blue and fuscous-purplish; marg. pulverulent, granular or determinate, paler or white.

Dead birch, elm; rare. Oct.-June. 5 in.

1864. C. atrovirens Fr. (from the colour, blackish, shaded greenish; *ater*, black, *vireo*, to be green) *a.*

Irregularly effused, tomentose beneath. *Hym.* waxy, smooth, pruinose, somewhat pale dull indigo with a greenish shade; marg. tomentose.

Rotten wood, leaves, sticks; rare. 3 in.

1865. C. nigrescens Fr. (from its blackish colour; *nigrico*, to be blackish) *a c.*

Effused, interrupted, waxy, agglutinate. *Hym.* here and there papillose, cracking when dry, yellowish then blackish or brown-whitish to brown-slate, densely powdered with large spores.

Branches, sticks, under the bark, oak. 3 in.

1866. C. comedens Fr. (from its habit of destroying the bark, when very thin, under which it grows; *comedo*, to consume) *a b c.*

Effused, innate, exposed only when the bark splits. *Hym.* even, sometimes papillose or finely granular, slightly viscid when moist, cracking when dry and peeling from matrix, flesh-colour, lilac or variegated pale dull buff and pale to somewhat dark purplish.

Sticks, branches, hazel, alder, hawthorn, chestnut, pine, oak, hornbeam; very common. Aug.-Mar. 5½ in. Never found above the bark.

LXXXVI. PENIOPHORA Cooke.

(From the shuttle-like setæ, metuloides, or modified cystidia borne on the hymenium; Gr. *penion*, a shuttle, *phero*, to bear.)

Resupinate, effused, coriaceous or subcarnose. *Hymenium*, as seen under a pocket lens, setulose, cystidia projecting, fusiform, colourless, hyaline at first smooth then verruculose above with minute particles of oxalate of lime, which are derived from water containing this substance in solution. The cystidia give the hymenium a velvety appearance, but they sometimes break away and leave the hymenium smooth; in this condition care must be taken not to confuse the species with *Corticium*. *Basidia* tetrasporous. *Spores* white, hyaline. (Fig. 101.)

Growing on wood, bark, leaves, etc. Young examples of some species may be readily confused with *Pezizæ*. Species **1867—1882**

Margin free, more or less incurved. **1867—1870**

Margin adpressed, often indeterminate. **1871—1881**

Setæ aggregate in fascicles. **1882**

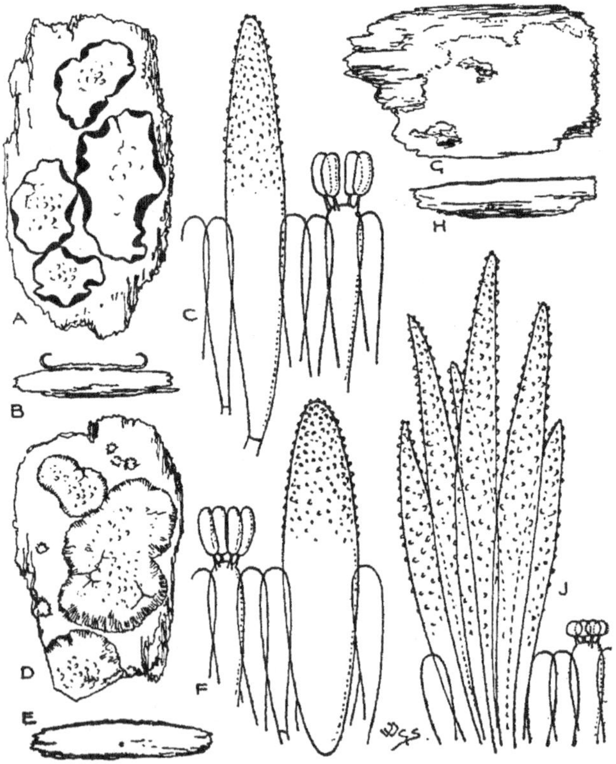

Fig. 101.—A, B, *Peniophora quercina* Cooke, entire and in section, one-half natural size. C, basidium, spores and cystidium of ditto, × 500. D, E, *P. rosea* Mass., entire and in section, one-half natural size. F, basidium, spores and cystidium of ditto, × 500. G, H, *P. hydnoides* Cooke & Mass., entire and in section, one-half natural size. J, basidium, spores and cystidium of ditto, × 500.

1867. P. quercina Cooke (from the usual habitat, oak, *quercus*) *a b c*. Cartilaginous, membranous, at first agglutinate, then fixed at centre, rigid, smooth, slightly cracking when dry, turning black beneath. *Hym.* papillose, or irregularly rugose, sometimes rugged all over, flesh-colour, pale purplish-rose, pale yellow-buff, dull salmon or liver-whitish.

Oak, birch, lilac, lime, hazel, ash, chestnut. Aug.–April. 7 in.

1868. P. pezizoides Mass. (from its resemblance to *Peziza;* Gr. *eidos*, resemblance) *a.*

Somewhat leathery, cup-shaped, then flattened, fixed at centre, externally pallid villous. *Hym.* continuous, ochreous or somewhat pale umber; mid. darker; marg. white.

Branches, horse-chestnut. ¼–⅜ in.

1869. P. gigantea Mass. (from its large size, *giganteus*) *a b c.*

Broadly effused, swelling when moist, fleshy-tough, waxy, cartilaginous when dry, slightly cracking; marg. free, incurved. *Hym.* even, continuous, partly granular, white to ivory, shaded greyish; marg. rugose, grey-ivory beneath.

Stumps, trunks, branches, fir, larch, sometimes investing fir-leaves, moss and grass; common. Oct.–April. 8¼ in.

1870. P. Crosslandii Mass. (after Charles Crossland of Halifax).

Effused, thin, soft when moist, wholly separating from matrix when dry. *Hym.* pale grey, faintly ochreous when dry.

Wood and bark, fir. Differing from **1869** in the smaller spores and shorter cystidia.

1871. P. limitata Cooke (from the margin limited by a black line) *a.*

Somewhat orbicular, grumous, indurated, smooth. *Hym.* lurid, lavender-brown, shaded brown.

Wood and bark, broom. 3¾ in.

1872. P. rosea Mass. (from the rose-coloured hymenium) *a b c.*

Effused, very thin, slightly cracking, showing grain of wood beneath. *Hym.* minutely papillose in middle, rose or rose-salmon becoming pale, sometimes clouded pale brown; marg. often finely fimbriate or rugoso-radiate and paler, sometimes tinted pale sulphur.

Wood, bark, birch, poplar, willow, walnut, rose, often spreading over grass, lichens, etc. Oct.–Mar. 4¾ in. Becoming rufous-salmon when dry. *Illosporium roseum* is probably a young condition of this or of **1873.**

1873. P. incarnata Mass. (from the flesh-coloured hymenium; *incarnatus*, flesh-coloured) *a b c.*

Somewhat waxy, agglutinate, thin, cracking when dry. *Hym.* dull pale red, pale vermilion, red-orange, salmon- or vermilion-white, or salmon, when young sometimes white; marg. radiate, varying darker or paler.

Wood, bark, bramble, gorse, birch, beech, laburnum, alder, plane, fir. Dec.–April. 5¼ in. *Illosporium carneum* is probably a young state of this or of **1872.**

1874. P. ochracea Mass. (from the ochreous-yellow hymenium) *a c.*

Broadly effused, cracking when dry. *Hym.* pallid ochreous, at first sprinkled with scattered shining gold-like atoms, sometimes pale vinous.

Wood, bark, ash. 5⅜ in. This must not be confounded with **1895.**

1875. P. cinerea Cooke (from the ash-coloured hymenium; *cinis*, ashes) *a b c*.

Waxy, becoming rigid, confluent, cracking when dry. *Hym.* minutely papillose or granular, cinereous or lurid, umber-white or white-lavender, often clouded umber, here and there pale salmon-white; marg. usually definite.

Wood, bark, herbaceous stems, pine, oak, poplar, lime, ash, honeysuckle, ivy, bramble, *Lycium*, *Viburnum*. Aug.–June. 6 in.

1876. P. velutina Cooke (from the velvety hymenium; *vellus*, fleece) *a c*.

Broadly effused, adnate, very delicate, cracking when dry. *Hym.* even, flesh-colour, rose-salmon, ivory, scarlet-white, sometimes rose-lavender, paler at the somewhat broad, fleecy, usually fimbriate marg., extending in radiate strands of mycelium.

Wood, bark, pine, bracken, sometimes involving moss. Oct.–May. 6 in.

1877. P. rimosa Cooke (from the cracked hymenium; *rima*, a cleft) *a*.

Broadly effused, agglutinate, cracking into areolæ with silky interstices. *Hym.* cinereous, pallid fawn, or somewhat pale brown-sienna-ochre.

Bark, wood. 4 in. Differing from 1876 principally in the larger spores.

1878. P. scotica Mass. (from the first place of finding, Scotland) *a*.

Broadly effused. *Hym.* cinnamon or pale brownish-ochre, obscurely vinous and clouded opaque cinnamon-brown; marg. fibrillosely-radiate.

Inside bark. 3½ in. The irregularities of the inner side of the bark show through the substance.

1879. P. phyllophila Mass. (from its habitat, leaves; Gr. *phullon*, a leaf, *philos*, loving) *a*.

Broadly effused, membranous. *Hym.* continuous, ivory, sometimes faintly sulphurous or ochreous; marg. fibrillose or indeterminate.

Dead leaves. 4⅜ in. Must not be confounded with the non-British *Corticium epiphyllum*, from which it is distinguished by its setæ.

1880. P. pubera Sacc. (from the downy hymenium; *puber*, downy) *a c*.

Broadly effused, closely adnate, very delicate, cracking. *Hym.* even, white, clay-colour or colour of milky coffee; marg. determinate to indeterminate but minutely ragged, sometimes slightly pulverulent.

Oak, beech, mountain ash, rough pieces of rotten wood. Jan.–Feb. 3½ in.

1881. P. terrestris Mass. (from its occasional place of growth, the ground, *terra*) *a b*.

Effused, very thin. *Hym.* cinereous, pallid fawn or lavender-slate-white.

Earth, twigs, branches, leaves. 5 in. Resembles pale dull lavender clouds.

1881a. P. Chrysanthemi W. G. Sm., *Corticium Chrysanthemi* Plowr. (from the habitat, base of living chrysanthemum stems) *a*.

Effused, incrusting, indeterminate, uneven from the irregularities of the matrix. *Sub.* filamentous. *Hym.* indeterminate, white.

Autumn. 5 in.

1882. P. hydnoides Cooke & Mass. (from the resemblance of the hymenium to certain resupinate species of *Hydnum ; eidos*, appearance) *a*.

Broadly effused, thin, somewhat innate. *Hym.* cinereous or dull lavender-white, clouded.

Bark. Nov. 5½ in. Somewhat resembles *Grandinia*.

LXXXVII. CONIOPHORA DC.

(From the dust or spore-powdered hymenium ; Gr. *konis*, dust, *phoreo*, to carry.)

Resupinate, broadly effused, margin defined or indeterminate, closely attached to the matrix. *Hymenium* even, powdered with

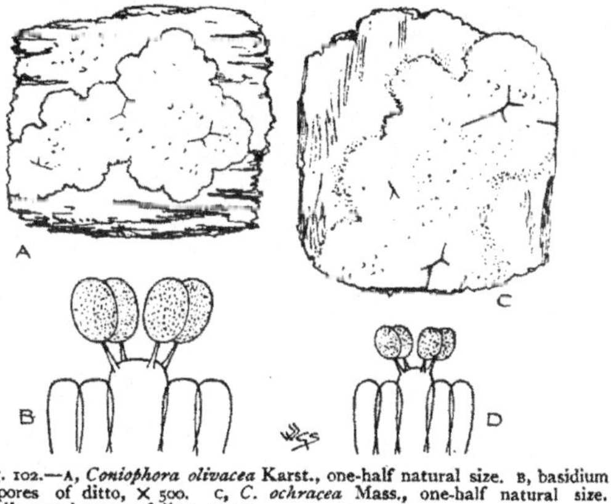

Fig. 102.—A, *Coniophora olivacea* Karst., one-half natural size. B, basidium and spores of ditto, × 500. C, *C. ochracea* Mass., one-half natural size. D, basidium and spores of ditto, × 500.

spores, without setæ. *Basidia* tetrasporous. *Spores* smooth, pale brownish or pale dull ochreous. (Fig. 102.).

Growing in broadly expanded patches on dead wood or bark.

Species **1883—1898**

Spores large.	**1883—1893**
Spores small.	**1894—1890**
Substance soft, subgelatinous.	**1897, 1898**

1883. C. olivacea Karst. (from the olivaceous colour) *a b c*.

Membranous, broadly effused, orbicular or elongate. *Hym.* dull olive, sparkling with oxalate of lime crystals when fresh; marg. fimbriate, whitish. *Spores* pale dull ochre.

Trunks and wood, pine; uncommon. Oct.–Nov. 5½ in.

1884. C. pulverulenta Sacc. (from the pulverulent hymenium) *a.*

Effused, dry, cracking. *Hym.* brown, ferruginous- or yellowish-brown, clouded and mottled darker; marg. membranous, white or pale yellowish-brown. *Spores* pale dull yellow.

Wood. 7 in. Has been considered a form of **1880**, var. *pulverulentus.*

1885. C. puteana Karst. (from its often growing in cellars, humid greenhouses, etc.; *puteanum*, a well) *a c.*

Broadly effused, fleshy, fragile, somewhat thick, compact, almost waxy, usually with open spaces exposing the matrix, sometimes cracked and showing the fibrillose substratum. *Hym.* subundulate, pallid, yellowish or buff to olivaceous-brown or brown-sienna; marg. mucedinous, white, sometimes indefinite. *Spores* tawny olive.

Bark and wood, cypress, dead wood, rotten trunks, running over dead leaves, damp walls in cellars, greenhouses, etc.; frequent. Oct.–Nov. 5¾ in.

1886. C. cinnamomea Mass. (from the cinnamon colour) *a b c.*

Effused, confluent, irregular, fleshy, soft, somewhat thick, cracking and exhibiting the fibrilloso-strigose Sub. *Hym.* pale cinnamon to dark rich sienna-umber or chocolate, spotted or mottled darker. *Spores* pale dull yellow or very pale cinnamon.

Wood, bark, plum, hazel, hornbeam, birch, oak; rare. Nov. 4 in.

1887. C. umbrina Karst. (from the brown colour, *umber*) *a b c.*

Effused, confluent, fleshy-soft, somewhat thick. *Sub.* villous. *Hym.* tuberculose then collapsing, varying pale to dark brown or sooty- or yellowish-umber, often slightly olive; marg. shortly fimbriate. *Spores* pale, dull brown.

Dead wood, branches, sticks, etc. Sometimes on the ground. Feb. 6 in. Not easily separating from the matrix. Surface sometimes rugoso-papillose all over.

1888. C. incrustans Mass. (from the incrusting habit) *a b.*

Effused, thin, indeterminate, almost waxy when perfectly developed. *Hym.* waxy-pulverulent, pallid to pale yellowish. *Spores* pale yellow.

Enveloping leaves, twigs, etc., with a film. 6 in. Thinner and less nodulose than **1898.**

1889. C. arida Karst. (from its dry substance; *aridus*, dry) *a b c.*

Effused, membranous, adnato-agglutinate, continuous, cracking. *Hym.* even to slightly tuberculose, sulphur-yellow, becoming ferruginous-umber, clouded; marg. broadly radiato-byssoid, sometimes pulverulent, sulphur-white, sometimes zoned ochre and sulphur near marg. *Spores* pale dull yellow.

Logs, woods, cellars, pine, fir, juniper; common. Oct.–Feb. 7 in.

1890. C. sulphurea Mass. (from the sulphur colour) *a b c.*

Effused, fibrilloso-flaxy, somewhat thick, when perfect thick and waxy-soft, cracking, often imperfect and spongy, passing into radiate, cord-like, branching sulphur-coloured strands. *Hym.* minutely papilloso-rugose or granulose, bright sulphur-yellow, or clouded and mottled sulphur and orange on an ivory ground, fading to yellow-umber, sometimes rich sienna-orange with a pale somewhat byssoid sulphur marg. *Spores* olive.

Dead wood, bark, twigs, leaves, stipules, etc., pine, oak, beech, sometimes on earth. Oct.–Jan. 5½ in. Var. *ochroidea* Mass. *Hym.* ochreous with a cream or sulphur margin. *Spores* larger than type, probably specifically distinct.

1891. C. subdealbata Mass. (from the whitish colour ; *sub*, somewhat, *dealbatus*, whitened) *a b.*

Broadly effused, thin, determinate, cracking. *Hym.* shining-white, thickly powdered, pallid fawn, or pale ochreous-whitish, mottled brownish-ochre or shaded olive, often with paler barren patches. *Spores* ochreous or pale dull yellow.

Bark, pine. Dec. 4¾ in.

1892. C. stabularis Karst. (from its being first recorded from an ox-stall ; *stabulum*, a stall) *a c.*

Effused, flaxy when young, fleshy-soft. *Hym.* tubercular, tubercles collapsing, white, whitish-yellow, pale yellowish-brown or vinous-fuscous, white-pruinose ; marg. definite flaxy or pulverulent, white.

Odour fetid from its habitat. Squared wood, fir. 4¼ in.

1893. C. Berkeleyi Mass. (after the Rev. M. J. Berkeley) *a b c.*

Effused, thick, cracked ; marg. determinate, sometimes minutely byssoid. *Hym.* vinous-brown, somewhat yellowish, somewhat paler at the pallid marg. *Spores* tawny.

Bark, decorticated wood. 4 in. Superficially resembles **1850.**

1894. C. Cookei Mass. (after Dr. M. C. Cooke) *a.*

Effused, fibrilloso-membranous. *Hym.* ferruginous-olive or yellowish-brown, clouded darker ; marg. byssoid, pallid. *Spores* pale dull yellow.

Rotting wood. 4½ in.

1895. C. ochracea Mass. (from the ochreous colour) *a b.*

Broadly effused, submembranous, indeterminate, cracking. *Hym.* with a woolly appearance, subpapillose, sulphur-white to ochre, salmon-white when dry, or pale vinous-brown with a yellowish tinge ; marg. broad, pulverulent, pale yellowish. *Spores* yellowish or pale dull yellow.

Wood, bark, elm. July. 5 in. Resembles a thin form of **1890.** Must not be confounded with **1874** or the var. *ochroidea* of **1890.**

1896. C. membranacea DC. (from the somewhat membranous substance) *a.*

Somewhat orbicular or effused, fragile, cracking and peeling off the host. *Hym.* dull ferruginous or opaque-yellowish, sienna-umber ; mid. somewhat dark, constantly paler at the fibrillose marg. *Spores* ochre.

Forming patches a foot or more in diam. on walls, wood, paper, etc. 9¾ in.

1897. C. gelatinosa W. G. Sm., *Aldridgea gelatinosa* Mass. (from its gelatinous substance) *a*.

Broadly effused, fleshy, subgelatinous, sometimes ⅛ in. thick, becoming collapsed and rigid. *Hym.* pallid becoming purple-brown; marg. determinate. *Spores* olive.

Sawdust, extending for several inches, the irregularities of the hymenium due to the substratum. 5 in.

1898. C. byssoidea Karst. (from its, at first, flaxy condition; Gr. *byssos*, fine flax, *eidos*, appearance) *a c*.

Irregularly effused, at first grey, then compact; mid. fleshy. *Hym.* light to ferruginous yellow, or colour of coffee and milk; marg. narrow, somewhat arachnoid or flaxy, whitish.

Forming a biscuit-coloured film sometimes a foot broad on fir-leaves, moss, earth, etc.; frequent. 4½ in.

LXXXVIII. EXOBASIDIUM Woron.

(From the habit, the spores growing from basidia only with no true sporophore; *ex*, out of.)

Entophytes parasitic on living leaves and stems, which they deform and hypertrophy. *Basidia* crowded, cylindrico-clavate,

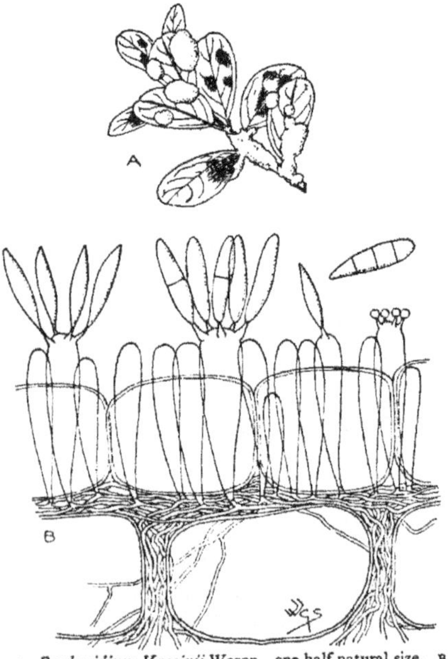

Fig. 103.—A, *Exobasidium Vaccinii* Woron., one-half natural size. B, section through hymenium of ditto, showing basidia and spores, × 500.

normally tetrasporous, but spores sometimes varying in number from one to five. The hymenial cells are borne on vegetative hyphæ which grow in the cellular tissue of the host and form a loose stratum on its external surface. *Spores* ovate to elongate-oval, continuous, rarely becoming septate, colourless. (Fig. 103.)

Most nearly allied to *Corticium*, but differing in the absence of a sporophore or subiculum and in being parasitic. *Exoascus*, an Ascomycetous fungus, sometimes grows in company with *Exobasidium*.

Species **1899, 1900**

1899. E. Vaccinii Woron. (from its growth on *Vaccinium* and its allies) *a b c.*

Innate, effused, usually orbicular or elongate, often confluent. *Hym.* on under surface of leaves and round stems, pale rose, covered with a white bloom.

June–Aug. ½ in. Red or purple patches occur on the upper surface of the leaves, opposite to the portion occupied by the fungus below.

1900. E. Rhododendri Cramer (from its growth on *Rhododendron*) *a c.*

Forming gall-like bullate excrescences, at first pale, then reddish and shining on stems, petioles and under surface of leaves.

On *Rhododendron ferrugineum*, *R. hirsutum*, *R. Wilsonii*. July. 1 in. Insects, mites, etc., are often found in the gall-like swellings of *Exobasidium*.

LXXXIX. CYPHELLA Fr.

(From the usual shape, like a small goblet, Gr. *kuphos*.)

Usually small, sometimes almost minute, somewhat membranous, cup shaped, rarely plane, adnate from a central or oblique point, sometimes with a stem and pendulous. *Hymenium* within the cup or disc. *Basidia* tetrasporous, rarely bisporous. *Spores* subovate. (Fig. 104.)

Closely allied to *Solenia*. *C. perpusilla* is intermediate. Owing to the strong external resemblance in many *Cyphellæ* to certain small species of *Peziza*, a genus of Ascomycetous fungi, it is necessary to examine microscopically the hymenium of some of the species for the decisive characters of basidia or asci. *C. faginea* Lib. is British, but it is an Ascomycete.

Species **1901—1922**

Stipitate.	**1901—1905**
Sessile or subsessile.	
Spores hyaline.	**1906—1918**
Spores coloured.	**1919—1922**

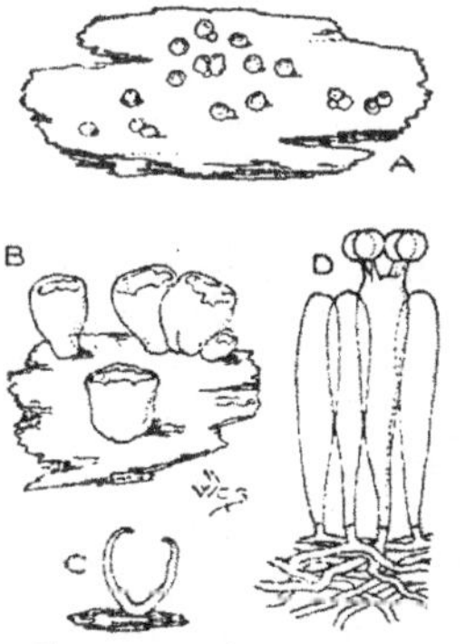

Fig. 104.—A, *Cyphella brunnea* Phill., × 2. B, ditto with C, section, × 10. D, basidium and spores, × 500.

1901. C. Pimii Phill. (after Greenwood Pim, mycologist) *a b*.
Fasciculate, cyathiform, erect or pendent, membranous, soft, fragile, pubescent, white or pale yellow. *Stem* somewhat slender, crooked, enlarged upwards ; marg. somewhat incised.

Dead herbaceous stems in water, nettle, butter-bur. Feb. $\frac{1}{8} \times \frac{1}{8}$ in.

1902. C. capula Fr. (from its resemblance to a small bowl, *capula*) *a b c*.
Membranous, obliquely campanulate, smooth; marg. sinuate, irregular. *Hym.* even, whitish to yellowish, rarely greyish, brownish, or wholly white.

Dead herbaceous stems, twigs, branches, nettles, fennel, moss, strawberry, comfrey ; frequent. Oct.–Aug. $\frac{1}{4}$ in.

1903. C. cernua Mass. (from the habit, bending downwards; *cernuus*, nodding) *a*.
Thin, obliquely campanulate, glabrous, stem elongate, equal, clear primrose-yellow, yellowish-white, somewhat sulphur or very pale yellow.

Bark, elder. $\frac{1}{4}$ in. Perhaps the same as **1902**, but Massee considers it distinct in its more erect habit and larger spores.

1904. C. cuticulosa Fr. ex Berk. (from the membranous substance; *cuticula*, the outer skin) *a b*.
Membranous, at first oblong, then cup-shaped, smooth externally, diaphanous white.

Dead grass stems ; rare. Feb. $\frac{1}{12}$–$\frac{1}{8}$ in.

1905. C. lacera Fr. (from the cleft margin ; *lacerus*, torn) *a b c*.
Membranous, cup-shaped, vertical or pendulous, torn into clefts, slightly striate and black fibrillose above. *Hym.* slightly wrinkled and whitish, grey externally.

Dead twigs. $\frac{1}{4} \times \frac{1}{4}$ in.

1906. C. galeata Fr. (from its resemblance to a metal skull-cap or helmet ; *galea*, a helmet) *a c*.
Membranous soft, subsessile, even ; marg. entire, then torn. *Hym.* slightly wrinkled, white to grey-white or pale rose-brown.

Mosses ; uncommon. Dec. $\frac{1}{4}$–$\frac{1}{12}$ in.

1907. C. Goldbachii Weinm. (after Carl Ludwig Goldbach) *a b c*.
Membranous, cup-shaped, sessile, lobed or cleft, externally villous. *Hym.* even, pallid, white, ivory, or white-ochre.

Dead leaves of *Aira cæspitosa, Carex paniculata*. Feb. $\frac{1}{8} \times \frac{1}{12}$ in.

1908. C. muscigena Fr. (from its habit, borne on mosses; *muscus*, moss, *gigno*, to bear) *a b c*.
Membranous soft, becoming plane, subdimidiate, externally silky under a lens. *Hym.* slightly wrinkled, shining-white, becoming pale brown.

Various mosses ; uncommon. Sept.–Jan. $\frac{3}{8}$ in. Varied in form, fan-shaped, spathulate, etc. Sometimes stipitate, sometimes $\frac{3}{4}$ in. across. This is *Cantharellus lævis* Fr.

1909. C. catilla W. G. Sm. (from its shape, like a small bowl, *catillus*) *a*.

Submembranous, expanded; marg. crisped, undulate. *Hym.* veined, pale grey or faint claret-brown.

Often imbricate. Moss and dead leaves. Nov. $\frac{3}{4}$ in. Closely allied to **1908**.

1909a. C. griseopallida Weinm. (from the pallid grey colour; *griseus*, grey) *a*.

Somewhat membranous, globose, then campanulate, sessile, floccose externally. *Hym.* even, smooth.

Elm, elder, honeysuckle, lime, pine. Dec. $\frac{1}{16}$ in. A supposed variety of this, found on dead *Carex paniculata* and described by Berkeley, is **1916**.

1910. C. fulva Berk. & Rav. (from the tawny colour; *fulvus*, tawny) *a b c*.

Membranous, cup-shaped, deflexed, externally hispid-tomentose, rich tawny-brown.

Dead bark, alder, elm. Jan. $\frac{1}{12}$ in.

1911. C. brunnea Phill. (from the brown colour; *brunneus*, brown) *a b*.

Subgelatinous, sessile, cupulate, grey-white-pruinate at the incurved lacerate marg., dull brown or somewhat vinous umber, mouth oblique, base very dark. *Hym.* smooth, discoloured brown.

Wood and bark, old elder trees. $\frac{1}{24}$ in.

1912. C. stuppea B. & Br. (from the coarsely hispid coat, as if clad with tow; *stuppeus*, made of tow) *a*.

Erumpent, sessile, cupulate, brownish, becoming white. *Hym.* fuscous.

Broom. Mar. $\frac{1}{24}$ in.

1913. C. alboviolascens Karst. (from the pallid violet hymenium; *albus*, white) *a b c*.

Cupulate, sessile or nearly so, densely villous, white. *Hym.* even, pallid, tinged with violet.

Gregarious, often proliferous. Wood, bark, twigs, burdock, nettle, thistle, holly, broom, lilac, sloe, oak, rose, poplar, ferns, chips. Oct.–June. $\frac{1}{24}$ in.

1914. C. villosa Karst. (from the villose coating) *a b c*.

Sessile, spherical with a contracted mouth, whitish.

Stems, leaves, branches, potato, vine, elder, burdock, sow-thistle, butter-bur, dog-wood, fennel, strawberry, *Lathyrus pratensis*, *Chenopodium*. Oct.–June. $\frac{1}{84}$–$\frac{1}{24}$ in. This and **1913** are much alike, both very white-villous externally, with hymenium often wholly hidden. Distinguished by the colour of the hymenium.

1915. C. dochmiospora B. & Br. (from the oblique spores; Gr. *dochmos*, oblique).

Minute, cupulate, villous, mouth subconnivent, white.

Twigs, stems of herbacious plants, etc. Oct. $\frac{1}{24}$–$\frac{1}{48}$ in.

1916. C. Berkeleyi Mass. (after the Rev. M. J. Berkeley) *a.*

Sessile, globose, then expanded, at length campanulate, minutely pilose, reddish-grey as is the even hymenium.

Dead *Carex paniculata.* Jan. $\frac{1}{12}$–$\frac{1}{24}$ in. Berkeley considered this to be a form of **1909a.** Massee has made it a species, owing to a difference in size of the spores and the different habitat.

1917. C. Bloxamii Berk. & Phill. (after the Rev. A. Bloxam) *a b.*

Floccoso-membranous, crenato-lobed, white villous externally. *Hym.* white, then pale yellow.

Furze. Mar.–April. Minute to $\frac{1}{16}$ in.

1917a. C. cyclas Cooke & Phill. (from the shape, like the bivalve shell, *Cyclas*) *a b.*

Conchiform, dimidiate, attached on one side, pendulous, clad with flexuous hairs, whitish. *Hym.* smooth, very pale rose-colour.

Dead wood. $\frac{1}{2}$ in.

1918. C. pallida B. & Br. (from the pallid colour) *a b c.*

Sessile, at first orbicular, then irregularly lobed, plane, densely tomentose to slightly hispid. *Hym.* at length wrinkled, pallid ochreous.

Old stems of *Clematis Vitalba.* Nov.–April. $\frac{1}{12}$ to $\frac{1}{24}$ in.

1919. C. fraxinicola B. & Br. (from its habitat, bark of ash, *Fraxinus; colo*, to inhabit) *a b c.*

Orbicular, shortly villous, white. *Hym.* light yellow, becoming fuscous. *Spores* pale olive.

Proliferous. Dec.–Feb. Minute to $\frac{1}{80}$ in.

1920. C. muscicola Fr. (from its growing on mosses; *muscus*, moss, *colo*, to inhabit) *a b.*

Membranous, subsessile, cupulate, nodding, externally slightly fibrilloso-striate; marg. slightly downy, repand, torn. *Hym.* even, whitish, rarely cinereous.

Mosses; rare. Dec.–Mar. $\frac{1}{4}$ in.

1921. C. ochroleuca B. & Br. (from the yellow-white colour; Gr. *ochros*, pale yellow, *leukos*, white) *a b.*

Membranous, cupulate, villous above; marg. at length split. *Hym.* even, pale ochre, brighter than exterior.

Sticks, bramble; rare. July–Oct. $\frac{1}{8}$ in. Rarely stipitate.

1922. C. fuscospora Curr. ex Cooke (from the colour of the spores; *fuscus*, dusky).

Sessile, white tomentose; marg. connivent. *Hym.* becoming yellow. *Spores* subglobose, punctulate.

Bark. $\frac{1}{4}$–$\frac{1}{5}$ mm.

XC. **SOLENIA** Hoffm.

(From the short pipe-like shape; Gr. *solen*, a pipe.)

Cylindrical, more or less contracted at the mouth, the external hairs in some species rough with particles of oxalate of lime, cavity covered with the hymenium. *Basidia* tetrasporous. (Fig. 105.)

The species are all minute, rarely exceeding $\frac{1}{12}$ in. in height. They grow, usually in crowded masses, on rotten wood. They are closely allied to *Cyphella*, but are distinguished by the densely crowded habit and by the cylinders never expanding.

Fries has placed this genus in the *Polyporaceæ*, but each cylinder in *Solenia* is a distinct Peziza-like plant and not comparable with a mere tube of one of the *Polyporaceæ*. To be allied to the *Polyporaceæ* each cylinder should be lined with minute basidia-bearing tubes.

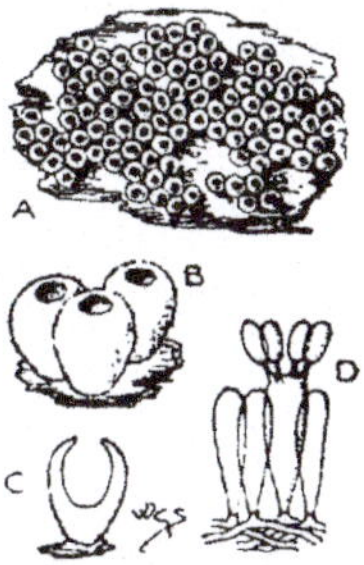

Fig. 105.—A, *Solenia anomala* Fr., × 5. B, ditto with C, section, × 20. D, basidium and spores, × 500.

Species **1923—1925**

Externally white. **1923, 1924**
Externally ochreous. **1925**

1923. S. maxima Mass. (from the comparatively large size of the spores; *maximus*, largest) *a*.

Gregarious or subfasciculate, not crowded, subcylindrical, slightly contracted at base, externally villous, with slender aseptate hyphæ rough with minute particles of oxalate of lime, whitish or pale buff, ochreous when dry.

Forming patches $\frac{1}{2} \times \frac{1}{4}$ in. May. $\frac{1}{12} \times \frac{1}{24}$ in. Spores $5 \times 3\ \mu$. This is **1925**, var. *cinerea* Cooke.

1924. S. fasciculata Pers. (from its growing in groups or fascicles) *a b c*.

Gregarious, usually fasciculate, cylindrico-clavate, externally minutely silky and almost smooth, sometimes faintly shaded, brownish-yellow.

Willow, pine. Nov.–April. $\frac{1}{24} \times \frac{1}{24}$ in. Spores $4\ \mu$.

1925. S. anomala Fr. (from its superficially anomalous or perhaps scattered character) *a b c*.

Usually densely crowded, pyriform, forming effused scattered patches; marg. incurved, pilose externally, varying from dull ochreous to ferruginous.

Birch, willow, vine, oak, beech, dog-wood. Aug.–May. $\frac{1}{24} \times \frac{1}{24}$ in. Spores $7 \times 4\ \mu$. Var. *ochracea* Mass. is somewhat smaller in size than type.

Fam. V. CLAVARIACEÆ

Hymenium not distinct from the hymenophore. Amphigenous somewhat fleshy fungi, vertical, branched or simple, never coriaceous or incrusting. The branched species of *Clavaria* must not be confounded with certain coriaceous species of *Thelephora*.

Species **1926—2000**

Key to the Genera.

Branched, sponge-like. Branches compressed, flat or leaf-like	91	Sparassis.
Branched to simple.		
Terete.		
Stem not distinct from club	92	Clavaria.
Stem distinct from club, not usually springing from a sclerotium	96	Pistillaria.
Usually springing from a sclerotium	95	Typhula.
Gelatinoso-cartilaginous, viscid	93	Calocera.
Hair-like	94	Pterula.

XCI. SPARASSIS Fr.

(From its appearance, as if torn to pieces; Gr. *sparasso*, to tear to pieces.)

Fleshy, with flat leaf-like branches, composed of two plates, fertile on both sides. *Basidia* tetrasporous. *Spores* continuous, hyaline. (Fig. 106.) Species **1926, 1926a**

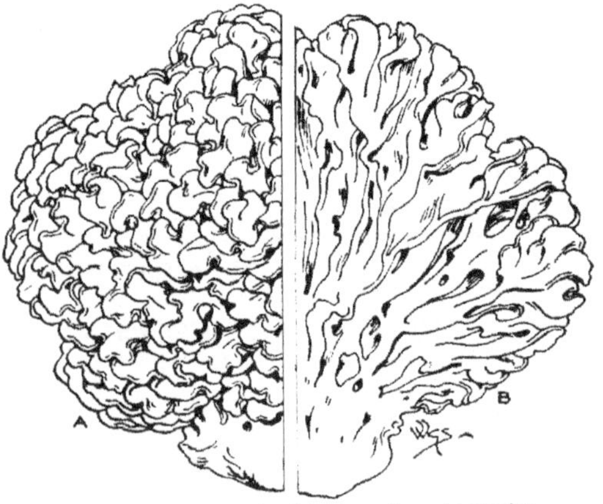

Fig. 106.—A, B, *Sparassis crispa* Fr., entire and in section. One-half natural size.

1926. S. crispa Fr. (from its crisped or curled appearance) *a b c*.
Whitish to pale yellowish. Very much branched, branches

recurved, intricate at apex, the whole forming a large sponge-like mass; stemless or with a short stout stem.

Edible. Odour very sweet of anise. Beside fir-trees and stumps, on dead fir-leaves, terrestrial on heaths. Aug.–Nov. 1 ft. 1 in. × 1 ft. 1 in. Sometimes attaining a diameter of 18 in. and a weight of 2 or 3 lbs.

1926a. S. laminosa Fr. (from the leaf-like plates or layers of the entire plant; *lamina*, a thin leaf) *a.*

Very much branched, branches laminar, spreading, yellowish straw-colour.

Woods. Sept. 10¼ × 7½ in. Stronger smelling and softer than **1926**, never fimbriate at edges, laminæ large, loose.

XCII. CLAVARIA L.

(From the club-like shape of some typical species; *clava*, a club.)

Fleshy, erect, branched or simple, branches somewhat round, with or without a distinct stem. *Hymenium* continuous, dry, even, homogeneous. *Basidia* tetrasporous. *Spores* coloured or uncoloured. For the most part terrestrial. (Fig. 107.) Species **1927—1975**

RAMARIÆ. Branched, branches attenuate upwards.

a. Leucosporæ. Spores white or nearly white.

Red, yellow or violet. **1927—1933**

White or grey. **1934—1942**

b. Ochrosporæ. Spores pale yellow, ochreous or cinnamon.

Yellow, pinkish-yellow or dull ochreous. **1943—1949**

Whitish, grey or violet. **1950, 1951**

Growing on wood. **1952, 1953**

SYNCORYNÆ. Clubs almost simple, tufted at the base.

Reddish. **1954—1956**

Yellowish or white. **1957—1963**

Smoky or blackish. **1964**

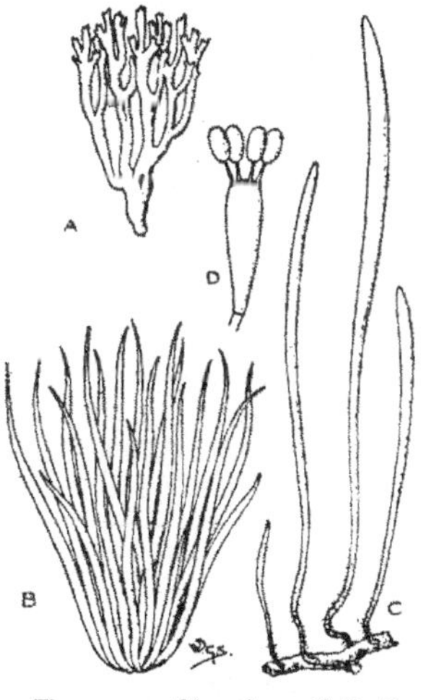

Fig. 107.—A, *Clavaria coralloides* L.; B, *C. vermicularis* Fr.; C, *C. fistulosa* Holmsk., one-third natural size. D, *C. pistillaris* L., basidium and spores, × 330.

HOLOCORYNÆ. Clubs entire or simple or nearly so, distinct at the base.

Whitish. **1965—1967**

Yellow or pallid, becoming darker. **1968—1973**

Brownish or flesh-colour. **1974, 1975**

There are analogous species in both sections of *Ramariæ* which resemble each other in form and colour, but the *Leucosporæ* are somewhat brittle-soft and several of the species are considered edible, whilst the *Ochrosporæ* are spongy-tough and mostly bitter and unfit for food. The spore-colour, the taste and the substance, whether soft or tough, are salient characters.

Ramariæ.

a. Leucosporæ.

1927. C. flava Schæff. (from the frequent light yellow colour; *flavus*, light yellow) *a c.*

Very much branched; branches cylindrical, vertical, even, obtuse, yellow. Trunk thick, fleshy, white.

Edible. Woods, pine. 6⅛ × 5¾ in. Fragile. Must not be confounded with **1943.**

1928. C. botrytis Pers. (from the deep vinous purple colour of the ultimate branchlets; Gr. *botrus*, a cluster of grapes) *a b c.*

Wholly white, buff, salmon, yellow, brownish, etc., sometimes more or less claret-colour. Much branched; branches swollen, unequal, somewhat wrinkled, rose or brown-crimson at the apices. Trunk thick, unequal.

Odour very pleasant. Woods, mixed, pine, beech; uncommon. Aug.–Nov. 5 × 5¼ in.

1929. C. amethystina Pers. (from the amethyst colour) *a c.*

Very much branched; branches round, even, obtuse, equal, purple, bright purple with a bluish shade or purple-lavender. Said to produce violent colic.

Grassy woods and pastures; uncommon. Aug.–Oct. 2 × 2¾ in. Myc. lavender-whitish or pale cobalt.

1930. C. fastigiata L. (from the parallel clustered and erect branches; in some cases the branches decrease in height outwardly, giving a sloping, obconic or gable-shape to the plant; *fastigium*, a slope, a gable) *a b c.*

Very much branched; branches short, divaricate, slightly tough, brittle towards base, egg-yellow, apices light yellow to fuscous.

Usually cæspitose. Slightly fragrant. Pastures; very common. Aug.–Mar. 1⅞ × 2¾ in.

1931. C. muscoides L. (from a fancied resemblance to a tuft of moss; *muscus*, moss; Gr. *eidos*, appearance) *a b c.*

Twice or thrice forked, becoming slender, branchlets crescent-shaped, acute, slightly tough, base tomentose, yellow.

Solitary. Pastures; frequent. Aug.–Oct. 3 × 3⅛ in.

1932. C. kewensis Mass. (from its first observed place of growth, Royal Gardens, Kew) *a.*

Branches numerous, repeatedly and dichotomously divided, axils rounded, trunk very short, dull ochreous.

Odour strong, spicy, resembling **168** or **1437.** Rotten stumps, under holly. April. 1⅞ × 2¾ in.

1933. C. curta Fr. (from its short stature; *curtus*, short) *a b*.

Very much branched, adpressed; branches short, crowded, obtuse; stem none, but there is usually a fleshy base, greenish-yellow.

On the ground, on fir-leaves. Oct. ¼ to ¾ in.

1934. C. coralloides L. (from its appearance, like white coral; Gr. *eidos*, appearance) *a b c*.

Branches unequal, dilated upwards; branchlets crowded, acute; trunk somewhat thick, repeatedly and irregularly much branched, base not downy, somewhat fragile, hollow within.

Commonly cæspitose. Edible; sold for food in markets of Italy. Shady woods; rare. Oct.–Nov. 5¼ × 5⅜ in. Sometimes faintly shaded buff.

1935. C. cinerea Pers. (from its ashy colour; *cinis*, ashes) *a b c*.

Branches and branchlets thickened, irregularly shaped, somewhat wrinkled, subobtuse; trunk somewhat thick, short, very much branched, fragile, stuffed.

Woods, mixed, fir; frequent. Aug.–Nov. 5⅝ × 4⅞ in.

1936. C. umbrina Berk. (from the colour, usually pale brown, *umbrinus*) *a*.

Slightly branched; branches and branchlets cylindrical, obtuse, forked, pale umber to salmon-yellowish.

Woods, lawns, etc. July–Oct. 2⅝ in.

1937. C. cristata Pers. (from the crested apices of the branchlets; *crista*, a crest) *a b c*.

Branches dilated above, acutely incised, crested, tough, even, stuffed, white, crests at length becoming pale buff.

Somewhat cæspitose. Woods, mixed, pine; frequent. June–Nov. 3⅛ × 3⅛ in.

1937a. C. gigaspora Cotton (from the large spores; Gr. *gigas*, a giant) *a*.

Irregularly branched; branches sometimes palmate, erect, terete or compressed, much compressed at the acute axils, ultimate branches attenuate, apices blunt, greyish with a shade of yellow.

Cæspitose, but distinct at the base, or solitary. Amongst moss on rocky heathy slopes. Nov. 1½ × ⅞.

1938. C. rugosa Bull. (from the wrinkled stem; *ruga*, a wrinkle) *a b c*.

Sparingly branched to simple, thickened upwards, irregularly shaped, tough, white.

Edible. Woods, beech, pastures; very common. Aug.–Dec. 5 in. Var. *macrospora* W. G. Sm., *C. macrospora* Britz. Spores longer and somewhat larger than type. Type 7–10 μ. Var. 8–10 × 12–14 μ.

1939. C. Krombholzii Fr. (after Julius Vicenz von Krombholz) *a b c*.

Slightly branched, even; branches somewhat compressed, obtuse, not wrinkled, fragile, white.

Cæspitose. Old pastures. Sept. 2¾ in.

1940. C. crassa Britz. (from the thickness of the clubs; *crassus*, thick) *a*.

Much branched; branches obtuse, club-shaped, compressed. Base attenuate downwards, violet.

Solitary. Not unlike **1938** and **1939** in habit, but differing in colour and in the attenuate base. 3 in.

1941. C. Kunzei Fr. (after Karl Sebastian Heinrich Kunze) *a b c*.

Very much branched from the thin cæspitose base; branches elongate, crowded, repeatedly forked, fastigiate, even, equal, axils compressed, somewhat fragile, white.

Woods; rare. Aug.–Sept. 2⅛ in.

1941a. C. subtilis Pers. (from its small and slender habit; *subtilis*, fine, slender) *a*.

Tough, thin, glabrous at base, of equal thickness throughout; branches few, dichotomously forked and somewhat fastigiate, white becoming yellowish.

Amongst grass. Sept. 1⅛ in.

1942. C. pyxidata Pers. (from the shape of the palmate ends of the branchlets in typical examples, like a small box, *pyxis*) *a c*.

Trunk thin, smooth, branched in an umbellate manner in typical examples, pallid to tan, somewhat rufescent, varying ivory and sulphur-white; branchlets taking a cup-like shape at the cuspo-radiate apices.

On the ground, on sticks, pine. 2¾ in. This has a tendency to form primary and secondary umbels, the latter with from 5 to 9 rays; when the branchlets are short the apices have a fancied resemblance to little boxes.

b. Ochrosporæ.

1943. C. aurea Schæff. (from the golden colour; *aurum*, gold) *a b c*.

Trunk thick, elastic, soft, divided into stout tense straight very dichotomous round obtuse somewhat toothed branches, buff, pale yellow, pale reddish, or pale salmon. Fleshy stem sometimes absent.

Woods, mixed, fir, beech. Aug.–Oct. 5 × 5¾ in. Must not be confounded with **1927.**

1944. C. rufescens Schæff. (from the red apices of the branchlets; *rufesco*, to become red) *a b c*.

Trunk thick, elastic, tan or pale salmon, much branched; branches multifid, crowded, somewhat fastigiate, even, tan, pale salmon or pale vermilion, with red, crimson or vermilion apices.

On wood. Sept. 4½ × 4 in.

1945. C. formosa Pers. (from its beauty; *formosus*, beautiful) *a b c*.

Trunk thick, elastic, ivory, pale buff or pale olive-brown, darker below, much elongato-branched; branchlets obtuse, luminous yellow, salmon or salmon-buff, apices sometimes orange or sulphur.

Odour pleasant. Downs, commons, woods, pine; rare. July–Oct. 4¾ × 6 in.

1946. C. spinulosa Pers. (from the attenuate spine-like branchlets; *spinulosus*, spined) *a b c*.

Trunk short, somewhat thick, pale yellow; branches crowded, tense, straight, attenuato-pointed, yellow-orange, dull yellow or warm brown.

Woods, fir, beech, mixed. Sept. 5 × 6 in.

1947. C. abietina Pers. (from the habitat, under fir, *abies*) *a b c*.

Trunk short, somewhat thick, very much branched; branches crowded, branchlets tense, straight, acute, olive-ochre or olive-sienna-ochreous, becoming green where touched.

Taste bitter. Under fir, larch; common. Aug.–Dec. 3 × 3 in.

1948. C. flaccida Fr. (from its flaccid habit) *a b c*.

Stem thin, very short, very much branched; branches crowded, even, unequal, acute, flaccid, often drooping, ochreous or buff.

Amongst leaves and moss in woods, pine; rare. Oct. 3 × 2¼ in.

1949. C. crocea Pers. (from the saffron colour; *croceus*, saffron) *a b*.

Minute, thin; stem naked, pallid or yellow; branches and branchlets orange-scarlet.

Downs, gardens, on peat. Sept.–Feb. ⅜ × ½ in. The general appearance is not that of a true Basidiomycete; it has somewhat the appearance of a slender minute branched *Calocera*.

1950. C. grisea Pers. (from the grey colour; *griseus*, grey) *a b c*.

Trunk somewhat thick, firm, often rooting, whitish; branches and branchlets attenuate, somewhat wrinkled, fuliginous-cinereous.

Woods; rare. Sept.–Oct. 4 × 5 in. Sometimes confounded with **1935**, in which the spores are white.

1951. C. condensata Fr. (from the closely adpressed branchlets; *condenso*, to press close together) *a*.

Tufts very dense, usually without a trunk, very much branched from the base, smooth, tan-rufescent or salmon; branches tense and straight, crowded, even, fastigiate, twice or thrice-toothed, yellow.

On the ground, under trees. 3½ × 4 in.

1952. C. stricta Pers. (from the tense, straight habit; *stringo*, to draw tight) *a b c*.

Trunk short, thin, very much branched; branches and branchlets tense and straight, even, crowded, adpressed, somewhat acute, pallid yellowish or whitish-yellow to buff-tan, fuscous where bruised.

Woods, pine, on and about stumps, sawdust; uncommon. Aug.–Nov. 4½ × 3½ in. Often springing from cord-like mycellium.

1953. C. crispula Fr. (from the slightly curled branchlets; *crispulus*, slightly curled) *a b c*.

Trunk very thin or obsolete, very much branched; branches flexuous, multifid, branchlets divaricate, tan to ochreous, buff or ochre, almost white below.

Base of trees, ash, elder; rare. Oct.–Dec. 3½ × 2⅝ in.

SYNCORYNÆ.

1954. C. purpurea Müll. (from the colour; *purpureus*, purple) *a c*.

Cl. elongate, hollow, then compressed, acute, fragile, opaque, claret-brown and slightly purplish to purple.

Cæspitose. Under pine; rare. Aug.–Oct. 3⅝ in.

1955. C. rufa Pers. (from the reddish colour; *rufus*, red) *a c*.

Cl. stuffed, thickened, fusiform, sometimes bifid, rarely trifid, pale warm brown, rarely bright red or orange-vermilion, apices darker with age.

Cæspitose. Odour somewhat strong. Sept. 2¼ in.

1956. C. rosea Dalman (from the rose colour) *a b c*.

Cl. stuffed, fragile, pale to claret-pink or full rose-colour, apices attenuate or rounded, becoming yellowish with age.

Somewhat fasciculate. Pastures, amongst moss; rare. 2¼ in.

1957. C. fusiformis Sow. (from the fusiform clubs) *a b c*.

Cl. somewhat fusiform, apex acute, sometimes bifid, trifid or toothed, firm, soon hollow, bright yellow, apices fuscous. *Spores* even.

Cæspitoso-connate. Amongst grass, woods, pastures, etc.; common. Sept.–Nov. 5 in. Sometimes intensely bitter. Often confounded with **1958** and **1959**. Var. *ceranoides* W. G. Sm., *Clavaria ceranoides* Pers., unequal, divided above, branched, upper part often antler-like.

1957a. C. luteoalba Rea (from colour of the club, yellow below, white above; *luteus*, yellow, *albus*, white) *a*.

Cl. simple, stuffed, attenuate to base, apex acute to blunt, deep rich yellow, apex constantly white.

Fasciculate or scattered. Amongst grass and bracken. Sept.–Oct. 2⅛ in.

1958. C. dissipabilis Britz. (from the scattered habit; *dissipabilis*, dissipated, scattered) *a*.

Cl. elongato-clavate, apex rounded, rarely subacute, round or compressed and channelled, solid, collapsing, fragile, even, glabrous, deep orange yellow. *Spores* spinulose.

Solitary or in small clusters. Short grass in pastures, roads, woods; common. Oct. 2⅜ in. Often confounded with **1957** and **1959**.

1959. C. inæqualis Müll. (from the clubs being often unequal in diameter and more or less unevenly compressed; *inæqualis*, unequal) *a b c*.

Cl. variable, usually simple and rounded above, narrow or

inflated, fragile, stuffed, rarely bifid, yellow, usually paler than **1957** and **1958.** *Spores* even.

Somewhat fasciculate or gregarious. Amongst grass, woods, pastures; very common. Aug.–Nov. 4⅝ in. Often confounded with **1957** and **1958.**

1960. C. argillacea Pers. (from the colour of clay, *argilla*) *a b c.*

Cl. usually simple, variable, sometimes short and obese, rarely with short rounded obtuse branchlets, apex rounded, sometimes in hammer-form, whitish-yellow to brownish-buff or buff-ivory, rarely sulphur, base yellowish.

Usually fasciculate. Woods, amongst moss, under firs. Aug.–Nov. 4⅝ in.

1961. C. vermicularis Fr. (from a fancied worm-like shape; *vermis*, a worm) *a b.*

Cl. stuffed, cylindrical, flexuous, acutely pointed, fragile, white, apices at length somewhat yellowish, base often faintly brownish.

Cæspitose. Amongst grass, pastures, roadsides; very common. May–Oct. 4¾ in.

1962. C. fragilis Holmsk. (from its brittleness; *fragilis*, fragile) *a b c.*

Cl. soon hollow, obtuse and rounded above, rarely forked, white, sometimes ivory, yellowish or pale brownish, white below.

Fasciculate. Pastures, roadsides; common. Aug.–Nov. 3 in.

1962a. C. Michelii Rea (after Piero Antonio Micheli) *a.*

Cl. thin, cylindrical, hollow, apex acute, very fragile, yellow, white at the base.

Fasciculate. On the ground under trees, cherry. Sept. This plant should not have been named *Michelii*, as Micheli's plant, *Nova Plant. Genera*, t. 87, figs. 6, 10, is stout, obtuse and white. According to Fries this is *C. fragilis.*

1963. C. striata Pers. (from the striate clubs) *a.*

Cl. elongate, compressed, somewhat tortuoso-flexuous, somewhat longitudinally striate, subfuliginous or pale dull to dark brown.

Cæspitose. Woods, amongst grass; rare. Sept. 3⅝ in.

1964. C. fumosa Pers. (from the smoky colour; *fumus*, smoke) *a b.*

Cl. even, usually tense and straight, somewhat compressed, fistulose, fragile, apices rounded or acute, fuliginous or yellowish dull umber.

Fasciculate. Pastures; uncommon. Aug.–Sept. 4¾ in.

Holocorynæ.

1965. C. canaliculata Fr. (from the small longitudinal channels on the clubs; *canaliculus*, a small channel) *a.*

Cl. rarely somewhat branched, slightly tough, hollow, very smooth, at length compressed or split longitudinally, white or shaded faintest salmon.

Solitary, sometimes in pairs, rarely in threes. On the ground. 3¾ in.

1965a. C. tenerima Mass. & Crossl. (from its tender substance; *tener*, tender).

Cl. cylindrical, flexuous, somewhat attenuate above, base not narrowed, hyaline white.

Gregarious but distinct at base. Amongst short grass. 1½ in. Resembles 1962 in habit.

1966. C. acuta Sow. (from the sharp pointed clubs; *acutus*, sharp) *a b c.*

Tense, straight with a translucent stem distinct from the opaque club, white.

Solitary. Sides of damp ditches, in garden pots; rare. Sept.–Nov. 1 in. Hardly a *Clavaria*. Sowerby's original drawing has the appearance of a *Torrubia* growing on a dead chrysalis, perhaps a *Pistillaria*.

1967. C. uncialis Grev. (from its usual length, about one inch, *uncia*) *a b c.*

Cl. stuffed, tough, tense and straight, obtuse, smooth, white, sometimes shaded faintest rose, becoming pale amber.

Dead stems of herbaceous plants, umbellifers; uncommon. May–Nov. 1 in.

1968. C. pistillaris L. (from the pestle shape; *pistillum*, a pestle) *a b c.*

Fleshy, stuffed, ovato-clavate, obtuse, light yellow to brownish-tan or rufescent.

Solitary. Taste bitter; odour not unpleasant. Woods, amongst ferns and leaves, roadsides under hedges; uncommon. Sept.–Nov. 11⅝ in.

1969. C. ligula Schaeff. (from the shape, like a small tongue or strap, *ligula*) *a c.*

Spongy-fleshy, elongato-clavate, obtuse, rarely slightly forked, villous at base, yellowish to pallid rufescent, rarely pale red.

Gregarious. Woods, fallen twigs and branches, amongst fir-leaves, moss. 4¼ in.

1970. C. contorta Holmsk. (from the usually twisted shape; *contortus*, twisted) *a b c.*

Contorted, wrinkled, obtuse, sometimes attenuate upwards or broadly spathulate, rarely bifid, spongy-fleshy, stuffed, watery-yellowish to somewhat pale olive-yellow.

Solitary or cæspitose. Wood, twigs, alder; rare. Sept.–Nov. ¾ in.

1971. C. fistulosa Holmsk. (from the hollow clubs; *fistula*, a pipe) *a c.*

Slender, tense, straight, hollow, light yellowish, yellow, ivory-buff or buff to orange and rufescent, rooting and villous below.

Twigs, branches, etc. 9¾ in.

1972. C. tuberosa Sow. (from the tuberous base) *a b.*

Tough, even, aculeate, hollow, tuberous and solid below, held to the matrix by flocci, ochre to somewhat pale salmon-umber.

Cæspitose. Sticks, birch, bursting through the bark. Nov. 3 in.

1973. C. juncea Fr. (from the rush-like shape; *juncus*, a rush) *a b c.*

Filiform, flaccid, hollow, acute, pallid to rufescent or yellow, pale ochre or pale orange, base creeping, fibrillose.

Gregarious. Woods, amongst leaves, twigs and moss, fir, poplar, beech, oak. Sept.–Nov. 5 in. Sometimes hair-like in thinness, the base spreading over leaves. Sometimes certain individuals spring from a small orange-brown sclerotium as in *Typhula* or a white base $\frac{1}{16}$ in. in diam. Must not be confounded with **1986** or **1992.**

1974. C. ardenia Sow. (after Lady Arden) *a b c.*

Apex of the very hollow inflated Cl. acute or rounded, then more or less collapsing and becoming depressed, brown-ochre with an olive shade or date-brown, base creeping, white tomentose.

Fallen branches, amongst fir-leaves, under laurels, on rotting hazel sticks; rare. Oct.–Dec. $8\frac{1}{8}$ in.

1975. C. incarnata Weinm. (from the flesh colour; *incarnatus*, flesh-coloured).

Solid, cylindrical, flesh-colour, internally purple.

Gregarious. On the ground.

XCIII. CALOCERA Fr.

(From the shape and colour, like beautiful horns; Gr. *kalos*, beautiful, *keras*, a horn.)

Gelatinoso-cartilaginous, horny when dry, somewhat cylindrical, simple or branched, viscid, without a distinct stem. *Hymenium*

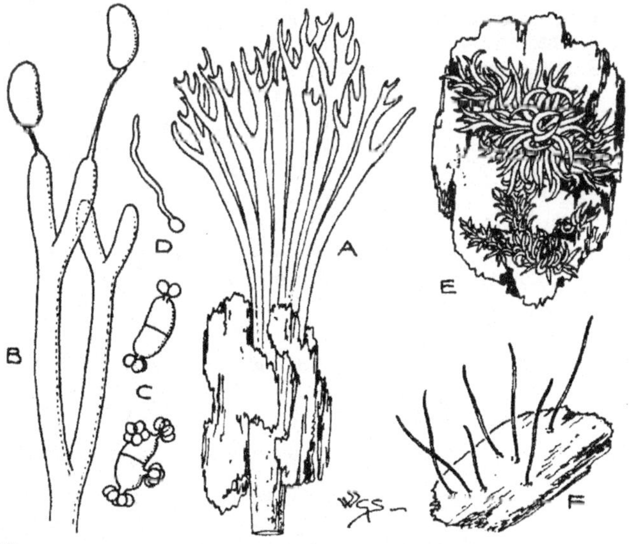

Fig. 108.—A, *Calocera viscosa* Fr., two-thirds natural size. B, basidia and spores of ditto; C, spores germinating and producing sporidiola; D, sporidiolum germinating; × 600. E, *C. cornea* Weinm.; F, *C. stricta* Fr.; two-thirds natural size.

amphigenous. *Basidia* furcate or bilobed, each lobe bearing a single one-spored sterigma. *Spores* oblong, curved, septate on germination, producing heads of ellipsoid sporidiola as in the *Tremellinaceæ*. (Fig. 108.)

Growing on wood.	Species **1976—1982**
Branched.	**1976, 1977**
Cæspitose.	**1978, 1979**
Simple, solitary.	**1980—1982**

1976. C. viscosa Fr. (from its viscidity) *a b c.*
Branches more or less vertical, repeatedly dichotomous, orange, the rooting base paler.

Rotten wood, stumps, chiefly pine; very common. July–Dec. 2½ in. Base 2½ in.

1977. C. palmata Fr. (from the palmate branching) *a c.*
Tremelloso-tough, compressed, dilated upwards and more or less horizontally palmato-divided, orange-yellow or orange.

Wood. ¾ in.

1978. C. cornea Weinm. (from the horn-like shape; *cornu*, a horn) *a b c.*
Simple, awl-shaped, sometimes bifid rarely trifid, connate at base, rooted, even, orange.

Dead wood, pine, oak, beech, willow, birch, old deal boards; frequent. June–Feb. ⅝ in.

1979. C. corticalis Fr. (from the habit, piercing through bark, *cortex*) *a c.*
Awl-shaped. *Cl.* somewhat distinct, soft, pellucid, pallid flesh-colour.

Dead bark. Dec.–Jan. ⅛ in.

1980. C. stricta Fr. (from the habit, tense and straight, *strictus*) *a b c.*
Linear, sometimes slightly spathulate, yellow, sometimes brownish below.

Ash, naked wood. Oct. 1 in.

1981. C. striata Fr. (from the striæ on the clubs) *a b.*
Lanceolato-clavate, tough, acute, striate when dry, yellow.

Wood, prostrate trunks. March. ¼ in.

1982. C. glossoides Fr. (from the tongue-like shape; Gr. *glossa*, the tongue, *eidos*, appearance) *a b c.*
Clavate, subtremellose, obtuse, compressed, stem round, light yellow.

Oak. Dec. ½ in.

XCIV. PTERULA Fr.

(From a fancied resemblance to a little wing; Gr. *pteron*, a wing.)

Subcartilaginous, filiform, arid, stemmed or stemless, branched in the two British species, subulate, apices attenuato-acute; branches

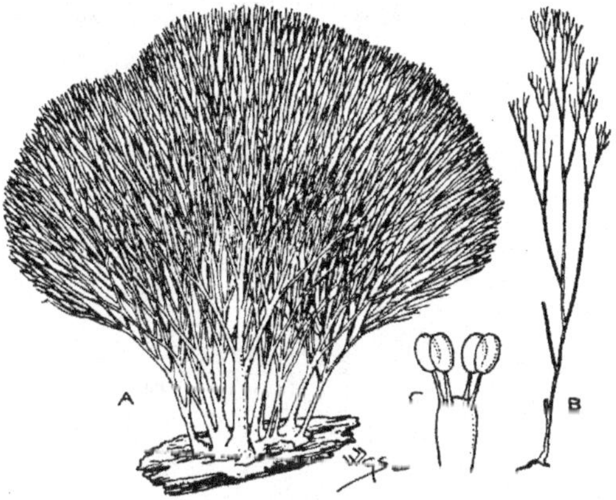

Fig. 109.—A, *Pterula subulata* Fr., natural size; B, single branch of ditto. C, basidium and spores of *P. multifida* Fr., × 500.

somewhat anastomosing, pubescent then smooth. *Basidia* tetrasporous. When young the stems and branches are considerably thicker than when mature. (Fig. 109.)

The cartilaginous substance indicates a relationship with *Tremellinaceæ* and *Calocera;* the basidia and spores are typical of the *Clavariaceæ*. Species **1983, 1984**

1983. P. subulata Fr. (from the awl-shaped branches; *subula*, an awl) *a c.*

Densely crowded, sparingly branched, tense, equal, the thread-like smooth branches multifid at the acute apices, whitish-cinereous, becoming yellowish.

On wood, amongst pines, gardens. 1¾ in.

1984. P. multifida Fr. (from the many branchlets; *multus*, many, *findo*, to cleave) *a b c.*

Very much branched, very delicate, flaccid, branches tense, straight, usually hair-like, variable in thickness, apices attenuate, pointed, base sometimes thick or with several confluent stems, at first pallid whitish then dull pale yellowish or sulphury-brown.

Branches, leaves, bases of dead ferns, etc. Sept.–Nov. 2 in.

XCV. **TYPHULA** Fr.

(From a fancied resemblance in some of the species to a minute *Typha* or reed-mace.)

Small thread-like fungi with a long or short filiform-stem, heterogeneous from the linear or slightly inflated club, often springing from a small sclerotium. *Hymenium* waxy. (Fig. 110.)

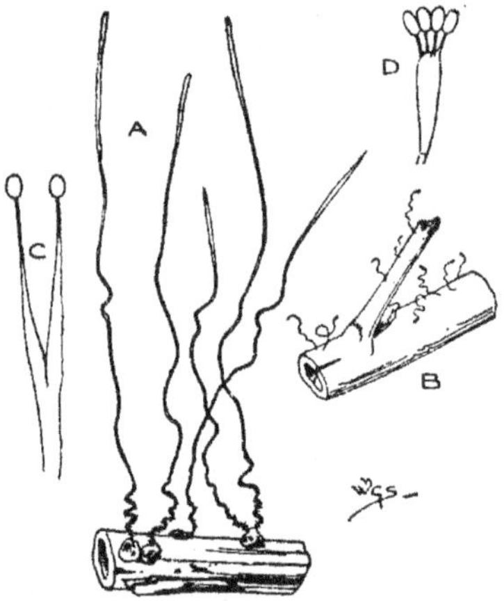

Fig. 110.—A, *Typhula phacorrhiza* Fr., one-half natural size. B, *T. gracillima* White, one-half natural size. C, basidium and spores of *T. gracilis* Berk. & Desm.; D, ditto of *T. erythropus* Fr.; × 500.

Growing on twigs, leaves, etc., rarely terrestrial, but the sclerotia will readily grow in earth if planted. Species **1985—1995**

a. Phacorrhizæ. Springing from a sclerotium. **1985—1989**

b. Leptorrhizæ. Sclerotium absent. **1990—1995**

The characters as given by Fries are far from permanent; all the species of *Typhula* may grow with or without a sclerotium. A sclerotium sometimes occurs in the allied genera, *Clavaria* and *Pistillaria*.

a. Phacorrhizæ.

1985. T. erythropus Fr. (from the red stem; Gr. *eruthros*, red, *pous*, a foot) *a b c*.

Cl. linear or nearly so, cylindrical, smooth, white. *St.* elongate, filiform, usually bent or variously twisted, finely pubescent under a lens, crimson-brown. *Scl.* $\frac{1}{20}$ in. or less, depressed, sometimes hidden beneath a thin bark or membrane and easily overlooked, brown, dark-brown when dry, sometimes absent.

Leaves, sticks, poplar, alder, ash, sunflower; common. Sept.–Nov. 1 in. The sclerotium is *Sclerotium scutellatum* A. & S.

1986. T. phacorrhiza Fr. (from the lentil-like base; Gr. *phakos*, a lentil, *rhiza*, a root) *a b c.*

Cl. linear, cylindrical, pallid to ivory and orange. *St.* linear, confluent with club, elongate, flexuous, yellowish, pilose at the sometimes fuscous base. *Scl.* $\frac{1}{5}$ in. or less, pale brown or buff, reddish-brown when dry.

Woods, hedgerows, gardens, herbaceous stems, leaves, moss, plane, hazel, alder, ash, apple, sometimes on earth; uncommon, but sometimes found in considerable abundance in patches. Oct.–Nov. 4¼ in. Must not be confounded with **1973** or **1992**. The sclerotium is *Sclerotium scutellatum* A. & S. and *S. complanatum* Tode.

1987. T. incarnata Fr. (from the flesh colour; *incarnatus*, flesh-coloured) *a c.*

Cl. cylindrical, obtuse, rarely forked, flesh-colour or white, more or less tinged with salmon-pink above. *St.* continuous with the club, pilose downwards. *Scl.* $\frac{1}{8}$ in. or less, compressed, fuscous.

Dead herbaceous stems; uncommon. 3¼ in. Closely allied to **1986**. The sclerotium is *Sclerotium fulvum* A. & S.

1988. T. gyrans Fr. (from the twisted habit; *gyro*, to turn in a circle) *a b c.* White.

Cl. cylindrical. *St.* filiform, gyroso-undulate, pubescent. *Scl.* $\frac{1}{10}$ in., pallid to orange and brown, becoming fuscous.

Straw, twigs, leaves, alder, poplar; uncommon. Oct. ½ in. The sclerotium is *Sclerotium complanatum* Tode.

1989. T. muscicola Fr. (from its habit, living on mosses; *muscus*, moss, *colo*, to inhabit) *a c.* White.

Cl. elongate, cylindrical, obtuse. *St.* thin, smooth. *Scl.* $\frac{1}{16}$ in., even, pallid.

Larger mosses; uncommon. 1½ in.

b. Leptorrhizæ.

1990. T. Grevillei Fr. (after Dr. Robert Kaye Greville) *a b c.* White.

Cl. thickened, obtuse. *St.* capillary, pilose white.

Dead leaves; frequent. Oct. ½ in. Only distinguished from **1988** by the absence of a sclerotium.

1991. T. gracillima White ex B. & Br. (from the very slender habit; *gracilis*, slender). White.

Cl. elongate. *St.* very slender, curved, smooth.

Various herbaceous plants. ½ in.

1992. T. filiformis Fr. (from the thread-like stem ; *filum*, a thread, *forma*, form) *a b c.*
Cl. elongate, thickened, smooth, whitish. *St.* often branched, decumbent, date-brown.

Dead leaves, holly, *Comptonia asplenifolia*, matting ; rare. Nov. 2⅝ in. Sometimes springing from a very small sclerotium. Must not be confounded with **1973** or **1986**.

1992a. T. tenuis Fr. (from its attenuation ; *tenuis*, thin) *a.*
Cl. simple, glabrous, hair-like, thickening upwards, brown-black.

Gregarious. On rotting wood in cellar. ⅛ to ¼ in. "It resembles a little black hair thickening upwards," Sowerby.

1993. T. gracilis Berk. & Desm. (from the slender habit ; *gracilis*, slender) *a b.* Pallid.
Cl. simple or forked, acute, pubescent. *St.* short, distinct, smooth or as club.

Putrid leaves ; uncommon. Dec. $\frac{3}{16}$ in. Has the appearance of *Isaria puberula*, but bears bisporous basidia.

1994. T. pusilla Schrœt. (from its very small size ; *pusillus*, very small) *a.* White.
Cl. linear, usually drooping. *St.* linear.

Horse-tail, plane, willow, wych-elm, alder. ¼ in. Leaves often covered with minute sclerotia, from which the *Typhula* sometimes springs.

1995. T. translucens B. & Br. (from its translucent substance) *a.* Pellucid white.
Cl. irregular, somewhat obovate. *St.* short, thickened upwards.

Terrestrial. Oct. "Not a fungus," Massee, *Brit. Fung. Fl.* vol. i. p. 91. "Resembling somewhat a prematurely dried Myxogast, but a true Basidiomycete," B. & Br. in *Ann. Mag. Nat. Hist.* ser. 4, vol. xvii. p. 138.

XCVI. PISTILLARIA Fr.

(From the form, like a pestle, *pistillum.*)

Small, club-shaped or attenuate downwards into a short continuous stem, rarely forked, cartilaginous and rigid when dry, sometimes springing from a sclerotium. (Fig. 111.) Species **1996—2000**

1996. P. tenuipes Mass. (from the thin stem ; *tenuis*, thin, *pes*, a foot) *a.*
Cl. inflated, obovate, wrinkled, pallid clay-colour or ochre. *St.* slender, flexuous, somewhat distinct from the club, ivory.

Bare heathy ground, charcoal heaps. Nov.–Jan. 1 in. *Clavaria tenuipes* B. & Br.

1997. P. micans Fr. (from its glistening surface ; *mico*, to glisten) *a b c.*
Cl. inflated, obovate, obtuse, smooth or slightly pruinose, glistening rose-colour, rarely purplish. *St.* short, attenuate, white.

Dead stems and leaves, thistles, nettles, spurge, *Artemisia* ; rare. July. ⅛ in.

1997a. P. fulgida Fr. (from its shining colour ; *fulgeo*, to shine) *a*.
Cl. lanceolato-clavate, subacute, white or pale orange below, yellow or orange above.

"On the bracteæ of *Dipsacus pilosus*," Sowerby. ⅛ in.

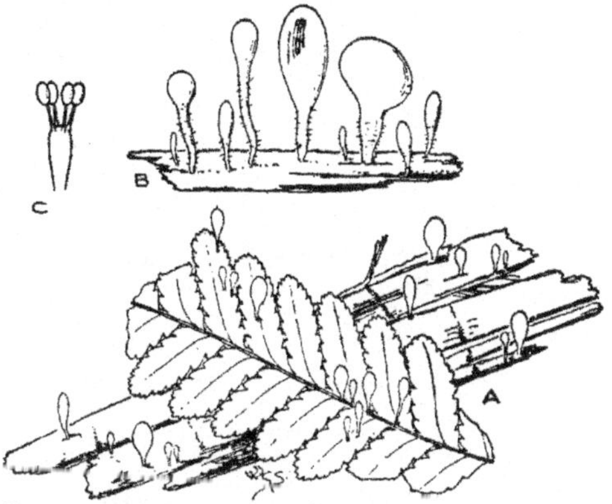

Fig. 111.—A, *Pistillaria puberula* Berk., natural size. B, ditto, × 4. C, basidium and spores, × 500.

1998. P. culmigena Fr. ex Mont. (from its habitat, grass-stems ; *culmus*, a haulm, *gigno*, to bear) *a b c*. White.
Cl. ovato-clavate, obtuse, hyaline-pellucid. *St.* distinct, very short, cartilaginous when dry.

Stems and leaves of grasses, *Secale*, *Triticum*, *Dactylis glomerata*. Oct.–Nov. ⅛ in.

1999. P. quisquiliaris Fr. (from its habitat, sweepings, rubbish, etc., *quisquiliæ*) *a b c*. Whitish.
Cl. ovato-clavate, somewhat compressed, rarely bifid, soft when fresh, rigid when dry. *St.* short, often springing from a sclerotium.

Fern-stems, oak, wych-elm, grass, willow-herb, *Chærophyllum*. April–Nov. ⅛ in.

1999a. P. ovata Fr. (from the ovate shape) *a*.
Cl. obovate, ventricose, subdepressed, hollow, white. *St.* very short, glabrous, pellucid.

Dead stems and leaves, birch, bramble, poplar, elm, willow. ⅛ in.

2000. P. puberula Berk. (from the tomentose stem ; *pubes*, downy) *a b c*. White.
Cl. ovate, ventricose. *St.* short, distinct, attenuate upwards, pellucid, fibrous, tomentose.

Ferns, stems and leaves of herbaceous plants ; rare. Sept.–Oct. 1/12 in.

2000a. P. pusilla Fr. (from its very small size ; *pusillus*, small) *a*.
Small, linear, attenuate upwards, stem scarcely distinct, white.

Dead leaves. Feb. ¼–⅜ in.

FAM. VI. TREMELLINACEÆ

Entire fungus homogeneous, gelatinous, collapsing when dry, retaining its form when moistened, traversed internally by branched hyphæ which terminate in basidia at the periphery. Basidia variable in form, elongate or fusoid, transversely septate or continuous, undivided or forked at the apex, or subglobose and longitudinally divided and bearing 2 to 4 sterigmata. Spores hyaline, from globose to sausage-shaped and curved, continuous or septate, often becoming variously septate on germination and producing sporidiola of different forms. Species **2001—2041**

KEY TO THE SUBFAMILIES AND GENERA.

Basidia elongate or fusoid, simple, transversely septate ...	Subfamily I.	AURICULARIOIDEÆ.
Attached broadly	97	AURICULARIA.
Attached by a point	98	HIRNEOLA.
Basidia subglobose at maturity, longitudinally quadripartite	Subfamily II.	EUTREMELLINOIDEÆ.
Cup-shaped, truncate or irregularly lobed ...	99	EXIDIA.
Pulvinate, gyrose	100	ULOCOLLA.
Brain-like or lobed	101	TREMELLA.
Convex, with a central hard nucleus	102	NÆMATELIA.
Erect, spathulate	103	GYROCEPHALUS.
Basidia cylindrical or clavate	Subfamily III.	DACRYOMYCETOIDEÆ.
Small, pulvinate, gyrose	104	DACRYOMYCES.
Irregularly cup-shaped, sessile	104A	FEMSJONIA.
Versiform, substipitate	105	GUEPINIA.
Stipitate, hymenium on expanded apex	106	DITIOLA.

SUBFAMILY I. AURICULARIOIDEÆ

Basidia elongate or fusoid, simple, transversely septate. Species **2001—2003**

XCVII. AURICULARIA Bull.

(From a fancied resemblance in some of the species to an ear, *auricula*.)

Subcoriaceous or gelatinous when moist, broadly attached. *Hymenium* remotely and reticulately costate or plicate, collapsing when dry. *Basidia* cylindrical, transversely tri- quadri- or quinque-septate, each joint or cell producing a single, slender sterigma from its apex. *Spores* oblong, curved, producing on germination a branched promycelium bearing several strongly curved sporidiola. (Fig. 112.)

A genus of fungi greatly resembling *Stereum* in habit, and perhaps more nearly allied to the *Thelephoraceæ* than the *Tremel-*

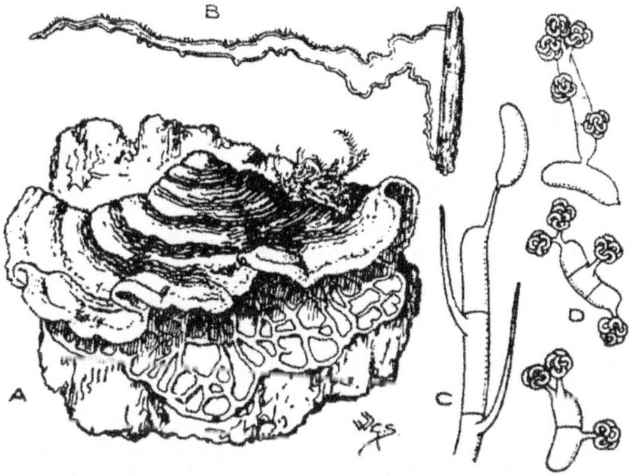

Fig. 112.—A, B, *Auricularia mesenterica* Pers., entire and in section. One-half natural size. C, basidium and spore; D, spores germinating and producing sporidiola. × 500.

linaceæ; the fructification is, however, very different from that of *Thelephora*, while the strigose cartilaginous pileus differs from typical *Tremellinaceæ*. The next genus *Hirneola* is clearly allied to *Auricularia* in habit, substance and fructification. Species 2001, 2002

2001. A. mesenterica Pers. (from a fancied resemblance to the mesenteric membrane) *a b c.*

P. resupinate, then reflexed, entire or slightly lobed, villous, yellowish, often green at base from algal growths; z. brownish. *Hym.* costato-plicate, brownish-violet, livid-purple or grey-liver.

Taste insipid; odour none. Trunks, stumps, beech, elm, walnut, vine, chestnut, furze, broom, yew; frequent. Jan.–Nov. 8½ in. Sometimes terrestrial, springing from buried wood, and then sometimes not unlike *Sticta pulmonaria* in appearance.

2002. A. lobata Sommerf. (from the lobed pileus) *a b c.*

P. effuso-reflexed, much lobed, strigose or velvety, warm yellowish-brown; z. darker, often green at base from algal growths. *Hym.* broadly reticulato-ribbed, fuscous-livid.

Bark of trees, elm; not uncommon. 4⅝ in. Closely allied to, if not identical with the last; the fruit is the same.

XCVIII. **HIRNEOLA** Fr.

(From the jug-shape when young; *hirnea*, a jug.)

Gelatinous or cartilagineo-gelatinous, soft and tremelloid when moist, cup- or ear-shaped. *Hymenium* gelatinous, even or plicate, barren under-surface minutely velvety, attached by a point, coriaceous

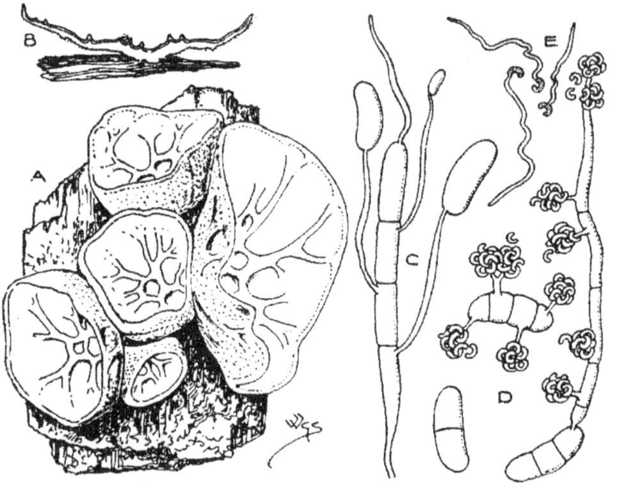

Fig. 113.—A, B, *Hirneola Auricula-Judæ* Berk., entire and in section. One-half natural size. C, basidium and spores; D, spores germinating and producing sporidiola; E, sporidiola germinating. × 500.

and membranous when dry, reviving with moisture. *Basidia* rod-shaped or fusoid, transversely septate, each cell bearing a single monosporous sterigma. *Spores* oblong, curved, becoming 1–3 septate and on germination producing minute sporidiola, sometimes on septate promycelium. (Fig. 113.)

2003. H. Auricula-Judæ Berk. ("Jew's-ear") *a b c*.

P. thin, concave, flexuous, somewhat veined, tomentose, brown flesh-colour to reddish-brown or blackish. *Hym.* venoso-plicate, glossy salmon, warm brown or reddish, purple-black when dry.

Cæspitose. Trunks, elder, elm, willow, walnut, oak, holly, *Robinia*; frequent. Jan.–Dec. 3 in. Much like *Peziza vesiculosa* in appearance when young. Formerly used as a remedy for dropsy. Eaten by inhabitants of Ukraine.

SUBFAMILY II. EUTREMELLINOIDEÆ

Basidia subglobose, at maturity longitudinally quadripartite and producing at the apex two to four elongate sterigmata.

Species **2004—2029**

XCIX. EXIDIA Fr.

(From the exuding receptacle; Gr. *exidio*, to exude.)

Gelatinously distended, tremelloid, irregularly gyrose and lobed, submarginate or effused, often papillose, when young cup-shaped and truncate. *Basidia* globose or ovoid, typically longitudinally four-

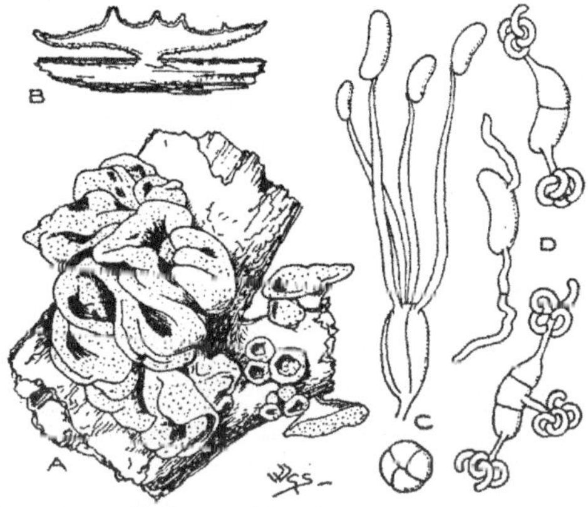

Fig. 114.—A, B, *Exidia glandulosa* Fr., entire and in section. One-half natural size. C, basidium and spores; D, spores germinating and producing sporidiola. × 500.

partite, bearing four elongate stout sterigmata. *Spores* reniform, for a long time continuous, becoming 1- to many-septate on germination, each cell or loculus of the spore giving origin to a very short promycelium bearing a crown of strongly-curved sporidiola. (Fig. 114.)

Species **2004—2007**

a. *Auriculinæ*. Pezizoid. **2004, 2005**

b. *Spiculariæ*. Somewhat plane, not lustrous beneath, glandular, here and there gyrose. **2006, 2007**

a. Auriculinæ.

2004. E. truncata Fr. (from its appearance, as if cut short; *trunco*, to maim, shorten) *a c.* Soft, distended, shining black. *Disc* truncato-plane, glandular, at length cavernous, rough with dots beneath. *St.* very short.

Bursting through bark, lime, willow. Dec. 1⅜ in. Like **2006** on a small scale.

2005. E. recisa Fr. (from its appearance as if cut back; *recido*, to cut short) *a b c.*

Very soft, warm yellow-brownish or brownish-vinous. *Disc* truncato-plane, ribbed, papillose, somewhat repand, rough with dots beneath. *St.* very short excentric.

Branches, willow, poplar, plum; common. Sept.-Dec. 1¼ in.

b. Spiculariæ.

2006. E. glandulosa Fr. (from the glandular hymenium) *a b c.*

Effused, thick, somewhat plane, undulate, sometimes pendulous, and 8 to 10 inches long, the upper hymenial surface papillose and somewhat veined, the lower surface next to the matrix crape-like to the touch and lustreless, black, blackish, purple-slate or olive-blackish.

Tasteless. Branches, oak, willow, *Ailanthus*, lime, alder, walnut. Aug.-Mar. 5½ in. When young somewhat cup-shaped and like one of the *Pezizæ* or *Bulgaria inquinans* in appearance. Translucent when quite fresh. "Witch's Butter."

2007. E. albida Bref. (from the whitish colour; *albus*, dead white) *a b c.*

Tough, undulate, subgyrose, pruinose, white or faintly shaded ivory or lavender.

Branches, birch, beech, alder, pine, hawthorn. Sept.-May. 1½ in.

C. **ULOCOLLA** Bref.

(From its glue-like consistency and appearance; Gr. *oulos*, shaggy, *kolla*, glue.)

Sporophore convex, pulvinate, gyrose, cerebriform, gelatinous, rather large. *Basidia* globose, soon longitudinally quadri-partite.

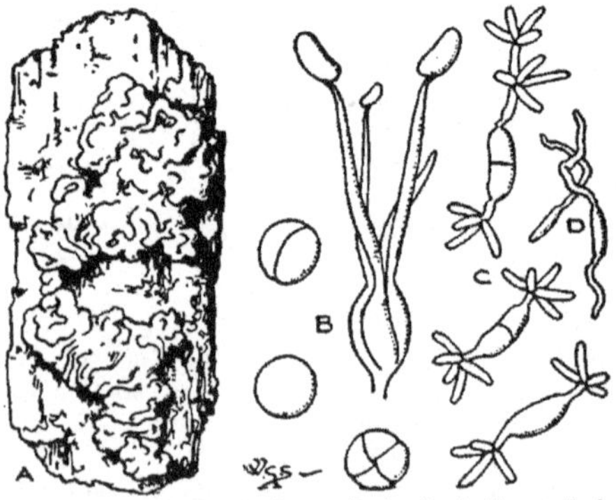

Fig. 115.—A, *Ulocolla saccharina* Bref. One-half natural size. B, basidia and spores; C, spores germinating and producing sporidiola; D, sporidiola germinating. × 500.

Sterigmata elongate, thick. *Spores* for a long time continuous, then uniseptate, reniform, each loculus on germinating giving origin to a very short promycelium bearing at its apex straight rod-like sporidiola. (Fig. 115.) Species **2008, 2009**

2008. U. saccharina Bref. (from the colour, like one kind of sugar, *saccharon*) *a b c.*

Effused, tubercular, gyroso-undulate, thick, sprinkled with a few minute obtuse evanescent papillæ, tawny cinnamon.

Branches, larch, pine; rare. Nov. 3 in.

2009. U. foliacea Bref. (from a fancied leafy or lamellate appearance; *folium*, a leaf) *a b c.*

Flaccid, even, diaphanous, undulate, plicate at base, cinnamon-salmon or vinous-buff, sometimes shaded yellowish.

Cæspitose. Stumps, logs, rails, etc., pine, oak, amongst fir-leaves, spreading to the ground. Aug.–Jan. 3½ in.

CI. TREMELLA Dill.

(From the jelly-like tremulous substance.)

Pulvinate or effused, brain-like or lobed, gelatinous, tremelloid, immarginate, generally smooth. *Basidia* globose, sometimes bifid or

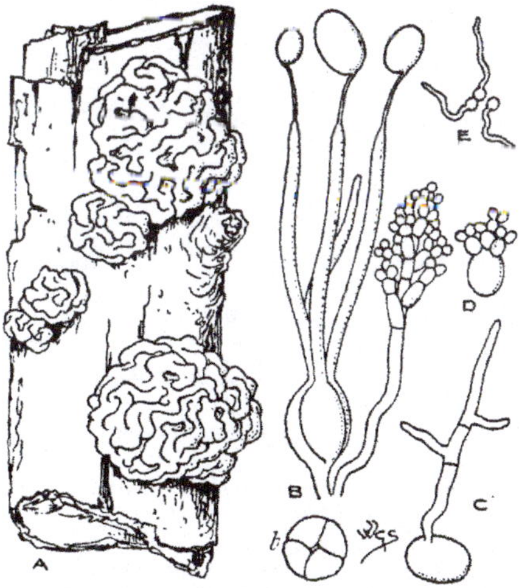

Fig. 116.—A, *Tremella lutescens* Pers. One-half natural size. B, basidium, with surface view, *b*, spores and conidia; C, spore germinating and producing hyphæ; D, ditto, producing sporidiola; E, sporidiola germinating. × 500.

trifid, but usually longitudinally quadripartite, each quadrant of the basidium elongating into a long, stout sterigma. *Spores* subglobose, continuous; on germination a tube is formed that is covered with broadly elliptical sporidiola. *Conidia* when present, as in **2012**, **2013** and **2021**, subglobose, racemose in sporiferous conceptacles in the sporophore. (Fig. 116.) Species **2010—2024**

a. Mesenteriformes. Cartilagineo-gelatinous; lobes large, foliaceous, not pruinose with spores, cæspitose. **2010—2012**

b. Cerebrinæ. Lobes short, contorted, brain-like, pruinose with white spores. **2013—2015**

c. Crustaceæ. Effused, applanate. **2016—2018**

d. Tuberculiformes. Small, suberumpent. **2019—2024**

a. Mesenteriformes.

2010. T. fimbriata Pers. (from the somewhat fimbriate edges of the lobes) *a b*.

Erect, corrugate, lobes flaccid, blackish-olive to brown-black; marg. undulato-fimbriate or incised.

Dead branches; rare. 4¾ in.

2011. T. frondosa Fr. (from a fancied leaf- or lamella-like appearance in the lobes; *frondosus*, leafy) *a b c*.

Even, gyroso-undulate, plicate at base, diaphanous, pale yellowish-buff, shaded salmon or vinous in places, base brownish-vinous.

Stumps, roots of living trees, oak, beech, hornbeam. July–Nov. 9 in.

2012. T. lutescens Pers. (from the yellowish colour; *luteus*, yellow) *a c*.

Tremulous, soft, inclining to fluid, undulato-gyrose, lobes crowded, entire, at first whitish then pale yellowish.

Stumps, oak; frequent. Aug.–Feb. 1⅝ in.

b. Cerebrinæ.

2013. T. mesenterica Retz. (from a fancied resemblance to the mesenteric membrane) *a b c*.

Expanded, somewhat tough, plicato-undulate, gyrose, changeable in form, smooth, vivid yellow to rich orange.

Dead branches, sticks, rails, oak, alder, beech, furze, broom, ivy; common. Jan.–Dec. 3½ in.

2014. T. intumescens Sm. (from the swollen habit; *intumesco*, to swell up) *a b c*.

Rounded or conglomerate, somewhat tortuous and lobed, obsoletely punctate, soft, brown.

Subcæspitose. Fallen trunks in a wet state, beech; rare. Jan. 2⅜ in. Becoming a thin black membrane when dry.

2015. T. vesicaria Sm. (from the swollen, bladder-like lobes; *vesica*, a bladder) *a b c*.

Erect, firm, gelatinous, gyrose, much waved and wrinkled, gelatinous within, dull biscuit.

On the ground; rare. Sept. 4⅝ in. Often much inflated and soon falling over.

c. Crustaceæ.

2016. T. viscosa Berk. (from the viscid substance) *a c.*

Effused, thin, resupinate, obscurely veined, undulato-uneven, following inequalities of matrix, white, becoming hyaline or translucent yellowish-white.

Dead branches, rotten wood, oak; common. Jan.–Dec. 4 in. Collapses completely in drying, leaving only a purple-grey stain on the bark.

2017. T. epigæa B. & Br. (from its habitat, the earth; Gr. *epi*, upon, *ge*, the earth) *a b c.*

Effused, thin, gelatinous, obscurely gyroso-plicate, hyaline, then white.

Covering twigs, moss, beech-nuts, etc., and spreading on to earth; rare. Aug.–Nov. 4 in.

2018. T. violacea Relh. (from the violet colour) *a.*

Erumpent, firm, somewhat compressed, gyrose, violaceous.

Gregarious. Trunks, pear; rare. ⅝ in. Sometimes confounded with **2020.**

d. Tuberculiformes.

2019. T. indecorata Sommf. (from its dull colour; *indecorus*, ugly) *a.*

Sessile, rounded, convex, plicate, moist, opaque dull livid or olivaceous, becoming dark brownish or slate-colour.

Branches, birch, oak, willow, hazel, poplar; rare. Oct.–Nov. Length of connate group 3¼ in.; width 1⅛ in.

2020. T. moriformis Sm. (from a fancied resemblance in the shape and colour to a mulberry; *morus*, a mulberry, *forma*, form) *a b c.*

Conlobate, sinuous, somewhat gelatinous but firm, mulberry-black. *Flesh* purple-crimson-blackish, base black.

Maple, chestnut, *Robinia*, elm, plane; rare. June–Feb. Common size about 1 in. × ⅝ in. Sometimes confounded with **2018.**

2021. T. tubercularia Berk. (from its appearance, like a little tuber, *tuberculum*) *a b.*

Erumpent, substipitate, pileate, stem round, dull white.

Bark, wood, oak, on old decayed *Corticium*. Oct.–Dec. $\frac{1}{16}$ in. *Tubercularia albida* B. & Br., *Brit. Fl.* vol. v. p. 354. "Coryne sp." Berk MS. on original example. "Certainly not a *Tremella*, but the conidial form of some species," Massee, *Brit. Fung. Fl.* vol. i. p. 64.

2022. T. versicolor B. & Br. (*versicolor*, changeable in colour) *a b c.*

Orbicular, tear-like, pallid when young, then orange, at length brownish-ochre.

On **1845**, ash, briar; uncommon. Feb.–May. $\frac{1}{16}$ in.

2023. T. atrovirens Sacc. (from the blackish-green colour; *ater*, black, *virens*, green) *a b c.*

Erumpent, disciform, very minutely papillate and rugose.

Gregarious or confluent, growing through cracks in bark. Dead branches, broom. Oct. $\frac{1}{16}$ to ⅛ in.

2024. T. clavata Pers. (from its somewhat clavate shape) *a*.
Incrassate, fleshy-red, becoming black at base.

Solitary, simple or connate. Branches. Average size of single growth $\frac{5}{8} \times \frac{1}{4}$ in. Possibly a form of *T. sarcoides* With., and, if so, a condition of *Ombrophila sarcoides* Jacq., one of the Discomycetes.

[*T. sarcoides* With. is an early condition of *Ombrophila scarcoides* Jacq. *T. foliicola* Fuck. is not a *Tremella*; it forms small gelatinous tufts or spots on the leaves of *Rubus fruticosus* in company with *Phragmidium*. *T. brassicæcola* B. & Br. is probably a form of *Hypocrea rufa* Fr.; it grows on and in cabbage stumps.]

CII. NÆMATELIA Fr.

(From the gelatinous substance which surrounds the nucleus; Gr. *naima*, gelatine, *eileo*, to roll or wrap round.)

Subgelatinous, firm, convex, with a firm fleshy central solid portion called the nucleus, not shrivelling when dry. *Basidia* subglobose, longitudinally quadripartite. *Spores* broadly elliptical, continuous. (Fig. 117.)

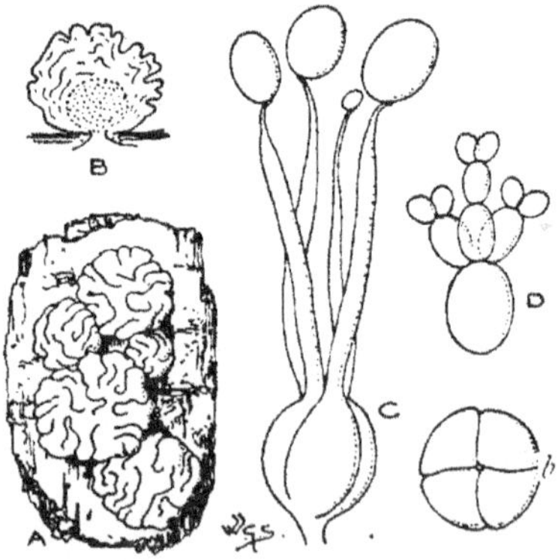

Fig. 117.—A, B, *Næmatelia encephala* Fr., entire and in section. One-half natural size. C, basidium with surface view, *b*, and spores; D, spore germinating and producing sporidiola. × 500.

Readily distinguished by the presence of the central solid nucleus, which is composed of interwoven hyphæ and covered with a thick gelatinous spore-bearing stratum. Growing on wood, sometimes on moss, rarely terrestrial. Species **2025—2028**

2025. N. encephala Fr. (from the contained nucleus; Gr. *en*, within, *kephale*, the head) *a b c*.

Somewhat sessile, pulvinate, plicato-rugose, pallid flesh-colour, salmon or salmon-whitish, nucleus hard, large, white.

Solitary to connate. Branches, rails, pine, larch; frequent. Sept.–Mar. 1 in. Connate for 4 in.

2026. N. rubiformis Fr. (from a fancied resemblance to a small fruit of *Rubus*, bramble) *a b.*

Somewhat sessile, orbicular, gyroso-tuberculose, yellow.

Dead branches, twigs, sometimes on moss, spreading to the ground. Oct. $\frac{3}{16}$ in.

2027. N. nucleata Fr. (from the nucleus) *a b c.*

Sessile, flattened, somewhat gyrose, whitish, then tawny yellow, nucleus white, about $\frac{1}{50}$ in. in diam.

Rotten wood, sticks, lime, plane, oak; rare. Mar. Single plants $\frac{3}{16}$ in. in diam. Connate to $1\frac{1}{4}$ in. Must not be confounded with **2007.**

2028. N. virescens Corda (from the greenish colour; *vireo*, to be green) *a b.*

Sessile, suborbicular, depressed, gyroso-tubercular, somewhat pale leaf-green. Forming green gelatinous masses.

Furze, ivy; common. Oct.–June. Single plant $\frac{3}{16}$ in. Connate chains $1\frac{1}{2}$ in.

CIII. **GYROCEPHALUS** Pers.

(From the globular shape; Gr. *guros*, round, *kephale*, the head.)

Sporophore erect, substipitate, flattened, subspathulate, irregular, tremelloso-cartilaginous. *Hymenium* confined to one surface.

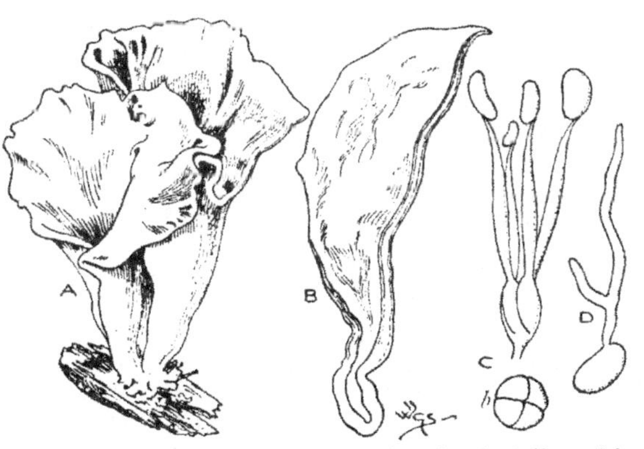

Fig. 118.—A, B, *Gyrocephalus rufus* Bref., entire and in section. One-half natural size. C, basidium with surface view, *b*, and spores; D, spore germinating. × 500.

Basidia globose, soon longitudinally quadripartite. *Sterigmata* 2–4, elongate, thick. *Spores* ovato-pyriform, continuous. (Fig. 118.)

2029. G. rufus Bref. (from the red colour; *rufus*, red) *a b c.*

Erect, cartilagineo-gelatinous, substipitate, subspathulate, sometimes stemless, variable, orange to red or dull scarlet above and orange below. *Hymenium* inferior, smooth.

Terrestrial, under pines, on rotten trunks, fragments of wood. $3\frac{3}{4} \times 3$ in.

Subfamily III. DACRYOMYCETOIDEÆ

Basidia cylindrical or clavate, divided at the apex into two long sterigmata, not septate, except in *Femsjonia*, where the basidia and sterigmata become septate with age. Species **2030—2041**

CIV. DACRYOMYCES Nees.

(From the tear-like habit; Gr. *dakru*, a tear, *mukes*, a fungus.)

Small, pulvinate, gelatinous, homogeneous, somewhat plicato-gyrose, everywhere covered with the hymenium. *Basidia* terminating normal hyphæ, tereti-clavate, bifurcate at the apex when mature.

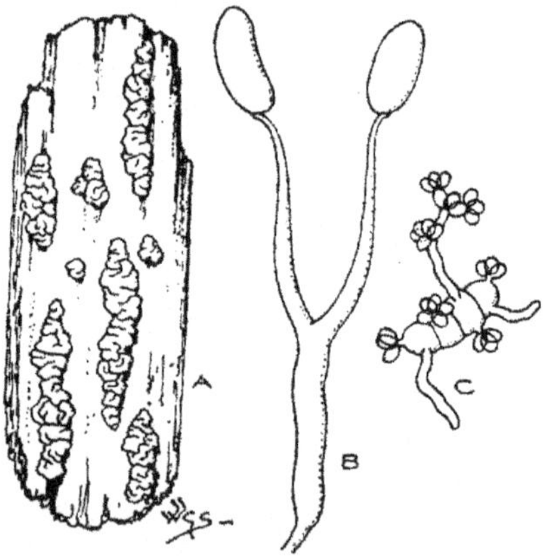

Fig. 119.—A, *Dacryomyces deliquescens* Duby. One-half natural size. B, basidium and spores; C, spore germinating and producing sporidiola. × 500.

Spores cylindric-oblong, curved, at maturity or during germination variously septate or muriform. In the conidia-bearing sporophores the conidia, which more or less resemble the spores in size and form, are produced in chains. (Fig. 119.)

The species grow on dead wood. Species **2030—2036**

Rose-coloured. **2030**
Orange, yellow or yellowish. **2031—2033**
Pallid or fuscous. **2034**

2030. D. macrosporus B. & Br. (from the long spores; Gr. *makros*, long) *a b c*.

Gelatinous, tuberculate, minutely gyroso-granular.

Branches, in cracks, on remains of *Sphæria stigma*. Dec.–Apr. Connate chains 2 × ⅛ in.

2031. **D. deliquescens** Duby (from its appearance as if dissolving; *deliquesco*, to dissolve) *a b c.*

Somewhat round, rooted, convex, immarginate, at length twisted, hyaline-yellowish, varying to sulphur, rarely to orange.

Dead wood, cut wood, pine, holly, ash, oak. Dec.–Mar. Connate chains $1\frac{1}{8} \times \frac{1}{8}$ in.

2032. **D. stillatus** Nees (from the drop-like habit; *stillo*, to drop) *a b c.*

Somewhat round, convex, at length plicate, yellow to orange or scarlet.

Rotten wood, fir-poles, cut wood, pine, fir. Jan.–Dec. Connate chains $1\frac{3}{4} \times \frac{1}{8}$ in. Distinguished from **2031** by being usually smaller in size and brighter and more persistent in colour.

2033. **D. chrysocomus** Tul. (from the colour as of golden hair; Gr. *chrusos*, gold, *kome*, hair) *a b c.*

Spherical, orbicular, immarginate, soon collapsing, pezizoid, at length flattened, even, bright golden-yellow, varying greenish- or buff-yellow or orange.

Rotten wood, fir; rare. $\frac{1}{16}$ in. Small in size, usually in large numbers scattered over wood or bark.

2034. **D. sebaceus** B. & Br. (from a fancied resemblance to droppings of tallow, *sebum*) *a b c.*

Somewhat round, cup-shaped, steel-grey.

Branches, ash, willow, pea-sticks. Jan.–April. Single plants about $\frac{1}{8}$ in. in diam. Curved chains $1\frac{1}{4}$ in.

2035. **D. tortus** Fr. (from the twisted habit; *torqueo*, to twist) *a b c.*

Round, depressed, gyroso-tuberculate, substipitate, yellow to orange, sometimes vermilion.

Dead wood, rotten powdery bark, oak. Oct.–June. Single plant $\frac{1}{8}$ in. Connate groups $\frac{7}{8}$ in.

2036. **D. vermiformis** B. & Br. (from a fancied resemblance to certain small worms; *vermis*, a worm, *forma*, form) *a c.*

Minute, worm-shaped, grey.

Rotten wood. April–Sept. Single plants $\frac{1}{24}$ in. Vermiform chains $\frac{3}{8}$ in.

[*D. succinius* Fr., appearing as minute dots on pine leaves has been shown by Phillips and Plowright to be the conidial stage of *Hymenoscypha electrina* Phill. & Plowr.]

CIVa. FEMSJONIA Fr.

(From Femsjo, the name of a place.)

Cup-shaped, distended with firm jelly, different on the two sides, disc thick, heterogeneous, even, not glandular. *Sporophores* immersed, globose. *Spores* oblong, curved, becoming septate, at first colourless, then orange-brown. (Fig. 120.)

Growing on wood, erumpent, brightly coloured.

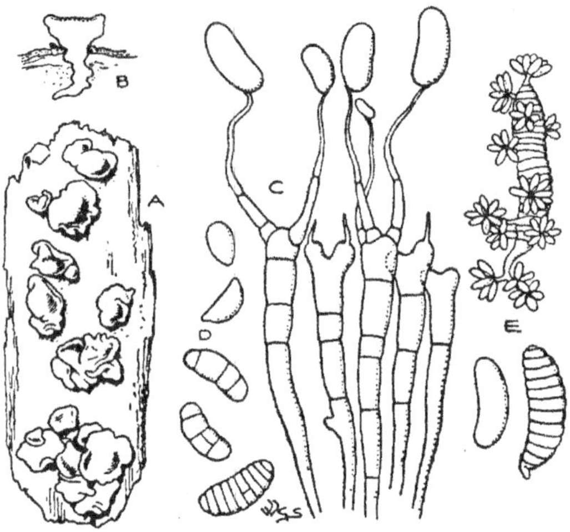

Fig. 120.—A, B, *Femsjonia luteoalba* Fr., entire and in section One-half natural size. C, basidia and spores; D, detached spores, the mature spores are septate. × 500. E, *Guepinia Femsjoniana* Olsen. Three spores, one germinating and producing sporidiola. × 500.

2036a. F. luteoalba Fr. (from the yellow disc and white exterior; *luteus*, yellow, *albus*, white) *a b*.

At first cup-shaped, becoming disc-shaped and plane, at length convex and folded, disc bright brassy-yellow, with a narrow white marg., under surface at first white-tomentose to brassy-white, the tomentum soon passing away, rooting, erumpent.

Branches, birch, oak. Sept. $\frac{5}{8}$ in. in diam. Looks like **2035** on a large scale. *Guepinia Femsjoniana* Olsen is perhaps the same as this.

CV. **GUEPINIA** Fr.

(After Jean Pierre Guepin.)

Unequally cup-shaped, substipitate, versiform, cartilagineo-gelatinous, the two surfaces diverse in structure. *Hymenium* unilateral. *Basidia* linear, deeply bifurcate. *Spores* curved. Chains of conidia produced on the tips of hyphæ springing from the surface opposed to the hymenium. (Fig. 121.) Species **2037, 2038**

2037. G. Peziza Tul. (from the *Peziza* shape) *a b.*
Cup-shaped, cup oblique, sessile or stipitate, yellow.
Dead branches, wood. ⅜ × ½ in.

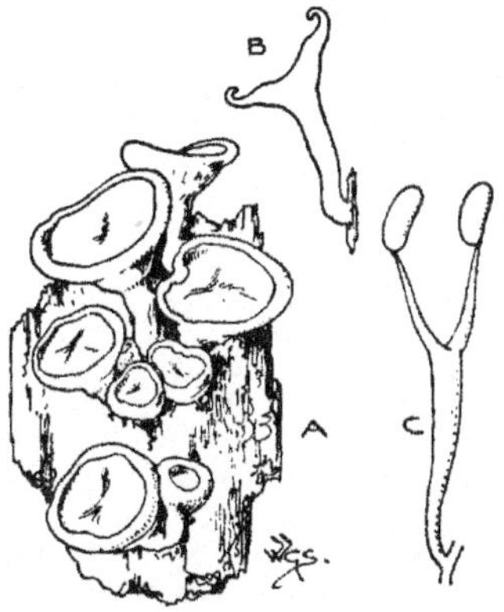

Fig. 121.—A, B, *Guepinia Peziza* Tul., entire and in section. × 2. C, basidium and spores. × 500.

2038. G. obliqua Mass. (from the oblique habit) *a.*
Minutely hairy on reverse side of Hym. *Hym.* slightly concave, oblique, passing into a very short stem-like base, glabrous, deep bright orange when moist, tinged with pink when dry.
Gregarious. Nov. ⅛ in. high.

CVI. DITIOLA Fr.

(From the down-like universal veil; Gr. *dittos*, double, *ioulos*, down.)

Veil universal, down-like. *Stroma* firm, more or less stem-like. *Hymenium* discoid, gelatinous. *Basidia* furcate. *Spores* at first continuous then septate. (Fig. 122.)

Massee has removed **2040** and made it the sole type of a new genus, *Dacryopsis*. It is undesirable to break up such a small and natural genus as *Ditiola*. *Dacryopsis* is said to be distinguished from *Ditiola* by having the hymenial surface at first covered with conidiophores and conidia as in *Tubercularia*; but this is hardly a distinction of generic value as conidia are present in some species of *Tremella* and not in others. Species **2039—2041**

2039. D. radicata Fr. (from the rooting base; *radix*, a root) *a c.*
Stipitate, rooting, cups nearly plane, disc golden-yellow. *St.* thick, white.
Wood, fir, amongst pine-leaves. ⅛ × ¼ in. No conidia recorded.

2040. D. nuda B. & Br. (from the absence of a veil; *nudus*, naked) *a.*

Head subpileiform, corrugate, sometimes umbilicate, golden-yellow. *St.* somewhat thick, not rooting, cylindrical, white-tomentose. Fruiting surface at first covered with conidiophores and conidia.

Stumps, fir. Sept. $\frac{1}{8} \times \frac{1}{4}$ in.

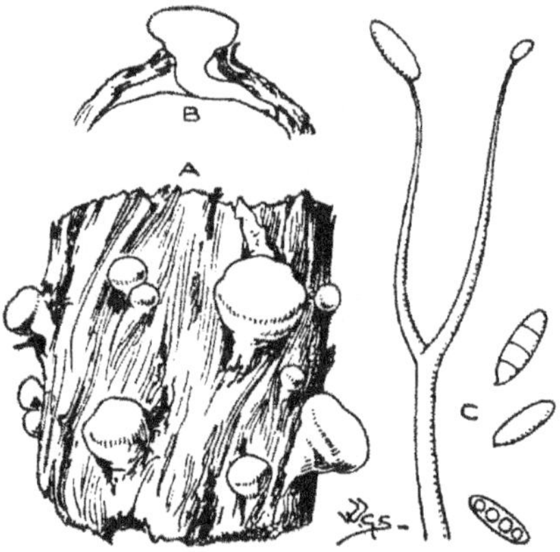

Fig. 122.—A, B, *Ditiola Ulicis* Plowr., entire and in section. × 2. C, basidium and spores. × 500.

2041. D. Ulicis Plowr. (from the habitat, gorse, *Ulex europæus*) *a c.*

Head globose, then flattened and wrinkled, pale lemon-yellow, slightly white-villous. *St.* firm, paler than head, sometimes absent or extending only from the wood through the bark.

Dead stems; very common. Jan. $\frac{1}{2} \times \frac{1}{4}$ in. No conidia recorded.

[The genus *Apyrenium* Fr. is omitted. *A. lignatile* Fr. is the conidial stage of *Hypocrĕa rufa* Fr. *A. armeniacum* B. & Br. is the same condition of *H. gelatinosa* Fr. *Hymenula punctiformis* B. & Br. is an imperfect form of some fungus not belonging to the *Tremellinaceæ.*]

ORDER II. GASTEROMYCETES

FAM. VII. **PHALLOIDACEÆ**

Pileus and stem or latticed receptacle at first enclosed in a universal volva of two membranes with a middle gelatinous stratum; volva at length opening at the apex. Gleba mucilaginous, at length elevated from the volva on a stem or lattice. Spores minute, elliptic-oblong, continuous, at first immersed in olive-brown mucilage.

Species **2042—2045**

KEY TO THE GENERA.

Pileus free, reticulate	107	ITHYPHALLUS.
Pileus adnate	108	MUTINUS.
Pileus divided in vertical lobes	110	LYSURUS.
Receptacle forming a hollow lattice of large meshes	109	CLATHRUS.

CVII. **ITHYPHALLUS** Fisch.

(From a fancied resemblance to the *Ithuphallos*, carried at the festivals of Bacchus.)

Pileus reticulate, attached to the apex of a hollow, perforate spongy, stem-like receptacle. *Veil* forming a thin film—sometimes a feeble white membrane—over the sporiferous mucous and slightly attached to the stem. (Fig. 123.)

2042. I. impudicus Fisch. (from its phallic resemblance; *impudicus*, shameless) *a b c*.

P. white, at first covered with olive-brown sporiferous mucous. *Rec.* elongato-fusiform, spongy-cellular, white. *Myc.* cord-like.

Highly fetid. Woods, thickets, gardens, generally amongst rotten tree-roots and stumps, by gates, in and by summer-houses, sheds, porches, often near to and in dwellings. May–Aug. Height 10 in. Var. *iosmus* Sacc. Borders of reticulations and pileus serrate, sometimes reddish-grey in colour, said to be violet-scented.

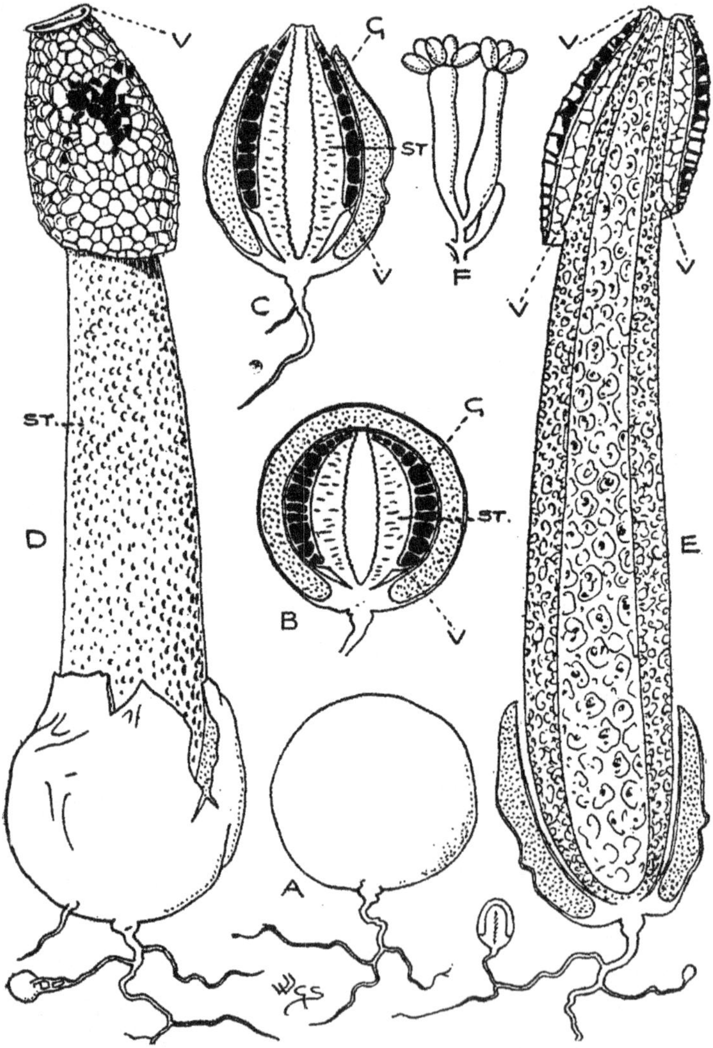

Fig. 123.—*Ithyphallus impudicus* Fisch. Two-thirds natural size.
A, young plant. B, section of ditto, showing veil at V. C, ditto, showing pileus emerging from volva, veil at V. D, mature plant, remains of veil at V. E, section of ditto. F, basidium and spores, × 1000. G, gleba; ST, stem or receptacle.

CVIII. **MUTINUS** Fr.

(An appellation of Priapus.)

Pileus adnate to the hollow, perforate or imperforate spongy stem, at first immersed in olive-brown sporiferous mucus. *Veil*

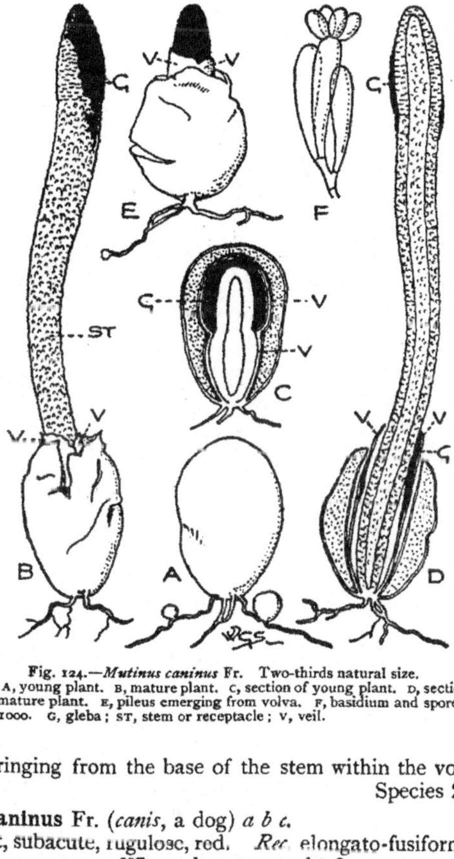

Fig. 124.—*Mutinus caninus* Fr. Two-thirds natural size. A, young plant. B, mature plant. C, section of young plant. D, section of mature plant. E, pileus emerging from volva. F, basidium and spores, × 1000. G, gleba; ST, stem or receptacle; V, veil.

distinct, springing from the base of the stem within the volva, white. (Fig. 124.) Species **2043, 2044**

2043. M. caninus Fr. (*canis*, a dog) *a b c*.

P. short, subacute, rugulose, red. *Rec.* elongato-fusiform, cellular, white or rosy. When the receptacle first emerges from the volva it is covered with the white veil.

Woods, bushy places, rotten stumps, pine, ash, decayed leaves in woods and hedges. June–Dec. Height 5¼ in.

2044. M. bambusinus Fisch. (from its original place of growth, at the base of bamboos, *bambusa*) *a.*

P. long, acutely conical, tuberculose or rugose, dull purple. *Rec.* elongato-fusiform, imperforate, spongy-cellular, pallid rubiginous or rosy.

Very fetid. Height 5¼ in.

CIX. CLATHRUS Mich.

(From the latticed receptacle; *clathrus*, a lattice.)

Receptacle hollow, globose, latticed, the meshes at first filled and covered inside by olive-brown sporiferous mucus. (Fig. 125.)

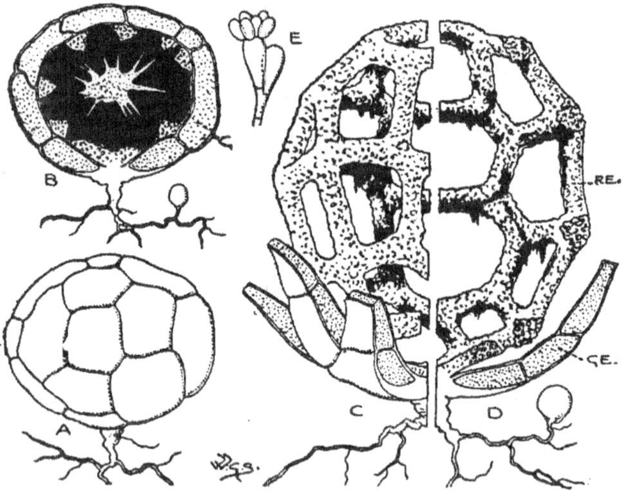

Fig. 125.—*Clathrus cancellatus* Tournef. One-half natural size.
A, young plant. B, ditto in section. C, mature plant, left half entire. D, ditto in section. E, basidium and spores, × 750. G, gleba; GE, gelatinous stratum of volva; RE, receptacle.

2045. C. cancellatus L. (from the lattice-like receptacle; *cancellatus*, latticed) *a b c.*

Rec. sessile, globose or ovate, the branches triangular in section with a flat face outwards, orange-vermilion outside, vermilion-crimson inside, inner sporiferous mucus dark olive, enclosing a central mass of soft dark gelatine. *Vo.* of two white membranes, marked internally and externally with lines corresponding with the lattice within. *Myc.* cord-like.

Single or cæspitose. Highly fetid. Plantations, gardens, stoves. Oct.-Nov. 4¾ × 3½ in. The interior of the receptacle at length drips with olive-brown mucus.

CX. **LYSURUS** Fr.

(From the free tail-like lobes of the mature receptacle ; Gr. *luo*, to loosen, *oura*, a tail.)

Stipitate, pileus divided in vertical lobes, free at the apex, distinct from the stem. Sporiferous mucus on the interior of the lobes. (Fig. 126.)

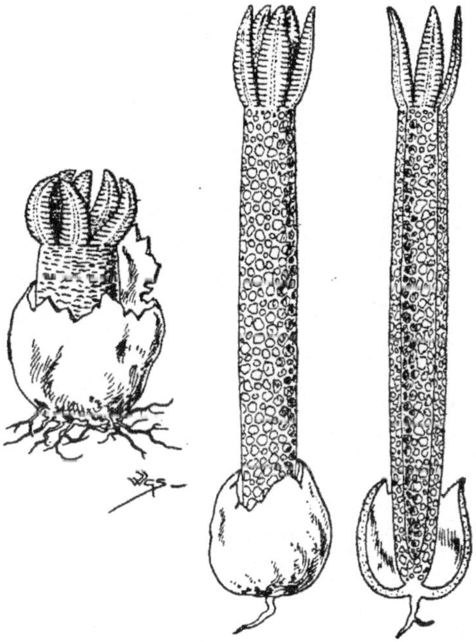

Fig. 126.—*Lysurus australiensis* Cooke & Mass. One-half natural size. Young plant on left, mature plant and section on right.

2045a. L. AUSTRALIENSIS Cooke & Mass. (from its native country—Australia).

P. for the most part 5-lobed, lobes attenuate upwards, at first connivent, then somewhat reflexed, with a longitudinal depression along the middle, transversely rugose, tawny. *Rec.* cylindrical, hollow cellular, whitish. *Vo.* globose, lobed, white.

Refuse of wheat, probably Australian, Kidderminster, Worcester. 14 Nov., 1902. Carleton Rea. 6¾ × ⅞ in.

Fam. VIII. **LYCOPERDACEÆ**

Peridium double, rarely single, gelatinous stratum absent. Spores forming a powdery mass at maturity, mixed with a well-developed capillitium. A very feeble membrane separating the gleba from the double outer peridium, may be detected in immature examples.

Species **2046—2084**

Key to the Genera.

Peridium vertically compressed. Stoma apical	111	Battarrea.
Peridium subglobose.		
Stoma apical	112	Tulostoma.
Stoma none	113	Queletia.
Exoperidium splitting in a stellate manner	114	Geaster.
Exoperidium breaking up in the form of warts or spines. Base sterile	115	Lycoperdon.
Exoperidium flaking off in patches. Sterile base absent ...	116	Bovista.

CXI. **BATTARREA** Pers.

(After Antonio Battarra.)

Volva universal, central stratum gelatinous. *Peridium* vertically compressed, pileus-like, bursting through the volva and at length raised on a tall stem. (Fig. 127.)

2046. B. phalloides Pers. (from a fancied resemblance to *Ithyphallus*; Gr. *eidos*, appearance) *a b*.

Pe. convex, bearing the spores and capillitium, orange-brown. *St.* hollow with a central pith of silky threads, wall of three layers, the outer broken up into coarse fibres pointing downwards, orange-brown. *Vo.* ovate formed of two white membranes with an internal gelatinous olive-yellow layer.

In and near decayed trees, ash, on sandhills. Dec. 12½ × 2 in. When in the young state sometimes buried several inches deep in sand or decayed wood. Sometimes 14 in. high with a pileus 2½ in. in diam.

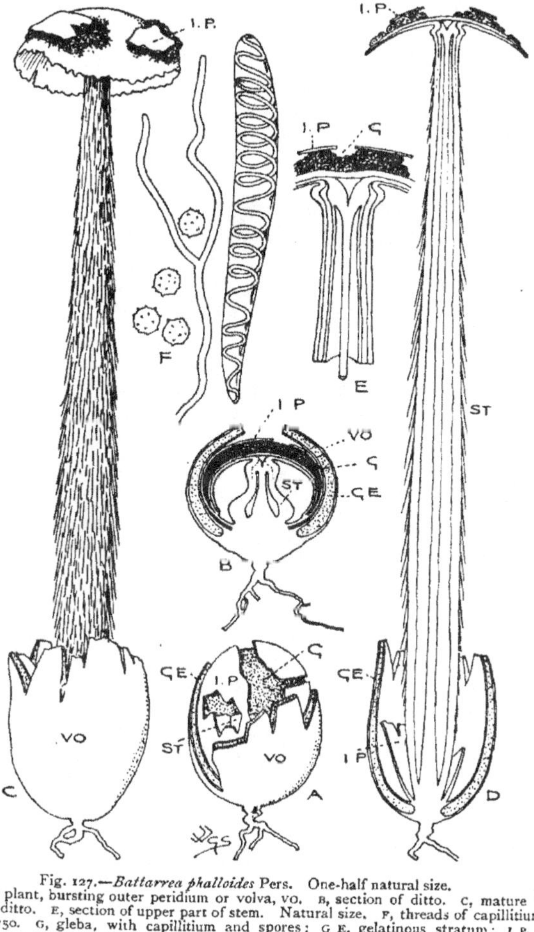

Fig. 127.—*Battarrea phalloides* Pers. One-half natural size.

A, young plant, bursting outer peridium or volva, VO. B, section of ditto. C, mature plant. D, section of ditto. E, section of upper part of stem. Natural size. F, threads of capillitium and spores. × 750. G, gleba, with capillitium and spores; G E, gelatinous stratum; I P, inner peridium; ST, stem; VO, volva.

CXII. **TULOSTOMA** Pers.

(From the structure of the mouth of the peridium; Gr. *tulos*, cartilage, *stoma*, a mouth.)

Peridium subglobose, dehiscing by an apical stoma, outer coat, veil or cortex, deciduous, distinct from the at length elongate stem. Threads of the capillitium swollen at the septa. (Fig. 128.)

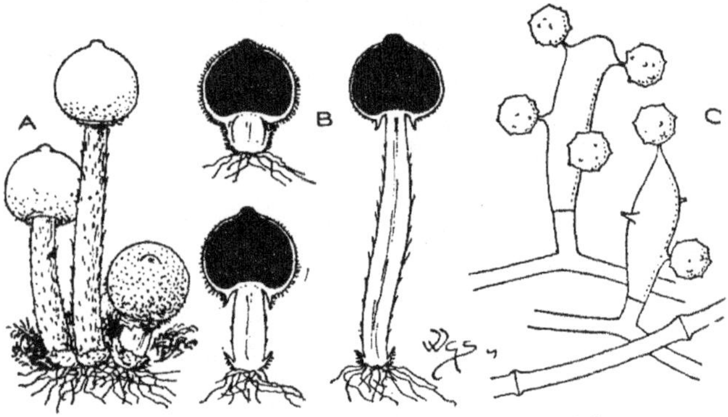

Fig. 128.—*Tulostoma mammosum* Fr. Two-thirds natural size. A, entire plant. B, sections. C, basidia, spores and thread of capillitium. × 1000.

2047. T. mammosum Fr. (from the breast-like peridium; *mamma*, a breast) *a b c*.

V. universal, distinct at the base of Pe. and at the lower part of St. *Pe.* globose, at first sessile or nearly so, stoma mammiform, entire, ochre. *St.* equal, squamoso-fibrillose, ragged at the base with remains of V., somewhat pale vinous-brown.

Amongst moss on old walls, dry banks, sandhills, etc. Oct.–Feb. $2\frac{1}{2} \times \frac{3}{8}$ in. There are two forms, brown, with a stout stem, and yellowish-white, with a slender stem. Intermediate forms occur.

CXIII. **QUELETIA** Fr.

(After Dr. Lucien Quelet.)

Peridium subglobose, simple, stoma none, the wall breaking up irregularly at maturity. *Stem* at length long, squarrose, the base of the peridium breaking at the apex and forming a collar or veil. *Spores* borne on tetrasporous basidia. (Fig. 129.)

2048. Q. mirabilis Fr. (from its unusual appearance; *mirabilis*, wonderful) *a*.

Pe. thin, even, glabrous, rigid, fragile, breaking away in irregular patches at maturity, whitish, then dull greyish. *St.* subclavate, fibroso-carnose, externally densely squarrose, whitish, brownish towards base.

Gregarious. 6¼ × 2¼ in. At first wholly subterranean.

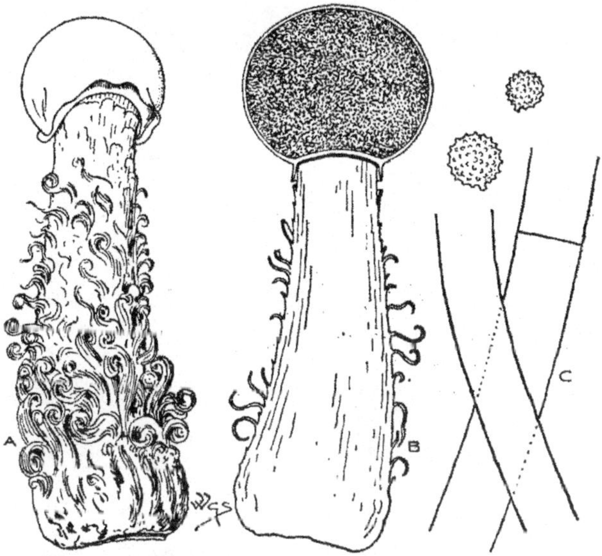

Fig. 129.—*Queletia mirabilis* Fr. One-half natural size. A, entire plant. B, section. C, threads of capillitium and spores. × 750.

CXIV. GEASTER Mich.

(From the star-like appearance of mature plants; Gr. *ge*, the earth, *aster*, a star.)

Peridium at first continuous, consisting of three layers, the two outermost—*exoperidium*—splitting from the apex in a stellate manner, the inner layer—*endoperidium*—sessile or pedicellate, furnished at the apex with one orifice, rarely more. *Columella* prominent, rudimentary or obsolete. *Capillitium* dense. Terrestrial. (Fig. 130.) Species **2049—2062**

a. Myriostomatæ. *Endoperidium* with many orifices. **2049**

b. Monostomatæ. *Endoperidium* with a single orifice.

Endoperidium pedicellate. **2050—2054**

Endoperidium sessile. **2055—2062**

Fig. 130.—A, B, *Geaster coliformis* Pers., entire and in section. One-half natural size. C, threads of capillitium and spores. × 750. D, E, *G. Berkeleyi* Mass., entire and in section. One-half natural size. F, thread of capillitium and spores. × 750. G, H, *G. fornicatus* Fr., entire and in section. One-half natural size. J, thread of capillitium and spores. × 750. K, L, *G. fimbriatus* Fr., entire and in section; M, young plant in section. One-half natural size. N, threads of capillitium and spores. × 750. O, basidium with spores and thread of capillitium of *G. rufescens* Pers.

a. Myriostomatæ.

2049. G. coliformis Pers. (from a fancied resemblance in the endoperidium to a colander, *colum*) *a b c.*

Ex. 7- to 10-segmented, unequal, pale umber. *End.* subglobose, scabrous, supported on numerous pedicels, pale slate. *Col.* usually obsolete or occurring as continuations of the pedicels. *Gl.* umber.

Sandy banks. Oct. Diam. of expanded Ex. 5½ in.

b. Monostomatæ.

2050. G. Bryantii Berk. (after Charles Bryant) *a b c.*

Ex. 8- to 12-segmented, becoming involute, pale umber to warm brown. *End.* subglobose or obpyriform, with a distinct groove between End. and Ped., dark purple-slate to warm umber, more or less whitish-farinose. *Per.* elongate, conical, sulcate. *Col.* hemispherical. *Gl.* dark brown.

Amongst leaves, under hedges, sandy places, on wood, dry marly banks, on the ground. July–Feb. Diam. of expanded Ex. 3¾ in. Var. *minor* Mass., under one inch when expanded.

2051. G. Schmidelii Vitt. (after Casimir Schmidel) *a c.*

Ex. 5- to 8-segmented, revolute, white to ivory. *End.* subglobose, lead-colour to pale salmon. *Per.* prominent, plicate, apex fimbriate, whitish. *Col.* large.

On the ground in open places. Diam. of expanded Ex. 2 in.

2051a. G. umbilicatus Fr., var. **Smithii** W. G. Sm., *G. Smithii* C. G. Lloyd (after Worthington George Smith) *a.*

Ex. 7- to 8-segmented, unequal, sublanceolate, pale umber-ochre, shaded darker, limbs slightly vinous-sienna, externally pale umber. *End.* sessile, globose, or depressed-globose, colour as Ex. with a tinge of olive. *Per.* conical plicato-striate in a circular depressed marginate umbilicus. *Gl.* deep vinous-umber.

Nov. Diam. of expanded Ex. 1⅝ in. Per. not like **2061**, but more like **2058** or **2057.**

2052. G. Berkeleyi Mass. (after the Rev. M. J. Berkeley) *a.*

Ex. 7- to 9-segmented, unequal, vinous-brown when dry, externally buff and umber. *End.* broadly ovate, shortly and stoutly pedicellate, coarsely farinose, pale brown. *Per.* prominent, conical, coarsely sulcate inside and out, surrounded by a smooth slightly depressed silky zone. *Col.* short, hemispherical. *Gl.* umber.

On the ground, amongst fir-leaves. Oct. Diam. of expanded Ex. 3¾ in. According to Massee this species was confused by Berkeley with **2051** and **2055**—see the dried examples in the Kew herbarium.

2053. G. limbatus Fr. (from the fringed or silky, fimbriato-ciliate peristome; *limbatus*, bordered, as with another colour) *a b c.*

Ex. multifid, unequal, vinous, scurfy and umber beneath. *End.* subpyriform to subglobose, shortly pedicellate, blackish-brown to deep vinous-umber. *Per.* slightly depressed, subacute, fimbriato-ciliate, surrounded by a small pale silky disc. *Col.* almost obsolete. *Gl.* dark umber.

Woods, hedgebanks, amongst firs, leaves. Sept.–Nov. Diam. of expanded Ex. 4 in.

2054. G. fornicatus Fr. (from the arched inner layer of the exoperidium ; *fornicatus*, arched over) *a b c.*

Ex. 4- (rarely 5-) segmented, the two layers separating, outer cup-like, scurfy and brown externally, tan within, inner arched and attached to outer by tips of segments, umber, blackish-umber or rufous, tan within. *End.* obpyriform, shortly pedicellate, dark brown. *Per.* conical, ciliato-striate. *Col.* slender, somewhat clavate.

Meadows, pastures, heaths, amongst firs. Mar.–Nov. Diam. of expanded Ex. 2¾ in. 5¾ in. high.

2055. G. striatus DC. (from the sulcato-striate peristome) *a b c.*

Ex. 6- to 9-segmented, coriaceous, thin, tan, darker externally. *End.* subsessile, globose, often scabrous, umber-white to lead-colour. *Per.* prominent, conical, sulcato-striate. *Gl.* brown.

On the ground, sandhills amongst dead leaves, amongst trees, firs. Mar.–Nov. Diam. of expanded Ex. 2 in.

2056. G. Michelianus W. G. Sm. (after Piero Antonio Micheli) *a b c.*

Ex. 4- to 6-segmented, unequal, thick, warm or pale vinous, outside much cracked, brown. *End.* subglobose, tan-umber. *Per.* mammiform, surrounded by a silky zone. *Col.* clavate. *Gl.* brown.

On the ground, amongst rhododendrons. Oct. Diam. of expanded Ex. 5 in. Differs from **2061a**, for which it has sometimes been mistaken. At maturity this plant will sometimes throw itself elastically some 9 inches from its place of growth.

2057. G. lagenæformis Vitt. (from a fancied resemblance to a narrow-mouthed bottle or jar ; *lagena*, a flask, *forma*, shape) *a b c.*

Ex. 7- to 9-segmented, attenuate, subequal, salmon-white to ivory, pale rose or vinous. *End.* subglobose or ovate, pale buff, umber-white or faintly vinous. *Per.* broadly-conical, sides incurved, silky, striate, surrounded by a disc. *Col.* clavate. *Gl.* brown, somewhat olive.

On the ground. April–Nov. Diam. of expanded Ex. 2¼ in.

2058. G. saccatus Fr. (from the sac or pouch-like shape when young ; *saccus*, a bag) *a b c.*

Ex. 7- to 9-segmented, pale dove-colour, very pale yellowish-lavender or pale buff, externally umber and ochre. *End.* globose, colour as Ex. *Per.* broadly-conical, sides incurved, stoma fimbriate, silky, surrounded by a silky disc. *Gl.* dark brown.

On the ground, amongst stones and earth, in hedges, sometimes on sticks with profuse Myc. Aug. Diam. of expanded Ex. 3¼ in.

2059. G. mammosus Chev. (from the breast-like endoperidium; *mammosus*, full-breasted) *a b c.*

Ex. 7- to 10-segmented, hygrometric, strongly incurved when dry, vinous-brown to umber, ochreous or glossy yellow-white-umber outside. *End.* globose, white-brown. *Per.* broadly and acutely conical, fimbriato-ciliate to striate, surrounded by a pale silky disc. *Gl.* vinous or dark brown.

On the ground. Feb.–Dec. Diam. of expanded Ex. 1⅜ in.

2060. G. rufescens Pers. (from the at length rufescent colour) *a b c.*

Ex. 6- to 9-segmented, thick, becoming deeply cracked, brownish flesh-colour or red-salmon, clouded rufescent, externally dark sienna or sienna-buff. *End.* globose, pale to dark umber, stoma slightly mammiform, dentate. *Col.* obsolete or short. *Gl.* dark brown.

Woods, pastures. Sept.–Dec. Diam. of expanded Ex. 4½ in. Var. *minor* Pers., 5–8 segmented, less than half the size of type.

2061. G. fimbriatus Fr. (from the fimbriate mouth of the peristome) *a b c.*

Ex. 5- to 15-segmented, flaccid, somewhat thick, buff- or umber-white with white cracks. *End.* subglobose, pale umber, stoma indeterminate, piloso-fimbriate. *Col.* obsolete. *Gl.* brown.

Woods, pine, on the ground amongst leaves. Aug.–Dec. Diam. of expanded Ex. 2⅜ in.

2061a. G. triplex Jungh. (from the superficial appearance, as of three peridia) *a.*

Ex. 5- to 6-segmented, the fleshy inner layer often becoming detached and forming a spurious peridium at the base of the End., vinous-brown. *End.* globose, with a definite even stoma, yellow-brown. *Col.* prominent, persistent.

Woods; fields. Diam. of expanded Ex. 4½ in. When young this plant is pouch-like with an acute apex as in **2058**, and greatly differs in this respect from **2056**, where also the colours of the peridia are the reverse of **2061a.**

2062. G. hygrometricus Pers. (from its absorption of moisture from the air; Gr. *hugros*, moisture, *metron*, a measure) *a b c.*

Ex. 7- to 20-segmented, harsh and horny, vinous-brown, yellowish-brown outside, rigidly inflexed when dry, and often cracked all over. *End.* ovato-globose in section, floccose, crimson-brown or purple-lavender, stoma irregularly torn. *Col.* obsolete. *Gl.* crimson-brown.

On the ground. Woods, pine. Feb.–Dec. Diam. of expanded Ex. 2¼ in.

CXV. LYCOPERDON L.

(From an old belief that puff-balls grew from the dung of the wolf; Gr. *lukos*, a wolf, *perdon*, dung.)

Peridium double, external layer or cortex becoming broken up into spine-like tufts or warts, inner layer smooth dehiscing by a small apical stoma; the whole of the upper portion being sometimes evanescent. Sterile base usually stem-like. (Fig. 131.)

The species are remarkable for their variability in size, very small mature examples of all the species being frequent. They are popularly named "Puff-balls" and "Devil's snuff-boxes."

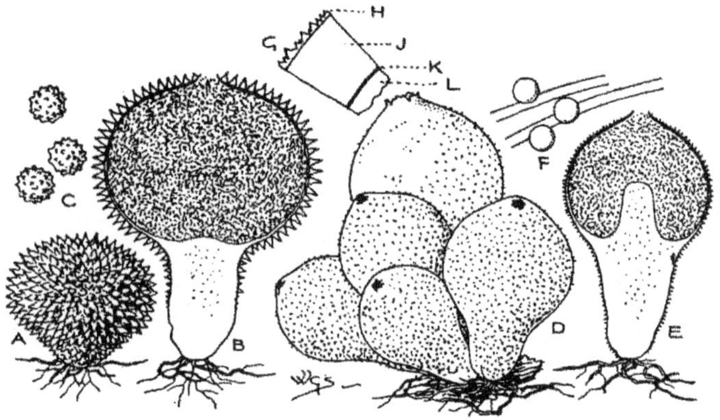

Fig. 131.—A, B, *Lycoperdon echinatum* Pers., young plant entire and mature plant in section. One-half natural size. C, spores. × 750. D, E, *L. pyriforme* Schæff., entire and in section. One-half natural size. F, spores and threads of capillitium. × 750. G, section through cortex of *L. gemmatum* Batsch × 5; H, outer scurfy coat; J, inner coat; K, thin membrane over gleba; L, gleba.

Some of the species are dangerous if eaten in a raw state; rapid inflammation of the throat and a greatly swollen tongue have been known to ensue. This is probably also true of species of *Bovista*.

Species **2063—2077**

Spores rough. **2063—2068**
Spores smooth. **2069—2077**

2063. L. echinatum Pers. (from the hedgehog-like spines of the peridium; *echinus*, a hedgehog) *a b c.*

Pe. globose, usually passing into a stout stem, sometimes stemless, at first covered with large pyramidal floccose spines, vinous-brown varying buff, stoma small, irregularly torn, base usually clad with white thread-like Myc. *St.* spongy-cellular within. *Cap.* and *Spores* deep vinous-brown.

Usually solitary. Amongst trees, beech. Mar.–Oct. Diam. 2¼ in. Sometimes made a var. of **2068**.

2064. L. Hoylei B. & Br. (after Mr. Hoyle, who first brought the plant to the notice of the Rev. M. J. Berkeley) *a b c.*

Pe. subglobose, passing into a short stem, at first covered with pyramidal floccose spines, purple-brown, stoma small, irregularly torn, base usually clad with white thread-like Myc. *St.* lacunose, spongy-cellular and olivaceous within. *Cap.* and *Spores* light in colour, pale purplish-brown, but described by Berkeley as lilac.

Woods, amongst leaves. Oct.–Dec. Diam. 1¼ in. Perhaps a form of **2068.**

2064a. L. velatum Vitt. (from the external layer of the peridium forming a universal veil; *velum*, a covering) *a b.*

Pe. pyriform or subglobose, slightly umbonate, outer layer tomentose, breaking up into evanescent star-shaped rosettes and forming a false ring-like appearance at the apex of the stem, at length almost completely disappearing, snow-white, inner layer furfuraceous and furnished with very thin short spines, white then delicate flesh-colour, finally greyish. *St.* with white cord-like Myc., white. *Cap.* and *Spores* white, then fulvous, finally ashy.

Amongst leaves. Oct. 2½ × 1⅞ in.

2065. L. atropurpureum Vitt. (from the blackish-purple capillitium and spores; *ater*, black) *a b c.*

Pe. subglobose or pyriform, at first covered with slender spines, warm yellowish-brown or rufous, stoma small, irregular. *St.* plicate below, spongy-cellular and pale dull purple within. *Cap.* and *Spores* dull- to blackish-purple.

Downs, woods, under oaks. Sept.–Nov. Diam. 1¾ in.

2066. L. excipuliforme Pers. (from a fancied resemblance to a basin, *excipula*) *a b c.*

Pe. subglobose, usually passing into a long, stout stem, rarely almost stemless, at first covered with slender spinose warts, white to ivory. *St.* even, somewhat plicate below, sometimes lacunose, spongy-cellular and pale olive within. *Cap.* and *Spores* dull olive or yellowish-brown.

Pastures, woods, under firs. May–Nov. Diam. 3½ in. Sometimes 7½ in. high. Sometimes confused with or made a var. of **2068, 2070** or **2071**, the two latter have smooth spores.

2067. L. saccatum Fr. (from a fancied resemblance to an inverted sac or bag, *saccus*) *a b c.*

Pe. subglobose, passing into a long stout stem, plicate below, at first covered with spinulose warts, brownish-ivory, stoma small, irregular. *St.* often more or less lacunose, spongy-cellular and yellowish-olive within, brownish below. *Cap.* and *Spores* yellowish-brown.

Often in rings of very large size, individuals a foot apart. Open places, near pines, in and near thickets. Aug.–Oct. Diam. 4 in. Often 8 in. or more high.

2068. L. gemmatum Batsch (from the peridium covered with pointed warts; *gemma*, a young bud) *a b c.*

Pe. subglobose, usually passing into a stem, rarely stemless, at first covered with small pointed spines, pale brown. *St.*

tapering downwards, coarsely spongy cellular and faintly olive within. *Cap.* and *Spores* olivaceous- to dark-umber.

Woods, shady places, pastures, from the decayed interior of an acorn. Sept.–Nov. 2⅝ × 3⅞ in. Sometimes confused with **2071**, of which **2068** has been considered a variety, but in **2071** the spores are smooth, while in this they are minutely warted or spiny.

2069. L. lacunosum Bull. (from the pits or depressions on the stem; *lacuna*, a hole or cavity) *a*.

Pe. subglobose, passing into a long stout stem, covered with small pointed spines, pale buff. *St.* tapering downwards covered with small lacunæ. *Cap.* and *Spores* olive-brown.

Woods. Autumn. 1¾ × 4¼ in.

2070. L. pyriforme Schæff. (from its pear-like shape; *pyrus*, a pear, *forma*, shape) *a b c*. Subumbonate.

Pe. thin, flaccid, at first covered with minute pointed warts, brown, stoma small, torn, base sometimes forming a Col., slightly cellular. *Cap.* and *Spores* olive.

Commonly cæspitose, often held together by coarse strands of white Myc. On the ground in fields and woods, on rotten stumps and branches, pollards, oak. July–Nov. 1⅝ × 2¼ in. Col. not a permanent character.

2071. L. perlatum Pers. (from the great width of the peridium in comparison with the slender stem; *perlatus*, very wide) *a b c*.

Pe. subglobose or flattened, often passing into a somewhat slender stem, umbonate, often plicate below, at first covered with stout obtuse short spines, base sometimes forming a Col., pale brownish-salmon, stoma small. *St.* slightly spongy-cellular and in section convex above. *Cap.* and *Spores* olivaceous.

Woods, thickets, amongst trees, firs. Sept.–Nov. 2¼ × 5 in. Col. not a permanent character. Sometimes made a var. of **2068**, or **2068** is sometimes made a var. of this, but the spores are different.

2072. L. molle Pers. (from its softness or soft scurfy surface; *mollis*, soft).

Turbinate, base broad, abrupt. *Pe.* papyraceous, at first furfuraceous, collapsing, umber to olive, stoma small, irregular. *Cap.* collapsing, ochreous-olive. Sterile base well developed, slightly spongy-cellular.

On the ground, woods, oak. 1⅝ × 2¼ in. Differing from **2068** in the smooth spores.

2072a. L. cruciatum Rost. (from the somewhat cross-shaped spines of the cortex) *a*.

Pe. subglobose to broadly pyriform, narrowed below into a short stout sterile stem-like base, 1¼–1½ in. high by ¾–1¼ in. broad. *Cor.* whitish, formed of groups of minute, more or less pyramidal spines, breaking away in large flakes and exposing the inner yellowish-brown minutely granulated peridium; dehiscence by a small irregularly torn apical stoma. *Gl.* umber. *Spores* globose, smooth, almost hyaline, 5–6 μ in diam.;

threads of capillitium mostly unbranched, slender, 4–5 μ thick, ends tapering, very slightly coloured yellow-brown. The lacunose sterile base not projecting into the Gl. as a Col.

On the ground among hazel bushes. $1\frac{3}{4} \times 1\frac{1}{4}$ in. Whitish outer wall of peridium peeling off in large flakes as in species of *Bovista*, exposing the brown, minutely granulated inner Pe.

2073. L. hiemale Bull. (from its time of growth, winter, *hiems*) *a b c.*

Pyriform, rarely almost sessile and globoso-depressed. *Pe.* membranous, flaccid, white then grey, yellowish or rufescent, stoma irregularly lacerate. *Cap.* and *Spores* smoky-yellow to olive-brownish. Sterile base well developed, at first verrucoso-aculeate, elastic-tough with a thick bark.

Fields, downs, amongst dead leaves. Sept.–Jan. $2\frac{1}{4} \times 3\frac{3}{8}$ in.

2074. L. cælatum Bull. (from the large conical warts of the peridium as if embossed or carved in bas-relief; *cælum*, a sculptor's chisel) *a b c.*

At first subglobose or pyriform, then depressed, flaccid and ruptured above, contracted below into a short stout tapering stem and more or less covered with large conical warts, whitish or ivory. *Cap.* and *Spores* ochreous-olive or opaque yellow-green. Sterile base spongy-cellular.

Odour disagreeable, musty, mouse-like. Fields, woods, grassy roadsides. May–Oct. $7\frac{1}{2} \times 6\frac{3}{4}$ in.

2074a. L. depressum Bonard. (from the depressed peridium) *a.*

Pe. obconical, at first rounded above then flattened, base somewhat plicate, yellowish-white, then greyish yellow, finally brownish, covered with spinulose warts. Stoma at first well defined, greatly extending with age. Sterile base well developed, separated from the olive Gl. by a diaphragm.

Commons, woods. Aug.–Oct. $1\frac{1}{4} \times 1\frac{1}{4}$ in.

2075. L. Bovista L. (from the resemblance to *Bovista*) *a b c.*

Globose, sessile, usually plicate at base, at first subtomentose then smooth and glossy, white, stoma large, irregular. *Cap.* and *Spores* dull olive or olive-umber.

Sometimes grows in rings 30–50 ft. in diam. Edible when young and perfectly white within. Rich pastures, orchards, gardens, roadsides, grassy places. Aug.–Oct. 12×11 in. Popular names "Puff-ball" and "Devil's snuff-box." Modern Greek name "kranion," like a skull. A drawing in British Museum shows a diam. of 18 in. A weight has been recorded of 20 lbs. An American example has been described with a larger diam. of 5 ft. 4 in. and a lesser of 4 ft. 6 in., height $9\frac{1}{2}$ in.

2076. L. furfuraceum Schæff. (from the branny or scurfy peridium; *furfur*, bran) *a b c.*

Pe. subglobose, passing into a short, stout stem, flaccid above, floccose, white then pale olive-yellowish, base white. *Cap.* and *Spores* olivaceous. Sterile base spongy-cellular, white.

Fields, fallow ground, old walls, about stumps. Sept. $1\frac{3}{4} \times 1\frac{5}{8}$ in.

2077. L. Cookei Mass. (after Dr. M. C. Cooke) *a c.*
Subglobose or flattened below and hemispherical, abruptly contracted into a short stem, minutely furfuraceous, smoky-brown, umber-salmon or umber with a faintly purplish shade. *St.* pale spongy-cellular within. *Cap.* and *Spores* yellow- to olive-umber.
Gregarious. Sept. 1⅛ × 1 in. Like a small form of **2071.**

CXVI. **BOVISTA** Pers.

(From the German popular name, *bofist*, a fairy ball.)

Peridium double, external layer or cortex smooth, fragile, deciduous, not remaining as adnate warts, except in **2C81** and **2083**,

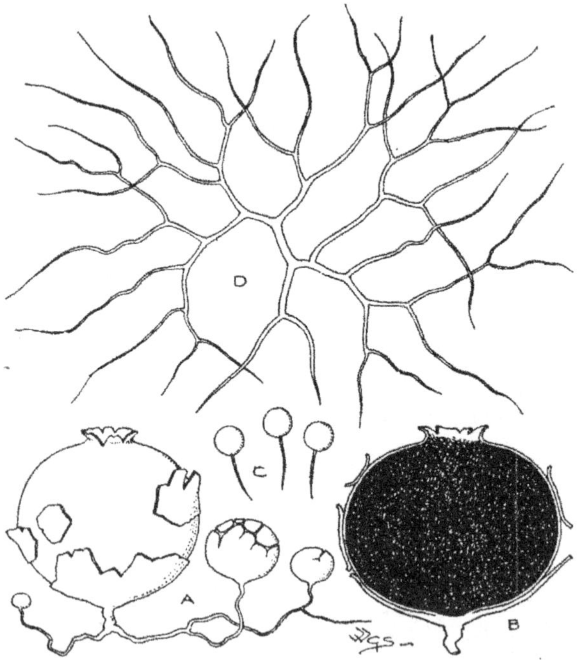

Fig. 132.—A, B, *Bovista nigrescens* Pers., entire and in section. One-half natural size. C, spores. × 750. D, thread of capillitium. × 100.

usually falling away in flakes, inner layer smooth, dehiscing by a definite or irregularly torn apical stoma. Sterile base absent. *Spores* globose or elliptical. (Fig. 132.) Species **2078—2084**

Spores globose. **2078—2083**
Spores elliptical. **2084**

2078. B. plumbea Pers. (from the lead-coloured inner layer of the peridium ; *plumbeus*, leaden) *a b c*.

Globose. *Cor*. thin, whitish, persistent towards base, inner layer tough, thin, dehiscing by a small irregular somewhat prominent stoma, lead-colour, slate-whitish or purplish-slate. *Cap*. and *Spores* brown.

Dry grassy and heathy places, downs, rarely on stumps. July–Nov. $1\frac{5}{8} \times 1\frac{5}{8}$ in.

2079. B. nigrescens Pers. (from the at length black or blackish colour of the inner layer of the peridium ; *nigrescens*, turning black) *a b c*.

Subglobose. *Cor*. papery, whitish, inner layer thin, tough, dehiscing by an irregular somewhat prominent stoma, blackish, brown-, slate-, olive-brown- or blue-black. *Cap*. and *Spores* olive-brown then purplish-brown.

Dry pastures, heathy places. Jan.–Dec. $2\frac{1}{2} \times 2\frac{1}{4}$ in.

2080. B. olivacea Cooke & Mass. (from the at length olive-coloured capillitium and spores) *a c*.

Globose. *Cor*. thin, fugacious, whitish or ochreous, inner layer thick, soft, becoming brittle and breaking away in patches above, white or ochreous. *Cap*. and *Spores* citrine then olive.

On the ground. Sept. 2×2 in.

2081. B. ammophila Lév. (from the habitat, sandy places ; Gr. *ammos*, sand, *phileo*, to love) *a*.

Broadly obovate, plicate below and tapering into a long stout root-like growth. *Cor*. broken up into tomentose warts, whitish, inner layer thin, dehiscing by a small irregularly torn stoma, pallid. *Cap*. and *Spores* olive.

Sandy places. Sept. $1\frac{3}{8} \times 1\frac{3}{4}$ in. without root.

2082. B. cepæformis Mass. (from the shape, like a flat onion ; *cepa*, an onion, *forma*, shape) *a*.

Subglobose, flattish, somewhat plicate below. *Cor*. minutely furfuraceous, whitish to very pale buff, inner layer dehiscing by a small torn stoma, whitish or pale buff. *Cap*. and *Spores* citrine. *Myc*. rooting, cord-like.

On the ground. $1\frac{5}{8} \times 1\frac{1}{8}$ in.

2083. B. pusilla Pers. (from the very small size ; *pusillus*, small) *a b*.

Subglobose. *Cor*. breaking up into minute scurfy squamules, pale yellowish-brown, inner layer flaccid, dehiscing by a small stoma, pale yellowish-brown or olivaceous ochre. *Myc*. cord-like, rooting.

Pastures, hedgebanks. $\frac{3}{4} \times \frac{5}{8}$ in.

2084. B. ovalispora Cooke & Mass. (from the oval spores) *a b*.

Subglobose. *Cor*. very thin, subpersistent towards base, whitish or ochreous, inner layer thin, flaccid, smooth, dull lead-colour, dehiscing by an irregular stoma. *Cap*. and *Spores* brown.

On the ground. Diam. 2 in. Formerly included with **2078** and **2079**, but differing in the oval spores and thinner cortex.

Fam. IX. **SCLERODERMACEÆ**

Peridium thick with a well-defined base, sometimes forming a stem, opening at the apex in an irregular manner, feebly developed with a sterile base. Capillitium absent or scanty. Species **2085—2089**

Key to the Genera.

Gleba composed of numerous distinct cells containing peridiola	117	Polysaccum.
Gleba cellular, cells without peridiola	118	Scleroderma.

CXVII. **POLYSACCUM** DC.

(From the many small cells within the peridium ; Gr. *polus*, many, *saccos*, a bag.)

Peridium irregularly globose, corky, attenuate downwards into a more or less elongated stem-like base, dehiscing by an irregular

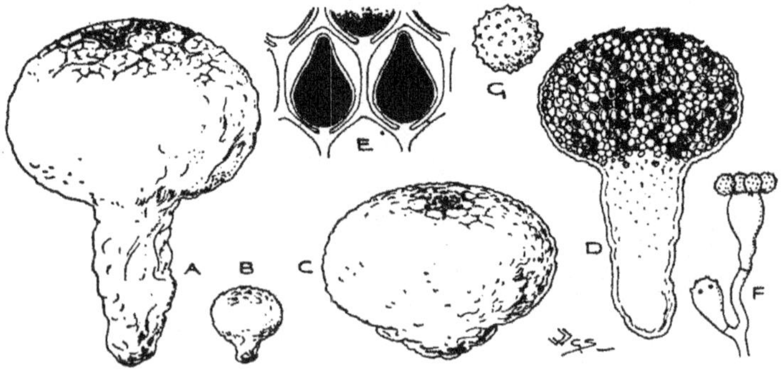

Fig. 133.—A, B, C, D, *Polysaccum pisocarpium* Fr., entire and in section. One-half natural size. E, section through part of gleba, showing cavities and enclosed peridiola. × 5. F, basidia and spores. × 250. G, spore. × 750.

rupture of the apex. *Gleba* consisting of numerous cells, each containing a peridiolum. *Peridiola* at first saccate or irregularly ovoid, sometimes filling the cell and becoming irregularly polygonal. (Fig. 133.)

2085. P. pisocarpium Fr. (from the pea-like peridiola ; *pisum*, a pea, *carpellum*, a fruit) *a b*.

Pe. irregularly globose or ovate, smooth, subtuberculose, corky, attenuate downwards into a more or less stem-like base, or sessile or nearly so, pale to olive-umber, base spongy-cellular within. *Gl.* composed of numerous cells, each containing an olive-umber Peri. *Spores* dark brown.

On the ground. May. $2\frac{5}{8} \times 3\frac{5}{8}$ in.

CXVIII. SCLERODERMA Pers.

(From the hard skin or peridium; Gr. *skleros*, hard, *derma*, skin.)

Peridium firm, cortex persistent in the form of warts, scales or granules, indehiscent or breaking up in an irregular or more or less

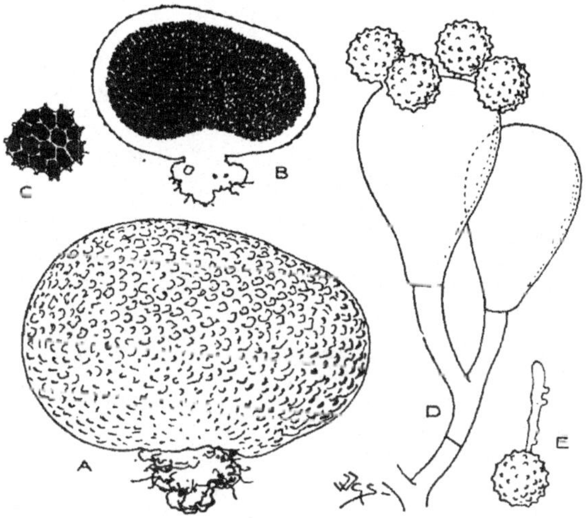

Fig. 134.—A, B, *Scleroderma vulgare* Hornem., entire and in section. One-half natural size. C, cells of gleba. × 5. D, basidia and spores; E, spore germinating. × 750.

stellate manner, convex internally at base. *Gleba* cellular, subpersistent, without peridiola. *Trama* in the cell walls. (Fig. 134.)

The peridium is frequently and extensively perforated by larvæ which eat the spore-mass and help to disseminate the spores.

Species **2086—2089**

2086. S. vulgare Hornem. (from its commonness; *vulgaris*, common) *a b c.*

Subsessile, often lobed below and usually slightly depressed above, ovate, reniform in vertical section. *Pe.* thick, corky, verrucose or adpressed-scaly, pale warm-brown, scales darker. *Gl.* dark slate-colour. *Trama* white.

Often cæspitose. Under the name of "vegetable tripe" it has been eaten when quite young and found not poisonous. Used for the adulteration of pâté de foie gras, sausage and poulard truffé in place of the French truffle, *Tuber macrosporum*. Odour rank, strong, disagreeable, sulphureous-alliaceous. Open places near trees. Aug.–Nov. Diam. 3¾ in. Often closely resembles horse-droppings. Sometimes attacked by **1481**, which in turn is not infrequently attacked by *Hypomyces luteovirens*. Pigs in search for truffles and pig-nuts will not eat *Scleroderma*. Sometimes 5½ inches in diam. Var. *cepa* W. G. Sm., *Scleroderma cepa* Pers. Flattish, onion-shaped. Var. *lævigatum* Fück. nomen. Even, smooth, often confounded with var. *cervinum* (Bolt. *Hist. Fung.* t. 116). Var. *aurantiacum* W. G. Sm., *Lycoperdon aurantiacum* Bull. Brassy-yellow. Var. *spadiceum* W. G. Sm., *Scleroderma spadiceum* Pers. Date-brown. Var. *cervinum* W. G. Sm., *Scleroderma cervinum* Pers. Usually less than an inch in diam., hard and granular all over, resembling *Elaphomyces granulatus*, of which species it is sometimes, without examination, placed in herbaria as a var. Fir woods. Probably a distinct species.

2087. S. verrucosum Pers. (from the warted peridium; *verrucosus*, full of warts) *a b c*.

Subsessile to stipitate. *Pe.* subglobose, thin, fragile and breaking up irregularly above, externally finely squarrose or minutely warted, when stipitate continued into a more or less elongate and sometimes lacunose stem, ochreous or dull brownish. *Gl.* at first umber, then slate-colour to blackish. *Trama* whitish. Often furnished with a mass of Myc. at base.

Sandy ground, commons, open places, near bushes. July–Nov. 3¼ × 5 in. Sometimes attacked by **1481**.

2088. S. Bovista Fr. (from its resemblance to a *Bovista*) *a b c*.

Subsessile, globose. *Pe.* thin, pliant, breaking up irregularly above, becoming nearly smooth, yellowish. *Gl.* yellowish- or olive- to purplish-brown. *Trama* yellow. Usually furnished with a mass of Myc. at base.

Sandy ground near bushes. Sept. Diam. 2¼ in.

2089. S. Geaster Fr. (from a fancied resemblance to a *Geaster*) *a b c*.

Sessile, globose. *Pe.* thick, almost smooth, splitting and expanding in an irregularly stellate manner, dull yellowish-brown. *Gl.* umber to pale and then dark slate. *Trama* whitish. Sometimes with a mass of Myc. at base.

Sandy places, stubble fields, woods, under oaks. Aug.–Oct. Diam. 3 in.

Fam. X. NIDULARIACEÆ

Spores produced in the interior of one or more indehiscent globose or compressed peridiola or sporangia which are at first contained within a peridium. Species **2090—2098**

Key to the Genera.

Periditiola several.

Peridiola attached by a funiculus or cord to the wall of the peridium.

Peridium of three layers, cup-shaped .... 119 CYATHUS.

Peridium of two layers, outer thick and felt-like, inner thin; cup-shaped .... 120 CRUCIBULUM.

Peridiola not attached by a funiculus to the wall of the peridium. Peridium of a single membrane, globose .... 121 NIDULARIA.

Peridiolum solitary, ejected from the peridium. Peridium globose, subglobose or urceolate, sessile .... 122 SPHÆROBOLUS.

CXIX. CYATHUS Haller.

(From the cup-shape; Gr. *kuathos*, a wine cup.)

Peridium cup-shaped, composed of three distinct, closely connected layers, apex at first closed with a white membrane or epiphragm,

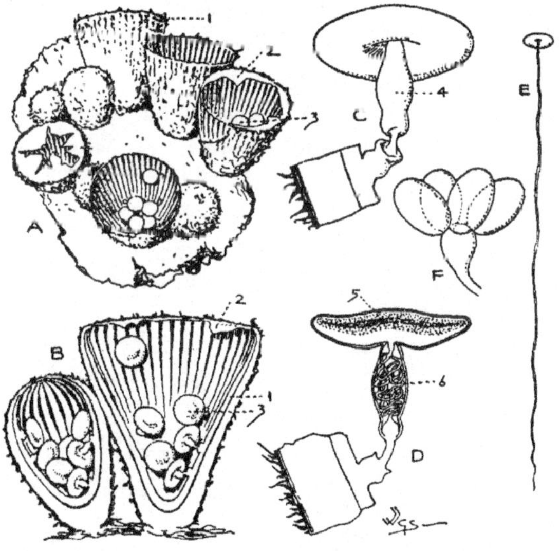

Fig. 135.—*Cyathus striatus* Hoffm.

A, group of plants. Natural size. B, section of young and mature plants. × 3. C, peridiolum; D, section of ditto. × 10. E, peridiolum with the filiform appendage (6 of fig. D) uncoiled and drawn out of the funiculus. × 3. F, basidia and spores. × 500. 1, peridium; 2, epiphragm; 3, peridiola; 4, funiculus; 5, hymenial layer, lining peridiolum; 6, filiform appendage coiled within the funiculus.

becoming broadly open. *Peridiola* compressed, umbilicate, attached to inner wall of peridium by an elastic cord or funiculus. (Fig. 135.)

The British species grow in fields and gardens, generally on chips, twigs and rotten wood; some foreign species grow on dung of horse and cow, and on bagging, matting, coffee-chaff, etc. Species **2090, 2091**

2090. C. striatus Hoffm. (from the striate inner surface of the peridium) *a b c.*

Obconical, truncate at base and apex, striate and shining lead-colour within, hirto-tomentose and ferruginous without. *Peri.* subcircular, biconvex, lead-colour. *Fun.* whitish, containing a long filiform appendage within. Springing from coarse brown Myc.

Fasciculate. Fields, woods, gardens, decaying twigs, wood, old willow, fir-cones, coco-nut fibre. Feb.–Nov. $\frac{3}{4} \times \frac{5}{8}$ in.

2091. C. vernicosus DC. (from the silky-shining outer surface of the peridium; *vernicosus*, shiny as though varnished) *a b c.*

Campanulate, undulate, tapering downwards to a narrow base, inside smooth, lead-colour or brownish, outside minutely silky, ferruginous, ochreous or greyish. *Peri.* circular, biconvex, blackish or blackish-brown. *Fun.* white.

On the ground, stubble fields, gardens, flower-pots, flower-boxes, rotten wood, chips. Feb.–Sept. $\frac{1}{4} \times \frac{1}{4}$ in.

CXX. **CRUCIBULUM** Tul.

(From a fancied resemblance to a small melting-pot, *crucibulum.*)

Peridium cup-shaped, consisting of two layers, the outer thick and felt-like, the inner a thin membrane which is at first continuous over

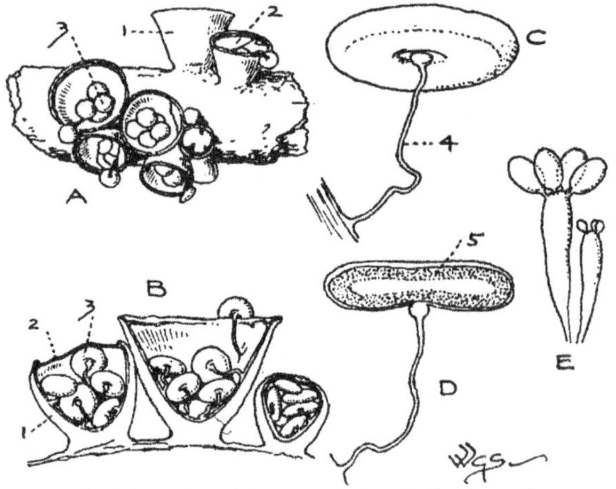

Fig. 136.—A, *Crucibulum vulgare* Tul. × $\frac{4}{5}$. B, section showing young and mature examples. × 3. C, peridiolum; D, section of ditto. × 12. E, basidia and spores. × 660. 1, peridium; 2, epiphragm; 3, peridiola; 4, funiculus; 5, hymenial layer lining peridiolum.

the apex as a flat epiphragm. *Peridiola* numerous, attached to the inner wall of the peridium by a long cord which originates from a depression in a central nipple-like tubercle on the under side of the peridiolum. (Fig. 136.)

2092. C. vulgare Tul. (from its commonness; *vulgaris*, common) *a b c*.

Pe. thick, sienna-white, cinnamon or greyish, externally minutely tomentose, internally smooth and shining. *Peri.* circular, biconvex, pale ochreous or sienna-white.

Rotten branches, planks, twigs, chips, cones, herbaceous stems, grass, rotten straw, dead fern-fronds. Oct.–Nov. $\frac{3}{8} \times \frac{1}{4}$ in.

CXXI. **NIDULARIA** Bull.

(From a fancied resemblance to a nest, *nidus*.)

Peridium globose, consisting of a single membrane, at first closed then opening by a more or less regular mouth or becoming irregularly

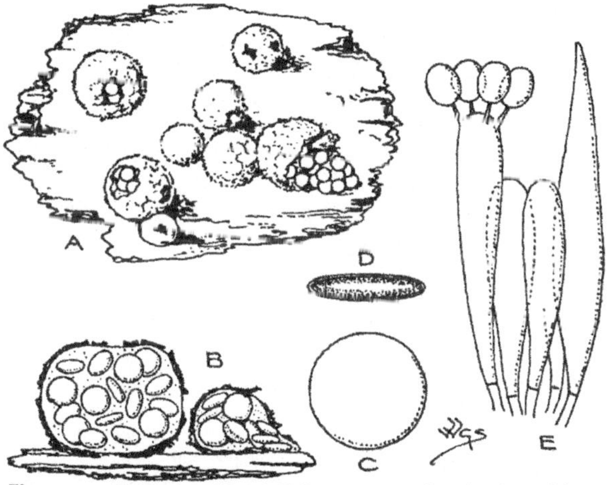

Fig. 137.—A, *Nidularia pisiformis* Tul. × $\frac{3}{2}$. B, sections showing peridiola embedded in mucus. × 3. C, D, peridiolum and section. × 12. E, basidium with spores, cystidium and the smaller cells of hymenial surface. × 660.

ruptured, containing numerous peridiola enveloped in mucus, not attached to inside by a funiculus. The mucus arises from disintegration of hyphæ. (Fig. 137.) Species **2093—2095**

2093. N. pisiformis Tul. (from the pea-like shape; *pisum*, a pea, *forma*, shape) *a b c*.

Pe. turbinate, strigose, tuberculose, splitting irregularly, cinnamon, pale buff or salmon. *Peri.* subrotund, biconvex, smooth, shining, brown.

Gregarious or solitary. Dead branches, chips, leaves, pine, sometimes on the ground, woods. May–Oct. $\frac{5}{16}$ in. Spores 7–8 × 8–9 μ. Var. *Broomei* Mass., on pine wood. Spores 10 × 4–5 μ.

2094. N. Berkeleyi Mass. (after the Rev. M. J. Berkeley) *a*.

Pe. subglobose, thick, hirto-tomentose, bright cinnamon. *Peri.* orbicular, biconvex, smooth, shining, bright brown.

Solitary or 2–3 together. Twigs, woods. $\frac{5}{16}$ in.

2095. N. confluens Fr. (from the confluent habit) *a b c*.

Pe. subglobose, thin, villous, whitish or buff-white. *Peri.* orbicular, compressed, pale yellowish-umber to chestnut.

Putrid leaves, twigs, chips, wood, rarely on the ground. Oct.

CXXII. SPHÆROBOLUS Tode.

(In reference to the ejection of the peridiolum from the peridium; Gr. *sphaira*, a ball, *ballo*, to throw.)

Peridium globose, subglobose or urceolate, sessile; *peridiolum* solitary, ejected from the peridium. (Fig. 138.) Species **2097, 2098**

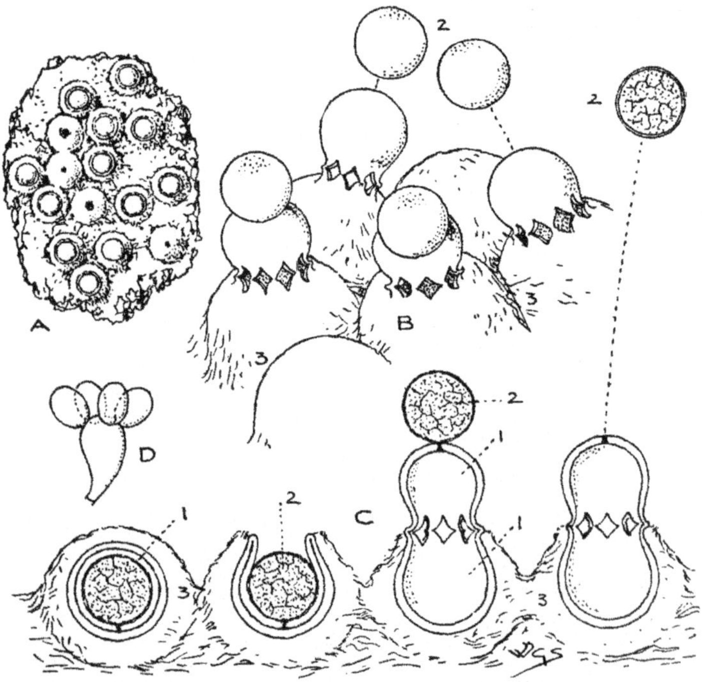

Fig. 138.—A, *Sphærobolus stellatus* Tode. × 3. B, group of ditto; C, sections of young and mature examples. × 12. D, basidium and spores. × 660. 1, peridium of two coats: 2, peridiola: 3, mycelium.

Peridium globose, of two layers, the inner at length extruded elastically and ejecting a single peridiolum. **2096, 2097**

Peridium homogeneous, sessile on a broad base, subrotund then urceolato-ventricose; *peridiolum* extruded from the apex of the peridium. **2098**

2096. S. stellatus Tode (from the stellate mouth of the peridium) *a b c*.

Pe. minutely tomentose, at first immersed in a thick web of white Myc., opening above in a 5- to 7-stellate manner, elastically extruding the inner layer and ejecting a globular yellowish or orange peri.; pale yellow or pale yellowish-buff, darker or orange within. The peri. which encloses basidia and spores is ejected for 8 or more inches with a slight clicking sound.

Usually crowded in troops. Rotten wood, sawdust, earth of old stumps, twigs, leaves, rotten grass. Often in stoves and greenhouses. Jan.–Dec. Diam. $\frac{1}{10}$ in.

2097. S. dentatus W. G. Sm. (from the toothed peridium; *dentatus*, toothed) *a*.

Pe. finely pilose with erect hairs, springing from scanty brownish Myc., opening above in a 4- to 7-stellate manner, the rays clad with long white hairs, which at first converge over the opening, outside pale livid-buff to vinous-brownish, pale brown or brown, inside white, ivory, white-greyish or faint olive-ivory, ejecting a reddish-brown peri.

Scattered. Dead elder. Failand, Bristol, Aug. 1888, R. Baker. Diam. $\frac{1}{18}$ in.

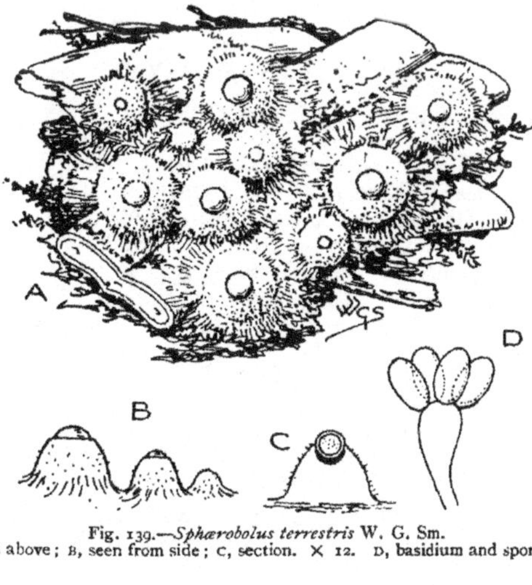

Fig. 139.—*Sphærobolus terrestris* W. G. Sm.
A, seen from above; B, seen from side; C, section. × 12. D, basidium and spores. × 660.

2098. S. terrestris W. G. Sm., *Thelebolus terrestris* A. & S.—not of Tode—(from the frequent habitat, the earth; *terrestris*, pertaining to the earth) *a b c*.

Pe. at first hemispherical, saffron-yellow honey-colour or sienna, seated on a dense tomentose Sub. of buff-white here and there brownish Myc. *Peri.* spherical, colour as Pe.

Crowded in troops. Earth, leaves, pine, moss, rotten wood, oak. Mar.-Nov. Diam. $\frac{1}{25}$ in.

FAM. XI. HYMENOGASTRACEÆ

Subterranean. Peridium indehiscent. Capillitium absent. Spores 2–7. Species **2099—2122**

KEY TO THE GENERA.

Spores globose warted.		
Sterile base of peridium well developed	123	OCTAVIANIA.
Sterile base of peridium absent	125	HYDNANGIUM.
Spores smooth.		
Sterile base of peridium absent. Spores elliptical ...	124	MELANOGASTER.
Gleba cartilagineo - glutinous. Spores minute, elliptical	126	HYSTERANGIUM.
Peridium with root-like fibres on the surface	127	RHIZOPOGON.
Spores large, elliptical or fusiform, rugulose or nodulose.		
Sterile base of peridium well developed	128	HYMENOGASTER.

CXXIII. OCTAVIANIA Vitt.

(After Dr. Vincent Ottaviani.)

Peridium continuous or cracked, cottony, running down into a well developed sterile base. *Trama* byssoid, easily divisible, cells at first empty, peripheral cells small, spherical, central cells much larger and irregular. *Spores* 2–7, globose, warted. (Fig. 140.)

Distinguished from *Hydnangium* by the presence of a sterile basal stratum and from *Melanogaster* in the cavities of the gleba being at first empty and the trama byssoid. Species **2099—2101**

2099. O. asterosperma Vitt. (from the star-shaped spores; Gr. *aster*, a star, *sperma*, a seed) *a b c*.

Globose, regular or irregular. *Pe.* continuous or cracked into areolæ, minutely downy, at first whitish-buff, then rosy-salmon, clouded umber. *Gl.* salmon, slightly buff, cells umber.

Generally near the surface, often partially exposed. Odour sweet, of *Ocymum basilicum*, sometimes of new cheese or pungent. Adhering to buried twigs and leaves. Oct. Diam. 1½ in. Changing to turquoise-blue and black where bruised. Commonly attacked by a bright orange mould, *Sepedonium chrysospermum*.

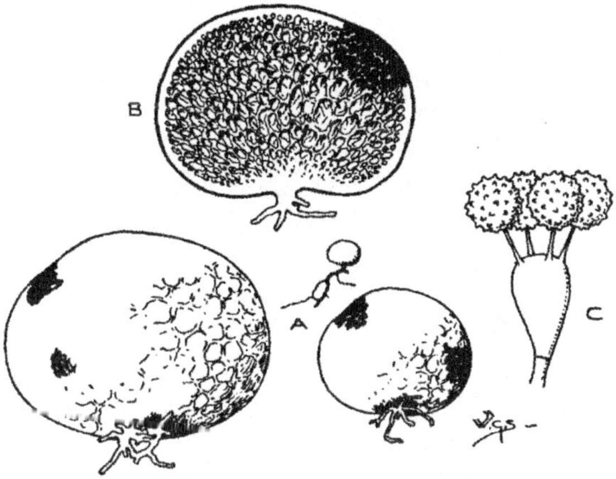

Fig 140.—A, B, *Octaviania asterosperma* Vitt., entire at different ages, and in section. Natural size. C, basidium and spores. × 500.

2100. O. Stephensii Tul. (after H. O. Stephens) *a b c*.

Irregularly globose or nodular-oblong, usually irregular in shape, base rugoso-plicate, cribrose, furnished with branched fibrous root-like growths. *Pe.* continuous, pale red to rufous, clouded darker. *Gl.* white then pale salmon, cells small, pale sienna.

Odour disagreeable like **1280**. About one-quarter of the plant is usually seen above ground, but often wholly superficial, covered with dead leaves, etc. Woods under lime-trees with *Nostoc*. Aug.–Dec. Diam. 1½ in. Exuding a white aromatic milky juice.

2101. O. compacta Tul. (from the compact substance) *a b c*.

Irregularly subglobose, minutely cottony. *Pe.* continuous or slightly granular, white to ochreous-white. *Gl.* salmon-white, cells pale salmon-buff.

Gregarious. Wholly or half-buried, involved in rooting masses of white Myc. Amongst roots, grass, earth and mycelium. Oct.–Nov. Diam. ¾ in.

CXXIV. **MELANOGASTER** Corda.

(From the black interior; Gr. *melas*, black, *gaster*, the belly.)

Peridium without a distinct base, branched fibres springing from every part of its surface. Cells of gleba more or less rounded, large in the centre, smaller towards the periphery, walls of cells thick, continuous with the peridium. *Spores* 3–5, subsessile on the basidia, smooth, usually coloured. (Fig. 141.)

The absence of a distinct rooting base, the branched fibres

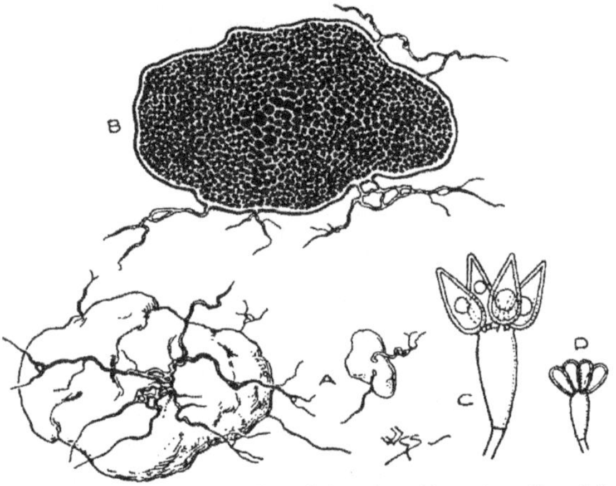

Fig. 141.—A, B, *Melanogaster ambiguus* Tul., entire and in section. Natural size. C, basidium and spores. × 500. D, basidium and spores of *M. variegatus* Tul. × 500.

traversing the whole surface of the peridium, and the smooth spores are characteristic. Species **2102, 2103**

2102. M. variegatus Tul. (from the variegated gleba) *a b c.*

Irregularly globose or nodular-oval. *Pe.* adpressedly tomentose, at first yellow or ochreous, then dark reddish-ferruginous. *Gl.* blackish with indigo-black cells. *Trama* white, whitish or bright orange.

More or less exposed above ground during growth, covered with leaves and twigs. Taste sweet and approved by many when cooked; odour sharp, strong, aromatic, agreeable, not unlike bitter almonds. Under beech, poplar, oak. June–Nov. 2¼ in. Spores 10 × 5 μ, elliptic-oblong. Eaten on the Continent, harmless but hardly pleasant. Sold half a century or more ago as an esculent in Bath under the name of Red Truffle. Var. *Broomeianus* Tul. *Trama* dull white or yellowish, never bright yellow or orange. Usually gregarious, some two or three inches beneath the surface or half exposed. Scentless when young but soon acquiring a sweet, treacley odour, or a scent of decaying pears.

2103. M. ambiguus Tul. (from its being doubtfully a true species; *ambiguus*, doubtful) *a b c.*

Irregularly globose or oval-nodular. *Pe.* nearly even, pale olive then reddish-brown. *Gl.* jet- or slate-black. *Trama* buff-white or white.

Highly fetid of assafœtida. Under beech, poplar, oak, fir. April–Oct. 1⅝ in. Spores 13–15 × 7–8 μ ovate, papillate. Var. *intermedius* Tul. has a yellowish trama which becomes red in drying.

CXXV. **HYDNANGIUM** Wallr. ex Klotzsch.

(From the tuberous form; Gr. *hudnon*, a tuber or truffle, *aggeion*, a vessel.)

Peridium fleshy or membranous, smooth or silky; sterile base not developed. *Trama* vesicular, cells minute, unequal, at first empty then filled with globose or ovate, echinulate spores. *Spores* 2–4. (Fig. 142.)

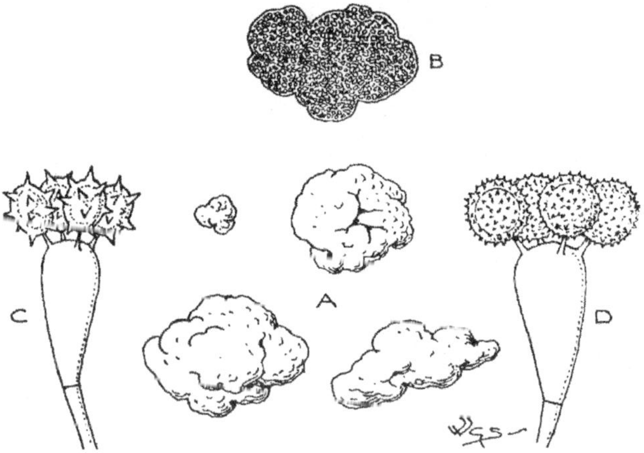

Fig. 142. A, B, *Hydnangium carotæcolor* Berk., entire and in section. × ⅓. C, basidium and spores. × 660. D, basidium and spores of *H. carneum* Wallr. × 660.

The globose or ovate, usually small spores and the absence of a sterile base separate *Hydnangium* from its nearest ally *Octaviania*.

Species **2104, 2105**

2104. H. carotæcolor Berk. (from the carrot-like colour; *carota*, carrot) *a b c.*

Globoso-oblong or oblong-rounded, nodular, rootless. *Pe.* thin, slightly tomentose, carrot-red, becoming dark orange-brownish when dry. *Gl.* orange.

Sometimes half-exposed, often superficial. Woods, downs, under trees. Aug.–Jan. Diam. 1⅛ in.

2105. H. carneum Wallr. ex Klotzsch (from the flesh-colour; *carneus*, of flesh) *a b c.*

Irregularly subglobose or rounded-elongate, rootless; *Pe.* very thin, at first covered with white fugacious down, flesh-colour, becoming ochreous-salmon when dry.

Subterranean or half-exposed, often superficial. Above the roots of *Eucalyptus*, etc. Oct.–Dec. Diam. ⅝ in.

CXXVI. **HYSTERANGIUM** Vitt.

(From the shape; Gr. *hustera*, the womb, *aggeion*, a vessel.)

Peridium more or less separating from the gleba in drying. *Gleba* cartilagineo-glutinous, cavities at first empty. *Spores* 2–3, seldom more, smooth. (Fig. 143.)

Characterised by the elongate elliptical smooth spores and the cartilaginous-mucous consistency of the gleba, which resembles

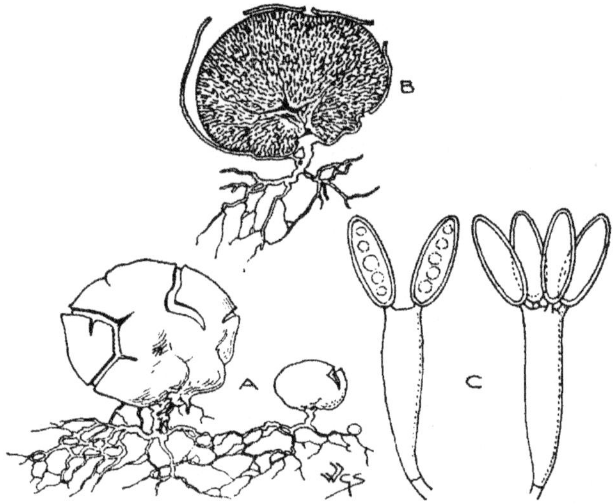

Fig. 143.—A, B, *Hysterangium nephriticum* Berk., entire and in section. × $\frac{4}{3}$. C, basidia and spores. × 660.

that of *Ithyphallus* and *Clathrus* in the young state. There is frequently a copious development of mycelium. Species **2106, 2107**

2106. H. nephriticum Berk. (from a fancied resemblance to a kidney when seen in section; Gr. *nephros*, the kidneys) *a b c.*

Irregularly globose, more or less indented. *Pe.* somewhat thick, firm, elastic, tomentose, white, becoming pale rufous when bruised. *Gl.* pale blue or grey, here and there greenish, cells irregular, minute, radiating from the base.

Commonly 8–9 in. deep in the ground. Odour at first slight, then like *Hypericum*, at last like decaying puff-balls. Woods. Sept.–Feb. ¾ in.

2107. H. Thwaitesii B. & Br. (after George Henry Kendrick Thwaites) *a b c.*

Irregularly globose, more or less indented. *Pe.* membranous, slightly silky, white, becoming rufous when bruised.

Commonly 8–9 in. deep in the ground. Woods. Aug.–Oct. ⅞ in.

CXXVII. **RHIZOPOGON** Fr.

(From the tuberous growth with the attached fibres; Gr. *rhiza*, a root, *pogon*, a beard.)

Peridium thick, subcoriaceous or submembranous, continuous or cracked and somewhat evanescent, with vein-like branched fibres

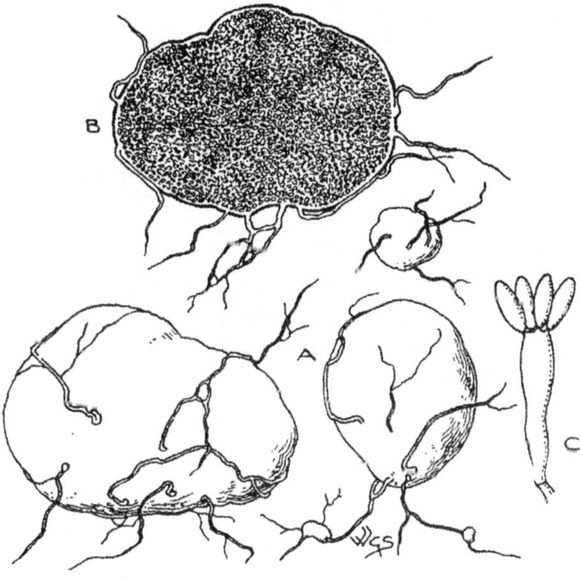

Fig. 144.—A, B, *Rhizopogon rubescens* Tul., entire and in section. × $\frac{4}{3}$. C, basidium and spores. × 660.

traversing and partially adnate to the surface; substance of gleba firm, cavities at first empty. *Spores* 2–7. (Fig. 144.)

Species **2108**, **2109**

2108. R. rubescens Tul. (from its becoming vinous rose-colour when bruised; *rubescens*, becoming red) *a b c*.

Irregularly subglobose or ovate. *Pe.* silky, white, when mature yellow, olive or warm-brown. *Gl.* at first white, then yellowish, greenish or brownish.

Somewhat superficial but often some inches deep. Odour acid, when young like sour ham, at length nauseous-fetid. Sandy fir-woods, plantations in very light peaty or fibrous soil, amongst pines. Oct.–Dec. Diam. $1\frac{3}{8}$ in. Var. *Berkeleyanus* W. G. Sm., *Melanogaster Berkeleyanus* Br., smaller and highly fetid.

2109. R. luteolus Fr. (from the dull yellowish colour when half-mature; *luteolus*, yellowish) *a b c*.

Irregularly subglobose or ovate. *Pe.* thick, subcoriaceous, whitish to dull yellow and olive- or nut-brown. *Gleba* black with white septa.

Solitary or gregarious. Subterranean or partly exposed. Taste insipid; odour at first weak then strong and offensive. Sandy fir-woods, under beech. Sept.-Oct. Diam. 1⅝ in.

CXXVIII. **HYMENOGASTER** Vitt.

(From the usually thin peridium; Gr. *humen*, a membrane, *gaster*, the belly.)

Globose or irregular. *Peridium* fleshy or thin, simple, homogeneous, running down into a sterile base, except **2116**. Cavities of gleba at first empty, radiating from the base or irregularly scattered.

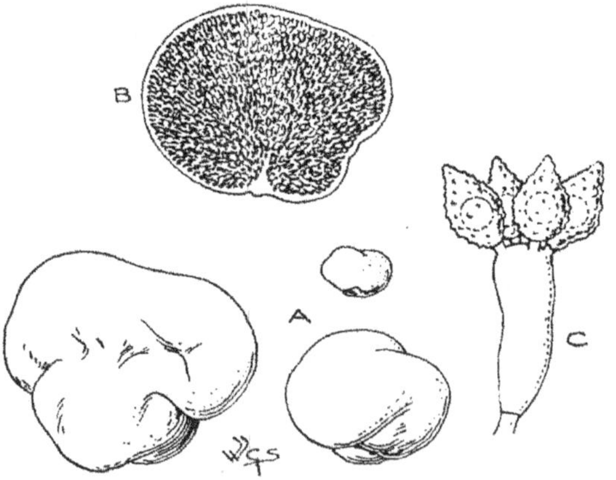

Fig. 145.—A, B, *Hymenogaster tener* Berk., entire and in section. × ⅓. C, basidium and spores. × 660.

Trama composed of elongate cells, not of byssoid flocci, not easily separable. *Spores* 1–4. (Fig. 145.)

All the species are nearly superficial in growth, none are edible.

Species **2110—2122**

2110. H. Klotzschii Tul. (after Johann Friedrich Klotzsch) *a b c*.

Irregularly globose to kidney-shaped. *Pe.* adpressedly downy, fibrillose at base, dull white becoming yellowish. *Gl.* pallid, becoming rufous-ochre.

Sandy soil. Dec. Diam. ⅝ in.

2111. H. muticus B. & Br. (from the absence of a base; *muticus*, curtailed, blunt) *a b.*

Globose, scarcely lobed. *Pe.* at length much cracked, white then brownish. *Gl.* almost destitute of a sterile base, somewhat firm and dry, cells loose, pale yellowish-brown.

Odour slight. Nov. Diam. $\frac{5}{8}$ in.

2112. H. luteus Vitt. (from the yellow gleba; *luteus*, yellow) *a b c.*

Irregularly globose or ovate, sometimes deformed. *Pe.* very thin, soft, silky, white to brownish. *Gl.* bright yellow.

Almost scentless or slightly sweet, becoming fetid. Sept.–Mar. Diam. 1 in.

2113. H. lycoperdineus Vitt. (from the shape, like a small puff-ball, *Lycoperdon*) *a b c.*

Subglobose, somewhat deformed, varied in form, plicate at base. *Pe.* smooth, white then pale whitish-salmon to brownish. *Gl.* soft, elastic, with somewhat large, irregular cavities, pallid to pale fuliginous.

Gregarious. Odour strong, onion-like, unpleasant when dry. In earth and clay Nov. Diam. 1$\frac{3}{8}$ in.

2114. H. decorus Bull. (*decorus*, seemly) *a b c.*

Irregularly globose or oval, sometimes slightly deformed. *Pe.* dull white becoming yellowish in places. *Gl.* lilac-brown at length blackish.

Woods, tan-pits. Oct.–Nov. Diam. 1$\frac{3}{8}$ in.

2115. H. vulgaris Tul. ex B. & Br. (from its commonness; *vulgaris*, common) *a b c.*

Subrotund, irregular, often indented, somewhat soft. *Pe.* white, becoming discoloured. *Gl.* with rather large irregular cavities, dull white then dark brown.

Gregarious or subcæspitose. Loose soil. July–Oct. Diam. $\frac{5}{8}$ in.

2116. H. pallidus B. & Br. (from the pallid peridium) *a b.*

Rounded, depressed. *Pe.* nearly smooth, somewhat soft, white, then dull tan-colour. *Gl.* without a sterile base, white, yellow, then pale brown.

Odour weak. Fir plantations. Oct. Diam. $\frac{3}{8}$ in.

2117. H. citrinus Vitt. (from the colour; *citreus*, lemon-yellow) *a b c.*

Irregularly globose or ovate, lobed, indented, deformed. *Pe.* shining as if silky, greenish-lemon when young, then lemon, golden-yellow, bright brassy-yellow or sulphur-cream, becoming rufous-black.

Used as food in past times by country folk in the truffle districts of England. Odour, which communicates itself to anything near, strong of cheese. May–Dec. Diam. 1$\frac{1}{4}$ in.

2118. H. olivaceus Vitt. (from the colour of the gleba, like a ripe olive) *a b c.*

Irregularly globose or ovate, lobed, indented, deformed. *Pe.* at first silky, whitish or faintly lemon, becoming rufescent when touched. *Gl.* whitish, passing through buff to rufous olive. *Trama* white.

Odour spicy. Woods. May–Nov. Diam. 1¼ in. A form between this and 2117 has been named var. *modestus* B. & Br. This is *H. populetorum* B. & Br. non Tul.

2119. **H. tener** Berk. (from the tender substance; *tener*, tender) *a b c*. Globose, irregular, more or less indented, silky or obscurely scaly, soft, tender. *Pe.* thin, dull white. *Gl.* with a conspicuous sterile base, pale rose, then greyish-brown.

Odour strong, pungent, like the odour of acrid *Lactarii*, in decay like sewer-gas. Woods, under oak-leaves. Sept.–Jan. Diam. 1 in. This is *H. lilacinus* B. & Br. non Vitt.

2120. **H. Thwaitesii** B. & Br. (after George Henry Kendrick Thwaites) *a c*.
Irregularly ovate or elongate to globose, firm. *Pe.* dull white, here and there discoloured. *Gl.* brown.

Sept.–Oct. Diam. ½ in.

2121. **H. griseus** Vitt. (from the grey gleba; *griseus*, grey) *a*. Globose, somewhat irregular or depressed. *Pe.* brown, at first covered with evanescent whitish down. *Gl.* with minute grey cells which become mottled-blackish.

Odour sweet of lily-of-the-valley. In leaf soil. Diam. ½ in.

2122. **H. pusillus** B. & Br. (from the small size; *pusillus*, small) *a b*. Obovate or subdepressed, white. *Gl.* with a comparatively large sterile base, cells comparatively large, dull white.

Almost scentless. Mossy ground. Oct. Diam. ⅛ in. Yellowish-brown when dry and then closely resembling *Sclerotium complanatum*.

RECENT ADDITIONS

52a. **Lepiota serena** Gill. (*serenus*, clear, bright). White. *P.* campanulate, glabrous; mid. sometimes slightly darkened. *St.* slender, subbulbous. *A.* deciduous.

Amongst grass. Carlisle, Miss Decima Graham, Sept. 1908. 1⅛ × 2⅜ × ⅛ in.

121a. **Tricholoma carneolum** Karst. (from the colour, *caro*, flesh). *P.* hemispherical to convexo-plane, pale flesh-colour. *St.* partly stuffed, rigid, subpruinose. *G.* finally very broad, very crowded, white.

Grassy places in woods. Yorkshire, Sept. 1908. 1½ × 1¼ × ⅛ in.

154b. **Clitocybe obscurata** Cooke in litt. (*obscurus*, obscure). *P.* plane, then infundibuliform, moist, greyish-umber, paler towards marg. *St.* solid, smooth or slightly striate, slightly attenuate downwards and never clavate, colour as P. *G.* decurrent, subdistant, white.

Odour of meal. Amongst grass and dead leaves in company with **154**. Carlisle, Miss Decima Graham, Sept. 1908. 2 × 1½ × ¼ in.

187a. Clitocybe connata Gill. (from the connate stems at the tuberous base). White.

P. fleshy, thin, conical, then expanded with an evanescent umbo, unequal, naked. *St.* hollow, thick, smooth. *G.* unequally decurrent, narrow, crowded, subarcuate.

Moist woods. Carlisle, Miss Decima Graham, Sept. 1908. 2¼ × 3 × ½ in.

262. Collybia dryophila Quél. Var. *alvearis* Cooke in litt. Large, coarse.

P. shaped like a bee-hive, rufescent at mid., 3 in. diam. *St.* very hollow.

Amongst dead leaves. Carlisle, Miss Decima Graham, Sept. 1908.

593a. Pholiota sororia Karst. (from its relationship to **593**, *soror*, a sister).

P. fleshy, convexo-expanded, striate, squamulose, fulvous-cinnamon. *St.* fistulose, equal, flexuous, with white fibrillose sq., apex furfuraceous. *G.* sinuato-adnate, crowded, pallid-cinnamon.

On wood-shavings. Yorkshire, Sept. 1908.

640a. Inocybe commixta Bres. (from being confused with **640**, *commixtus*, mixed together).

P. fleshy, conico-campanulate, expanso-umbonate, white or white-cinereous, silky-fibrillose, dry; marg. often fissile. *St.* solid, white, slightly pruinose, apex furfuraceous, base turbinato-bulbous. *G.* crowded, free, white to grey-cinnamon.

Pine-woods. Yorkshire, Sept. 1908.

GENERA OMITTED

Cenococcum Fr., one of the *Tuberaceæ*.

Pilacre Fr., placed by Lindau (in Engler and Prantl, *P. flanzenfamilien*, vol. i. pt. 1**, p. 86) near *Auricularia*, is of doubtful position.

Polyangium Link, an insect production.

Thelebolus Tode, one of the Ascomycetes.

CORRIGENDA

Page 110, No. **505a**, *for* **C. Sarnicus** *read* **C. sarnicus.**
,, 171, No. **788**, *for* **P. Elvensis** *read* **P. elvensis.**
,, 317, after **BOLETUS** *for* Dill. *read* L.
,, 333, after **POLYPORUS** *for* Mich. *read* Haller.
,, 344, No. **1563**, *after* **fumosus** *add* Fr.
,, 368, after **MERULIUS** *for* Halle *read* Haller.
,, 436, No. **1965a**, *for* **tenerima** *read* **tenerrima.**
,, 449, after **TREMELLA** *for* Dill. *read* L.
,, 462, after **CLATHRUS** *for* Mich. *read* L.
,, 467, after **GEASTER** *for* Mich. *read* Pers.

GLOSSARY

Adnate (*adnascor*, to grow to), of gills broadly attached to the stem.

Adnexed (*ad*, to, *necto*, to join), of gills slightly joined to the stem.

Æcidia (Gr. *oikidion*, a little house), small cup-shaped fungi.

Æruginous (*ærugo*, rust of copper), the blue-green colour of verdigris.

Æstival (*æstivalis*), belonging or peculiar to summer.

Agglutinate (*agglutino*, to glue), glued to a surface.

Algal (*alga*, a sea-weed), relating to *Algæ*, a group to which belong the sea-weeds and many fresh-water plants, including microscopic green plants which grow in water and damp places.

Alveolate (*alveus*, a hollow), in the form of small depressions like a shallow honeycomb.

Amphigenous (Gr. *amphi*, around, *gennao*, to beget), not confined to one surface, all round, as in the hymenium of *Clavaria*.

Anastomosing (Gr. *anastomosis*, a bringing to a point), united by running together irregularly.

Annular, in the form of a ring.

Annulate, bearing a ring on the stem.

Annulus (*annulus*, a ring), the ring round the stem.

Apiculate (*apex*, the top), terminating in a small point.

Appendiculate (*appendicula*, a small appendage), hanging in small fragments.

Applanate (*ad*, to, *planatus*, made flat), flattened.

Approximate (*ad*, to, *proximo*, to approach), of gills which approach but do not touch the stem.

Arachnoid (Gr. *arachne*, a spider's web, *eidos*, resemblance), like a cobweb.

Arcuate (*arcuatus*, bent like a bow), of gills, bow-shaped.

Areolate (*area*, a plot, a space), divided into small areas or patches.

Ascus (Gr. *askos*, a wine-skin), the swollen end of a hyphal branch, in which spores are borne in *Ascomycetes*.

Atomate, sprinkled with small particles.

Basidium (*basidium*, a little pedestal), the mother-cell which bears the spores in *Hymenomycetes* and *Gasteromycetes*.

Bifid (*bifidus*, twice-cleft), divided half-way into two.

Bullate (*bulla*, a bubble), furnished with a boss or stud.

Byssoid (*byssus*, fine flax), composed of fine threads.

Cæspitose (*cæspes*, a clump), growing in tufts.

Calyptra (*calyptra*, a hood), any extinguisher-shaped covering.

Campanulate (*campana*, a bell), bell-shaped.

Cancellate (*cancellatus*), latticed, as in *Clathrus*.

Capillitium (*capillus*, a hair), the dense mass of sterile fibres mixed with the spores in the gleba of the *Gasteromycetes*.

Cerebriform (*cerebrum*, the brain), having an irregular brain-like appearance, like the kernel of a walnut.

Ciliate (*cilium*, an eye-lash), fringed with hairs.

Cinereous (*cinis*, ashes), ash-coloured.

Cingulate (*cingulum*, a girdle), bearing a ring or annulus on the stem.

Clavate (*clava*, a club), club-shaped.

Columella (*columella*, a small pillar), the small columnar base of barren tissue in *Lycoperdon*, *Geaster*, etc.

Conceptacle (*conceptaculum*, a receptacle), a hollow case enclosing reproductive bodies.

Conchiform (*concha*, a shell), shell-shaped.

Concrete (*concretus*, growing together), as when the scales adhere to the flesh of the pileus.

Conidia (Gr. *konis*, dust), simple asexual spores.

Connate (*con*, together, *nascor*, to be born), united by growing together.

Connivent (*conniveo*, to wink), arching over to meet.

Coriaceous (*corium*, leather), leathery.

Cortex (bark or rind), the periderm in the *Gasteromycetes*.

Corticate, with a bark-like covering.

Cortina (*cortina*, a veil), a veil of spider-web structure as the veil in *Cortinarius*.

Cortinate, furnished with a veil.

Costate (*costatus*, ribbed), ribbed or veined.

Crenate (*crena*, a notch), with rounded notches on the margin.

Crenulate, with small rounded notches.

Cribrose (*cribrum*, a sieve), pierced with many holes.

Cupulate (*cupula*, a little cup), cup-shaped.

Cuspidate (*cuspis*, a point, spear), with a sharp spear-like point.

Cuticle (*cuticula*, the outer skin), the outermost skin.

Cyathiform (Gr. *kuathos*, a wine-cup), cup-shaped.

Decumbent (*decumbo*, to recline), reclining but with summit ascending.

Decurrent (*decurro*, to run down), of gills and tubes running down the stem in the *Agaricaceæ*, *Polyporaceæ*, etc.

Dehiscing (*dehisco*, to yawn), opening when ripe.

Deliquescent (*deliquescens*, melting away), becoming fluid when mature.

Denticulate (*denticulatus*, minutely toothed), as in the minute teeth on the edge of the gills in certain *Agaricaceæ*.

Depressed (*depressus*, sunk down), slightly hollowed.

Determinate (*determino*, to limit), ending definitely, as in the margin of some species of *Corticium*, etc., not powdered or clouded off indefinitely.

Diaphanous (Gr. *dia*, through, *phaino*, to appear), more or less transparent.

Dichotomous (Gr. *dichotomeo*, to cut in two), divided into two, forked.

Dimidiate (*dimidio*, to halve), as when one part of an organ is smaller than the other as in the pilei of many fungi.

Disciform (*discus*, a quoit), circular and flat.

Dissepiments (*dissepimentum*, a partition), dividing walls.

Distant, of gills in the *Agaricaceæ* with a wide distance between them.

Divaricate (*divarico*, to spread asunder), very divergent, separating at an obtuse angle.

Echinulate (*echinus*, a hedgehog), with short bristles.

Effused (*effundo*, to pour out), spread over without regular form.

Emarginate (*e*, out of, *margo*, the margin), of gills with a sudden curve as if scooped out at the point of attachment to the stem.

Endoperidium (Gr. *endon*, within), the inner layer of the peridium.

Endophyte, Entophyte (Gr. *entos*, within, *phuton*, a plant), growing within another plant.

Epiphyte (Gr. *epi*, upon, *phuton*, a plant), growing upon another plant.

Erumpent (*erumpens*, breaking through), bursting through the epidermis.

Evanescent (*evanescens*, disappearing), lasting only a short time.

Excipulus (*excipula*, a basin), the rim round the base of the apothecium.

Exoperidium (Gr. *exo*, without), the outer layer of the peridium.

Farinose (*farina*, meal), mealy.

Fasciate (*fascia*, a band or girth), zoned as with coloured bands.

Fasciculate (*fascis*, a bundle), growing in small bundles.

Fastigiate (*fastigium*, a slope or gable), with branches parallel, clustered and erect, sometimes decreasing in height outwards and giving a gable shape.

Ferruginous (*ferrum*, iron), of the colour of iron-rust.

Fibrillose (*fibra*, a fibre), clothed with small fibres.

Filiform (*filum*, a thread), thread-like.

Fimbriate (*fimbria*, a fringe), fringed.

Fissile (*fissilis*, tending to split), capable of being split.

Fistulose (*fistula*, a pipe), hollow like a pipe.

Flocci (*floccus*, a lock of wool), woolly locks.

Floccose, woolly.

Flocculent, diminutive of floccose.

Flocculose, covered with small flocci.

Free, of gills which reach the stem, but are not attached to it.

Fuliginous (*fuligineus*, sooty), soot-coloured.

Funiculus (*funis*, a rope), a cord of hyphæ in *Nidularia* which attaches the peridiolum to the inner surface of the wall of the peridium.

Furcate (*furcatus*, forked), as in the branching gills common in *Russula*.

Furfuraceous (*furfur*, bran), with branny scales or scurf.

Fuscous (*fuscus*, dusky), of a dingy brownish colour.

Fusiform (*fusus*, a spindle), somewhat thick but tapering towards each end.

Fusoid, somewhat spindle-shaped.

Gibbous (*gibbus*, hump-backed), of a pileus with a swelling or broad umbo, or convex above and flat beneath.

Gills, the plates of an agaric on which the hymenium is situated.

Glabrous (*glaber*, without hair), not bearing hairs.

Glaucous (Gr. *glaukos*, bluish-grey), covered with a bluish-grey bloom.

Gleba (*gleba*, a clod), the contents of the peridium in the *Gasteromycetes*.

Greaved, of a stem clothed as if with a greave.

Gregarious (*gregarius*, belonging to a flock), growing in company, but not cæspitose.

Grumous (*grumus*, a little heap), clotted; of flesh composed of little clustered grains.

Guttate (*gutta*, a tear), with tear-like spots.

Gyrose (Gr. *guros*, round), curved backward and forward in turn.

Habit (*habitus*, appearance), the general appearance.

Habitat (*habitatio*, dwelling), the locality.

Heterogeneous (Gr. *heteros*, one of two, *gennao*, to beget), not uniform in structure.

Hispid (*hispidus*, bristly), beset with rough hairs or bristles.

Homogeneous (Gr. *homos*, one and the same, *gennao*, to beget), uniform in structure.

Hyaline (*hualos*, clear), colourless, translucent.

Hybernaculum, hibernaculum (*hiberno*, to winter), applied to forms in which certain fungi pass the winter, sclerotia.

Hygrometric (Gr. *hugros*, moist, *metron*, a measure), influenced by moisture.

Hygrophanous (Gr. *hugros*, moist, *phaino*, to appear), of a watery appearance when moist.

Hymenium (Gr. *humen*, a membrane), the spore-bearing surface.

Hymenophore (Gr. *humen*, a membrane, *phero*, to bear), the structure which bears the hymenium, as the under surface of the pileus to which the gills are attached in *Agaricaceæ*, the sporophore.

Hypertrophy (Gr. *huper*, above, *trophe*, food), an abnormal enlargement of an organ by disease.

Hypha (Gr. *huphe*, a web), pl. **hyphæ**, the thread-like element of which a fungus is made up.

Imbricate (*imbricatus*, covered with tiles), overlapping as the tiles on a roof.

Immarginate (*in*, negative, *margo*, a margin), without a distinct border.

Immersed, sunk in the matrix.

Inferior (*inferior*, lower), of the annulus on the stem, far down.

Infundibuliform (*infundibulum*, a funnel), funnel-shaped.

Innate (*in*, into, *nascor*, to be born), adhering by growing into.

Involute (*in*, in, *volvo*, to roll), rolled inwards.

Isabelline or Isabella colour, dirty-tawny, as of soiled linen.

Laciniate (*lacinia*, a lappet, flap), divided into flaps.

Lacunæ (*lacuna*, a hole or cavity), holes or depressions.

Lacunose, marked with small hollows.

Lamellæ (*lamella*, a thin plate), the gills in *Agaricaceæ*.

Lanceolate (*lancea*, a lance or spear), lance-shaped; narrow and tapering to both ends.

Linear (*linea*, a line), narrow and straight, several times longer than wide.

Loculus (*loculus*, a little compartment), a cell of a septate spore.

Mammiform (*mamma*, a breast), breast-shaped.

Marginate (*margo*, a margin), having a distinct border.

Matrix (*matrix*, the womb), the substance on which a fungus grows.

Medial (*medium*, the middle), of the annulus, when at the middle of the stem.

Merismoid (Gr. *merizo*, to divide, *eidos*, form), having a branched or laciniate pileus.

Merulioid, resembling the hymenium of *Merulius*.

Mitrate (Gr. *mitra*, a head-dress), with thick rounded pileus.

Mucedinous (*mucedus*, mouldy), of the nature of the group *Mucedines*, naked-spored moulds.

Multifid (*multifidus*, many cleft), divided half-way into many lobes.

Multipartite (*multipartitus*, many times divided), divided many times and more deeply than multifid.

Muricate (*murex*, a sharp stone, a bit with sharp spikes, a spiked ball), rough with short hard tubercular excrescences.

Muriform (*murus*, a wall), with cells resembling bricks in a wall.

Mycelium (Gr. *mukes*, a fungus), spawn of fungi, the vegetative portion, composed of hyphæ.

Mycology (Gr. *mukes*, a fungus, *logos*, a discourse), the study of fungi.

Nuclear (*nucleus*, a kernel), pertaining to a kernel-like growth as in *Næmatelia*.

Obconic (*ob*, inversely, *conus*, a cone), inversely conical.

Obovate (*ob*, inversely, *ovum*, an egg), inversely egg-shaped.

Obtuse (*obtusus*), blunt or rounded at the end.

Ochreous (*ochra*, yellow-earth), yellow, not red-ochre.

Operculum (*operculum*, a lid), a cover.

Ostiolum (*ostiolum*, a little door), the aperture through which the spores escape in the *Gasteromycetes*.

Ovate (*ovum*, an egg), egg-shaped.

Papilla (*papilla*, a nipple), a nipple-like elevation.

Papillate, Papillose, with nipple-like elevations.

Paraphyses (Gr. *para*, beside, *phusis*, growth), sterile filaments in a hymenium.

Partial, of a veil, clothing the stem and reaching to the edge of the pileus, but not extending beyond it.

Patelliform (*patella*, a small dish), dish-shaped.

Pectinate (*pecten*, a comb), like the teeth of a comb.

Pedicel (*pediculus*, a small foot), a support or stalk, as the small column which supports the inner peridium in some species of *Geaster*.

Pedicellate, born on a pedicel.

Pellicle (*pellicula*, a small skin), a delicate superficial membrane.

Pelliculose, furnished with a pellicle or delicate distinct skin.

Peltate (*pelta*, a target), target-shaped.

Pencilled, with pencil-like hairs either on the tip or border.

Penicillate (*penicillum*, a little brush), pencil-shaped.

Peridiola, diminutive of peridium.

Peridium (Gr. *peridion*, a little pouch), the outer enveloping coat enclosing the gleba in the *Gasteromycetes*.

Peristome (Gr. *peri*, about, *stoma*, a mouth), the fringe or its homologue round the orifice of the peridium in *Geaster*, *Tulostoma*, etc., amongst the *Gasteromycetes*.

Perithecium (Gr. *peri*, about, *theke*, a case), a receptacle enclosing spores.

Peronate (*pero*, a kind of high boot), sheathed, *e.g.* of a stem which has a woolly covering like a legged boot.

Pervious (*pervius*, passable), having an open passage-way; hollow to the base, as in some stems.

Pezizoid (*eidos*, resemblance), cup-shaped, like a *Peziza*, a genus of *Ascomycetes*.

Pileate, with a cap.

Pileoli, secondary pilei, arising from the division of a primary pileus.

Pileus (*pileus*, a cap), a part of the receptacle of a fungus, *e.g.* the cap-like head in *Agaricaceæ*.

Pilose (*pilus*, a hair), covered with hairs.

Plane (*planus*), level, even, flat.

Plasmodium (Gr. *plasma*, formed substance), the naked nucleated protoplasm of the *Mycetozoa*.

Plicate (*plico*, to fold), folded into plaits.

Plumose (*plumosus*, feathered), downy.

Polymorphic (Gr. *polus*, many, *morphe*, a change), with several or various forms.

Poriform, in the form of pores.

Porous, furnished with pores.

Potential (*potentia*, a force), existing in possibility, rather than in actuality.

Proliferous (*proles*, offspring, *fero*, to bear), applied to an organ which gives rise to secondary organs of the same kind.

Promycelium (*pro*, for, *mycelium*, the vegetative portion of the thallus composed of hyphæ), the short-lived product of germination of a spore which bears spores of a different nature from the mother-spore.

Protoplasm (Gr. *protos*, first, *plasma*, formed substance), the living substance of plants and animals.

Pruinose (*pruina*, hoar-frost), covered with frost-like bloom.

Pruniform (*prunus*, a plum), plum-shaped.

Pubescent (*puber*, downy), slightly hairy.

Pulverulent (*pulverulentus*, dusty), powdered as if dusted over.

Pulvinate (*pulvinus*, a cushion), cushion-shaped.

Punctate (*punctus*, a point), dotted with points.

Pyriform (*pyrus*, a pear), pear-shaped.

Racemose (*racemus*, a bunch of grapes), borne in a bunch.

Radiate (*radius*, the spoke of a wheel), spreading from a centre.

Receptacle (*receptaculum*, a reservoir), an axis bearing one or more organs, as the stem upon which the hymenium is elevated in the *Phalloidaceæ*.

Remote, of gills which do not reach the stem, but leave a free space round it.

Reniform (*renes*, the kidneys), kidney-shaped.

Repand (*repando*, to throw open), bent backwards.

Resupinate (*resupino*, to throw on the back), with hymenium upwards.

Reticulate (*rete*, a net), netted.

Revolute (*revolvo*, to roll back), rolled backwards; of the margin of a pileus—the opposite of involute.

Rhizomorphoid, rhizomorphous (Gr. *rhiza*, a root, *morphe*, shape), like a rhizomorph, a root-like branched strand of mycelial hyphæ.

Rimose (*rima*, a crack), cracked.

Rimulose, diminutive of rimose.

Ring, a part of the veil adhering to the stem in the form of a ring or annulus.

Rivulose (*rivula*, a small stream), marked with lines like rivulets.

Rounded, applied to gills and tubes, rounded near point of contact with the stem.

Rubiginous (*rubigo*, rust), colour of iron-rust.

Rufescent (*rufescens*), becoming reddish.

Rufous (*rufus*), reddish.

Rugose (*ruga*, a wrinkle), wrinkled.

Rugulose, somewhat wrinkled.

Scabrous (*scaber*, rough), rough on the surface.

Scales, applied to the broken up epidermis of the pileus and stem.

Scarious (Gr. *skaros*, a kind of sea-fish), applied to a shrivelled membrane.

Scissile (*scindo*, to cleave), capable of being cleft asunder; of two plates, *e.g.* lying together, but capable of being separated.

Sclerotioid, sclerotium-like.

Sclerotium (Gr. *skleros*, hard), a compact mass of hyphæ in a dormant state.

Scrobiculate (*scrobis*, a trench), marked with small pits.

Separating, becoming detached, as gills from a stem, or resupinate fungi from the matrix.

Septate, divided.

Septum (*septum*, a hedge), a division.

Serrate (*serra*, a saw), toothed like a saw.

Serrulate, minutely toothed.

Sessile (*sedeo*, to sit), seated without a stem.

Setæ (*seta*, a bristle), bristles.

Setulose, resembling a fine bristle.

Simple, applied to a stem without ring or volva.

Sinuate, waved; of a pileus with a wavy or curved margin; of gills with a sudden wave, curve or sinus where they reach the stem.

Sinus (*sinus*), a curve.

Smooth, applied to a surface which is destitute of hairs or other protuberances.

Spathulate (Gr. *spathe*, a spatula), spoon-like.

Spawn, see **Mycelium**.

Sphæriæ, small spherical black fungi common on branches.

Spiculæ (*spiculum*, a small needle), fine erect acutely pointed spines.

Spinulose (*spina*, a thorn), bearing fine spines.

Sporangium (Gr. *spora*, a seed, *aggeion*, a vessel), a spore-bearing receptacle.

Spores (Gr. *spora*, a seed), the analogues of seeds of flowering plants.

Sporidia, diminutive of spores—small spores.

Sporophore (Gr. *spora*, a seed, *phoreo*, to carry), the part which bears the hymenial surface.

Squamose (*squama*, a scale), scaly.

Squamule, a small scale.

Squamulose, covered with small scales.

Squarrose (*squarrosus*, rough, scurfy), rough with scales.

Sterigma (Gr. *sterigma*, a prop), pl. **sterigmata**, the slender thread which carries a spore.

Stipes (*stipes*), a stalk.

Stipitate, stemmed.

Stoloniferous (*stolo*, a short sucker or runner, *fero*, to bear), stolon-bearing.

Stoma (Gr. *stoma*, a mouth), the mouth through which the spores escape in certain fungi, as *Geaster*, *Tulostoma*, etc.

Striate (*stria*, a furrow, flute of a column), marked with lines in the form of channels.

Strigæ (*striga*, a swathe), small straight hair-like scales.

Strigose, rough with sharp-pointed hairs, hispid.

Stroma (Gr. *stroma*, a mattress), a cushion-like body in which the perithecia of some fungi are immersed.

Stuffed, of a stem filled with substance of a different texture from its walls.

Subiculum (*subiculum*, an under layer), the felted or byssoid mycelium of a resupinate fungus, forming the under layer bearing the hymenium.

Sulcate (*sulcus*, a furrow), marked with grooves.

Superior (*super*, above), applied to the annulus when it is near the apex of the stem.

Terete (*teres*, rounded), circular in transverse section.

Tessellate (*tessellatus*, squared stones), checquered.

Tetrasporous (Gr. *tetras*, four, *spora*, a seed), bearing four spores.

Thallus (Gr. *thallos*, a sprout), the whole of the body which does not serve for reproduction.

Tomentose (*tomentum*, wool, etc.), downy.

Trama (*trama*, the weft or filling of a web), the substance between the hymenial layers in *Agaricaceæ* and *Polyporaceæ*.

Tremelloid (*tremo*, to tremble), of the consistency of jelly.

Truncate (*trunco*, to maim), ending abruptly as if cut short.

Tubercle (*tuberculum*, a little tuber), a small wart-like excrescence.

Tuberculate, with small warty projections or excrescences.

Tuberculiform, resembling a little tuber.

Tubiform (*tuba*, a trumpet), trumpet-shaped.

Turbinate (*turbinatus*, cone-shaped), top-shaped.

Umbilicate (*umbilicus*, the navel), with a small central depression.

Umbonate (*umbo*, the boss of a shield), with a central boss-like elevation.

Uncinate (*uncinus*, a hook), hooked.

Universal, of the veil or volva which entirely envelops the fungus when young.

Urceolate (*urceola*, a pitcher), pitcher-like, hollow and contracted at the mouth like an urn or pitcher.

Veil (*velum*, a covering), a covering of various texture more or less completely enwrapping a fungus when young.

Veliform (*velum*, a covering), of a thin veil-like covering.

Venose (*venosus*, veiny), having veins.

Ventricose (*venter*, the belly), swollen in the middle.

Vermiform (*vermis*, a worm), worm-shaped.

Verrucose (*verruca*, a wart), warty.

Versiform (*versiformis*, changing shape), of different shapes, or altering in shape with age.

Villous (*villus*, a tuft of hair), downy with soft hairs.

Vinous, of the colour of claret, dark or dilute.

Virgate (*virga*, a twig, stripe), streaked.

Volva (*volva*, a wrapper), a covering or sac enveloping the young fungus.

Volvate, having a volva.

Zone, a circle of hairs, asperities, colour, etc.

μ, a micromillimetre or $\frac{1}{1000}$ of a millimetre ; used for spore measurements.

INDEX

LONDON:
PRINTED BY WILLIAM CLOWES AND SONS, LIMITED,
DUKE STREET, STAMFORD STREET, S.E., AND GREAT WINDMILL STREET, W.

SER. I. LEUCOSPORI.

\+ Hymenophore distinct from the fleshy stem.

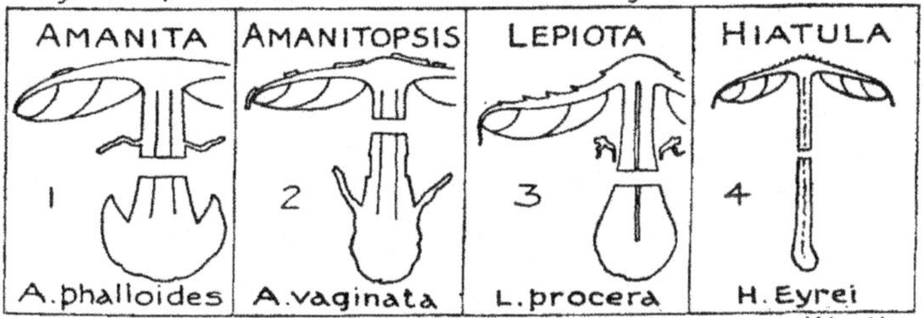

\+ + Hymenophore confluent and homogeneous with the fleshy stem.

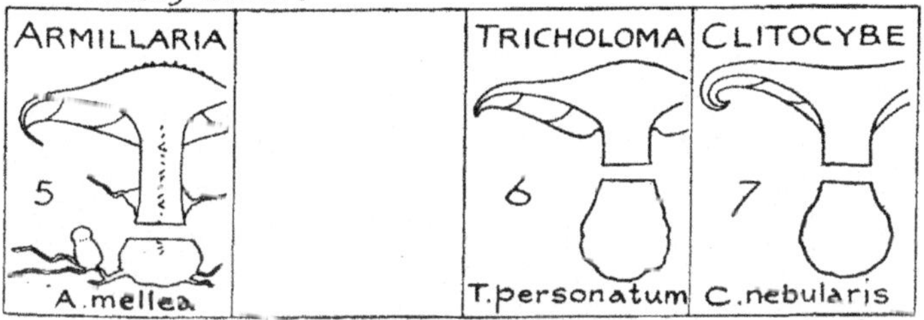

\+ + + Hymenophore confluent with but heterogeneous from the cartilaginous stem.

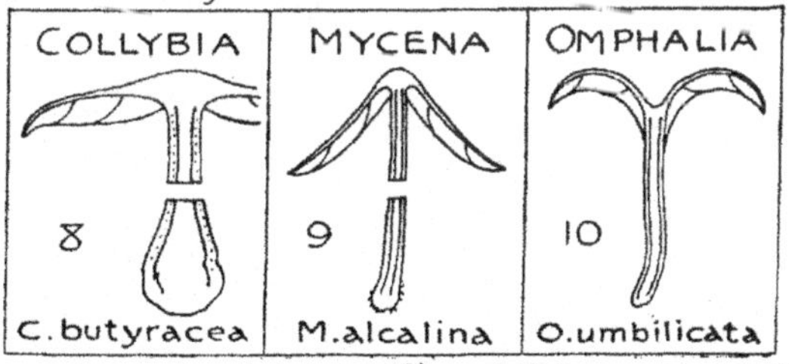

\+ + + + Hymenophore confluent & homogeneous with the fleshy excentric stem.

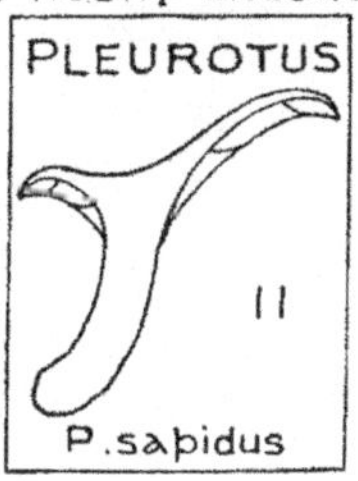

SER. II. HYPORHODII.

+ Hymenophore distinct from the fleshy stem.

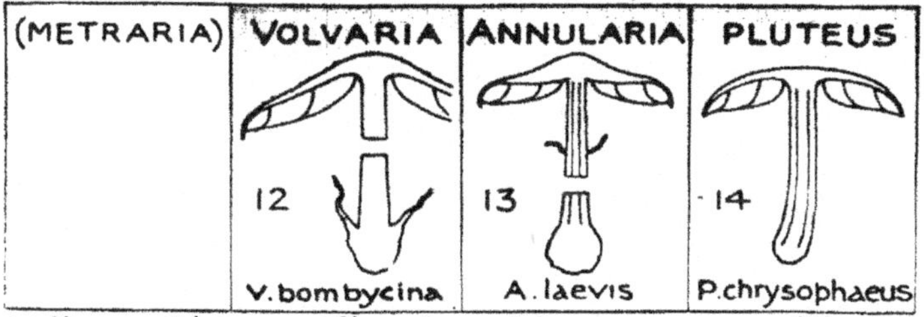

++ Hymenophore confluent and homogeneous with the fleshy stem.

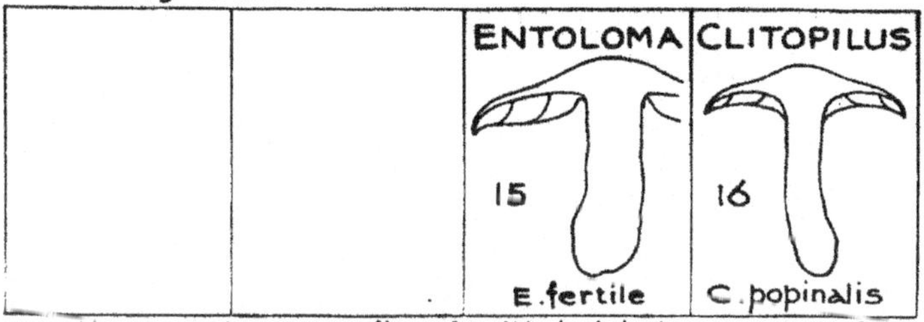

+++ Hymenophore confluent with but heterogeneous from the cartilaginous stem.

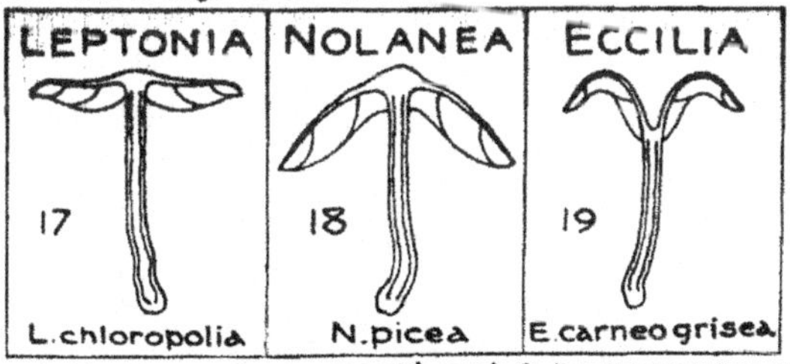

++++ Hymenophore confluent & homogeneous with the fleshy excentric stem.

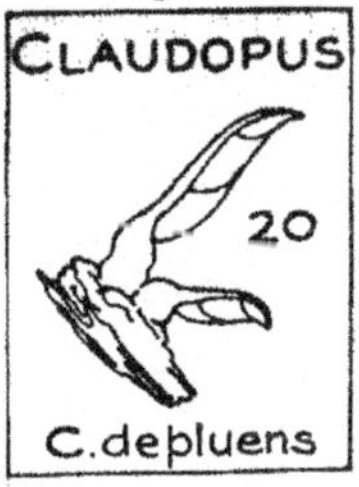

Ser. III. Dermini.

+ Hymenophore distinct from the fleshy stem.

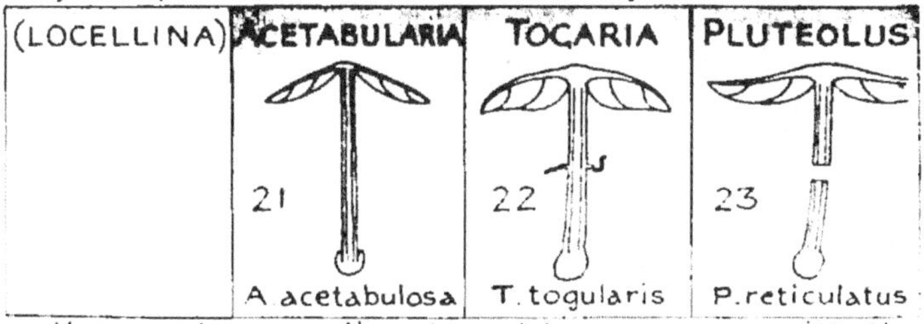

++ Hymenophore confluent and homogeneous with the fleshy stem.

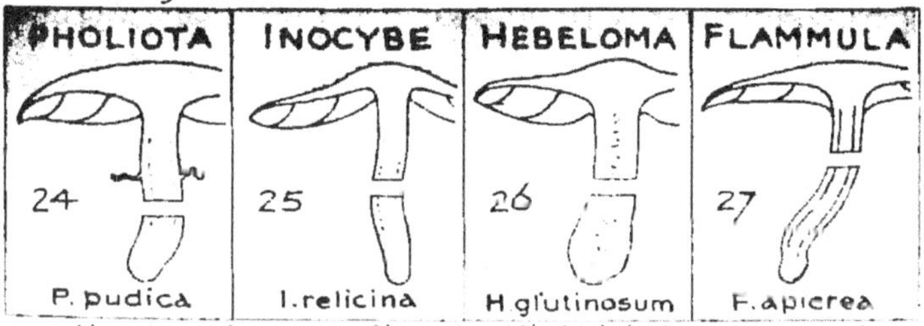

+++ Hymenophore confluent with but heterogeneous from the cartilaginous stem.

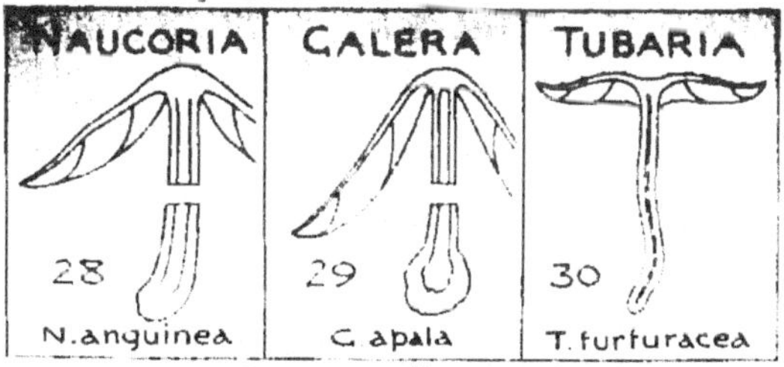

++++ Hymenophore confluent & homogeneous with the fleshy excentric stem.

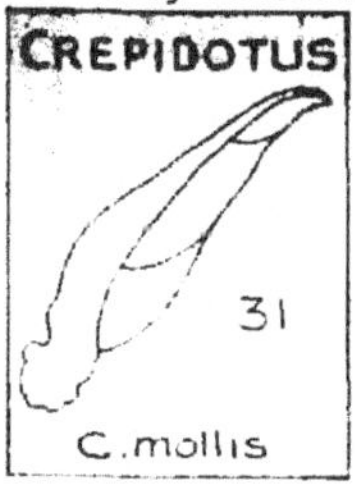

SER. IV. PRATELLI.

\+ Hymenophore distinct from the fleshy stem.

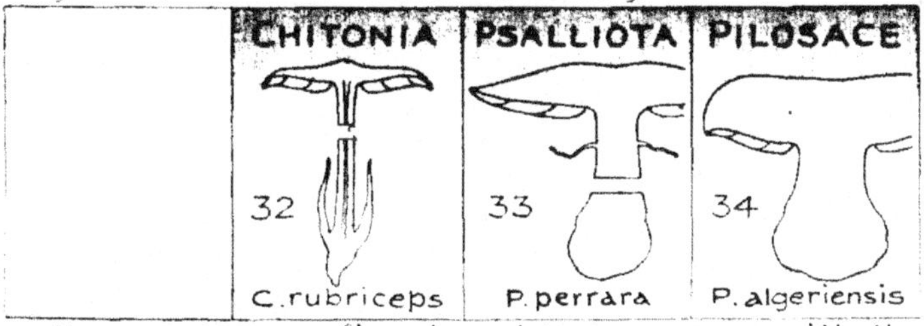

++ Hymenophore confluent and homogeneous with the fleshy stem.

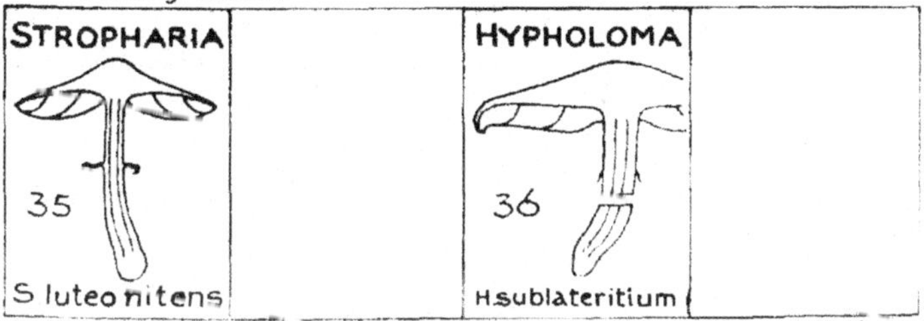

+++ Hymenophore confluent with but heterogeneous from the cartilaginous stem.

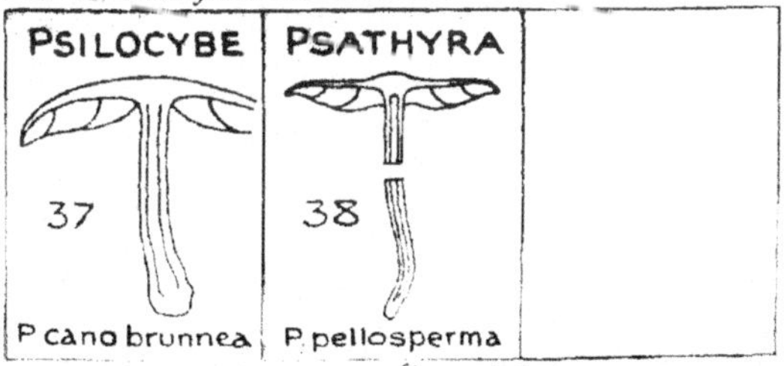

++++ Hymenophore confluent & homogeneous with the fleshy excentric stem.

SER. V. COPRINARII

* Hymenophore [illegible] the fleshy stem.

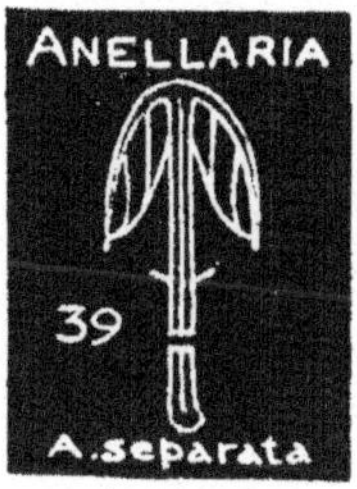

** Hymenophore [illegible] and [illegible] with the fleshy stem.

*** Hymenophore [illegible] the [illegible] stem.

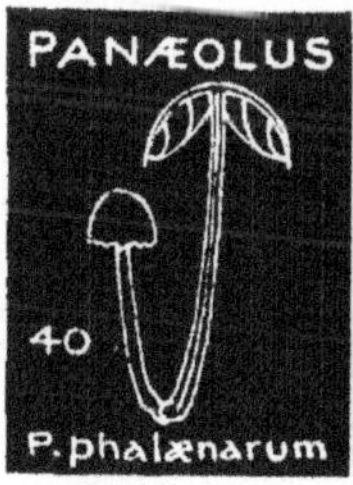

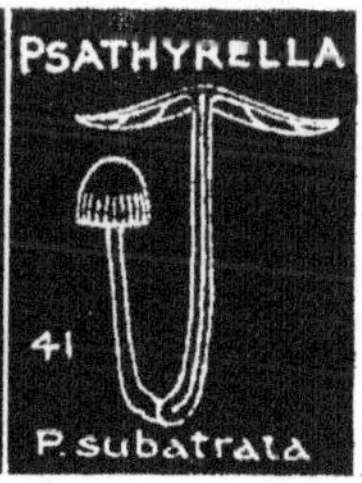

**** Hymenophore confluent & [illegible] with the fleshy excentric stem.

SER. V. COPRINARII.

+ Hymenophore distinct from the fleshy stem.

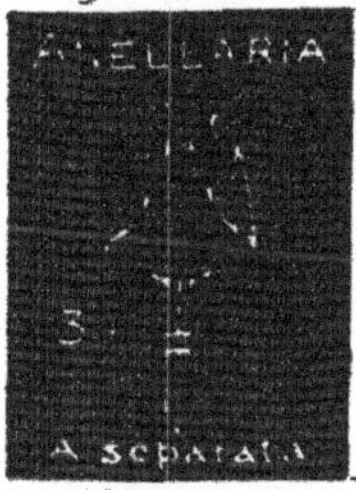

++ Hymenophore confluent and homogeneous with the fleshy stem.

+++ Hymenophore confluent with but heterogeneous from the cartilaginous stem.

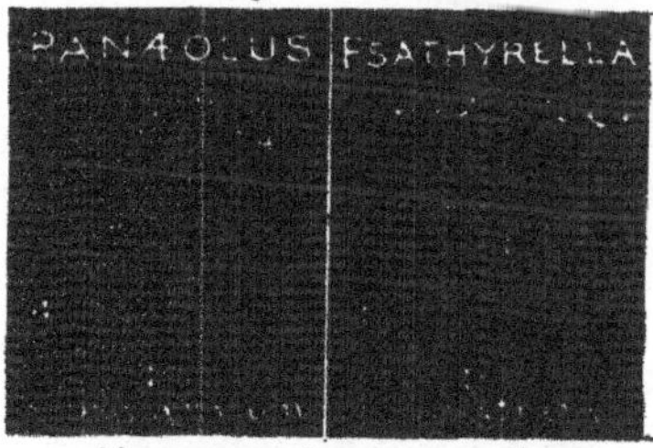

++++ Hymenophore confluent & homogeneous with the fleshy excentric stem.

Printed in Dunstable, United Kingdom

91231746R10312